Exponential function: $y = P_0 a^x$

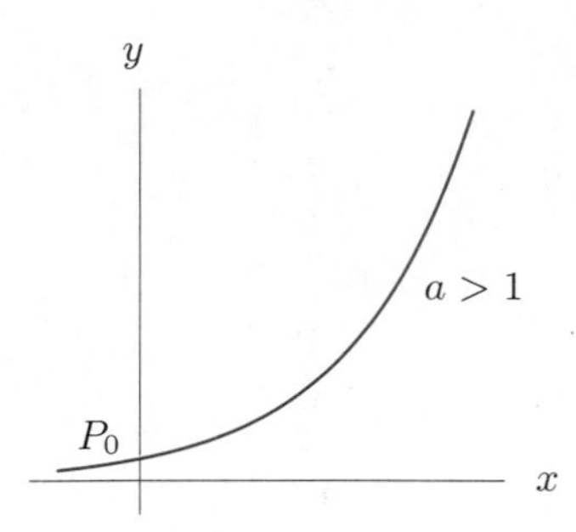

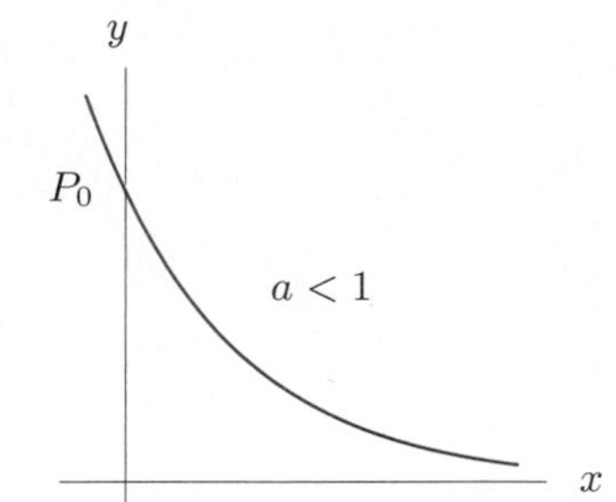

Logarithm function: $y = \ln x$

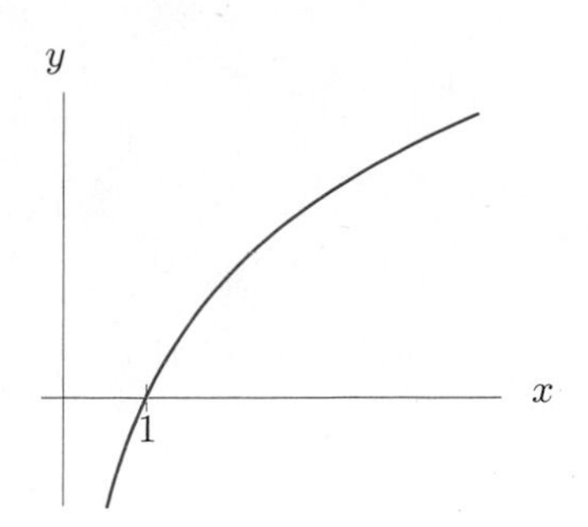

Periodic functions

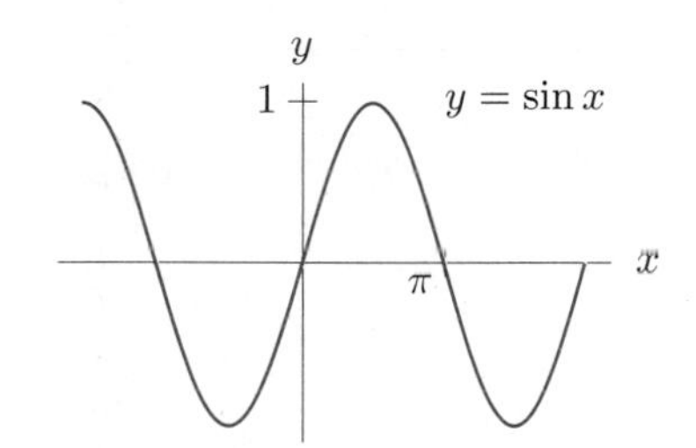

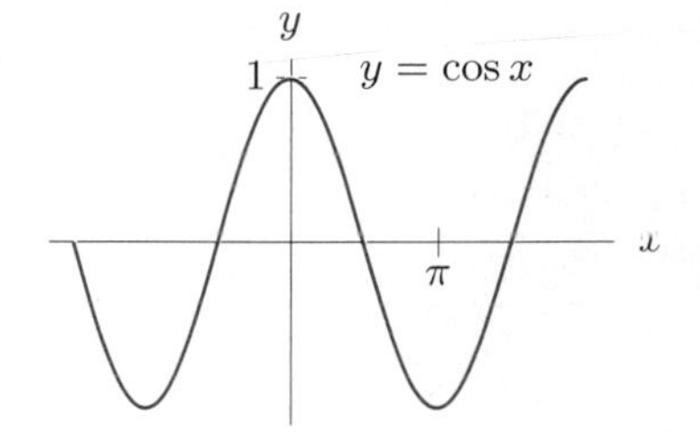

Logistic function: $y = \dfrac{L}{1 + Ce^{-kx}}$

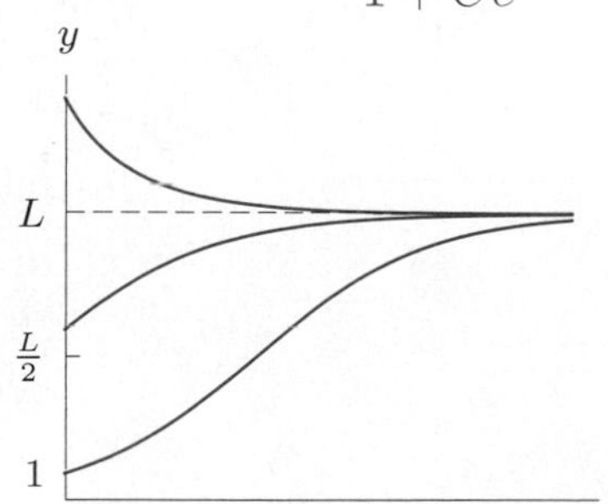

Surge function: $y = axe^{-bx}$

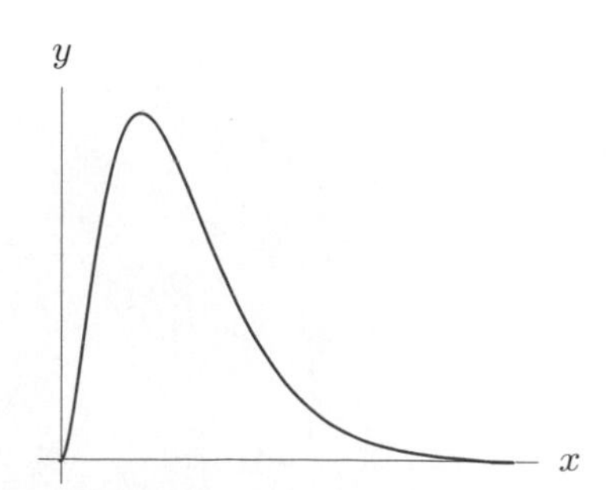

APPLIED CALCULUS

Third Edition

Produced by the Calculus Consortium and initially funded by a National Science Foundation Grant.

Deborah Hughes-Hallett
University of Arizona

William G. McCallum
University of Arizona

Andrew M. Gleason
Harvard University

Brad G. Osgood
Stanford University

Patti Frazer Lock
St. Lawrence University

Douglas Quinney
University of Keele

Daniel E. Flath
Macalester College

Karen Rhea
University of Michigan

David O. Lomen
University of Arizona

Jeff Tecosky-Feldman
Haverford College

David Lovelock
University of Arizona

Thomas W. Tucker
Colgate University

with the assistance of

Otto K. Bretscher
Colby College

Eric Connally
Harvard Extension School

Richard D. Porter
Northeastern University

Sheldon P. Gordon
SUNY at Farmingdale

Andrew Pasquale
Chelmsford High School

Joe B. Thrash
University of Southern Mississippi

Coordinated by

Elliot J. Marks

John Wiley & Sons, Inc.

ACQUISITIONS EDITOR	Mary K. Kittell
VICE PRESIDENT AND PUBLISHER	Laurie Rosatone
DEVELOPMENTAL EDITOR	Anne Scanlan-Rohrer/Two Ravens Editorial
ASSOCIATE EDITOR	Kelly Boyle
MEDIA EDITOR	Stefanie Liebman
SENIOR PRODUCTION EDITOR	Ken Santor
COVER DESIGNER	Hope Miller
COVER AND CHAPTER OPENING PHOTO	©Mike Yamashita/Taxi/Getty Images

Problems from Calculus: The Analysis of Functions, by Peter D. Taylor (Toronto: Wall & Emerson, Inc., 1992). Reprinted with permission of the publisher.

This book was set in Times Roman by the Consortium using TeX, Mathematica, and the package AsTeX, which was written by Alex Kasman. It was printed and bound by Von Hoffmann Press. The cover was printed by Von Hoffmann Press.

This book is printed on acid-free paper.

This material is based upon work supported by the National Science Foundation under Grant No. DUE-9352905. Opinions expressed are those of the authors and not necessarily those of the Foundation.

ISBN-13: 978-0-471-68121-2
ISBN-10: 0-471-68121-0

Printed in the United States of America

10 9 8 7 6 5 4 3 2 1

Dedicated to Robin, Kari, Eric, and Dennis,
and
Laura, Pearl, Felix, Natt, Isaac, and Matthias.

PREFACE

Calculus is one of the greatest achievements of the human intellect. Inspired by problems in astronomy, Newton and Leibniz developed the ideas of calculus 300 years ago. Since then, each century has demonstrated the power of calculus to illuminate questions in mathematics, the physical sciences, engineering, and the social and biological sciences.

Calculus has been so successful because of its extraordinary power to reduce complicated problems to simple rules and procedures. Therein lies the danger in teaching calculus: it is possible to teach the subject as nothing but the rules and procedures—thereby losing sight of both the mathematics and of its practical value. This third edition of *Applied Calculus* continues our effort to refocus the teaching of calculus on concepts as well as procedures.

A Focused Vision: Conceptual Understanding

Our goal is to provide students with a clear understanding of the ideas of calculus as a solid foundation for subsequent courses. We began work on this book by talking to faculty in business, economics, biology, and a wide range of other fields, as well as to many mathematicians who teach applied calculus. As a result of these discussions we included some new topics and omitted some traditional topics whose inclusion we could not justify. In the process, we also changed the focus of certain topics.

The Third Edition: Flexibility and the Rule of Four

Since our First Edition struck a new balance between concepts, modeling, and skills, we have helped widen the range of choices for calculus instructors. We have found that the Rule of Four, which emphasizes multiple representations of functions, encourages students to reflect on the meaning of the material.

Students usually learn most when they are active, so we feel the exercises in a text are of central importance. Our problems probe student understanding in areas often taken for granted. The influence of these problems, praised for their creativity and variety, has extended far beyond the users of our textbooks. We have been guided by the following principles.

- Our problems are varied. Some are straightforward and some are challenging. Many problems require students to understand the concepts and cannot be done by following a template in the text.
- The Rule of Four: Where appropriate, topics are presented geometrically, numerically, analytically, and verbally.
- The Key Concept chapters on the derivative and the definite integral (Chapters 2 and 5) can be covered at the outset of the course, right after Chapter 1.
- Because different users often choose very different topics to cover in a one-semester applied calculus course, we have designed this book for either a one-semester course (with much flexibility in choosing topics) or a two-semester course. Sample syllabi are provided in the Instructor's Manual.

Changes in the Third Edition

The third edition has the same vision as the first two editions. Previous editions were used by a large and diverse group of schools in semester and quarter systems, in large lectures and small classes, in computer labs, small groups, and traditional settings, and with a number of different technologies. In preparing this edition, we solicited comments from a large number of mathematicians who had used the text. We continued to discuss with our colleagues in client disciplines the mathematical needs of their students. We were offered many valuable suggestions, which we have tried to incorporate, while maintaining our original commitment to a focused treatment of a limited number of topics. The changes we have made include:

- Data throughout the book has been updated. More problems have been added throughout; new spreadsheet projects have been added.

- The material in Section 1.4 on taxes has been simplified and easier problems added.
- The Focus on Modeling section on fitting formulas to data now contains an introduction to correlation.
- The subsection on maximizing profit has been moved from Section 2.5 to Section 4.4.
- Section 3.3 on the chain rule has been expanded to provide more practice, especially with functions given graphically.
- Sections 4.3 and 4.4 contain more mid-level problems. The sections on economics in Chapter 4 have been rewritten for clarity.
- In Chapter 9 there is a new project introducing the concept of duality, important in economics.
- Chapter 10 contains a new Focus on Theory section on separation of variables and a new project on the 2003 SARS outbreak.
- There are two new spreadsheet projects: one on whether to buy a new or used car, and one on the outbreak of flu during the First World War.
- A pre-test is included for students whose skills may need a refresher prior to taking the course.

What Student Background is Expected?

This book is intended for students in business, the social sciences, and the life sciences. We have found the material to be thought-provoking for well-prepared students while still accessible to students with weak algebra backgrounds. Providing numerical and graphical approaches as well as the algebraic gives students several ways of mastering the material. This approach encourages students to persist, thereby lowering failure rates. A pre-test over background material is available in the appendix to the book; an algebra refresher is avalable at the student book companion site at www.wiley.com/college/hugheshallett.

Technology

We take advantage of computers and graphing calculators to help students learn to think mathematically. For example, using a graphing calculator to zoom in on functions is an excellent way of seeing local linearity. The ability to use technology effectively as a tool is important. Students are expected to use their own judgment to determine where technology is useful.

However, the book does not require any specific software or technology. Instructors have used the materials with graphing calculators, graphing software, and computer algebra systems. Any technology with the ability to graph functions and perform numerical integration will suffice.

Content

This content represents our vision of how applied calculus can be taught. It is flexible enough to accommodate individual course needs and requirements. Topics can easily be added or deleted, or the order changed.

Chapter 1: Functions and Change

Chapter 1 introduces the concept of a function and the idea of change, including the distinction between total change and rate of change. All elementary functions are introduced here. Although the functions are probably familiar, the graphical, numerical, verbal, and modeling approach to them is likely to be new. We introduce exponential functions early, since they are fundamental to the understanding of real-world processes.

Focus on Modeling: The first section introduces the student to fitting formulas to data and the second section provides further discussion of compound interest and the definition of the number e.

Focus on Theory: This section discusses limits to infinity and end behavior.

Chapter 2: Rate of Change: The Derivative

Chapter 2 presents the key concept of the derivative according to the Rule of Four. The purpose of this chapter is to give the student a practical understanding of the meaning of the derivative and its interpretation as an instantaneous rate of change. After finishing this chapter, a student will be able to approximate derivatives numerically by taking difference quotients, visualize derivatives graphically as the slope of the graph, and interpret the meaning of first and second derivatives in various applications. The student will also understand the concept of marginality and recognize the derivative as a function in its own right.

Focus on Theory: This section discusses limits and continuity and presents the symbolic definition of the derivative.

Chapter 3: Short-Cuts to Differentiation

The derivatives of all the functions in Chapter 1 are introduced, as well as the rules for differentiating products, quotients, and composite functions.

Focus on Theory: This section uses the definition of the derivative to obtain the differentiation rules.

Focus on Practice: This section provides a collection of differentiation problems for skill-building.

Chapter 4: Using the Derivative

The aim of this chapter is to enable the student to use the derivative in solving problems, including optimization and graphing. It is not necessary to cover all the sections.

Chapter 5: Accumulated Change: The Definite Integral

Chapter 5 presents the key concept of the definite integral, in the same spirit as Chapter 2.

The purpose of this chapter is to give the student a practical understanding of the definite integral as a limit of Riemann sums, and to bring out the connection between the derivative and the definite integral in the Fundamental Theorem of Calculus. We use the same method as in Chapter 2, introducing the fundamental concept in depth without going into technique. The student will finish the chapter with a good grasp of the definite integral as a limit of Riemann sums, with the ability to approximate a definite integral numerically, and with an understanding of how to interpret the definite integral in various contexts.

Chapter 5 can be covered immediately after Chapter 2 without difficulty.

Focus on Theory: This section presents the Second Fundamental Theorem of Calculus and the properties of the definite integral.

Chapter 6: Using the Definite Integral

This chapter presents applications of the definite integral. It is not meant to be comprehensive and it is not necessary to cover all the sections.

Chapter 7: Antiderivatives

This chapter covers antiderivatives from a graphical, numerical, and analytical point of view. Integration by substitution is included. The Fundamental Theorem of Calculus is used to evaluate definite integrals and to analyze antiderivatives.

Focus on Practice: This section provides a collection of integration problems for skill-building.

Chapter 8: Probability

This chapter covers probability density functions, cumulative distribution functions, the median, and the mean.

Chapter 9: Functions of Several Variables

Chapter 9 introduces functions of two variables from several points of view, using contour diagrams, formulas, and tables. It gives students the skills to read contour diagrams and think graphically, to read tables and think numerically, and to apply these skills, along with their algebraic skills, to modeling. The idea of the partial derivative is introduced from graphical, numerical, and symbolic viewpoints. Partial derivatives are then applied to optimization problems, ending with a discussion of constrained optimization using Lagrange multipliers.

Focus on Theory: This section uses optimization to derive the formula for the regression line.

Chapter 10: Mathematical Modeling Using Differential Equations

This chapter introduces differential equations. The emphasis is on modeling, qualitative solutions, and interpretation. This chapter includes applications of systems of differential equations to population models, the spread of disease, and predator-prey interactions.

Focus on Theory: This section explains the technique of separation of variables.

Chapter 11: Geometric Series

This chapter covers geometric series and their applications to business, economics, and the life sciences.

Supplementary Materials

Supplements for the instructor can be obtained by sending a request on your institutional letterhead to Mathematics Marketing Manager, John Wiley & Sons Inc., 111 River Street, Hoboken, NJ 07030-5774, or by contacting your local Wiley representative. The following supplementary materials are available.

- **Instructor's Manual and Test Bank** containing teaching tips, sample syllabii, calculator programs, overhead transparency masters, and test questions and solutions arranged by section.
- **Instructor's Solution Manual** with complete solutions to all problems.
- **Student's Solution Manual** with complete solutions to half the odd-numbered problems.
- **Student Study Guide** with additional study aids for students that are tied directly to the book.
- **Additional Material for Instructors**, elaborating specially marked points in the text, as well as password protected electronic versions of the instructor ancillaries, can be found on the web at www.wiley.com/college/hugheshallett.
- **Additional Material for Students**, at the student book companion site at www.wiley.com/college/hugheshallett, includes an algebra refresher and web quizzes.

Getting Started Technology Manual Series:

- **Getting Started with Mathematica**, 2nd cdn, by C-K. Cheung, G.E. Keough, Robert H. Gross, and Charles Landraitis of Boston College
- **Getting Started with Maple**, 2nd edn, by C-K. Cheung, G.E. Keough, both of Boston College, and Michael May of St. Louis University
- **Graphing Calculator Guide for the TI84/83** by Carl Swenson of Seattle University
- **Graphing Calculator Guide for the TI-89** by Carl Swenson of Seattle University

ConcepTests

ConcepTests, modeled on the pioneering work of Harvard physicist Eric Mazur, are questions designed to promote active learning during class, particularly (but not exclusively) in large lectures. Our evaluation data show students taught with ConcepTests outperformed students taught by traditional lecture methods 73% versus 17% on conceptual questions, and 63% versus 54% on computational problems. A new supplement to *Applied Calculus*, 3rd edn, containing ConcepTests by section, is available from your Wiley representative.

Wiley Faculty Network

A peer-to-peer network of academic faculty dedicated to the effective use of technology in the classroom. This group can help you apply innovative classroom techniques and implement specific software packages. Visit www.wherefacultyconnect.com or ask your Wiley representative for details.

eGrade Plus

eGrade Plus is a powerful online tool that provides a completely integrated suite of teaching and learning resources in one easy-to-use Web site. *eGrade Plus* includes an online version of the text, with fully integrated electronic versions of all student supplements, including the Student Solutions Manual, the Student Study Guide, and an Algebra and Trigonometry Refresher. Instructors have additional access to electronic versions of the Instructor's Manual, the Instructor's Solutions Manual, additional projects, as well as other valuable resources. *eGrade Plus* also offers an online assessment system with full gradebook capabilities, which contains over 1000 skill-building questions from the Exercise sections in each chapter. Please view our online demo at www.wiley.com/college/egradeplus. Here you will find additional information about the features and benefits of eGrade Plus, how to request a "test drive" of eGrade Plus, and how to adopt it for class use.

Acknowledgements

First and foremost, we want to express our appreciation to the National Science Foundation for their faith in our ability to produce a revitalized calculus curriculum and, in particular, to Louise Raphael, John Kenelly, John Bradley, Bill Haver, and James Lightbourne. We also want to thank the members of our Advisory Board, Benita Albert, Lida Barrett, Bob Davis, Lovenia DeConge-Watson, John Dosscy, Ron Douglas, Don Lewis, Seymour Parter, John Prados, and Steve Rodi for their ongoing guidance and advice.

In addition, we want to thank all the people across the country who encouraged us to write this book and who offered so many helpful comments. We would like to thank the following people, for all that they have done to help our project succeed: Ruth Baruth, Alon Ben-David, Jeffery Bergen, Ted Bick, Graeme Bird, Kelly Boyle, Kelly Brooks, Lucille Buonocore, J. Curtis Chipman, Dipa Choudhury, Larry Crone, Jane Devoe, Jeff Edmunds, Gail Ferrell, Joe Fiedler, Holland Filgo, Sally Fischbeck, David Flath, Ron Frazer, Lynn Garner, David Graser, Ole Hald, Jenny Harrison, John Hennessey, Yvette Hester, David Hornung, Richard Iltis, Adrian Iovita, Jerry Johnson, Thomas Judson, Bonnie Kelly, Mary Kittell, Donna Krawczyk, Theodore Laetsch, T.-Y. Lam, Sylvain Laroche, Kurt Lemmert, Suzanne Lenhart, Madelyn Lesure, Janny Leung, Ben Levitt, Thomas Lucas, Alfred Manaster, Peter McClure, Georgia Kamvosoulis Mederer, Kurt Mederer, David Meredith, Nolan Miller, Mohammad Moazzam, Saadat Moussavi, Patricia Oakley, Mary Ellen O'Leary, Jim Osterburg, Mary Parker, Ruth Parsons, Greg Peters, Laura Piscitelli, Kim Presser, Sarah Richardson, Laurie Rosatone, Daniel Rovey, Harry Row, Kenneth Santor, Anne Scanlan-Rohrer, Alfred Schipke, Virginia Stallings, Brian Stanley, Marian Stas, Mary Jane Sterling, Robert Styer, "Suds" Sudholz, Thomas Timchek, Jake Thomas, Praja Trivedi, J. Jerry Uhl, Nicola Viegi, Tilaka Vijithakumara, Alan Weinstein, Rachel Deyette Werkema, Aaron Wootton, Hung-Hsi Wu, and Sam Xu.

Reports from the following reviewers were most helpful in shaping the third edition:

Victor Akatsa, Carol Blumberg, Mary Ann Collier, Murray Eisenberg, Donna Fatheree, Dan Fuller, Ken Hannsgen, Marek Kossowski, Sheri Lehavi, Deborah Lurie, Jan Mays, Jeffery Meyer, Bobra Palmer, Barry Peratt, Russ Potter, Ken Price, Maijian Qian, Emily Roth, Lorenzo Traldi, Joan Weiss, Christos Xenophontos.

Reports from the following reviewers were most helpful in shaping the second edition:

Victor Akatsa, Carol Joyce Blumberg, Jennifer Fowler, Helen Hancock, Ken Hannsgen, John Haverhals, Mako E. Haruta, Linda Hill, Thom Kline, Jill Messer Lamping, Dennis Lewandowski, Lige Li, William O. Martin, Ted Marsden, Michael Mocciola, Maijian Qian, Joyce Quella, Peter Penner, Barry Peratt, Emily Roth, Jerry Schuur, Barbara Shabell, Peter Sternberg, Virginia Stover, Bruce Yoshiwara, Katherine Yoshiwara.

With particular thanks to Scott Clark for all his work on the data sets.

Deborah Hughes-Hallett
Andrew M. Gleason
Patti Frazer Lock
Daniel E. Flath
David O. Lomen
David Lovelock
William G. McCallum
Brad G. Osgood
Douglas Quinney
Karen Rhea
Jeff Tecosky-Feldman
Thomas W. Tucker

APPLICATIONS INDEX

Business and Economics

Life Sciences and Ecology

Social Sciences

Physical Sciences

To Students: How to Learn from this Book

- This book may be different from other math textbooks that you have used, so it may be helpful to know about some of the differences in advance. At every stage, this book emphasizes the *meaning* (in practical, graphical or numerical terms) of the symbols you are using. There is much less emphasis on "plug-and-chug" and using formulas, and much more emphasis on the interpretation of these formulas than you may expect. You will often be asked to explain your ideas in words or to explain an answer using graphs.
- The book contains the main ideas of calculus in plain English. Success in using this book will depend on reading, questioning, and thinking hard about the ideas presented. It will be helpful to read the text in detail, not just the worked examples.
- There are few examples in the text that are exactly like the homework problems, so homework problems can't be done by searching for similar–looking "worked out" examples. Success with the homework will come by grappling with the ideas of calculus.
- For many problems in the book, there is more than one correct approach and more than one correct solution. Sometimes, solving a problem relies on common sense ideas that are not stated in the problem explicitly but which you know from everyday life.
- Some problems in this book assume that you have access to a graphing calculator or computer. There are many situations where you may not be able to find an exact solution to a problem, but you can use a calculator or computer to get a reasonable approximation.
- This book attempts to give equal weight to four methods for describing functions: graphical (a picture), numerical (a table of values), algebraic (a formula), and verbal (words). Sometimes it's easier to translate a problem given in one form into another. For example, you might replace the graph of a parabola with its equation, or plot a table of values to see its behavior. It is important to be flexible about your approach: if one way of looking at a problem doesn't work, try another.
- Students using this book have found discussing these problems in small groups helpful. There are a great many problems which are not cut-and-dried; it can help to attack them with the other perspectives your colleagues can provide. If group work is not feasible, see if your instructor can organize a discussion session in which additional problems can be worked on.
- You are probably wondering what you'll get from the book. The answer is, if you put in a solid effort, you will get a real understanding of one of the crowning achievements of human creativity—calculus—as well as a real sense of the power of mathematics in the age of technology.

CONTENTS

3 SHORT-CUTS TO DIFFERENTIATION 141

4 USING THE DERIVATIVE 175

5 ACCUMULATED CHANGE: THE DEFINITE INTEGRAL 235

6 USING THE DEFINITE INTEGRAL 275

7 ANTIDERIVATIVES 299

8 PROBABILITY 323

9 FUNCTIONS OF SEVERAL VARIABLES 343

10 MATHEMATICAL MODELING USING DIFFERENTIAL EQUATIONS 397

11 GEOMETRIC SERIES 445

Chapter One

FUNCTIONS AND CHANGE

Functions are truly fundamental to mathematics. In everyday language we say, "The performance of the stock market is a function of consumer confidence" or "The patient's blood pressure is a function of the drugs prescribed." In each case, the word *function* expresses the idea that knowledge of one fact tells us another. In mathematics, the most important functions are those in which knowledge of one number tells us another number. If we know the length of the side of a square, its area is determined. If the circumference of a circle is known, its radius is determined.

Calculus starts with the study of functions. This chapter lays the foundation for calculus by surveying the behavior of some common functions. We also see ways of handling the graphs, tables, and formulas that represent these functions.

Calculus enables us to study change. In this chapter we see how to measure change and average rate of change.

1.1 WHAT IS A FUNCTION?

In mathematics, a *function* is used to represent the dependence of one quantity upon another.

Let's look at an example. In December 2004 the temperatures in International Falls, Minnesota, were unusually low over winter vacation. The daily low temperatures for December 17–26 are given in Table 1.1.

Table 1.1 *Daily low temperature in International Falls, December 17–26, 2004*

Date	17	18	19	20	21	22	23	24	25	26
Low temperature (°F)	−14	−21	−30	5	−9	−23	−31	−35	−14	−22

Although you may not have thought of something so unpredictable as temperature as being a function, the temperature *is* a function of date, because each day gives rise to one and only one low temperature. There is no formula for temperature (otherwise we would not need the weather bureau), but nevertheless the temperature does satisfy the definition of a function: Each date, t, has a unique low temperature, L, associated with it.

We define a function as follows:

A **function** is a rule that takes certain numbers as inputs and assigns to each a definite output number. The set of all input numbers is called the **domain** of the function and the set of resulting output numbers is called the **range** of the function.

The input is called the *independent variable* and the output is called the *dependent variable.* In the temperature example, the set of dates $\{17, 18, 19, 20, 21, 22, 23, 24, 25, 26\}$ is the domain and the set of temperatures $\{-35, -31, -30, -23, -22, -21, -14, -9, 5\}$ is the range. We call the function f and write $L = f(t)$. Notice that a function may have identical outputs for different inputs (December 17 and 25, for example).

Some quantities, such as date, are *discrete*, meaning they take only certain isolated values (dates must be integers). Other quantities, such as time, are *continuous* as they can be any number. For a continuous variable, domains and ranges are often written using interval notation:

$$\text{The set of numbers } t \text{ such that } a \leq t \leq b \text{ is written } [a, b].$$
$$\text{The set of numbers } t \text{ such that } a < t < b \text{ is written } (a, b).$$

The Rule of Four: Tables, Graphs, Formulas, and Words

Functions can be represented by tables, graphs, formulas, and descriptions in words. For example, the function giving the daily low temperatures in International Falls can be represented by the graph in Figure 1.1, as well as by Table 1.1.

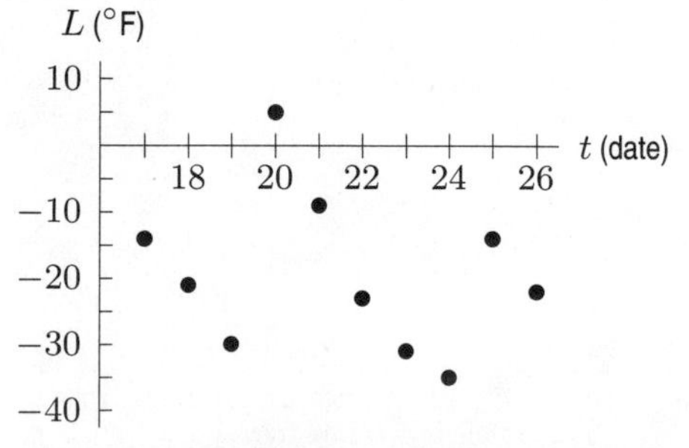

Figure 1.1: International Falls temperatures, December 2004

Other functions arise naturally as graphs. Figure 1.2 contains electrocardiogram (EKG) pictures showing the heartbeat patterns of two patients, one normal and one not. Although it is possible to construct a formula to approximate an EKG function, this is seldom done. The pattern of repetitions is what a doctor needs to know, and these are more easily seen from a graph than from a formula. However, each EKG gives electrical activity as a function of time.

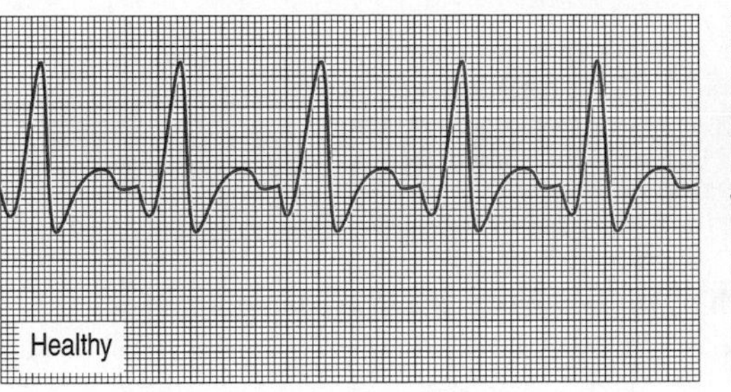

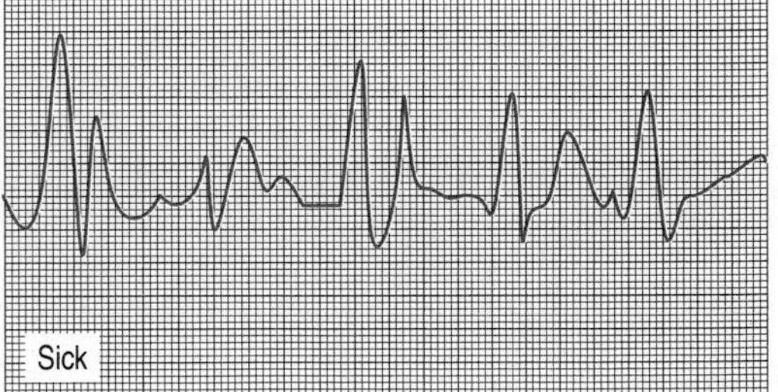

Figure 1.2: EKG readings on two patients

Consider the snow tree cricket. Surprisingly enough, all such crickets chirp at essentially the same rate if they are at the same temperature. That means that the chirp rate is a function of temperature. In other words, if we know the temperature, we can determine the chirp rate. Even more surprisingly, the chirp rate, C, in chirps per minute, increases steadily with the temperature, T, in degrees Fahrenheit, and can be computed, to a fair degree of accuracy, using the formula

$$C = f(T) = 4T - 160.$$

The graph of this function is in Figure 1.3.

Since $C = f(T)$ increases with T, we say that f is an *increasing function.*

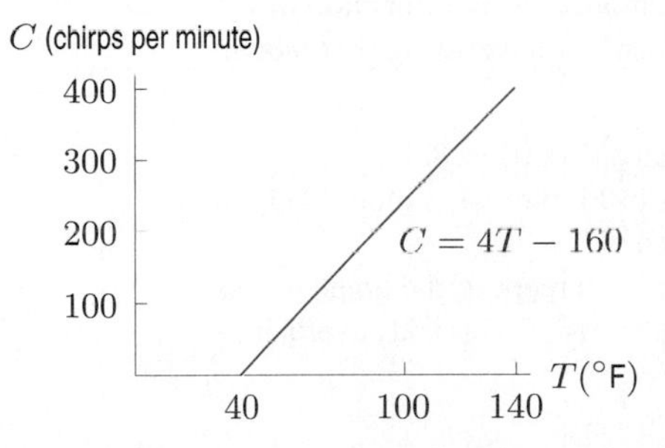

Figure 1.3: Cricket chirp rate as a function of temperature

Function Notation and Intercepts

We write $y = f(t)$ to express the fact that y is a function of t. The independent variable is t, the dependent variable is y, and f is the name of the function. The graph of a function has an *intercept* where it crosses the horizontal or vertical axis. Horizontal intercepts are also called the *zeros* of the function.

Example 1 The value of a car, V, is a function of the age of the car, a, so $V = f(a)$.

(a) Interpret the statement $f(5) = 9$ in terms of the value of a car if V is in thousands of dollars and a is in years.

(b) In the same units, the value of a Honda[1] is approximated by $f(a) = 13.25 - 0.9a$. Find and interpret the vertical and horizontal intercepts of the graph of this depreciation function f.

[1]From data obtained from the Kelley Blue Book, www.kbb.com.

Solution

(a) Since $V = f(a)$, the statement $f(5) = 9$ means $V = 9$ when $a = 5$. This tells us that the car is worth \$9000 when it is 5 years old.

(b) Since $V = f(a)$, a graph of the function f has the value of the car on the vertical axis and the age of the car on the horizontal axis. The vertical intercept is the value of V when $a = 0$. It is $V = f(0) = 13.25$, so the Honda was valued at \$13,250 when new. The horizontal intercept is the value of a such that $V(a) = 0$, so

$$13.25 - 0.9a = 0$$
$$a = \frac{13.25}{0.9} = 14.7.$$

At age 15 years, the Honda has no value.

Since $V = f(a)$ decreases with a, we say that f is a *decreasing function.*

Problems for Section 1.1

1. The time T in minutes that it takes Dan to run x kilometers is a function $T = f(x)$. Explain the meaning of the statement $f(5) = 23$ in terms of running.

2. The population of a city, P, in millions, is a function of t, the number of years since 1970, so $P = f(t)$. Explain the meaning of the statement $f(35) = 12$ in terms of the population of this city.

3. Let $W = f(t)$ represent wheat production in Argentina, in millions of metric tons, where t is years since 1990. Interpret the statement $f(12) = 9$ in terms of wheat production.

4. The number of sales per month, S, is a function of the amount, a, (in dollars) spent on advertising that month, so $S = f(a)$.

(a) Interpret the statement $f(1000) = 3500$.
(b) Which of the graphs in Figure 1.4 is more likely to represent this function?
(c) What does the vertical intercept of the graph of this function represent, in terms of sales and advertising?

(I) S ... a (II) S ... a

Figure 1.4

5. Describe what Figure 1.5 tells you about an assembly line whose productivity is represented as a function of the number of workers on the line.

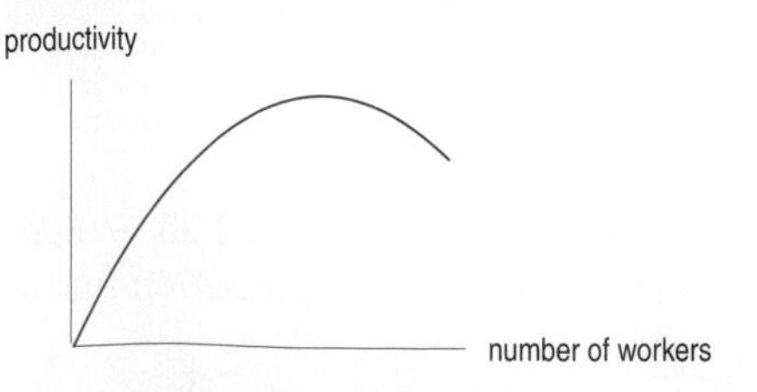

Figure 1.5

For the functions in Problems 6–10, find $f(5)$.

6. $f(x) = 2x + 3$

7. $f(x) = 10x - x^2$

8. 

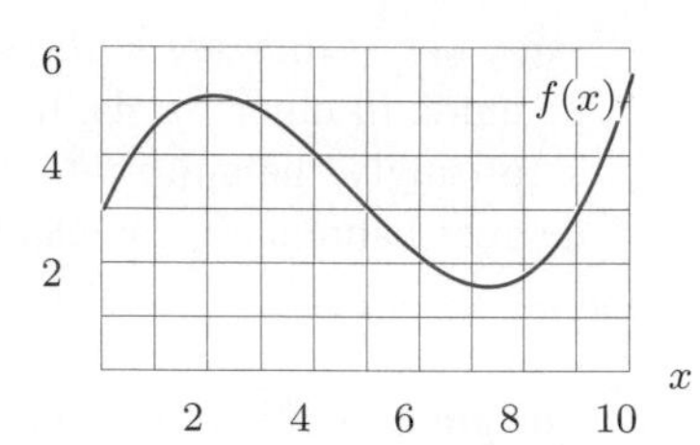

9. 

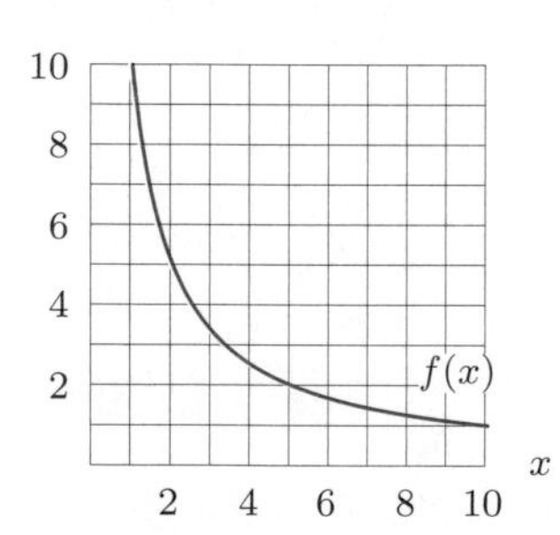

10.

x	1	2	3	4	5	6	7	8
$f(x)$	2.3	2.8	3.2	3.7	4.1	5.0	5.6	6.2

11. A filter-based clean-up program is adopted for a very polluted lake. The filters gradually clog, impairing their effectiveness. This problem is dealt with by removing the filters once a week for two days for cleaning after which they are replaced as good as new. Graph the quantity of pollution in the lake as a function of time for a three week time period.

12. Financial investors know that, in general, the higher the expected rate of return on an investment, the higher the corresponding risk.

(a) Graph this relationship, showing expected return as a function of risk.
(b) On the figure from part (a), mark a point with high expected return and low risk. (Investors hope to find such opportunities.)

13. In tide pools on the New England coast, snails eat algae. Describe what Figure 1.6 tells you about the effect of snails on the diversity of algae.[2] Does the graph support the statement that diversity peaks at intermediate predation levels?

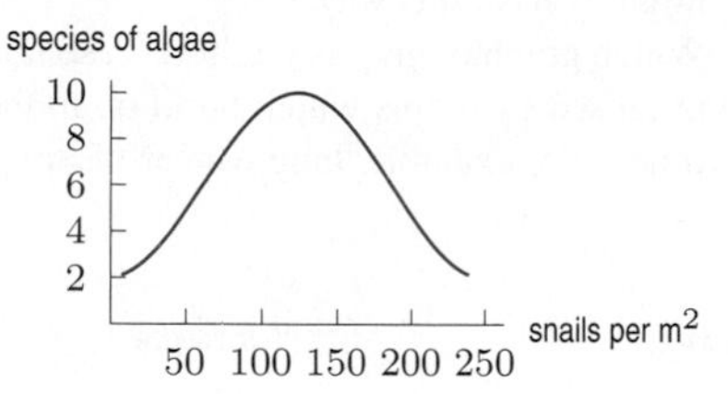

Figure 1.6

14. An object is put outside on a cold day at time $t = 0$. Its temperature, $H = f(t)$, in °C, is graphed in Figure 1.7.

(a) What does the statement $f(30) = 10$ mean in terms of temperature? Include units for 30 and for 10 in your answer.
(b) Explain what the vertical intercept, a, and the horizontal intercept, b, represent in terms of temperature of the object and time outside.

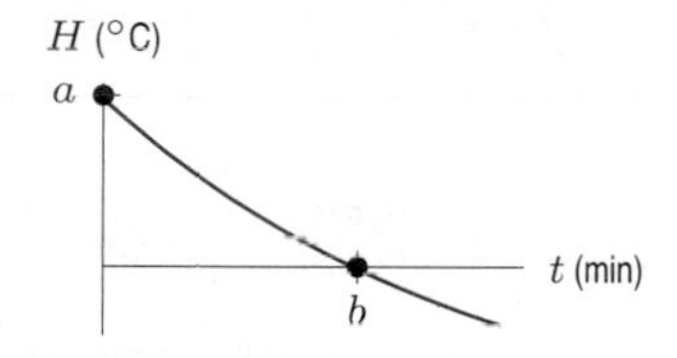

Figure 1.7

15. In the Andes mountains in Peru, the number, N, of species of bats is a function of the elevation, h, in feet above sea level, so $N = f(h)$.

(a) Interpret the statement $f(500) = 100$ in terms of bat species.
(b) What are the meanings of the vertical intercept, k, and horizontal intercept, c, in Figure 1.8?

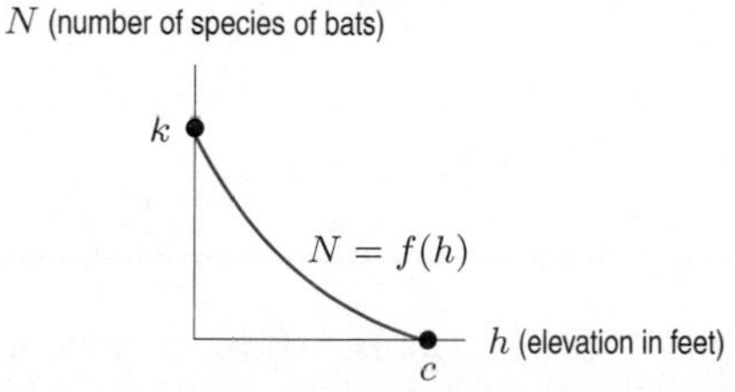

N (number of species of bats)

k

$N = f(h)$

h (elevation in feet)

c

Figure 1.8

16. After an injection, the concentration of a drug in a patient's body increases rapidly to a peak and then slowly decreases. Graph the concentration of the drug in the body as a function of the time since the injection was given. Assume that the patient has none of the drug in the body before the injection. Label the peak concentration and the time it takes to reach that concentration.

17. Figure 1.9 shows the amount of nicotine, $N = f(t)$, in mg, in a person's bloodstream as a function of the time, t, in hours, since the person finished smoking a cigarette.

(a) Estimate $f(3)$ and interpret it in terms of nicotine.
(b) About how many hours have passed before the nicotine level is down to 0.1 mg?
(c) What is the vertical intercept? What does it represent in terms of nicotine?
(d) If this function had a horizontal intercept, what would it represent?

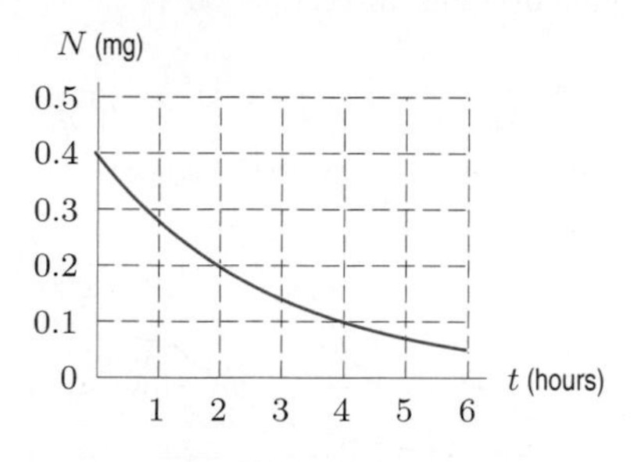

Figure 1.9

18. **(a)** A potato is put in an oven to bake at time $t = 0$. Which of the graphs in Figure 1.10 could represent the potato's temperature as a function of time?
(b) What does the vertical intercept represent in terms of the potato's temperature?

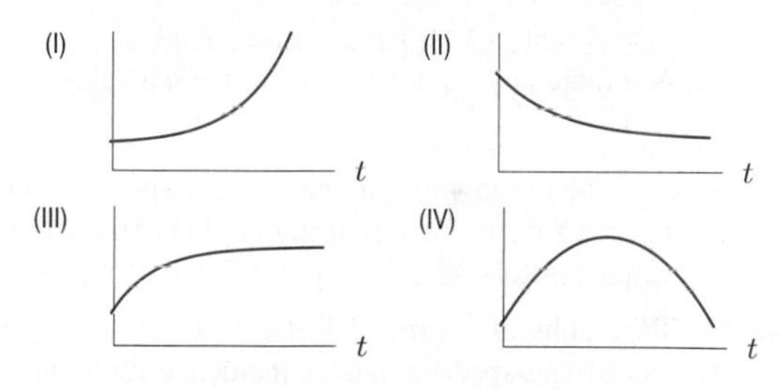

Figure 1.10

19. A deposit is made into an interest-bearing account. Figure 1.11 shows the balance, B, in the account t years later.

(a) What was the original deposit?
(b) Estimate $f(10)$ and interpret it.
(c) When does the balance reach \$5000?

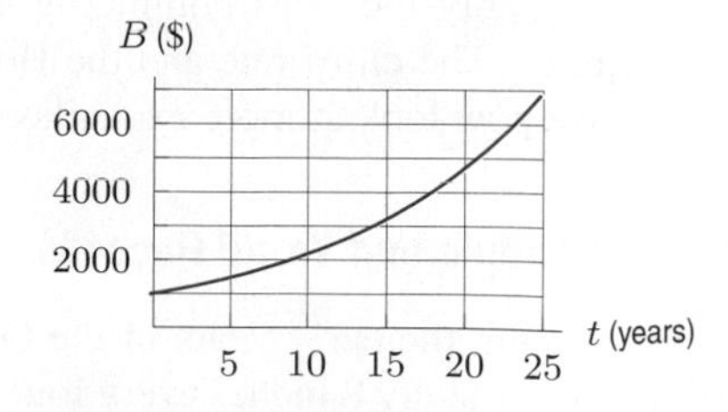

Figure 1.11

[2] Rosenzweig, M.L., *Species Diversity in Space and Time*, p. 343, (Cambridge: Cambridge University Press, 1995).

20. When a patient with a rapid heart rate takes a drug, the heart rate plunges dramatically and then slowly rises again as the drug wears off. Sketch the heart rate against time from the moment the drug is administered.

21. Figure 1.12 shows fifty years of fertilizer use in the US, India, and the former Soviet Union.[3]

(a) Estimate fertilizer use in 1970 in the US, India, and the former Soviet Union.

(b) Write a sentence for each of the three graphs describing how fertilizer use has changed in each region over this 50-year period.

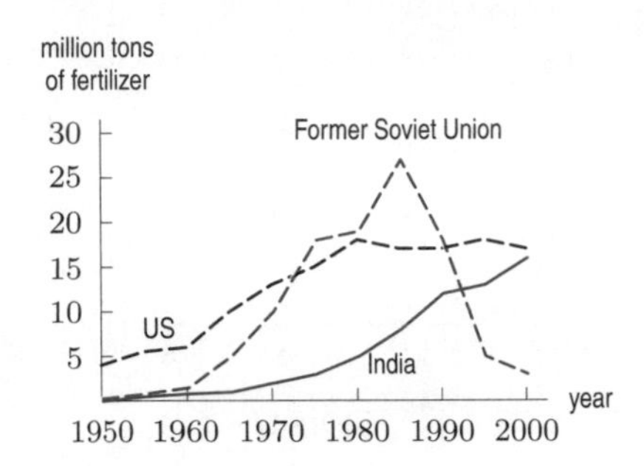

Figure 1.12

22. Let $y = f(x) = x^2 + 2$.

(a) Find the value of y when x is zero.

(b) What is $f(3)$?

(c) What values of x give y a value of 11?

(d) Are there any values of x that give y a value of 1?

23. The gas mileage of a car (in miles per gallon) is highest when the car is going about 45 miles per hour and is lower when the car is going faster or slower than 45 mph. Graph gas mileage as a function of speed of the car.

24. The six graphs in Figure 1.13 show frequently observed patterns of age-specific cancer incidence rates, in number of cases per 1000 people, as a function of age.[4] The scales on the vertical axes are equal.

(a) For each of the six graphs, write a sentence explaining the effect of age on the cancer rate.

(b) Which graph shows a relatively high incidence rate for children? Suggest a type of cancer that behaves this way.

(c) Which graph shows a brief decrease in the incidence rate at around age 50? Suggest a type of cancer that might behave this way.

(d) Which graph or graphs might represent a cancer that is caused by toxins which build up in the body over time? (For example, lung cancer.) Explain.

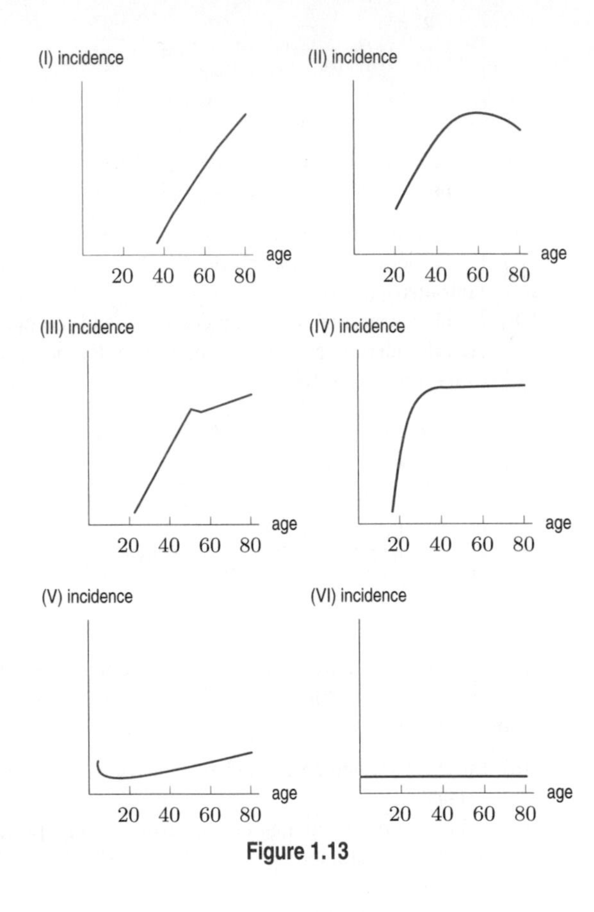

Figure 1.13

1.2 LINEAR FUNCTIONS

Probably the most commonly used functions are the *linear functions*, whose graphs are straight lines. The chirp-rate and the Honda depreciation functions in the previous section are both linear. We now look at more examples of linear functions.

Olympic and World Records

During the early years of the Olympics, the height of the men's winning pole vault increased approximately 8 inches every four years. Table 1.2 shows that the height started at 130 inches in 1900,

[3]The Worldwatch Institute, *Vital Signs 2001*, p. 32, (New York: W.W. Norton, 2001).

[4]Abraham M. Lilienfeld, *Foundations of Epidemiology*, p. 155, (New York: Oxford University Press, 1976).

and increased by the equivalent of 2 inches a year between 1900 and 1912. So the height was a linear function of time.

Table 1.2 *Winning height (approximate) for Men's Olympic pole vault*

Year	1900	1904	1908	1912
Height (inches)	130	138	146	154

If y is the winning height in inches and t is the number of years since 1900, we can write

$$y = f(t) = 130 + 2t.$$

Since $y = f(t)$ increases with t, we see that f is an increasing function. The coefficient 2 tells us the rate, in inches per year, at which the height increases. This rate is the *slope* of the line in Figure 1.14. The slope is given by the ratio

$$\text{Slope} = \frac{\text{Rise}}{\text{Run}} = \frac{146 - 138}{8 - 4} = \frac{8}{4} = 2 \text{ inches/year}.$$

Calculating the slope (rise/run) using any other two points on the line gives the same value.

What about the constant 130? This represents the initial height in 1900, when $t = 0$. Geometrically, 130 is the intercept on the vertical axis.

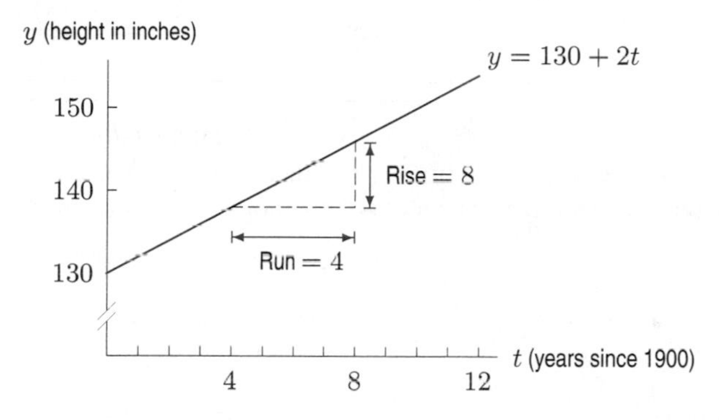

Figure 1.14: Olympic pole vault records

You may wonder whether the linear trend continues beyond 1912. Not surprisingly, it does not exactly. The formula $y = 130+2t$ predicts that the height in the 2004 Olympics would be 338 inches or 28 feet 2 inches, which is considerably higher than the actual value of 19 feet 6.25 inches.[5] There is clearly a danger in *extrapolating* too far from the given data. You should also observe that the data in Table 1.2 is *discrete*, because it is given only at specific points (every four years). However, we have treated the variable t as though it were *continuous*, because the function $y = 130 + 2t$ makes sense for all values of t. The graph in Figure 1.14 is of the continuous function because it is a solid line, rather than four separate points representing the years in which the Olympics were held.

Example 1 If y is the world record time to run the mile, in seconds, and t is the number of years since 1900, then records show that, approximately,

$$y = g(t) = 260 - 0.4t.$$

Explain the meaning of the intercept, 260, and the slope, -0.4, in terms of the world record time to run the mile and sketch the graph.

[5] *The World Almanac and Book of Facts 2005*, p. 866 (New York).

Solution The intercept, 260, tells us that the world record was 260 seconds in 1900 (at $t = 0$). The slope, -0.4, tells us that the world record decreased at a rate of about 0.4 seconds per year. See Figure 1.15.

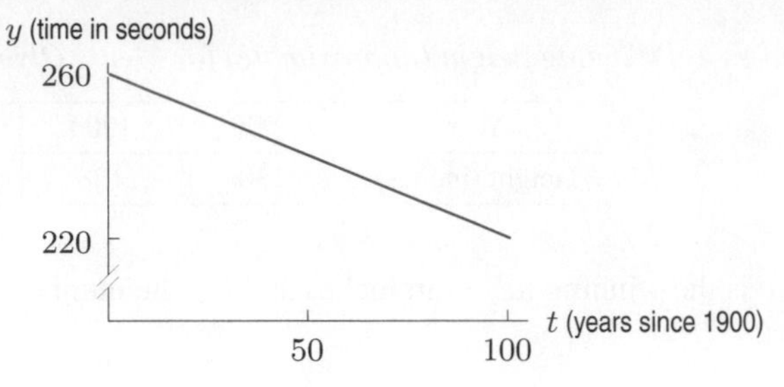

Figure 1.15: World record time to run the mile

Slope and Rate of Change

We use the symbol Δ (the Greek letter capital delta) to mean "change in," so Δx means change in x and Δy means change in y.

The slope of a linear function $y = f(x)$ can be calculated from values of the function at two points, given by x_1 and x_2, using the formula

$$\text{Slope} = \frac{\text{Rise}}{\text{Run}} = \frac{\Delta y}{\Delta x} = \frac{f(x_2) - f(x_1)}{x_2 - x_1}.$$

The quantity $(f(x_2) - f(x_1))/(x_2 - x_1)$ is called a *difference quotient* because it is the quotient of two differences. (See Figure 1.16). Since slope $= \Delta y/\Delta x$, the slope represents the *rate of change* of y with respect to x. The units of the slope are y-units over x-units.

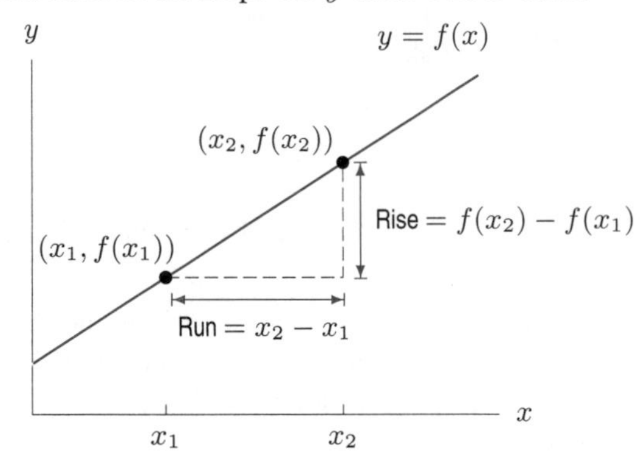

Figure 1.16: Difference quotient $= \dfrac{f(x_2) - f(x_1)}{x_2 - x_1}$

Linear Functions in General

A **linear function** has the form

$$y = f(x) = b + mx.$$

Its graph is a line such that

- m is the **slope**, or rate of change of y with respect to x.
- b is the **vertical intercept** or value of y when x is zero.

Notice that if the slope, m, is zero, we have $y = b$, a horizontal line. For a line of slope m through the point (x_0, y_0), we have

$$\text{Slope} = m = \frac{y - y_0}{x - x_0}.$$

Therefore we can write the equation of the line in the *point-slope form*:

The equation of a line of slope m through the point (x_0, y_0) is

$$y - y_0 = m(x - x_0).$$

Example 2 The solid waste generated each year in the cities of the US is increasing. The solid waste generated,[6] in millions of tons, was 205.2 in 1990 and 229.2 in 2001.

(a) Assuming that the amount of solid waste generated by US cities is a linear function of time, find a formula for this function by finding the equation of the line through these two points.

(b) Use this formula to predict the amount of solid waste generated in the year 2020.

Solution (a) We are looking at the amount of solid waste, W, as a function of year, t, and the two points are $(1990, 205.2)$ and $(2001, 229.2)$. The slope of the line is

$$m = \frac{\Delta W}{\Delta t} = \frac{229.2 - 205.2}{2001 - 1990} = \frac{24}{11} = 2.182 \text{ million tons/year.}$$

To find the equation of the line, we find the vertical intercept. We substitute the point $(1990, 205.2)$ and the slope $m = 2.182$ into the equation for W:

$$\begin{aligned} W &= b + mt \\ 205.2 &= b + (2.182)(1990) \\ 205.2 &= b + 4342.18 \\ -4136.98 &= b. \end{aligned}$$

The equation of the line is $W = -4136.98 + 2.182t$. Alternately, we could use the point-slope form of a line, $W - 205.2 = 2.182(t - 1990)$.

(b) To calculate solid waste predicted for the year 2020, we substitute $t = 2020$ into the equation of the line, $W = -4136.98 + 2.182t$, and calculate W:

$$W = -4136.98 + (2.182)(2020) = 270.66.$$

The formula predicts that in the year 2020, there will be 270.66 million tons of solid waste.

Recognizing Data from a Linear Function: Values of x and y in a table could come from a linear function $y = b + mx$ if differences in y-values are constant for equal differences in x.

[6] *Statistical Abstracts of the US*, 2004–2005, Table 363.

Example 3 Which of the following tables of values could represent a linear function?

x	0	1	2	3
$f(x)$	25	30	35	40

x	0	2	4	6
$g(x)$	10	16	26	40

t	20	30	40	50
$h(t)$	2.4	2.2	2.0	1.8

Solution Since $f(x)$ increases by 5 for every increase of 1 in x, the values of $f(x)$ could be from a linear function with slope $= 5/1 = 5$.

Between $x = 0$ and $x = 2$, the value of $g(x)$ increases by 6 as x increases by 2. Between $x = 2$ and $x = 4$, the value of y increases by 10 as x increases by 2. Since the slope is not constant, $g(x)$ could not be a linear function.

Since $h(t)$ decreases by 0.2 for every increase of 10 in t, the values of $h(t)$ could be from a linear function with slope $= -0.2/10 = -0.02$.

Example 4 The data in the following table lie on a line. Find formulas for each of the following functions, and give units for the slope in each case:

(a) q as a function of p
(b) p as a function of q

p(dollars)	5	10	15	20
q(tons)	100	90	80	70

Solution (a) If we think of q as a linear function of p, then q is the dependent variable and p is the independent variable. We can use any two points to find the slope. The first two points give

$$\text{Slope} = m = \frac{\Delta q}{\Delta p} = \frac{90-100}{10-5} = \frac{-10}{5} = -2.$$

The units are the units of q over the units of p, or tons per dollar.

To write q as a linear function of p, we use the equation $q = b+mp$. We know that $m = -2$, and we can use any of the points in the table to find b. Substituting $p = 10$, $q = 90$ gives

$$\begin{aligned} q &= b + mp \\ 90 &= b + (-2)(10) \\ 90 &= b - 20 \\ 110 &= b. \end{aligned}$$

Thus, the equation of the line is

$$q = 110 - 2p.$$

(b) If we now consider p as a linear function of q, then p is the dependent variable and q is the independent variable. We have

$$\text{Slope} = m = \frac{\Delta p}{\Delta q} = \frac{10-5}{90-100} = \frac{5}{-10} = -0.5.$$

The units of the slope are dollars per ton.

Since p is a linear function of q, we have $p = b + mq$ and $m = -0.5$. To find b, we substitute any point from the table, such as $p = 10$, $q = 90$, into this equation:

$$\begin{aligned} p &= b + mq \\ 10 &= b + (-0.5)(90) \\ 10 &= b - 45 \\ 55 &= b. \end{aligned}$$

Thus, the equation of the line is

$$p = 55 - 0.5q.$$

Alternatively, we could take our answer to part (a), that is $q = 110 - 2p$, and solve for p.

Families of Linear Functions

Formulas such as $f(x) = b + mx$, in which the constants m and b can take on various values, represent a *family of functions*. All the functions in a family share certain properties—in this case, the graphs are lines. The constants m and b are called *parameters*. Figures 1.17 and 1.18 show graphs with several values of m and b. Notice the greater the magnitude of m, the steeper the line.

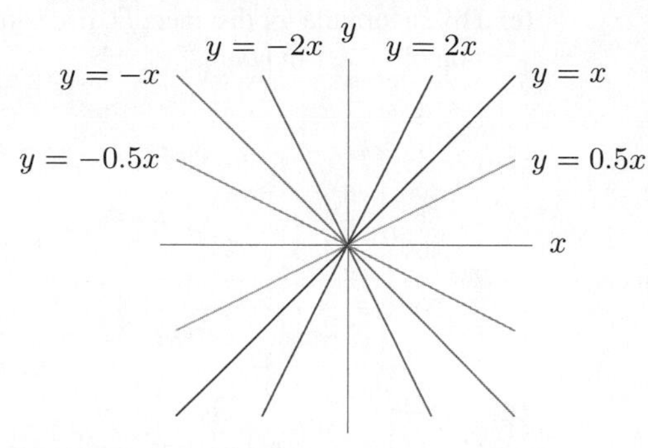

Figure 1.17: The family $y = mx$ (with $b = 0$)

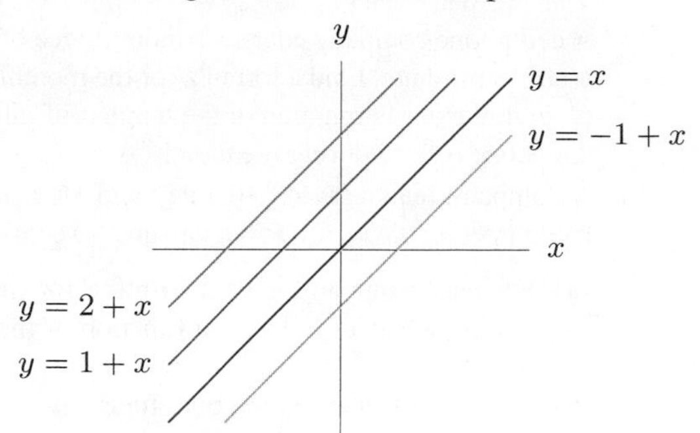

Figure 1.18: The family $y = b + x$ (with $m = 1$)

Problems for Section 1.2

For Problems 1–4, determine the slope and the y-intercept of the line whose equation is given.

1. $7y + 12x - 2 = 0$

2. $3x + 2y = 8$

3. $12x = 6y + 4$

4. $-4y + 2x + 8 = 0$

For Problems 5–8, find the equation of the line that passes through the given points.

5. $(0, 2)$ and $(2, 3)$

6. $(0, 0)$ and $(1, 1)$

7. $(-2, 1)$ and $(2, 3)$

8. $(4, 5)$ and $(2, -1)$

9. Figure 1.19 shows four lines given by equation $y = b + mx$. Match the lines to the conditions on the parameters m and b.

(a) $m > 0, b > 0$ **(b)** $m < 0, b > 0$
(c) $m > 0, b < 0$ **(d)** $m < 0, b < 0$

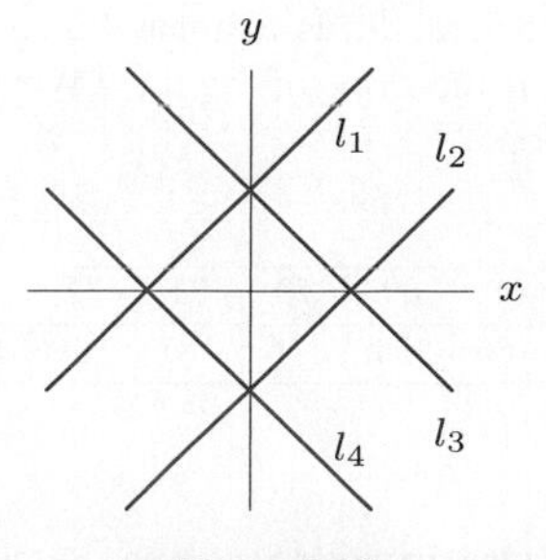

Figure 1.19

10. **(a)** Which two lines in Figure 1.20 have the same slope? Of these two lines, which has the larger y-intercept?
(b) Which two lines have the same y-intercept? Of these two lines, which has the larger slope?

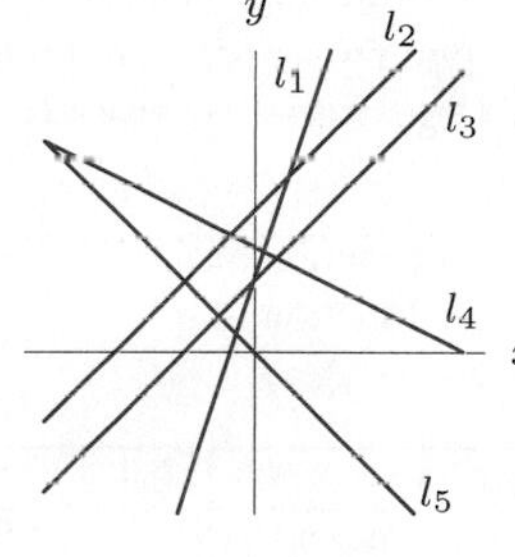

Figure 1.20

11. Match the graphs in Figure 1.21 with the following equations. (Note that the x and y scales may be unequal.)

(a) $y = x - 5$ **(b)** $-3x + 4 = y$
(c) $5 = y$ **(d)** $y = -4x - 5$
(e) $y = x + 6$ **(f)** $y = x/2$

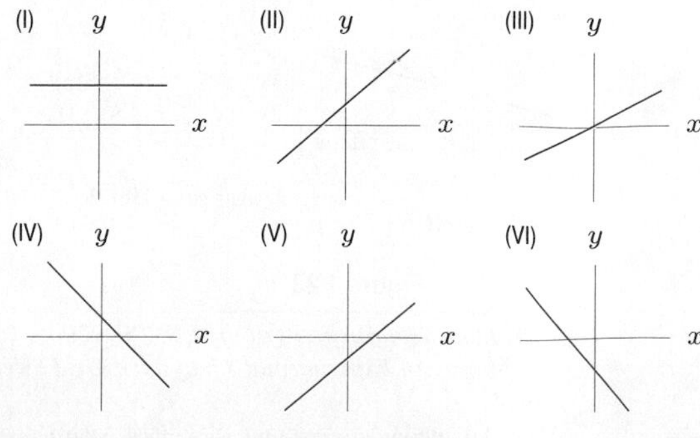

Figure 1.21

12. A city's population was 30,700 in the year 2000 and is growing by 850 people a year.

(a) Give a formula for the city's population, P, as a function of the number of years, t, since 2000.
(b) What is the population predicted to be in 2010?
(c) When is the population expected to reach 45,000?

13. A cell phone company charges a monthly fee of \$25 plus \$0.05 per minute. Find a formula for the monthly charge, C, in dollars, as a function of the number of minutes, m, the phone is used during the month.

14. A company rents cars at \$40 a day and 15 cents a mile. Its competitor's cars are \$50 a day and 10 cents a mile.

(a) For each company, give a formula for the cost of renting a car for a day as a function of the distance traveled.
(b) On the same axes, graph both functions.
(c) How should you decide which company is cheaper?

15. Which of the following tables could represent linear functions?

(a)

x	0	1	2	3
y	27	25	23	21

(b)

t	15	20	25	30
s	62	72	82	92

(c)

u	1	2	3	4
w	5	10	18	28

16. For each table in Problem 15 that could represent a linear function, find a formula for that function.

17. A company's pricing schedule in Table 1.3 is designed to encourage large orders. (A gross is 12 dozen.) Find a formula for:

(a) q as a linear function of p.
(b) p as a linear function of q.

Table 1.3

q (order size, gross)	3	4	5	6
p (price/dozen)	15	12	9	6

18. World milk production rose at an approximately constant rate between 1997 and 2003.[7] See Figure 1.22.

(a) Estimate the vertical intercept and interpret it in terms of milk production.
(b) Estimate the slope and interpret it in terms of milk production.
(c) Give an approximate formula for milk production, M, as a function of t.

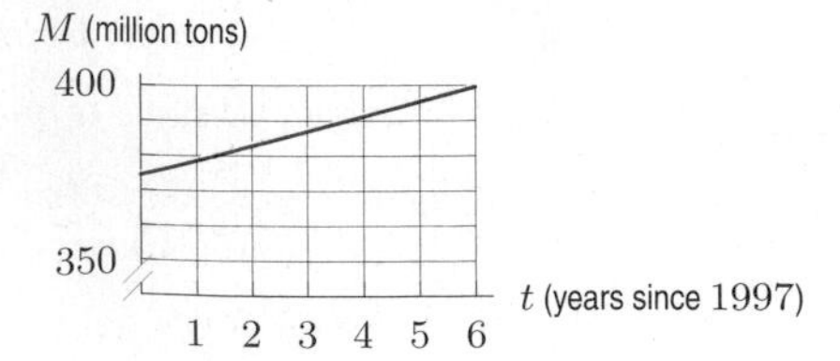

Figure 1.22

19. Figure 1.23 shows the distance from home, in miles, of a person on a 5-hour trip.

(a) Estimate the vertical intercept. Give units and interpret it in terms of distance from home.
(b) Estimate the slope of this linear function. Give units, and interpret it in terms of distance from home.
(c) Give a formula for distance, D, from home as a function of time, t in hours.

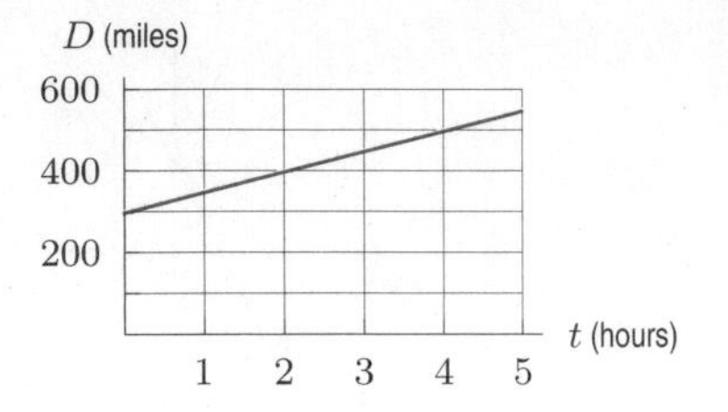

Figure 1.23

20. Search and rescue teams work to find lost hikers. Members of the search team separate and walk parallel to one another through the area to be searched. Table 1.4 shows the percent, P, of lost individuals found for various separation distances, d, of the searchers.[8]

Table 1.4

Separation distance d (ft)	20	40	60	80	100
Approximate percent found, P	90	80	70	60	50

(a) Explain how you know that the percent found, P, could be a linear function of separation distance, d.
(b) Find P as a linear function of d.
(c) What is the slope of the function? Give units and interpret the answer.
(d) What are the vertical and horizontal intercepts of the function? Give units and interpret the answers.

21. Table 1.5 gives the average weight, w, in pounds, of American men in their sixties for various heights, h, in inches.[9]

(a) How do you know that the data in this table could represent a linear function?
(b) Find weight, w, as a linear function of height, h. What is the slope of the line? What are the units for the slope?
(c) Find height, h, as a linear function of weight, w. What is the slope of the line? What are the units for the slope?

Table 1.5

h (inches)	68	69	70	71	72	73	74	75
w (pounds)	166	171	176	181	186	191	196	201

[7] *Statistical Abstracts of the US 2004–2005*, Table 1355.
[8] From *An Experimental Analysis of Grid Sweep Searching*, by J. Wartes (Explorer Search and Rescue, Western Region, 1974).
[9] Adapted from "Average Weight of Americans by Height and Age," *The World Almanac* (New Jersey: Funk and Wagnalls, 1992), p. 956.

22. Sales of music compact discs (CDs) have declined since 1999. Sales were 938.2 million in 1999 and 745.9 million in 2003.[10]

(a) Find a formula for sales, S, of music CDs, in millions of units, as a linear function of the number of years, t, since 1999.
(b) Give units for and interpret the slope and the vertical intercept of this function.
(c) Use the formula to predict music CD sales in 2010.

23. The monthly charge for a waste collection service is $32 for 100 kg of waste and is $48 for 180 kg of waste.

(a) Find a linear formula for the cost, C, of waste collection as a function of the number of kilograms of waste, w.
(b) What is the slope of the line found in part (a)? Give units and interpret your answer in terms of the cost of waste collection.
(c) What is the vertical intercept of the line found in part (a)? Give units with your answer and interpret it in terms of the cost of waste collection.

24. The number of species of coastal dune plants in Australia decreases as the latitude, in °S, increases. There are 34 species at 11°S and 26 species at 44°S.[11]

(a) Find a formula for the number, N, of species of coastal dune plants in Australia as a linear function of the latitude, l, in °S.
(b) Give units for and interpret the slope and the vertical intercept of this function.
(c) Graph this function between $l = 11$°S and $l = 44$°S. (Australia lies entirely within these latitudes.)

25. A controversial 1992 Danish study[12] reported that men's average sperm count has decreased from 113 million per milliliter in 1940 to 66 million per milliliter in 1990.

(a) Express the average sperm count, S, as a linear function of the number of years, t, since 1940.
(b) A man's fertility is affected if his sperm count drops below about 20 million per milliliter. If the linear model found in part (a) is accurate, in what year will the average male sperm count fall below this level?

Problems 26–31 concern the maximum heart rate (MHR), which is the maximum number of times a person's heart can safely beat in one minute. If a is age in years, the formulas used to estimate MHR, both in the medical profession and in fitness training, are

$$\text{For females: MHR} = 226 - a \text{ beats/minute,}$$

$$\text{For males: MHR} = 220 - a \text{ beats/minute.}$$

26. Which of the following is the correct statement?

(a) As you age, your maximum heart rate decreases by one beat per year.
(b) As you age, your maximum heart rate decreases by one beat per minute.
(c) As you age, your maximum heart rate decreases by one beat per minute per year.

27. Which of the following is the correct statement for a male and female of the same age?

(a) Their maximum heart rates are the same.
(b) The male's maximum heart rate exceeds the female's.
(c) The female's maximum heart rate exceeds the male's.

28. What can be said about the ages of a male and a female with the same maximum heart rate?

29. Recently[13] it has been suggested that a more accurate predictor of MHR for both males and females is given by

$$\text{MHR} = 208 - 0.7a.$$

(a) At what age do the old and new formulas give the same MHR for females? For males?
(b) Which of the following is true?
  (i) The new formula predicts a higher MHR for young people and a lower MHR for older people than the old formula.
  (ii) The new formula predicts a lower MHR for young people and a higher MHR for older people than the old formula.
(c) When testing for heart disease, doctors ask patients to walk on a treadmill while the speed and incline are gradually increased until their heart rates reach 85 percent of the MHR. For a 65 year old male, what is the difference in beats per minute between the heart rate reached if the old formula is used and the heart rate reached if the new formula is used?

30. Experiments[14] suggest that the female MHR decreases by 12 beats per minute by age 21, and by 19 beats per minute by age 33. Is this consistent with MHR being approximately linear with age?

31. Experiments[15] suggest that the male MHR decreases by 9 beats per minute by age 21, and by 26 beats per minute by age 33. Is this consistent with MHR being approximately linear with age?

---

[10] *The World Almanac and Book of Facts 2005*, p. 309 (New York).
[11] Rosenzweig, M.L., *Species Diversity in Space and Time,* p. 292, (Cambridge: Cambridge University Press, 1995).
[12] "Investigating the Next Silent Spring," *US News and World Report*, p. 50-52, (March 11, 1996).
[13] www.physsportsmed.com/issues/2001/07_01/jul01news.htm, accessed January 4, 2005
[14] www.css.edu/users/tboone2/asep/May2002JEPonline.html, accessed January 4, 2005
[15] www.css.edu/users/tboone2/asep/May2002JEPonline.html, accessed January 4, 2005

1.3 RATES OF CHANGE

In the previous section, we saw that the height of the winning Olympic pole vault increased at an approximately constant rate of 2 inches/year between 1900 and 1912. Similarly, the world record for the mile decreased at an approximately constant rate of 0.4 seconds/year. We now see how to calculate rates of change when they are not constant.

Example 1 Table 1.6 shows the height of the winning pole vault at the Olympics[16] during the 1960s and 1990s. Find the rate of change of the winning height between 1960 and 1968, and between 1992 and 2000. In which of these two periods did the height increase faster than during the period 1900–1912?

Table 1.6 *Winning height in men's Olympic pole vault (approximate)*

Year	1960	1964	1968	...	1992	1996	2000
Height (inches)	185	201	213	...	228	233	232

Solution From 1900 to 1912, the height increased by 2 inches/year. To compare the 1960s and 1990s, we calculate

$$\text{Average rate of change of height 1960 to 1968} = \frac{\text{Change in height}}{\text{Change in time}} = \frac{213 - 185}{1968 - 1960} = 3.5 \text{ inches/year.}$$

$$\text{Average rate of change of height 1992 to 2000} = \frac{\text{Change in height}}{\text{Change in time}} = \frac{232 - 228}{2000 - 1992} = 0.5 \text{ inches/year.}$$

Thus, the height was increasing more quickly during the 1960s than from 1900 to 1912. During the 1990s, the height was increasing more slowly than from 1900 to 1912.

In Example 1, the function does not have a constant rate of change (it is not linear). However, we can compute an *average rate of change* over any interval. The word average is used because the rate of change may vary within the interval. We have the following general formula.

> If y is a function of t, so $y = f(t)$, then
>
> $$\textbf{Average rate of change} \text{ of } y \text{ between } t = a \text{ and } t = b = \frac{\Delta y}{\Delta t} = \frac{f(b) - f(a)}{b - a}.$$
>
> The units of average rate of change of a function are units of y per unit of t.

The average rate of change of a linear function is the slope, and a function is linear if the rate of change is the same on all intervals.

Example 2 Using Figure 1.24, estimate the average rate of change of the number of farms[17] in the US between 1950 and 1970.

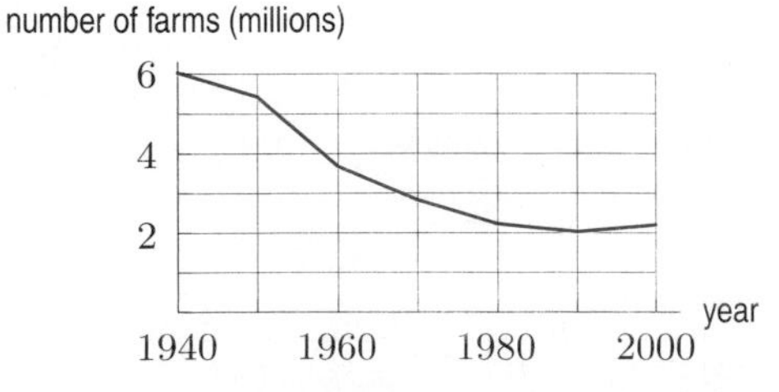

Figure 1.24: Number of farms in the US (in millions)

[16] *The World Almanac and Book of Facts, 2005*, p. 866 (New York).
[17] *The World Almanac and Book of Facts, 2005*, p. 136 (New York).

Solution Figure 1.24 shows that the number, N, of farms in the US was approximately 5.4 million in 1950 and approximately 2.8 million in 1970. If time, t, is in years, we have

$$\text{Average rate of change} = \frac{\Delta N}{\Delta t} = \frac{2.8 - 5.4}{1970 - 1950} = -0.13 \text{ million farms per year.}$$

The average rate of change is negative because the number of farms is decreasing. During this period, the number of farms decreased at an average rate of 0.13 million, or 130,000, per year.

Increasing and Decreasing Functions

Since the rate of change of the height of the winning pole vault is positive, we know that the height is increasing. The rate of change of number of farms is negative, so the number of farms is decreasing. See Figure 1.25. In general:

> A function f is **increasing** if the values of $f(x)$ increase as x increases.
> A function f is **decreasing** if the values of $f(x)$ decrease as x increases.
>
> The graph of an *increasing* function *climbs* as we move from left to right.
> The graph of a *decreasing* function *descends* as we move from left to right.

Figure 1.25: Increasing and decreasing functions

We have looked at how an Olympic record and the number of farms change over time. In the next example, we look at a rate of change with respect to a quantity other than time.

Example 3 High levels of PCB (polychlorinated biphenyl, an industrial pollutant) in the environment affect pelicans' eggs. Table 1.7 shows that as the concentration of PCB in the eggshells increases, the thickness of the eggshell decreases, making the eggs more likely to break.[18]

Find the average rate of change in the thickness of the shell as the PCB concentration changes from 87 ppm to 452 ppm. Give units and explain why your answer is negative.

Table 1.7 *Thickness of pelican eggshells and PCB concentration in the eggshells*

Concentration, c, in parts per million (ppm)	87	147	204	289	356	452
Thickness, h, in millimeters (mm)	0.44	0.39	0.28	0.23	0.22	0.14

Solution Since we are looking for the average rate of change of thickness with respect to change in PCB concentration, we have

$$\text{Average rate of change of thickness} = \frac{\text{Change in the thickness}}{\text{Change in the PCB level}} = \frac{\Delta h}{\Delta c} = \frac{0.14 - 0.44}{452 - 87}$$
$$= -0.00082 \frac{\text{mm}}{\text{ppm}}.$$

The units are thickness units (mm) over PCB concentration units (ppm), or millimeters over parts per million. The average rate of change is negative because the thickness of the eggshell decreases as the PCB concentration increases. The thickness of pelican eggs decreases by an average of 0.00082 mm for every additional part per million of PCB in the eggshell.

[18] Risebrough, R. W., "Effects of environmental pollutants upon animals other than man." *Proceedings of the 6th Berkeley Symposium on Mathematics and Statistics, VI*, p. 443–463, (Berkeley: University of California Press, 1972).

Visualizing Rate of Change

For a function $y = f(x)$, the change in the value of the function between $x = a$ and $x = c$ is $\Delta y = f(c) - f(a)$. Since Δy is a difference of two y-values, it is represented by the vertical distance in Figure 1.26. The average rate of change of f between $x = a$ and $x = c$ is represented by the slope of the line joining the points A and C in Figure 1.27. This line is called the *secant line* between $x = a$ and $x = c$.

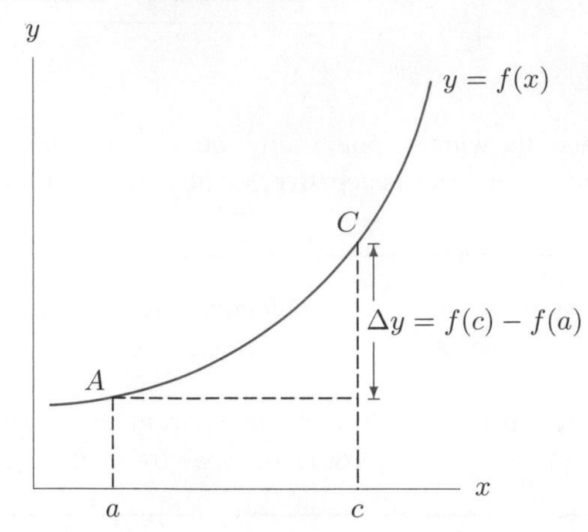

Figure 1.26: The change in a function is represented by a vertical distance

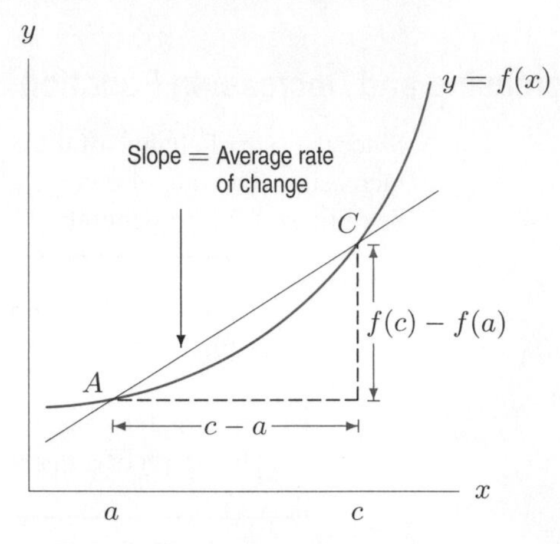

Figure 1.27: The average rate of change is represented by the slope of the line

Example 4

(a) Find the average rate of change of $y = f(x) = \sqrt{x}$ between $x = 1$ and $x = 4$.
(b) Graph $f(x)$ and represent this average rate of change as the slope of a line.
(c) Which is larger, the average rate of change of the function between $x = 1$ and $x = 4$ or the average rate of change between $x = 4$ and $x = 5$? What does this tell us about the graph of the function?

Solution

(a) Since $f(1) = \sqrt{1} = 1$ and $f(4) = \sqrt{4} = 2$, between $x = 1$ and $x = 4$, we have

$$\text{Average rate of change} = \frac{\Delta y}{\Delta x} = \frac{f(4) - f(1)}{4 - 1} = \frac{2 - 1}{3} = \frac{1}{3}.$$

(b) A graph of $f(x) = \sqrt{x}$ is given in Figure 1.28. The average rate of change of f between 1 and 4 is the slope of the secant line between $x = 1$ and $x = 4$.
(c) Since the secant line between $x = 1$ and $x = 4$ is steeper than the secant line between $x = 4$ and $x = 5$, the average rate of change between $x = 1$ and $x = 4$ is larger than it is between $x = 4$ and $x = 5$. The rate of change is decreasing. This tells us that the graph of this function is bending downward.

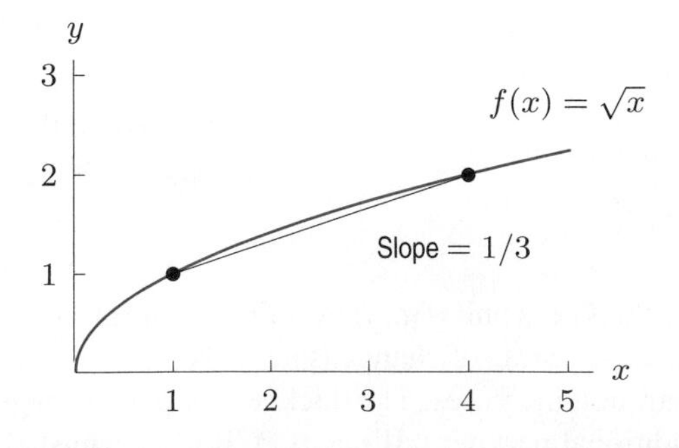

Figure 1.28: Average rate of change = Slope of secant line

Concavity

Figure 1.28 shows a graph that is bending downward because the rate of change is decreasing. The graph in Figure 1.26 bends upward because the rate of change of the function is increasing. We make the following definitions.

> The graph of a function is **concave up** if it bends upward as we move left to right; the graph is **concave down** if it bends downward. (See Figure 1.29.) A line is neither concave up nor concave down.

Figure 1.29: Concavity of a graph

Example 5 Using Figure 1.30, estimate the intervals over which:

(a) The function is increasing; decreasing. (b) The graph is concave up; concave down.

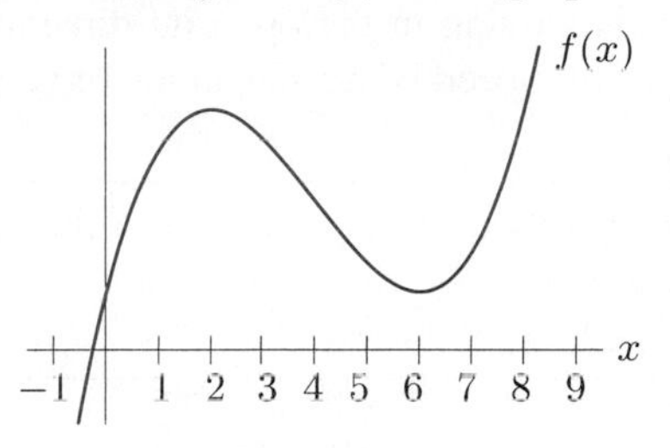

Figure 1.30

Solution

(a) The graph suggests that the function is increasing for $x < 2$ and for $x > 6$. It appears to be decreasing for $2 < x < 6$.

(b) The graph is concave down on the left and concave up on the right. It is difficult to tell exactly where the graph changes concavity, although it appears to be about $x = 4$. Approximately, the graph is concave down for $x < 4$ and concave up for $x > 4$.

Example 6 From the following values of $f(t)$, does f appear to be increasing or decreasing? Do you think its graph is concave up or concave down?

t	0	5	10	15	20	25	30
$f(t)$	12.6	13.1	14.1	16.2	20.0	29.6	42.7

Solution Since the given values of $f(t)$ increase as t increases, f appears to be increasing. As we read from left to right, the change in $f(t)$ starts small and gets larger (for constant change in t), so the graph is climbing faster. Thus, the graph appears to be concave up. Alternatively, plot the points and notice that a curve through these points bends up.

Distance, Velocity, and Speed

A grapefruit is thrown up in the air. The height of the grapefruit above the ground first increases and then decreases. See Table 1.8.

Table 1.8 *Height, y, of the grapefruit above the ground t seconds after it is thrown*

t (sec)	0	1	2	3	4	5	6
y (feet)	6	90	142	162	150	106	30

Example 7 Find the change and average rate of change of the height of the grapefruit during the first 3 seconds. Give units and interpret your answers.

Solution The change in height during the first 3 seconds is $\Delta y = 162 - 6 = 156$ ft. This means that the grapefruit goes up a total of 156 feet during the first 3 seconds. The average rate of change during this 3 second interval is $156/3 = 52$ ft/sec. During the first 3 seconds, the grapefruit is rising at an average rate of 52 ft/sec.

The average rate of change of height with respect to time is *velocity*. You may recognize the units (feet per second) as units of velocity.

$$\text{Average velocity} = \frac{\text{Change in distance}}{\text{Change in time}} = \begin{array}{c}\text{Average rate of change of distance}\\ \text{with respect to time}\end{array}$$

There is a distinction between *velocity* and *speed*. Suppose an object moves along a line. If we pick one direction to be positive, the velocity is positive if the object is moving in that direction and negative if it is moving in the opposite direction. For the grapefruit, upward is positive and downward is negative. Speed is the magnitude of velocity, so it is always positive or zero.

Example 8 Find the average velocity of the grapefruit over the interval $t = 4$ to $t = 6$. Explain the sign of your answer.

Solution Since the height is $y = 150$ feet at $t = 4$ and $y = 30$ feet at $t = 6$, we have

$$\text{Average velocity} = \frac{\text{Change in distance}}{\text{Change in time}} = \frac{\Delta y}{\Delta t} = \frac{30 - 150}{6 - 4} = -60 \text{ ft/sec.}$$

The negative sign means the height is decreasing and the grapefruit is moving downward.

Example 9 A car travels away from home on a straight road. Its distance from home at time t is shown in Figure 1.31. Is the car's average velocity greater during the first hour or during the second hour?

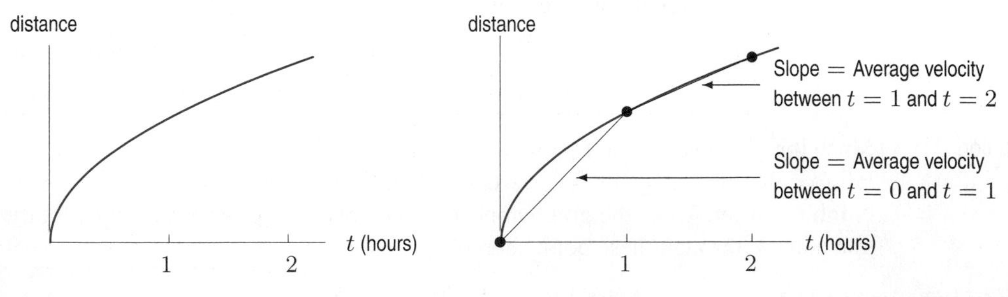

Figure 1.31: Distance of car from home

Figure 1.32: Average velocities of the car

Solution Average velocity is represented by the slope of a secant line. Figure 1.32 shows that the secant line between $t = 0$ and $t = 1$ is steeper than the secant line between $t = 1$ and $t = 2$. Thus, the average velocity is greater during the first hour.

Problems for Section 1.3

In Problems 1–4, decide whether the graph is concave up, concave down, or neither.

1.

2.

3.

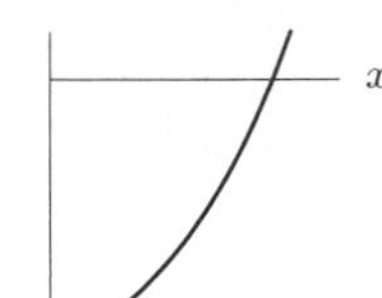

4.

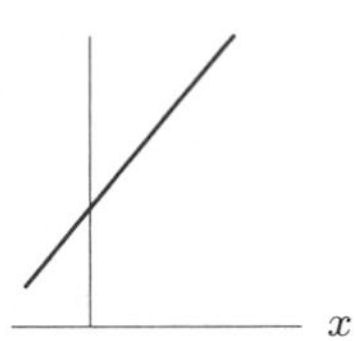

5. Table 1.9 gives values of a function $w = f(t)$. Is this function increasing or decreasing? Is the graph of this function concave up or concave down?

Table 1.9

t	0	4	8	12	16	20	24
w	100	58	32	24	20	18	17

6. Identify the x-intervals on which the function graphed in Figure 1.33 is:

(a) Increasing and concave up
(b) Increasing and concave down
(c) Decreasing and concave up
(d) Decreasing and concave down

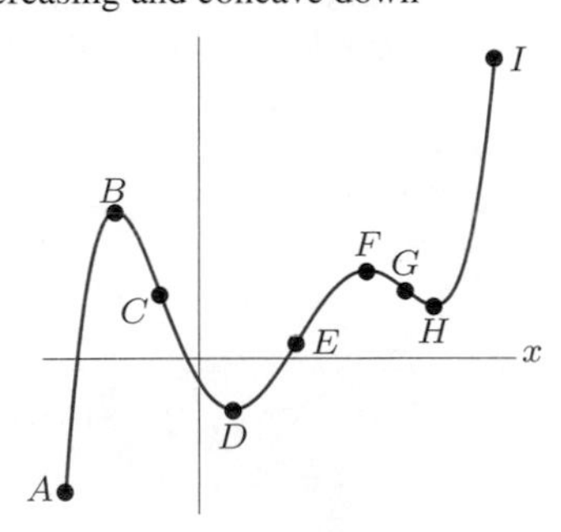

Figure 1.33

7. When a new product is advertised, more and more people try it. However, the rate at which new people try it slows as time goes on.

(a) Graph the total number of people who have tried such a product against time.
(b) What do you know about the concavity of the graph?

8. Graph a function $f(x)$ which is increasing everywhere and concave up for negative x and concave down for positive x.

9. Find the average rate of change of $f(x) = 2x^2$ between $x = 1$ and $x = 3$.

10. When a deposit of \$1000 is made into an account paying 8% interest, compounded annually, the balance, \$$B$, in the account after t years is given by $B = 1000(1.08)^t$. Find the average rate of change in the balance over the interval $t = 0$ to $t = 5$. Give units and interpret your answer in terms of the balance in the account.

11. Table 1.10 shows the production of tobacco in the US.[19]

(a) What is the average rate of change in tobacco production between 1996 and 2003? Give units and interpret your answer in terms of tobacco production.
(b) During this seven-year period, is there any interval during which the average rate of change was positive? If so, when?

Table 1.10 *Tobacco production, in millions of pounds*

Year	1996	1997	1998	1999	2000	2001	2002	2003
Production	1517	1787	1480	1293	1053	991	879	831

12. Do you expect the average rate of change (in units per year) of each of the following to be positive or negative? Explain your reasoning.

(a) Number of acres of rain forest in the world.
(b) Population of the world.
(c) Number of polio cases each year in the US, since 1950.
(d) Height of a sand dune that is being eroded.
(e) Cost of living in the US.

13. Figure 1.34 shows the length, L, in cm, of a sturgeon (a type of fish) as a function of the time, t, in years.[20]

(a) Is the function increasing or decreasing? Is the graph concave up or concave down?
(b) Estimate the average rate of growth of the sturgeon between $t = 5$ and $t = 15$. Give units and interpret your answer in terms of the sturgeon.

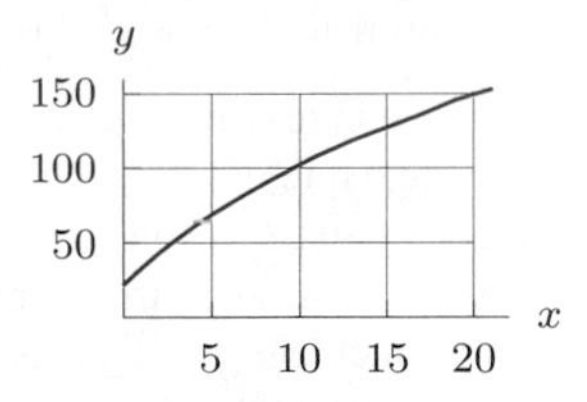

Figure 1.34

14. Table 1.11 shows the total US labor force, L. Find the average rate of change between 1940 and 2000; between 1940 and 1960; between 1980 and 2000. Give units and interpret your answers in terms of the labor force.[21]

Table 1.11 *US labor force, in thousands of workers*

Year	1940	1960	1980	2000
L	47,520	65,778	99,303	136,891

[19] *The World Almanac and Book of Facts 2005*, pp. 138–139 (New York).
[20] Data from von Bertalanffy, L., *General System Theory*, p. 177, (New York: Braziller, 1968).
[21] *The World Almanac and Book of Facts 2005*, p. 144 (New York).

15. The total world marine catch[22] of fish, in metric tons, was 17 million in 1950 and 99 million in 2001. What was the average rate of change in the marine catch during this period? Give units and interpret your answer.

16. Figure 1.35 shows the total value of US imports, in billions of dollars.[23]

(a) Was the value of the exports higher in 1985 or in 2003? Approximately how much higher?
(b) Estimate the average rate of change between 1985 and 2003. Give units and interpret your answer in terms of the value of US imports.

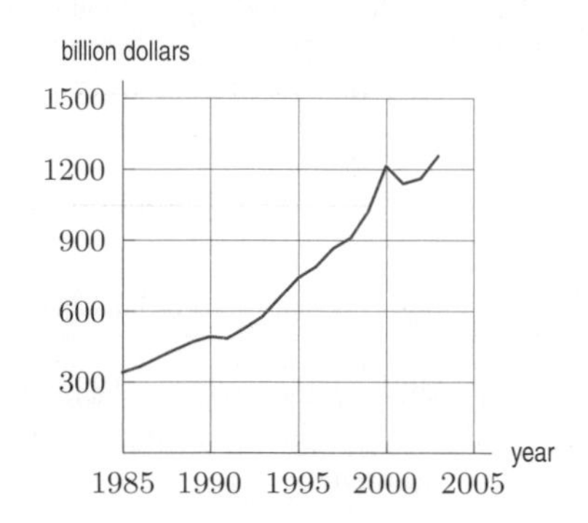

Figure 1.35

17. Table 1.12 gives sales of Pepsico, which operates two major businesses: beverages (including Pepsi) and snack foods.[24]

(a) Find the change in sales between 1999 and 2004.
(b) Find the average rate of change in sales between 1998 and 2004. Give units and interpret your answer.

Table 1.12 *Pepsico sales, in millions of dollars*

Year	1999	2000	2001	2002	2003	2004
Sales	20,367	20,438	23,512	25,112	26,971	29,261

18. Table 1.13 gives the revenues, R, of General Motors, the world's largest auto manufacturer.[25]

(a) Find the change in revenues between 1999 and 2004.
(b) Find the average rate of change in revenues between 1999 and 2004. Give units and interpret your answer.
(c) From 1999 to 2004, were there any one-year intervals during which the average rate of change was negative? If so, which?

Table 1.13 *GM revenues, billions of dollars*

Year	1999	2000	2001	2002	2003	2004
R	176.6	183.3	177.3	177.3	185.5	193.0

19. The number of US households with cable television[26] was 12,168,450 in 1977 and 73,365,880 in 2003. Estimate the average rate of change in the number of US households with cable television during this 26-year period. Give units and interpret your answer.

20. Figure 1.9 on page 5 shows the amount of nicotine $N = f(t)$, in mg, in a person's bloodstream as a function of the time, t, in hours, since the last cigarette.

(a) Is the average rate of change in nicotine level positive or negative? Explain.
(b) Find the average rate of change in the nicotine level between $t = 0$ and $t = 3$. Give units and interpret your answer in terms of nicotine.

21. Table 1.14 shows the concentration, c, of creatinine in the bloodstream of a dog.[27]

(a) Including units, find the average rate at which the concentration is changing between the
(i) 6th and 8th minutes. (ii) 8th and 10th minutes.
(b) Explain the sign and relative magnitudes of your results in terms of creatinine.

Table 1.14

t (minutes)	2	4	6	8	10
c (mg/ml)	0.439	0.383	0.336	0.298	0.266

Problems 22–23 refer to Figure 1.36 which shows the contraction velocity of a muscle as a function of the load it pulls against.

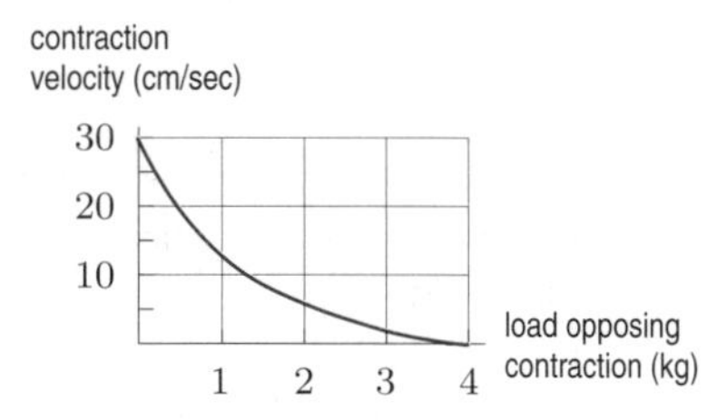

Figure 1.36

22. In terms of the muscle, interpret the

(a) Vertical intercept **(b)** Horizontal intercept

23. **(a)** Find the change in muscle contraction velocity when the load changes from 1 kg to 3 kg. Give units.
(b) Find the average rate of change in the contraction velocity between 1 kg and 3 kg. Give units.

[22] *The World Almanac and Book of Facts 2005*, p. 143 (New York).
[23] www.ita.doc.gov/td/industry/otea/usfth/aggregate/H03t26.pdf, accessed April 19, 2005.
[24] www.pepsico.com, accessed February 20, 2005.
[25] www.gm.com/company/investor_information/earnings/hist_earnings/index.html, accessed February 20, 2005.
[26] *The World Almanac and Book of Facts 2005*, p. 310 (New York).
[27] From Cullen, M.R., *Linear Models in Biology*, (Chichester: Ellis Horwood, 1985).

24. Figure 1.37 shows the age-adjusted death rates from different types of cancer among US males.[28]

(a) Discuss how the death rate has changed for the different types of cancers.
(b) For which type of cancer has the average rate of change between 1930 and 1967 been the largest? Estimate the average rate of change for this cancer type. Interpret your answer.
(c) For which type of cancer has the average rate of change between 1930 and 1967 been the most negative? Estimate the average rate of change for this cancer type. Interpret your answer.

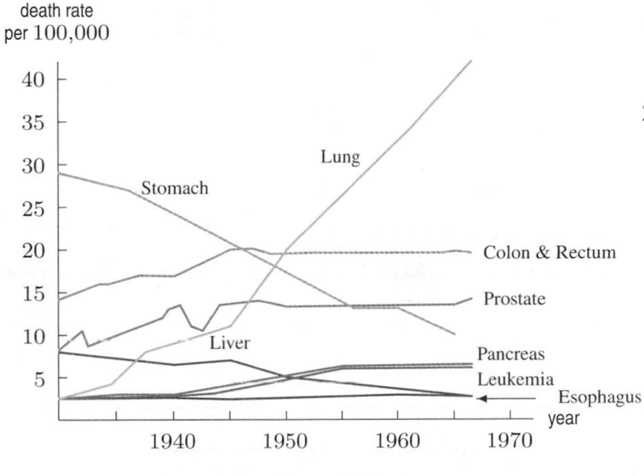

Figure 1.37

25. The volume of water in a pond over a period of 20 weeks is shown in Figure 1.38.

(a) Is the average rate of change of volume positive or negative over the following intervals?
(i) $t = 0$ and $t = 5$ (ii) $t = 0$ and $t = 10$
(iii) $t = 0$ and $t = 15$ (iv) $t = 0$ and $t = 20$
(b) During which of the following time intervals was the average rate of change larger?
(i) $0 \leq t \leq 5$ or $0 \leq t \leq 10$
(ii) $0 \leq t \leq 10$ or $0 \leq t \leq 20$
(c) Estimate the average rate of change between $t = 0$ and $t = 10$. Interpret your answer in terms of water.

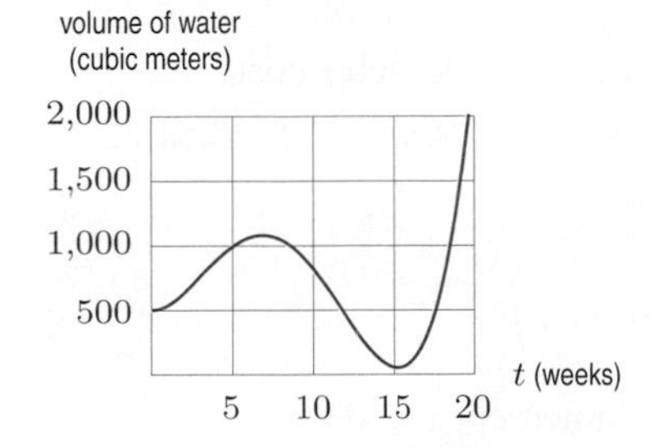

Figure 1.38

26. Table 1.15 gives the sales, S, of Intel Corporation, a leading manufacturer of integrated circuits.[29]

(a) Find the change in sales between 1998 and 2003.
(b) Find the average rate of change in sales between 1998 and 2003. Give units and interpret your answer.
(c) If the average rate of change continues at the same rate as between 2001 and 2003, in which year will the sales first reach 40,000 million dollars?

Table 1.15 *Intel sales, in millions of dollars*

Year	1998	1999	2000	2001	2002	2003
S	26,273	29,389	33,726	26,539	26,764	30,141

27. In an experiment, a lizard is encouraged to run as fast as possible. Figure 1.39 shows the distance run in meters as a function of the time in seconds.[30]

(a) If the lizard were running faster and faster, what would be the concavity of the graph? Does this match what you see?
(b) Estimate the average velocity of the lizard during this 0.8 second experiment.

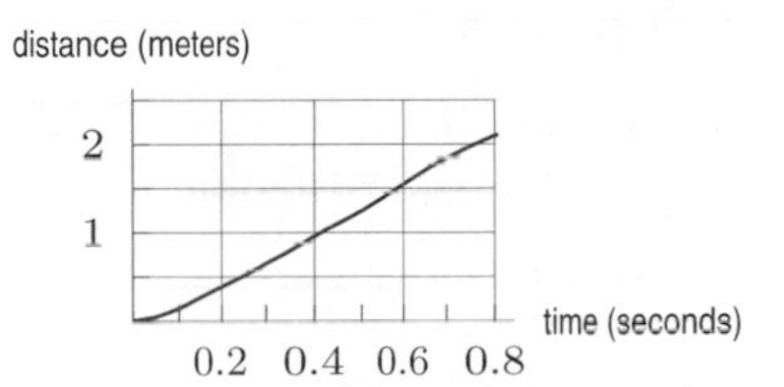

Figure 1.39

28. Values of $F(t)$, $G(t)$, and $H(t)$ are in Table 1.16. Which graph is concave up and which is concave down? Which function is linear?

Table 1.16

t	$F(t)$	$G(t)$	$H(t)$
10	15	15	15
20	22	18	17
30	28	21	20
40	33	24	24
50	37	27	29
60	40	30	35

[28] Abraham M. Lilienfeld, *Foundations of Epidemiology*, p. 67, (New York; Oxford University Press, 1976).
[29] www.intel.com/intel/finance/pastfin/10yrFinancials.xls, accessed April 19, 2005.
[30] Data from Huey, R.B. and Hertz, P.E., "Effects of Body Size and Slope on the Acceleration of a Lizard," *J. Exp. Biol.*, Volume 110, 1984, p. 113-123.

29. Experiments suggest that the male maximum heart rate (the most times a male's heart can safely beat in a minute) decreases by 9 beats per minute during the first 21 years of his life, and by 26 beats per minute during the first 33 years.[31] If you model the maximum heart rate as a function of age, should you use a function that is increasing or decreasing, concave up or concave down?

30. Draw a graph of distance against time with the following properties: The average velocity is always positive and the average velocity for the first half of the trip is less than the average velocity for the second half of the trip.

31. A car starts slowly and then speeds up. Eventually the car slows down and stops. Graph the distance that the car has traveled against time.

32. Figure 1.40 shows the position of an object at time t.

(a) Draw a line on the graph whose slope represents the average velocity between $t = 2$ and $t = 8$.

(b) Is average velocity greater between $t = 0$ and $t = 3$ or between $t = 3$ and $t = 6$?

(c) Is average velocity positive or negative between $t = 6$ and $t = 9$?

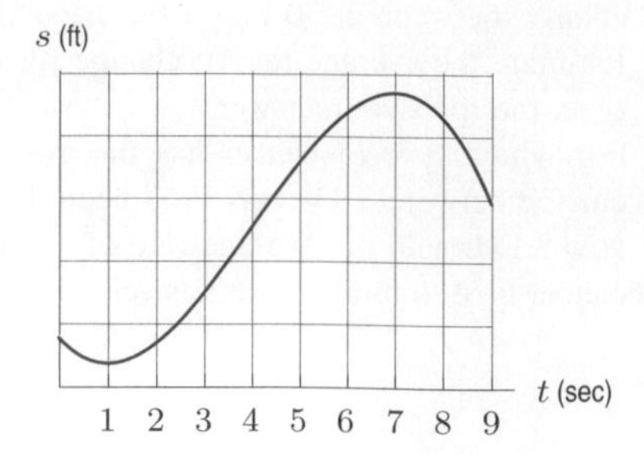

Figure 1.40

1.4 APPLICATIONS OF FUNCTIONS TO ECONOMICS

In this section, we look at some of the functions of interest to decision-makers in a firm or industry.

The Cost Function

The **cost function**, $C(q)$, gives the total cost of producing a quantity q of some good.

What sort of function do you expect C to be? The more goods that are made, the higher the total cost, so C is an increasing function. Costs of production can be separated into two parts: the *fixed costs,* which are incurred even if nothing is produced, and the *variable cost,* which depends on how many units are produced.

An Example: Manufacturing Costs

Let's consider a company that makes radios. The factory and machinery needed to begin production are fixed costs, which are incurred even if no radios are made. The costs of labor and raw materials are variable costs since these quantities depend on how many radios are made. The fixed costs for this company are \$24,000 and the variable costs are \$7 per radio. Then,

$$\begin{aligned}\text{Total costs for the company} &= \text{Fixed cost} + \text{Variable cost}\\ &= 24{,}000 + 7 \cdot \text{Number of radios},\end{aligned}$$

so, if q is the number of radios produced,

$$C(q) = 24{,}000 + 7q.$$

This is the equation of a line with slope 7 and vertical intercept 24,000.

Example 1 Graph the cost function $C(q) = 24{,}000 + 7q$. Label the fixed costs and variable cost per unit.

[31] www.css.edu/users/tboone2/asep/May2002JEPonline.html, accessed January 4, 2005.

Solution The graph of $C(q)$ is the line in Figure 1.41. The fixed costs are represented by the vertical intercept of 24,000. The variable cost per unit is represented by the slope of 7, which is the change in cost corresponding to a unit change in production.

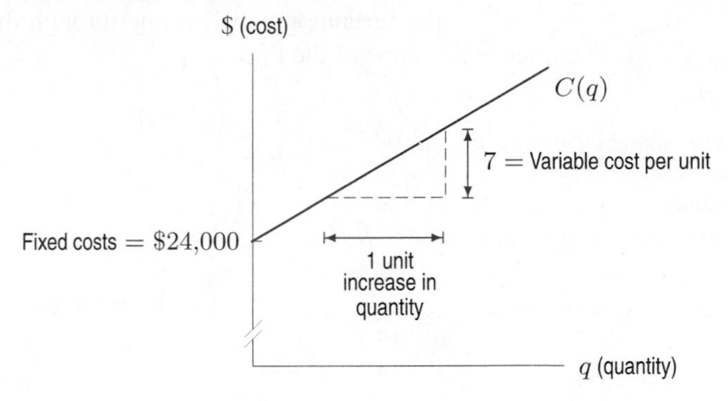

Figure 1.41: Cost function for the radio manufacturer

If $C(q)$ is a linear cost function,

- Fixed costs are represented by the vertical intercept.
- Variable cost per unit is represented by the slope.

Example 2 In each case, draw a graph of a linear cost function satisfying the given conditions:

(a) Fixed costs are large but variable cost per unit is small.
(b) There are no fixed costs but variable cost per unit is high.

Solution

(a) The graph is a line with a large vertical intercept and a small slope. See Figure 1.42.
(b) The graph is a line with a vertical intercept of zero (so the line goes through the origin) and a large positive slope. See Figure 1.43. Figures 1.42 and 1.43 have the same scales.

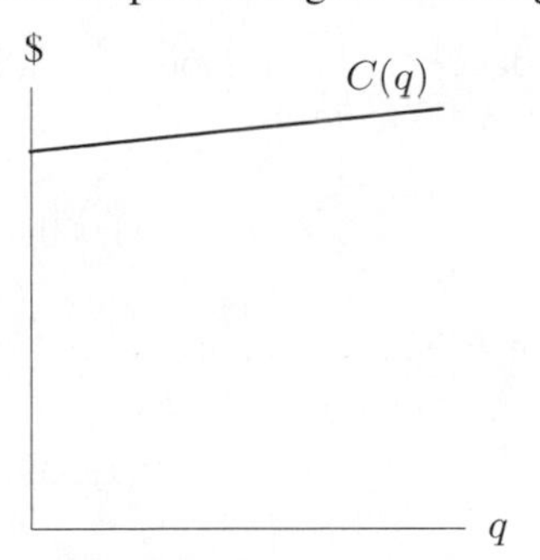

Figure 1.42: Large fixed costs, small variable cost per unit

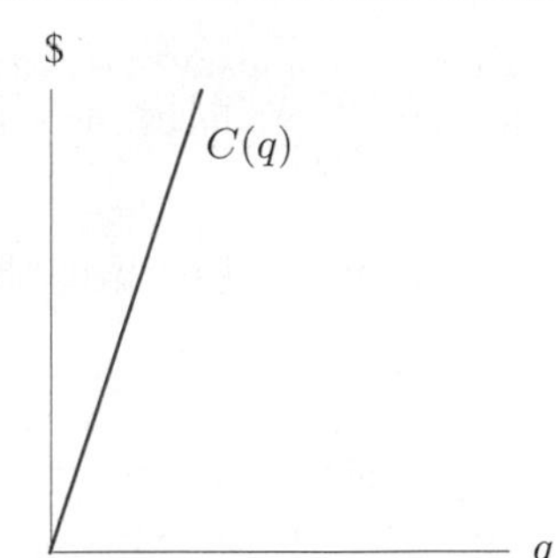

Figure 1.43: No fixed costs, high variable cost per unit

The Revenue Function

The **revenue function**, $R(q)$, gives the total revenue received by a firm from selling a quantity, q, of some good.

If the good sells for a price of p per unit, and the quantity sold is q, then

$$\text{Revenue} = \text{Price} \cdot \text{Quantity}, \quad \text{so} \quad R = pq.$$

If the price does not depend on the quantity sold, so p is a constant, the graph of revenue as a function of q is a line through the origin, with slope equal to the price p.

Example 3 If radios sell for \$15 each, sketch the manufacturer's revenue function. Show the price of a radio on the graph.

Solution Since $R(q) = pq = 15q$, the revenue graph is a line through the origin with a slope of 15. See Figure 1.44. The price is the slope of the line.

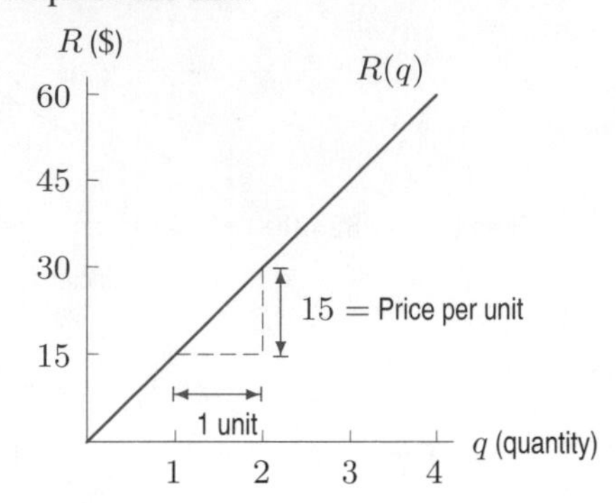

Figure 1.44: Revenue function for the radio manufacturer

Example 4 Graph the cost function $C(q) = 24{,}000 + 7q$ and the revenue function $R(q) = 15q$ on the same axes. For what values of q does the company make money?

Solution The company makes money whenever revenues are greater than costs, so we find the values of q for which the graph of $R(q)$ lies above the graph of $C(q)$. See Figure 1.45.

We find the point at which the graphs of $R(q)$ and $C(q)$ cross:

$$\begin{aligned}
\text{Revenue} &= \text{Cost} \\
15q &= 24{,}000 + 7q \\
8q &= 24{,}000 \\
q &= 3000.
\end{aligned}$$

Thus, the company makes a profit if it produces and sells more than 3000 radios. The company loses money if it produces and sells fewer than 3000 radios.

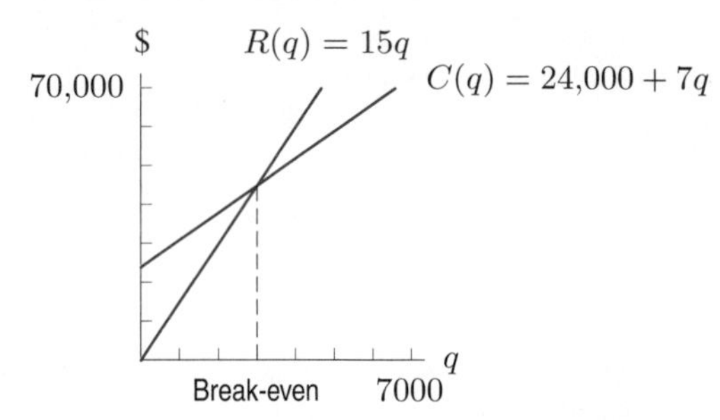

Figure 1.45: Cost and revenue functions for the radio manufacturer: What values of q generate a profit?

The Profit Function

Decisions are often made by considering the profit, usually written[32] as π to distinguish it from the price, p. We have

$$\text{Profit} = \text{Revenue} - \text{Cost} \quad \text{so} \quad \pi = R - C.$$

The *break-even point* for a company is the point where the profit is zero and revenue equals cost.

[32] This π has nothing to do with the area of a circle, and merely stands for the Greek equivalent of the letter "p."

Example 5 Find a formula for the profit function of the radio manufacturer. Graph it, marking the break-even point.

Solution Since $R(q) = 15q$ and $C(q) = 24{,}000 + 7q$, we have

$$\pi(q) = R(q) - C(q) = 15q - (24{,}000 + 7q) = -24{,}000 + 8q.$$

Notice that the negative of the fixed costs is the vertical intercept and the break-even point is the horizontal intercept. See Figure 1.46.

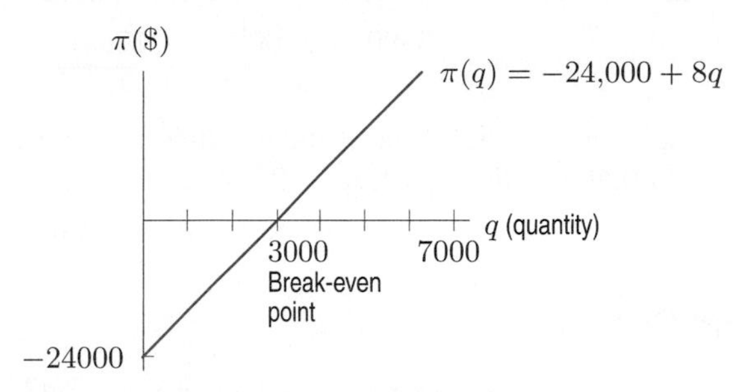

Figure 1.46: Profit for radio manufacturer

Example 6

(a) Using Table 1.17, estimate the break-even point for this company.
(b) Find the company's profit if 1000 units are produced.
(c) What price do you think the company is charging for its product?

Table 1.17 *Company's estimates of cost and revenue for a product*

q	500	600	700	800	900	1000	1100
$C(q)$, in $	5000	5500	6000	6500	7000	7500	8000
$R(q)$, in $	4000	4800	5600	6400	7200	8000	8800

Solution

(a) The break-even point is the value of q for which revenue equals cost. Since revenue is below cost at $q = 800$ and revenue is greater than cost at $q = 900$, the break-even point is between 800 and 900. The values in the table suggest that the break-even point is closer to 800, as the cost and revenue are closer there. A reasonable estimate for the break-even point is $q = 830$.
(b) If the company produces 1000 units, the cost is \$7500 and the revenue is \$8000, so the profit is $8000 - 7500 = 500$ dollars.
(c) From the table, it appears that $R(q) = 8q$. This indicates the company is selling the product for \$8 each.

The Marginal Cost, Marginal Revenue, and Marginal Profit

The terms marginal cost, marginal revenue, and marginal profit are used to mean the rate of change, or slope, of cost, revenue, and profit, respectively. Marginal cost is also the variable cost for one additional unit. The term *marginal* is used because we are looking at how the cost, revenue, or profit change "at the margin," that is, by the addition of one more unit. For example, for the radio manufacturer, the marginal cost is 7 dollars/item (the additional cost of producing one more item is \$7), the marginal revenue is 15 dollars/item (the additional revenue from selling one more item is \$15), and the marginal profit is 8 dollars/item (the additional profit from selling one more item is \$8).

The Depreciation Function

Suppose that the radio manufacturer has a machine that costs \$20,000 and is sold ten years later for \$3000. We say the value of the machine *depreciates* from \$20,000 today to a resale value of \$3000 in ten years. The depreciation formula gives the value, $V(t)$, in dollars, of the machine as a function of the number of years, t, since the machine was purchased. We assume that the value of the machine depreciates linearly.

The value of the machine when it is new ($t = 0$) is \$20,000, so $V(0) = 20{,}000$. The resale value at time $t = 10$ is \$3000, so $V(10) = 3000$. We have

$$\text{Slope} = m = \frac{3000 - 20{,}000}{10 - 0} = \frac{-17{,}000}{10} = -1700 \text{ dollars per year.}$$

This slope tells us that the value of the machine is decreasing at a rate of \$1700 per year. Since $V(0) = 20{,}000$, the vertical intercept is 20,000, so

$$V(t) = 20{,}000 - 1700t \text{ dollars.}$$

Supply and Demand Curves

The quantity, q, of an item that is manufactured and sold, depends on its price, p. As the price increases, manufacturers are usually willing to supply more of the product, whereas the quantity demanded by consumers falls.

> The **supply curve**, for a given item, relates the quantity, q, of the item that manufacturers are willing to make per unit time to the price, p, for which the item can be sold.
> The **demand curve** relates the quantity, q, of an item demanded by consumers per unit time to the price, p, of the item.

Economists often think of the quantities supplied and demanded as functions of price. However, for historical reasons, the economists put price (the independent variable) on the vertical axis and quantity (the dependent variable) on the horizontal axis. (The reason for this state of affairs is that economists originally took price to be the dependent variable and put it on the vertical axis. Later, when the point of view changed, the axes did not.) Thus, typical supply and demand curves look like those shown in Figure 1.47.

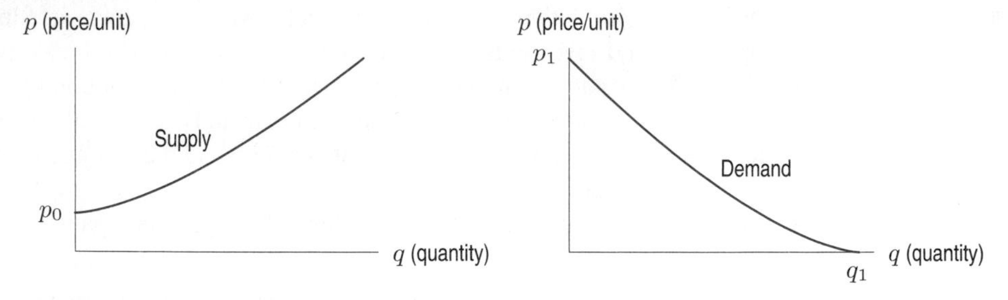

Figure 1.47: Supply and demand curves

Example 7 What is the economic meaning of the prices p_0 and p_1 and the quantity q_1 in Figure 1.47?

Solution The vertical axis corresponds to a quantity of zero. Since the price p_0 is the vertical intercept on the supply curve, p_0 is the price at which the quantity supplied is zero. In other words, for prices below p_0, the suppliers will not produce anything. The price p_1 is the vertical intercept on the demand curve, so it corresponds to the price at which the quantity demanded is zero. In other words, for prices above p_1, consumers buy none of the product.

The horizontal axis corresponds to a price of zero, so the quantity q_1 on the demand curve is the quantity demanded if the price were zero—the quantity that could be given away free.

Equilibrium Price and Quantity

If we plot the supply and demand curves on the same axes, as in Figure 1.48, the graphs cross at the *equilibrium point*. The values p^* and q^* at this point are called the *equilibrium price* and *equilibrium quantity*, respectively. It is assumed that the market naturally settles to this equilibrium point. (See Problem 20.)

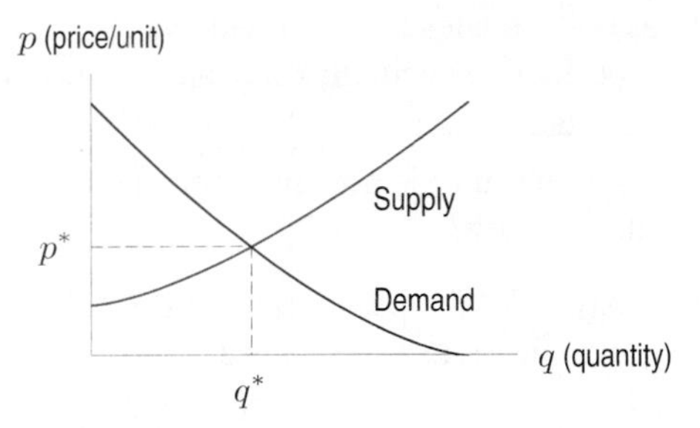

Figure 1.48: The equilibrium price and quantity

Example 8 Find the equilibrium price and quantity if

$$\text{Quantity supplied } = 3p - 50 \quad \text{and} \quad \text{Quantity demanded } = 100 - 2p.$$

Solution To find the equilibrium price and quantity, we find the point at which

$$\begin{aligned} \text{Supply } &= \text{Demand} \\ 3p - 50 &= 100 - 2p \\ 5p &= 150 \\ p &= 30. \end{aligned}$$

The equilibrium price is \$30. To find the equilibrium quantity, we use either the demand curve or the supply curve. At a price of \$30, the quantity produced is $100 - 2 \cdot 30 = 40$ items. The equilibrium quantity is 40 items. In Figure 1.49, the demand and supply curves intersect at $p^* = 30$ and $q^* = 40$.

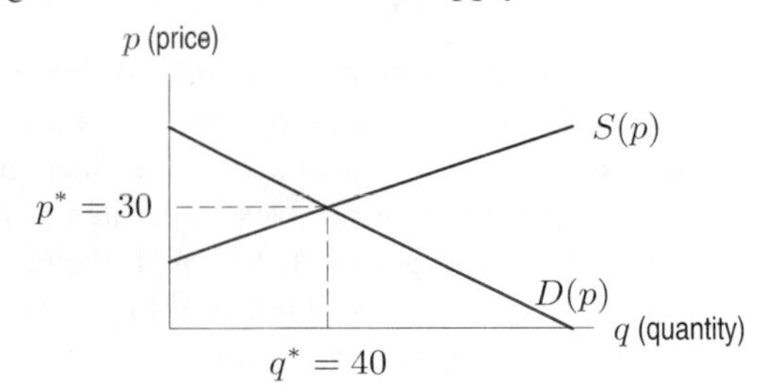

Figure 1.49: Equilibrium: $p^* = 30$, $q^* = 40$

Notice that the supply equation shows how suppliers react to the amount of money they obtain for each unit sold. In Example 8,

$$\text{Quantity supplied } = 3\,(\text{Amount per unit received by suppliers}) - 50.$$

In this case, the amount received is the price, p, so the equation of the supply curve is

$$q = 3p - 50.$$

Similarly, the demand equation shows how consumers react to the amount paid per unit. In Example 8,

$$\text{Quantity demanded } = 100 - 2\,(\text{Amount per unit paid by consumers})\,.$$

Here again, the amount paid is the price, p, so the equation of the demand curve is

$$q = 100 - 2p.$$

The Effect of Taxes on Equilibrium

What effect do taxes have on the equilibrium price and quantity for a product? We distinguish between two types of taxes.[33] A *specific tax* is a fixed amount per unit of a product sold regardless of the selling price. This is the case with such items as gasoline, alcohol, and cigarettes. A specific tax is usually imposed on the producer. A *sales tax* is a fixed percentage of the selling price. Many cities and states collect sales tax on a wide variety of items. A sales tax is usually imposed on the consumer. We consider a specific tax now; a sales tax is considered in Problems 36 and 37.

Example 9 A specific tax of $5 per unit is now imposed upon suppliers in Example 8. What are the new equilibrium price and quantity?

Solution The consumers pay p dollars per unit, but the suppliers receive only $p-5$ dollars per unit because \$5 goes to the government as taxes. Since

$$\text{Quantity supplied } = 3(\text{Amount per unit received by suppliers}) - 50,$$

the new supply equation is

$$\text{Quantity supplied } = 3(p-5) - 50 = 3p - 65;$$

the demand equation is unchanged:

$$\text{Quantity demanded } = 100 - 2p.$$

At the equilibrium price, we have

$$\begin{aligned} \text{Demand} &= \text{Supply} \\ 100 - 2p &= 3p - 65 \\ 165 &= 5p \\ p &= 33. \end{aligned}$$

The equilibrium price is \$33. The equilibrium quantity is 34 units, since the quantity demanded is $q = 100 - 2 \cdot 33 = 34$.

In Example 8, the equilibrium price was \$30; with the imposition of a \$5 tax in Example 9, the equilibrium price is \$33. Thus the equilibrium price increases by \$3 as a result of the tax. Notice that this is less than the amount of the tax. The consumer ends up paying \$3 more than if the tax did not exist. However the government receives \$5 per item. The producer pays the other \$2 of the tax, retaining \$28 of the price paid per item. Although the tax was imposed on the producer, some of the tax is passed on to the consumer in terms of higher prices. The tax has increased the price and reduced the number of items sold. See Figure 1.50. Notice that the taxes have the effect of moving the supply curve up by \$5 because suppliers have to be paid \$5 more to produce the same quantity.

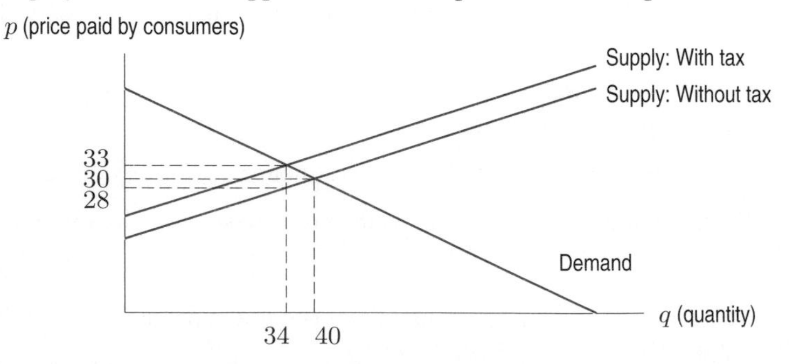

Figure 1.50: Specific tax shifts the supply curve, altering the equilibrium price and quantity

[33] Adapted from Barry Bressler, *A Unified Approach to Mathematical Economics*, p. 81–88, (New York: Harper & Row, 1975).

A Budget Constraint

An ongoing debate in the federal government concerns the allocation of money between defense and social programs. In general, the more that is spent on defense, the less that is available for social programs, and vice versa. Let's simplify the example to guns and butter. Assuming a constant budget, we show that the relationship between the number of guns and the quantity of butter is linear. Suppose that there is \$12,000 to be spent and that it is to be divided between guns, costing \$400 each, and butter, costing \$2000 a ton. Suppose the number of guns bought is g, and the number of tons of butter is b. Then the amount of money spent on guns is \$400g, and the amount spent on butter is \$2000b. Assuming all the money is spent,

$$\text{Amount spent on guns} + \text{Amount spent on butter} = \$12{,}000$$

or

$$400g + 2000b = 12{,}000.$$

Thus, dividing both sides by 400,

$$g + 5b = 30.$$

This equation is the budget constraint. Since the budget constraint can be written as

$$g = 30 - 5b,$$

the graph of the budget constraint is a line. See Figure 1.51.

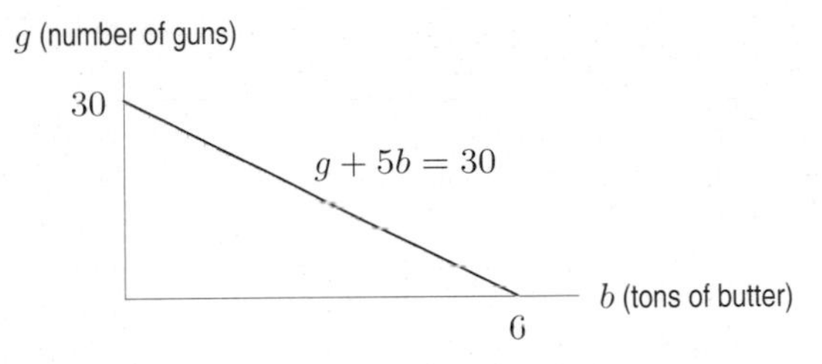

Figure 1.51: Budget constraint

Problems for Section 1.4

1. The table shows the cost of manufacturing various quantities of an item and the revenue obtained from their sale.

Quantity	0	10	20	30	40	50	60	70	80
Cost (\$)	120	400	600	780	1000	1320	1800	2500	3400
Revenue (\$)	0	300	600	900	1200	1500	1800	2100	2400

(a) What range of production levels appears to be profitable?

(b) Calculate the profit or loss for each of the quantities shown. Estimate the most profitable production level.

2. A company has cost and revenue functions, in dollars, given by $C(q) = 6000 + 10q$ and $R(q) = 12q$.

(a) Find the cost and revenue if the company produces 500 units. Does the company make a profit? What about 5000 units?

(b) Find the break-even point and illustrate it graphically.

3. A company has cost function $C(q) = 4000 + 2q$ dollars and revenue function $R(q) = 10q$ dollars.

(a) What are the fixed costs for the company?

(b) What is the variable cost per unit?

(c) What price is the company charging for its product?

(d) Graph $C(q)$ and $R(q)$ on the same axes and label the break-even point, q_0. Explain how you know the company makes a profit if the quantity produced is greater than q_0.

(e) Find the break-even point q_0.

4. Values of a linear cost function are in Table 1.18. What are the fixed costs and the variable cost per unit (the marginal cost)? Find a formula for the cost function.

Table 1.18

q	0	5	10	15	20
$C(q)$	5000	5020	5040	5060	5080

5. **(a)** Estimate the fixed costs and the variable cost per unit for the cost function in Figure 1.52.
(b) Estimate $C(10)$ and interpret it in terms of cost.

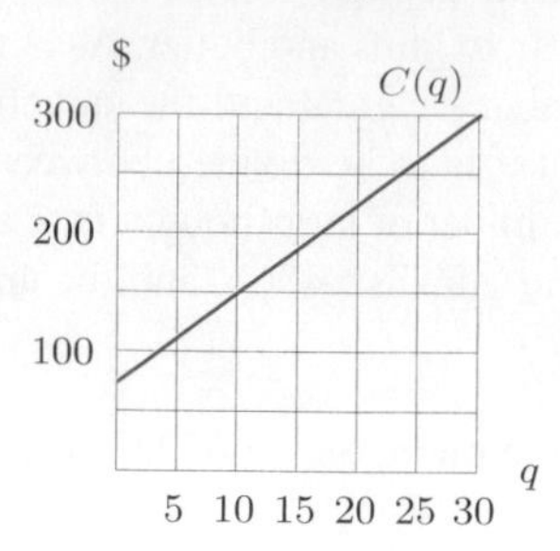

Figure 1.52

6. **(a)** What are the fixed costs and the variable cost per unit for the cost function in Figure 1.53?
(b) Explain what $C(100) = 2500$ tells you about costs.

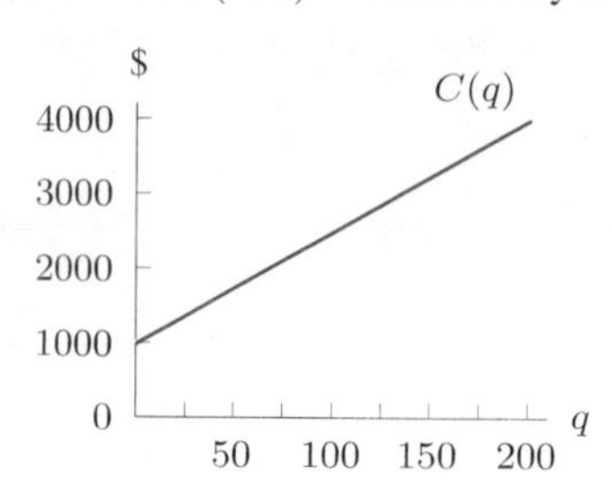

Figure 1.53

7. Figure 1.54 shows cost and revenue for a company.

(a) Approximately what quantity does this company have to produce to make a profit?
(b) Estimate the profit generated by 600 units.

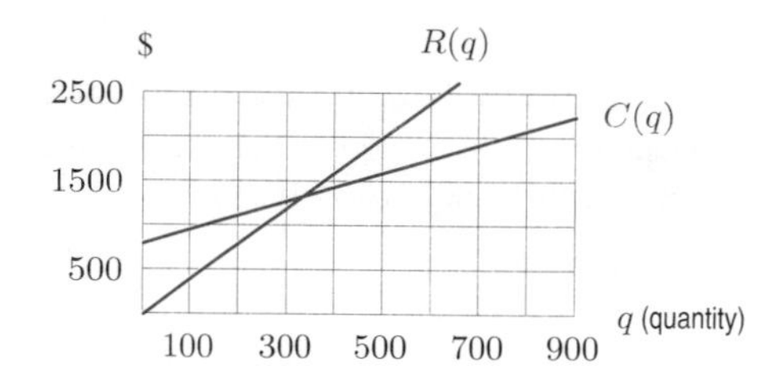

Figure 1.54

8. In Figure 1.55, which shows the cost and revenue functions for a product, label each of the following:

(a) Fixed costs **(b)** Break-even quantity
(c) Quantities at which the company:
(i) Makes a profit (ii) Loses money

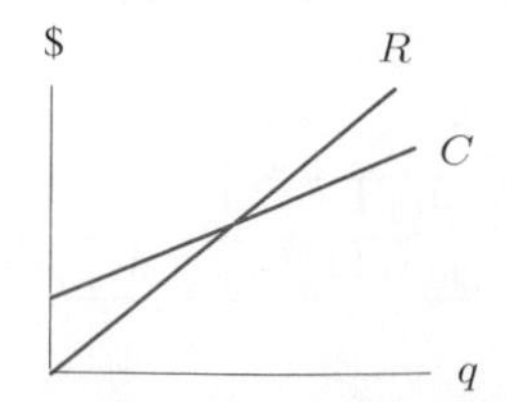

Figure 1.55

9. A movie theater has fixed costs of $5000 per day and variable costs averaging $2 per customer. The theater charges $7 per ticket.

(a) How many customers per day does the theater need in order to make a profit?
(b) Find the cost and revenue functions and graph them on the same axes. Mark the break-even point.

10. A company producing jigsaw puzzles has fixed costs of $6000 and variable costs of $2 per puzzle. The company sells the puzzles for $5 each.

(a) Find formulas for the cost function, the revenue function, and the profit function.
(b) Sketch a graph of $R(q)$ and $C(q)$ on the same axes. What is the break-even point, q_0, for the company?

11. The Quick-Food company provides a college meal-service plan. Quick-Food has fixed costs of $350,000 per term and variable costs of $400 per student. Quick-Food charges $800 per student per term. How many students must sign up with the Quick-Food plan in order for the company to make a profit?

12. Production costs for manufacturing running shoes consist of a fixed overhead of $650,000 plus variable costs of $20 per pair of shoes. Each pair of shoes sells for $70.

(a) Find the total cost, $C(q)$, the total revenue, $R(q)$, and the total profit, $\pi(q)$, as a function of the number of pairs of shoes produced, q.
(b) Find the marginal cost, marginal revenue, and marginal profit.
(c) How many pairs of shoes must be produced and sold for the company to make a profit?

13. **(a)** Give an example of a possible company where the fixed costs are zero (or very small).
(b) Give an example of a possible company where the variable cost per unit is zero (or very small).

14. For tax purposes, you may have to report the value of your assets, such as cars or refrigerators. The value you report drops with time. "Straight-line depreciation" assumes that the value is a linear function of time. If a $950 refrigerator depreciates completely in seven years, find a formula for its value as a function of time.

15. A $15,000 robot depreciates linearly to zero in 10 years.

(a) Find a formula for its value as a function of time.
(b) How much is the robot worth three years after it is purchased?

16. A $50,000 tractor has a resale value of $10,000 twenty years after it was purchased. Assume that the value of the tractor depreciates linearly from the time of purchase.

(a) Find a formula for the value of the tractor as a function of the time since it was purchased.
(b) Graph the value of the tractor against time.
(c) Find the horizontal and vertical intercepts, give units, and interpret them.

17. Suppose that $q = f(p)$ is the demand curve for a product, where p is the selling price in dollars and q is the quantity sold at that price.

(a) What does the statement $f(12) = 60$ tell you about demand for this product?
(b) Do you expect this function to be increasing or decreasing? Why?

18. The cost, C, in millions of dollars of producing q items is given by $C = 5.7 + 0.002q$. Interpret the 5.7 and the 0.002 in terms of production. Give units.

19. The demand curve for a quantity q of a product is $q = 5500 - 100p$ where p is price in dollars. Interpret the 5500 and the 100 in terms of demand. Give units.

20. Figure 1.56 shows supply and demand for a product.

(a) What is the equilibrium price for this product? At this price, what quantity is produced?
(b) Choose a price above the equilibrium price—for example, $p = 12$. At this price, how many items are suppliers willing to produce? How many items do consumers want to buy? Use your answers to these questions to explain why, if prices are above the equilibrium price, the market tends to push prices lower (toward the equilibrium).
(c) Now choose a price below the equilibrium price—for example, $p = 8$. At this price, how many items are suppliers willing to produce? How many items do consumers want to buy? Use your answers to these questions to explain why, if prices are below the equilibrium price, the market tends to push prices higher (toward the equilibrium).

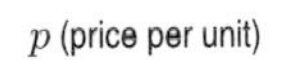

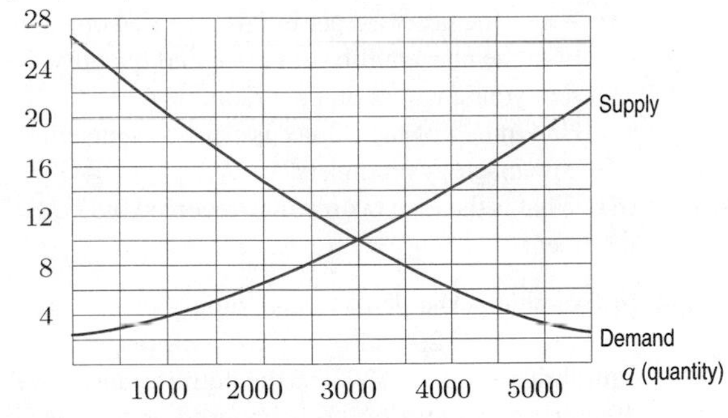

Figure 1.56

21. A company produces and sells shirts. The fixed costs are \$7000 and the variable costs are \$5 per shirt.

(a) Shirts are sold for \$12 each. Find cost and revenue as functions of the quantity of shirts, q.
(b) The company is considering changing the selling price of the shirts. Demand is $q = 2000 - 40p$, where p is price in dollars and q is the number of shirts. What quantity is sold at the current price of \$12? What profit is realized at this price?
(c) Use the demand equation to write cost and revenue as functions of the price, p. Then write profit as a function of price.
(d) Graph profit against price. Find the price that maximizes profits. What is this profit?

22. One of the graphs in Figure 1.57 is a supply curve, and the other is a demand curve. Which is which? Explain how you made your decision using what you know about the effect of price on supply and demand.

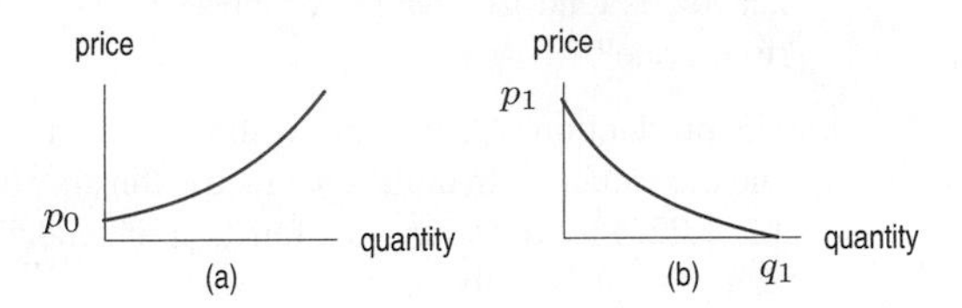

Figure 1.57

23. One of Tables 1.19 and 1.20 represents a supply curve; the other represents a demand curve.

(a) Which table represents which curve? Why?
(b) At a price of \$155, approximately how many items would consumers purchase?
(c) At a price of \$155, approximately how many items would manufacturers supply?
(d) Will the market push prices higher or lower than \$155?
(e) What would the price have to be if you wanted consumers to buy at least 20 items?
(f) What would the price have to be if you wanted manufacturers to supply at least 20 items?

Table 1.19

p (\$/unit)	182	167	153	143	133	125	118
q (quantity)	5	10	15	20	25	30	35

Table 1.20

p (\$/unit)	6	35	66	110	166	235	316
q (quantity)	5	10	15	20	25	30	35

24. A demand curve is given by $75p + 50q = 300$, where p is the price of the product, in dollars, and q is the quantity demanded at that price. Find p- and q-intercepts and interpret them in terms of consumer demand.

25. Table 1.21 gives data for the linear demand curve for a product, where p is the price of the product and q is the quantity sold every month at that price. Find formulas for the following functions. Interpret their slopes in terms of demand.

(a) q as a function of p. **(b)** p as a function of q.

Table 1.21

p (dollars)	16	18	20	22	24
q (tons)	500	460	420	380	340

26. The demand curve for a product is given by $q = 120{,}000 - 500p$ and the supply curve is given by $q = 1000p$ for $0 \le q \le 120{,}000$, where price is in dollars.

(a) At a price of \$100, what quantity are consumers willing to buy and what quantity are producers willing to supply? Will the market push prices up or down?

(b) Find the equilibrium price and quantity. Does your answer to part (a) support the observation that market forces tend to push prices closer to the equilibrium price?

27. The US production, Q, of zinc in thousands of metric tons and the value, P, in dollars per metric ton are given[34] in Table 1.22. Plot the value as a function of production. Sketch a possible supply curve.

Table 1.22 *US zinc production*

Year	1998	1999	2000	2001	2002
Q	368	372	371	311	295
P	1130	1180	1230	969	852

28. When the price, p, charged for a boat tour was \$25, the average number of passengers per week, N, was 500. When the price was reduced to \$20, the average number of passengers per week increased to 650. Find a formula for the demand curve, assuming that it is linear.

29. You have a budget of \$1000 for the year to cover your books and social outings. Books cost (on average) \$40 each and social outings cost (on average) \$10 each. Let b denote the number of books purchased per year and s denote the number of social outings in a year.

(a) What is the equation of your budget constraint?

(b) Graph the budget constraint. (It does not matter which variable you put on which axis.)

(c) Find the vertical and horizontal intercepts, and give a financial interpretation for each.

30. A company has a total budget of \$500,000 and spends this budget on raw materials and personnel. The company uses m units of raw materials, at a cost of \$100 per unit, and hires r employees, at a cost of \$25,000 each.

(a) What is the equation of the company's budget constraint?

(b) Solve for m as a function of r.

(c) Solve for r as a function of m.

31. Linear supply and demand curves are shown in Figure 1.58, with price on the vertical axis.

(a) Label the equilibrium price p_0 and the equilibrium quantity q_0 on the axes.

(b) Explain the effect on equilibrium price and quantity if the slope of the supply curve increases. Illustrate your answer graphically.

(c) Explain the effect on equilibrium price and quantity if the slope of the demand curve becomes more negative. Illustrate your answer graphically.

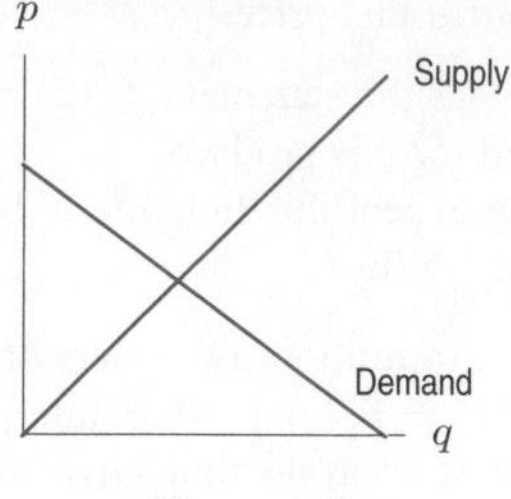

Figure 1.58

32. A demand curve has equation $q = 100 - 5p$, where p is price in dollars. A \$2 tax is imposed on consumers. Find the equation of the new demand curve. Sketch both curves.

33. A supply curve has equation $q = 4p - 20$, where p is price in dollars. A \$2 tax is imposed on suppliers. Find the equation of the new supply curve. Sketch both curves.

34. A tax of \$8 per unit is imposed on the supplier of an item. The original supply curve is $q = 0.5p - 25$ and the demand curve is $q = 165 - 0.5q$, where p is price in dollars. Find the equilibrium price and quantity before and after the tax is imposed.

35. The demand and supply curves for a product are given in terms of price, p, by

$$q = 2500 - 20p \quad \text{and} \quad q = 10p - 500.$$

(a) Find the equilibrium price and quantity. Represent your answers on a graph.

(b) A specific tax of \$6 per unit is imposed on suppliers. Find the new equilibrium price and quantity. Represent your answers on the graph.

(c) How much of the \$6 tax is paid by consumers and how much by producers?

(d) What is the total tax revenue received by the government?

36. In Example 8, the demand and supply curves are given by $q = 100 - 2p$ and $q = 3p - 50$, respectively; the equilibrium price is \$30 and the equilibrium quantity is 40 units. A sales tax of 5% is imposed on the consumer.

(a) Find the equation of the new demand and supply curves.

(b) Find the new equilibrium price and quantity.

(c) How much is paid in taxes on each unit? How much of this is paid by the consumer and how much by the producer?

(d) How much tax does the government collect?

37. Answer the questions in Problem 36, assuming that the 5% sales tax is imposed on the supplier instead of the consumer.

[34] http://minerals.usgs.gov/minerals/pubs/of01-006/zinc.xls, accessed April 23, 2005.

1.5 EXPONENTIAL FUNCTIONS

The function $f(x) = 2^x$, where the power is variable, is an *exponential function*. The number 2 is called the base. Exponential functions of the form $f(x) = a^x$, where a is a positive constant, are used to represent many phenomena in the natural and social sciences.

Population Growth

Table 1.23 contains approximate population data for the McAllen, Texas, metropolitan area.[35] McAllen was in the news on June 26, 1954, when Hurricane Alice dropped 27 inches of rainfall in 24 hours—one of the highest ever one-day rainfalls in the US. To see how the population of McAllen is growing, we look at the yearly increases in population in the third column of Table 1.23. If the population had been growing linearly, all the numbers in the third column would be the same. But populations usually grow faster as they get bigger, so it is not surprising that the numbers in the third column increase.

Table 1.23 *Population of McAllen, Texas*

Year	Population (thousands)	Increase in population (thousands)
2000	570	
		21
2001	591	
		22
2002	613	
		23
2003	636	

Suppose we divide each year's population by the previous year's population. We get, approximately,

$$\frac{\text{Population in 2001}}{\text{Population in 2000}} = \frac{591 \text{ thousand}}{570 \text{ thousand}} \approx 1.037,$$
$$\frac{\text{Population in 2002}}{\text{Population in 2001}} = \frac{613 \text{ thousand}}{591 \text{ thousand}} \approx 1.037,$$
$$\frac{\text{Population in 2003}}{\text{Population in 2002}} = \frac{636 \text{ thousand}}{613 \text{ thousand}} \approx 1.037.$$

The fact that all calculations are near 1.037 shows the population grew by about 3.7% between 2000 and 2001, between 2001 and 2002, and between 2002 and 2003. Whenever we have a constant percent increase (here 3.7%), we have *exponential growth*. If t is the number of years since 2000 and population is in thousands,

$$\text{When } t = 0, \quad \text{population} = 570 = 570(1.037)^0.$$
$$\text{When } t = 1, \quad \text{population} = 591 = 570(1.037)^1.$$
$$\text{When } t = 2, \quad \text{population} = 613 = 591(1.037) = 570(1.037)^2.$$
$$\text{When } t = 3, \quad \text{population} = 636 = 613(1.037) = 570(1.037)^3.$$

So P, the population in thousands t years after 2000, is given by

$$P = 570(1.037)^t \text{ thousand.}$$

Since the variable t is in the exponent, this is an exponential function. The base, 1.037, represents the factor by which the population grows each year and is called the *growth factor*. Assuming that the formula holds for 50 years, the population graph has the shape in Figure 1.59. The population is growing, so the function is increasing. Since the population grows faster as time passes, the graph is concave up. This behavior is typical of an exponential function. Even exponential functions that climb slowly at first, such as this one, eventually climb extremely quickly.

[35] *Statistical Abstracts of the US 2004–2005*, Table 24.

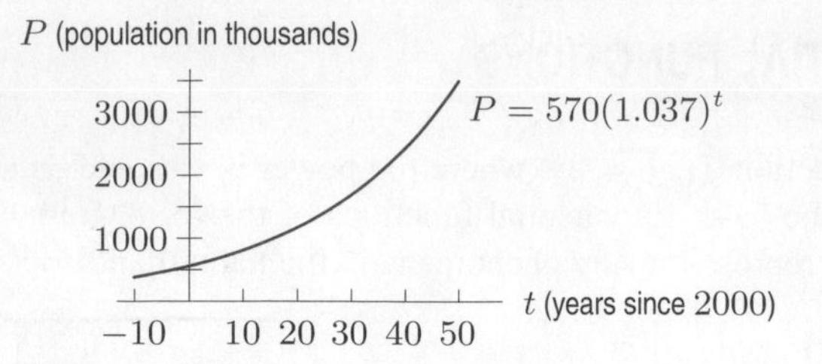

Figure 1.59: Population of McAllen, Texas (estimated): Exponential growth

Elimination of a Drug from the Body

Now we look at a quantity that is decreasing instead of increasing. When a patient is given medication, the drug enters the bloodstream. The rate at which the drug is metabolized and eliminated depends on the particular drug. For the antibiotic ampicillin, approximately 40% of the drug is eliminated every hour. A typical dose of ampicillin is 250 mg. Suppose $Q = f(t)$, where Q is the quantity of ampicillin, in mg, in the bloodstream at time t hours since the drug was given. At $t = 0$, we have $Q = 250$. Since the quantity remaining at the end of each hour is 60% of the quantity remaining the hour before, we have

$$
\begin{aligned}
f(0) &= 250\\
f(1) &= 250(0.6)\\
f(2) &= 250(0.6)(0.6) = 250(0.6)^2\\
f(3) &= 250(0.6)^2(0.6) = 250(0.6)^3.
\end{aligned}
$$

So, after t hours,

$$Q = f(t) = 250(0.6)^t.$$

This function is called an *exponential decay* function. As t increases, the function values get arbitrarily close to zero. The t-axis is a *horizontal asymptote* for this function.

Notice the way the values in Table 1.24 are decreasing. Each additional hour a smaller quantity of drug is removed than the previous hour (100 mg the first hour, 60 mg the second, and so on). This is because as time passes, there is less of the drug in the body to be removed. Thus, the graph in Figure 1.60 bends upward. Compare this to the exponential growth in Figure 1.59, where each step upward is larger than the previous one. Notice that both graphs are concave up.

Table 1.24 *Value of decay function*

t (hours)	Q (mg)
0	250
1	150
2	90
3	54
4	32.4
5	19.4

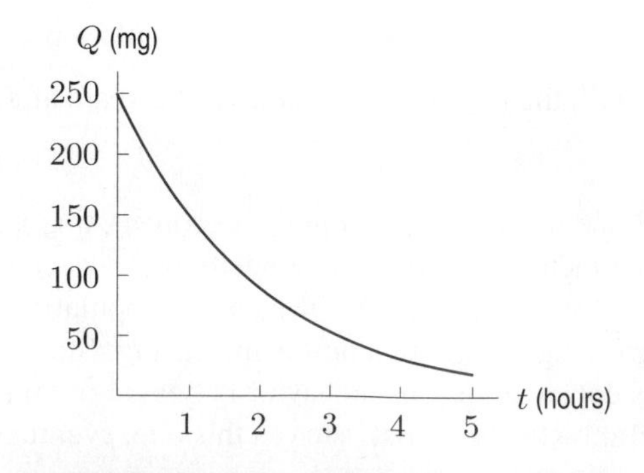

Figure 1.60: Drug elimination: Exponential decay

The General Exponential Function

Exponential growth is often described in terms of percent growth rates. The population of McAllen is growing at 3.7% per year, so it increases by a factor of $a = 1 + 0.037 = 1.037$ every year. Similarly, 40% of the ampicillin is removed every hour, so the quantity remaining decays by a factor of $a = 1 - 0.40 = 0.6$ each hour. We have the following general formulas.

> We say that P is an **exponential function** of t with base a if
>
> $$P = P_0 a^t,$$
>
> where P_0 is the initial quantity (when $t = 0$) and a is the factor by which P changes when t increases by 1. If $a > 1$, we have **exponential growth**; if $0 < a < 1$, we have **exponential decay**. The factor a is given by
>
> $$a = 1 + r$$
>
> where r is the decimal representation of the percent rate of change; r may be positive (for growth) or negative (for decay).

The largest possible domain for the exponential function is all real numbers,[36] provided $a > 0$.

Comparison Between Linear and Exponential Functions

Every exponential function changes at a constant percent, or *relative*, rate. For example, the population of McAllen increased approximately 3.7% per year. Every linear function changes at a constant (absolute) rate. For example, the Olympic pole vault record increased by 2 inches per year.

> A **linear** function has constant absolute rate of change.
> An **exponential** function has constant relative (or percent) rate of change.

Example 1 The amount of adrenaline in the body can change rapidly. Suppose the initial amount is 15 mg. Find a formula for A, the amount in mg, at a time t minutes later if A is:

(a) Increasing by 0.4 mg per minute.
(b) Decreasing by 0.4 mg per minute.
(c) Increasing by 3% per minute.
(d) Decreasing by 3% per minute.

Solution

(a) This is a linear function with initial quantity 15 and slope 0.4, so

$$A = 15 + 0.4t.$$

(b) This is a linear function with initial quantity 15 and slope -0.4, so

$$A = 15 - 0.4t.$$

(c) This is an exponential function with initial quantity 15 and base $1 + 0.03 = 1.03$, so

$$A = 15(1.03)^t.$$

(d) This is an exponential function with initial quantity 15 and base $1 - 0.03 = 0.97$, so

$$A = 15(0.97)^t.$$

[36]The reason we do not want $a \leq 0$ is that, for example, we cannot define $a^{1/2}$ if $a < 0$. Also, we do not usually have $a = 1$, since $P = P_0 a^t = P_0 1^t = P_0$ is then a constant function.

Example 2 Sales[37] at Borders Books and Music stores increased from \$2503 million in 1997 to \$3699 million in 2003. Assuming that sales have been increasing exponentially, find an equation of the form $P = P_0a^t$, where P is Borders sales in millions of dollars and t is the number of years since 1997. What is the percent growth rate?

Solution We know that $P = 2503$ when $t = 0$, so $P_0 = 2503$. To find a, we use the fact that $P = 3699$ when $t = 6$. Substituting gives

$$P = P_0a^t$$
$$3699 = 2503a^6.$$

Dividing both sides by 2503, we get

$$\frac{3699}{2503} = a^6$$
$$1.478 = a^6.$$

Taking the sixth root of both sides gives

$$a = (1.478)^{1/6} = 1.07.$$

Since $a = 1.07$, the equation for Borders sales as a function of the number of years since 1997 is

$$P = 2503(1.07)^t.$$

During this period, sales increased by 7% per year.

> **Recognizing Data from an Exponential Function**: Values of t and P in a table could come from an exponential function $P = P_0a^t$ if ratios of P values are constant for equally spaced t values.

Example 3 Which of the following tables of values could correspond to an exponential function, a linear function, or neither? For those which could correspond to an exponential or linear function, find a formula for the function.

(a)

x	$f(x)$
0	16
1	24
2	36
3	54
4	81

(b)

x	$g(x)$
0	14
1	20
2	24
3	29
4	35

(c)

x	$h(x)$
0	5.3
1	6.5
2	7.7
3	8.9
4	10.1

Solution (a) We see that f cannot be a linear function, since $f(x)$ increases by different amounts ($24-16 = 8$ and $36 - 24 = 12$) as x increases by one. Could f be an exponential function? We look at the ratios of successive $f(x)$ values:

$$\frac{24}{16} = 1.5 \qquad \frac{36}{24} = 1.5 \qquad \frac{54}{36} = 1.5 \qquad \frac{81}{54} = 1.5.$$

[37] http://phx.corporate-ir.net/phoenix.zhtml?c=65380&p=irol-annualreports, accessed April 14, 2005.

Since the ratios are all equal to 1.5, this table of values could correspond to an exponential function with a base of 1.5. Since $f(0) = 16$, a formula for $f(x)$ is

$$f(x) = 16(1.5)^x.$$

Check by substituting $x = 0, 1, 2, 3, 4$ into this formula; you get the values given for $f(x)$.

(b) As x increases by one, $g(x)$ increases by 6 (from 14 to 20), then 4 (from 20 to 24), so g is not linear. We check to see if g could be exponential:

$$\frac{20}{14} = 1.43 \quad \text{and} \quad \frac{24}{20} = 1.2.$$

Since these ratios (1.43 and 1.2) are different, g is not exponential.

(c) For h, notice that as x increases by one, the value of $h(x)$ increases by 1.2 each time. So h could be a linear function with a slope of 1.2. Since $h(0) = 5.3$, a formula for $h(x)$ is

$$h(x) = 5.3 + 1.2x.$$

The Family of Exponential Functions and the Number e

The formula $P = P_0a^t$ gives a family of exponential functions with parameters P_0 (the initial quantity) and a (the base). The base tells us whether the function is increasing ($a > 1$) or decreasing ($0 < a < 1$). Since a is the factor by which P changes when t is increased by 1, large values of a mean fast growth; values of a near 0 mean fast decay. (See Figures 1.61 and 1.62.) All members of the family $P = P_0a^t$ are concave up if $P_0 > 0$.

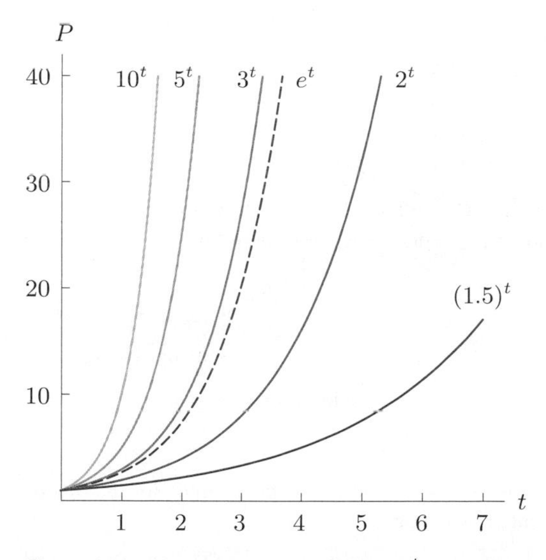

Figure 1.61: Exponential growth: $P = a^t$, for $a > 1$

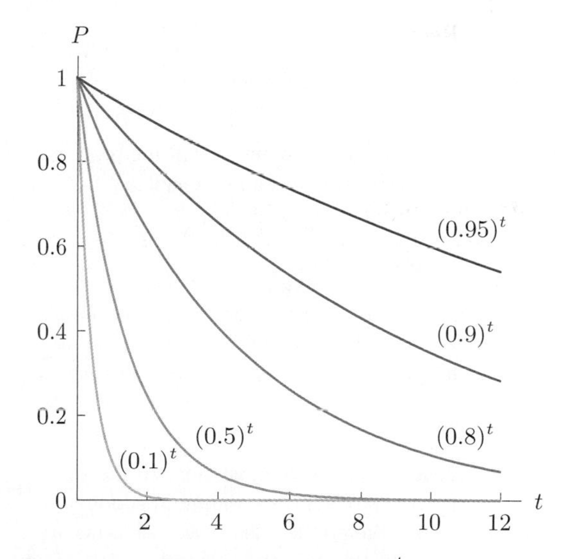

Figure 1.62: Exponential decay: $P = a^t$, for $0 < a < 1$

In practice the most commonly used base is the number $e = 2.71828\ldots$. The fact that most calculators have an e^x button is an indication of how important e is. Since e is between 2 and 3, the graph of $y = e^t$ in Figure 1.61 is between the graphs of $y = 2^t$ and $y = 3^t$.

The base e is used so often that it is called the natural base. At first glance, this is somewhat mysterious: What could be natural about using 2.71828... as a base? The full answer to this question must wait until Chapter 3, where you will see that many calculus formulas come out more neatly when e is used as the base.

Problems for Section 1.5

1. The following functions give the populations of four towns with time t in years.

(i) $P = 600(1.12)^t$ (ii) $P = 1{,}000(1.03)^t$
(iii) $P = 200(1.08)^t$ (iv) $P = 900(0.90)^t$

(a) Which town has the largest percent growth rate? What is the percent growth rate?
(b) Which town has the largest initial population? What is that initial population?
(c) Are any of the towns decreasing in size? If so, which one(s)?

2. Each of the following functions gives the amount of a substance present at time t. In each case, give the amount present initially (at $t = 0$), state whether the function represents exponential growth or decay, and give the percent growth or decay rate.

(a) $A = 100(1.07)^t$ **(b)** $A = 5.3(1.054)^t$
(c) $A = 3500(0.93)^t$ **(d)** $A = 12(0.88)^t$

3. A town has a population of 1000 people at time $t = 0$. In each of the following cases, write a formula for the population, P, of the town as a function of year t.

(a) The population increases by 50 people a year.
(b) The population increases by 5% a year.

4. The gross domestic product, G, of Switzerland was 240 billion dollars in 2003. Give a formula for G (in billions of dollars) t years after 2003 if G increases by
(a) 3% per year **(b)** 8 billion dollars per year

5. A product costs \$80 today. How much will the product cost in t days if the price is reduced by

(a) \$4 a day **(b)** 5% a day

6. An air-freshener starts with 30 grams and evaporates. In each of the following cases, write a formula for the quantity, Q grams, of air-freshener remaining t days after the start and sketch a graph of the function. The decrease is:

(a) 2 grams a day **(b)** 12% a day

7. The world's economy has been expanding. In the year 2001, the gross world product (total output in goods and services) was 32.4 trillion dollars and was increasing at 3.6% a year.[38] Assume this growth rate continues.

(a) Find a formula for the gross world product, W (in trillions of dollars), as a function of t, the number of years since the year 2001.
(b) What is the predicted gross world product in 2010?
(c) Graph W as a function of t.
(d) Use the graph to estimate when the gross world product will pass 50 trillion dollars.

8. A 50 mg dose of quinine is given to a patient to prevent malaria. Quinine leaves the body at a rate of 6% per hour.

(a) Find a formula for the amount, A (in mg), of quinine in the body t hours after the dose is given.
(b) How much quinine is in the body after 24 hours?
(c) Graph A as a function of t.
(d) Use the graph to estimate when 5 mg of quinine remains.

9. World population is approximately $P = 6.4(1.0126)^t$, with P in billions and t in years since 2004.

(a) What is the yearly percent rate of growth of the world population?
(b) What was the world population in 2004? What does this model predict for the world population in 2010?
(c) Use part (b) to find the average rate of change of the world population between 2004 and 2010.

10. Figure 1.63 shows graphs of several cities' populations against time. Match each of the following descriptions to a graph and write a description to match each of the remaining graphs.

(a) The population increased at 5% per year.
(b) The population increased at 8% per year.
(c) The population increased by 5000 people per year.
(d) The population was stable.

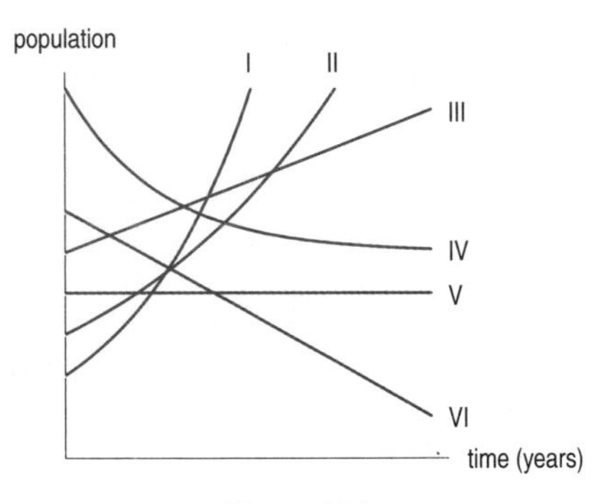

Figure 1.63

For Problems 11–12, find a possible formula for the function represented by the data.

11.

x	0	1	2	3
$f(x)$	4.30	6.02	8.43	11.80

12.

t	0	1	2	3
$g(t)$	5.50	4.40	3.52	2.82

[38] www.eia.doe.gov/oiaf/ieo/pdf/appb1_b8.pdf, Table B 3, accessed April 10, 2005.

Give a possible formula for the functions in Problems 13–14.

13.

14.

15. The number of passengers using a railway fell from 190,205 to 174,989 during a 5-year period. Find the annual percentage decrease over this period.

16. The company that produces Cliffs Notes (abridged versions of classic literature) was started in 1958 with \$4000 and sold in 1998 for \$14,000,000. Find the annual percent increase in the value of this company, over the 40 years.

17. **(a)** Which (if any) of the functions in the following table could be linear? Find formulas for those functions.
(b) Which (if any) of these functions could be exponential? Find formulas for those functions.

x	$f(x)$	$g(x)$	$h(x)$
-2	12	16	37
-1	17	24	34
0	20	36	31
1	21	54	28
2	18	81	25

18. Determine whether each of the following tables of values could correspond to a linear function, an exponential function, or neither. For each table of values that could correspond to a linear or an exponential function, find a formula for the function.

(a)

x	$f(x)$
0	10.5
1	12.7
2	18.9
3	36.7

(b)

t	$s(t)$
-1	50.2
0	30.12
1	18.072
2	10.8432

(c)

u	$g(u)$
0	27
2	24
4	21
6	18

19. The worldwide carbon dioxide emission[39], C, from consumption of fossil fuels was 6.03 billion tons in 1995 and 6.69 billion tons in 2002. Find a formula for the emission C in t years after 1995 if:

(a) C is a linear function of t. What is the absolute yearly rate of increase in carbon dioxide emission?
(b) C is an exponential function of t. What is the relative yearly rate of increase in carbon dioxide emission?

20. During the 1980s, Costa Rica had the highest deforestation rate in the world, at 2.9% per year. (This is the rate at which land covered by forests is shrinking.) Assuming the rate continues, what percent of the land in Costa Rica covered by forests in 1980 will be forested in 2015?

21. Aircrafts require longer takeoff distances, called takeoff rolls, at high altitude airports because of diminished air density. The table shows how the takeoff roll for a certain light airplane depends on the airport elevation. (Takeoff rolls are also strongly influenced by air temperature; the data shown assume a temperature of 0° C.) Determine a formula for this particular aircraft that gives the takeoff roll as an exponential function of airport elevation.

Elevation (ft)	Sea level	1000	2000	3000	4000
Takeoff roll (ft)	670	734	805	882	967

22. **(a)** Make a table of values for $y = e^x$ using $x = 0, 1, 2, 3$.
(b) Plot the points found in part (a). Does the graph look like an exponential growth or decay function?
(c) Make a table of values for $y = e^{-x}$ using $x = 0, 1, 2, 3$.
(d) Plot the points found in part (c). Does the graph look like an exponential growth or decay function?

23. Graph $y = 100e^{-0.4x}$. Describe what you see.

24. **(a)** Niki invested \$10,000 in the stock market. The investment was a loser, declining in value 10% per year each year for 10 years. How much was the investment worth after 10 years?
(b) After 10 years, the stock began to gain value at 10% per year. After how long will the investment regain its initial value (\$10,000)?

25. A photocopy machine can reduce copies to 80% of their original size. By copying an already reduced copy, further reductions can be made.

(a) If a page is reduced to 80%, what percent enlargement is needed to return it to its original size?
(b) Estimate the number of times in succession that a page must be copied to make the final copy less than 15% of the size of the original.

26. Match the functions $h(s)$, $f(s)$, and $g(s)$, whose values are in Table 1.25, with the formulas

$$y = a(1.1)^s, \quad y = b(1.05)^s, \quad y = c(1.03)^s,$$

assuming a, b, and c are constants. Note that the function values have been rounded to two decimal places.

Table 1.25

s	$h(s)$	s	$f(s)$	s	$g(s)$
2	1.06	1	2.20	3	3.47
3	1.09	2	2.42	4	3.65
4	1.13	3	2.66	5	3.83
5	1.16	4	2.93	6	4.02
6	1.19	5	3.22	7	4.22

[39] *Statistical Abstracts of the United States 2004–2005*, Table 1333.

27. The 2004 US presidential debates questioned whether the minimum wage has kept pace with inflation. Decide the question using the following information:[40] In 1938, the minimum wage was 25¢; in 2004, it was \$5.15. During the same period, inflation averaged 4.3%.

28. Whooping cough was thought to have been almost wiped out by vaccinations. It is now known that the vaccination wears off, leading to an increase in the number of cases, w, from 1248 in 1981 to 18,957 in 2004.

(a) With t in years since 1980, find an exponential function that fits this data.

(b) What does your answer to part (a) give as the average annual percent growth rate of the number of cases?

(c) On May 4, 2005, the *Arizona Daily Star* reported (correctly) that the number of cases had more than doubled between 2000 and 2004. Does your model confirm this report? Explain.

1.6 THE NATURAL LOGARITHM

In Section 1.5, we projected the population of McAllen, Texas (in thousands), by the function

$$P = f(t) = 570(1.037)^t,$$

where t is the number of years since 2000. Now suppose that instead of calculating the population at time t, we ask when the population will reach 900,000. We want to find the value of t for which

$$900 = f(t) = 570(1.037)^t.$$

We use logarithms to solve for a variable in an exponent.

Definition and Properties of the Natural Logarithm

We define the natural logarithm of x, written $\ln x$, as follows:

The **natural logarithm** of x, written $\ln x$, is the power of e needed to get x. In other words,

$$\ln x = c \quad \text{means} \quad e^c = x.$$

The natural logarithm is sometimes written $\log_e x$.

For example, $\ln e^3 = 3$ since 3 is the power of e needed to give e^3. Similarly, $\ln(1/e) = \ln e^{-1} = -1$. A calculator gives $\ln 5 = 1.6094$, because $e^{1.6094} = 5$. However if we try to find $\ln(-7)$ on a calculator, we get an error message because e to any power is never negative or 0. In general

$\ln x$ is not defined if x is negative or 0.

To work with logarithms, we use the following properties:

Properties of the Natural Logarithm

1. $\ln(AB) = \ln A + \ln B$
2. $\ln\left(\dfrac{A}{B}\right) = \ln A - \ln B$
3. $\ln(A^p) = p \ln A$
4. $\ln e^x = x$
5. $e^{\ln x} = x$

In addition, $\ln 1 = 0$ because $e^0 = 1$, and $\ln e = 1$ because $e^1 = e$.

[40] http://www.dol.gov/esa/minwage/chart.htm#5

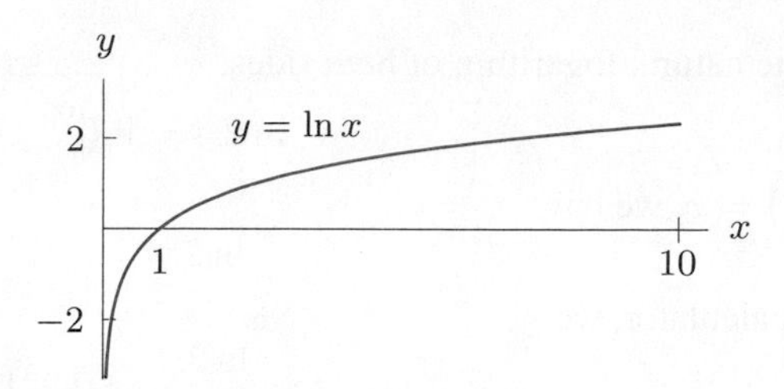

Figure 1.64: The natural logarithm function climbs very slowly

Using the LN button on a calculator, we get the graph of $f(x) = \ln x$ in Figure 1.64. Observe that, for large x, the graph of $y = \ln x$ climbs very slowly as x increases. The x-intercept is $x = 1$, since $\ln 1 = 0$. For $x > 1$, the value of $\ln x$ is positive; for $0 < x < 1$, the value of $\ln x$ is negative.

Solving Equations Using Logarithms

Natural logs can be used to solve for unknown exponents.

Example 1 Find t such that $3^t = 10$.

Solution First, notice that we expect t to be between 2 and 3, because $3^2 = 9$ and $3^3 = 27$. To find t exactly, we take the natural logarithm of both sides and solve for t:

$$\ln(3^t) = \ln 10.$$

The third property of logarithms tells us that $\ln(3^t) = t\ln 3$, so we have

$$t\ln 3 = \ln 10$$
$$t = \frac{\ln 10}{\ln 3}.$$

Using a calculator to find the natural logs gives

$$t = 2.096.$$

Example 2 We return to the question of when the population of McAllen reaches 900 thousand. To get an answer, we solve $900 = 570(1.037)^t$ for t, using logs.

Solution Dividing both sides of the equation by 570, we get

$$\frac{900}{570} = (1.037)^t.$$

Now take natural logs of both sides:

$$\ln\left(\frac{900}{570}\right) = \ln(1.037^t).$$

Using the fact that $\ln(1.037^t) = t\ln 1.037$, we get

$$\ln\left(\frac{900}{570}\right) = t\ln(1.037).$$

Solving this equation using a calculator to find the logs, we get

$$t = \frac{\ln(900/570)}{\ln(1.037)} = 12.57 \text{ years}.$$

Since $t = 0$ in 2000, this value of t corresponds to the year 2012.

Example 3 Find t such that $12 = 5e^{3t}$.

Solution It is easiest to begin by isolating the exponential, so we divide both sides of the equation by 5:

$$2.4 = e^{3t}.$$

Now take the natural logarithm of both sides:

$$\ln 2.4 = \ln(e^{3t}).$$

Since $\ln(e^x) = x$, we have

$$\ln 2.4 = 3t,$$

so, using a calculator, we get

$$t = \frac{\ln 2.4}{3} = 0.2918.$$

Exponential Functions with Base e

An exponential function with base a has formula

$$P = P_0 a^t.$$

For any positive number a, we can write $a = e^k$ where $k = \ln a$. Thus, the exponential function can be rewritten as

$$P = P_0 a^t = P_0 (e^k)^t = P_0 e^{kt}.$$

If $a > 1$, then k is positive, and if $0 < a < 1$, then k is negative. We conclude:

> Writing $a = e^k$, so $k = \ln a$, any exponential function can be written in two forms
>
> $$P = P_0 a^t \quad \text{or} \quad P = P_0 e^{kt}.$$
>
> - If $a > 1$, we have exponential growth; if $0 < a < 1$, we have exponential decay.
> - If $k > 0$, we have exponential growth; if $k < 0$, we have exponential decay.
> - k is called the *continuous* growth or decay rate.

The word continuous in continuous growth rate is used in the same way to describe continuous compounding of interest earned on money. See the Focus on Modeling Section on page 78.

Example 4 (a) Convert the function $P = 1000e^{0.05t}$ to the form $P = P_0 a^t$.
(b) Convert the function $P = 500(1.06)^t$ to the form $P = P_0 e^{kt}$.

Solution (a) Since $P = 1000e^{0.05t}$, we have $P_0 = 1000$. We want to find a so that

$$1000a^t = 1000e^{0.05t} = 1000(e^{0.05})^t.$$

We take $a = e^{0.05} = 1.0513$, so the following two functions give the same values:

$$P = 1000e^{0.05t} \quad \text{and} \quad P = 1000(1.0513)^t.$$

So a continuous growth rate of 5% is equivalent to a growth rate of 5.13% per unit time.

(b) We have $P_0 = 500$ and we want to find k with

$$500(1.06)^t = 500(e^k)^t,$$

so we take

$$1.06 = e^k$$
$$k = \ln(1.06) = 0.0583.$$

The following two functions give the same values:

$$P = 500(1.06)^t \quad \text{and} \quad P = 500e^{0.0583t}.$$

So a growth rate of 6% per unit time is equivalent to a continuous growth rate of 5.83%.

Example 5 Sketch graphs of $P = e^{0.5t}$, a continuous growth rate of 50%, and $Q = 5e^{-0.2t}$, a continuous decay rate of 20%.

Solution The graph of $P = e^{0.5t}$ is in Figure 1.65. Notice that the graph is the same shape as the previous exponential growth curves: increasing and concave up. The graph of $Q = 5e^{-0.2t}$ is in Figure 1.66; it has the same shape as other exponential decay functions.

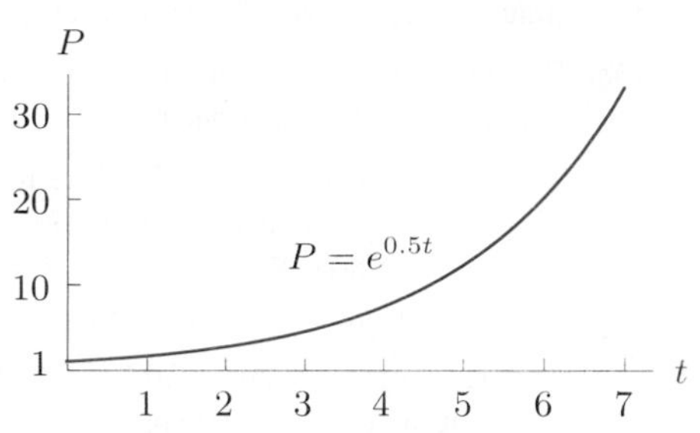

Figure 1.65: Continuous exponential growth function

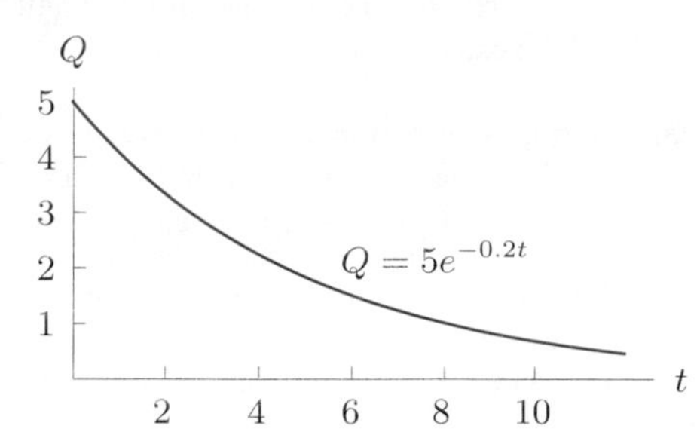

Figure 1.66: Continuous exponential decay function

Problems for Section 1.6

For Problems 1–16, solve for t using natural logarithms.

1. $5^t = 7$

2. $10 = 2^t$

3. $2 = (1.02)^t$

4. $130 = 10^t$

5. $50 = 10 \cdot 3^t$

6. $100 = 25(1.5)^t$

7. $a = b^t$

8. $10 = e^t$

9. $5 = 2e^t$

10. $e^{3t} = 100$

11. $10 = 6e^{0.5t}$

12. $40 = 100e^{-0.03t}$

13. $B = Pe^{rt}$

14. $2P = Pe^{0.3t}$

15. $5e^{3t} = 8e^{2t}$

16. $7 \cdot 3^t = 5 \cdot 2^t$

The functions in Problems 17–20 represent exponential growth or decay. What is the initial quantity? What is the growth rate? State if the growth rate is continuous.

17. $P = 5(1.07)^t$

18. $P = 7.7(0.92)^t$

19. $P = 15e^{-0.06t}$

20. $P = 3.2e^{0.03t}$

21. A city's population is 1000 and growing at 5% a year.

(a) Find a formula for the population at time t years from now assuming that the 5% per year is an:

(i) Annual rate (ii) Continuous annual rate

(b) In each case in part (a), estimate the population of the city in 10 years.

22. The following formulas give the populations of four different towns, A, B, C, and D, with t in years from now.

$$P_A = 600e^{0.08t} \qquad P_B = 1000e^{-0.02t}$$
$$P_C = 1200e^{0.03t} \qquad P_D = 900e^{0.12t}$$

(a) Which town is growing fastest (that is, has the largest percentage growth rate)?

(b) Which town is the largest now?

(c) Are any of the towns decreasing in size? If so, which one(s)?

Write the functions in Problems 23–26 in the form $P = P_0 a^t$. Which represent exponential growth and which represent exponential decay?

23. $P = 15e^{0.25t}$

24. $P = 2e^{-0.5t}$

25. $P = P_0 e^{0.2t}$

26. $P = 7e^{-\pi t}$

In Problems 27–30, put the functions in the form $P = P_0 e^{kt}$.

27. $P = 15(1.5)^t$

28. $P = 10(1.7)^t$

29. $P = 174(0.9)^t$

30. $P = 4(0.55)^t$

31. A fishery stocks a pond with 1000 young trout. The number of trout t years later is given by $P(t) = 1000e^{-0.5t}$.

(a) How many trout are left after six months? After 1 year?
(b) Find $P(3)$ and interpret it in terms of trout.
(c) At what time are there 100 trout left?
(d) Graph the number of trout against time, and describe how the population is changing. What might be causing this?

32. During a recession a firm's revenue declines continuously so that the revenue, R (measured in millions of dollars), in t years time is given by $R = 5e^{-0.15t}$.

(a) Calculate the current revenue and the revenue in two years' time.
(b) After how many years will the revenue decline to \$2.7 million?

33. (a) What is the continuous percent growth rate for $P = 100e^{0.06t}$, with time, t, in years?
(b) Write this function in the form $P = P_0a^t$. What is the annual percent growth rate?

34. (a) What is the annual percent decay rate for $P = 25(0.88)^t$, with time, t, in years?
(b) Write this function in the form $P = P_0e^{kt}$. What is the continuous percent decay rate?

35. The gross world product is $W = 32.4(1.036)^t$, where W is in trillions of dollars and t is years since 2001. Find a formula for gross world product using a continuous growth rate.

36. The population, P, in millions, of Nicaragua was 5.4 million in 2004 and growing at an annual rate of 3.4%. Let t be time in years since 2004.

(a) Express P as a function in the form $P = P_0a^t$.
(b) Express P as an exponential function using base e.
(c) Compare the annual and continuous growth rates.

37. What annual percent growth rate is equivalent to a continuous percent growth rate of 8%?

38. What continuous percent growth rate is equivalent to an annual percent growth rate of 10%?

39. The population of the world can be represented by $P = 6.4(1.0126)^t$, where P is in billions of people and t is years since 2004. Find a formula for the population of the world using a continuous growth rate.

40. The population of a city is 50,000 in 2001 and is growing at a continuous yearly rate of 4.5%.

(a) Give the population of the city as a function of the number of years since 2001. Sketch a graph of the population against time.
(b) What will be the population of the city in the year 2011?
(c) Calculate the time for the population of the city to reach 100,000. This is called the doubling time of the population.

41. In 1980, there were about 170 million vehicles (cars and trucks) and about 227 million people in the United States. The number of vehicles has been growing at 4% a year, while the population has been growing at 1% a year. When was there, on average, one vehicle per person?

1.7 EXPONENTIAL GROWTH AND DECAY

Many quantities in nature change according to an exponential growth or decay function of the form $P = P_0e^{kt}$, where P_0 is the initial quantity and k is the continuous growth or decay rate.

Example 1 The Environmental Protection Agency (EPA) recently investigated a spill of radioactive iodine. The radiation level at the site was about 2.4 millirems/hour (four times the maximum acceptable limit of 0.6 millirems/hour), so the EPA ordered an evacuation of the surrounding area. The level of radiation from an iodine source decays at a continuous hourly rate of $k = -0.004$.

(a) What was the level of radiation 24 hours later?
(b) Find the number of hours until the level of radiation reached the maximum acceptable limit, and the inhabitants could return.

Solution (a) The level of radiation, R, in millirems/hour, at time t, in hours since the initial measurement, is given by

$$R = 2.4e^{-0.004t},$$

so the level of radiation 24 hours later was

$$R = 2.4e^{(-0.004)(24)} = 2.18 \text{ millirems per hour.}$$

(b) A graph of $R = 2.4e^{-0.004t}$ is in Figure 1.67. The maximum acceptable value of R is 0.6 millirems per hour, which occurs at approximately $t = 350$. Using logarithms, we have

$$\begin{aligned} 0.6 &= 2.4e^{-0.004t} \\ 0.25 &= e^{-0.004t} \\ \ln 0.25 &= -0.004t \\ t &= \frac{\ln 0.25}{-0.004} = 346.57. \end{aligned}$$

The inhabitants will not be able to return for 346.57 hours, or about 15 days.

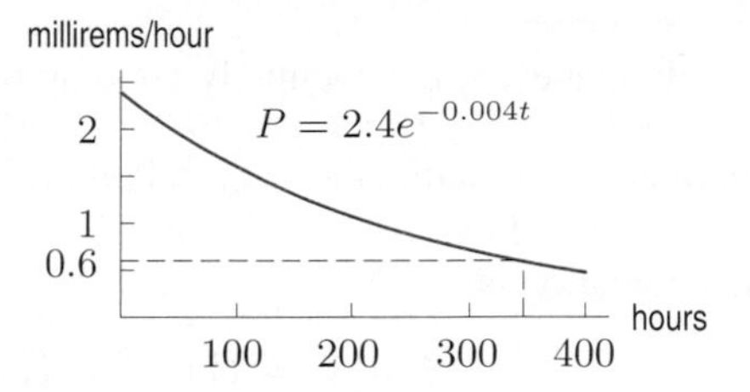

Figure 1.67: The level of radiation from radioactive iodine

Example 2 The population of Kenya[41] was 19.5 million in 1984 and 32.0 million in 2004. Assuming the population increases exponentially, find a formula for the population of Kenya as a function of time.

Solution If we measure the population, P, in millions and time, t, in years since 1984, we can say

$$P = P_0 e^{kt} = 19.5e^{kt},$$

where $P_0 = 19.5$ is the initial value of P. We find k using the fact that $P = 32.0$ when $t = 20$:

$$32.0 = 19.5e^{k\cdot 20}.$$

Divide both sides by 19.5, giving

$$\frac{32.0}{19.5} = e^{20k}.$$

Take natural logs of both sides:

$$\ln\left(\frac{32.0}{19.5}\right) = \ln(e^{20k}).$$

Since $\ln(e^{20k}) = 20k$, this becomes

$$\ln\left(\frac{32.0}{19.5}\right) = 20k.$$

Using a calculator, we get

$$k = \frac{1}{20}\ln\left(\frac{32.0}{19.5}\right) = 0.025,$$

and therefore

$$P = 19.5e^{0.025t}.$$

Since $k = 0.025 = 2.5\%$, the population of Kenya was growing at a continuous rate of 2.5% per year.

---

[41]The World Almanac and Book of Facts 2005, p. 792 (New York).

Doubling Time and Half-Life

Every exponential growth function has a constant doubling time and every exponential decay function has a constant half-life.

> The **doubling time** of an exponentially increasing quantity is the time required for the quantity to double.
> The **half-life** of an exponentially decaying quantity is the time required for the quantity to be reduced by a factor of one half.

Example 3 Show algebraically that every exponentially growing function has a fixed doubling time.

Solution Consider the exponential function $P = P_0 a^t$. For any base a with $a > 1$, there is a positive number d such that $a^d = 2$. We show that d is the doubling time. If the population is P at time t, then at time $t + d$, the population is

$$P_0 a^{t+d} = P_0 a^t a^d = (P_0 a^t)(2) = 2P.$$

So, no matter what the initial quantity and no matter what the initial time, the size of the population is doubled d time units later.

Example 4 The release of chlorofluorocarbons used in air conditioners and household sprays (hair spray, shaving cream, etc.) destroys the ozone in the upper atmosphere. The quantity of ozone, Q, is decaying exponentially at a continuous rate of 0.25% per year. What is the half-life of ozone? In other words, at this rate, how long will it take for half the ozone to disappear?

Solution If Q_0 is the initial quantity of ozone and t is in years, then

$$Q = Q_0 e^{-0.0025t}.$$

We want to find the value of t making $Q = Q_0/2$, so

$$\frac{Q_0}{2} = Q_0 e^{-0.0025t}.$$

Dividing both sides by Q_0 and taking natural logs gives

$$\ln\left(\frac{1}{2}\right) = -0.0025t,$$

so

$$t = \frac{\ln(1/2)}{-0.0025} = 277 \text{ years}.$$

Half the present atmospheric ozone will be gone in 277 years.

Financial Applications: Compound Interest

We deposit \$100 in a bank paying interest at a rate of 8% per year. How much is in the account at the end of the year? This depends on how often the interest is compounded. If the interest is paid into the account *annually*, that is, only at the end of the year, then the balance in the account after one year is \$108. However, if the interest is paid twice a year, then 4% is paid at the end of the first six months and 4% at the end of the year. Slightly more money is earned this way, since the interest paid early in the year will earn interest during the rest of the year. This effect is called *compounding*.

In general, the more often interest is compounded, the more money is earned (although the increase may not be large). What happens if interest is compounded more frequently, such as every minute or every second? The benefit of increasing the frequency of compounding becomes negligible beyond a certain point. When that point is reached, we find the balance using the number e and we say that the interest per year is *compounded continuously*. If we have deposited $\$100$ in an account paying 8% interest per year compounded continuously, the balance after one year is $100e^{0.08} = \$108.33$. Compounding is discussed further in the Focus on Modeling section on page 78. In general:

> An amount P_0 is deposited in an account paying interest at a rate of r per year. Let P be the balance in the account after t years.
> - If interest is compounded annually, then $P = P_0(1+r)^t$.
> - If interest is compounded continuously, then $P = P_0e^{rt}$, where $e = 2.71828...$.

We write P_0 for the initial deposit because it is the value of P when $t = 0$. Note that for a 7% interest rate, $r = 0.07$. If a rate is continuous, we will say so explicitly.

Example 5 A bank advertises an interest rate of 8% per year. If you deposit \$5000, how much is in the account 3 years later if the interest is compounded (a) Annually? (b) Continuously?

Solution

(a) For annual compounding, $P = P_0(1+r)^t = 5000(1.08)^3 = \$6298.56$.
(b) For continuous compounding, $P = P_0e^{rt} = 5000e^{0.08\cdot 3} = \$6356.25$. As expected, the amount in the account 3 years later is larger if the interest is compounded continuously (\$6356.25) than if the interest is compounded annually (\$6298.56).

Example 6 If \$10,000 is deposited in an account paying interest at a rate of 5% per year, compounded continuously, how long does it take for the balance in the account to reach \$15,000?

Solution

Since interest is compounded continuously, we use $P = P_0e^{rt}$ with $r = 0.05$ and $P_0 = 10{,}000$. We want to find the value of t for which $P = 15{,}000$. The equation is

$$15{,}000 = 10{,}000e^{0.05t}.$$

Now divide both sides by 10,000, then take logarithms and solve for t:

$$\begin{aligned} 1.5 &= e^{0.05t} \\ \ln(1.5) &= \ln(e^{0.05t}) \\ \ln(1.5) &= 0.05t \\ t &= \frac{\ln(1.5)}{0.05} = 8.1093. \end{aligned}$$

It takes about 8.1 years for the balance in the account to reach \$15,000.

Example 7

(a) Calculate the doubling time, D, for interest rates of 2%, 3%, 4%, and 5% per year, compounded annually.
(b) Use your answers to part (a) to check that an interest rate of $i\%$ gives a doubling time approximated for small values of i by

$$D \approx \frac{70}{i} \text{ years.}$$

This is the "Rule of 70" used by bankers: To compute the approximate doubling time of an investment, divide 70 by the percent annual interest rate.

Solution (a) We find the doubling time for an interest rate of 2% per year using the formula $P = P_0(1.02)^t$ with t in years. To find the value of t for which $P = 2P_0$, we solve

$$\begin{aligned}2P_0 &= P_0(1.02)^t\\ 2 &= (1.02)^t\\ \ln 2 &= \ln(1.02)^t\\ \ln 2 &= t\ln(1.02) \qquad \text{(using the third property of logarithms)}\\ t &= \frac{\ln 2}{\ln 1.02} = 35.003 \text{ years.}\end{aligned}$$

With an annual interest rate of 2%, it takes about 35 years for an investment to double in value. Similarly, we find the doubling times for 3%, 4%, and 5% in Table 1.26.

Table 1.26 *Doubling time as a function of interest rate*

i (% annual growth rate)	2	3	4	5
D (doubling time in years)	35.003	23.450	17.673	14.207

(b) We compute $(70/i)$ for $i = 2, 3, 4, 5$. The results are shown in Table 1.27.

Table 1.27 *Approximate doubling time as a function of interest rate: Rule of* 70

i (% annual growth rate)	2	3	4	5
$(70/i)$ (Approximate doubling time in years)	35.000	23.333	17.500	14.000

Comparing Tables 1.26 and Table 1.27, we see that the quantity $(70/i)$ gives a reasonably accurate approximation to the doubling time, D, for the small interest rates we considered.

Present and Future Value

Many business deals involve payments in the future. For example, when a car is bought on credit, payments are made over a period of time. Being paid \$100 in the future is clearly worse than being paid \$100 today for many reasons. If we are given the money today, we can do something else with it—for example, put it in the bank, invest it somewhere, or spend it. Thus, even without considering inflation, if we are to accept payment in the future, we would expect to be paid more to compensate for this loss of potential earnings.[42] The question we consider now is, how much more?

To simplify matters, we consider only what we would lose by not earning interest; we do not consider the effect of inflation. Let's look at some specific numbers. Suppose we deposit \$100 in an account that earns 7% interest per year compounded annually, so that in a year's time we have \$107. Thus, \$100 today is worth \$107 a year from now. We say that the \$107 is the *future value* of the \$100, and that the \$100 is the *present value* of the \$107. In general, we say the following:

- The **future value**, B, of a payment, P, is the amount to which the P would have grown if deposited today in an interest-bearing bank account.
- The **present value**, P, of a future payment, B, is the amount that would have to be deposited in a bank account today to produce exactly B in the account at the relevant time in the future.

Due to the interest earned, the future value is larger than the present value. The relation between the present and future values depends on the interest rate, as follows.

[42]This is referred to as the time value of money.

Suppose B is the *future value* of P and P is the *present value* of B.
If interest is compounded annually at a rate r for t years, then

$$B = P(1+r)^t, \quad \text{or equivalently,} \quad P = \frac{B}{(1+r)^t}.$$

If interest is compounded continuously at a rate r for t years, then

$$B = Pe^{rt}, \quad \text{or equivalently,} \quad P = \frac{B}{e^{rt}} = Be^{-rt}.$$

The rate, r, is sometimes called the *discount rate*. The present value is often denoted by PV, and the future value by FV.

Example 8 You win the lottery and are offered the choice between \$1 million in four yearly installments of \$250,000 each, starting now, and a lump-sum payment of \$920,000 now. Assuming a 6% interest rate per year, compounded continuously, and ignoring taxes, which should you choose?

Solution We assume that you pick the option with the largest present value. The first of the four \$250,000 payments is made now, so

$$\text{Present value of first payment} = \$250{,}000.$$

The second payment is made one year from now and so

$$\text{Present value of second payment} = \$250{,}000e^{-0.06(1)}.$$

Calculating the present value of the third and fourth payments similarly, we find:

$$\begin{aligned}\text{Total present value} &= \$250{,}000 + \$250{,}000e^{-0.06(1)} + \$250{,}000e^{-0.06(2)} + \$250{,}000e^{-0.06(3)} \\ &= \$250{,}000 + \$235{,}441 + \$221{,}730 + \$208{,}818 \\ &= \$915{,}989.\end{aligned}$$

Since the present value of the four payments is less than \$920,000, you are better off taking the \$920,000 now.

Alternatively, we can compare the future values of the two pay schemes. We calculate the future value of both schemes three years from now, on the date of the last \$250,000 payment. At that time,

$$\text{Future value of the lump-sum payment} = \$920{,}000e^{0.06(3)} = \$1{,}101{,}440.$$

The future value of the first \$250,000 payment is $\$250{,}000e^{0.06(3)}$. Calculating the future value of the other payments similarly, we find:

$$\begin{aligned}\text{Total future value} &= \$250{,}000e^{0.06(3)} + \$250{,}000e^{0.06(2)} + \$250{,}000e^{0.06(1)} + \$250{,}000 \\ &= \$299{,}304 + \$281{,}874 + \$265{,}459 + \$250{,}000 \\ &= \$1{,}096{,}637.\end{aligned}$$

As we expect, the future value of the \$920,000 payment is greater, so you are better off taking the \$920,000 now.[43]

[43] If you read the fine print, you will find that many lotteries do not make their payments right away, but often spread them out, sometimes far into the future. This is to reduce the present value of the payments made, so that the value of the prizes is less than it might first appear!

Problems for Section 1.7

1. The half-life of nicotine in the blood is 2 hours. A person absorbs 0.4 mg of nicotine by smoking a cigarette. Fill in the following table with the amount of nicotine remaining in the blood after t hours. Estimate the length of time until the amount of nicotine is reduced to 0.04 mg.

t (hours)	0	2	4	6	8	10
Nicotine (mg)	0.4					

2. If you deposit \$10,000 in an account earning interest at an 8% annual rate compounded continuously, how much money is in the account after five years?

3. If you need \$20,000 in your bank account in 6 years, how much must be deposited now? The interest rate is 10%, compounded continuously.

4. If a bank pays 6% per year interest compounded continuously, how long does it take for the balance in an account to double?

5. You invest \$5000 in an account which pays interest compounded continuously.

(a) How much money is in the account after 8 years, if the annual interest rate is 4%?
(b) If you want the account to contain \$8000 after 8 years, what yearly interest rate is needed?

6. Suppose \$1000 is invested in an account paying interest at a rate of 5.5% per year. How much is in the account after 8 years if the interest is compounded

(a) Annually? **(b)** Continuously?

7. Find the doubling time of a quantity that is increasing by 7% per year.

8. A cup of coffee contains 100 mg of caffeine, which leaves the body at a continuous rate of 17% per hour.

(a) Write a formula for the amount, A mg, of caffeine in the body t hours after drinking a cup of coffee.
(b) Graph the function from part (a). Use the graph to estimate the half-life of caffeine.
(c) Use logarithms to find the half-life of caffeine.

9. A population, currently 200, is growing at 5% per year.

(a) Write a formula for the population, P, as a function of time, t, years in the future.
(b) Graph P against t.
(c) Estimate the population 10 years from now.
(d) Use the graph to estimate the doubling time of the population.

10. The antidepressant fluoxetine (or Prozac) has a half-life of about 3 days. What percentage of a dose remains in the body after one day? After one week?

11. You need \$10,000 in your account 3 years from now and the interest rate is 8% per year, compounded continuously. How much should you deposit now?

12. A firm decides to increase output at a constant rate from its current level of 20,000 to 30,000 units during the next five years. Calculate the annual percent rate of increase required to achieve this growth.

13. The quantity, Q, of radioactive carbon-14 remaining t years after an organism dies is given by the formula

$$Q = Q_0 e^{-0.000121t},$$

where Q_0 is the initial quantity.

(a) A skull uncovered at an archeological dig has 15% of the original amount of carbon-14 present. Estimate its age.
(b) Calculate the half-life of carbon-14.

14. **(a)** Figure 1.68 shows exponential growth. Starting at $t = 0$, estimate the time for the population to double.
(b) Repeat part (a), but this time start at $t = 3$.
(c) Pick any other value of t for the starting point, and notice that the doubling time is the same no matter where you start.

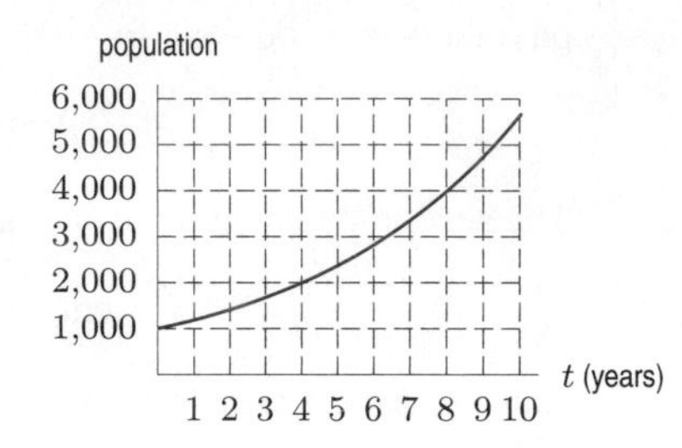

Figure 1.68

15. An exponentially growing animal population numbers 500 at time $t = 0$; two years later, it is 1500. Find a formula for the size of the population in t years and find the size of the population at $t = 5$.

16. If the quantity of a substance decreases by 4% in 10 hours, find its half-life.

17. Figure 1.69 shows the balances in two bank accounts. Both accounts pay the same interest rate, but one compounds continuously and the other compounds annually. Which curve corresponds to which compounding method? What is the initial deposit in each case?

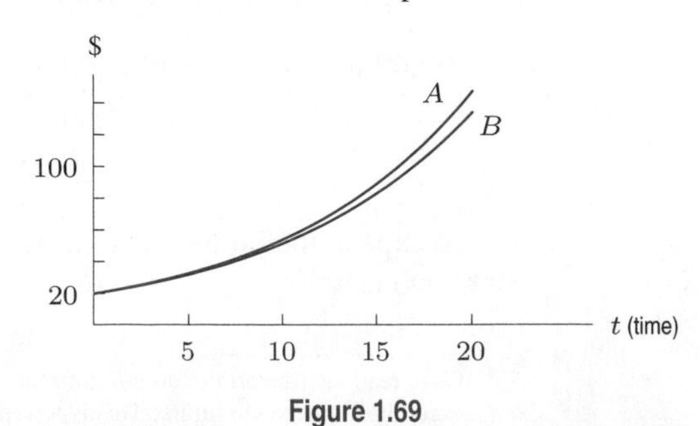

Figure 1.69

18. Pregnant women metabolize some drugs at a slower rate than the rest of the population. The half-life of caffeine is about 4 hours for most people. In pregnant women, it is 10 hours.[44] (This is important because caffeine, like all psychoactive drugs, crosses the placenta to the fetus.) If a pregnant woman and her husband each have a cup of coffee containing 100 mg of caffeine at 8 am, how much caffeine does each have left in the body at 10 pm?

19. The half-life of radioactive strontium-90 is 29 years. In 1960, radioactive strontium-90 was released into the atmosphere during testing of nuclear weapons, and was absorbed into people's bones. How many years does it take until only 10% of the original amount absorbed remains?

20. If $12,000 is deposited in an account paying 8% interest per year, compounded continuously, how long will it take for the balance to reach $20,000?

21. You want to invest money for your child's education in a certificate of deposit (CD). You want it to be worth $12,000 in 10 years. How much should you invest if the CD pays interest at a 9% annual rate compounded **(a)** Annually? **(b)** Continuously?

22. When you rent an apartment, you are often required to give the landlord a security deposit which is returned if you leave the apartment undamaged. In Massachusetts the landlord is required to pay the tenant interest on the deposit once a year, at a 5% annual rate, compounded annually. The landlord, however, may invest the money at a higher (or lower) interest rate. Suppose the landlord invests a $1000 deposit at a yearly rate of

(a) 6%, compounded continuously
(b) 4%, compounded continuously.

In each case, determine the net gain or loss by the landlord at the end of the first year. (Give your answer to the nearest cent.)

23. In 1923, koalas were introduced on Kangaroo Island off the coast of Australia. In 1996, the population was 5000. By 2005, the population had grown to 27,000, prompting a debate on how to control their growth and avoid koalas dying of starvation.[45] Assuming exponential growth, find the (continuous) rate of growth of the koala population between 1996 and 2005. Find a formula for the population as a function of the number of years since 1996, and estimate the population in the year 2020.

24. The total world marine catch in 1950 was 17 million tons and in 2001 was 99 million tons.[46] If the marine catch is increasing exponentially, find the (continuous) rate of increase. Use it to predict the total world marine catch in the year 2020.

25. **(a)** Use the Rule of 70 to predict the doubling time of an investment which is earning 8% interest per year.
(b) Find the doubling time exactly, and compare your answer to part (a).

26. The island of Manhattan was sold for $24 in 1626. Suppose the money had been invested in an account which compounded interest continuously.

(a) How much money would be in the account in the year 2005 if the yearly interest rate was
(i) 5%? (ii) 7%?
(b) If the yearly interest rate was 6%, in what year would the account be worth one million dollars?

27. Owing to an innovative rural public health program, infant mortality in Senegal, West Africa, is being reduced at a rate of 10% per year. How long will it take for infant mortality to be reduced by 50%?

28. In 2004, the world's population was 6.4 billion, and the population was projected to reach 8.5 billion by the year 2030. What annual growth rate is projected?

29. A picture supposedly painted by Vermeer (1632–1675) contains 99.5% of its carbon-14 (half-life 5730 years). From this information decide whether the picture is a fake. Explain your reasoning.

30. A business associate who owes you $3000 offers to pay you $2800 now, or else pay you three yearly installments of $1000 each, with the first installment paid now. If you use only financial reasons to make your decision, which option should you choose? Justify your answer, assuming a 6% interest rate per year, compounded continuously.

31. Big Tree McGee is negotiating his rookie contract with a professional basketball team. They have agreed to a three-year deal which will pay Big Tree a fixed amount at the end of each of the three years, plus a signing bonus at the beginning of his first year. They are still haggling about the amounts and Big Tree must decide between a big signing bonus and fixed payments per year, or a smaller bonus with payments increasing each year. The two options are summarized in the table. All values are payments in millions of dollars.

	Signing bonus	Year 1	Year 2	Year 3
Option #1	6.0	2.0	2.0	2.0
Option #2	1.0	2.0	4.0	6.0

(a) Big Tree decides to invest all income in stock funds which he expects to grow at a rate of 10% per year, compounded continuously. He would like to choose the contract option which gives him the greater future value at the end of the three years when the last payment is made. Which option should he choose?
(b) Calculate the present value of each contract offer.

[44]From Robert M. Julien, *A Primer of Drug Action*, 7th ed., p. 159, (New York: W. H. Freeman, 1995).
[45]news.yahoo.com/s/afp/australiaanimalskoalas, accessed June 1, 2005.
[46]*The World Almanac and Book of Facts 2005*, p. 143 (New York).

32. A company is considering whether to buy a new machine, which costs \$97,000. The cash flows (adjusted for taxes and depreciation) that would be generated by the new machine are given in the following table:

Year	1	2	3	4
Cash flow	\$50,000	\$40,000	\$25,000	\$20,000

(a) Find the total present value of the cash flows. Treat each year's cash flow as a lump sum at the end of the year and use an interest rate of 7.5% per year, compounded annually.

(b) Based on a comparison of the cost of the machine and the present value of the cash flows, would you recommend purchasing the machine?

33. You win \$38,000 in the state lottery to be paid in two installments—\$19,000 now and \$19,000 one year from now. A friend offers you \$36,000 in return for your two lottery payments. Instead of accepting your friend's offer, you take out a one-year loan at an interest rate of 8.25% per year, compounded annually. The loan will be paid back by a single payment of \$19,000 (your second lottery check) at the end of the year. Which is better, your friend's offer or the loan?

34. You are considering whether to buy or lease a machine whose purchase price is \$12,000. Taxes on the machine will be \$580 due in one year, \$464 due in two years, and \$290 due in three years. If you buy the machine, you expect to be able to sell it after three years for \$5,000. If you lease the machine for three years, you make an initial payment of \$2650 and then three payments of \$2650 at the end of each of the next three years. The leasing company will pay the taxes. The interest rate is 7.75% per year, compounded annually. Should you buy or lease the machine? Explain.

35. You are buying a car that comes with a one-year warranty and are considering whether to purchase an extended warranty for \$375. The extended warranty covers the two years immediately after the one-year warranty expires. You estimate that the yearly expenses that would have been covered by the extended warranty are \$150 at the end of the first year of the extension and \$250 at the end of the second year of the extension. The interest rate is 5% per year, compounded annually. Should you buy the extended warranty? Explain.

36. You have the option of renewing the service contract on your three-year old dishwasher. The new service contract is for three years at a price of \$200. The interest rate is 7.25% per year, compounded annually, and you estimate that the costs of repairs if you do not buy the service contract will be \$50 at the end of the first year, \$100 at the end of the second year, and \$150 at the end of the third year. Should you buy the service contract? Explain.

1.8 NEW FUNCTIONS FROM OLD

We have studied linear and exponential functions, and the logarithm function. In this section, we learn how to create new functions by composing, stretching, and shifting functions we already know.

Composite Functions

A drop of water falls onto a paper towel. The area, A of the circular damp spot is a function of r, its radius, which is a function of time, t. We know $A = f(r) = \pi r^2$; suppose $r = g(t) = t + 1$. By substitution, we express A as a function of t:

$$A = f(g(t)) = \pi(t+1)^2.$$

The function $f(g(t))$ is a "function of a function," or a *composite function*, in which there is an *inside function* and an *outside function*. To find $f(g(2))$, we first add one ($g(2) = 2 + 1 = 3$) and then square and multiply by π. We have

$$f(g(2)) = \pi(2+1)^2 \underset{\text{First calculation}}{=} \pi 3^2 \underset{\text{Second calculation}}{=} 9\pi.$$

The inside function is $t + 1$ and the outside function is squaring and multiplying by π. In general, the inside function represents the calculation that is done first and the outside function represents the calculation done second.

Example 1 If $f(t) = t^2$ and $g(t) = t + 2$, find

(a) $f(t+1)$ (b) $f(t)+3$ (c) $f(t+h)$ (d) $f(g(t))$ (e) $g(f(t))$

Solution
(a) Since $t+1$ is the inside function, $f(t+1)=(t+1)^2$.
(b) Here 3 is added to $f(t)$, so $f(t)+3=t^2+3$.
(c) Since $t+h$ is the inside function, $f(t+h)=(t+h)^2$.
(d) Since $g(t)=t+2$, substituting $t+2$ into f gives $f(g(t))=f(t+2)=(t+2)^2$.
(e) Since $f(t)=t^2$, substituting t^2 into g gives $g(f(t))=g(t^2)=t^2+2$.

Example 2 If $f(x)=e^x$ and $g(x)=5x+1$, find (a) $f(g(x))$ (b) $g(f(x))$

Solution
(a) Substituting $g(x)=5x+1$ into f gives $f(g(x))=f(5x+1)=e^{5x+1}$.
(b) Substituting $f(x)=e^x$ into g gives $g(f(x))=g(e^x)=5e^x+1$.

Example 3 Using the following table, find $g(f(0))$, $f(g(0))$, $f(g(1))$, and $g(f(1))$.

x	0	1	2	3
$f(x)$	3	1	−1	−3
$g(x)$	0	2	4	6

Solution To find $g(f(0))$, we first find $f(0)=3$ from the table. Then we have $g(f(0))=g(3)=6$. For $f(g(0))$, we must find $g(0)$ first. Since $g(0)=0$, we have $f(g(0))=f(0)=3$. Similar reasoning leads to $f(g(1))=f(2)=-1$ and $g(f(1))=g(1)=2$.

We can write a composite function using a new variable u to represent the value of the inside function. For example

$$y=(t+1)^4 \quad \text{is the same as} \quad y=u^4 \quad \text{with} \quad u=t+1.$$

Other expressions for u, such as $u=(t+1)^2$, with $y=u^2$, are also possible.

Example 4 Use a new variable u for the inside function to express each of the following as a composite function:
(a) $y=\ln(3t)$ (b) $w=5(2r+3)^2$ (c) $P=e^{-0.03t}$

Solution
(a) We take the inside function to be $3t$, so $y=\ln u$ with $u=3t$.
(b) We take the inside function to be $2r+3$, so $w=5u^2$ with $u=2r+3$.
(c) We take the inside function to be $-0.03t$, so $P=e^u$ with $u=-0.03t$.

Stretches of Graphs

The graph of $y=f(x)$ is in Figure 1.70. What does the graph of $y=3f(x)$ look like? The factor 3 in the function $y=3f(x)$ stretches each $f(x)$ value by multiplying it by 3. What does the graph of $y=-2f(x)$ look like? The factor -2 in the function $y=-2f(x)$ stretches $f(x)$ by multiplying by 2 and reflecting it about the x-axis. See Figure 1.71.

> Multiplying a function by a constant, c, stretches the graph vertically (if $c>1$) or shrinks the graph vertically (if $0<c<1$). A negative sign (if $c<0$) reflects the graph about the x-axis, in addition to shrinking or stretching.

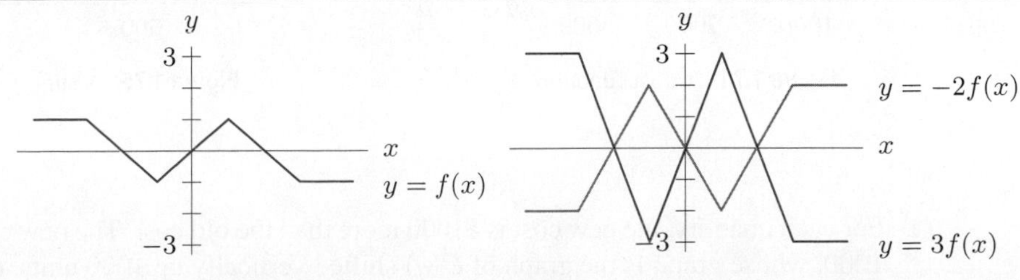

Figure 1.70: Graph of $f(x)$

Figure 1.71: Multiples of the function $f(x)$

Shifted Graphs

Consider the function $y = x^2 + 4$. The y-coordinates for this function are exactly 4 units larger than the corresponding y-coordinates of the function $y = x^2$. So the graph of $y = x^2 + 4$ is obtained from the graph of $y = x^2$ by adding 4 to the y-coordinate of each point; that is, by moving the graph of $y = x^2$ up 4 units. (See Figure 1.72.)

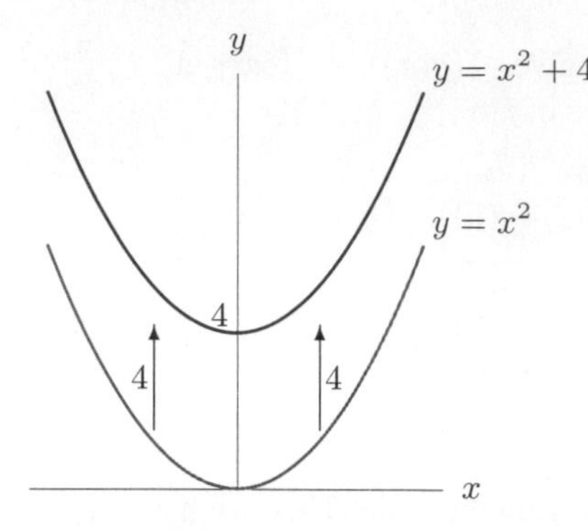

Figure 1.72: Vertical shift: Graphs of $y = x^2$ and $y = x^2 + 4$

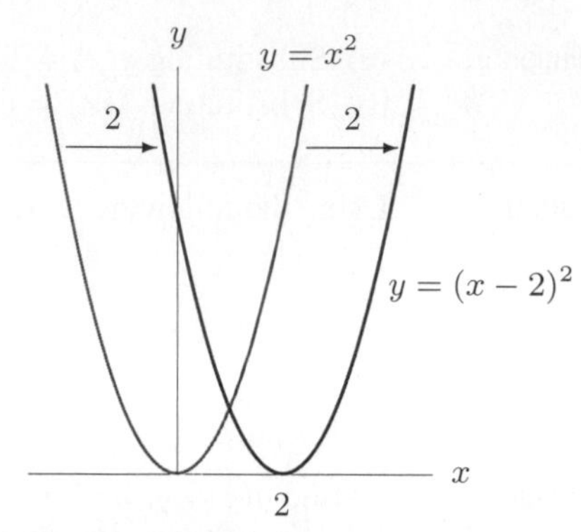

Figure 1.73: Horizontal shift: Graphs of $y = x^2$ and $y = (x - 2)^2$

A graph can also be shifted to the left or to the right. In Figure 1.73, we see that the graph of $y = (x - 2)^2$ is the graph of $y = x^2$ shifted to the right 2 units. In general,

- The graph of $y = f(x) + k$ is the graph of $y = f(x)$ moved up k units (down if k is negative).
- The graph of $y = f(x - k)$ is the graph of $y = f(x)$ moved to the right k units (to the left if k is negative).

Example 5 (a) A cost function, $C(q)$, for a company is shown in Figure 1.74. The fixed cost increases by \$1000. Sketch a graph of the new cost function.

(b) A supply curve, S, for a product is given in Figure 1.75. A new factory opens and produces 100 units of the product no matter what the price. Sketch a graph of the new supply curve.

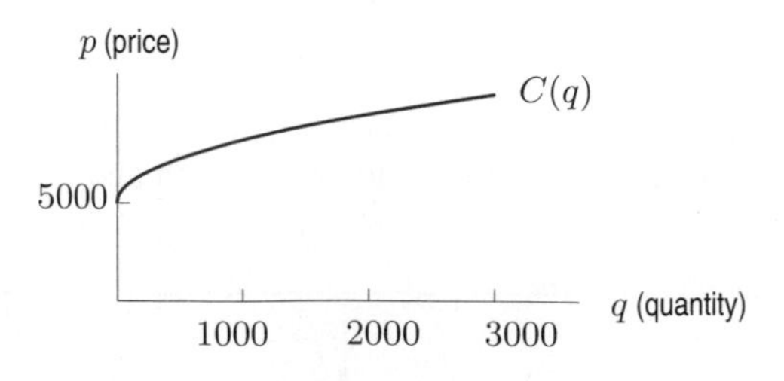

Figure 1.74: A cost function

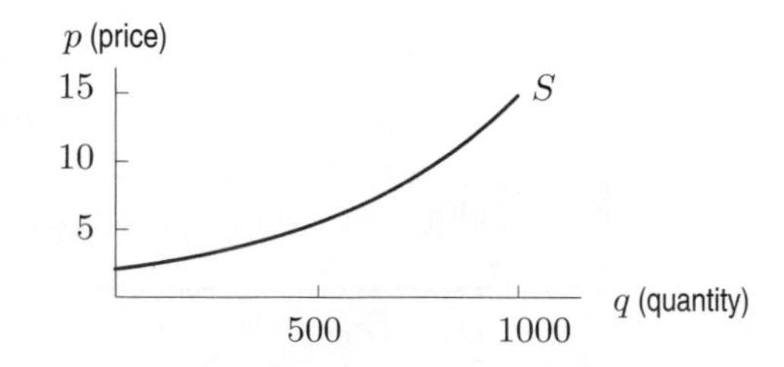

Figure 1.75: A supply function

Solution (a) For each quantity, the new cost is \$1000 more than the old cost. The new cost function is $C(q) + 1000$, whose graph is the graph of $C(q)$ shifted vertically up 1000 units. (See Figure 1.76.)

(b) To see the effect of the new factory, look at an example. At a price of 10 dollars, approximately 800 units are currently produced. With the new factory, this amount increases by 100 units, so the new amount produced is 900 units. At each price, the quantity produced increases by 100, so the new supply curve is S shifted horizontally to the right by 100 units. (See Figure 1.77.)

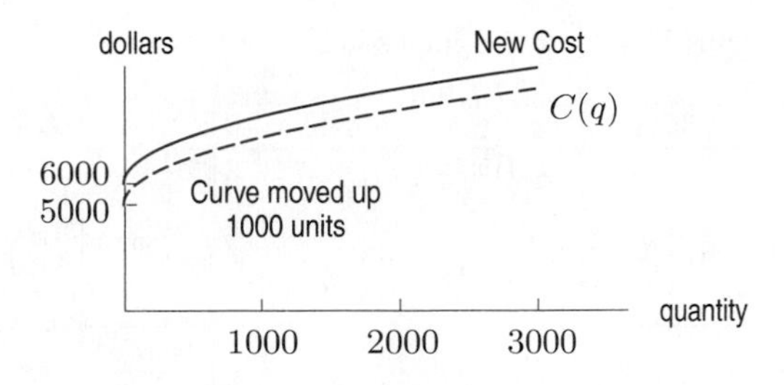

Figure 1.76: New cost function (original curve dashed)

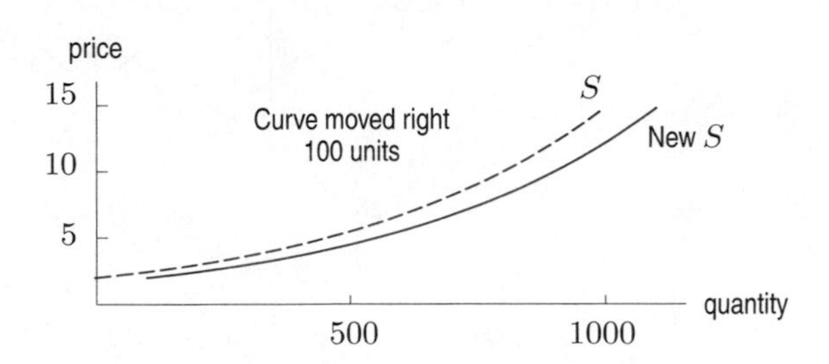

Figure 1.77: New supply curve (original curve dashed)

Problems for Section 1.8

1. For $g(x) = x^2 + 2x + 3$, find and simplify:

(a) $g(2+h)$ **(b)** $g(2)$
(c) $g(2+h) - g(2)$

2. If $f(x) = x^2 + 1$, find and simplify:

(a) $f(t+1)$ **(b)** $f(t^2+1)$ **(c)** $f(2)$
(d) $2f(t)$ **(e)** $[f(t)]^2 + 1$

For the functions f and g in Problems 3–6, find

(a) $f(g(1))$ **(b)** $g(f(1))$ **(c)** $f(g(x))$
(d) $g(f(x))$ **(e)** $f(t)g(t)$

3. $f(x) = x^2$, $g(x) = x + 1$

4. $f(x) = \sqrt{x+4}$, $g(x) = x^2$

5. $f(x) = e^x$, $g(x) = x^2$

6. $f(x) = 1/x$, $g(x) = 3x + 4$

7. Let $f(x) = x^2$ and $g(x) = 3x - 1$. Find the following:

(a) $f(2) + g(2)$ **(b)** $f(2) \cdot g(2)$
(c) $f(g(2))$ **(d)** $g(f(2))$

In Problems 8–10, find the following

(a) $f(g(x))$ **(b)** $g(f(x))$ **(c)** $f(f(x))$

8. $f(x) = 2x^2$ and $g(x) = x + 3$

9. $f(x) = 2x + 3$ and $g(x) = 5x^2$

10. $f(x) = x^2 + 1$ and $g(x) = \ln x$

11. Use the variable u for the inside function to express each of the following as a composite function:

(a) $y = 2^{3x-1}$ **(b)** $P = \sqrt{5t^2 + 10}$
(c) $w = 2\ln(3r + 4)$

12. Use the variable u for the inside function to express each of the following as a composite function:

(a) $y = (5t^2 - 2)^6$ **(b)** $P = 12e^{-0.6t}$
(c) $C = 12\ln(q^3 + 1)$

In Problems 13–16, use Figure 1.78 to graph the functions.

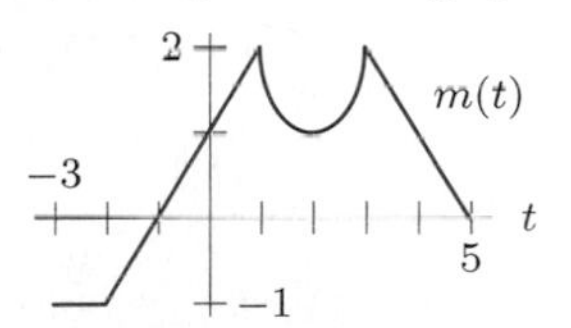

Figure 1.78

13. $n(t) = m(t) + 2$

14. $p(t) = m(t - 1)$

15. $k(t) = m(t + 1.5)$

16. $w(t) = m(t - 0.5) - 2.5$

Graph the functions in Problems 17–22 using Figure 1.79.

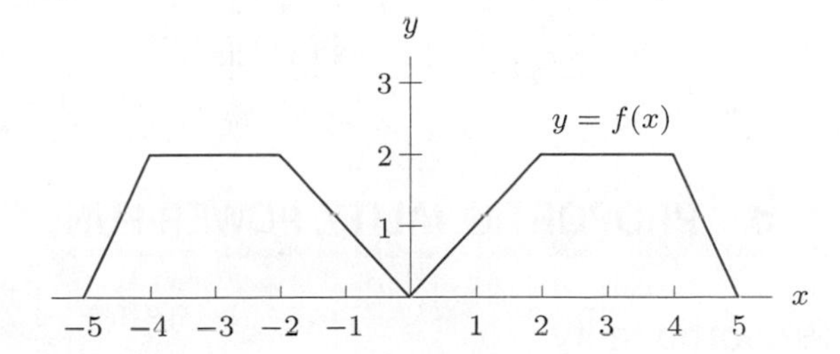

Figure 1.79

17. $y = f(x) + 2$ **18.** $y = 2f(x)$

19. $y = f(x - 1)$ **20.** $y = -3f(x)$

21. $y = 2f(x) - 1$ **22.** $y = 2 - f(x)$

For the functions $f(x)$ in Problems 23–26, graph:

(a) $y = f(x) + 2$ **(b)** $y = f(x - 1)$
(c) $y = 3f(x)$ **(d)** $y = -f(x)$

23.

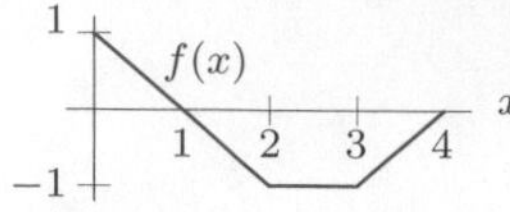

24.

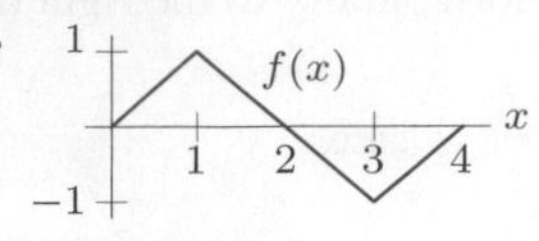

25.

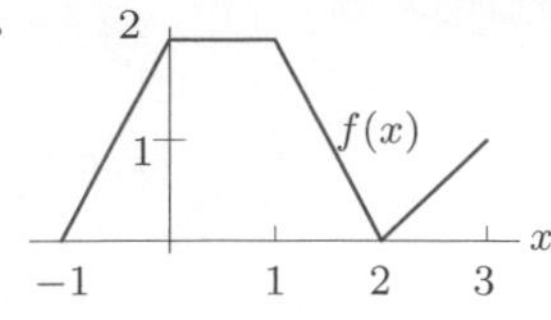

26.

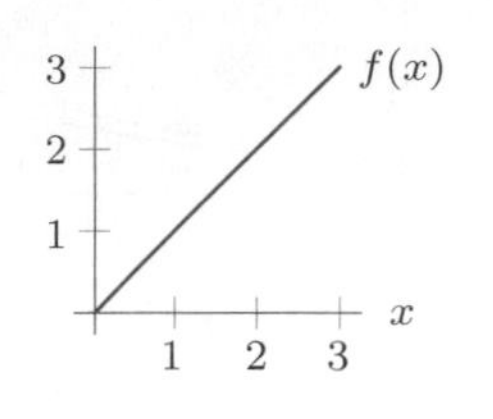

In Problems 27–29, use Figure 1.80 to graph the function.

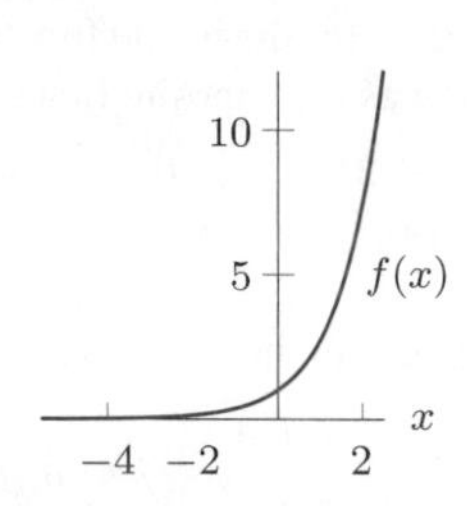

Figure 1.80

27. $5f(x)$ **28.** $f(x+5)$ **29.** $f(x) + 5$

30. Use Table 1.28 to find:

(a) $f(g(0))$ **(b)** $f(g(1))$ **(c)** $f(g(2))$
(d) $g(f(2))$ **(e)** $g(f(3))$

Table 1.28

x	0	1	2	3	4	5
$f(x)$	10	6	3	4	7	11
$g(x)$	2	3	5	8	12	15

31. Make a table of values for each of the following functions using Table 1.28.

(a) $f(x) + 3$ **(b)** $f(x - 2)$ **(c)** $5g(x)$
(d) $-f(x) + 2$ **(e)** $g(x - 3)$ **(f)** $f(x) + g(x)$

For Problems 32–34, use the graphs in Figure 1.81.

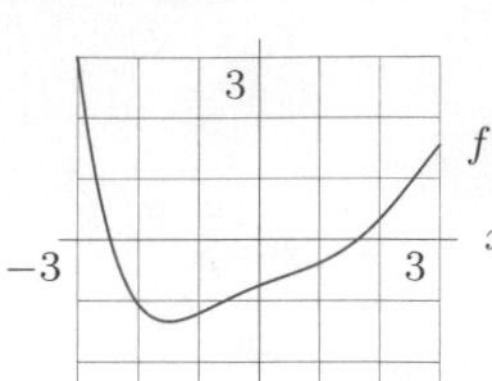

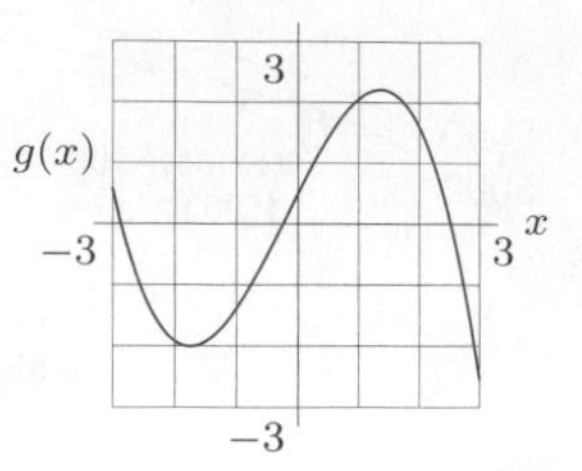

Figure 1.81

32. Estimate $f(g(1))$. **33.** Estimate $g(f(2))$.

34. Estimate $f(f(1))$.

35. **(a)** Write an equation for a graph obtained by vertically stretching the graph of $y = x^2$ by a factor of 2, followed by a vertical upward shift of 1 unit. Sketch it.
(b) What is the equation if the order of the transformations (stretching and shifting) in part (a) is interchanged?
(c) Are the two graphs the same? Explain the effect of reversing the order of transformations.

36. Using Table 1.29, create a table of values for $f(g(x))$ and for $g(f(x))$.

Table 1.29

x	−3	−2	−1	0	1	2	3
$f(x)$	0	1	2	3	2	1	0
$g(x)$	3	2	2	0	−2	−2	−3

37. A plan is adopted to reduce the pollution in a lake to the legal limit. The quantity Q of pollutants in the lake after t weeks of clean-up is modeled by the function $Q = f(t)$ where $f(t) = A + Be^{Ct}$.

(a) What are the signs of A, B and C?
(b) What is the initial quantity of pollution in the lake?
(c) What is the legal limit of pollution in the lake?

1.9 PROPORTIONALITY, POWER FUNCTIONS, AND POLYNOMIALS

Proportionality

A common functional relationship occurs when one quantity is *proportional* to another. For example, if apples are $1.40 a pound, we say the price you pay, p dollars, is proportional to the weight you buy, w pounds, because

$$p = f(w) = 1.40w.$$

As another example, the area, A, of a circle is proportional to the square of the radius, r:

$$A = f(r) = \pi r^2.$$

We say y is (directly) **proportional** to x if there is a nonzero constant k such that

$$y = kx.$$

This k is called the constant of proportionality.

We also say that one quantity is *inversely proportional* to another if one is proportional to the reciprocal of the other. For example, the speed, v, at which you make a 50-mile trip is inversely proportional to the time, t, taken, because v is proportional to $1/t$:

$$v = 50\left(\frac{1}{t}\right) = \frac{50}{t}.$$

Notice that if y is directly proportional to x, then the magnitude of one variable increases (decreases) when the magnitude of the other increases (decreases). If, however, y is inversely proportional to x, then the magnitude of one variable increases when the magnitude of the other decreases.

Example 1 The heart mass of a mammal is proportional to its body mass.[47]

(a) Write a formula for heart mass, H, as a function of body mass, B.
(b) A human with a body mass of 70 kilograms has a heart mass of 0.42 kilograms. Use this information to find the constant of proportionality.
(c) Estimate the heart mass of a horse with a body mass of 650 kg.

Solution (a) Since H is proportional to B, for some constant k, we have

$$H = kB.$$

(b) We use the fact that $H = 0.42$ when $B = 70$ to solve for k:

$$\begin{aligned} H &= kB \\ 0.42 &= k(70) \\ k &= \frac{0.42}{70} = 0.006. \end{aligned}$$

(c) Since $k = 0.006$, we have $H = 0.006B$, so the heart mass of the horse is given by

$$H = 0.006(650) = 3.9 \text{ kilograms.}$$

Example 2 The period of a pendulum, T, is the amount of time required for the pendulum to make one complete swing. For small swings, the period, T, is approximately proportional to the square root of l, the pendulum's length. So

$$T = k\sqrt{l} \quad \text{where } k \text{ is a constant.}$$

Notice that T is not directly proportional to l, but T is proportional to $\sqrt{l}$.

Example 3 An object's weight, w, is inversely proportional to the square of its distance, r, from the earth's center. So, for some constant k,

$$w = \frac{k}{r^2}.$$

Here w is not inversely proportional to r, but to r^2.

[47]K. Schmidt-Nielson: *Scaling-Why is Animal Size So Important?* (Cambridge: CUP, 1984).

Power Functions

In each of the previous examples, one quantity is proportional to the power of another quantity. We make the following definition:

> We say that $Q(x)$ is a **power function** of x if $Q(x)$ is proportional to a constant power of x. If k is the constant of proportionality, and if p is the power, then
>
> $$Q(x) = k \cdot x^p.$$

For example, the function $H = 0.006B$ is a power function with $p = 1$. The function $T = k\sqrt{l} = kl^{1/2}$ is a power function with $p = 1/2$, and the function $w = k/r^2 = kr^{-2}$ is a power function with $p = -2$.

Example 4 Which of the following are power functions? For those which are, write the function in the form $y = kx^p$, and give the coefficient k and the exponent p.

(a) $y = \dfrac{5}{x^3}$ (b) $y = \dfrac{2}{3x}$ (c) $y = \dfrac{5x^2}{2}$

(d) $y = 5 \cdot 2^x$ (e) $y = 3\sqrt{x}$ (f) $y = (3x^2)^3$

Solution

(a) Since $y = 5x^{-3}$, this is a power function with $k = 5$ and $p = -3$.
(b) Since $y = (2/3)x^{-1}$, this is a power function with $k = 2/3$ and $p = -1$.
(c) Since $y = (5/2)x^2$, this is a power function with $k = 5/2 = 2.5$ and $p = 2$.
(d) This is not a power function. It is an exponential function.
(e) Since $y = 3x^{1/2}$, this is a power function with $k = 3$ and $p = 1/2$.
(f) Since $y = 3^3 \cdot (x^2)^3 = 27x^6$, this is a power function with $k = 27$ and $p = 6$.

Graphs of Power Functions

The graph of $y = x^2$ is shown in Figure 1.82. It is decreasing for negative x and increasing for positive x. Notice that it is bending upward, or concave up, for all x. The graph of $y = x^3$ is shown in Figure 1.83. Notice that it is bending downward, or concave down for negative x and bending upward, or concave up for positive x. The graph of $y = \sqrt{x} = x^{1/2}$ is shown in Figure 1.84. Notice that the graph is increasing and concave down.

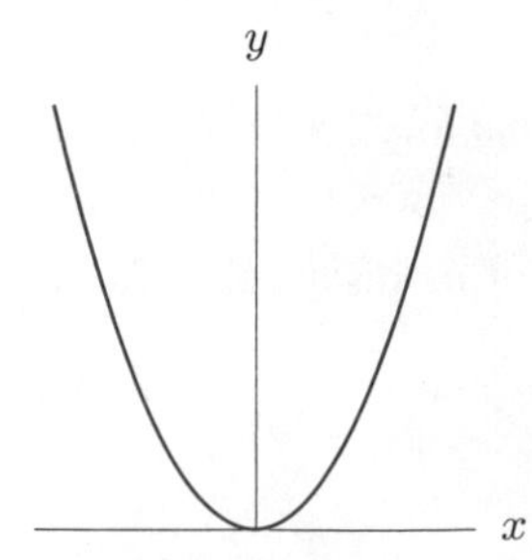

Figure 1.82: Graph of $y = x^2$

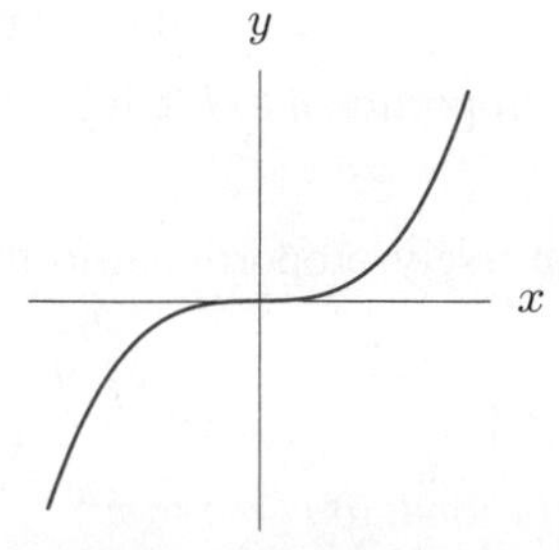

Figure 1.83: Graph of $y = x^3$

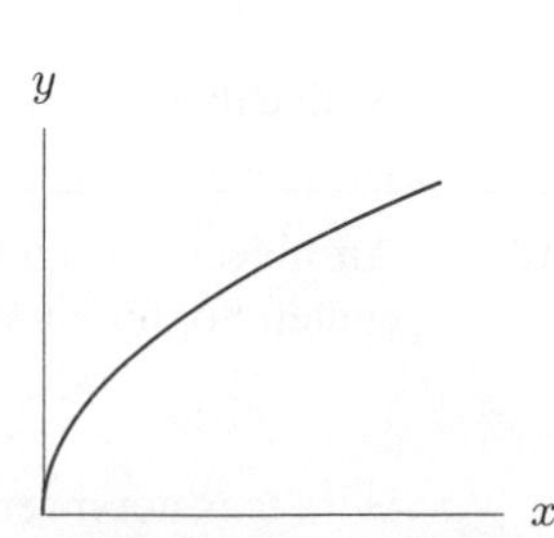

Figure 1.84: Graph of $y = x^{1/2}$

Example 5 If N is the average number of species found on an island and A is the area of the island, observations have shown[48] that N is approximately proportional to the cube root of A. Write a formula for N as a function of A and describe the shape of the graph of this function.

Solution For some positive constant k, we have

$$N = k\sqrt[3]{A} = kA^{1/3}.$$

It turns out that the value of k depends on the region of the world in which the island is found. The graph of N against A (for $A > 0$) has a shape similar to the graph in Figure 1.84. It is increasing and concave down. Thus, larger islands have more species on them (as we would expect), but the increase slows as the island gets larger.

The function $y = x^0 = 1$ has a graph that is a horizontal line. For negative powers, rewriting

$$y = x^{-1} = \frac{1}{x} \quad \text{and} \quad y = x^{-2} = \frac{1}{x^2}$$

makes it clear that as $x > 0$ increases, the denominators increase and the functions decrease. The graphs of $y = x^{-1}$ and $y = x^{-2}$ have both the x- and y-axes as asymptotes. (See Figure 1.85.)

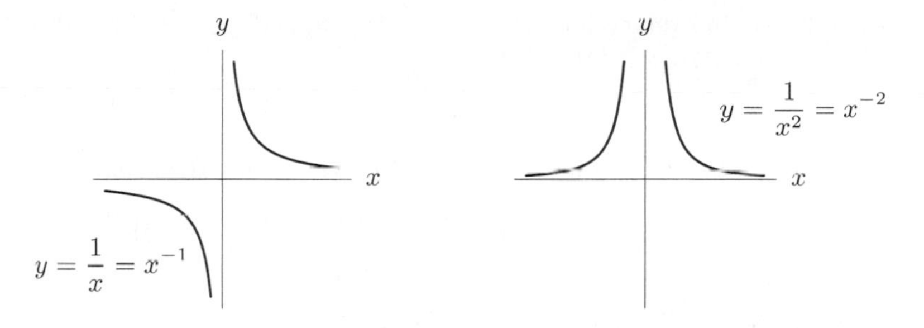

Figure 1.85: Graphs of negative powers of x

Polynomials

Sums of power functions with non-negative integer exponents are called *polynomials*, which are functions of the form

$$y = p(x) = a_n x^n + a_{n-1}x^{n-1} + \cdots + a_1 x + a_0.$$

Here, n is a nonnegative integer, called the *degree* of the polynomial, and a_n is a non-zero number called the *leading coefficient*. We call $a_n x^n$ the *leading term*. If $n = 2$, the polynomial is called quadratic; if $n = 3$, the polynomial is called cubic.

The shape of the graph of a polynomial depends on its degree. See Figure 1.86. The graph of a quadratic polynomial is a parabola. It opens up if the leading coefficient is positive (as in Figure 1.86) and opens down if the leading coefficient is negative. A cubic polynomial may have the shape of the graph of $y = x^3$, or the shape shown in Figure 1.86, or it may be a reflection of one of these about the x-axis.

Notice in Figure 1.86 that the graph of the quadratic "turns around" once, the cubic "turns around" twice, and the quartic (fourth degree) "turns around" three times. An nth degree polynomial "turns around" at most $n - 1$ times (where n is a positive integer), but there may be fewer turns.

[48] *Scientific American*, p. 112, (September, 1989).

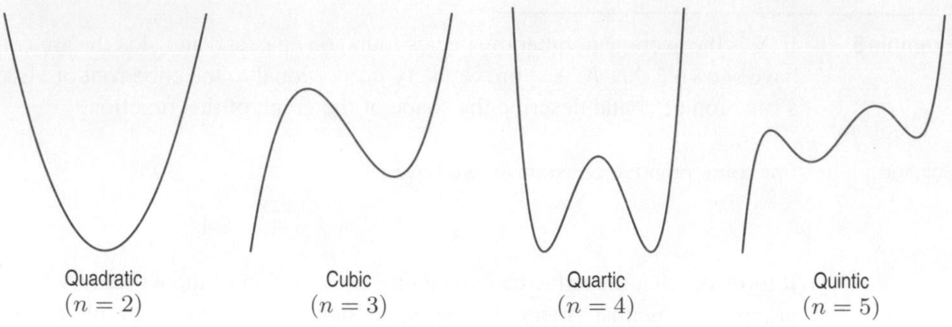

Quadratic ($n = 2$) Cubic ($n = 3$) Quartic ($n = 4$) Quintic ($n = 5$)

Figure 1.86: Graphs of typical polynomials of degree n

Example 6 A company finds that the average number of people attending a concert is 75 if the price is \$50 per person. At a price of \$35 per person, the average number of people in attendance is 120.

(a) Assume that the demand curve is a line. Write the demand, q, as a function of price, p.
(b) Use your answer to part (a) to write the revenue, R, as a function of price, p.
(c) Use a graph of the revenue function to determine what price should be charged to obtain the greatest revenue.

Solution (a) Two points on the line are $(p, q) = (50, 75)$ and $(p, q) = (35, 120)$. The slope of the line is

$$m = \frac{120 - 75}{35 - 50} = \frac{45}{-15} = -3 \text{ people/dollar.}$$

To find the vertical intercept of the line, we use the slope and one of the points:

$$75 = b + (-3)(50)$$
$$225 = b$$

The demand function is $q = 225 - 3p$.

(b) Since $R = pq$ and $q = 225 - 3p$, we see that $R = p(225 - 3p) = 225p - 3p^2$.

(c) The revenue function is the quadratic polynomial graphed in Figure 1.87. The maximum revenue occurs at $p = 37.5$. Thus, the company maximizes revenue by charging \$37.50 per person.

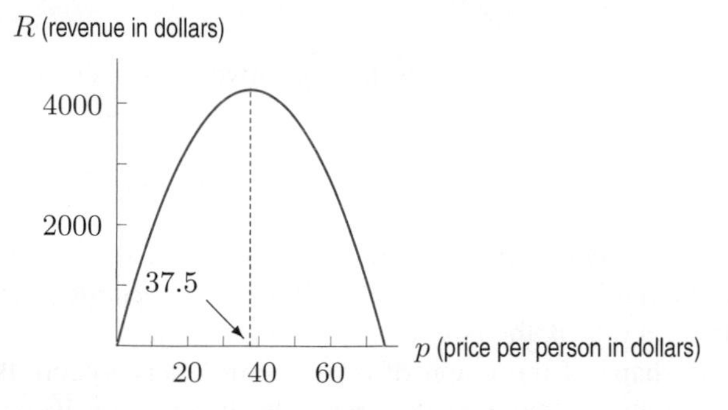

Figure 1.87: Revenue function for concert ticket sales

Example 7 Using a calculator or computer, sketch $y = x^4$ and $y = x^4 - 15x^2 - 15x$ for $-20 \le x \le 20$ and $0 \le y \le 200{,}000$. What do you observe?

Solution From Figure 1.88, we see that the two graphs look indistinguishable. The reason is that the leading term of each polynomial (the one with the highest power of x) is the same, namely x^4, and for the large values of x in this window, the leading term dominates the other terms.

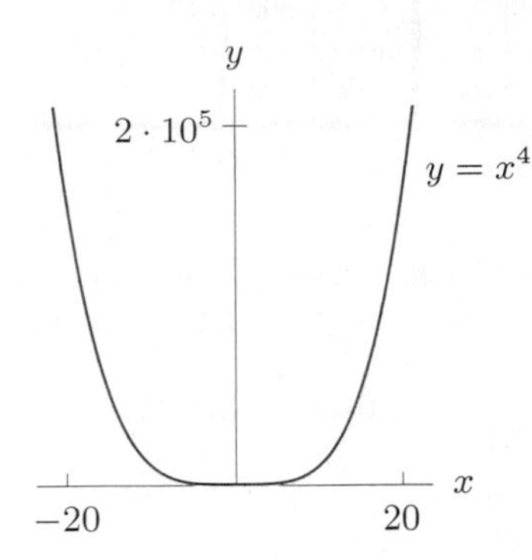

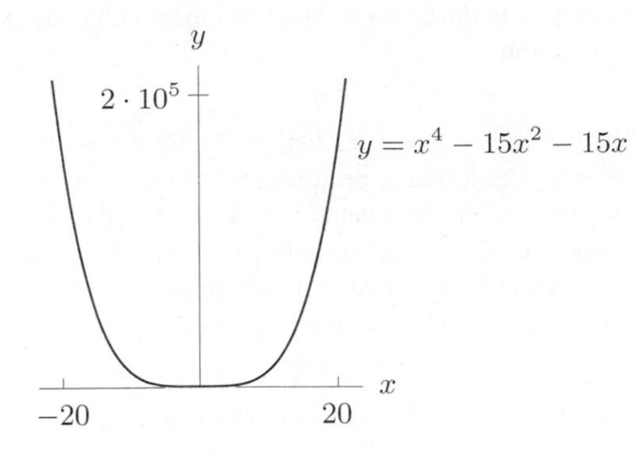

Figure 1.88: Graphs of $y = x^4$ and $y = x^4 - 15x^2 - 15x$ look almost indistinguishable in a large window

Problem 41 compares the graphs of these two functions in a smaller window with Figure 1.88.

We see in Example 7 that, from a distance, the polynomial $y = x^4 - 15x^2 - 15x$ looks like the power function $y = x^4$. In general, if the graph of a polynomial of degree n

$$y = a_n x^n + a_{n-1}x^{n-1} + \cdots + a_1 x + a_0$$

is viewed in a large enough window, it has approximately the same shape as the graph of the power function given by the leading term:

$$y = a_n x^n.$$

Problems for Section 1.9

In Problems 1–12, determine whether or not the function is a power function. If it is a power function, write it in the form $y = kx^p$ and give the values of k and p.

1. $y = 5\sqrt{x}$ **2.** $y = \dfrac{3}{x^2}$ **3.** $y = 2^x$

4. $y = \dfrac{3}{8x}$ **5.** $y = (3x^5)^2$ **6.** $y = \dfrac{5}{2\sqrt{x}}$

7. $y = 3 \cdot 5^x$ **8.** $y = \dfrac{2x^2}{10}$ **9.** $y = \dfrac{8}{x}$

10. $y = (5x)^3$ **11.** $y = 3x^2 + 4$ **12.** $y = \dfrac{x}{5}$

In Problems 13–16, write a formula representing the function.

13. The strength, S, of a beam is proportional to the square of its thickness, h.

14. The energy, E, expended by a swimming dolphin is proportional to the cube of the speed, v, of the dolphin.

15. The average velocity, v, for a trip over a fixed distance, d, is inversely proportional to the time of travel, t.

16. The gravitational force, F, between two bodies is inversely proportional to the square of the distance d between them.

17. The specific heat, s, of an element is the number of calories of heat required to raise the temperature of one gram of the element by one degree Celsius. Use the following table to decide if s is proportional or inversely proportional to the atomic weight, w, of the element. If so, find the constant of proportionality.

Element	Li	Mg	Al	Fe	Ag	Pb	Hg
w	6.9	24.3	27.0	55.8	107.9	207.2	200.6
s	0.92	0.25	0.21	0.11	0.056	0.031	.033

18. The following table gives values for a function $p = f(t)$. Could p be proportional to t?

t	0	10	20	30	40	50
p	0	25	60	100	140	200

19. The blood mass of a mammal is proportional to its body mass. A rhinoceros with body mass 3000 kilograms has blood mass of 150 kilograms. Find a formula for the blood mass of a mammal as a function of the body mass and estimate the blood mass of a human with body mass 70 kilograms.

20. The number of species of lizards, N, found on an island off Baja California is proportional to the fourth root of the area, A, of the island.[49] Write a formula for N as a function of A. Graph this function. Is it increasing or decreasing? Is the graph concave up or concave down? What does this tell you about lizards and island area?

21. The surface area of a mammal, S, satisfies the equation $S = kM^{2/3}$, where M is the body mass, and the constant of proportionality k depends on the body shape of the mammal. A human of body mass 70 kilograms has surface area 18,600 cm^2. Find the constant of proportionality for humans. Find the surface area of a human with body mass 60 kilograms.

22. Biologists estimate that the number of animal species of a certain body length is inversely proportional to the square of the body length.[50] Write a formula for the number of animal species, N, of a certain body length as a function of the length, L. Are there more species at large lengths or at small lengths? Explain.

23. The circulation time of a mammal (that is, the average time it takes for all the blood in the body to circulate once and return to the heart) is proportional to the fourth root of the body mass of the mammal.

(a) Write a formula for the circulation time, T, in terms of the body mass B.
(b) If an elephant of body mass 5230 kilograms has a circulation time of 148 seconds, find the constant of proportionality.
(c) What is the circulation time of a human with body mass 70 kilograms?

24. Allometry is the study of the relative size of different parts of a body as a consequence of growth. In this problem, you will check the accuracy of an allometric equation: the weight of a fish is proportional to the cube of its length.[51] Table 1.30 relates the weight, y, in gm, of plaice (a type of fish) to its length, x, in cm. Does this data support the hypothesis that (approximately) $y = kx^3$? If so, estimate the constant of proportionality, k.

Table 1.30

x	y	x	y	x	y
33.5	332	37.5	455	41.5	623
34.5	363	38.5	500	42.5	674
35.5	391	39.5	538	43.5	724
36.5	419	40.5	574		

25. The DuBois formula relates a person's surface area, s, in m^2, to weight w, in kg, and height h, in cm, by

$$s = 0.01w^{0.25}h^{0.75}.$$

(a) What is the surface area of a person who weighs 65 kg and is 160 cm tall?
(b) What is the weight of a person whose height is 180 cm and who has a surface area of 1.5 m^2?
(c) For people of fixed weight 70 kg, solve for h as a function of s. Simplify your answer.

26. According to the National Association of Realtors,[52] the minimum annual gross income, m, in thousands of dollars, needed to obtain a 30-year home loan of A thousand dollars at 9% is given in Table 1.31. Note that the larger the loan, the greater the income needed.

Table 1.31

A	50	75	100	150	200
m	17.242	25.863	34.484	51.726	68.968

Table 1.32

r	8	9	10	11	12
m	31.447	34.484	37.611	40.814	44.084

Of course, not every mortgage is financed at 9%. In fact, excepting for slight variations, mortgage interest rates are generally determined not by individual banks but by the economy as a whole. The minimum annual gross income, m, in thousands of dollars, needed for a home loan of \$100,000 at various interest rates, r, is given in Table 1.32. Note that obtaining a loan at a time when interest rates are high requires a larger income.

(a) Is the size of the loan, A, proportional to the minimum annual gross income, m?
(b) Is the percentage rate, r, proportional to the minimum annual gross income, m?

[49] Rosenzweig, M.L., *Species Diversity in Space and Time*, p. 143, (Cambridge: Cambridge University Press, 1995).
[50] *US News & World Report*, August 18, 1997, p. 79.
[51] Adapted from "On the Dynamics of Exploited Fish Populations" by R. J. H. Beverton and S. J. Holt, *Fishery Investigations*, Series II, 19, 1957.
[52] "Income needed to get a Mortgage," *The World Almanac 1992*, p. 720.

27. A standard tone of 20,000 dynes/cm^2 (about the loudness of a rock band) is assigned a value of 10. A subject listened to other sounds, such as a light whisper, normal conversation, thunder, a jet plane at takeoff, and so on. In each case, the subject was asked to judge the loudness and assign it a number relative to 10, the value of the standard tone. This is a "judgment of magnitude" experiment. The power law $J = al^{0.3}$ was found to model the situation well, where l is the actual loudness (measured in dynes/cm^2) and J is the judged loudness.

(a) What is the value of a?
(b) What is the judged loudness if the actual loudness is 0.2 dynes/cm^2 (normal conversation)?
(c) What is the actual loudness if judged loudness is 20?

For the functions in Problems 28–35:

(a) What is the degree of the polynomial? Is the leading coefficient positive or negative?
(b) What power function approximates $f(x)$ for large x? Without using a calculator or computer, sketch the graph of the function in a large window.
(c) Using a calculator or computer, sketch a graph of the function. How many turning points does the function have? How does the number of turning points compare to the degree of the polynomial?

28. $f(x) = 5x^3 - 17x^2 + 9x + 50$
29. $f(x) = x^2 + 10x - 5$
30. $f(x) = 8x - 3x^2$
31. $f(x) = 17 + 8x - 2x^3$
32. $f(x) = -9x^5 + 82x^3 + 12x^2$
33. $f(x) = 100 + 5x - 12x^2 + 3x^3 - x^4$
34. $f(x) = 0.01x^4 + 2.3x^2 - 7$
35. $f(x) = 0.2x^7 + 1.5x^4 - 3x^3 + 9x - 15$

36. Each of the graphs in Figure 1.89 is of a polynomial. The windows are large enough to show global behavior.

(a) What is the minimum possible degree of the polynomial?
(b) Is the leading coefficient of the polynomial positive or negative?

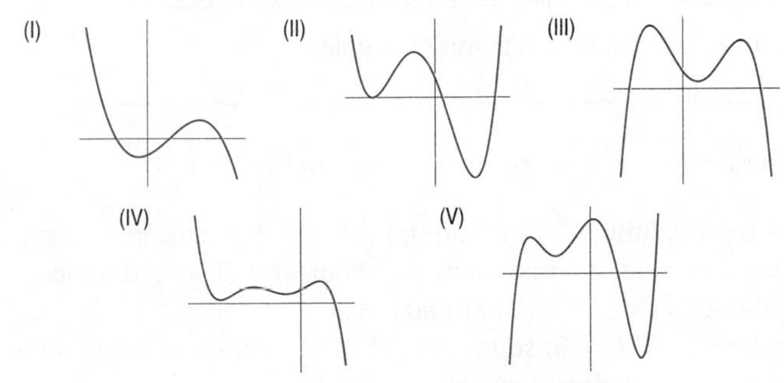

Figure 1.89

37. A sporting goods wholesaler finds that when the price of a product is \$25, the company sells 500 units per week. When the price is \$30, the number sold per week decreases to 460 units.

(a) Find the demand, q, as a function of price, p, assuming that the demand curve is linear.
(b) Use your answer to part (a) to write revenue as a function of price.
(c) Graph the revenue function in part (b). Find the price that maximizes revenue. What is the revenue at this price?

38. A health club has cost and revenue functions given by $C = 10{,}000 + 35q$ and $R = pq$, where q is the number of annual club members and p is the price of a one-year membership. The demand function for the club is $q = 3000 - 20p$.

(a) Use the demand function to write cost and revenue as functions of p.
(b) Graph cost and revenue as a function of p, on the same axes. (Note that price does not go above \$170 and that the annual costs of running the club reach \$120,000.)
(c) Explain why the graph of the revenue function has the shape it does.
(d) For what prices does the club make a profit?
(e) Estimate the annual membership fee that maximizes profit. Mark this point on your graph.

39. Use shifts of power functions to find a possible formula for each of the graphs:

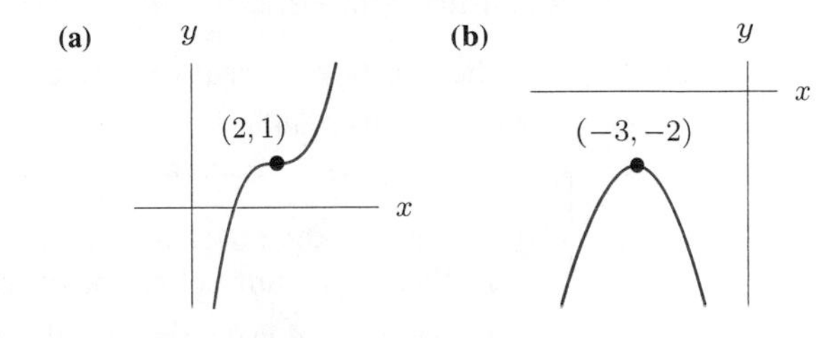

40. Find a calculator window in which the graphs of $f(x) = x^3 + 1000x^2 + 1000$ and $g(x) = x^3 - 1000x^2 - 1000$ appear indistinguishable.

41. Do the functions $y = x^4$ and $y = x^4 - 15x^2 - 15x$ look similar in the window $-4 \le x \le 4; -100 \le y \le 100$? Comment on the difference between your answer to this question and what you see in Figure 1.88.

1.10 PERIODIC FUNCTIONS

What Are Periodic Functions?

Many functions have graphs that oscillate, resembling a wave. Figure 1.90 shows the number of new housing construction starts (one-family units) in the US, 2002–2004, where t is time in quarter-years.[53] Notice that few new homes begin construction during the first quarter of a year (January, February, and March), whereas many new homes are begun in the second quarter (April, May, and June). We expect this pattern of oscillations to continue into the future.

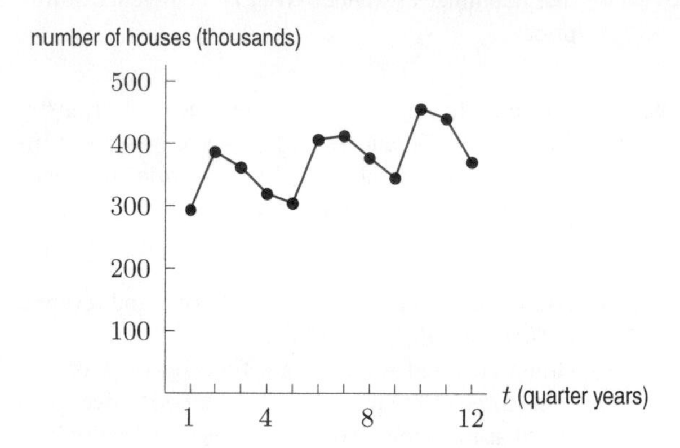

Figure 1.90: New housing construction starts, 2002–2004

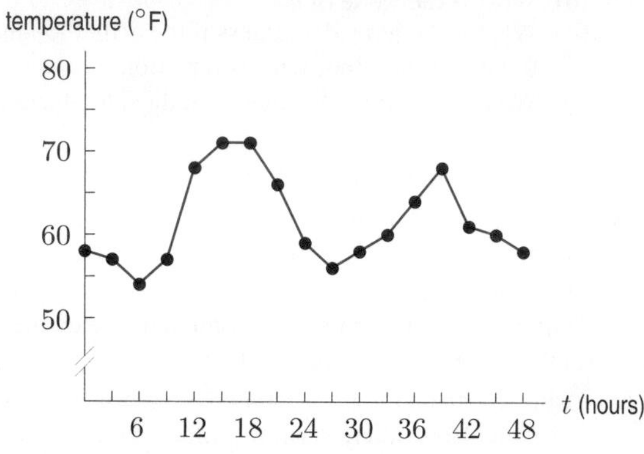

Figure 1.91: Temperature in Phoenix after midnight February 17, 2005

Let's look at another example. Figure 1.91 is a graph of the temperature (in °F) in Phoenix, Az, in hours after midnight, February 17, 2005. Notice that the maximum is in the afternoon and the minimum is in the early morning.[54] Again, the graph looks like a wave.

Functions whose values repeat at regular intervals are called *periodic*. Many processes, such as the number of housing starts or the temperature, are approximately periodic. The water level in a tidal basin, the blood pressure in a heart, retail sales in the US, and the position of air molecules transmitting a musical note are also all periodic functions of time.

Amplitude and Period

Periodic functions repeat exactly the same cycle forever. If we know one cycle of the graph, we know the entire graph.

For any periodic function of time:

- The **amplitude** is half the difference between its maximum and minimum values.
- The **period** is the time for the function to execute one complete cycle.

Example 1 Estimate the amplitude and period of the new housing starts function shown in Figure 1.90.

Solution Figure 1.90 is not exactly periodic, since the maximum and minimum are not the same for each cycle. Nonetheless, the minimum is about 300, and the maximum is about 450. The difference between them is 150, so the amplitude is about $\frac{1}{2}(150) = 75$ thousand houses.

The wave completes a cycle between $t = 1$ and $t = 5$, so the period is $t = 4$ quarter-years, or one year. The business cycle for new housing construction is one year.

[53] http://www.census.gov/const/www/quarterly_starts_completions.pdf, accessed April 15, 2005.
[54] http://www.weather.com, accessed February 20, 2005.

Example 2 Figure 1.92 shows the temperature in an unopened freezer. Estimate the temperature in the freezer at 12:30 and at 2:45.

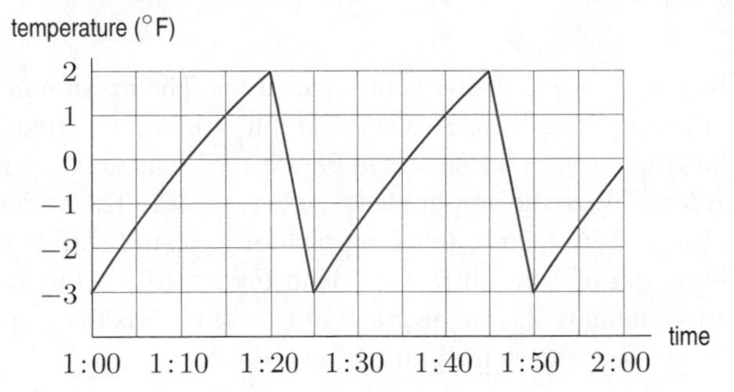

Figure 1.92: Oscillating freezer temperature. Estimate the temperature at 12:30 and 2:45

Solution The maximum and minimum values each occur every 25 minutes, so the period is 25 minutes. The temperature at 12:30 should be the same as at 12:55 and at 1:20, namely, 2°F. Similarly, the temperature at 2:45 should be the same as at 2:20 and 1:55, or about -1.5°F.

The Sine and Cosine

Many periodic functions are represented using the functions called *sine* and *cosine*. The keys for the sine and cosine on a calculator are usually labeled as sin and cos.

Warning: Your calculator can be in either "degree" mode or "radian" mode. For this book, always use "radian" mode.

Graphs of the Sine and Cosine

The graphs of the sine and the cosine functions are periodic; see Figures 1.93 and 1.94. Notice that the graph of the cosine function is the graph of the sine function, shifted $\pi/2$ to the left.

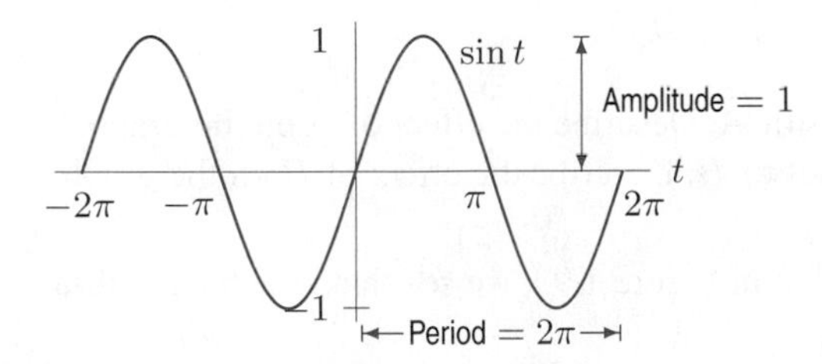

Figure 1.93: Graph of $\sin t$

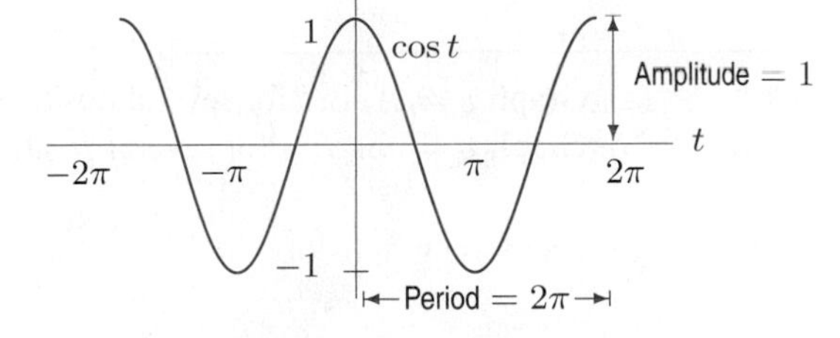

Figure 1.94: Graph of $\cos t$

The maximum and minimum values of $\sin t$ are $+1$ and -1, so the amplitude of the sine function is 1. The graph of $y = \sin t$ completes a cycle between $t = 0$ and $t = 2\pi$; the rest of the graph repeats this portion. The period of the sine function is 2π.

Example 3 Use a graph of $y = 3\sin 2t$ to estimate the amplitude and period of this function.

Solution In Figure 1.95, the waves have a maximum of $+3$ and a minimum of -3, so the amplitude is 3. The graph completes one complete cycle between $t = 0$ and $t = \pi$, so the period is π.

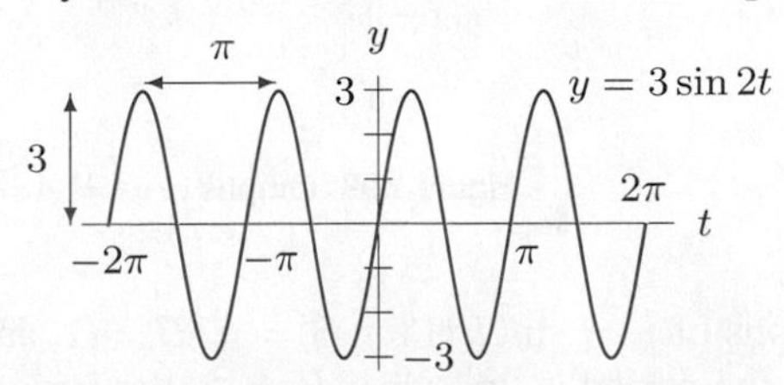

Figure 1.95: The amplitude is 3 and the period is π

Example 4 Explain how the graphs of each of the following functions differ from the graph of $y = \sin t$.

(a) $y = 6\sin t$ (b) $y = 5 + \sin t$ (c) $y = \sin(t + \frac{\pi}{2})$

Solution

(a) The graph of $y = 6\sin t$ is in Figure 1.96. The maximum and minimum values are $+6$ and -6, so the amplitude is 6. This is the graph of $y = \sin t$ stretched vertically by a factor of 6.

(b) The graph of $y = 5 + \sin t$ is in Figure 1.97. The maximum and minimum values of this function are 6 and 4, so the amplitude is $(6 - 4)/2 = 1$. The amplitude (or size of the wave) is the same as for $y = \sin t$, since this is a graph of $y = \sin t$ shifted up 5 units.

(c) The graph of $y = \sin(t + \frac{\pi}{2})$ is in Figure 1.98. This has the same amplitude, namely 1, and period, namely 2π, as the graph of $y = \sin t$. It is the graph of $y = \sin t$ shifted $\pi/2$ units to the left. (In fact, this is the graph of $y = \cos t$.)

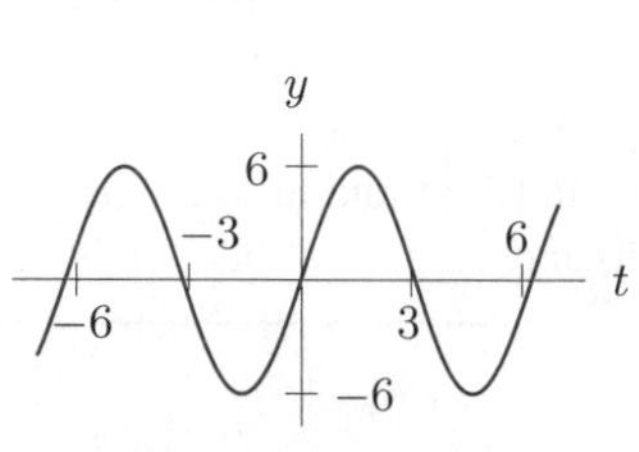

Figure 1.96: Graph of $y = 6\sin t$

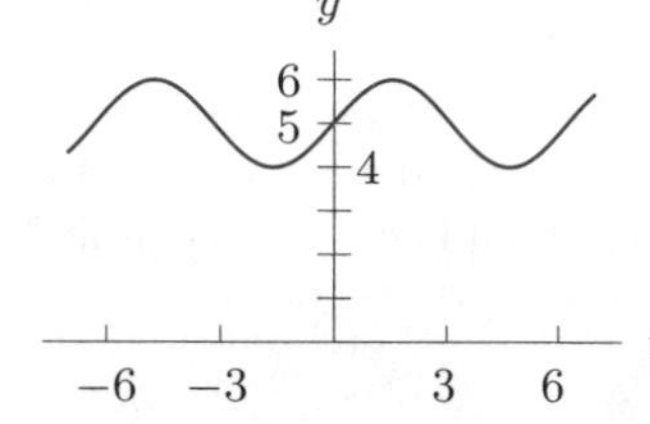

Figure 1.97: Graph of $y = 5 + \sin t$

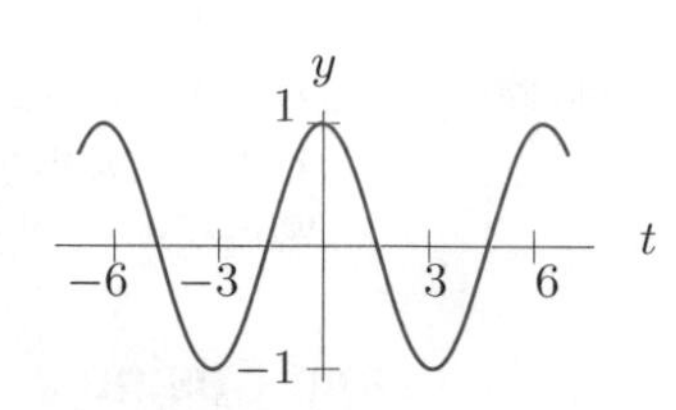

Figure 1.98: Graph of $y = \sin(t + \frac{\pi}{2})$

Families of Curves: The Graph of $y = A\sin(Bt)$

The constants A and B in the expression $y = A\sin(Bt)$ are called *parameters*. We can study families of curves by varying one parameter at a time and studying the result.

Example 5

(a) Graph $y = A\sin t$ for several positive values of A. Describe the effect of A on the graph.

(b) Graph $y = \sin(Bt)$ for several positive values of B. Describe the effect of B on the graph.

Solution

(a) From the graphs of $y = A\sin t$ for $A = 1, 2, 3$ in Figure 1.99, we see that A is the amplitude.

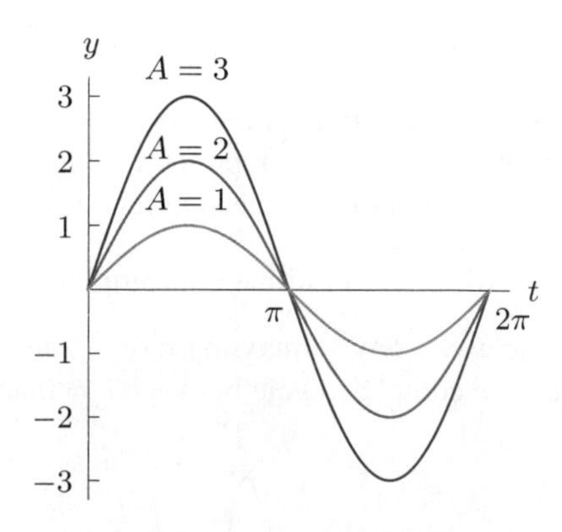

Figure 1.99: Graphs of $y = A\sin t$ with $A = 1,2,3$

(b) The graphs of $y = \sin(Bt)$ for $B = \frac{1}{2}$, $B = 1$, and $B = 2$ are shown in Figure 1.100. When $B = 1$, the period is 2π; when $B = 2$, the period is π; and when $B = \frac{1}{2}$, the period is 4π.

The parameter B affects the period of the function. The graphs suggest that the larger B is, the shorter the period. In fact, the period is $2\pi/B$.

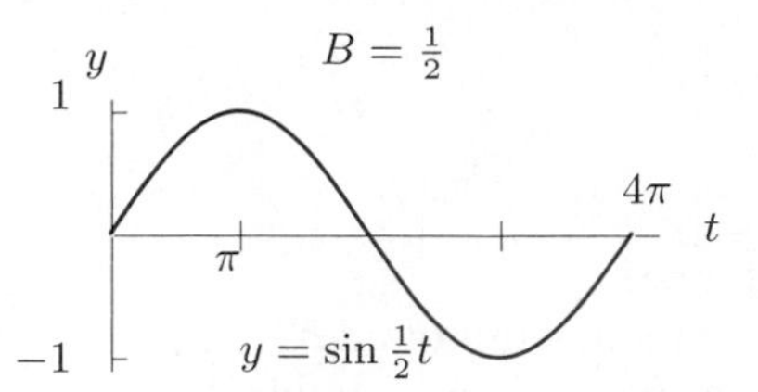

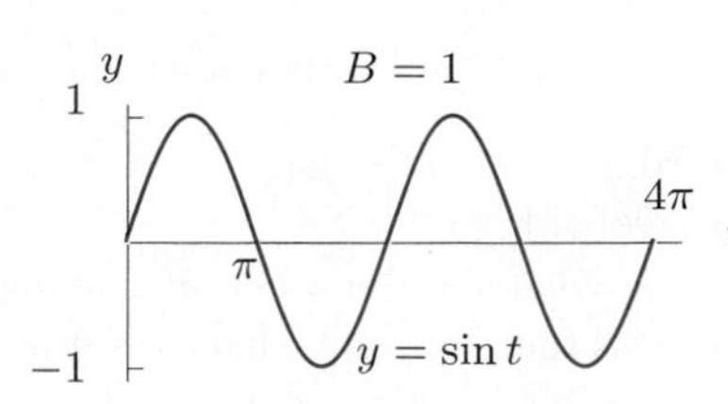

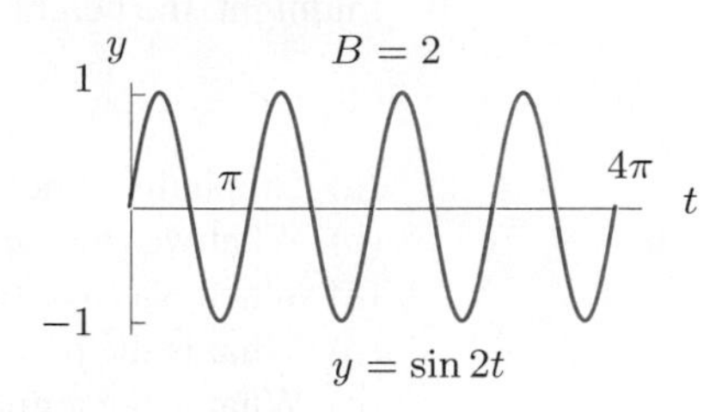

Figure 1.100: Graphs of $y = \sin(Bt)$ with $B = \frac{1}{2}, 1, 2$

In Example 5, the amplitude of $y = A\sin(Bt)$ was determined by the parameter A, and the period was determined by the parameter B. In general, we have

> The functions $y = A\sin(Bt) + C$ and $y = A\cos(Bt) + C$ are periodic with
>
> $$\text{Amplitude} = |A|, \qquad \text{Period} = \frac{2\pi}{|B|}, \qquad \text{Vertical shift} = C$$

Example 6 Find possible formulas for the following periodic functions.

(a)

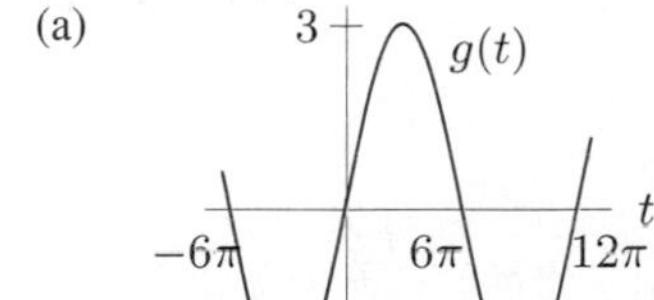

(b)

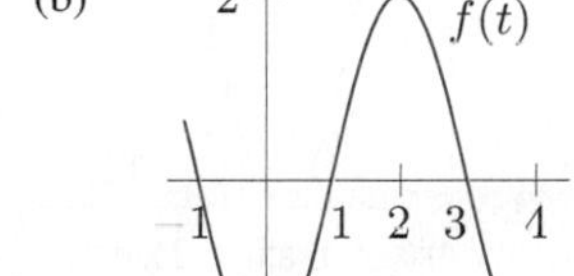

(c)

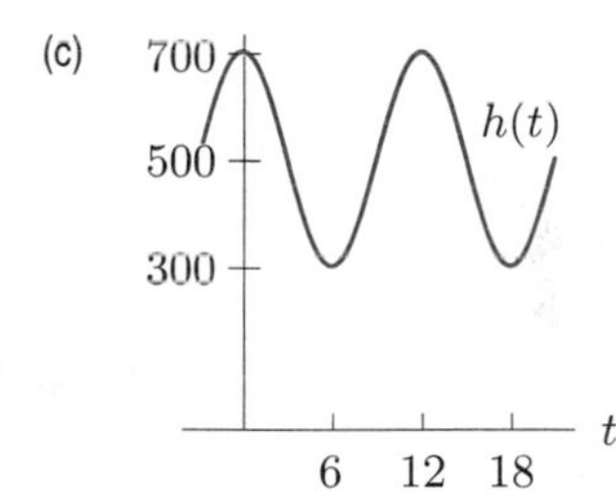

Solution

(a) This function looks like a sine function of amplitude 3, so $g(t) = 3\sin(Bt)$. Since the function executes one full oscillation between $t = 0$ and $t = 12\pi$, when t changes by 12π, the quantity Bt changes by 2π. This means $B \cdot 12\pi = 2\pi$, so $B = 1/6$. Therefore, $g(t) = 3\sin(t/6)$ has the graph shown.

(b) This function looks like an upside down cosine function with amplitude 2, so $f(t) = -2\cos(Bt)$. The function completes one oscillation between $t = 0$ and $t = 4$. Thus, when t changes by 4, the quantity Bt changes by 2π, so $B \cdot 4 = 2\pi$, or $B = \pi/2$. Therefore, $f(t) = -2\cos(\pi t/2)$ has the graph shown.

(c) This function looks like a cosine function. The maximum is 700 and the minimum is 300, so the amplitude is $\frac{1}{2}(700 - 300) = 200$. The height halfway between the maximum and minimum is 500, so the cosine curve has been shifted up 500 units, so $h(t) = 500 + 200\cos(Bt)$. The period is 12, so $B \cdot 12 = 2\pi$. Thus, $B = \pi/6$. The function $h(t) = 500 + 200\cos(\pi t/6)$ has the graph shown.

Example 7 On April 25, 2005, high tide in Portland, Maine was at midnight.[55] The height of the water in the harbor is a periodic function, since it oscillates between high and low tide. If t is in hours since midnight, the height (in feet) is approximated by the formula

$$y = 4.9 + 4.4\cos\left(\frac{\pi}{6}t\right).$$

(a) Graph this function from $t = 0$ to $t = 24$.
(b) What was the water level at high tide?
(c) When was low tide, and what was the water level at that time?
(d) What is the period of this function, and what does it represent in terms of tides?
(e) What is the amplitude of this function, and what does it represent in terms of tides?

Solution

(a) See Figure 1.101.
(b) The water level at high tide was 9.3 feet (given by the y-intercept on the graph).
(c) Low tide occurs at $t = 6$ (6 am) and at $t = 18$ (6 pm). The water level at this time is 0.5 feet.
(d) The period is 12 hours and represents the interval between successive high tides or successive low tides. Of course, there is something wrong with the assumption in the model that the period is 12 hours. If so, the high tide would always be at noon or midnight, instead of progressing slowly through the day, as it in fact does. The interval between successive high tides actually averages about 12 hours 39 minutes, which could be taken into account in a more precise mathematical model.
(e) The maximum is 9.3, and the minimum is 0.5, so the amplitude is $(9.3 - 0.5)/2$, which is 4.4 feet. This represents half the difference between the depths at high and low tide.

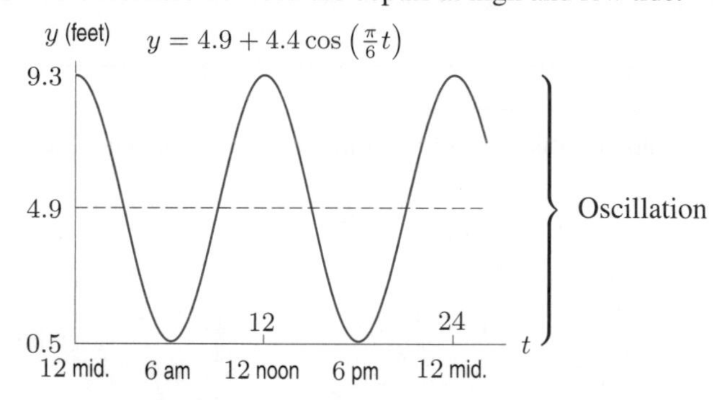

Figure 1.101: Graph of the function approximating the depth of the water in Portland, Maine on April 25, 2005

Problems for Section 1.10

1. A graduate student in environmental science studied the temperature fluctuations of a river. Figure 1.102 shows the temperature of the river (in °C) every hour, with hour 0 being midnight of the first day.

(a) Explain why a periodic function could be used to model these data.
(b) Approximately when does the maximum occur? The minimum? Why does this make sense?
(c) What is the period for these data? What is the amplitude?

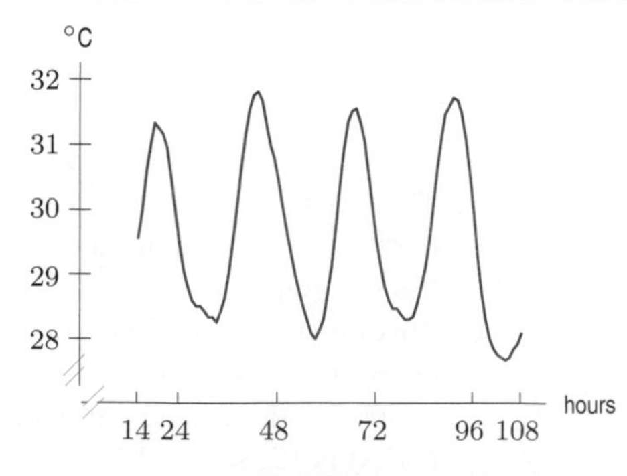

Figure 1.102

[55] www.maineharbors.com/aprpt05.htm, accessed April 15, 2005.

Find the period and amplitude in Problems 2–4.

2. $y = 7\sin(3t)$

3. $z = 3\cos(u/4) + 5$

4. $r = 0.1\sin(\pi t) + 2$

5. A person breathes in and out every three seconds. The volume of air in the person's lungs varies between a minimum of 2 liters and a maximum of 4 liters. Which of the following is the best formula for the volume of air in the person's lungs as a function of time?

(a) $y = 2 + 2\sin\left(\frac{\pi}{3}t\right)$ **(b)** $y = 3 + \sin\left(\frac{2\pi}{3}t\right)$

(c) $y = 2 + 2\sin\left(\frac{2\pi}{3}t\right)$ **(d)** $y = 3 + \sin\left(\frac{\pi}{3}t\right)$

6. Sketch a possible graph of sales of sunscreen in the northeastern US over a 3-year period, as a function of months since January 1 of the first year. Explain why your graph should be periodic. What is the period?

For Problems 7–12, sketch graphs of the functions. What are their amplitudes and periods?

7. $y = 3\sin x$

8. $y = 3\sin 2x$

9. $y = -3\sin 2\theta$

10. $y = 4\cos 2x$

11. $y = 4\cos(\frac{1}{2}t)$

12. $y = 5 - \sin 2t$

13. Delta Cephei is one of the most visible stars in the night sky. Its brightness has periods of 5.4 days, the average brightness is 4.0 and its brightness varies by ± 0.35. Find a formula that models the brightness of Delta Cephei as a function of time, t, with $t = 0$ at peak brightness.

14. Values of a function are given in the following table. Explain why this function appears to be periodic. Approximately what are the period and amplitude of the function? Assuming that the function is periodic, estimate its value at $t = 15$, at $t = 75$, and at $t = 135$.

t	20	25	30	35	40	45	50	55	60
$f(t)$	1.8	1.4	1.7	2.3	2.0	1.8	1.4	1.7	2.3

15. The following table shows values of a periodic function $f(x)$. The maximum value attained by the function is 5.

(a) What is the amplitude of this function?
(b) What is the period of this function?
(c) Find a formula for this periodic function.

x	0	2	4	6	8	10	12
$f(x)$	5	0	-5	0	5	0	-5

16. Figure 1.103 shows the levels of the hormones estrogen and progesterone during the monthly ovarian cycles in females.[56] Is the level of both hormones periodic? What is the period in each case? Approximately when in the monthly cycle is estrogen at a peak? Approximately when in the monthly cycle is progesterone at a peak?

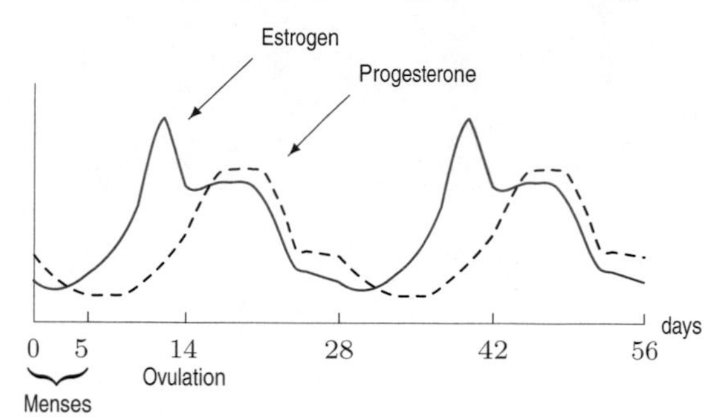

Figure 1.103

17. Figure 1.104 shows the number of reported[57] cases of mumps by month, in the US, for 1972–73.

(a) Find the period and amplitude of this function, and interpret each in terms of mumps.
(b) Predict the number of cases of mumps 30 months and 45 months after January 1, 1972.

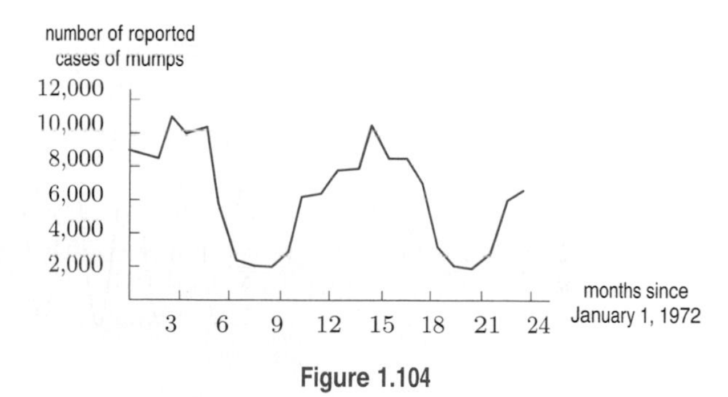

Figure 1.104

18. Most breeding birds in the northeast US migrate elsewhere during the winter. The number of bird species in an Ohio forest preserve oscillates between a high of 28 in June and a low of 10 in December.[58]

(a) Graph the number of bird species in this preserve as a function of t, the number of months since June. Include at least three years on your graph.
(b) What are the amplitude and period of this function?
(c) Find a formula for the number of bird species, B, as a function of the number of months, t since June.

[56] Robert M. Julien, *A Primer of Drug Action*, Seventh Edition, p. 360, (W. H. Freeman and Co., New York: 1995).

[57] Center for Disease Control, 1974, *Reported Morbidity and Mortality in the United States 1973*, Vol. 22, No. 53. Prior to the licensing of the vaccine in 1967, 100,000–200,000 cases of mumps were reported annually. Since 1995, fewer than 1000 cases are reported annually. Source: CDC.

[58] Rosenzweig, M.L., *Species Diversity in Space and Time*, p. 71, (Cambridge University Press, 1995).

For Problems 19–30, find a possible formula for each graph.

19.

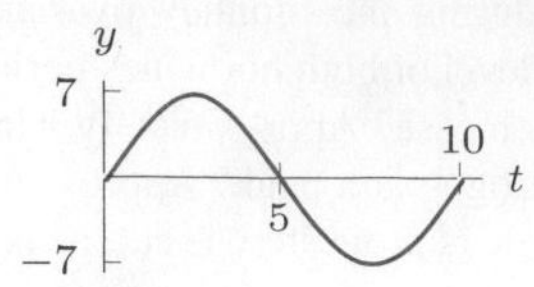

20.

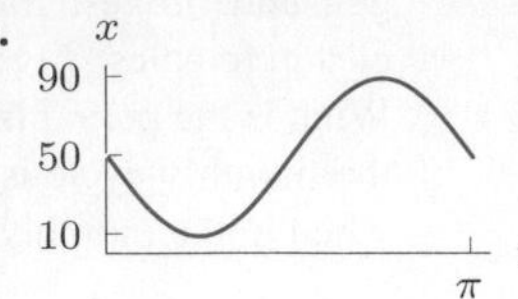

21.

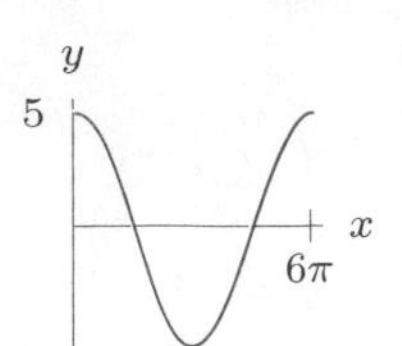

22.

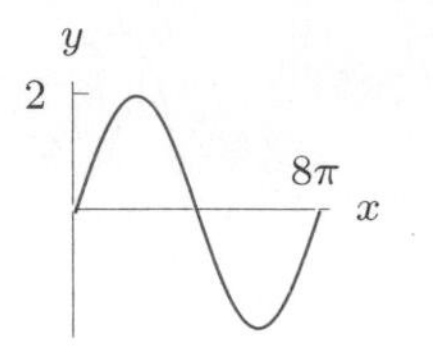

23.

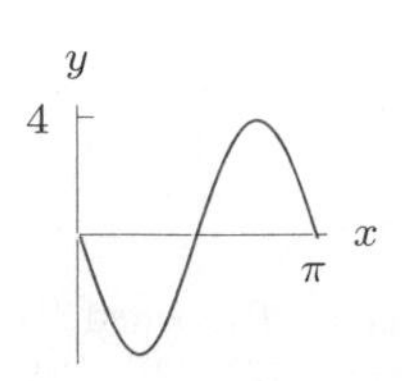

24.

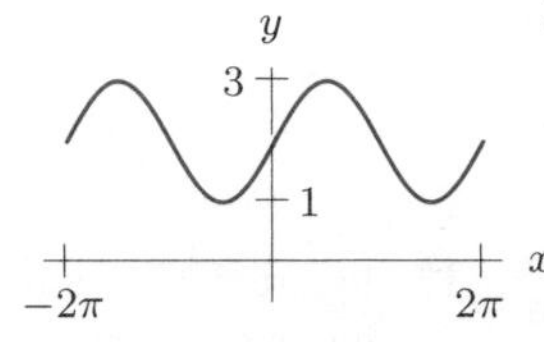

25.

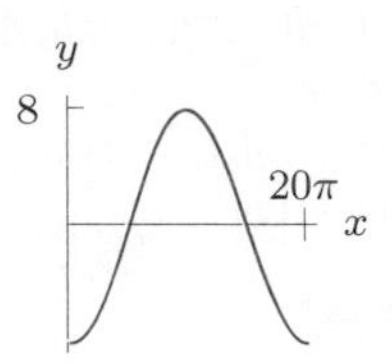

26.

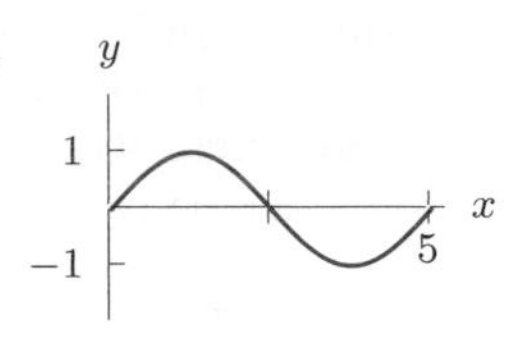

27.

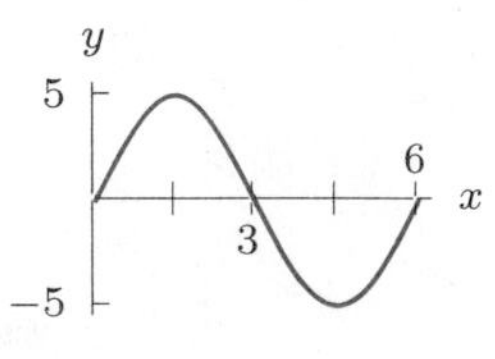

28.

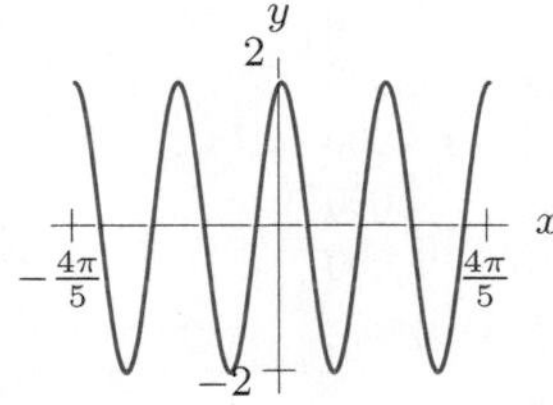

29.

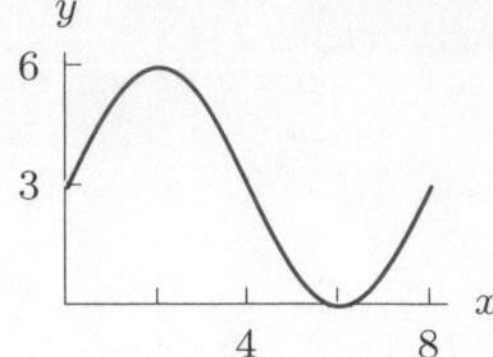

30.

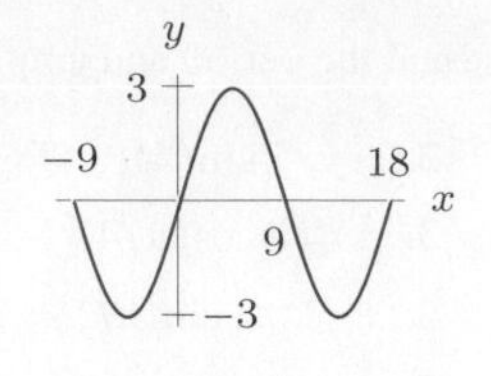

31. The depth of water in a tank oscillates once every 6 hours. If the smallest depth is 5.5 feet and the largest depth is 8.5 feet, find a possible formula for the depth in terms of time in hours.

32. The desert temperature, H, oscillates daily between 40° F at 5 am and 80° F at 5 pm. Write a possible formula for H in terms of t, measured in hours from 5 am.

33. Table 1.33 gives values for $g(t)$, a periodic function.

(a) Estimate the period and amplitude for this function.
(b) Estimate $g(34)$ and $g(60)$.

Table 1.33

t	0	2	4	6	8	10	12	14
$g(t)$	14	19	17	15	13	11	14	19
t	16	18	20	22	24	26	28	
$g(t)$	17	15	13	11	14	19	17	

34. The Bay of Fundy in Canada has the largest tides in the world. The difference between low and high water levels is 15 meters (nearly 50 feet). At a particular point the depth of the water, y meters, is given as a function of time, t, in hours since midnight by

$$y = D + A\cos\left(B(t - C)\right).$$

(a) What is the physical meaning of D?
(b) What is the value of A?
(c) What is the value of B? Assume the time between successive high tides is 12.4 hours.
(d) What is the physical meaning of C?

CHAPTER SUMMARY

- **Function terminology**
  Domain/range, increasing/decreasing, concavity, intercepts.
- **Linear functions**
  Slope, y-intercept. Grow by equal amounts in equal times.
- **Economic applications**
  Cost, revenue, and profit functions, break-even point. Supply and demand curves, equilibrium point. Depreciation function. Budget constraint. Present and future value.
- **Change, average rate of change, relative rate of change**
- **Exponential functions**
  Exponential growth and decay, growth rate, the number e, continuous growth rate, doubling time, half life, compound interest. Grow by equal percentages in equal times.
- **The natural logarithm function**
- **New functions from old**
  Composition, shifting, stretching.
- **Power functions and proportionality**
- **Polynomials**
- **Periodic functions**
  Sine, cosine, amplitude, period.

REVIEW PROBLEMS FOR CHAPTER ONE

1. Which graph in Figure 1.105 best matches each of the following stories?[59] Write a story for the remaining graph.

(a) I had just left home when I realized I had forgotten my books, and so I went back to pick them up.
(b) Things went fine until I had a flat tire.
(c) I started out calmly but sped up when I realized I was going to be late.

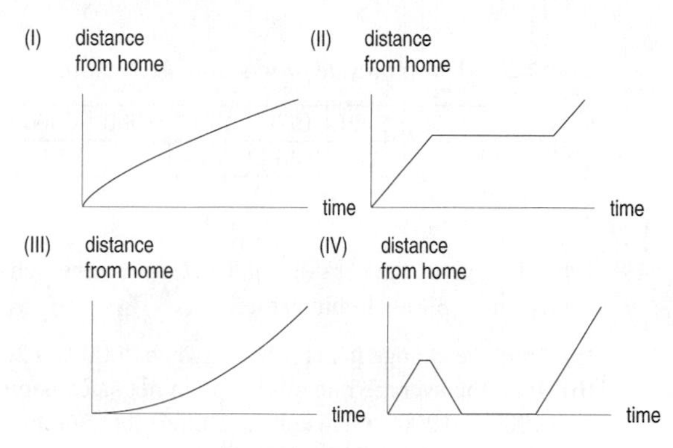

Figure 1.105

2. The population of Washington DC grew from 1900 to 1950, stayed approximately constant during the 1950s, and decreased from about 1960 to 2000. Graph the population as a function of years since 1900.

3. It warmed up throughout the morning, and then suddenly got much cooler around noon, when a storm came through. After the storm, it warmed up before cooling off at sunset. Sketch temperature as a function of time.

4. A gas tank 6 meters underground springs a leak. Gas seeps out and contaminates the soil around it. Graph the amount of contamination as a function of the depth (in meters) below ground.

5. **(a)** The graph of $r = f(p)$ is in Figure 1.106. What is the value of r when p is 0? When p is 3?
(b) What is $f(2)$?

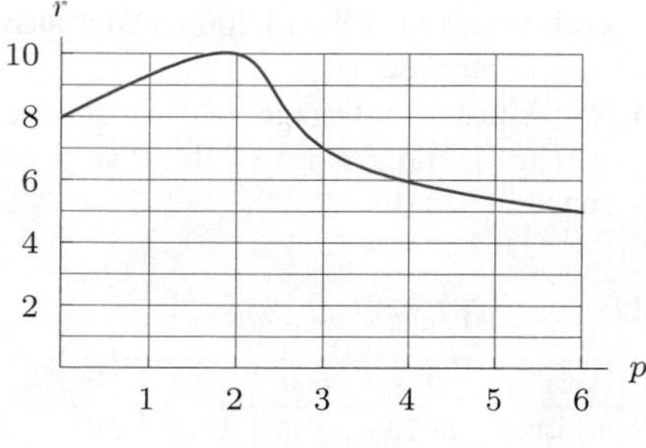

Figure 1.106

6. The yield, Y, of an apple orchard (in bushels) as a function of the amount, a, of fertilizer (in pounds) used on the orchard is shown in Figure 1.107.

(a) Describe the effect of the amount of fertilizer on the yield of the orchard.
(b) What is the vertical intercept? Explain what it means in terms of apples and fertilizer.
(c) What is the horizontal intercept? Explain what it means in terms of apples and fertilizer.
(d) What is the range of this function for $0 \leq a \leq 80$?
(e) Is the function increasing or decreasing at $a = 60$?
(f) Is the graph concave up or down near $a = 40$?

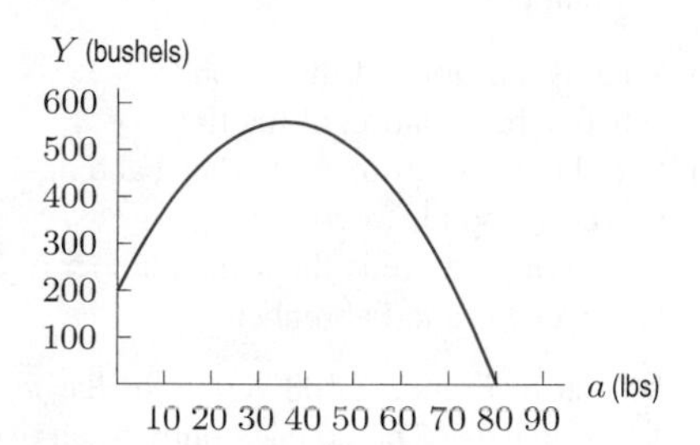

Figure 1.107

7. Let $y = f(x) = 3x - 5$.

(a) What is $f(1)$?
(b) Find the value of y when x is 5.
(c) Find the value of x when y is 4.
(d) Find the average rate of change of f between $x = 2$ and $x = 4$.

Find the equation of the line passing through the points in Problems 8–11.

8. $(0, -1)$ and $(2, 3)$

9. $(-1, 3)$ and $(2, 2)$

10. $(0, 2)$ and $(2, 2)$

11. $(-1, 3)$ and $(-1, 4)$

12. Find the linear equation used to generate the values in Table 1.34.

Table 1.34

x	5.2	5.3	5.4	5.5	5.6
y	27.8	29.2	30.6	32.0	33.4

13. The percentage of people, P, below the poverty level in the US[60] is given in Table 1.35. Find a formula for the percentage as a linear function of time since 2000.

Table 1.35

Year (since 2000)	0	1	2	3
P (percentage)	11.3	11.7	12.1	12.5

[59] Adapted from Jan Terwel, "Real Math in Cooperative Groups in Secondary Education." *Cooperative Learning in Mathematics*, ed. Neal Davidson, p. 234, (Reading: Addison Wesley, 1990).
[60] *The World Almanac and Book of Facts 2005*, p. 128 (New York).

14. Residents of the town of Maple Grove who are connected to the municipal water supply are billed a fixed amount yearly plus a charge for each cubic foot of water used. A household using 1000 cubic feet was billed \$90, while one using 1600 cubic feet was billed \$105.

(a) What is the charge per cubic foot?
(b) Write an equation for the total cost of a resident's water as a function of cubic feet of water used.
(c) How many cubic feet of water used would lead to a bill of \$130?

15. The graph of Fahrenheit temperature, °F, as a function of Celsius temperature, °C, is a line. You know that 212°F and 100°C both represent the temperature at which water boils. Similarly, 32°F and 0°C both represent water's freezing point.

(a) What is the slope of the graph?
(b) What is the equation of the line?
(c) Use the equation to find what Fahrenheit temperature corresponds to 20°C.
(d) What temperature is the same number of degrees in both Celsius and Fahrenheit?

16. The graphs in Figure 1.108 represent the temperature, H, of four loaves of bread each put into an oven at time $t = 0$.

(a) Which curve corresponds to the bread that was put into the hottest oven?
(b) Which curve corresponds to the bread that had the lowest temperature at the time that it was put into the oven?
(c) Which two curves correspond to loaves of bread that were at the same temperature when they were put into the oven?
(d) Write a sentence describing any differences between the curves shown in (II) and (III). In terms of bread, what might cause this difference?

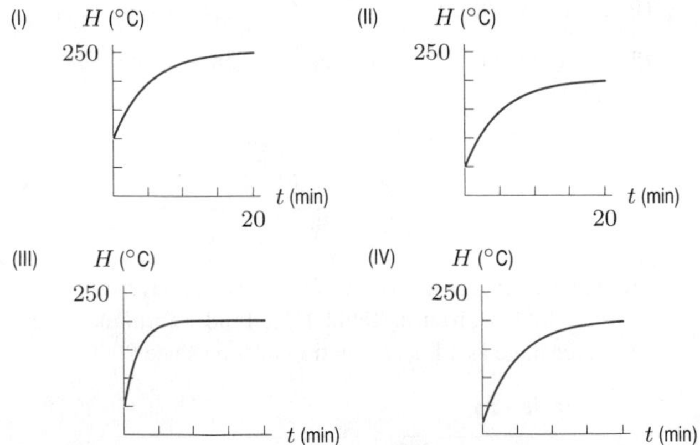

Figure 1.108

17. You drive at a constant speed from Chicago to Detroit, a distance of 275 miles. About 120 miles from Chicago you pass through Kalamazoo, Michigan. Sketch a graph of your distance from Kalamazoo as a function of time.

18. Table 1.36 shows world bicycle production.[61]

(a) Find the change in bicycle production between 1950 and 2000. Give units.
(b) Find the average rate of change in bicycle production between 1950 and 2000. Give units and interpret your answer in terms of bicycle production.

Table 1.36 *World bicycle production, in millions*

Year	1950	1960	1970	1980	1990	2000
Bicycles	11	20	36	62	92	101

19. Table 1.37 gives the net sales of The Gap, Inc, which operates nearly 3000 clothing stores.[62]

(a) Find the change in net sales between 2000 and 2003.
(b) Find the average rate of change in net sales between 2000 and 2003. Give units and interpret your answer.
(c) From 1998 to 2003, were there any one-year intervals during which the average rate of change was negative? If so, when?

Table 1.37 *Gap net sales, in millions of dollars*

Year	1998	1999	2000	2001	2002	2003
Sales	9054	11,635	13,673	13,848	14,455	15,854

20. Find the average rate of change of $f(x) = 3x^2 + 4$ between $x = -2$ and $x = 1$. Illustrate your answer graphically.

21. Table 1.38 shows attendance in millions at NFL football games.[63]

(a) Find the average rate of change in the number of games from 1999 to 2003. Give units.
(b) Find the annual increase in the number of games for each year from 1999 to 2003. (Your answer should be four numbers.)
(c) Show that the average rate of change found in part (a) is the average of the four yearly changes found in part (b).

Table 1.38 *NFL football games*

Year	1999	2000	2001	2002	2003
Attendance	20.76	20.95	20.59	21.51	21.64

[61] www.earth-policy.org/Indicators/indicator11_data1.htm, accessed April 19, 2005.
[62] www.gapinc.com/financmedia/AR_proxy.htm, accessed February 2, 2005.
[63] *Statistical Abstracts of the United States 2004–2005, Table 1239.*

22. Sketch reasonable graphs for the following. Pay particular attention to the concavity of the graphs.

(a) The total revenue generated by a car rental business, plotted against the amount spent on advertising.
(b) The temperature of a cup of hot coffee standing in a room, plotted as a function of time.

23. Each of the functions g, h, k in Table 1.39 is increasing, but each increases in a different way. Which of the graphs in Figure 1.109 best fits each function?

(a) (b) (c)

Figure 1.109

Table 1.39

t	$g(t)$	$h(t)$	$k(t)$
1	23	10	2.2
2	24	20	2.5
3	26	29	2.8
4	29	37	3.1
5	33	44	3.4
6	38	50	3.7

Find possible formulas for the graphs in Problems 24–29.

24.

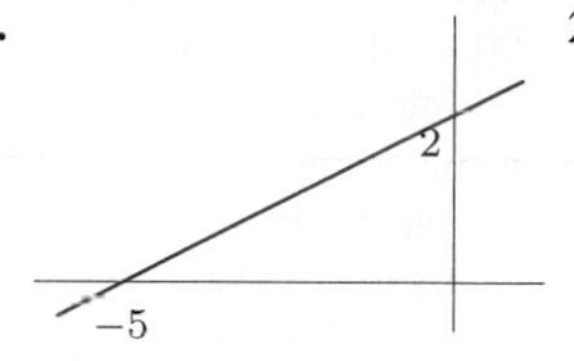

25.

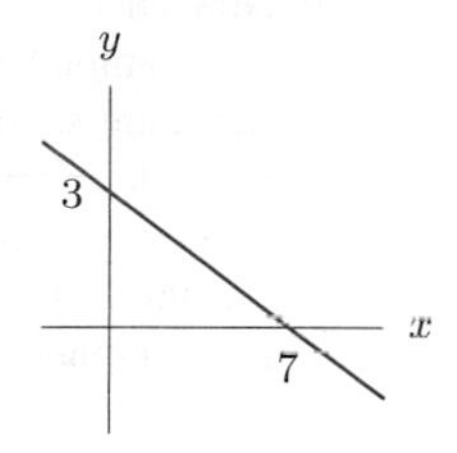

26.

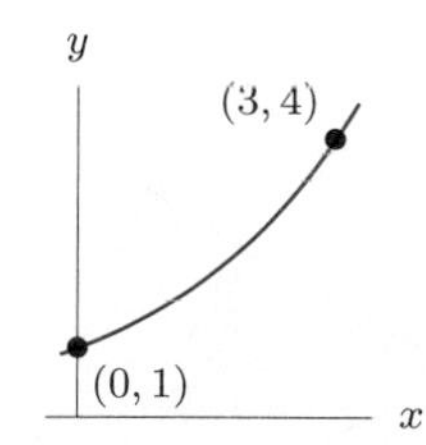

27.

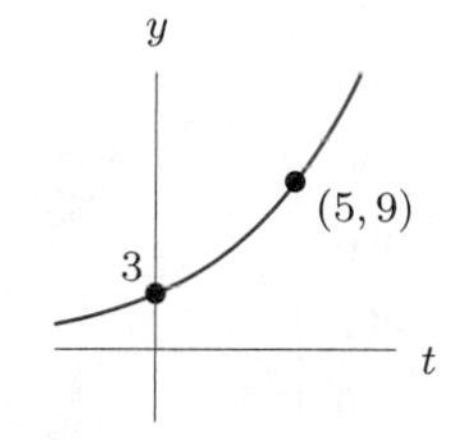

28.

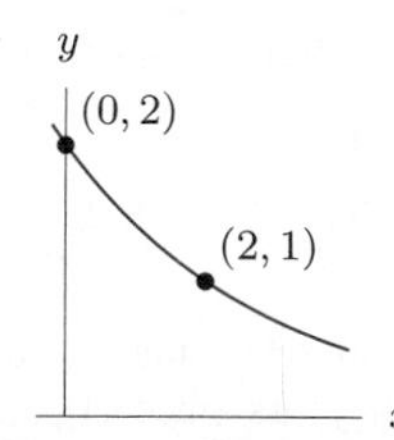

29.

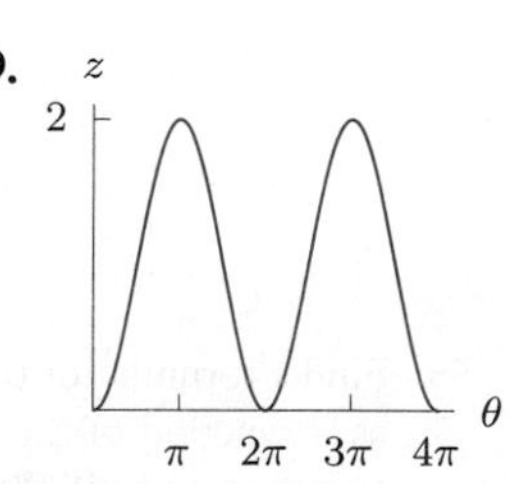

30. Find the average rate of change between $x = 0$ and $x = 10$ of each of the following functions: $y = x, y = x^2, y = x^3$, and $y = x^4$. Which has the largest average rate of change? Graph the four functions, and draw lines whose slopes represent these average rates of change.

31. The rate, R, at which a population in a confined space increases is proportional to the product of the current population, P, and the difference between the carrying capacity, L, and the current population. (The carrying capacity is the maximum population the environment can sustain.)

(a) Write R as a function of P.
(b) Sketch R as a function of P.

32. Table 1.40 gives values for three functions. Which functions could be linear? Which could be exponential? Which are neither? For those which could be linear or exponential, give a possible formula for the function.

Table 1.40

x	$f(x)$	$g(x)$	$h(x)$
0	25	30.8	15,000
1	20	27.6	9,000
2	14	24.4	5,400
3	7	21.2	3,240

In Problems 33–34, use shifts of a power function to find a possible formula for the graph.

33.

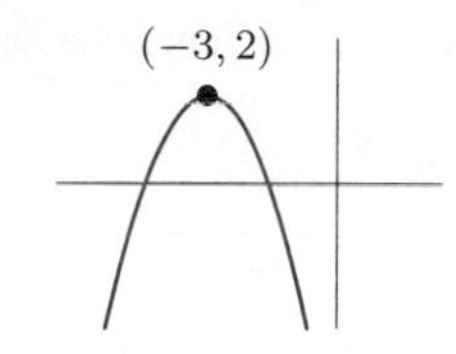

34.

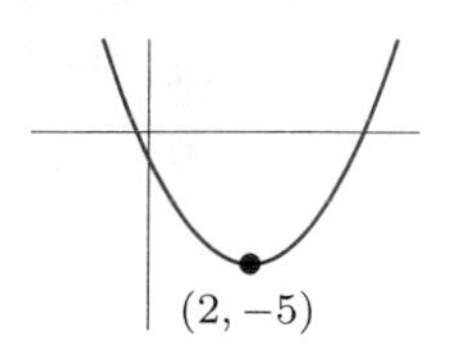

For Problems 35–38, solve for x using logs.

35. $3^x = 11$

36. $20 = 50(1.04)^x$

37. $e^{5x} = 100$

38. $25e^{3x} = 10$

39. Write the exponential functions $P = e^{0.08t}$ and $Q = e^{-0.3t}$ in the form $P = a^t$ and $Q = b^t$.

40. If $h(x) = x^3 + 1$ and $g(x) = \sqrt{x}$, find

(a) $g(h(x))$
(b) $h(g(x))$
(c) $h(h(x))$
(d) $g(x) + 1$
(e) $g(x + 1)$

41. Let $f(x) = 2x + 3$ and $g(x) = \ln x$. Find formulas for each of the following functions.

(a) $g(f(x))$ **(b)** $f(g(x))$ **(c)** $f(f(x))$

42. For $f(n) = 3n^2 - 2$ and $g(n) = n + 1$, find and simplify:

(a) $f(n) + g(n)$
(b) $f(n)g(n)$
(c) The domain of $f(n)/g(n)$
(d) $f(g(n))$
(e) $g(f(n))$

Simplify the quantities in Problems 43–46 using $m(z) = z^2$.

43. $m(z + 1) - m(z)$

44. $m(z + h) - m(z)$

45. $m(z) - m(z - h)$

46. $m(z + h) - m(z - h)$

For Problems 47–48, determine functions f and g such that $h(x) = f(g(x))$. [Note: There is more than one correct answer. Do not choose $f(x) = x$ or $g(x) = x$.]

47. $h(x) = (x + 1)^3$

48. $h(x) = x^3 + 1$

49. An amusement park charges an admission fee of $7 per person as well as an additional $1.50 for each ride.

(a) For one visitor, find the park's total revenue $R(n)$ as a function of the number of rides, n, taken.
(b) Find $R(2)$ and $R(8)$ and interpret your answers in terms of amusement park fees.

50. A company that makes Adirondack chairs has fixed costs of $5000 and variable costs of $30 per chair. The company sells the chairs for $50 each.

(a) Find formulas for the cost and revenue functions.
(b) Find the marginal cost and marginal revenue.
(c) Graph the cost and the revenue functions on the same axes.
(d) Find the break-even point.

51. A photocopying company has two different price lists. The first price list is $100 plus 3 cents per copy; the second price list is $200 plus 2 cents per copy.

(a) For each price list, find the total cost as a function of the number of copies needed.
(b) Determine which price list is cheaper for 5000 copies.
(c) For what number of copies do both price lists charge the same amount?

52. You have a budget of $\$k$ to spend on soda and suntan oil, which cost $\$p_1$ per liter and $\$p_2$ per liter respectively.

(a) Write an equation expressing the relationship between the number of liters of soda and the number of liters of suntan oil that you can buy if you exhaust your budget. This is your budget constraint.
(b) Graph the budget constraint, assuming that you can buy fractions of a liter. Label the intercepts.
(c) Suppose your budget is doubled. Graph the new budget constraint on the same axes.
(d) With a budget of $\$k$, the price of suntan oil doubles. Graph the new budget constraint on the same axes.

53. A corporate office provides the demand curve in Figure 1.110 to its ice cream shop franchises. At a price of $1.00 per scoop, 240 scoops per day can be sold.

(a) Estimate how many scoops could be sold per day at a price of 50¢ per scoop. Explain.
(b) Estimate how many scoops per day could be sold at a price of $1.50 per scoop. Explain.

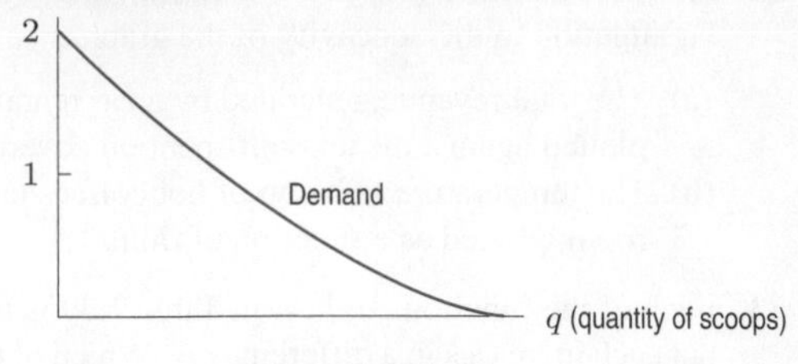

Figure 1.110

54. Figure 1.111 shows supply and demand curves.

(a) What is the equilibrium price for this product? At this price, what quantity is produced?
(b) Choose a price above the equilibrium price—for example, $p = 300$. At this price, how many items are suppliers willing to produce? How many items do consumers want to buy? Use your answers to these questions to explain why, if prices are above the equilibrium price, the market tends to push prices lower (toward the equilibrium).
(c) Now choose a price below the equilibrium price—for example, $p = 200$. At this price, how many items are suppliers willing to produce? How many items do consumers want to buy? Use your answers to these questions to explain why, if prices are below the equilibrium price, the market tends to push prices higher (toward the equilibrium).

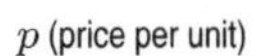

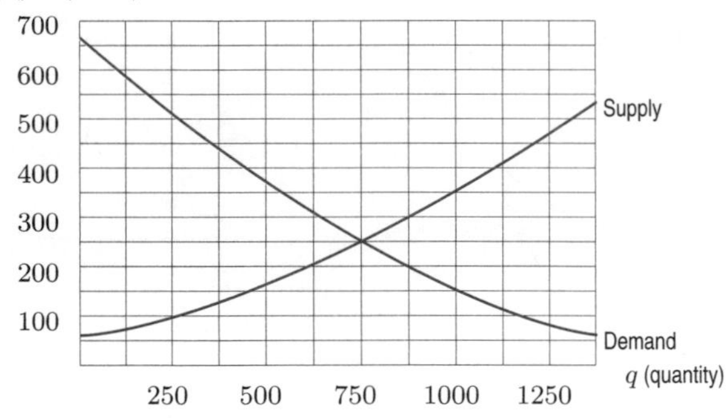

Figure 1.111

55. Find a formula for the number of zebra mussels in a bay as a function of the number of years since 2003, given that there were 2700 at the start of 2003 and 3186 at the start of 2004.

(a) Assume that the number of zebra mussels is growing linearly. Give units for the slope of the line and interpret it in terms of zebra mussels.
(b) Assume that the number of zebra mussels is growing exponentially. What is the percent rate of growth of the zebra mussel population?

56. Worldwide, wind energy[64] generating capacity, W, was 13,559 megawatts in 1998 and 39,151 megawatts in 2003.

(a) Use the values given to write W, in megawatts, as a linear function of t, the number of years since 1998.
(b) Use the values given to write W as an exponential function of t.
(c) Graph the functions found in parts (a) and (b) on the same axes. Label the given values.

57. If the world's population increased exponentially from 4.478 billion in 1980 to 5.423 billion in 1991 and continued to increase at the same percentage rate between 1991 and 2004, calculate what the world's population would have been in 2004. How does this compare to the actual population of 6.378 billion, and what conclusions, if any, can you draw?

58. When the Olympic Games were held outside Mexico City in 1968, there was much discussion about the effect the high altitude (7340 feet) would have on the athletes. Assuming air pressure decays exponentially by 0.4% every 100 feet, by what percentage is air pressure reduced by moving from sea level to Mexico City?

59. The median price, P, of a home rose from \$60,000 in 1980 to \$180,000 in 2000. Let t be the number of years since 1980.

(a) Assume the increase in housing prices has been linear. Give an equation for the line representing price, P, in terms of t. Use this equation to complete column (a) of Table 1.41. Use units of \$1000.
(b) If instead the housing prices have been rising exponentially, find an equation of the form $P = P_0a^t$ to represent housing prices. Complete column (b) of Table 1.41.
(c) On the same set of axes, sketch the functions represented in column (a) and column (b) of Table 1.41.
(d) Which model for the price growth do you think is more realistic?

Table 1.41

t	(a) Linear growth price in \$1000 units	(b) Exponential growth price in \$1000 units
0	60	60
10		
20	180	180
30		
40		

60. (a) A population, P, grows at a continuous rate of 2% a year and starts at 1 million. Write P in the form $P = P_0e^{kt}$, with P_0, k constants.
(b) Plot the population in part (a) against time.

61. (a) What is the continuous percent growth rate for the function $P = 10e^{0.15t}$?
(b) Write this function in the form $P = P_0a^t$.
(c) What is the annual (not continuous) percent growth rate for this function?
(d) Graph $P = 10e^{0.15t}$ and your answer to part (b) on the same axes. Explain what you see.

62. The half-life of a radioactive substance is 12 days. There are 10.32 grams initially.

(a) Write an equation for the amount, A, of the substance as a function of time.
(b) When is the substance reduced to 1 gram?

63. Air pressure, P, decreases exponentially with the height, h, in meters above sea level:

$$P = P_0e^{-0.00012h}$$

where P_0 is the air pressure at sea level.

(a) At the top of Mount McKinley, height 6194 meters (about 20,320 feet), what is the air pressure, as a percent of the pressure at sea level?
(b) The maximum cruising altitude of an ordinary commercial jet is around 12,000 meters (about 39,000 feet). At that height, what is the air pressure, as a percent of the sea level value?

64. A radioactive substance has a half-life of 8 years. If 200 grams are present initially, how much remains at the end of 12 years? How long until only 10% of the original amount remains?

65. One of the main contaminants of a nuclear accident, such as that at Chernobyl, is strontium-90, which decays exponentially at a rate of approximately 2.5% per year.

(a) Write the percent of strontium-90 remaining, P, as a function of years, t, since the nuclear accident. [Hint: 100% of the contaminant remains at $t = 0$.]
(b) Graph P against t.
(c) Estimate the half-life of strontium-90.
(d) After the Chernobyl disaster, it was predicted that the region would not be safe for human habitation for 100 years. Estimate the percent of original strontium-90 remaining at this time.

66. The number of people living with HIV infections increased worldwide approximately exponentially from 2.5 million in 1985 to 37.8 million in 2003.[65] (HIV is the virus that causes AIDS.)

(a) Give a formula for the number of HIV infections, H, (in millions) as a function of years, t, since 1985. Use the form $H = H_0e^{kt}$. Graph this function.
(b) What was the yearly continuous percent change in the number of HIV infections between 1985 and 2003?

[64] www.wwindea.org/wind_energy.htm, accessed April 24, 2005.
[65] *The World Almanac and Book of Facts 2005*, p. 89 (New York).

67. The size of an exponentially growing bacteria colony doubles in 5 hours. How long will it take for the number of bacteria to triple?

68. Interest is compounded annually. Consider the following choices of payments to you:

Choice 1: \$1500 now and \$3000 one year from now

Choice 2: \$1900 now and \$2500 one year from now

(a) If the interest rate on savings were 5% per year, which would you prefer?

(b) Is there an interest rate that would lead you to make a different choice? Explain.

69. A person is to be paid \$2000 for work done over a year. Three payment options are being considered. Option 1 is to pay the \$2000 in full now. Option 2 is to pay \$1000 now and \$1000 in a year. Option 3 is to pay the full \$2000 in a year. Assume an annual interest rate of 5% a year, compounded continuously.

(a) Without doing any calculations, which option is the best option financially for the worker? Explain.

(b) Find the future value, in one year's time, of all three options.

(c) Find the present value of all three options.

70. Using Figure 1.112, find $f(g(x))$ and $g(f(x))$ for $x = -3, -2, -1, 0, 1, 2, 3$. Then graph $f(g(x))$ and $g(f(x))$.

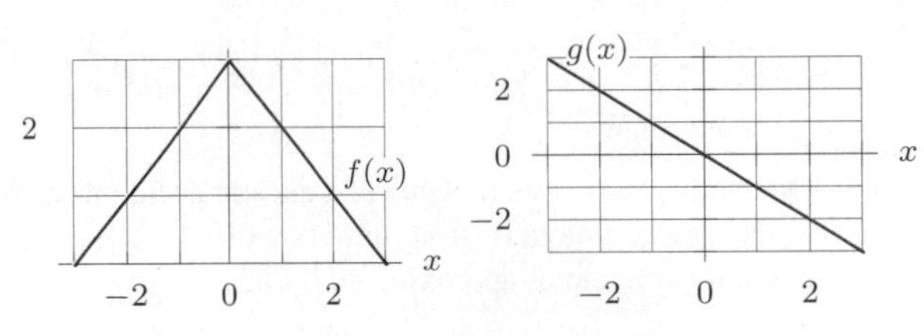

Figure 1.112

71. The Heaviside step function, H, is graphed in Figure 1.113. Graph the following functions.

(a) $2H(x)$ **(b)** $H(x) + 1$ **(c)** $H(x + 1)$
(d) $-H(x)$ **(e)** $H(-x)$

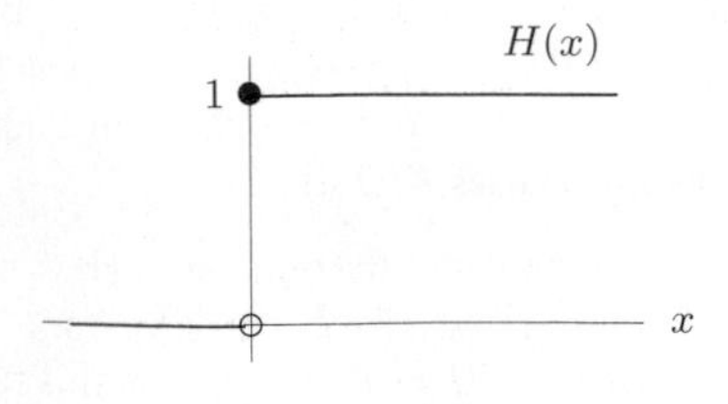

Figure 1.113

72. A population of animals varies periodically between a low of 700 on January 1 and a high of 900 on July 1. Graph the population against time.

73. When a car's engine makes less than about 200 revolutions per minute, it stalls. What is the period of the rotation of the engine when it is about to stall?

For Problems 74–75, find a possible formula for each graph.

74.

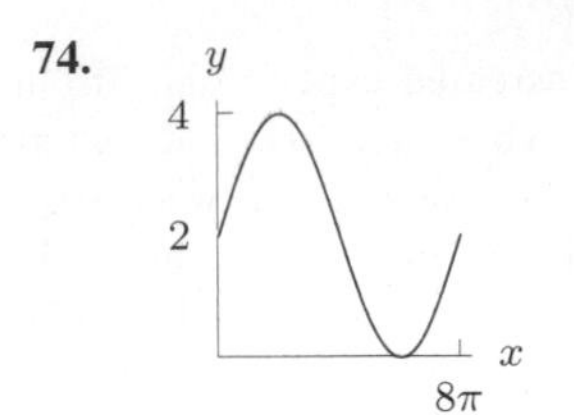

75.

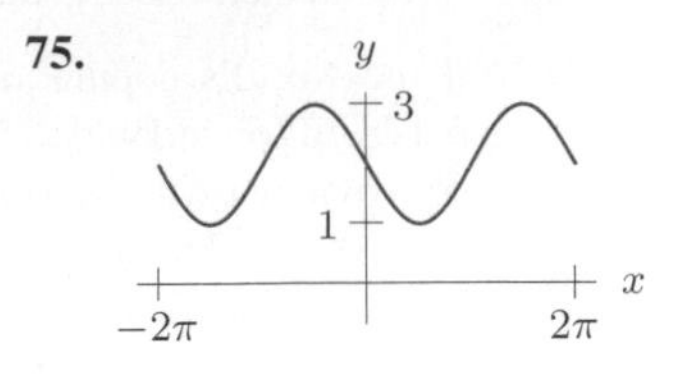

76. Figure 1.114 shows quarterly beer production during the period 1997 to 1999. Quarter 1 reflects production during the first three months of the year, etc.[66]

(a) Explain why a periodic function should be used to model these data.

(b) Approximately when does the maximum occur? The minimum? Why does this make sense?

(c) What are the period and amplitude for these data?

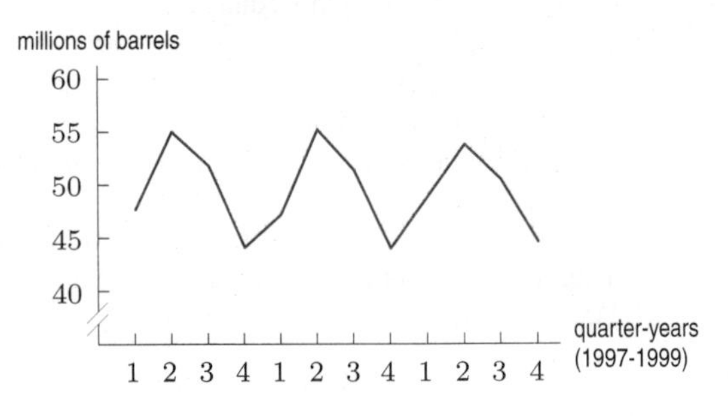

Figure 1.114

77. In a US household, the voltage in volts in an electric outlet is given by

$$V = 156 \sin(120\pi t),$$

where t is in seconds. However, in a European house, the voltage is given (in the same units) by

$$V = 339 \sin(100\pi t).$$

Compare the voltages in the two regions, considering the maximum voltage and number of cycles (oscillations) per second.

[66] www.beerinstitute.org/pdfs/PRODUCTION_AND_WITHDRAWALS_OF_MALT_BEVERAGES_1997_1999.pdf, accessed May 7, 2005.

PROJECTS FOR CHAPTER ONE

1. Compound Interest

The newspaper article below is from *The New York Times*, May 27, 1990. Fill in the three blanks. (For the first blank, assume that daily compounding is essentially the same as continuous compounding. For the last blank, assume the interest has been compounded yearly, and give your answer in dollars. Ignore the occurrence of leap years.)

213 Years After Loan, Uncle Sam Is Dunned

By LISA BELKIN

Special to The New York Times

SAN ANTONIO, May 26 — More than 200 years ago, a wealthy Pennsylvania merchant named Jacob DeHaven lent $450,000 to the Continental Congress to rescue the troops at Valley Forge. That loan was apparently never repaid.

So Mr. DeHaven's descendants are taking the United States Government to court to collect what they believe they are owed.

The total: ____ in today's dollars if the interest is compounded daily at 6 percent, the going rate at the time. If compounded yearly, the bill is only ____.

Family Is Flexible

The descendants say that they are willing to be flexible about the amount of a settlement and that they might even accept a heartfelt thank you or perhaps a DeHaven statue. But they also note that interest is accumulating at ____ a second.

2. Population Center of the US

Since the opening up of the West, the US population has moved westward. To observe this, we look at the "population center" of the US, which is the point at which the country would balance if it were a flat plate with no weight, and every person had equal weight. In 1790 the population center was east of Baltimore, Maryland. It has been moving westward ever since, and in 2000 it was in Edgar Springs, Missouri. During the second half of the 20th century, the population center has moved about 50 miles west every 10 years.

(a) Let us measure position westward from Edgar Springs along the line running through Baltimore. For the years since 2000, express the approximate position of the population center as a function of time in years from 2000.

(b) The distance from Baltimore to Edgar Springs is a bit over 1000 miles. Could the population center have been moving at roughly the same rate for the last two centuries?

(c) Could the function in part (a) continue to apply for the next four centuries? Why or why not? [Hint: You may want to look at a map. Note that distances are in air miles and are not driving distances.]

FOCUS ON MODELING

FITTING FORMULAS TO DATA

In this section we see how the formulas that are used in a mathematical model can be developed. Some of the formulas we use are exact. However, many formulas we use are approximations, often constructed from data.

Fitting a Linear Function To Data

A company wants to understand the relationship between the amount spent on advertising, a, and total sales, S. The data they collect might look like that found in Table 1.42.

Table 1.42 *Advertising and sales: Linear relationship*

a (advertising in \$1000s)	3	4	5	6
S (sales in \$1000s)	100	120	140	160

The data in Table 1.42 are linear, so a formula fits it exactly. The slope of the line is 20, and we can determine that the vertical intercept is 40, so the line is

$$S = 40 + 20a.$$

Now suppose that the company collected the data in Table 1.43. This time the data are not linear. In general, it is difficult to find a formula to fit data exactly. We must be satisfied with a formula that is a good approximation to the data.

Table 1.43 *Advertising and sales: Nonlinear relationship*

a (advertising in \$1000s)	3	4	5	6
S (sales in \$1000s)	105	117	141	152

Figure 1.115 shows the data in Table 1.43. Since the relationship is nearly, though not exactly, linear, it is well approximated by a line. Figure 1.116 shows the line $S = 40 + 20a$ and the data.

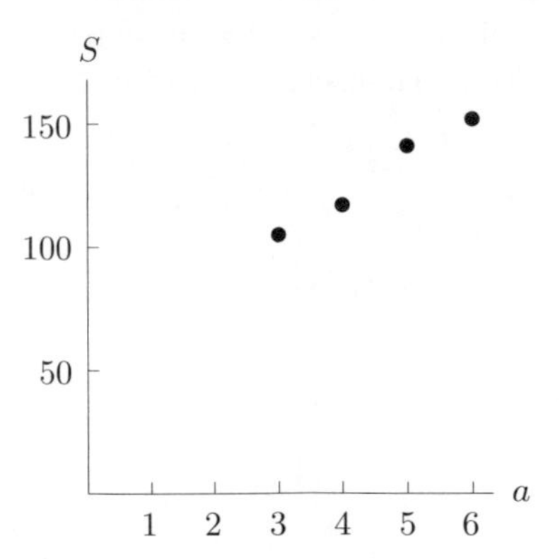

Figure 1.115: The sales data from Table 1.43

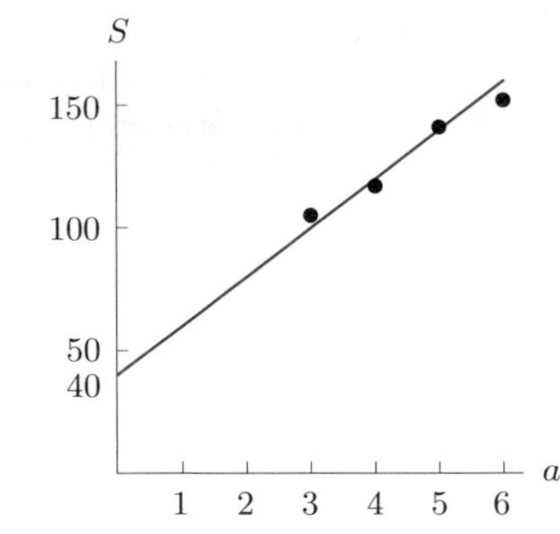

Figure 1.116: The line $S = 40 + 20a$ and the data from Table 1.43

The Regression Line

Is there a line that fits the data better than the one in Figure 1.116? If so, how do we find it? The process of fitting a line to a set of data is called *linear regression* and the line of best fit is called the *regression line.* (See page 80 for a discussion of what "best fit" means.) Many calculators and computer programs calculate the regression line from the data points. Alternatively, the regression line can be estimated by plotting the points on paper and fitting a line "by eye." In Chapter 9, we derive the formulas for the regression line. For the data in Table 1.43, the regression line is

$$S = 54.5 + 16.5a.$$

This line is graphed with the data in Figure 1.117.

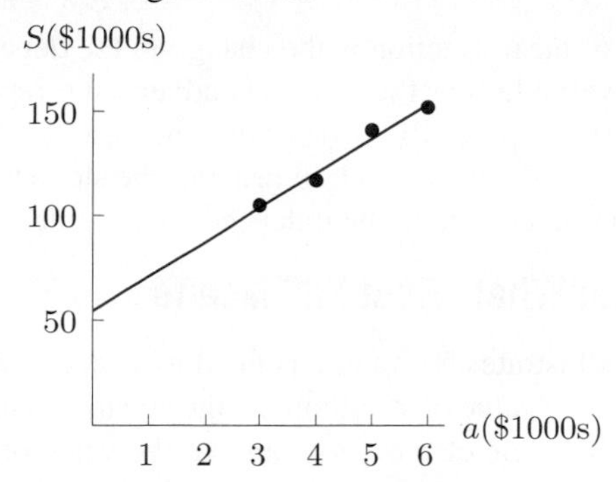

Figure 1.117: The regression line $S = 54.5 + 16.5a$ and the data from Table 1.43

Using the Regression Line to Make Predictions

We can use the formula for sales as a function of advertising to make predictions. For example, to predict total sales if \$3500 is spent on advertising, substitute $a = 3.5$ into the regression line:

$$S = 54.5 + 16.5(3.5) = 112.25.$$

The regression line predicts sales of \$112,250. To see that this is reasonable, compare it to the entries in Table 1.43. When $a = 3$, we have $S = 105$, and when $a = 4$, we have $S = 117$. Predicted sales of $S = 112.25$ when $a = 3.5$ makes sense because it falls between 105 and 117. See Figure 1.118. Of course, if we spent \$3500 on advertising, sales would probably not be exactly \$112,250. The regression equation allows us to make predictions, but does not provide exact results.

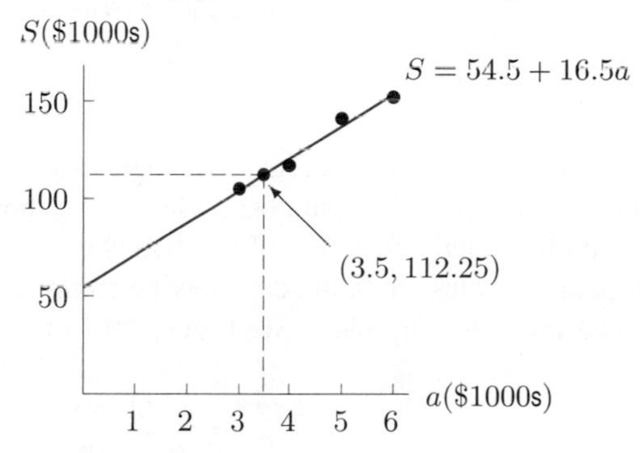

Figure 1.118: Predicting sales when spending \$3,500 on advertising

---

Example 1 Predict total sales given advertising expenditures of \$4800 and \$10,000.

Solution When \$4800 is spent on advertising, $a = 4.8$, so

$$S = 54.5 + 16.5(4.8) = 133.7.$$

Sales are predicted to be \$133,700. When \$10,000 is spent on advertising, $a = 10$, so

$$S = 54.5 + 16.5(10) = 219.5.$$

Sales are predicted to be \$219,500.

Consider the two predictions made in Example 1 at $a = 4.8$ and $a = 10$. We have more confidence in the accuracy of the prediction when $a = 4.8$, because we are *interpolating* within an interval we already know something about. The prediction for $a = 10$ is less reliable, because we are *extrapolating* outside the interval defined by the data values in Table 1.43. In general, interpolation is safer than extrapolation.

Interpreting the Slope of the Regression Line

The slope of a linear function is the change in the dependent variable divided by the change in the independent variable. For the sales and advertising regression line, the slope is 16.5. This tells us that S increases by about 16.5 whenever a increases by 1. If advertising expenses increase by \$1000, sales increase by about \$16,500. In general, the slope tells us the expected change in the dependent variable for a unit change in the independent variable.

How Regression Works: What "Best Fit" Means

Figure 1.119 illustrates how a line is fitted to a set of data. We assume that the value of y is in some way related to the value of x, although other factors could influence y as well. Thus, we assume that we can pick the value of x exactly but that the value of y may be only partially determined by this x-value.

A calculator or computer finds the line that minimizes the sum of the squares of the vertical distances between the data points and the line. See Figure 1.119. The regression line is also called a *least-squares line*, or the *line of best fit*.

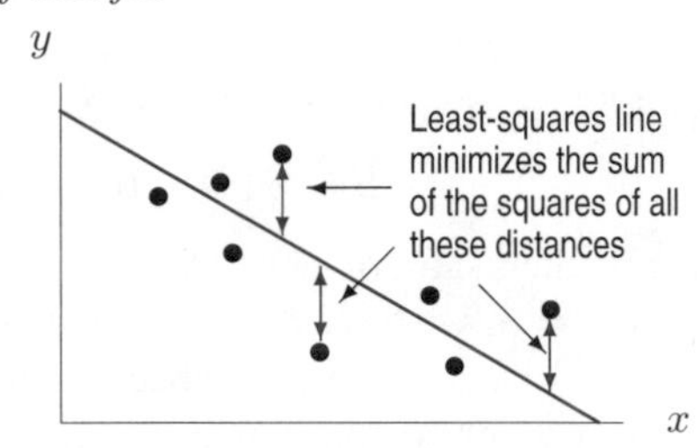

Figure 1.119: Data and the corresponding least-squares regression line

Correlation

When a computer or calculator calculates a regression line, it also gives a *correlation coefficient*, r. This number lies between -1 and $+1$ and measures how well the regression line fits the data. If $r = 1$, the data lie exactly on a line of positive slope. If $r = -1$, the data lie exactly on a line of negative slope. If r is close to 0, the data may be completely scattered, or there may be a non-linear relationship between the variables. (See Figure 1.120.)

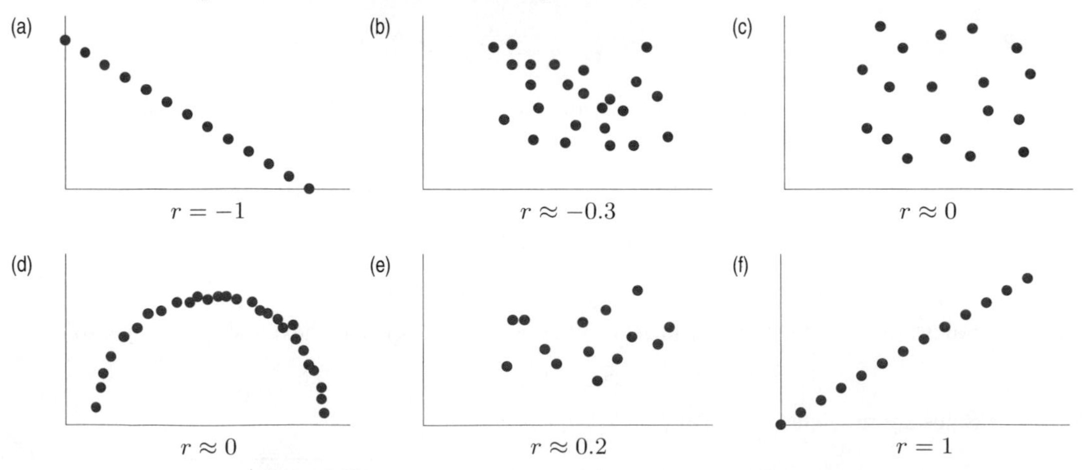

Figure 1.120: Various data sets and correlation coefficients

Example 2 The correlation coefficient for the sales data in Table 1.43 on page 78 is $r \approx 0.99$. The fact that r is positive tells us that the regression line has positive slope. The fact that r is close to 1 tells us that the regression line fits the data well.

The Difference between Relation, Correlation, and Causation

It is important to understand that a high correlation (either positive or negative) between two quantities does *not* imply causation. For example, there is a high correlation between children's reading level and shoe size.[67] However, large feet do not cause a child to read better (or vice versa). Larger feet and improved reading ability are both a consequence of growing older.

Notice also that a correlation of 0 does not imply that there is no relationship between x and y. For example, in Figure 1.120(d) there is a relationship between x and y-values, while Figure 1.120(c) exhibits no apparent relationship. Both data sets have a correlation coefficient of $r \approx 0$. Thus, a correlation of $r = 0$ usually implies there is no linear relationship between x and y, but this does not mean there is no relationship at all.

Regression When the Relationship Is Not Linear

Table 1.44 shows the population of the US (in millions) from 1790 to 1860. These points are plotted in Figure 1.121. Do the data look linear? Not really. It appears to make more sense to fit an exponential function than a linear function to this data. Finding the exponential function of best fit is called *exponential regression.* One algorithm used by a calculator or computer gives the exponential function that fits the data as

$$P = 3.9(1.03)^t,$$

where P is the US population in millions and t is years since 1790. Other algorithms may give different answers. See Figure 1.122.

Since the base of this exponential function is 1.03, the US population was increasing at the rate of about 3% per year between 1790 and 1860. Is it reasonable to expect the population to continue to increase at this rate? It turns out that this exponential model does not fit the population of the US well beyond 1860. In Section 4.7, we see another function that is used to model the US population.

Table 1.44 *US Population in millions, 1790–1860*

Year	1790	1800	1810	1820	1830	1840	1850	1860
Population	3.9	5.3	7.2	9.6	12.9	17.1	23.1	31.4

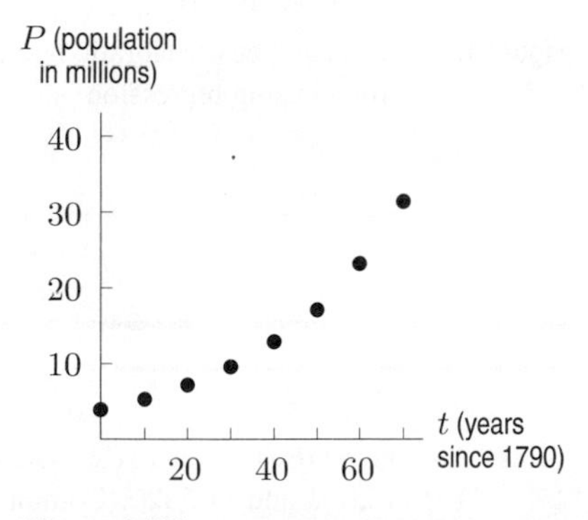

Figure 1.121: US Population 1790–1860

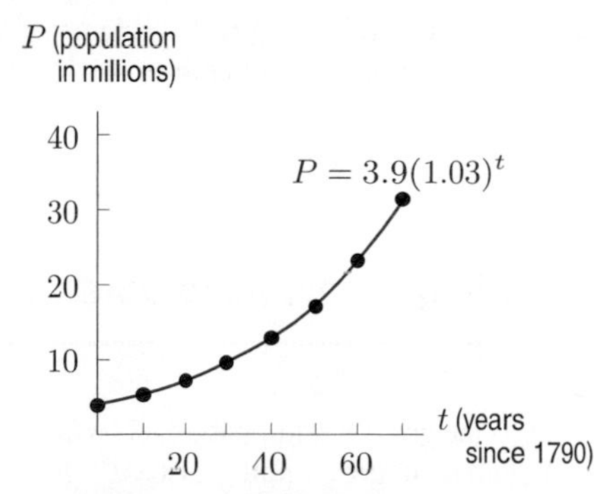

Figure 1.122: US population and an exponential regression function

Calculators and computers can do linear regression, exponential regression, logarithmic regression, quadratic regression, and more. To fit a formula to a set of data, the first step is to graph the data and identify the appropriate family of functions.

---

[67]From *Statistics*, 2ed, by David Freedman. Robert Pisani, Roger Purves, Ani Adhikari, p. 142 (New York: W.W.Norton, 1991).

Example 3 The average fuel efficiency (miles per gallon of gasoline) of US automobiles declined until the 1960s and then started to rise as manufacturers made cars more fuel efficient.[68] See Table 1.45.

(a) Plot the data. What family of functions should be used to model the data: linear, exponential, logarithmic, power function, or a polynomial? If a polynomial, state the degree and whether the leading coefficient is positive or negative.

(b) Use quadratic regression to fit a quadratic polynomial to the data; graph it with the data.

Table 1.45 *What function fits these data?*

Year	1940	1950	1960	1970	1980	1986
Average miles per gallon	14.8	13.9	13.4	13.5	15.5	18.3

Solution

(a) The data are shown in Figure 1.123, with time t in years since 1940. Miles per gallon decreases and then increases, so a good function to model the data is a quadratic (degree 2) polynomial. Since the parabola opens up, the leading coefficient is positive.

(b) If $f(t)$ is average miles per gallon, one algorithm for quadratic regression tells us that the quadratic polynomial that fits the data is

$$f(t) = 0.00617t^2 - 0.225t + 15.10.$$

In Figure 1.124, we see that this quadratic does fit the data reasonably well.

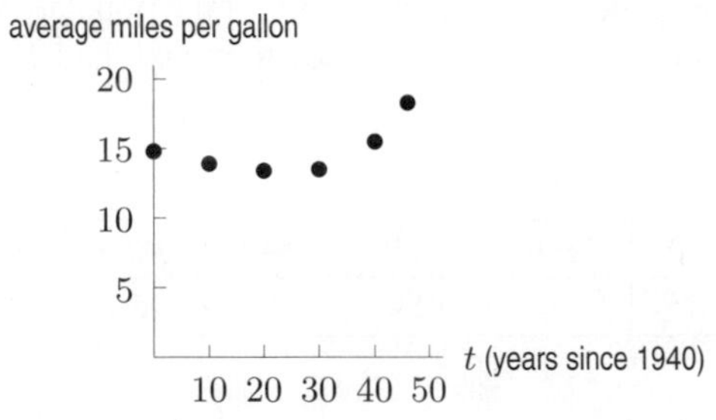

Figure 1.123: Data showing fuel efficiency of US automobiles over time

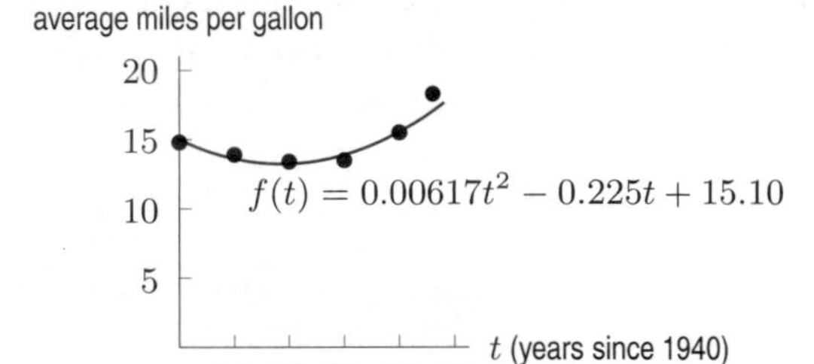

Figure 1.124: Data and best quadratic polynomial, found using regression

Problems on Fitting Formulas to Data

1. Table 1.46 gives the gross world product, G, which measures global output of goods and services.[69] If t is in years since 1950, the regression line for these data is

$$G = 3.543 + 0.734t.$$

(a) Plot the data and the regression line on the same axes. Does the line fit the data well?

(b) Interpret the slope of the line in terms of gross world product.

(c) Use the regression line to estimate gross world product in 2005 and in 2020. Comment on your confidence in the two predictions.

Table 1.46 *G, in trillions of 1999 dollars*

Year	1950	1960	1970	1980	1990	2000
G	6.4	10.0	16.3	23.6	31.9	43.2

[68] C. Schaufele and N. Zumoff, *Earth Algebra, Preliminary Version*, p. 91, (New York: Harper Collins, 1993).
[69] The Worldwatch Institute, *Vital Signs* 2001, p. 57, (New York: W.W. Norton, 2001).

2. Table 1.47 shows worldwide cigarette production as a function of t, the number of years since 1950.[70]

(a) Find the regression line for this data.
(b) Use the regression line to estimate world cigarette production in the year 2010.
(c) Interpret the slope of the line in terms of cigarette production.
(d) Plot the data and the regression line on the same axes. Does the line fit the data well?

Table 1.47 *Cigarette production, P, in billions*

t	0	10	20	30	40	50
P	1686	2150	3112	4388	5419	5564

3. Table 1.48 shows the US Gross National Product (GNP).[71]

(a) Plot GNP against years since 1970. Does a line fit the data well?
(b) Find the regression line and graph it with the data.
(c) Use the regression line to estimate the GNP in 1985 and in 2020. Which estimate do you have more confidence in? Why?

Table 1.48 *GNP in 2003 dollars*

Year	1970	1980	1990	2000
GNP (billions)	1045	2824	5838	9856

4. The acidity of a solution is measured by its pH, with lower pH values indicating more acidity. A study of acid rain was undertaken in Colorado between 1975 and 1978, in which the acidity of rain was measured for 150 consecutive weeks. The data followed a generally linear pattern and the regression line was determined to be

$$P = 5.43 - 0.0053t,$$

where P is the pH of the rain and t is the number of weeks into the study.[72]

(a) Is the pH level increasing or decreasing over the period of the study? What does this tell you about the level of acidity in the rain?
(b) According to the line, what was the pH at the beginning of the study? At the end of the study ($t = 150$)?
(c) What is the slope of the regression line? Explain what this slope is telling you about the pH.

5. In a 1977 study[73] of 21 of the best American female runners, researchers measured the average stride rate, S, at different speeds, v. The data are given in Table 1.49.

(a) Find the regression line for these data, using stride rate as the dependent variable.
(b) Plot the regression line and the data on the same axes. Does the line fit the data well?
(c) Use the regression line to predict the stride rate when the speed is 18 ft/sec and when the speed is 10 ft/sec. Which prediction do you have more confidence in? Why?

Table 1.49 *Stride rate, S, in steps/sec, and speed, v, in ft/sec*

v	15.86	16.88	17.50	18.62	19.97	21.06	22.11
S	3.05	3.12	3.17	3.25	3.36	3.46	3.55

6. Table 1.50 shows the atmospheric concentration of carbon dioxide, CO_2, (in parts per million, ppm) at the Mauna Loa Observatory in Hawaii.[74]

(a) Find the average rate of change of the concentration of carbon dioxide between 1980 and 2000. Give units and interpret your answer in terms of carbon dioxide.
(b) Plot the data, and find the regression line for carbon dioxide concentration against years since 1980. Use the regression line to predict the concentration of carbon dioxide in the atmosphere in the year 2020.

Table 1.50

Year	1980	1985	1990	1995	2000
CO_2	349.6	346.7	354.4	360.8	368.9

7. In Problem 6, carbon dioxide concentration was modeled as a linear function of time. However, if we include data for carbon dioxide concentration from as far back as 1900, the data appear to be more exponential than linear. (They looked linear in Problem 6 because we were only looking at a small piece of the graph.) If C is the CO_2 concentration in ppm and t is in years since 1900, an exponential regression function to fit the data is

$$C = 272.27(1.0026)^t.$$

(a) What is the annual percent growth rate during this period? Interpret this rate in terms of CO_2 concentration.
(b) What CO_2 concentration is given by the model for 1900? For 1980? Compare the 1980 estimate to the actual value in Table 1.50.

[70]The Worldwatch Institute, *Vital Signs* 2001, p. 77, (New York: W.W. Norton, 2001).
[71]*The World Almanac and Book of Facts 2005*, p. 111 (New York).
[72]William M. Lewis and Michael C. Grant, "Acid Precipitation in the Western United States," *Science* 207 (1980) pp. 176-177.
[73]R.C. Nelson, C.M. Brooks, and N.L. Pike, "Biomechanical Comparison of Male and Female Distance Runners." *The Marathon: Physiological, Medical, Epidemiological, and Psychological Studies*, ed. P. Milvy, pp. 793-807, (New York: New York Academy of Sciences, 1977).
[74]www.cmdl.noaa.gov/ccgg/iadv, accessed on February 20, 2005.

8. Table 1.51 shows the average yearly US health care expenditures per consumer unit for various years.[75] Does a linear or an exponential model appear to fit these data best? Find a formula for the regression function you decide is best. Graph the function with the data and assess how well it fits the data.

Table 1.51 *Health expenditures, $ per consumer unit*

Years since 1998, t	0	1	2	3	4
Expenditure, C	1903	1959	2066	2182	2350

9. **(a)** Fit an exponential function to the population data in Table 1.52. Plot the data and the exponential function on the same axes.
(b) At approximately what percentage rate was the population growing between 1960 and 2000?
(c) If the population continues to grow at the same percentage rate, what population is projected for 2020?

Table 1.52 *US Population 1960-2000*

t, years since 1960	0	10	20	30	40
population (m)	179.3	203.3	226.5	248.7	281.4

10. Table 1.53 shows the public debt, D, of the US[76] in billions of dollars, t years after 1998.
(a) Plot the public debt against the number of years since 1998.
(b) Does the data look more linear or more exponential?
(c) Fit an exponential function to the data and graph it with the data.
(d) What annual percentage growth rate does the exponential model show?
(e) Do you expect this model to give accurate predictions beyond 2004? Explain.

Table 1.53

t	0	1	2	3	4	5	6
D	5526	5656	5674	5808	6228	6783	7379

11. A company collects the data in Table 1.54. Find the regression line and interpret its slope. Sketch the data and the line. What is the correlation coefficient? Why is the value you get reasonable?

Table 1.54 *Cost to produce various quantities of a product*

q (quantity in units)	25	50	75	100	125
C (cost in dollars)	500	625	689	742	893

12. Match the r values with scatter plots in Figure 1.125.

$$r=-0.98,\quad r=-0.5,\quad r=-0.25,$$
$$r=0,\quad r=0.7,\quad r=1.$$

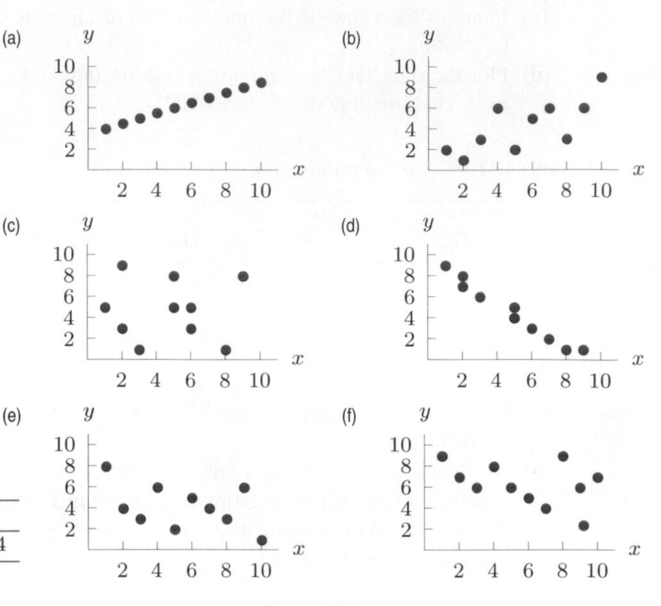

Figure 1.125

13. Table 1.55 shows the number of cars in the US.[77]
(a) Plot the data, with number of passenger cars as the dependent variable.
(b) Does a linear or exponential model appear to fit the data better?
(c) Use a linear model first: Find the regression line for these data. Graph it with the data. Use the regression line to predict the number of passenger cars in the year 2010 ($t=70$).
(d) Interpret the slope of the regression line found in part (c) in terms of passenger cars.
(e) Now use an exponential model: Find the exponential regression function for these data. Graph it with the data. Use the exponential function to predict the number of passenger cars in the year 2010 ($t=70$). Compare your prediction with the prediction obtained from the linear model.
(f) What annual percent growth rate in number of US passenger cars does your exponential model show?

Table 1.55 *Number of passenger cars, in millions*

t (years since 1940)	0	10	20	30	40	50	60
N (millions of cars)	27.5	40.3	61.7	89.2	121.6	133.7	133.6

[75] *Statistical Abstracts of the United States 2004–2005*, Table 125.
[76] *The World Almanac and Book of Facts 2005*, p. 119 (New York).
[77] *The World Almanac and Book of Facts 2005*, p. 237 (New York).

14. Table 1.56 gives the population of the world in billions.

(a) Plot these data. Does a linear or exponential model seem to fit the data best?
(b) Find the exponential regression function.
(c) What annual percent growth rate does the exponential function show?
(d) Predict the population of the world in the year 2010 and in the year 2050. Comment on the relative confidence you have in these two estimates.

Table 1.56 *World population*

Year (since 1950)	0	10	20	30	40	50	54
World population (billions)	2.6	3.1	3.7	4.5	5.4	6.1	6.4

15. In 1969, all field goal attempts were analyzed in the National Football League and American Football League. See Table 1.57. (The data has been summarized: all attempts between 10 and 19 yards from the goal post are listed as 14.5 yards out, etc.)

(a) Graph the data, with success rate as the dependent variable. Discuss whether a linear or an exponential model fits best.
(b) Find the linear regression function; graph it with the data. Interpret the slope of the regression line in terms of football.
(c) Find the exponential regression function; graph it with the data. What success rate does this function predict from a distance of 50 yards?
(d) Using the graphs in parts (b) and (c), decide which model seems to fit the data best.

Table 1.57 *Successful fraction of field goal attempts*

Distance from goal, x yards	14.5	24.5	34.5	44.5	52.0
Fraction successful, Y	0.90	0.75	0.54	0.29	0.15

16. Table 1.58 shows the number of Japanese cars imported into the US.[78]

(a) Plot the number of Japanese cars imported against the number of years since 1964.
(b) Does the data look more linear or more exponential?
(c) Fit an exponential function to the data and graph it with the data.
(d) What annual percentage growth rate does the exponential model show?
(e) Do you expect this model to give accurate predictions beyond 1971? Explain.

Table 1.58 *Imported Japanese cars, 1964–1971*

Year since 1964	0	1	2	3	4	5	6	7
Cars (thousands)	16	24	56	70	170	260	381	704

17. Figure 1.126 shows oil production in the Middle East.[79] If you were to model this function with a polynomial, what degree would you choose? Would the leading coefficient be positive or negative?

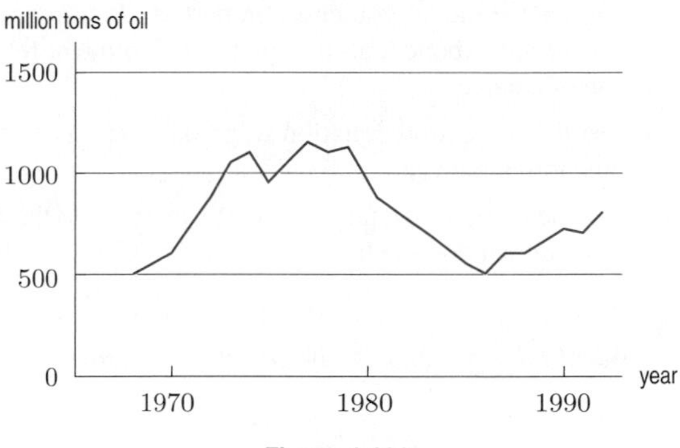

Figure 1.126

18. After the oil crisis in 1973, the average fuel efficiency, E, of cars increased until the early 1990s, when it started to decrease again.

(a) Plot the data[80] in Table 1.59, using t in years since 1975. If you were to fit a quadratic polynomial to the data, what would be the sign of the leading coefficient?
(b) Fit a quadratic polynomial and plot it with the data.

Table 1.59

Year	1975	1980	1985	1990	1995	2000
E, mpg	13.1	19.2	21.3	21.5	21.1	20.7

19. Table 1.60 gives the area of rain forest destroyed for agriculture and development.[81]

(a) Plot these data.
(b) Are the data increasing or decreasing? Concave up or concave down? In each case, interpret your answer in terms of rain forest.
(c) Use a calculator or computer to fit a logarithmic function to this data. Plot this function on the axes in part (a).
(d) Use the curve you found in part (c) to predict the area of rain forest destroyed in 2010.

Table 1.60 *Destruction of rain forest*

x (year)	1960	1970	1980	1988
y (million hectares)	2.21	3.79	4.92	5.77

[78] *The World Almanac 1995.*
[79] Lester R. Brown, et. al., *Vital Signs*, p. 49, (New York: W. W. Norton and Co., 1994).
[80] *The World Almanac and Book of Facts* (New York, 2005).
[81] C. Schaufele and N. Zumoff, *Earth Algebra, Preliminary Version*, p. 131, (New York: Harper Collins, 1993).

In Problems 20–22, tables of data are given.[82]

(a) Use a plot of the data to decide whether a linear, exponential, logarithmic, or quadratic model fits the data best.

(b) Use a calculator or computer to find the regression equation for the model you chose in part (a). If the equation is linear or exponential, interpret the absolute or relative rate of change.

(c) Use the regression equation to predict the value of the function in the year 2005.

(d) Plot the regression equation on the same axes as the data, and comment on the fit.

20. *World solar power, S, in megawatts; t in years since 1990*

t	0	1	2	3	4	5	6	7	8	9	10
S	46	55	58	60	69	79	89	126	153	201	288

21. *Nuclear warheads, N, in thousands; t in years since 1960*

t	0	5	10	15	20	25	30	35	40
N	20	39	40	52	61	69	60	43	32

22. *Carbon dioxide, C, in ppm; t in years since 1970*

t	0	5	10	15	20	25	30
C	325.5	331.0	338.5	345.7	354.0	360.9	369.40

23. For each graph in Figure 1.127, decide whether the best fit for the data appears to be a linear function, an exponential function, or a polynomial.

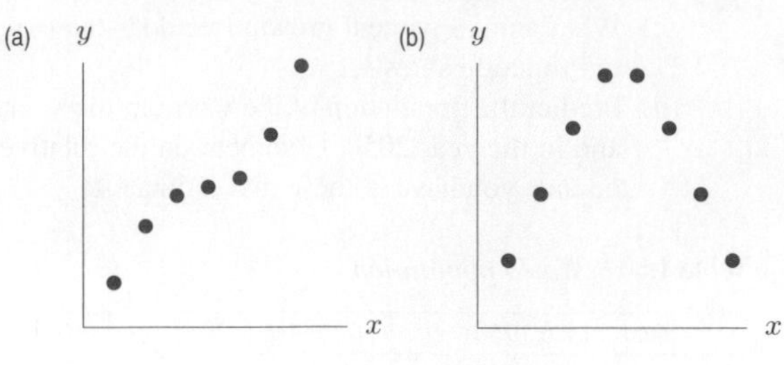

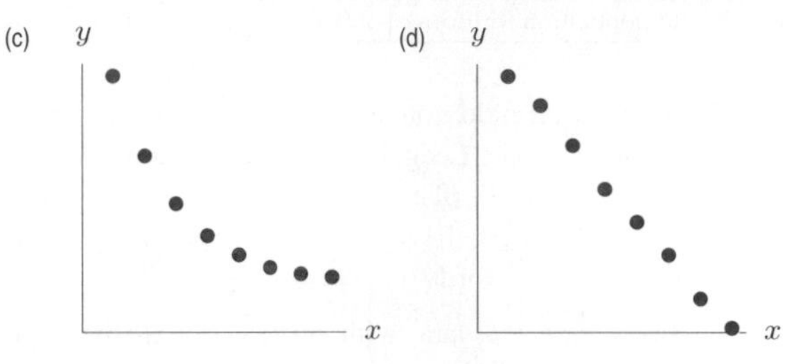

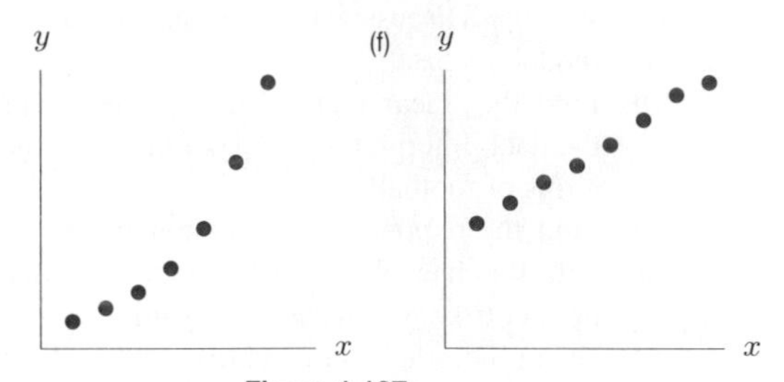

Figure 1.127

COMPOUND INTEREST AND THE NUMBER e

If you have some money, you may decide to invest it to earn interest. The interest can be paid in many different ways—for example, once a year or many times a year. If the interest is paid more frequently than once per year and the interest is not withdrawn, there is a benefit to the investor since the interest earns interest. This effect is called *compounding*. You may have noticed banks offering accounts that differ both in interest rates and in compounding methods. Some offer interest compounded annually, some quarterly, and others daily. Some even offer continuous compounding.

What is the difference between a bank account advertising 8% compounded annually (once per year) and one offering 8% compounded quarterly (four times per year)? In both cases 8% is an annual rate of interest. The expression 8% *compounded annually* means that at the end of each year, 8% of the current balance is added. This is equivalent to multiplying the current balance by 1.08. Thus, if \$100 is deposited, the balance, B, in dollars, will be

$$B = 100(1.08) \quad \text{after one year,}$$
$$B = 100(1.08)^2 \quad \text{after two years,}$$
$$B = 100(1.08)^t \quad \text{after } t \text{ years.}$$

The expression 8% *compounded quarterly* means that interest is added four times per year (every three months) and that $\frac{8}{4} = 2\%$ of the current balance is added each time. Thus, if \$100 is

[82] The Worldwatch Institute, *Vital Signs* 2001, (New York: W.W. Norton & Company, 2001), p. 47.

deposited, at the end of one year, four compoundings have taken place and the account will contain $\$100(1.02)^4$. Thus, the balance will be

$$\begin{aligned} B &= 100(1.02)^4 && \text{after one year,} \\ B &= 100(1.02)^8 && \text{after two years,} \\ B &= 100(1.02)^{4t} && \text{after } t \text{ years.} \end{aligned}$$

Note that 8% is *not* the rate used for each three month period; the annual rate is divided into four 2% payments. Calculating the total balance after one year under each method shows that

$$\begin{aligned} \text{Annual compounding:} &\quad B = 100(1.08) = 108.00, \\ \text{Quarterly compounding:} &\quad B = 100(1.02)^4 = 108.24. \end{aligned}$$

Thus, more money is earned from quarterly compounding, because the interest earns interest as the year goes by. In general, the more often interest is compounded, the more money will be earned (although the increase may not be very large).

We can measure the effect of compounding by introducing the notion of *effective annual yield*. Since $\$100$ invested at 8% compounded quarterly grows to $\$108.24$ by the end of one year, we say that the *effective annual yield* in this case is 8.24%. We now have two interest rates that describe the same investment: the 8% compounded quarterly and the 8.24% effective annual yield. Banks call the 8% the *annual percentage rate*, or APR. We may also call the 8% the *nominal rate* (nominal means "in name only"). However, it is the effective yield that tells you exactly how much interest the investment really pays. Thus, to compare two bank accounts, simply compare the effective annual yields. The next time you walk by a bank, look at the advertisements, which should (by law) include both the APR, or nominal rate, and the effective annual yield. We often abbreviate *annual percentage rate* to *annual rate*.

Using the Effective Annual Yield

Example 4 Which is better: Bank X paying a 7% annual rate compounded monthly or Bank Y offering a 6.9% annual rate compounded daily?

Solution We find the effective annual yield for each bank.

Bank X: There are 12 interest payments in a year, each payment being $0.07/12 = 0.005833$ times the current balance. If the initial deposit were $\$100$, then the balance B would be

$$\begin{aligned} B &= 100(1.005833) && \text{after one month,} \\ B &= 100(1.005833)^2 && \text{after two months,} \\ B &= 100(1.005833)^t && \text{after } t \text{ months.} \end{aligned}$$

To find the effective annual yield, we look at one year, or 12 months, giving $B = 100(1.005833)^{12} = 100(1.072286)$, so the effective annual yield $\approx 7.23\%$.

Bank Y: There are 365 interest payments in a year (assuming it is not a leap year), each being $0.069/365 = 0.000189$ times the current balance. Then the balance is

$$\begin{aligned} B &= 100(1.000189) && \text{after one day,} \\ B &= 100(1.000189)^2 && \text{after two days,} \\ B &= 100(1.000189)^t && \text{after } t \text{ days.} \end{aligned}$$

so at the end of one year we have multiplied the initial deposit by

$$(1.000189)^{365} = 1.071413$$

so the effective annual yield for Bank Y $\approx 7.14\%$.

Comparing effective annual yields for the banks, we see that Bank X is offering a better investment, by a small margin.

Example 5 If \$1000 is invested in each bank in Example 4, write an expression for the balance in each bank after t years.

Solution For Bank X, the effective annual yield $\approx 7.23\%$, so after t years the balance, in dollars, will be

$$B = 100\left(1+\frac{0.07}{12}\right)^{12t} = 1000(1.005833)^{12t} = 1000(1.0723)^t.$$

For Bank Y, the effective annual yield $\approx 7.14\%$, so after t years the balance, in dollars, will be

$$B = 1000\left(1+\frac{0.069}{365}\right)^{365t} = 1000(1.0714)^t.$$

(Again, we are ignoring leap years.)

If interest at an annual rate of r is compounded n times a year, then r/n times the current balance is added n times a year. Therefore, with an initial deposit of P, the balance t years later is

$$B = P\left(1+\frac{r}{n}\right)^{nt}.$$

Note that r is the nominal rate; for example, $r = 0.05$ when the annual rate is 5%.

Increasing the Frequency of Compounding: Continuous Compounding

Let us look at the effect of increasing the frequency of compounding. How much effect does it have?

Example 6 Find the effective annual yield for a 7% annual rate compounded
(a) 1000 times a year. (b) 10,000 times a year.

Solution (a) In one year, a deposit is multiplied by

$$\left(1+\frac{0.07}{1000}\right)^{1000} \approx 1.0725056,$$

giving an effective annual yield of about 7.25056%.

(b) In one year, a deposit is multiplied by

$$\left(1+\frac{0.07}{10{,}000}\right)^{10{,}000} \approx 1.0725079,$$

giving an effective annual yield of about 7.25079%.

You can see that there's not a great deal of difference between compounding 1000 times each year (about three times per day) and 10,000 times each year (about 30 times per day). What happens if we compound more often still? Every minute? Every second? You may be surprised to know that the effective annual yield does not increase indefinitely, but tends to a finite value. The benefit of increasing the frequency of compounding becomes negligible beyond a certain point.

For example, if you were to compute the effective annual yield on a 7% investment compounded n times per year for values of n larger than 100,000, you would find that

$$\left(1+\frac{0.07}{n}\right)^n \approx 1.0725082.$$

So the effective annual yield is about 7.25082%. Even if you take $n = 1{,}000{,}000$ or $n = 10^{10}$, the effective annual yield does not change appreciably. The value 7.25082% is an upper bound that is approached as the frequency of compounding increases.

When the effective annual yield is at this upper bound, we say that the interest is being *compounded continuously*. (The word *continuously* is used because the upper bound is approached by compounding more and more frequently.) Thus, when a 7% nominal annual rate is compounded so frequently that the effective annual yield is 7.25082%, we say that the 7% is compounded *continuously*. This represents the most one can get from a 7% nominal rate.

Where Does the Number e Fit In?

It turns out that e is intimately connected to continuous compounding. To see this, use a calculator to check that $e^{0.07} \approx 1.0725082$, which is the same number we obtained by compounding 7% a large number of times. So you have discovered that for very large n

$$\left(1+\frac{0.07}{n}\right)^n \approx e^{0.07}.$$

As n gets larger, the approximation gets better and better, and we write

$$\lim_{n\to\infty}\left(1+\frac{0.07}{n}\right)^n = e^{0.07},$$

meaning that as n increases, the value of $(1+0.07/n)^n$ approaches $e^{0.07}$.

If P is deposited at an annual rate of 7% compounded continuously, the balance, B, after t years, is given by

$$B = P(e^{0.07})^t = Pe^{0.07t}.$$

> If interest on an initial deposit of P is *compounded continuously* at an annual rate r, the balance t years later can be calculated using the formula
>
> $$B = Pe^{rt}.$$

In working with compound interest, it is important to be clear whether interest rates are nominal rates or effective yields, as well as whether compounding is continuous or not.

Example 7 Find the effective annual yield of a 6% annual rate, compounded continuously.

Solution In one year, an investment of P becomes $Pe^{0.06}$. Using a calculator, we see that

$$Pe^{0.06} = P(1.0618365).$$

So the effective annual yield is about 6.18%.

Example 8 You invest money in a certificate of deposit (CD) for your child's education, and you want it to be worth \$120,000 in 10 years. How much should you invest if the CD pays interest at a 9% annual rate compounded quarterly? Continuously?

Solution Suppose you invest P initially. A 9% annual rate compounded quarterly has an effective annual yield given by $(1 + 0.09/4)^4 = 1.0930833$, or 9.30833%. So after 10 years you have

$$P(1.0930833)^{10} = 120{,}000.$$

Therefore, you should invest

$$P = \frac{120{,}000}{(1.0930833)^{10}} = \frac{120{,}000}{2.4351885} = 49{,}277.50.$$

On the other hand, if the CD pays 9% per year, compounded continuously, after 10 years you have

$$Pe^{(0.09)10} = 120{,}000.$$

So you would need to invest

$$P = \frac{120{,}000}{e^{(0.09)10}} = \frac{120{,}000}{2.4596031} = 48{,}788.36.$$

Notice that to achieve the same result, continuous compounding requires a smaller initial investment than quarterly compounding. This is to be expected since the effective annual yield is higher for continuous than for quarterly compounding.

Problems on Compound Interest and the Number e

1. A department store issues its own credit card, with an interest rate of 2% per month. Explain why this is not the same as an annual rate of 24%. What is the effective annual rate?
2. A deposit of \$10,000 is made into an account paying a nominal yearly interest rate of 8%. Determine the amount in the account in 10 years if the interest is compounded:

   (a) Annually **(b)** Monthly **(c)** Weekly
   (d) Daily **(e)** Continuously
3. A deposit of \$50,000 is made into an account paying a nominal yearly interest rate of 6%. Determine the amount in the account in 20 years if the interest is compounded:

   (a) Annually **(b)** Monthly **(c)** Weekly
   (d) Daily **(e)** Continuously
4. Use a graph of $y = (1 + 0.07/x)^x$ to estimate the number that $(1 + 0.07/x)^x$ approaches as $x \to \infty$. Confirm that the value you get is $e^{0.07}$.
5. Find the effective annual yield of a 6% annual rate, compounded continuously.
6. What nominal annual interest rate has an effective annual yield of 5% under continuous compounding?
7. What is the effective annual yield, under continuous compounding, for a nominal annual interest rate of 8%?
8. **(a)** Find the effective annual yield for a 5% annual interest rate compounded n times/year if

(i) $n = 1000$ (ii) $n = 10{,}000$
(iii) $n = 100{,}000$

(b) Look at the sequence of answers in part (a), and predict the effective annual yield for a 5% annual rate compounded continuously.

(c) Compute $e^{0.05}$. How does this confirm your answer to part (b)?

9. **(a)** Find $(1 + 0.04/n)^n$ for $n = 10{,}000$, and $100{,}000$, and $1{,}000{,}000$. Use the results to predict the effective annual yield of a 4% annual rate compounded continuously.

(b) Confirm your answer by computing $e^{0.04}$.

10. A bank account is earning interest at 6% per year compounded continuously.

(a) By what percentage has the bank balance in the account increased over one year? (This is the effective annual yield.)

(b) How long does it take the balance to double?

(c) For an interest rate of r, find a formula giving the doubling time in terms of the interest rate.

11. Explain how you can match the interest rates (a)–(e) with the effective annual yields I–V without calculation.

(a) 5.5% annual rate, compounded continuously.
(b) 5.5% annual rate, compounded quarterly.
(c) 5.5% annual rate, compounded weekly.
(d) 5% annual rate, compounded yearly.
(e) 5% annual rate, compounded twice a year.

I. 5% II. 5.06% III. 5.61%
IV. 5.651% V. 5.654%

Countries with very high inflation rates often publish monthly rather than yearly inflation figures, because monthly figures are less alarming. Problems 12–13 involve such high rates, which are called *hyperinflation*.

12. In 1989, US inflation was 4.6% a year. In 1989 Argentina had an inflation rate of about 33% a month.

(a) What is the yearly equivalent of Argentina's 33% monthly rate?

(b) What is the monthly equivalent of the US 4.6% yearly rate?

13. Between December 1988 and December 1989, Brazil's inflation rate was 1290% a year. (This means that between 1988 and 1989, prices increased by a factor of $1 + 12.90 = 13.90$.)

(a) What would an article which cost 1000 cruzados (the Brazilian currency unit) in 1988 cost in 1989?

(b) What was Brazil's monthly inflation rate during this period?

FOCUS ON THEORY

LIMITS TO INFINITY AND END BEHAVIOR

Comparing Power Functions

As x gets large, how do different power functions compare? For positive powers, Figure 1.128 shows that the higher the power of x, the faster the function climbs. For large values of x (in fact, for all $x > 1$), $y = x^5$ is above $y = x^4$, which is above $y = x^3$, and so on. Not only are the higher powers larger, but they are *much* larger. This is because if $x = 100$, for example, 100^5 is one hundred times as big as 100^4 which is one hundred times as big as 100^3. As x gets larger (written as $x \to \infty$), any positive power of x completely swamps all lower powers of x. We say that, as $x \to \infty$, higher powers of x *dominate* lower powers.

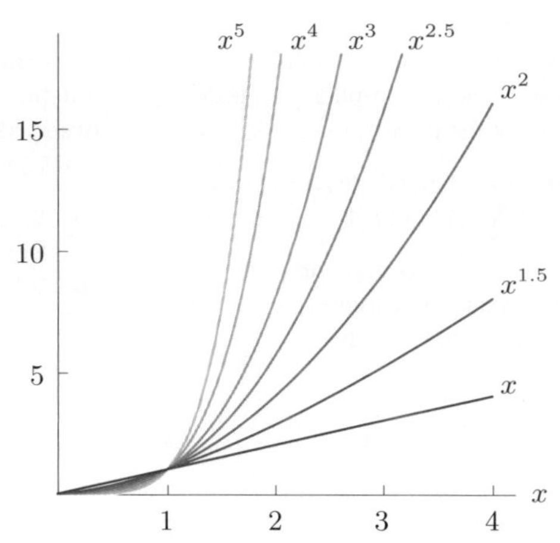

Figure 1.128: Powers of x: Which is largest for large values of x?

Limits to Infinity

When we consider the values of a function $f(x)$ as $x \to \infty$, we are looking for the *limit* as $x \to \infty$. This is abbreviated

$$\lim_{x\to\infty} f(x).$$

The notation $\lim_{x\to\infty} f(x) = L$ means that the values of the function approach L as the values of x get larger and larger. We have $f(x) \to L$ as $x \to \infty$. The behavior of a function as $x \to \infty$ and as $x \to -\infty$ is called the *end behavior* of the function.

Example 9 Find $\lim_{x\to\infty} f(x)$ and $\lim_{x\to-\infty} f(x)$ in each case.

(a) $f(x) = x^2$ (b) $f(x) = -x^3$ (c) $f(x) = e^x$

Solution

(a) As x gets larger and larger without bound, the function x^2 gets larger and larger without bound, so as $x \to \infty$, we have $x^2 \to \infty$. Thus,

$$\lim_{x\to\infty} (x^2) = \infty.$$

The square of a negative number is positive, so as $x \to -\infty$, we have $x^2 \to +\infty$. Thus,

$$\lim_{x\to-\infty} (x^2) = \infty.$$

To see this graphically, look at Figure 1.129. As $x \to \infty$ or as $x \to -\infty$, the function values get bigger and bigger and the "ends" of the graph go up.

(b) The graph of $f(x) = -x^3$ is in Figure 1.130. As $x \to \infty$, the function values get more and more negative; as $x \to -\infty$, the function values are positive and get larger and larger. We have

$$\lim_{x\to\infty} (-x^3) = -\infty \quad \text{and} \quad \lim_{x\to-\infty} (-x^3) = \infty.$$

(c) The graph of $f(x) = e^x$ is in Figure 1.131. As $x \to \infty$, the function values get larger without bound, and as $x \to -\infty$, the function values get closer and closer to zero. We have

$$\lim_{x\to\infty} (e^x) = \infty \quad \text{and} \quad \lim_{x\to-\infty} (e^x) = 0.$$

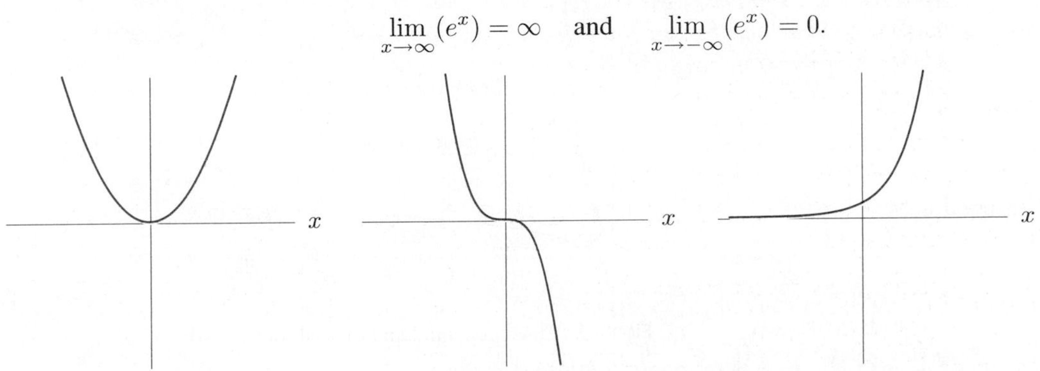

Figure 1.129: End behavior of $f(x) = x^2$

Figure 1.130: End behavior of $f(x) = -x^3$

Figure 1.131: End behavior of $f(x) = e^x$

Limits to finite values of x such as 0 are covered on page 136.

Exponential Functions and Power Functions: Which Dominate?

How does the growth of power functions compare to the growth of exponential functions, as x gets large? In everyday language, the word exponential is often used to imply very fast growth. But do exponential functions always grow faster than power functions? To determine what happens "in the long run," we often want to know which functions dominate as $x \to \infty$. Let's compare functions of the form $y = a^x$ for $a > 1$, and $y = x^n$ for $n > 0$.

First we consider $y = 2^x$ and $y = x^3$. The close-up, or local, view in Figure 1.132(a) shows that between $x = 2$ and $x = 4$, the graph of $y = 2^x$ lies below the graph of $y = x^3$. But Figure 1.132(b) shows that the exponential function $y = 2^x$ eventually overtakes $y = x^3$. The far-away, or global, view in Figure 1.132(c) shows that, for large x, the value of x^3 is insignificant compared to 2^x. Indeed, 2^x is growing so much faster than x^3 that the graph of 2^x appears almost vertical in comparison to the more leisurely climb of x^3.

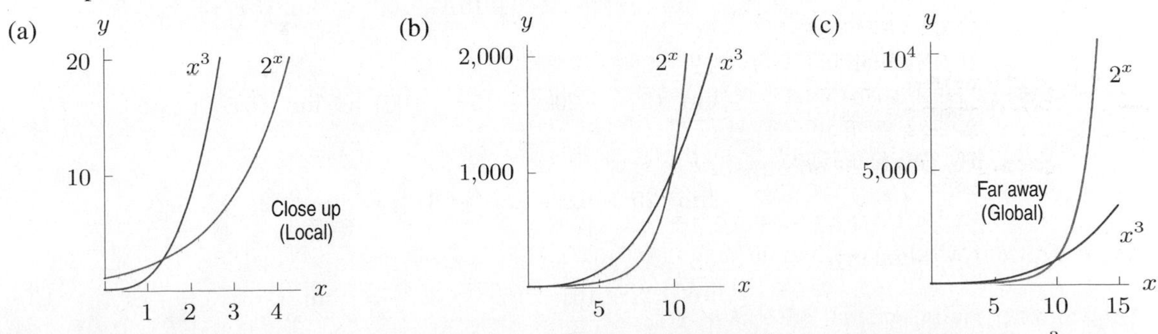

Figure 1.132: Local and global views of $y = 2^x$ and $y = x^3$: Notice that $y = 2^x$ eventually dominates $y = x^3$

In fact, *every* exponential growth function eventually dominates *every* power function. Although an exponential function may be below a power function for some values of x, if we look at large enough x-values, a^x eventually dominates x^n, no matter what n is (provided $a > 1$).

What about the logarithm function? We know that exponential functions grow very quickly and logarithm functions grow very slowly. See Figure 1.133. Logarithm functions grow so slowly, in fact, that every power function x^n eventually dominates $\ln x$ (provided n is positive).

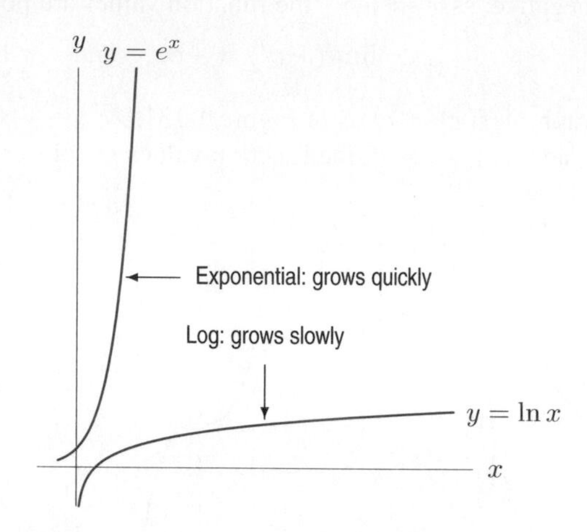

Figure 1.133: Exponential and logarithmic growth

End Behavior of Polynomials

We saw in Section 1.9 that if a polynomial is viewed in a large enough window, it has approximately the same shape as the power function given by the leading term. Provided $a_n \neq 0$, we have

$$\lim_{x\to\infty} (a_n x^n + a_{n-1}x^{n-1} + \cdots + a_1 x + a_0) = \lim_{x\to\infty} (a_n x^n).$$

A similar result holds as $x \to -\infty$. The end behavior of a polynomial is the same as the end behavior of its leading term.

Example 10 Find $\lim_{x\to\infty} f(x)$ and $\lim_{x\to-\infty} f(x)$ in each case. (a) $f(x) = 5x^3 - 20x^2 + 15x - 100$ (b) $f(x) = 25 + 10x - 15x^2 - 3x^4$

Solution (a) We have

$$\lim_{x\to\infty} (5x^3 - 20x^2 + 15x - 100) = \lim_{x\to\infty} (5x^3) = \infty.$$

Similarly,

$$\lim_{x\to-\infty} (5x^3 - 20x^2 + 15x - 100) = \lim_{x\to-\infty} (5x^3) = -\infty.$$

(b) We have

$$\lim_{x\to\infty} (25 + 10x - 15x^2 - 3x^4) = \lim_{x\to\infty} (-3x^4) = -\infty.$$

Similarly,

$$\lim_{x\to-\infty} (25 + 10x - 15x^2 - 3x^4) = \lim_{x\to-\infty} (-3x^4) = -\infty.$$

Problems on Limits to Infinity and End Behavior

1. As $x \to \infty$, which of the three functions $y = 1000x^2$, $y = 20x^3$, $y = 0.1x^4$ has the largest values? Which has the smallest values? Sketch a global picture of these three functions (for $x \geq 0$) on the same axes.
2. As $x \to \infty$, which of the two functions $y = 5000x^3$ and $y = 0.2x^4$ dominates? The two graphs intersect at the origin. Are there any other points of intersection? If so, find their x values.
3. Graph $y = x^{1/2}$ and $y = x^{2/3}$ for $x \geq 0$ on the same axes. Which function has larger values as $x \to \infty$?
4. By hand, graph $f(x) = x^3$ and $g(x) = 20x^2$ on the same axes. Which function has larger values as $x \to \infty$?
5. By hand, graph $f(x) = x^5$, $g(x) = -x^3$, and $h(x) = 5x^2$ on the same axes. Which has the largest positive values as $x \to \infty$? As $x \to -\infty$?

In Problems 6–7, graph $f(x)$ and $g(x)$ in two windows. Describe what you see.

(a) Window $-7 \leq x \leq 7$ and $-15 \leq y \leq 15$
(b) Window $-50 \leq x \leq 50$ and $-10{,}000 \leq y \leq 10{,}000$

6. $f(x) = 0.2x^3 - 5x + 3$ and $g(x) = 0.2x^3$
7. $f(x) = 3 - 5x + 5x^2 + x^3 - x^4$ and $g(x) = -x^4$

In Problems 8–10, draw a possible graph for $f(x)$.

8. $\lim_{x\to\infty} f(x) = -\infty$ and $\lim_{x\to-\infty} f(x) = -\infty$
9. $\lim_{x\to\infty} f(x) = -\infty$ and $\lim_{x\to-\infty} f(x) = +\infty$
10. $\lim_{x\to\infty} f(x) = 1$ and $\lim_{x\to-\infty} f(x) = +\infty$
11. A continuous function has $\lim_{x\to\infty} f(x) = 3$.
   (a) In words, explain what this limit means.
   (b) In each part, (i)–(iv), graph a function $f(x)$ with this limit and which is
   (i) Increasing (ii) Decreasing
   (iii) Concave up (iv) Oscillating
12. A continuous function has $\lim_{x\to-\infty} g(x) = 3$.
   (a) In words, explain what this limit means.
   (b) In each part, (i)–(iv), graph a function $g(x)$ with this limit and which is
   (i) Increasing (ii) Decreasing
   (iii) Concave up (iv) Oscillating
13. Estimate $\lim_{x\to\infty} \dfrac{1}{x}$. Explain your reasoning.
14. If $f(x) = -x^2$, what is $\lim_{x\to\infty} f(x)$? What is $\lim_{x\to-\infty} f(x)$?

In Problems 15–18, find $\lim_{x\to\infty} f(x)$ and $\lim_{x\to-\infty} f(x)$.

15. $f(x) = -10x^4$
16. $f(x) = 2^x$
17. $f(x) = 8(1 - e^{-x})$
18. $f(x) = 4x^5 - 25x^3 - 60x^2 + 1000x + 5000$

Which function in Problems 19–24 has larger values as $x \to \infty$?

19. $3x^5$ or $58x^4$
20. $12x^6$ or $(1.06)^x$
21. $x^{1/2}$ or $\ln x$
22. $x^3 + 2x^2 + 25x + 100$ or $10 - 6x^2 + x^4$
23. $5x^3 + 20x^2 + 150x + 200$ or $0.5x^4$
24. $5x^3 + 20x^2 + 150x + 200$ or $e^{0.2x}$
25. Match $y = 70x^2$, $y = 5x^3$, $y = x^4$, and $y = 0.2x^5$ with their graphs in Figure 1.134.

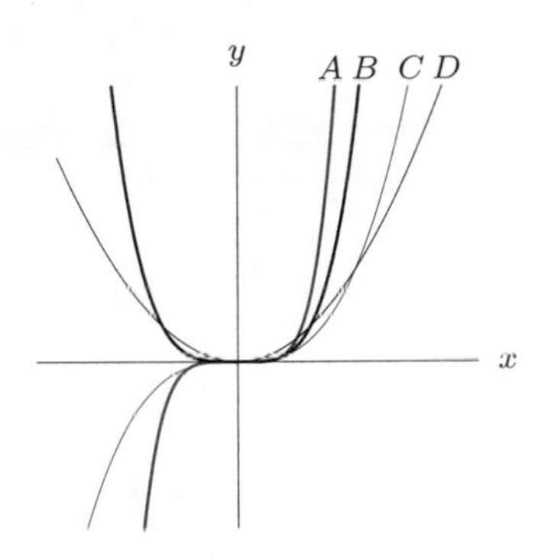

Figure 1.134

26. Match $y = e^x$, $y = \ln x$, $y = x^2$, and $y = x^{1/2}$ with their graphs in Figure 1.135.

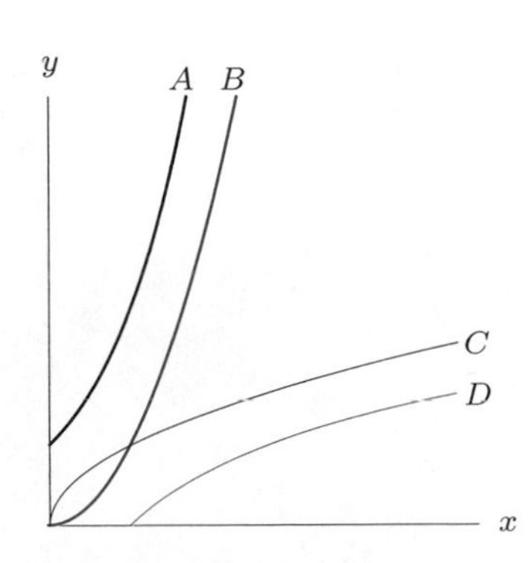

Figure 1.135

27. Global graphs of $y = x^5, y = 100x^2$, and $y = 3^x$ are in Figure 1.136. Which function corresponds to which curve?

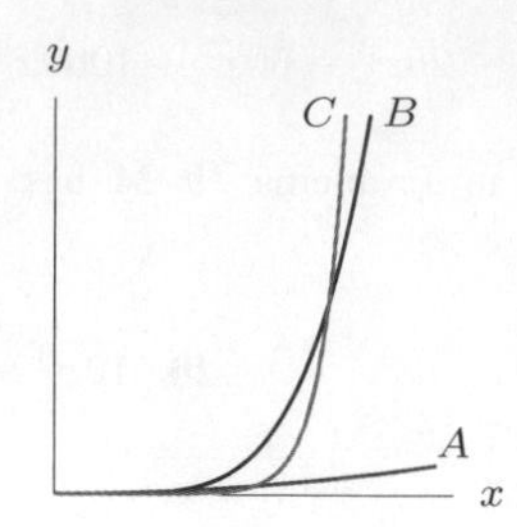

Figure 1.136

28. Use a graphing calculator or a computer to graph $y = x^4$ and $y = 3^x$. Determine approximate domains and ranges that give each of the graphs in Figure 1.137.

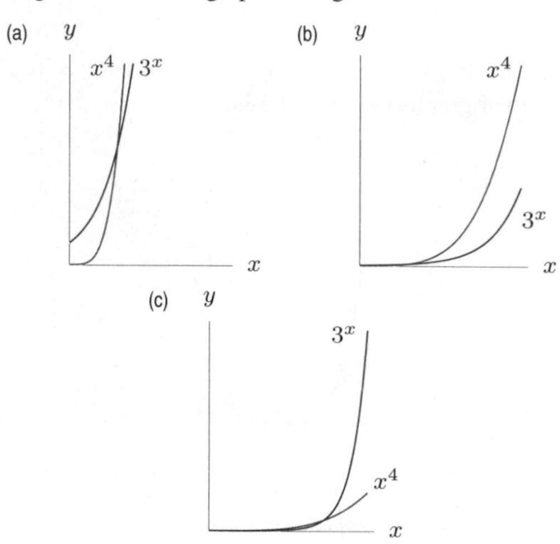

Figure 1.137

29. Graph $f(x) = e^{-x^2}$ using a window that includes both positive and negative values of x.

(a) For what values of x is f increasing? For what values is it decreasing?
(b) Is the graph of f concave up or concave down near $x = 0$?
(c) As $x \to \infty$, what happens to the value of $f(x)$? As $x \to -\infty$, what happens to $f(x)$?

30. Graph $f(x) = \ln(x^2 + 1)$ using a window that includes positive and negative values of x.

(a) For what values of x is f increasing? For what values is it decreasing?
(b) Is f concave up or concave down near $x = 0$?
(c) As $x \to \infty$, what happens to the value of $f(x)$? As $x \to -\infty$,what happens to $f(x)$?

Chapter Two

RATE OF CHANGE: THE DERIVATIVE

Chapter 1 introduced the average rate of change of a function on an interval. In this chapter, we investigate the *instantaneous* rate of change of a function at a point. The notion of rate of change at a given instant leads us to the concept of the *derivative*.

The derivative can be interpreted geometrically as the slope of a curve and physically as a rate of change. Derivatives can be used to represent everything from fluctuations in interest rates, to the rate at which fish are dying, to the rate of growth of a tumor.

2.1 INSTANTANEOUS RATE OF CHANGE

Chapter 1 introduced the average rate of change of a function over an interval. In this section, we consider the rate of change of a function at a point. We saw in Chapter 1 that when an object is moving along a straight line, the average rate of change of position with respect to time is the average velocity. If position is expressed as $y = f(t)$, where t is time, then

$$\text{Average rate of change in position between } t = a \text{ and } t = b = \frac{\Delta y}{\Delta t} = \frac{f(b) - f(a)}{b - a}.$$

If you drive 200 miles in 4 hours, your average velocity is $200/4 = 50$ miles per hour. Of course, this does not mean that you travel at exactly 50 mph the entire trip. Your velocity at a given instant during the trip is shown on your speedometer, and this is the quantity that we investigate now.

Instantaneous Velocity

We throw a grapefruit straight upward into the air. Table 2.1 gives its height, y, at time t. What is the velocity of the grapefruit at exactly $t = 1$? We use average velocities to estimate this quantity.

Table 2.1 *Height of the grapefruit above the ground*

t (sec)	0	1	2	3	4	5	6
$y = s(t)$ (feet)	6	90	142	162	150	106	30

The average velocity on the interval $0 \le t \le 1$ is 84 ft/sec and the average velocity on the interval $1 \le t \le 2$ is 52 ft/sec. Notice that the average velocity before $t = 1$ is larger than the average velocity after $t = 1$ since the grapefruit is slowing down. We expect the velocity *at* $t = 1$ to be between these two average velocities. How can we find the velocity at *exactly* $t = 1$? We look at what happens near $t = 1$ in more detail. Suppose that we find the average velocities on either side of $t = 1$ over smaller and smaller intervals, as in Figure 2.1. Then, for example,

$$\text{Average velocity between } t = 1 \text{ and } t = 1.01 = \frac{\Delta y}{\Delta t} = \frac{s(1.01) - s(1)}{1.01 - 1} = \frac{90.678 - 90}{0.01} = 67.8 \text{ ft/sec}.$$

We expect the instantaneous velocity at $t = 1$ to be between the average velocities on either side of $t = 1$. In Figure 2.1, the values of the average velocity before $t = 1$ and the average velocity after $t = 1$ get closer together as the size of the interval shrinks. For the smallest intervals in Figure 2.1, both velocities are 68.0 ft/sec (to one decimal place), so we say the velocity at $t = 1$ is 68.0 ft/sec (to one decimal place).

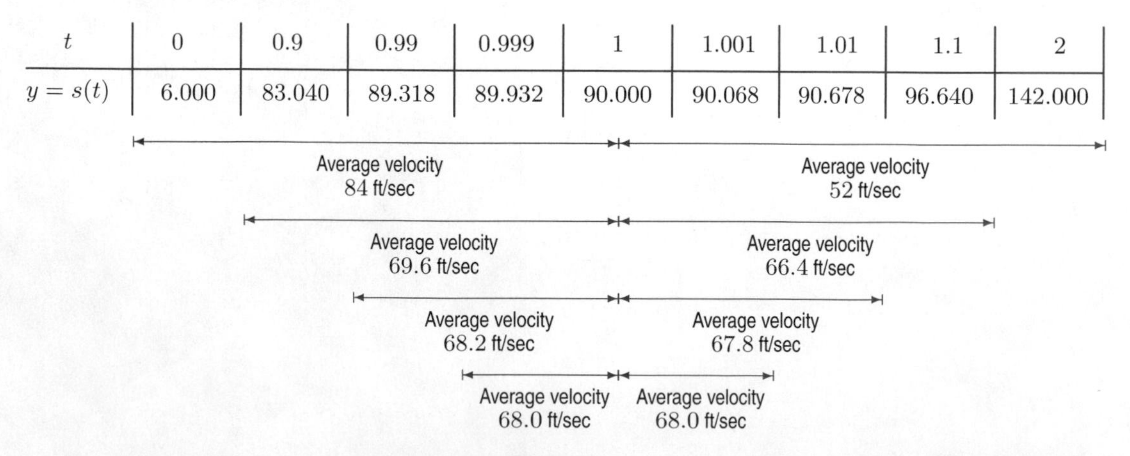

t	0	0.9	0.99	0.999	1	1.001	1.01	1.1	2
$y = s(t)$	6.000	83.040	89.318	89.932	90.000	90.068	90.678	96.640	142.000

Figure 2.1: Average velocities over intervals on either side of $t = 1$ showing successively smaller intervals

Of course, if we showed more decimal places, the average velocities before and after $t = 1$ would no longer agree. To calculate the velocity at $t = 1$ to more decimal places of accuracy, we take smaller and smaller intervals on either side of $t = 1$ until the average velocities agree to the number of decimal places we want. In this way, we can estimate the velocity at $t = 1$ to any accuracy.

Defining Instantaneous Velocity Using the Idea of a Limit

When we take smaller intervals near $t = 1$, it turns out that the average velocities for the grapefruit are always just above or just below 68 ft/sec. It seems natural, then, to define velocity at the instant $t = 1$ to be 68 ft/sec. This is called the *instantaneous velocity* at this point. Its definition depends on our being convinced that smaller and smaller intervals provide average velocities that come arbitrarily close to 68. This process is referred to as *taking the limit.*

> The **instantaneous velocity** of an object at time t is defined to be the limit of the average velocity of the object over shorter and shorter time intervals containing t.

Notice that the instantaneous velocity seems to be exactly 68, but what if it were 68.000001? How can we be sure that we have taken small enough intervals? Showing that the limit is exactly 68 requires more precise knowledge of how the velocities were calculated and of the limiting process; see the Focus on Theory section on page 135.

Instantaneous Rate of Change

We can define the *instantaneous rate of change* of any function $y = f(t)$ at a point $t = a$. We mimic what we did for velocity and look at the average rate of change over smaller and smaller intervals.

> The **instantaneous rate of change** of f at a, also called the **rate of change** of f at a, is defined to be the limit of the average rates of change of f over shorter and shorter intervals around a.

Since the average rate of change is a difference quotient of the form $\Delta y/\Delta t$, the instantaneous rate of change is a limit of difference quotients. In practice, we often approximate a rate of change by one of these difference quotients.

Example 1 The quantity (in mg) of a drug in the blood at time t (in minutes) is given by $Q = 25(0.8)^t$. Estimate the rate of change of the quantity at $t = 3$ and interpret your answer.

Solution We estimate the rate of change at $t = 3$ by computing the average rate of change over intervals near $t = 3$. We can make our estimate as accurate as we like by choosing our intervals small enough. Let's look at the average rate of change over the interval $3 \leq t \leq 3.01$:

$$\text{Average rate of change} = \frac{\Delta Q}{\Delta t} = \frac{25(0.8)^{3.01} - 25(0.8)^3}{3.01 - 3.00} = \frac{12.7715 - 12.80}{3.01 - 3.00} = -2.85.$$

A reasonable estimate for the rate of change of the quantity at $t = 3$ is -2.85. Since Q is in mg and t in minutes, the units of $\Delta Q/\Delta t$ are mg/minute. Since the rate of change is negative, the quantity of the drug is decreasing. After 3 minutes, the quantity of the drug in the body is decreasing at 2.85 mg/minute.

In Example 1, we estimated the rate of change using an interval to the right of the point ($t = 3$ to $t = 3.01$). We could use an interval to the left of the point, or we could average the rates of change to the left and the right. In this text, we usually use an interval to the right of the point.

The Derivative at a Point

The instantaneous rate of change of a function f at a point a is so important that it is given its own name, the *derivative of f at a*, denoted $f'(a)$ (read "f-prime of a"). If we want to emphasize that $f'(a)$ is the rate of change of $f(x)$ as the variable x increases, we call $f'(a)$ the derivative of f *with respect to x* at $x = a$. Notice that the derivative is just a new name for the rate of change of a function.

> The **derivative of f at a**, written $f'(a)$, is defined to be the instantaneous rate of change of f at the point a.

A definition of the derivative using a formula is given in the Focus on Theory section on page 135.

Example 2 Estimate $f'(2)$ if $f(x) = x^3$.

Solution Since $f'(2)$ is the derivative, or rate of change, of $f(x) = x^3$ at 2, we look at the average rate of change over intervals near 2. Using the interval $2 \le x \le 2.001$, we see that

$$\text{Average rate of change on } 2 \le x \le 2.001 = \frac{(2.001)^3 - 2^3}{2.001 - 2} = \frac{8.012 - 8}{0.001} = 12.0.$$

The rate of change of $f(x)$ at $x = 2$ appears to be approximately 12, so we estimate $f'(2) = 12$.

Visualizing the Derivative: Slope of the Graph and Slope of the Tangent Line

Figure 2.2 shows the average rate of change of a function represented by the slope of the secant line joining points A and B. The derivative is found by taking the average rate of change over smaller and smaller intervals. In Figure 2.3, as point B moves toward point A, the secant line becomes the tangent line at point A. Thus, the derivative is represented by the slope of the tangent line to the graph at the point.

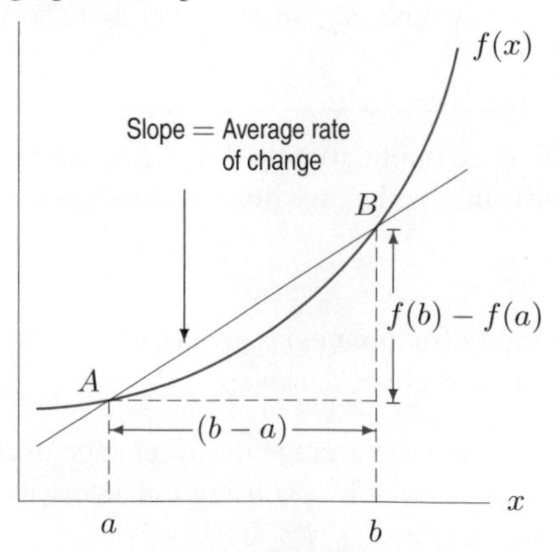

Figure 2.2: Visualizing the average rate of change of f between a and b

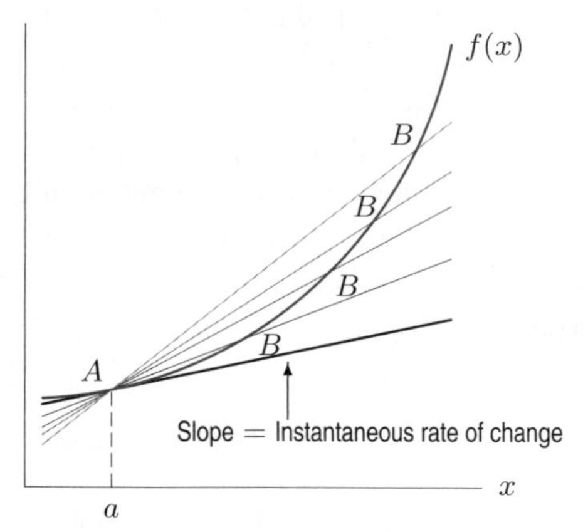

Figure 2.3: Visualizing the instantaneous rate of change of f at a

Alternatively, take the graph of a function around a point and "zoom in" to get a close-up view. (See Figure 2.4.) The more we zoom in, the more the graph appears to be straight. We call the slope of this line the *slope of the graph* at the point; it also represents the derivative.

> The derivative of a function at the point A is equal to
>
> - The slope of the graph of the function at A.
> - The slope of the line tangent to the curve at A.

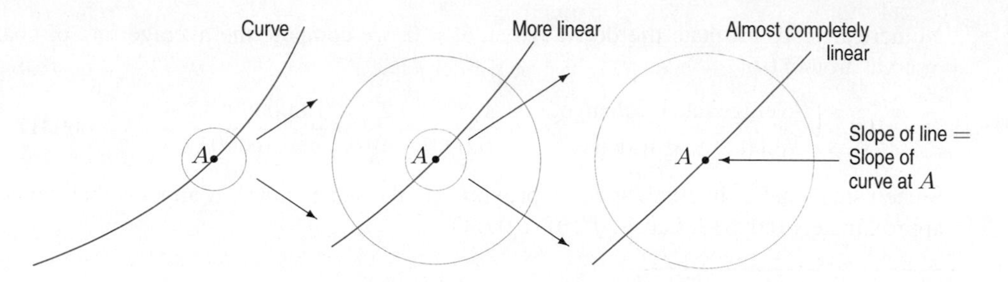

Figure 2.4: Finding the slope of a curve at a point by "zooming in"

The slope interpretation is often useful in gaining rough information about the derivative, as the following examples show.

Example 3 Use a graph of $f(x) = x^2$ to determine whether each of the following quantities is positive, negative, or zero: (a) $f'(1)$ (b) $f'(-1)$ (c) $f'(2)$ (d) $f'(0)$

Solution Figure 2.5 shows tangent line segments to the graph of $f(x) = x^2$ at the points $x = 1$, $x = -1$, $x = 2$, and $x = 0$. Since the derivative is the slope of the tangent line at the point, we have:

(a) $f'(1)$ is positive.
(b) $f'(-1)$ is negative.
(c) $f'(2)$ is positive (and larger than $f'(1)$).
(d) $f'(0) = 0$ since the graph has a horizontal tangent at $x = 0$.

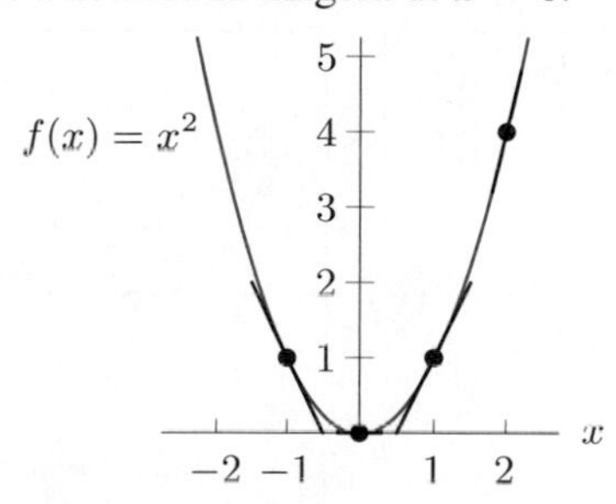

Figure 2.5: Tangent lines showing sign of derivative of $f(x) = x^2$

Example 4 Estimate the derivative of $f(x) = 2^x$ at $x = 0$ graphically and numerically.

Solution Graphically: If we draw a tangent line at $x = 0$ to the exponential curve in Figure 2.6, we see that it has a positive slope between 0.5 and 1.

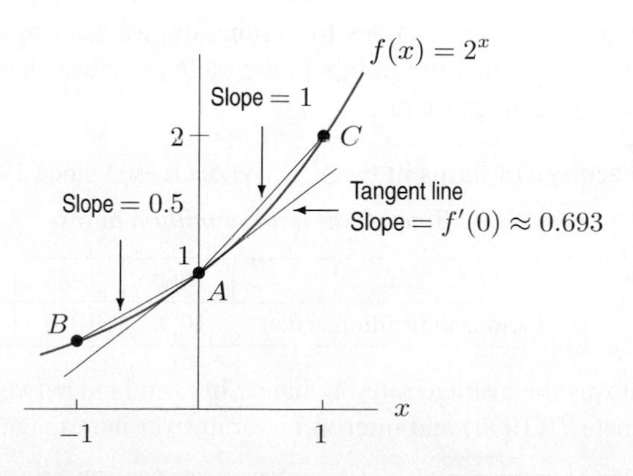

Figure 2.6: Graph of $f(x) = 2^x$ showing the derivative at $x = 0$

Numerically: To estimate the derivative at $x = 0$, we compute the average rate of change on an interval around 0.

$$\begin{array}{c}\text{Average rate of change} \\ \text{on } 0 \leq x \leq 0.0001\end{array} = \frac{2^{0.0001} - 2^0}{0.0001 - 0} = \frac{1.000069317 - 1}{0.0001} = 0.69317.$$

Since using smaller intervals gives approximately the same values, it appears that the derivative is approximately 0.69317; that is, $f'(0) \approx 0.693$.

Example 5 The graph of a function $y = f(x)$ is shown in Figure 2.7. Indicate whether each of the following quantities is positive or negative, and illustrate your answers graphically.

(a) $f'(1)$ (b) $\dfrac{f(3) - f(1)}{3 - 1}$ (c) $f(4) - f(2)$

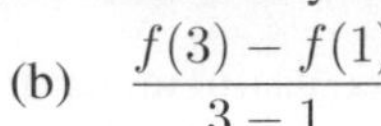

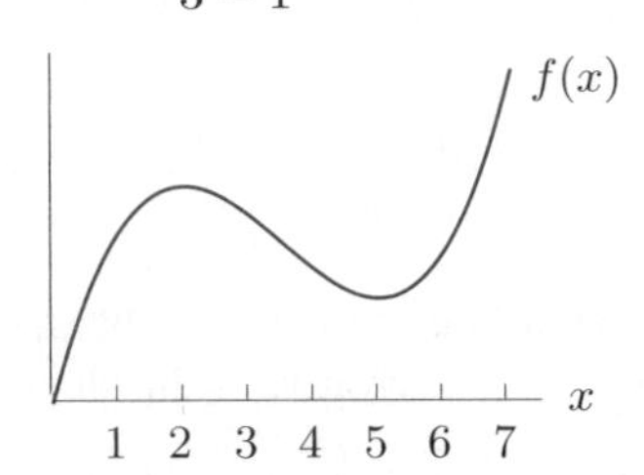

Figure 2.7

Solution (a) Since $f'(1)$ is the slope of the graph at $x = 1$, we see in Figure 2.8 that $f'(1)$ is positive.

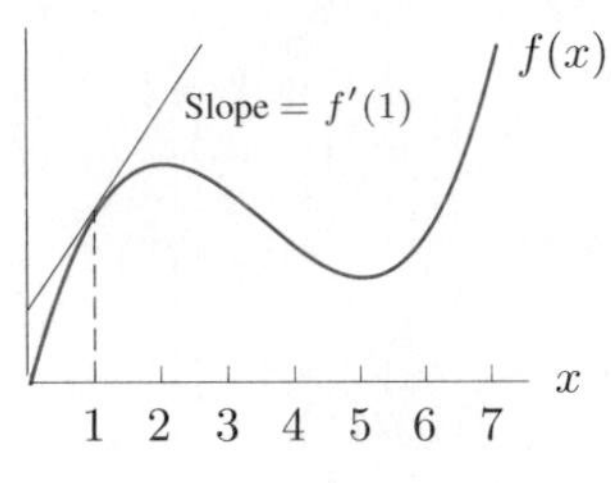

Figure 2.8

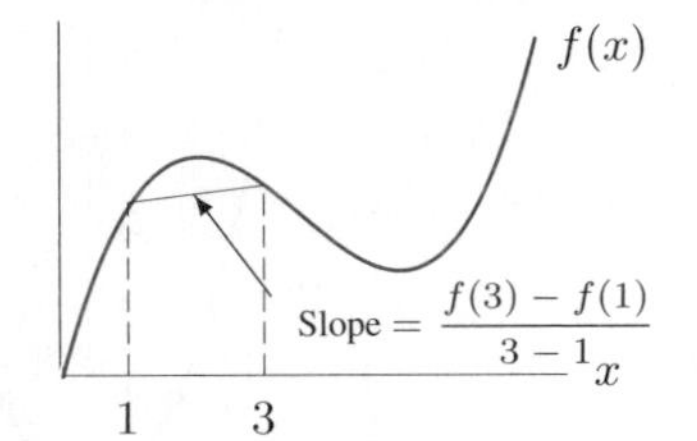

Figure 2.9

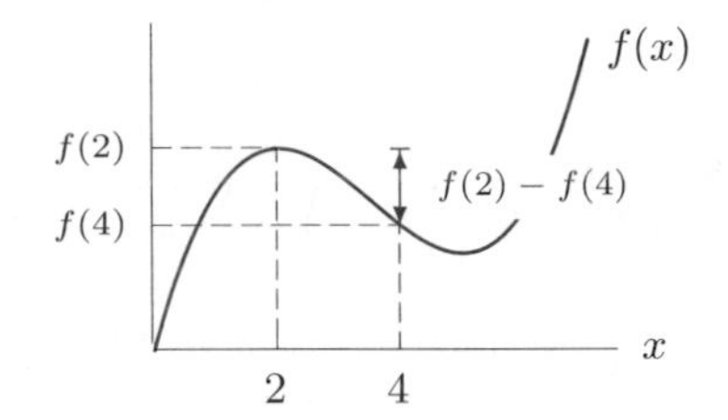

Figure 2.10

(b) The difference quotient $(f(3) - f(1))/(3 - 1)$ is the slope of the secant line between $x = 1$ and $x = 3$. We see from Figure 2.9 that this slope is positive.

(c) Since $f(4)$ is the value of the function at $x = 4$ and $f(2)$ is the value of the function at $x = 2$, the expression $f(4) - f(2)$ is the change in the function between $x = 2$ and $x = 4$. Since $f(4)$ lies below $f(2)$, this change is negative. See Figure 2.10.

Estimating the Derivative of a Function Given Numerically

If we are given a table of values for a function, we can estimate values of its derivative. To do this, we have to assume that the points in the table are close enough together that the function does not change wildly between them.

Example 6 The total acreage of farms in the US[1] has decreased since 1980. See Table 2.2.

Table 2.2 *Total farm land in million acres*

Year	1980	1985	1990	1995	2000
Farm land (million acres)	1039	1012	987	963	945

(a) What was the average rate of change in farm land between 1980 and 2000?
(b) Estimate $f'(1995)$ and interpret your answer in terms of farm land.

[1] *Statistical Abstracts of the United States 2004–2005*, Table 796.

Solution (a) Between 1980 and 2000,

$$\text{Average rate of change} = \frac{945 - 1039}{2000 - 1980} = \frac{-94}{20} = -4.7 \text{ million acres per year.}$$

Between 1980 and 2000, the amount of farm land was decreasing at an average rate of 4.7 million acres per year.

(b) We use the interval from 1995 to 2000 to estimate the instantaneous rate of change at 1995:

$$f'(1995) = \begin{matrix}\text{Rate of change}\\ \text{in 1995}\end{matrix} \approx \frac{945 - 963}{2000 - 1995} = \frac{-18}{5} = -3.6 \text{ million acres per year.}$$

In 1995, the amount of farm land was decreasing at a rate of approximately 3.6 million acres per year.

Problems for Section 2.1

1. The distance (in feet) of an object from a point is given by $s(t) = t^2$, where time t is in seconds.

(a) What is the average velocity of the object between $t = 3$ and $t = 5$?

(b) By using smaller and smaller intervals around 3, estimate the instantaneous velocity at time $t = 3$.

2. In a time of t seconds, a particle moves a distance of s meters from its starting point, where $s = 4t^2 + 3$.

(a) Find the average velocity between $t = 1$ and $t = 1 + h$ if:

(i) $h = 0.1$, (ii) $h = 0.01$, (iii) $h = 0.001$.

(b) Use your answers to part (a) to estimate the instantaneous velocity of the particle at time $t = 1$.

3. The size, S, of a tumor (in cubic millimeters) is given by $S = 2^t$, where t is the number of months since the tumor was discovered. Give units with your answers.

(a) What is the total change in the size of the tumor during the first six months?

(b) What is the average rate of change in the size of the tumor during the first six months?

(c) Estimate the rate at which the tumor is growing at $t = 6$. (Use smaller and smaller intervals.)

4. Match the points labeled on the curve in Figure 2.11 with the given slopes.

Slope	Point
-3	
-1	
0	
$1/2$	
1	
2	

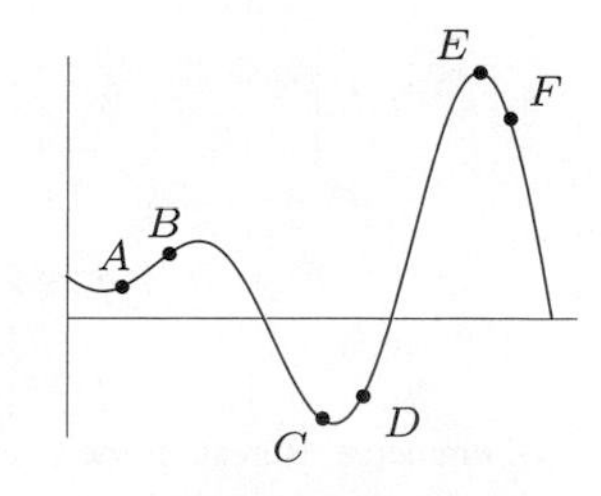

Figure 2.11

5. If t is in years since 2000, the population, in thousands, of the McAllen, Texas metropolitan area was given by $P(t) = 570(1.037)^t$. Estimate the rate of growth, in people per year, in 2006.

6. Figure 2.12 shows the cost, $y = f(x)$, of manufacturing x kilograms of a chemical.

(a) Is the average rate of change of the cost greater between $x = 0$ and $x = 3$, or between $x = 3$ and $x = 5$? Explain your answer graphically.

(b) Is the instantaneous rate of change of the cost of producing x kilograms greater at $x = 1$ or at $x = 4$? Explain your answer graphically.

(c) What are the units of these rates of change?

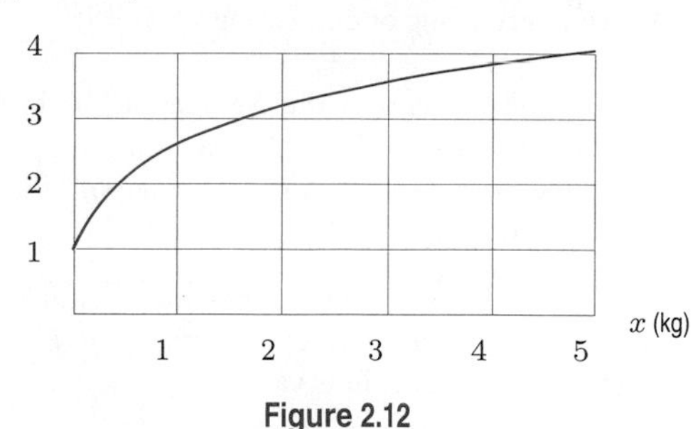

Figure 2.12

7. Find the average velocity over the interval $0 \leq t \leq 0.8$, and estimate the velocity at $t = 0.2$ of a car whose position, s, is given by the following table.

t (sec)	0	0.2	0.4	0.6	0.8	1.0
s (ft)	0	0.5	1.8	3.8	6.5	9.6

8. The following table gives the percent of the US population living in urban areas as a function of year.[2]

Year	1800	1830	1860	1890	1920
Percent	6.0	9.0	19.8	35.1	51.2
Year	1950	1980	1990	2000	
Percent	64.0	73.7	75.2	79.0	

(a) Find the average rate of change of the percent of the population living in urban areas between 1890 and 1990.
(b) Estimate the rate at which this percent is increasing at the year 1990.
(c) Estimate the rate of change of this function for the year 1830 and explain what it is telling you.
(d) Is this function increasing or decreasing?

9. **(a)** The function f is given in Figure 2.13. At which of the labeled points is $f'(x)$ positive? Negative? Zero?
(b) At which labeled point is f' largest? At which labeled point is f' most negative?

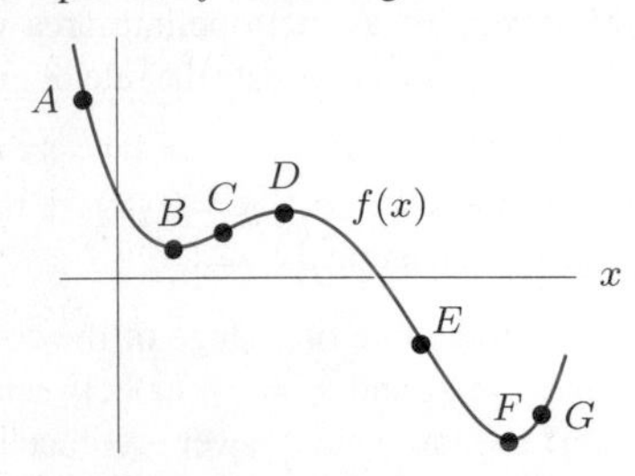

Figure 2.13

10. For $-3 \le x \le 7$, use a calculator or computer to graph

$$f(x) = (x^3 - 6x^2 + 8x)(2 - 3^x).$$

(a) How many zeros does f have in this interval?
(b) Is f increasing or decreasing at $x = 0$? At $x = 2$? At $x = 4$?
(c) On which interval is the average rate of change of f greater: $-1 \le x \le 0$ or $2 \le x \le 3$?
(d) Is the instantaneous rate of change of f greater at $x = 0$ or at $x = 2$?

11. Let $f(x) = 5^x$. Use a small interval to estimate $f'(2)$. Now improve your accuracy by estimating $f'(2)$ again, using an even smaller interval.

12. **(a)** Let $g(t) = (0.8)^t$. Use a graph to determine whether $g'(2)$ is positive, negative, or zero.
(b) Use a small interval to estimate $g'(2)$.

13. **(a)** Use a graph of $f(x) = 2 - x^3$ to decide whether $f'(1)$ is positive or negative. Give reasons.
(b) Use a small interval to estimate $f'(1)$.

14. Figure 2.14 shows the graph of f. Match the derivatives in the table with the points a, b, c, d, e.

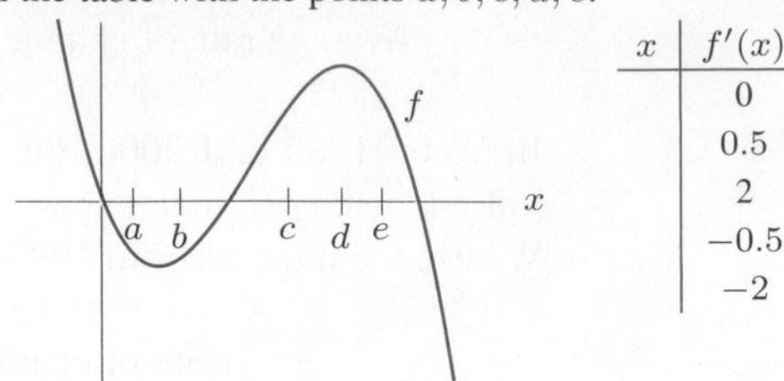

x	$f'(x)$
	0
	0.5
	2
	-0.5
	-2

Figure 2.14

15. Estimate $P'(0)$ if $P(t) = 200(1.05)^t$. Explain how you obtained your answer.

16. For the function $f(x) = 3^x$, estimate $f'(1)$. From the graph of $f(x)$, would you expect your estimate to be greater than or less than the true value of $f'(1)$?

17. Table 2.3 gives $P = f(t)$, the percent of households in the US with cable television t years since 1990.[3]

(a) Does $f'(6)$ appear to be positive or negative? What does this tell you about the percent of households with cable television?
(b) Estimate $f'(2)$. Estimate $f'(10)$. Explain what each is telling you, in terms of cable television.

Table 2.3

t (years since 1990)	0	2	4	6	8	10	12
P (% with cable)	59.0	61.5	63.4	66.7	67.4	67.8	68.9

18. Figure 2.15 shows $N = f(t)$, the number of farms in the US[4] between 1930 and 2000 as a function of year, t.

(a) Is $f'(1950)$ positive or negative? What does this tell you about the number of farms?
(b) Which is more negative: $f'(1960)$ or $f'(1980)$? Explain.

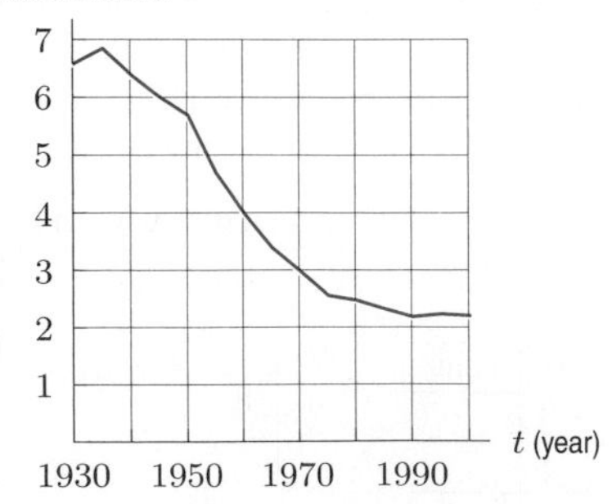

Figure 2.15

[2] *Statistical Abstracts of the US*, 1985, US Department of Commerce, Bureau of the Census, p. 22, and *World Almanac and Book of Facts 2005*, p. 624 (New York).
[3] *The World Almanac and Book of Facts 2005*, p. 310, (New York).
[4] www.nass.usda.gov:81/ipedb/farmnum.htm, accessed April 11, 2005.

19. Estimate the instantaneous rate of change of the function $f(x) = x \ln x$ at $x = 1$ and at $x = 2$. What do these values suggest about the concavity of the graph between 1 and 2?

20. Use the graph in Figure 2.7 on page 102 to decide if each of the following quantities is positive, negative or approximately zero. Illustrate your answers graphically.

(a) The average rate of change of $f(x)$ between $x = 3$ and $x = 7$.
(b) The instantaneous rate of change of $f(x)$ at $x = 3$.

21. Use Figure 2.16 to fill in the blanks in the following statements about the function g at point B.

(a) $g(\_\_) = \_\_$ **(b)** $g'(\_\_) = \_\_$

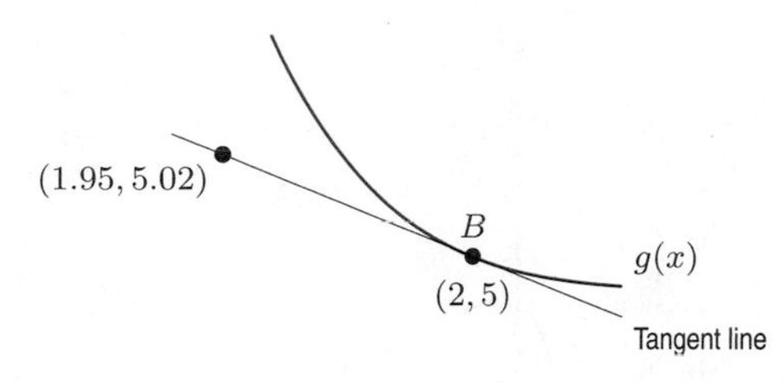

Figure 2.16

22. Use Figure 2.17 to fill in the blanks in the following statements about the function f at point A.
(a) $f(\_\_) = \_\_$ **(b)** $f'(\_\_) = \_\_$

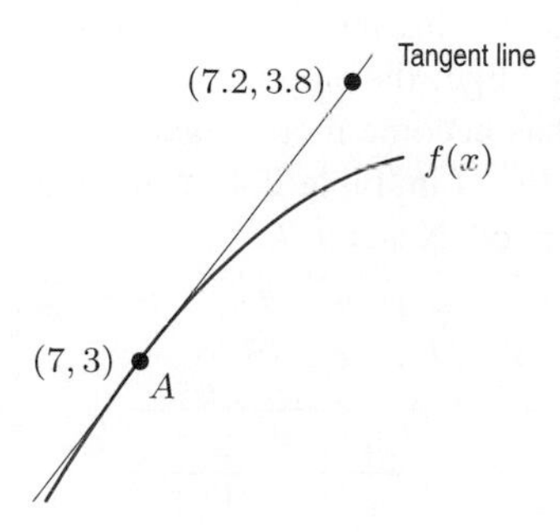

Figure 2.17

23. Show how to represent the following on Figure 2.18.

(a) $f(4)$ **(b)** $f(4) - f(2)$
(c) $\dfrac{f(5) - f(2)}{5 - 2}$ **(d)** $f'(3)$

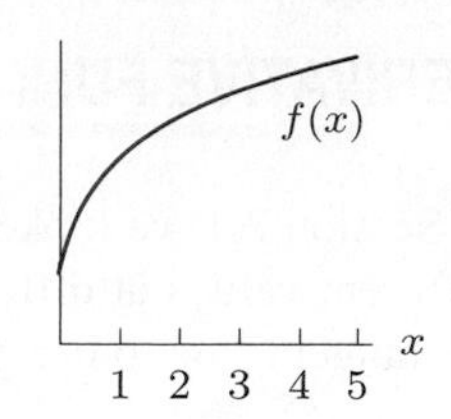

Figure 2.18

24. For each of the following pairs of numbers, use Figure 2.18 to decide which is larger. Explain your answer.

(a) $f(3)$ or $f(4)$?
(b) $f(3) - f(2)$ or $f(2) - f(1)$?
(c) $\dfrac{f(2) - f(1)}{2 - 1}$ or $\dfrac{f(3) - f(1)}{3 - 1}$?
(d) $f'(1)$ or $f'(4)$?

25. **(a)** Graph $f(x) = x^2$ and $g(x) = x^2 + 3$ on the same axes. What can you say about the slopes of the tangent lines to the two graphs at the point $x = 0$? $x = 1$? $x = 2$? $x = a$, where a is any value?
(b) Explain why adding a constant to any function will not change the value of the derivative at any point.

26. The following table shows the number of hours worked in a week, $f(t)$, hourly earnings, $g(t)$, in dollars, and weekly earnings, $h(t)$, in dollars, of production workers as functions of t, the year.[5]

(a) Indicate whether each of the following derivatives is positive, negative, or zero: $f'(t)$, $g'(t)$, $h'(t)$. Interpret each answer in terms of hours or earnings.
(b) Estimate each of the following derivatives, and interpret your answers:
(i) $f'(1970)$ and $f'(1995)$
(ii) $g'(1970)$ and $g'(1995)$
(iii) $h'(1970)$ and $h'(1995)$

t	1970	1975	1980	1985	1990	1995	2000
$f(t)$	37.0	36.0	35.2	34.9	34.3	34.3	34.3
$g(t)$	3.40	4.73	6.84	8.73	10.09	11.64	14.00
$h(t)$	125.80	170.28	240.77	304.68	349.29	399.53	480.41

[5] *The World Almanac and Book of Facts 2005*, p. 151 (New York). Production workers includes nonsupervisory workers in mining, manufacturing, construction, transportation, public utilities, wholesale and retail trade, finance, insurance, real estate, and services.

2.2 THE DERIVATIVE FUNCTION

In Section 2.1 we looked at the derivative of a function at a point. In general, the derivative takes on different values at different points and is itself a function. Recall that the derivative is the slope of the tangent line to the graph at the point.

Finding the Derivative of a Function Given Graphically

Example 1 Estimate the derivative of the function $f(x)$ graphed in Figure 2.19 at $x = -2, -1, 0, 1, 2, 3, 4, 5$.

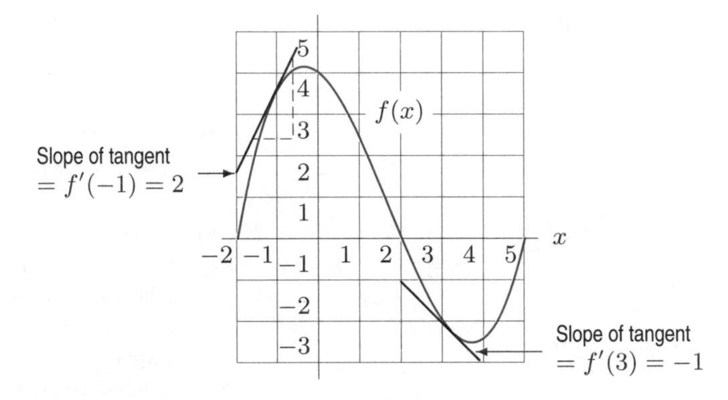

Figure 2.19: Estimating the derivative graphically as the slope of a tangent line

Solution From the graph, we estimate the derivative at any point by placing a straight edge so that it forms the tangent line at that point, and then using the grid to estimate the slope of the tangent line. For example, the tangent at $x = -1$ is drawn in Figure 2.19, and has a slope of about 2, so $f'(-1) \approx 2$. Notice that the slope at $x = -2$ is positive and fairly large; the slope at $x = -1$ is positive but smaller. At $x = 0$, the slope is negative, by $x = 1$ it has become more negative, and so on. Some estimates of the derivative, to the nearest integer, are listed in Table 2.4. You should check these values yourself. Is the derivative positive where you expect? Negative?

Table 2.4 *Estimated values of derivative of function in Figure 2.19*

x	-2	-1	0	1	2	3	4	5
Derivative at x	6	2	-1	-2	-2	-1	1	4

The important point to notice is that for every x-value, there is a corresponding value of the derivative. The derivative, therefore, is a function of x.

> For a function f, we define the **derivative function**, f', by
>
> $$f'(x) = \text{Instantaneous rate of change of } f \text{ at } x.$$

Example 2 Plot the values of the derivative function calculated in Example 1. Compare the graphs of f' and f.

Solution Graphs of f and f' are in Figures 2.20 and 2.21, respectively. Notice that f' is positive (its graph is above the x-axis) where f is increasing, and f' is negative (its graph is below the x-axis) where f is decreasing. The value of $f'(x)$ is 0 where f has a maximum or minimum value (at approximately $x = -0.5$ and $x = 3.7$).

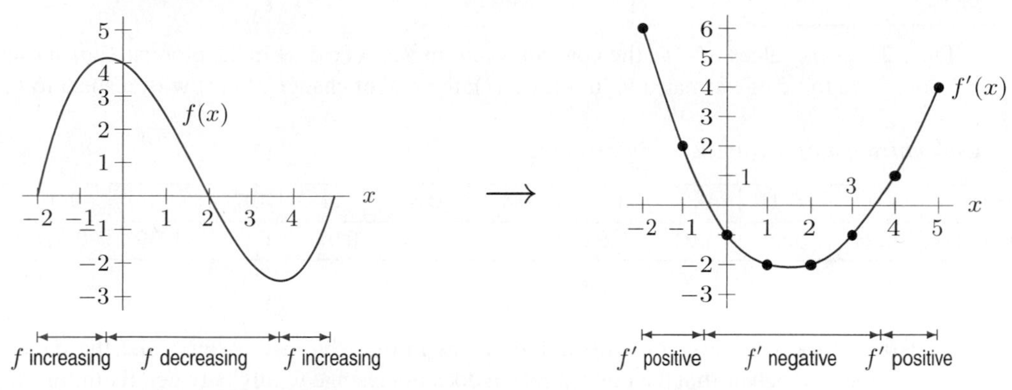

Figure 2.20: The function f

Figure 2.21: Estimates of the derivative, f'

Example 3 The graph of f is in Figure 2.22. Which of the graphs (a)–(c) is a graph of the derivative, f'?

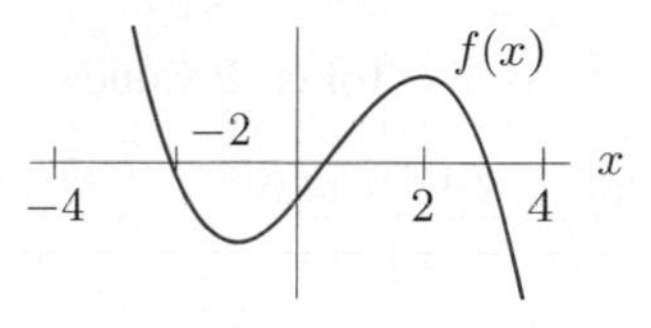

Figure 2.22

(a)

(b)

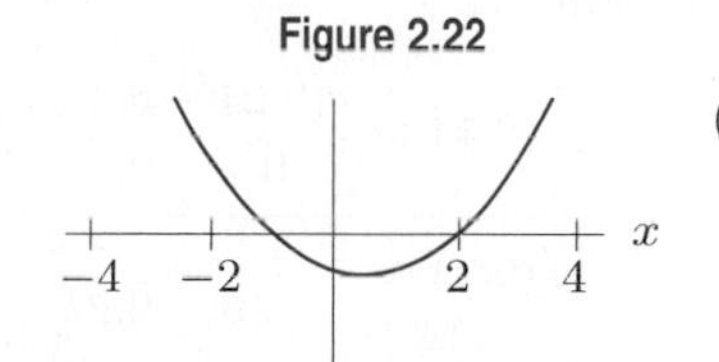

(c)

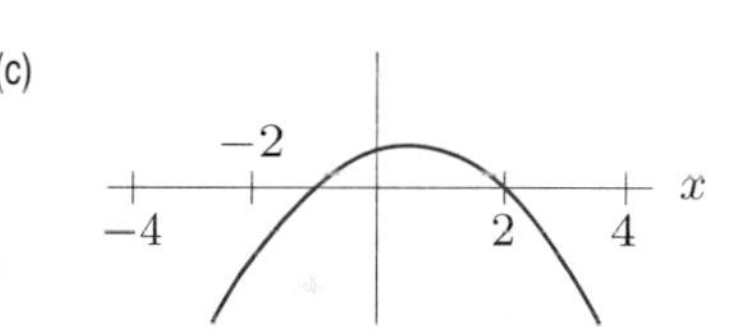

Solution Since the graph of $f(x)$ is horizontal at $x = -1$ and $x = 2$, the derivative is zero there. Therefore, the graph of $f'(x)$ has x-intercepts at $x = -1$ and $x = 2$.

The function f is decreasing for $x < -1$, increasing for $-1 < x < 2$, and decreasing for $x > 2$. The derivative is positive (its graph is above the x-axis) where f is increasing, and the derivative is negative (its graph is below the x-axis) where f is decreasing. The correct graph is (c).

What Does the Derivative Tell Us Graphically?

Where the derivative, f', of a function is positive, the tangent to the graph of f is sloping up; where f' is negative, the tangent is sloping down. If $f' = 0$ everywhere, then the tangent is horizontal everywhere and so f is constant. The sign of the derivative f' tells us whether the function f is increasing or decreasing.

> If $f' > 0$ on an interval, then f is *increasing* over that interval.
> If $f' < 0$ on an interval, then f is *decreasing* over that interval.
> If $f' = 0$ on an interval, then f is *constant* over that interval.

The magnitude of the derivative gives us the magnitude of the rate of change of f. If f' is large in magnitude, then the graph of f is steep (up if f' is positive or down if f' is negative); if f' is small in magnitude, the graph of f is gently sloping.

Estimating the Derivative of a Function Given Numerically

If we are given a table of function values instead of a graph of the function, we can estimate values of the derivative.

Example 4 Table 2.5 gives values of $c(t)$, the concentration (mg/cc) of a drug in the bloodstream at time t (min). Construct a table of estimated values for $c'(t)$, the rate of change of $c(t)$ with respect to t.

Table 2.5 *Concentration of a drug as a function of time*

t (min)	0	0.1	0.2	0.3	0.4	0.5	0.6	0.7	0.8	0.9	1.0
$c(t)$ (mg/cc)	0.84	0.89	0.94	0.98	1.00	1.00	0.97	0.90	0.79	0.63	0.41

Solution To estimate the derivative of c using the values in the table, we assume that the data points are close enough together that the concentration does not change wildly between them. From the table, we see that the concentration is increasing between $t = 0$ and $t = 0.4$, so we expect a positive derivative there. From $t = 0.5$ to $t = 1.0$, the concentration starts to decrease, and the rate of decrease gets larger and larger, so we would expect the derivative to be negative and of greater and greater magnitude.

We estimate the derivative for each value of t using a difference quotient. For example,

$$c'(0) \approx \frac{c(0.1) - c(0)}{0.1 - 0} = \frac{0.89 - 0.84}{0.1} = 0.5 \text{ (mg/cc) per minute.}$$

Similarly, we get the estimates

$$c'(0.1) \approx \frac{c(0.2) - c(0.1)}{0.2 - 0.1} = \frac{0.94 - 0.89}{0.1} = 0.5$$

$$c'(0.2) \approx \frac{c(0.3) - c(0.2)}{0.3 - 0.2} = \frac{0.98 - 0.94}{0.1} = 0.4$$

and so on. These values are tabulated in Table 2.6. Notice that the derivative has small positive values up until $t = 0.4$, and then it gets more and more negative, as we expected.

Table 2.6 *Derivative of concentration*

t	0	0.1	0.2	0.3	0.4	0.5	0.6	0.7	0.8	0.9
$c'(t)$	0.5	0.5	0.4	0.2	0.0	−0.3	−0.7	−1.1	−1.6	−2.2

Improving Numerical Estimates for the Derivative

In the previous example, our estimate for the derivative of $c(t)$ at $t = 0.2$ used the point to the right. We found the average rate of change between $t = 0.2$ and $t = 0.3$. However, we could equally well have gone to the left and used the rate of change between $t = 0.1$ and $t = 0.2$ to approximate the derivative at 0.2. For a more accurate result, we could average these slopes, getting the approximation

$$c'(0.2) \approx \frac{1}{2}\left(\begin{array}{c}\text{Slope to left}\\ \text{of } 0.2\end{array} + \begin{array}{c}\text{Slope to right}\\ \text{of } 0.2\end{array}\right) = \frac{0.5 + 0.4}{2} = 0.45.$$

Each of these methods of approximating the derivative gives a reasonable answer. We will usually estimate the derivative by going to the right.

Finding the Derivative of a Function Given by a Formula

If we are given a formula for a function f, can we come up with a formula for f'? Using the definition of the derivative, we often can. Indeed, much of the power of calculus depends on our ability to find formulas for the derivatives of all the familiar functions. This is explained in detail in Chapter 3. In the next example, we see how to guess a formula for the derivative.

Example 5 Guess a formula for the derivative of $f(x) = x^2$.

Solution We use difference quotients to estimate the values of $f'(1)$, $f'(2)$, and $f'(3)$. Then we look for a pattern in these values which we use to guess a formula for $f'(x)$.

Near $x = 1$, we have

$$f'(1) \approx \frac{1.001^2 - 1^2}{0.001} = \frac{1.002 - 1}{0.001} = \frac{0.002}{0.001} = 2.$$

Similarly,

$$f'(2) \approx \frac{2.001^2 - 2^2}{0.001} = \frac{4.004 - 4}{0.001} = \frac{0.004}{0.001} = 4$$

$$f'(3) \approx \frac{3.001^2 - 3^2}{0.001} = \frac{9.006 - 9}{0.001} = \frac{0.006}{0.001} = 6.$$

Knowing the value of f' at specific points cannot tell us the formula for f', but it can be suggestive: knowing $f'(1) \approx 2$, $f'(2) \approx 4$, $f'(3) \approx 6$ suggests that $f'(x) = 2x$. In Chapter 3, we show that this is indeed the case.

Problems for Section 2.2

1. The graph of $f(x)$ is given in Figure 2.23. Draw tangent lines to the graph at $x = -2$, $x = -1$, $x = 0$, and $x = 2$. Estimate $f'(-2)$, $f'(-1)$, $f'(0)$, and $f'(2)$.

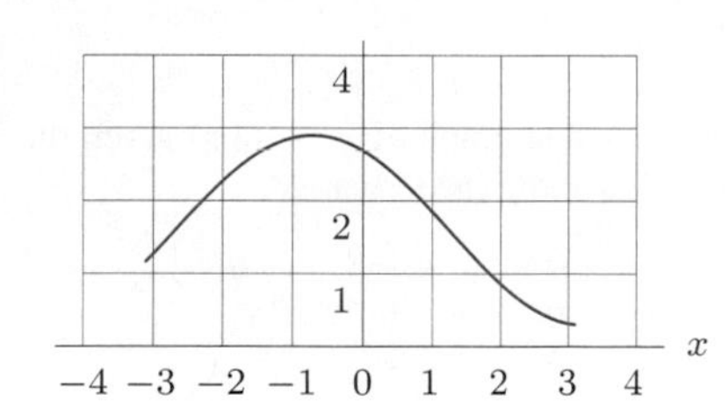

Figure 2.23

For Problems 2–7, graph the derivative of the given functions.

2.

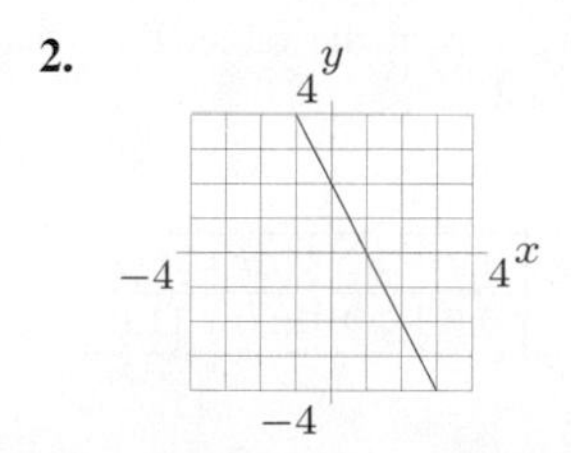

3.

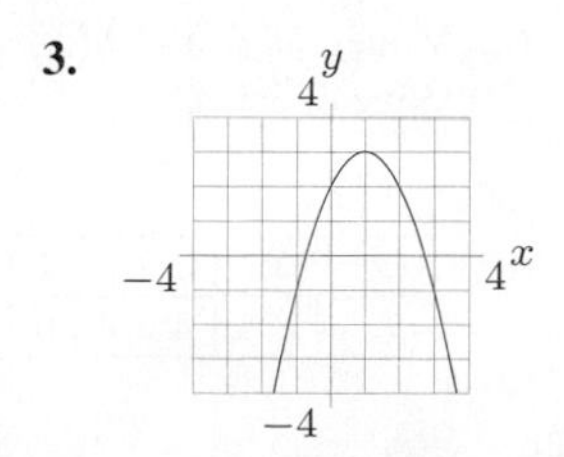

4.

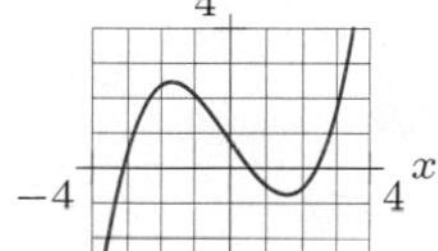

5.

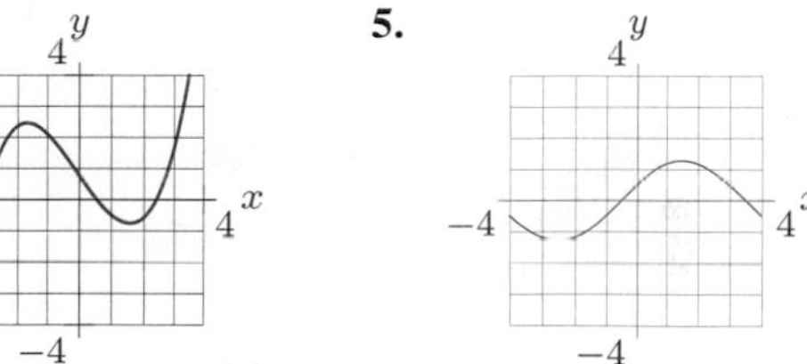

6.

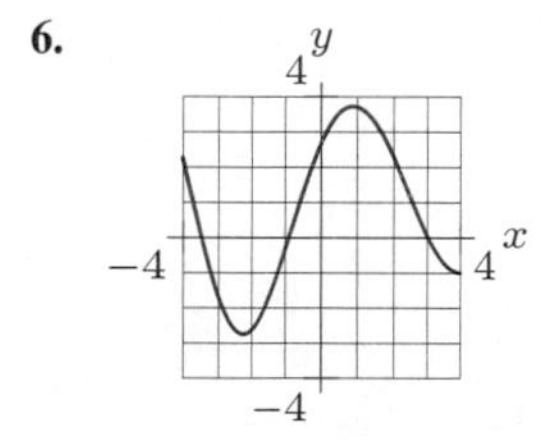

7.

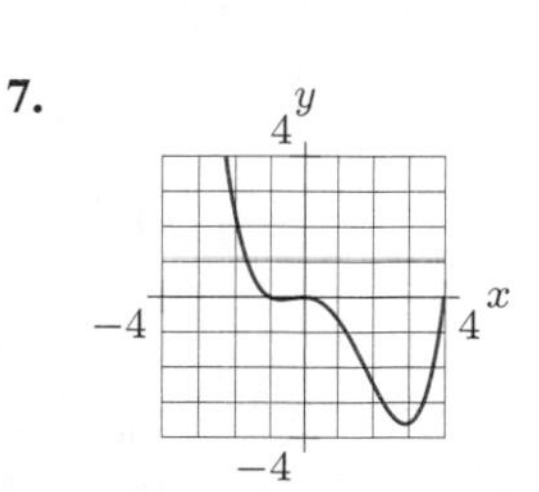

8. A city grew in population throughout the 1980s. The population was at its largest in 1990, and then shrank throughout the 1990s. Let $P = f(t)$ represent the population of the city t years since 1980. Sketch graphs of $f(t)$ and $f'(t)$, labeling the units on the axes.

Match the functions in Problems 9–12 with one of the derivatives in Figure 2.24.

(I)

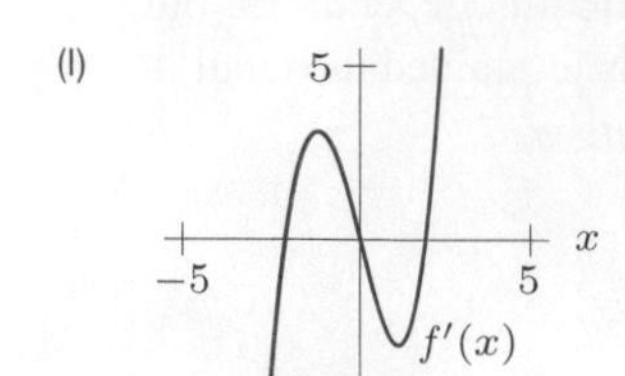

(II)

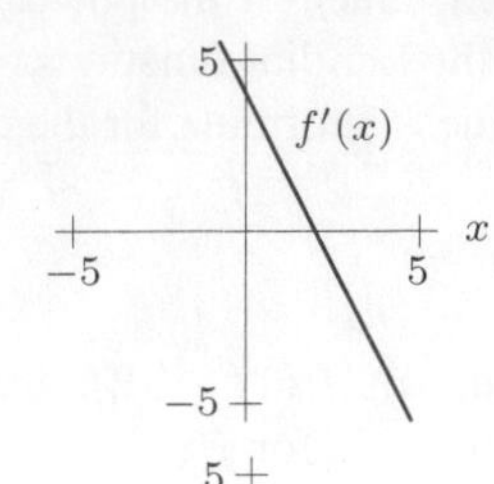

(III)

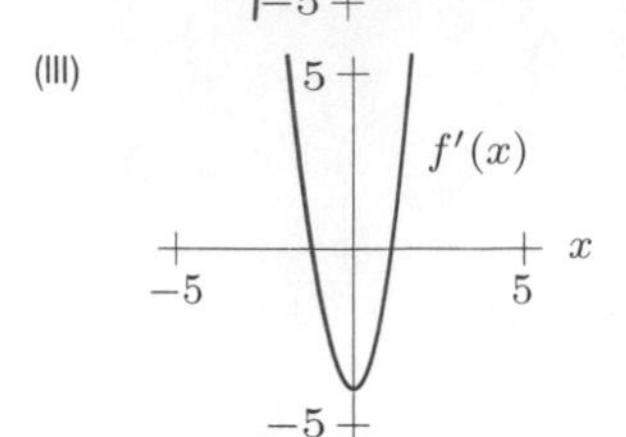

(IV)

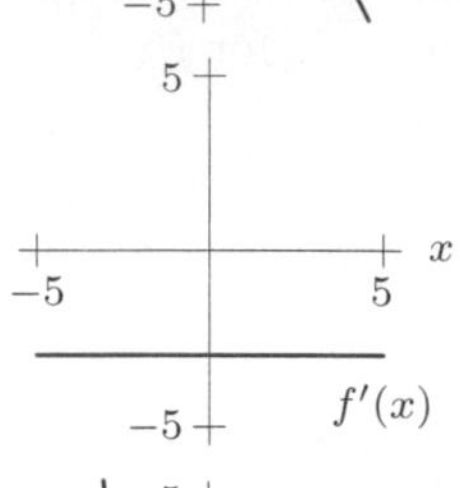

(V)

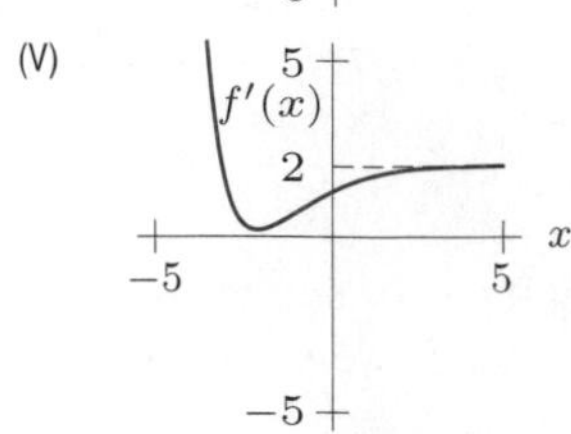

(VI)

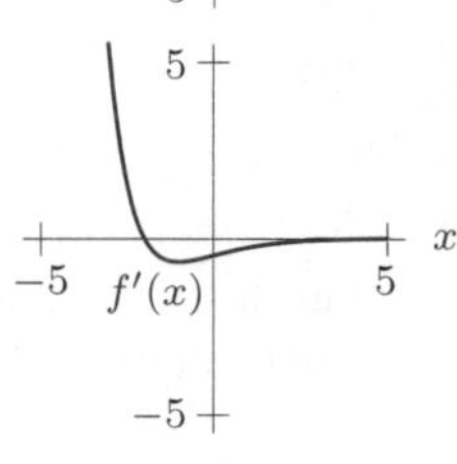

(VII)

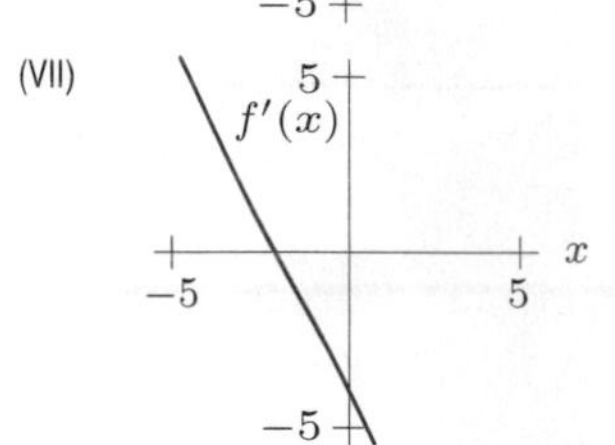

(VIII)

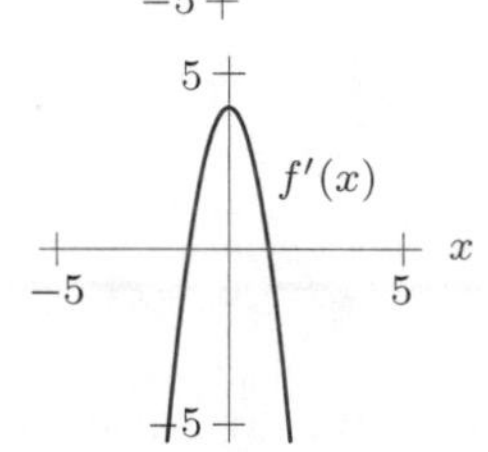

Figure 2.24

9.

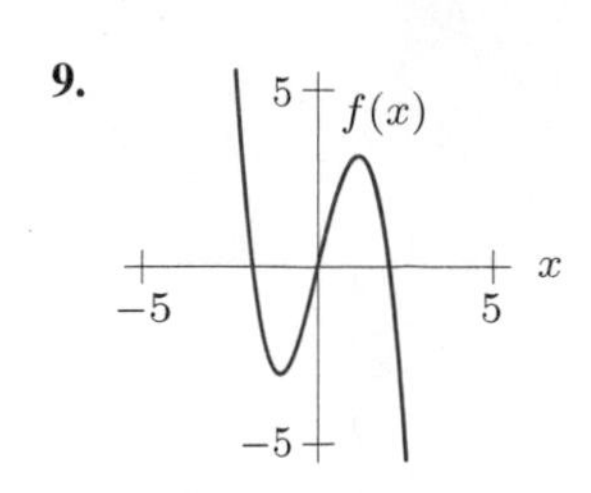

10.

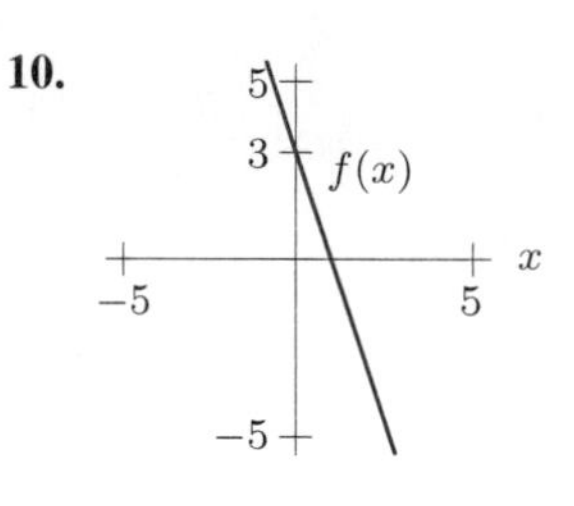

11.

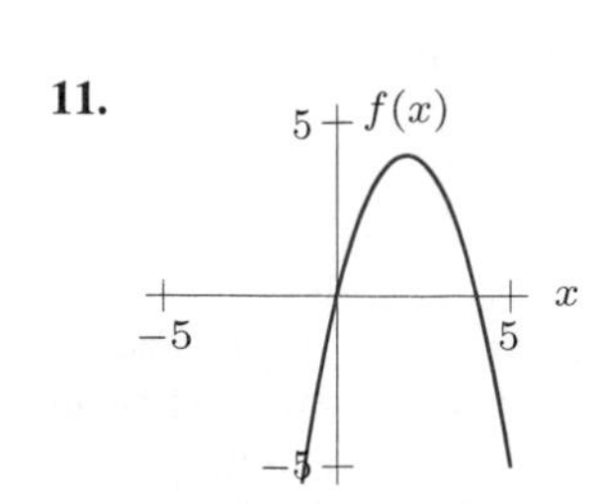

12.

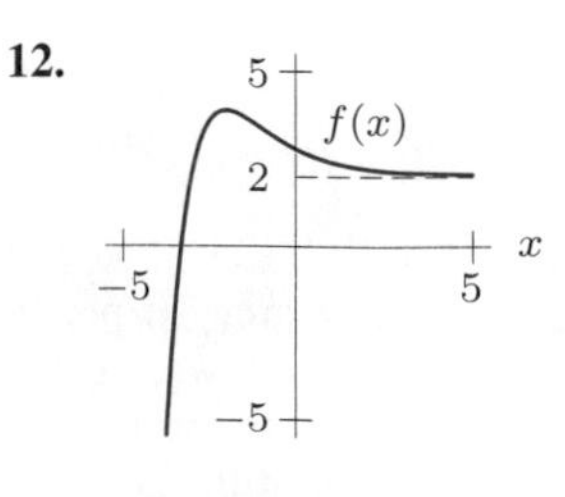

13. In the graph of f in Figure 2.25, at which of the labeled x-values is

(a) $f(x)$ greatest? **(b)** $f(x)$ least?
(c) $f'(x)$ greatest? **(d)** $f'(x)$ least?

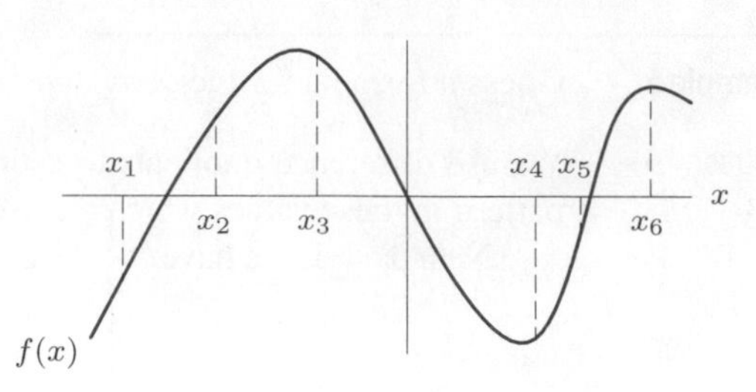

Figure 2.25

14. **(a)** Estimate $f'(2)$ using the values of f in the table.
(b) For what values of x does $f'(x)$ appear to be positive? Negative?

x	0	2	4	6	8	10	12
$f(x)$	10	18	24	21	20	18	15

15. Find approximate values for $f'(x)$ at each of the x-values given in the following table.

x	0	5	10	15	20
$f(x)$	100	70	55	46	40

16. Draw a possible graph of $y = f(x)$ given the following information about its derivative.

- $f'(x) > 0$ for $x < -1$
- $f'(x) < 0$ for $x > -1$
- $f'(x) = 0$ at $x = -1$

17. Draw a possible graph of $y = f(x)$ given the following information about its derivative.

- $f'(x) > 0$ on $1 < x < 3$
- $f'(x) < 0$ for $x < 1$ and $x > 3$
- $f'(x) = 0$ at $x = 1$ and $x = 3$

18. Values of x and $g(x)$ are given in the table. For what value of x is $g'(x)$ closest to 3?

x	2.7	3.2	3.7	4.2	4.7	5.2	5.7	6.2
$g(x)$	3.4	4.4	5.0	5.4	6.0	7.4	9.0	11.0

For Problems 19–26, sketch the graph of $f'(x)$.

19.

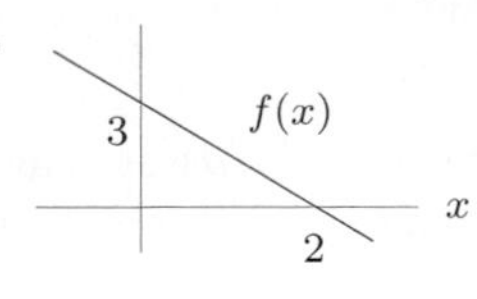

20.

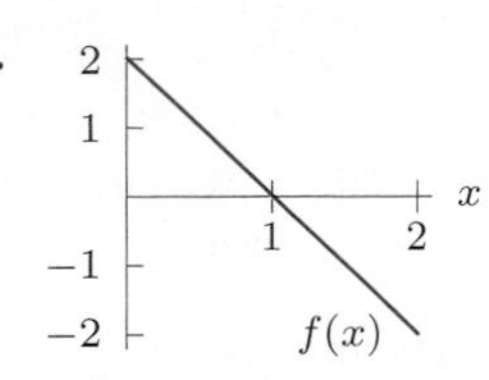

21.

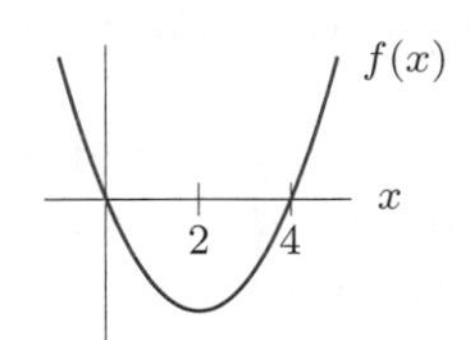

22.

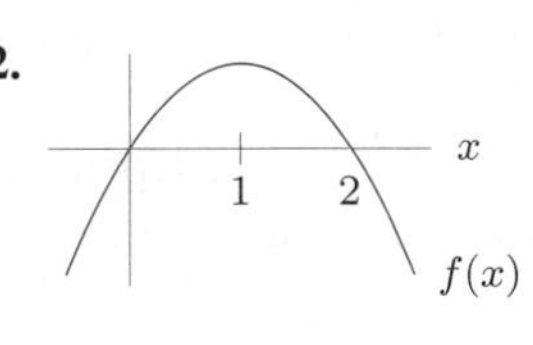

23.

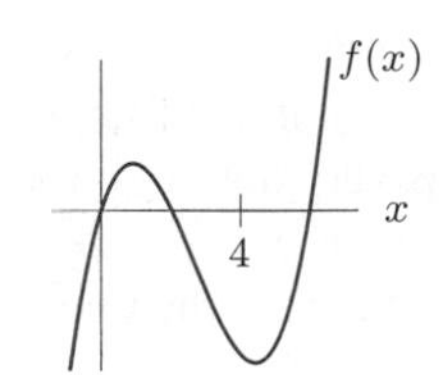

24.

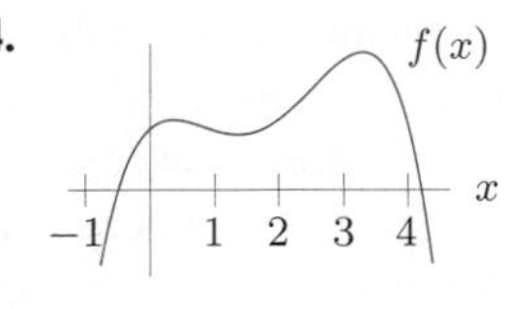

25.

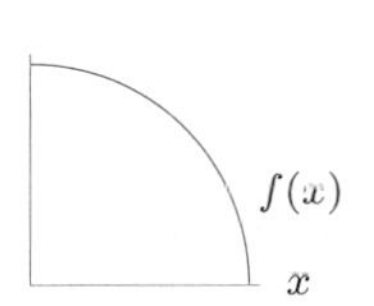

26.

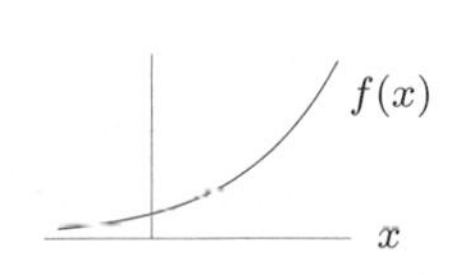

27. **(a)** Let $f(x) = \ln x$. Use small intervals to estimate $f'(1)$, $f'(2)$, $f'(3)$, $f'(4)$, and $f'(5)$.
(b) Use your answers to part (a) to guess a formula for the derivative of $f(x) = \ln x$.

28. Suppose $f(x) = \frac{1}{3}x^3$. Estimate $f'(2)$, $f'(3)$, and $f'(4)$. What do you notice? Can you guess a formula for $f'(x)$?

29. Figure 2.26 is the graph of f', the derivative of a function f. On what interval(s) is the function f

(a) Increasing? **(b)** Decreasing?

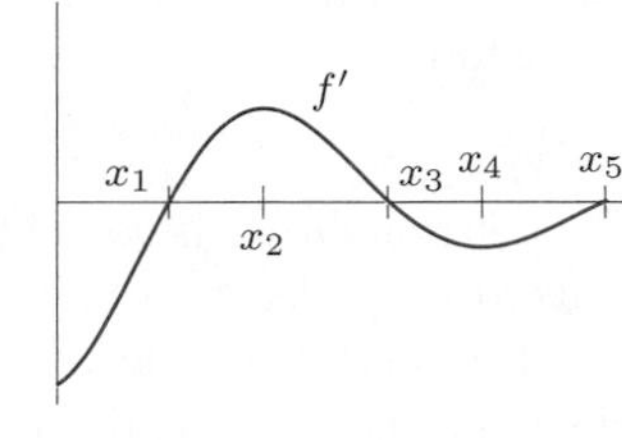

Figure 2.26: Graph of f', not f

30. A child inflates a balloon, admires it for a while and then lets the air out at a constant rate. If $V(t)$ gives the volume of the balloon at time t, then Figure 2.27 shows $V'(t)$ as a function of t. At what time does the child:

(a) Begin to inflate the balloon?
(b) Finish inflating the balloon?
(c) Begin to let the air out?
(d) What would the graph of $V'(t)$ look like if the child had alternated between pinching and releasing the open end of the balloon, instead of letting the air out at a constant rate?

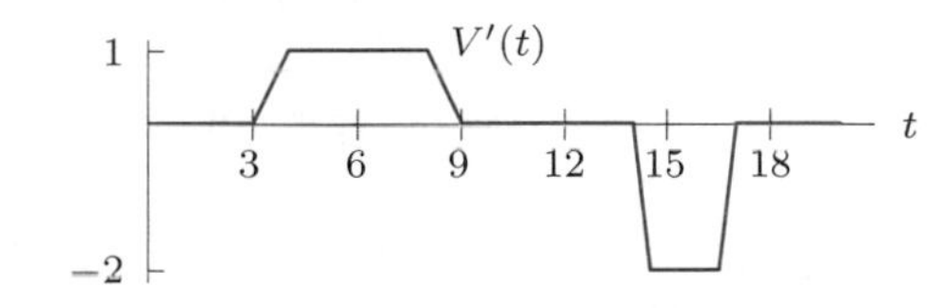

Figure 2.27

2.3 INTERPRETATIONS OF THE DERIVATIVE

We have seen the derivative interpreted as a slope and as a rate of change. In this section, we see other interpretations. The purpose of these examples is not to make a catalog of interpretations but to illustrate the process of obtaining them. There is another notation for the derivative that is often helpful.

An Alternative Notation for the Derivative

So far we have used the notation f' to stand for the derivative of the function f. An alternative notation for derivatives was introduced by the German mathematician Gottfried Wilhelm Leibniz (1646–1716) when calculus was first being developed. We know that $f'(x)$ is approximated by the average rate of change over a small interval. If $y = f(x)$, then the average rate of change is given by $\Delta y/\Delta x$. For small Δx, we have

$$f'(x) \approx \frac{\Delta y}{\Delta x}.$$

Leibniz's notation for the derivative, dy/dx, is meant to remind us of this. If $y = f(x)$, then we write

$$f'(x) = \frac{dy}{dx}.$$

Leibniz's notation is quite suggestive, especially if we think of the letter d in dy/dx as standing for "small difference in . . . ." The notation dy/dx reminds us that the derivative is a limit of ratios of the form

$$\frac{\text{Difference in } y\text{-values}}{\text{Difference in } x\text{-values}}.$$

The notation dy/dx is useful for determining the units for the derivative: the units for dy/dx are the units for y divided by (or "per") the units for x.

The separate entities dy and dx officially have no independent meaning: they are part of one notation. In fact, a good formal way to view the notation dy/dx is to think of d/dx as a single symbol meaning "the derivative with respect to x of . . .". Thus, dy/dx could be viewed as

$$\frac{d}{dx}(y), \quad \text{meaning "the derivative with respect to } x \text{ of } y\text{."}$$

On the other hand, many scientists and mathematicians really do think of dy and dx as separate entities representing "infinitesimally" small differences in y and x, even though it is difficult to say exactly how small "infinitesimal" is. It may not be formally correct, but it is very helpful intuitively to think of dy/dx as a very small change in y divided by a very small change in x.

For example, recall that if $s = f(t)$ is the position of a moving object at time t, then $v = f'(t)$ is the velocity of the object at time t. Writing

$$v = \frac{ds}{dt}$$

reminds us that v is a velocity since the notation suggests a distance, ds, over a time, dt, and we know that distance over time is velocity. Similarly, we recognize

$$\frac{dy}{dx} = f'(x)$$

as the slope of the graph of $y = f(x)$ by remembering that slope is vertical rise, dy, over horizontal run, dx.

The disadvantage of the Leibniz notation is that it is awkward to specify the x-value at which a derivative is evaluated. To specify $f'(2)$, for example, we have to write

$$\left.\frac{dy}{dx}\right|_{x=2}.$$

Using Units to Interpret the Derivative

Suppose $s = f(t)$ gives the position in meters of a body from a fixed point as a function of time, t, in seconds. Then, knowing that

$$\frac{ds}{dt} = f'(2) = 10 \text{ meters/sec}$$

tells us that when $t = 2$ sec, the body is moving at a velocity of 10 meters/sec. If the body continues to move at this velocity for a whole second (from $t = 2$ to $t = 3$), it would move an additional 10 meters. In general:

- The units of the derivative of a function are the units of the dependent variable divided by the units of the independent variable.
- If the derivative of a function is not changing rapidly near a point, then the derivative is approximately equal to the change in the function when the independent variable increases by 1 unit.

We define the derivative of velocity, dv/dt, as *acceleration.*

Example 1 If the velocity of a body at time t seconds is measured in meters/sec, what are the units of the acceleration?

Solution Since acceleration, dv/dt, is the derivative of velocity, the units of acceleration are units of velocity divided by units of time, or (meters/sec)/sec, written meters/sec^2.

The following examples illustrate how useful units can be in suggesting interpretations of the derivative.

Example 2 The cost C (in dollars) of building a house A square feet in area is given by the function $C = f(A)$. What is the practical interpretation of the function $f'(A)$?

Solution In the Leibniz notation,

$$f'(A) = \frac{dC}{dA}.$$

This is a cost divided by an area, so it is measured in dollars per square foot. You can think of dC as the extra cost of building an extra dA square feet of house. Thus, dC/dA is the additional cost per square foot. So if you are planning to build a house roughly A square feet in area, $f'(A)$ is the cost per square foot of the *extra* area involved in building a slightly larger house, and is called the *marginal cost*. The marginal cost is not necessarily the same thing as the average cost per square foot for the entire house, since once you are already set up to build a large house, the cost of adding a few square feet could be comparatively small.

Example 3 The cost of extracting T tons of ore from a copper mine is $C = f(T)$ dollars. What does it mean to say that $f'(2000) = 100$?

Solution In the Leibniz notation,

$$f'(2000) = \left.\frac{dC}{dT}\right|_{T=2000}.$$

Since C is measured in dollars and T is measured in tons, dC/dT must be measured in dollars per ton. So the statement

$$\left.\frac{dC}{dT}\right|_{T=2000} = 100$$

says that when 2000 tons of ore have already been extracted from the mine, the cost of extracting the next ton is approximately \$100. Another way of saying this is that it costs about \$100 to extract the 2001st ton. Note that this may well be different from the cost of extracting the tenth ton, which is likely to be more accessible.

Example 4 If $q = f(p)$ gives the number of pounds of sugar produced when the price per pound is p dollars, then what are the units and the meaning of

$$\left.\frac{dq}{dp}\right|_{p=3} = 50?$$

Solution The units of dq/dp are the units of q over the units of p, or pounds/dollar. The statement

$$\left.\frac{dq}{dp}\right|_{p=3} = f'(3) = 50 \text{ pounds/dollar}$$

tells us that the rate of change of q with respect to p is 50 when $p = 3$. This means that when the price is \$3, the quantity produced is increasing at 50 pounds for each dollar increase in price. This is an instantaneous rate of change, meaning that if the rate were to remain 50 pounds/dollar and if the price were to increase by a whole dollar, the quantity produced would increase by 50 pounds. In fact, the rate probably does not remain constant, so the quantity produced would probably not increase by exactly 50 pounds.

Example 5 The length of time, L, (in hours) that a drug stays in a person's system is a function of the quantity administered, q, in mg, so $L = f(q)$.

(a) Interpret the statement $f(10) = 6$. Give units for the numbers 10 and 6.
(b) Write the derivative of the function $L = f(q)$ in Leibniz notation. If $f'(10) = 0.5$, what are the units of the 0.5?
(c) Interpret the statement $f'(10) = 0.5$ in terms of dose and duration.

Solution

(a) We know that $f(q) = L$. In the statement $f(10) = 6$, we have $q = 10$ and $L = 6$, so the units are 10 mg and 6 hours. The statement $f(10) = 6$ tells us that a dose of 10 mg lasts 6 hours.
(b) Since $L = f(q)$, we see that L depends on q. The derivative of this function is dL/dq. Since L is in hours and q is in mg, the units of the derivative are hours per mg. In the statement $f'(10) = 0.5$, the 0.5 is the derivative and the units are hours per mg.
(c) The statement $f'(10) = 0.5$ tells us that, at a dose of 10 mg, the rate of change of duration is 0.5 hour per mg. In other words, if we increase the dose by 1 mg, the drug stays in the body approximately 30 minutes longer.

Example 6 You are told that water is flowing through a pipe at a rate of 10 cubic feet per second. Interpret this rate as the derivative of some function.

Solution You might think at first that the statement has something to do with the velocity of the water, but in fact a flow rate of 10 cubic feet per second could be achieved either with very slowly moving water through a large pipe, or with very rapidly moving water through a narrow pipe. If we look at the units—cubic feet per second—we realize that we are being given the rate of change of a quantity measured in cubic feet. But a cubic foot is a measure of volume, so we are being told the rate of change of a volume. If you imagine all the water that is flowing through ending up in a tank somewhere and let $V(t)$ be the volume of the tank at time t, then we are being told that the rate of change of $V(t)$ is 10, or that

$$V'(t) = \frac{dV}{dt} = 10.$$

Using the Derivative to Estimate Values of a Function

Since the derivative tells us how fast the value of a function is changing, we can use the derivative at a point to estimate values of the function at nearby points.

Example 7 The number of new subscriptions to a newspaper, y, in a month is a function of the amount, x, in dollars spent on advertising in that month, so $y = f(x)$.

(a) Interpret the statements $f(250) = 180$ and $f'(250) = 2$.
(b) Use the statements given in part (a) to estimate $f(251)$ and $f(260)$. Which estimate is more reliable?

Solution

(a) The statement $f(250) = 180$ tells us that $y = 180$ when $x = 250$. This means that if \$250 a month is spent on advertising, there are 180 new subscriptions a month. Since the derivative is dy/dx, the statement $f'(250) = 2$ tells us that

$$\frac{dy}{dx} = 2 \quad \text{when } x = 250.$$

This means that if the amount spent on advertising is \$250 and increases by \$1, the number of new subscriptions will go up by about 2.

(b) The statement $f(250) = 180$ says that when \$250 is spent on advertising, there are 180 new subscriptions. The statement $f'(250) = 2$ means that the number of new subscriptions increases at a rate of 2 subscriptions per additional dollar spent on advertising. If one more dollar is spent on advertising (so $x = 251$), we expect 2 more subscriptions in addition to the 180, so

$$f(251) \approx 180 + 2 = 182.$$

Similarly, if 10 dollars more were spent on advertising (so $x = 260$), we expect about $10(2) = 20$ new subscriptions, so

$$f(260) \approx 180 + 10(2) = 200.$$

Note that to estimate $f(260)$, we have to assume that the rate of 2 new subscriptions for each additional dollar continues all the way from $x = 250$ to $x = 260$. This means that the estimate of $f(251)$ is more reliable.

In Example 7, representing the change in y by Δy and the change in x by Δx, we used the following result:

Local Linear Approximation

$$\Delta y \approx f'(x)\Delta x \qquad \text{for } \Delta x \text{ near } 0.$$

Example 8 Climbing health care costs have been a source of concern for some time. Use the data[6] in Table 2.7 to estimate average (per consumer unit) expenditures in 2005 and 2020.

Table 2.7 *Average yearly health care costs (per consumer unit) for various years since 1990*

Year	1990	1995	1998	2000	2002
Per capita expenditure (\$)	1480	1732	1903	2066	2350

Solution Health care costs increased throughout the period shown. Between 2000 and 2002, they increased $(2350 - 2066)/2 = \$142$ per year. To make estimates beyond 2002 we assume that costs continue to climb at the same rate. Therefore, we estimate

$$\begin{aligned}\text{Costs in 2005} &= \text{Costs in 2002} + \text{Change in costs}\\ &\approx \$2350 + \$142 \cdot 3 = \$2776.\end{aligned}$$

Since 2020 is 18 years beyond 2002,

$$\text{Costs in 2020} \approx \$2350 + \$142 \cdot 18 = \$4906.$$

The estimate for 2005 in the preceding example is much more likely to be close to the true value than the estimate for 2020. The further we extrapolate from the given data, the less accurate we are likely to be. It is unlikely that the rate of change of health care costs will stay at \$142/year until 2020.

Graphically, what we have done is to extend the line joining the points for 2000 and 2002 to make projections for the future. See Figure 2.28. You might be concerned that we used only the last two pieces of data to make the estimates. Isn't there valuable information to be gained from the rest of the data? Yes, indeed—though there's no fixed way of taking this information into account. You might look at the rate of change for the years before 2000 and take an average, or you might use linear or exponential regression.

[6] *Statistical Abstracts of the United States 2004–2005*, Table 125.

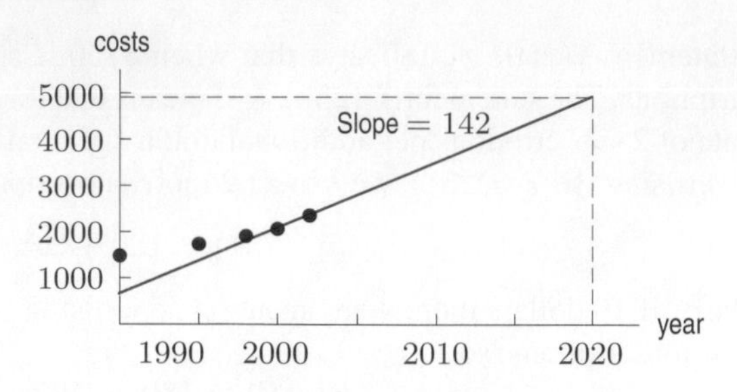

Figure 2.28: Graph of health care costs

Problems for Section 2.3

1. The average weight W of an oak tree in kilograms that is x meters tall is given by the function $W = f(x)$. What are the units of measurement of $f'(x)$?

2. Figure 2.29 shows world solar energy output, in megawatts, as a function of years since 1990.[7] Estimate $f'(6)$. Give units and interpret your answer.

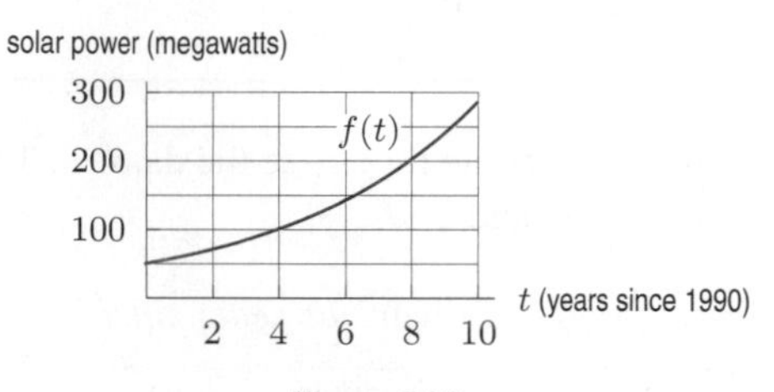

Figure 2.29

3. The cost, $C = f(w)$, in dollars of buying a chemical is a function of the weight bought, w, in pounds.

(a) In the statement $f(12) = 5$, what are the units of the 12? What are the units of the 5? Explain what this is saying about the cost of buying the chemical.
(b) Do you expect the derivative f' to be positive or negative? Why?
(c) In the statement $f'(12) = 0.4$, what are the units of the 12? What are the units of the 0.4? Explain what this is saying about the cost of buying the chemical.

4. The cost, C (in dollars) to produce g gallons of ice cream can be expressed as $C = f(g)$. Using units, explain the meaning of the following statements in terms of ice cream.

(a) $f(200) = 350$ **(b)** $f'(200) = 1.4$

5. The time for a chemical reaction, T (in minutes), is a function of the amount of catalyst present, a (in milliliters), so $T = f(a)$.

(a) If $f(5) = 18$, what are the units of 5? What are the units of 18? What does this statement tell us about the reaction?
(b) If $f'(5) = -3$, what are the units of 5? What are the units of -3? What does this statement tell us?

6. The percent, P, of US households with a personal computer is a function of the number of years, t, since 1982 (when the percent was essentially zero), so $P = f(t)$. Interpret the statements $f(20) = 57$ and $f'(20) = 3$.

7. A yam has just been taken out of the oven and is cooling off before being eaten. The temperature, T, of the yam (measured in degrees Fahrenheit) is a function of how long it has been out of the oven, t (measured in minutes). Thus, we have $T = f(t)$.

(a) Is $f'(t)$ positive or negative? Why?
(b) What are the units for $f'(t)$?

8. The quantity sold, q, of a certain product is a function of the price, p, so $q = f(p)$. Interpret each of the following statements in terms of demand for the product:

(a) $f(15) = 200$ **(b)** $f'(15) = -25$.

9. The weight, W, in lbs, of a child is a function of its age, a, in years, so $W = f(a)$.

(a) Do you expect $f'(a)$ to be positive or negative? Why?
(b) What does $f(8) = 45$ tell you? Give units for the numbers 8 and 45.
(c) What are the units of $f'(a)$? Explain what $f'(a)$ tells you in terms of age and weight.
(d) What does $f'(8) = 4$ tell you about age and weight?
(e) As a increases, do you expect $f'(a)$ to increase or decrease? Explain.

10. The thickness, P, in mm, of pelican eggshells depends on the concentration, c, of PCBs in the eggshell, measured in ppm (parts per million); that is, $P = f(c)$.

(a) The derivative $f'(c)$ is negative. What does this tell you?
(b) Give units and interpret $f(200) = 0.28$ and $f'(200) = -0.0005$ in terms of PCBs and eggs.

[7]The Worldwatch Institute, *Vital Signs* 2001, p. 47, (New York: W.W. Norton, 2001).

Problems 11–14 concern $g(t)$ in Figure 2.30, which gives the weight of a human fetus as a function of its age.

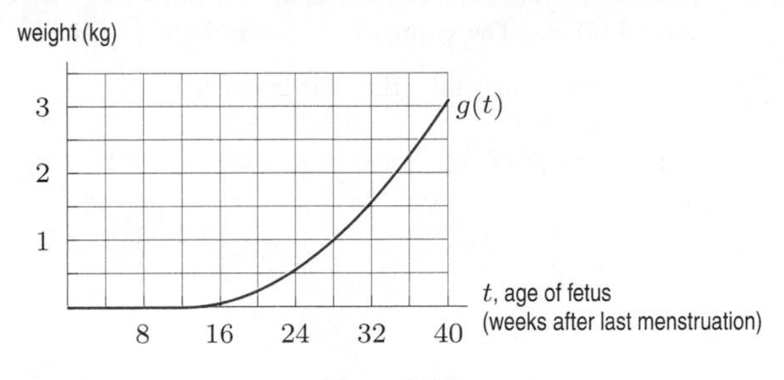

Figure 2.30

11. **(a)** What are the units of $g'(24)$?
(b) What is the biological meaning of $g'(24) = 0.096$?

12. **(a)** Which is greater, $g'(20)$ or $g'(36)$?
(b) What does your answer say about fetal growth?

13. Is the instantaneous weight growth rate greater or less than the average rate of change of weight over the 40 week period

(a) At week 20? **(b)** At week 36?

14. Estimate **(a)** $g'(20)$ **(b)** $g'(36)$
(c) The average rate of change of weight for the entire 40 week gestation.

15. The wind speed W in meters per second at a distance x kilometers from the center of a hurricane is given by the function $W = h(x)$. What does the fact that $h'(15) > 0$ tell you about the hurricane?

16. You drop a rock from a high tower. After it falls x meters its speed S in meters per second is $S = h(x)$. What is the meaning of $h'(20) = 0.5$?

17. If t is the number of years since 2003, the population, P, of China, in billions, can be approximated by the function

$$P = f(t) = 1.291(1.006)^t.$$

Estimate $f(6)$ and $f'(6)$, giving units. What do these two numbers tell you about the population of China?

18. Figure 2.31 shows the length, L, in cm, of a sturgeon (a type of fish) as a function of the time, t, in years.[8] Estimate $f'(10)$. Give units and interpret your answer.

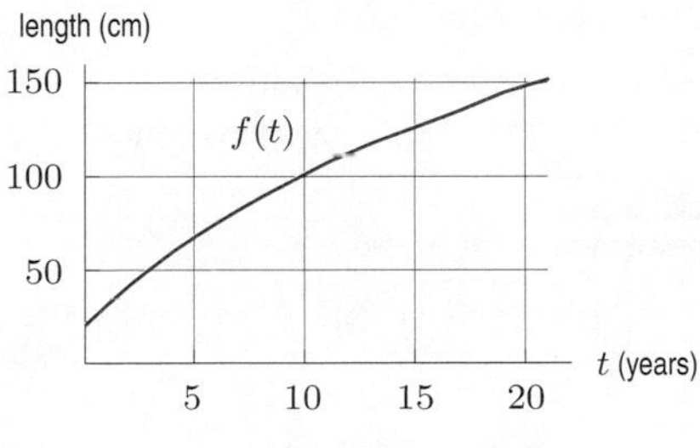

Figure 2.31

19. After investing \$1000 at an annual interest rate of 7% compounded continuously for t years, your balance is \$$B$, where $B = f(t)$. What are the units of dB/dt? What is the financial interpretation of dB/dt?

20. For some painkillers, the size of the dose, D, given depends on the weight of the patient, W. Thus, $D = f(W)$, where D is in milligrams and W is in pounds.

(a) Interpret the statements $f(140) = 120$ and $f'(140) = 3$ in terms of this painkiller.
(b) Use the information in the statements in part (a) to estimate $f(145)$.

21. For a function $f(x)$, we know that $f(20) = 68$ and $f'(20) = -3$. Estimate $f(21)$, $f(19)$ and $f(25)$.

22. Suppose that $f(x)$ is a function with $f(20) = 345$ and $f'(20) = 6$. Estimate $f(22)$.

23. The quantity, Q mg, of nicotine in the body t minutes after a cigarette is smoked is given by $Q = f(t)$.

(a) Interpret the statements $f(20) = 0.36$ and $f'(20) = -0.002$ in terms of nicotine. What are the units of the numbers 20, 0.36, and -0.002?
(b) Use the information given in part (a) to estimate $f(21)$ and $f(30)$. Justify your answers.

24. Table 2.8 shows world gold production,[9] $G = f(t)$, as a function of year, t.

(a) Does $f'(t)$ appear to be positive or negative? What does this mean in terms of gold production?
(b) In which time interval does $f'(t)$ appear to be greatest?
(c) Estimate $f'(2002)$. Give units and interpret your answer in terms of gold production.
(d) Use the estimated value of $f'(2002)$ to estimate $f(2003)$ and $f(2010)$, and interpret your answers.

Table 2.8 *World gold production*

t (year)	1990	1993	1996	1999	2002
G (mn troy ounces)	70.2	73.3	73.6	82.6	82.9

[8]Data from von Bertalanffy, L., *General System Theory*, p. 177, (New York: Braziller, 1968).
[9]*The World Almanac and Book of Facts 2005*, p. 135, (New York).

25. Figure 2.32 shows how the contraction velocity, $v(x)$, of a muscle changes as the load on it changes.

(a) Find the slope of the line tangent to the graph of contraction velocity at a load of 2 kg. Give units.
(b) Using your answer to part (a), estimate the change in the contraction velocity if the load is increased from 2 kg by adding 50 grams.
(c) Express your answer to part (a) as a derivative of $v(x)$.

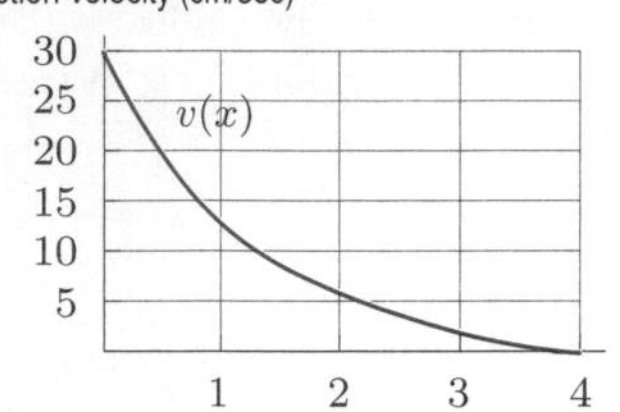

Figure 2.32

26. Suppose $C(r)$ is the total cost of paying off a car loan borrowed at an annual interest rate of $r\%$. What are the units of $C'(r)$? What is the practical meaning of $C'(r)$? What is its sign?

27. A climber on Mount Everest is 6000 meters from the start of his trail and at elevation 8000 meters above sea level. At x meters from the start, the elevation of the trail is $h(x)$ meters above sea level. If $h'(x) = 0.5$ for x near 6000, what is the approximate elevation another 3 meters along the trail?

28. Let $f(v)$ be the gas consumption (in liters/km) of a car going at velocity v (in km/hr). In other words, $f(v)$ tells you how many liters of gas the car uses to go one kilometer at velocity v. Explain what the following statements tell you about gas consumption:

$$f(80) = 0.05 \quad \text{and} \quad f'(80) = 0.0005.$$

29. Figure 2.33 shows how the pumping rate of a person's heart changes after bleeding.

(a) Find the slope of the line tangent to the graph at time 2 hours. Give units.
(b) Using your answer to part (a), estimate how much the pumping rate increases during the minute beginning at time 2 hours.
(c) Express your answer to part (a) as a derivative of $g(t)$.

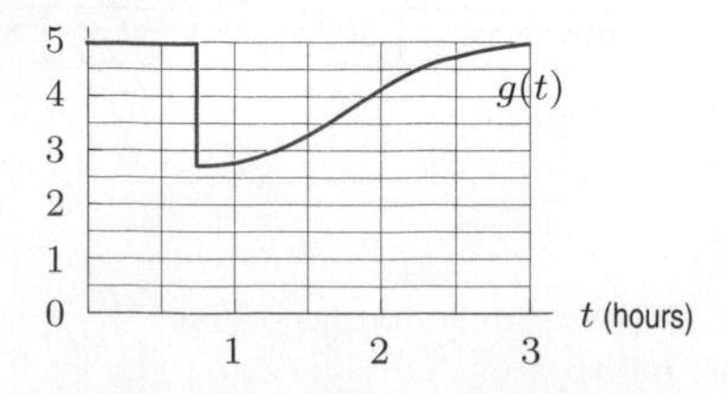

Figure 2.33

30. To study traffic flow, a city installs a device which records $C(t)$, the total number of cars that have passed by t hours after 4:00 am. The graph of $C(t)$ is in Figure 2.34.

(a) When is the traffic flow the greatest?
(b) Estimate $C'(2)$.
(c) What does $C'(2)$ mean in practical terms?

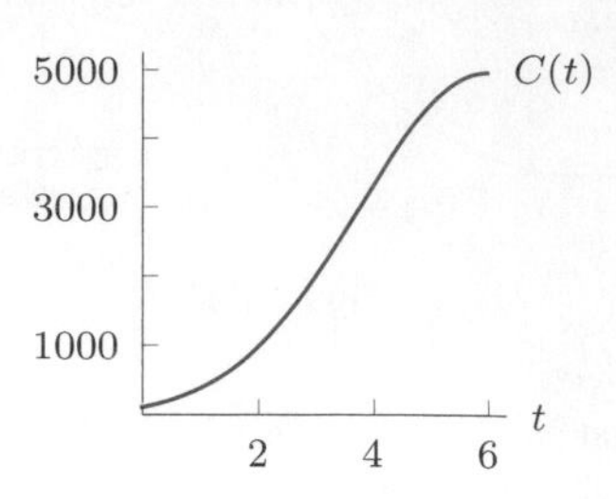

Figure 2.34

Problems 31–35 refer to Figure 2.35, which shows the depletion of food stores in the human body during starvation.

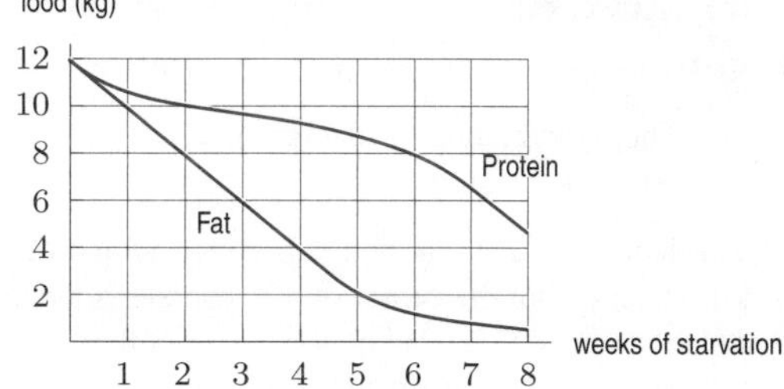

Figure 2.35

31. Which is being consumed at a greater rate, fat or protein, during the

(a) Third week? **(b)** Seventh week?

32. The fat storage graph is linear for the first four weeks. What does this tell you about the use of stored fat?

33. Estimate the rate of fat consumption after

(a) 3 weeks **(b)** 6 weeks **(c)** 8 weeks

34. What seems to happen during the sixth week? Why do you think this happens?

35. Figure 2.36 shows the derivatives of the protein and fat storage functions. Which graph is which?

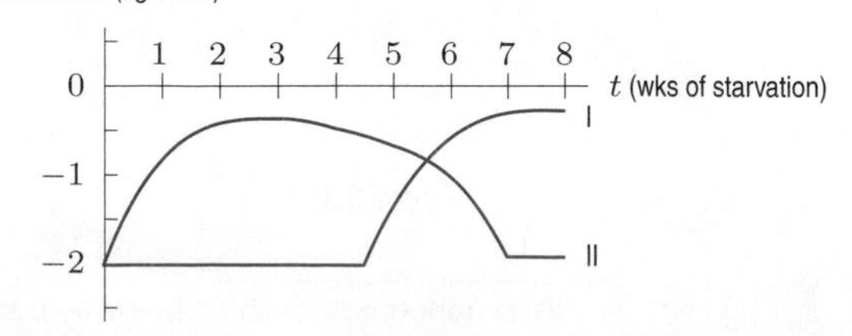

Figure 2.36

36. The table[10] shows $f(t)$, total sales of music compact discs (CDs), in millions, and $g(t)$, total sales of music cassettes, in millions, as a function of year t.

(a) Estimate $f'(2002)$ and $g'(2002)$. Give units with your answers and interpret each answer in terms of sales of CDs or cassettes.

(b) Use $f'(2002)$ to estimate $f(2003)$ and $f(2010)$. Interpret your answers in terms of sales of CDs.

Year, t	1994	1996	1998	2000	2002
CD sales, $f(t)$	662.1	778.9	847.0	942.5	803.3
Cassette sales, $g(t)$	345.4	225.3	158.5	76.0	31.1

37. When you breathe, a muscle (called the diaphragm) reduces the pressure around your lungs and they expand to fill with air. The table shows the volume of a lung as a function of the reduction in pressure from the diaphragm. Pulmonologists (lung doctors) define the *compliance* of the lung as the derivative of this function.[11]

(a) What are the units of compliance?

(b) Estimate the maximum compliance of the lung.

(c) Explain why the compliance gets small when the lung is nearly full (around 1 liter).

Pressure reduction (cm of water)	Volume (liters)
0	0.20
5	0.29
10	0.49
15	0.70
20	0.86
25	0.95
30	1.00

38. A person with a certain liver disease first exhibits larger and larger concentrations of certain enzymes (called SGOT and SGPT) in the blood. As the disease progresses, the concentration of these enzymes drops, first to the predisease level and eventually to zero (when almost all of the liver cells have died). Monitoring the levels of these enzymes allows doctors to track the progress of a patient with this disease. If $C = f(t)$ is the concentration of the enzymes in the blood as a function of time,

(a) Sketch a possible graph of $C = f(t)$.

(b) Mark on the graph the intervals where $f' > 0$ and where $f' < 0$.

(c) What does $f'(t)$ represent, in practical terms?

2.4 THE SECOND DERIVATIVE

Since the derivative is itself a function, we can calculate its derivative. For a function f, the derivative of its derivative is called the *second derivative*, and written f''. If $y = f(x)$, the second derivative can also be written as $\dfrac{d^2y}{dx^2}$, which means $\dfrac{d}{dx}\left(\dfrac{dy}{dx}\right)$, the derivative of $\dfrac{dy}{dx}$.

What Does the Second Derivative Tell Us?

Recall that the derivative of a function tells us whether the function is increasing or decreasing:

If $f' > 0$ on an interval, then f is increasing over that interval.
If $f' < 0$ on an interval, then f is decreasing over that interval.

Since f'' is the derivative of f', we have

If $f'' > 0$ on an interval, then f' is increasing over that interval.
If $f'' < 0$ on an interval, then f' is decreasing over that interval.

So the question becomes: What does it mean for f' to be increasing or decreasing? The case in which f' is increasing is shown in Figure 2.37, where the graph of f is bending upward, or is *concave up*. In the case when f' is decreasing, shown in Figure 2.38, the graph is bending downward, or is *concave down*.

> $f'' > 0$ on an interval means f' is increasing, so the graph of f is concave up there.
> $f'' < 0$ on an interval means f' is decreasing, so the graph of f is concave down there.

[10] *The World Almanac and Book of Facts 2005*, p. 309 (New York).

[11] Adapted from John B. West, *Respiratory Physiology* 4th Ed. (New York: Williams and Wilkins, 1990).

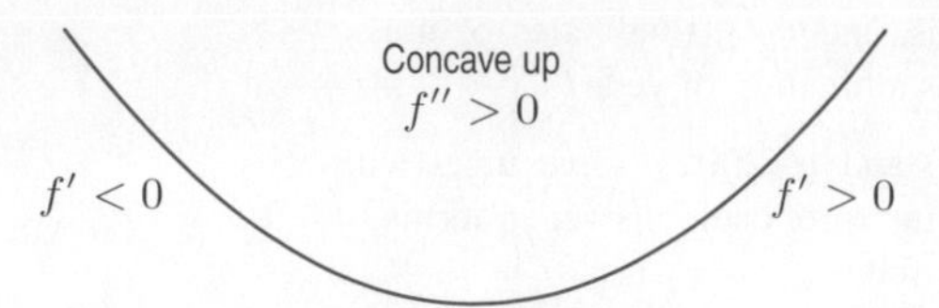

Figure 2.37: Meaning of f'': The slope increases from negative to positive as you move from left to right, so f'' is positive and f is concave up

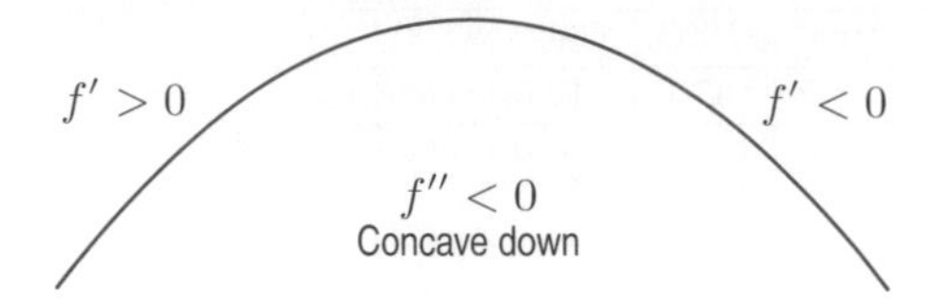

Figure 2.38: Meaning of f'': The slope decreases from positive to negative as you move from left to right, so f'' is negative and f is concave down

---

Example 1 For the functions whose graphs are given in Figure 2.39, decide where their second derivatives are positive and where they are negative.

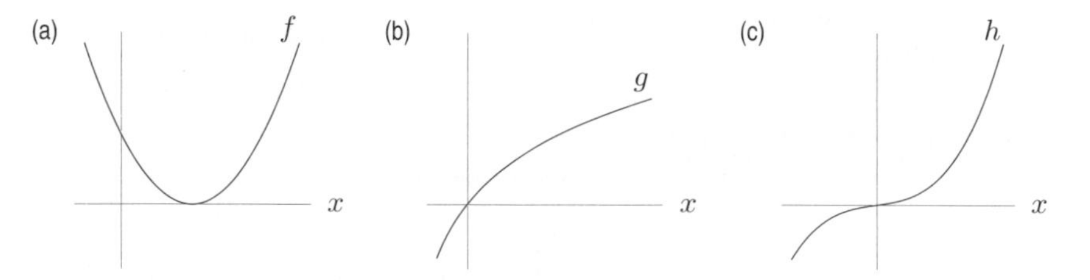

Figure 2.39: What signs do the second derivatives have?

Solution From the graphs it appears that

(a) $f'' > 0$ everywhere, because the graph of f is concave up everywhere.
(b) $g'' < 0$ everywhere, because the graph is concave down everywhere.
(c) $h'' > 0$ for $x > 0$, because the graph of h is concave up there; $h'' < 0$ for $x < 0$, because the graph of h is concave down there.

Interpretation of the Second Derivative as a Rate of Change

If we think of the derivative as a rate of change, then the second derivative is a rate of change of a rate of change. If the second derivative is positive, the rate of change is increasing; if the second derivative is negative, the rate of change is decreasing.

The second derivative is often a matter of practical concern. In 1985 a newspaper headline reported the Secretary of Defense as saying that Congress and the Senate had cut the defense budget. As his opponents pointed out, however, Congress had merely cut the rate at which the defense budget was increasing.[12] In other words, the derivative of the defense budget was still positive (the budget was increasing), but the second derivative was negative (the budget's rate of increase had slowed).

[12]In the *Boston Globe*, March 13, 1985, Representative William Gray (D–Pa.) was reported as saying: "It's confusing to the American people to imply that Congress threatens national security with reductions when you're really talking about a reduction in the increase."

Example 2 A population, P, growing in a confined environment often follows a *logistic* growth curve, like the graph shown in Figure 2.40. Describe how the rate at which the population is increasing changes over time. What is the sign of the second derivative d^2P/dt^2? What is the practical interpretation of t^* and L?

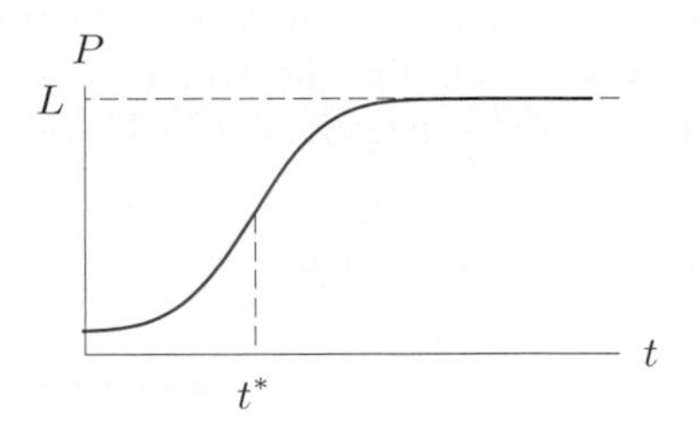

Figure 2.40: Logistic growth curve

Solution Initially, the population is increasing, and at an increasing rate. So, initially dP/dt is increasing and $d^2P/dt^2 > 0$. At t^*, the rate at which the population is increasing is a maximum; the population is growing fastest then. Beyond t^*, the rate at which the population is growing is decreasing, so $d^2P/dt^2 < 0$. At t^*, the graph changes from concave up to concave down and $d^2P/dt^2 = 0$.

The quantity L represents the limiting value of the population that is approached as t tends to infinity; L is called the *carrying capacity* of the environment and represents the maximum population that the environment can support.

Example 3 Table 2.9 shows the number of abortions per year, A, reported in the US[13] in the year t.

Table 2.9 *Abortions reported in the US (1972–2000)*

Year, t	1972	1975	1980	1985	1990	1995	2000
Thousands of abortions reported, A	587	1034	1554	1589	1609	1359	1313

(a) Calculate the average rate of change for the time intervals shown between 1972 and 2000.
(b) What can you say about the sign of d^2A/dt^2 during the period 1972–1995?

Solution (a) For each time interval we can calculate the average rate of change of the number of abortions per year over this interval. For example, between 1972 and 1975

$$\text{Average rate of change} = \frac{\Delta A}{\Delta t} = \frac{1034 - 587}{1975 - 1972} = \frac{447}{3} \approx 149.$$

Thus, between 1972 and 1975, there were approximately 149,000 more abortions reported each year. Values of $\Delta A/\Delta t$ are listed in Table 2.10:

Table 2.10 *Rate of change of number of abortions reported*

Time	1972–1975	1975–1980	1980–1985	1985–1990	1990–1995	1995–2000
Average rate of change, $\Delta A/\Delta t$ (1000s/year)	149	104	7	4	−50	−9.2

[13]*Statistical Abstracts of the United States 2004–2005*, Table 89.

(b) We assume the data lies on a smooth curve. Since the values of $\Delta A/\Delta t$ are decreasing dramatically for 1975–1995, we can be pretty certain that dA/dt also decreases, so d^2A/dt^2 is negative for this period. For 1972–1975, the sign of d^2A/dt^2 is less clear; abortion data from 1968 would help. Figure 2.41 confirms this; the graph appears to be concave down for 1975–1995. The fact that dA/dt is positive during the period 1972–1980 tells us that the number of abortions reported increased from 1972 to 1980. The fact that dA/dt is negative during the period 1990–2000 tells us that the number of abortions reported decreased from 1990 to 2000. The fact that d^2A/dt^2 is negative for 1975–1995 tells us that the rate of increase slowed over this period.

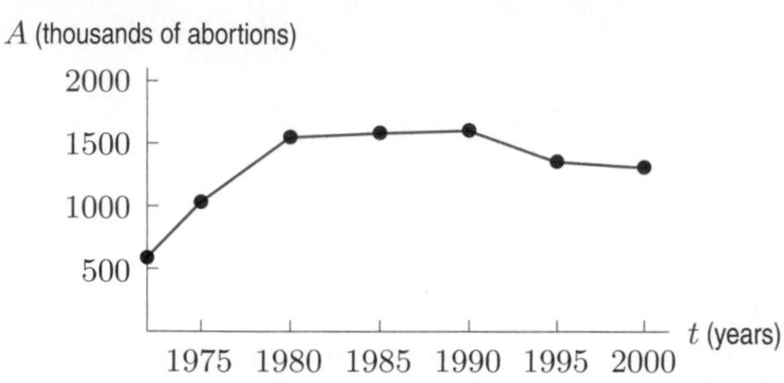

Figure 2.41: How the number of reported abortions in the US is changing with time

Problems for Section 2.4

1. For the function graphed in Figure 2.42, are the following nonzero quantities positive or negative?

(a) $f(2)$ **(b)** $f'(2)$ **(c)** $f''(2)$

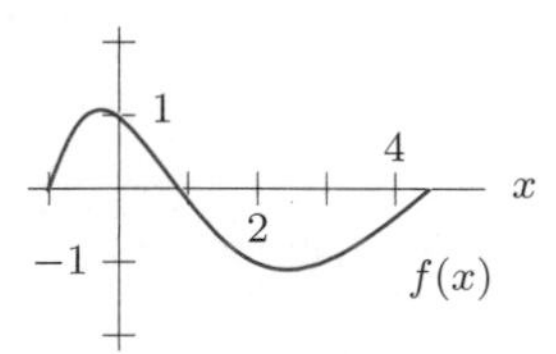

Figure 2.42

For Problems 2–7, give the signs of the first and second derivatives for each of the following functions. Each derivative is either positive everywhere, zero everywhere, or negative everywhere.

2.

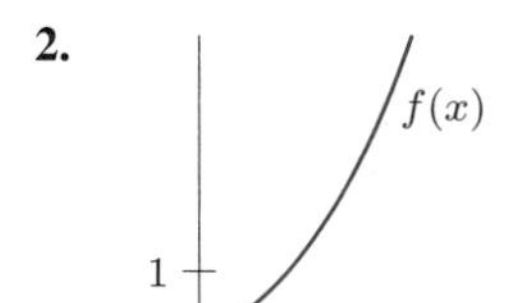

3.

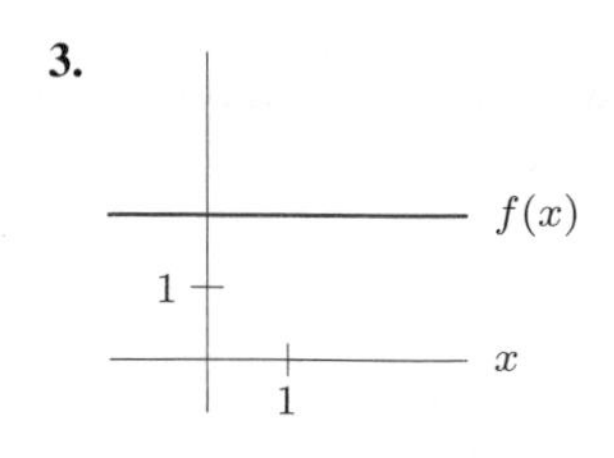

4.

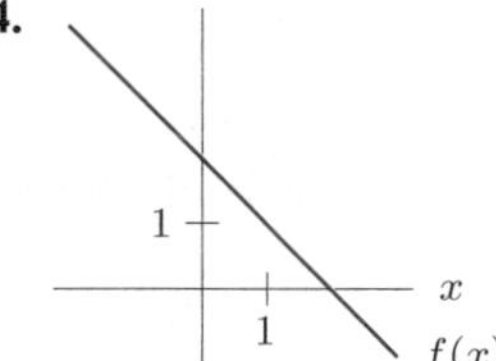

5.

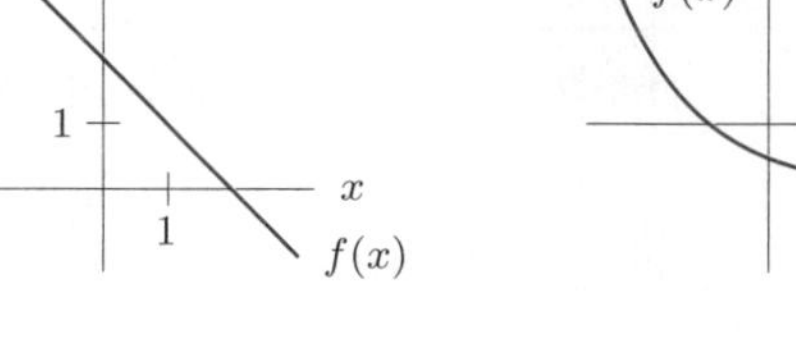

6.

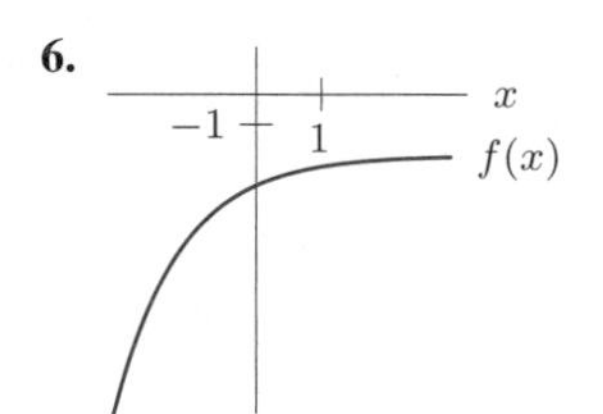

7.

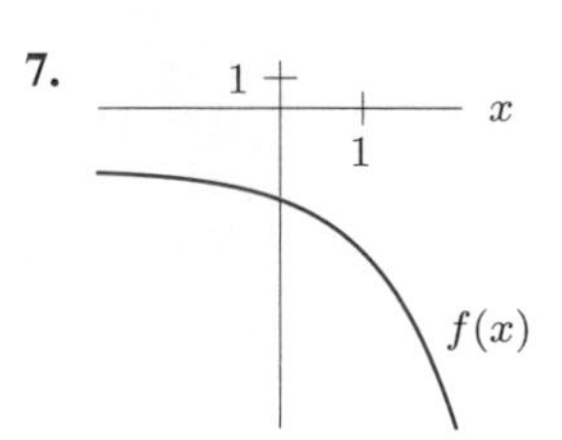

In Problems 8–9, use the values given for each function.

(a) Does the derivative of the function appear to be positive or negative over the given interval? Explain.

(b) Does the second derivative of the function appear to be positive or negative over the given interval? Explain.

8.

t	100	110	120	130	140
$w(t)$	10.7	6.3	4.2	3.5	3.3

9.

t	0	1	2	3	4	5
$s(t)$	12	14	17	20	31	55

In Problems 10–11, use the graph given for each function.

(a) Estimate the intervals on which the derivative is positive and the intervals on which the derivative is negative.

(b) Estimate the intervals on which the second derivative is positive and the intervals on which the second derivative is negative.

10.

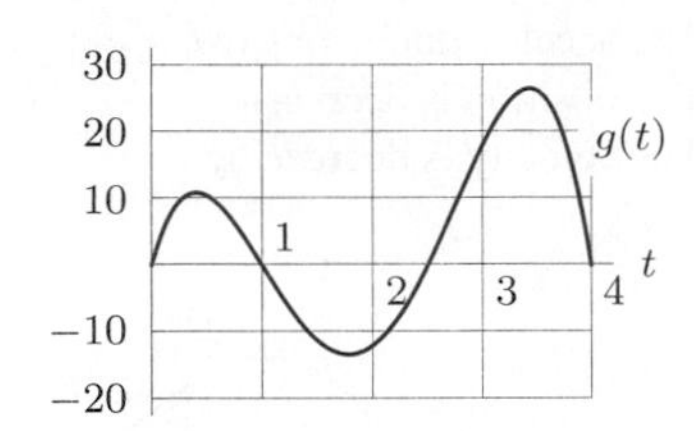

11.

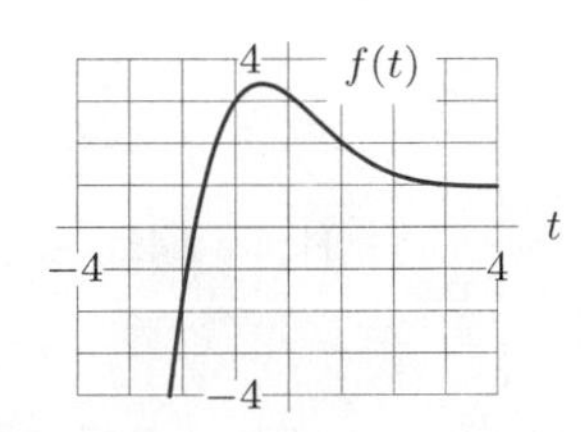

12. **(a)** Graph a function whose first and second derivatives are everywhere positive.

(b) Graph a function whose second derivative is everywhere negative but whose first derivative is everywhere positive.

(c) Graph a function whose second derivative is everywhere positive but whose first derivative is everywhere negative.

(d) Graph a function whose first and second derivatives are everywhere negative.

13. At exactly two of the labeled points in Figure 2.43, the derivative f' is 0; the second derivative f'' is not zero at any of the labeled points. On a copy of the table, give the signs of f, f', f'' at each marked point.

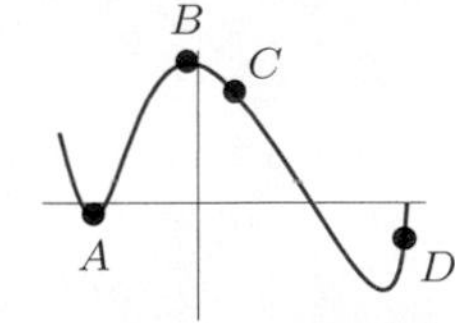

Point	f	f'	f''
A			
B			
C			
D			

Figure 2.43

14. The length L of the day in minutes (sunrise to sunset) x kilometers north of the equator on June 21 is given by $L = f(x)$. What are the units of

(a) $f'(3000)$? **(b)** $f''(3000)$?

15. For three minutes the temperature of a feverish person has had positive first derivative and negative second derivative. Which of the following is correct?

(a) The temperature rose in the last minute more than it rose in the minute before.

(b) The temperature rose in the last minute, but less than it rose in the minute before.

(c) The temperature fell in the last minute but less than it fell in the minute before.

(d) The temperature rose two minutes ago but fell in the last minute.

16. Yesterday's temperature at t hours past midnight was $f(t)$ °C. At noon the temperature was 20°C. The first derivative, $f'(t)$, decreased all morning, reaching a low of 2°C/hour at noon, then increased for the rest of the day. Which one of the following must be correct?

(a) The temperature fell in the morning and rose in the afternoon.

(b) At 1 pm the temperature was 18°C.

(c) At 1 pm the temperature was 22°C.

(d) The temperature was lower at noon than at any other time.

(e) The temperature rose all day.

17. Values of $f(t)$ are given in the following table.

(a) Does this function appear to have a positive or negative first derivative? Second derivative? Explain.

(b) Estimate $f'(2)$ and $f'(8)$.

t	0	2	4	6	8	10
$f(t)$	150	145	137	122	98	56

18. Sketch the graph of a function f such that $f(2) = 5$, $f'(2) = 1/2$, and $f''(2) > 0$.

19. Sketch a graph of a continuous function f with the following properties:

- $f'(x) > 0$ for all x
- $f''(x) < 0$ for $x < 2$ and $f''(x) > 0$ for $x > 2$.

20. At which of the marked x-values in Figure 2.44 can the following statements be true?

(a) $f(x) < 0$

(b) $f'(x) < 0$

(c) $f(x)$ is decreasing

(d) $f'(x)$ is decreasing

(e) Slope of $f(x)$ is positive

(f) Slope of $f(x)$ is increasing

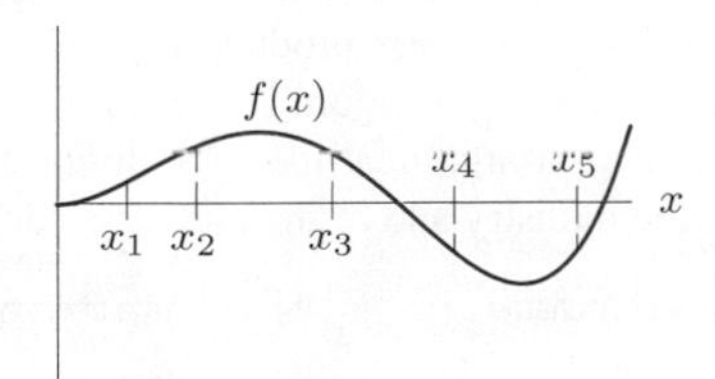

Figure 2.44

21. A function f has $f(5) = 20$, $f'(5) = 2$, and $f''(x) < 0$, for $x \geq 5$. Which of the following are possible values for $f(7)$ and which are impossible?

(a) 26 **(b)** 24 **(c)** 22

22. The table gives the number of passenger cars, $C = f(t)$, in millions,[14] in the US in the year t.

(a) Do $f'(t)$ and $f''(t)$ appear to be positive or negative during the period 1940–1980?
(b) Estimate $f'(1975)$. Using units, interpret your answer in terms of passenger cars.

t	1940	1950	1960	1970	1980	1990	2000
C	27.5	40.3	61.7	89.2	121.6	133.7	133.6

23. "Winning the war on poverty" has been described cynically as slowing the rate at which people are slipping below the poverty line. Assuming that this is happening:

(a) Graph the total number of people in poverty against time.
(b) If N is the number of people below the poverty line at time t, what are the signs of dN/dt and d^2N/dt^2? Explain.

24. Let $P(t)$ represent the price of a share of stock of a corporation at time t. What does each of the following statements tell us about the signs of the first and second derivatives of $P(t)$?

(a) "The price of the stock is rising faster and faster."
(b) "The price of the stock is close to bottoming out."

25. In economics, *total utility* refers to the total satisfaction from consuming some commodity. According to the economist Samuelson:[15]

> As you consume more of the same good, the total (psychological) utility increases. However, ... with successive new units of the good, your total utility will grow at a slower and slower rate because of a fundamental tendency for your psychological ability to appreciate more of the good to become less keen.

(a) Sketch the total utility as a function of the number of units consumed.
(b) In terms of derivatives, what is Samuelson saying?

26. An industry is being charged by the Environmental Protection Agency (EPA) with dumping unacceptable levels of toxic pollutants in a lake. Over a period of several months, an engineering firm makes daily measurements of the rate at which pollutants are being discharged into the lake. The engineers produce a graph similar to either Figure 2.45(a) or Figure 2.45(b). For each case, give an idea of what argument the EPA might make in court against the industry and of the industry's defense.

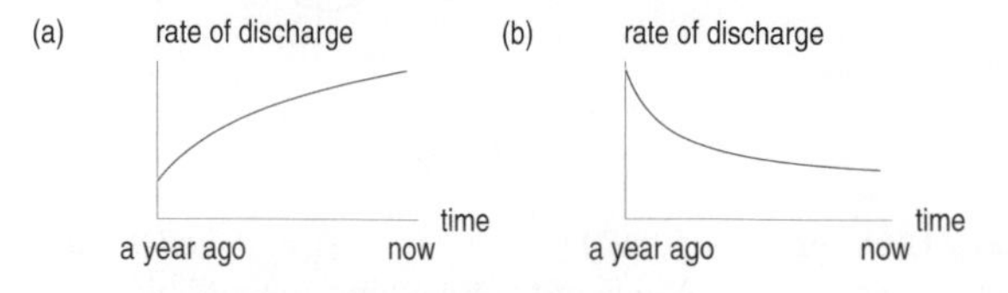

Figure 2.45

27. Sketch the graph of the height of a particle against time if velocity is positive and acceleration is negative.

28. Figure 2.46 gives the position, $f(t)$, of a particle at time t. At which of the marked values of t can the following statements be true?

(a) The position is positive
(b) The velocity is positive
(c) The acceleration is positive
(d) The position is decreasing
(e) The velocity is decreasing

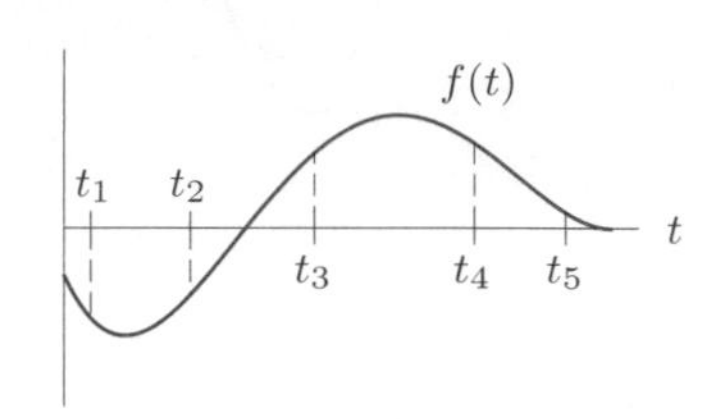

Figure 2.46

29. Each of the graphs in Figure 2.47 shows the position of a particle moving along the x-axis as a function of time, $0 \leq t \leq 5$. The vertical scales of the graphs are the same. During this time interval, which particle has

(a) Constant velocity?
(b) The greatest initial velocity?
(c) The greatest average velocity?
(d) Zero average velocity?
(e) Zero acceleration?
(f) Positive acceleration throughout?

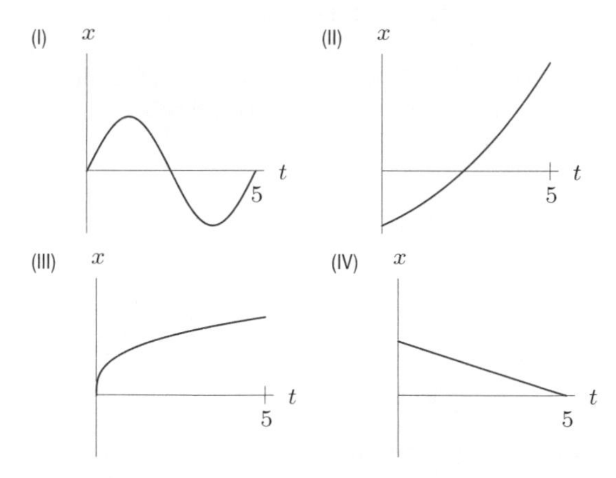

Figure 2.47

[14] *The World Almanac and Book of Facts 2005*, p. 237 (New York).
[15] From Paul A. Samuelson, *Economics*, 11th edition (New York: McGraw-Hill, 1981).

2.5 MARGINAL COST AND REVENUE

Management decisions within a particular firm or industry usually depend on the costs and revenues involved. In this section we look at the cost and revenue functions.

Graphs of Cost and Revenue Functions

The graph of a cost function may be linear, as in Figure 2.48, or it may have the shape shown in Figure 2.49. The intercept on the C-axis represents the fixed costs, which are incurred even if nothing is produced. (This includes, for instance, the cost of the machinery needed to begin production.) In Figure 2.49, the cost function increases quickly at first and then more slowly because producing larger quantities of a good is usually more efficient than producing smaller quantities—this is called *economy of scale*. At still higher production levels, the cost function increases faster again as resources become scarce; sharp increases may occur when new factories have to be built. Thus, the graph of a cost function, C, may start out concave down and become concave up later on.

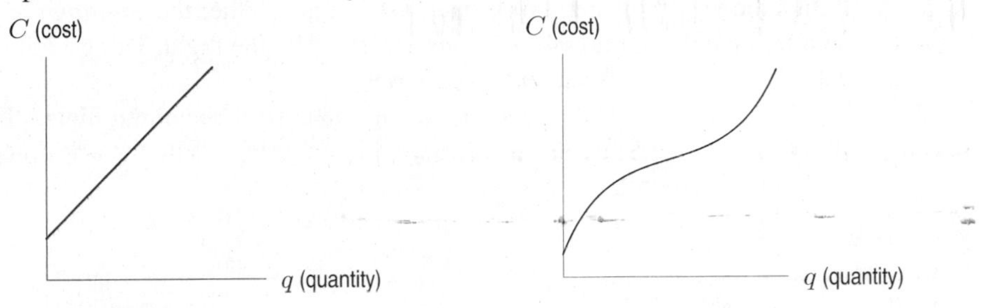

Figure 2.48: A linear cost function

Figure 2.49: A nonlinear cost function

The revenue function is $R = pq$, where p is price and q is quantity. If the price, p, is a constant, the graph of R against q is a straight line through the origin with slope equal to the price. (See Figure 2.50). In practice, for large values of q, the market may become glutted, causing the price to drop and giving R the shape in Figure 2.51.

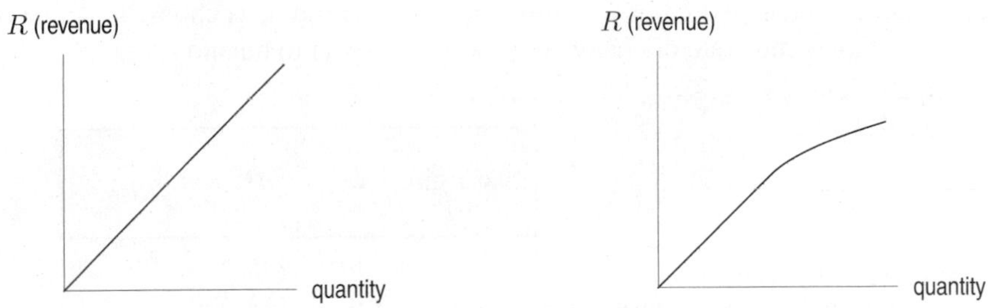

Figure 2.50: Revenue: Constant price

Figure 2.51: Revenue: Decreasing price

Example 1 If cost, C, and revenue, R, are given by the graph in Figure 2.52, for what production quantities does the firm make a profit?

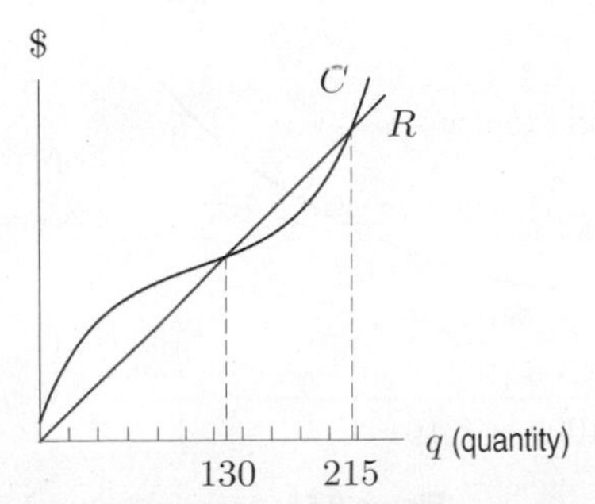

Figure 2.52: Costs and revenues for Example 1

Solution The firm makes a profit whenever revenues are greater than costs, that is, when $R > C$. The graph of R is above the graph of C approximately when $130 < q < 215$. Production between 130 units and 215 units will generate a profit.

Marginal Analysis

Many economic decisions are based on an analysis of the costs and revenues "at the margin." Let's look at this idea through an example.

Suppose you are running an airline and you are trying to decide whether to offer an additional flight. How should you decide? We'll assume that the decision is to be made purely on financial grounds: if the flight will make money for the company, it should be added. Obviously you need to consider the costs and revenues involved. Since the choice is between adding this flight and leaving things the way they are, the crucial question is whether the *additional costs* incurred are greater or smaller than the *additional revenues* generated by the flight. These additional costs and revenues are called *marginal costs* and *marginal revenues*.

Suppose $C(q)$ is the function giving the cost of running q flights. If the airline had originally planned to run 100 flights, its costs would be $C(100)$. With the additional flight, its costs would be $C(101)$. Therefore,

$$\text{Marginal cost} = C(101) - C(100).$$

Now

$$C(101) - C(100) = \frac{C(101) - C(100)}{101 - 100},$$

and this quantity is the average rate of change of cost between 100 and 101 flights. In Figure 2.53 the average rate of change is the slope of the secant line. If the graph of the cost function is not curving too fast near the point, the slope of the secant line is close to the slope of the tangent line there. Therefore, the average rate of change is close to the instantaneous rate of change. Since these rates of change are not very different, many economists choose to define marginal cost, MC, as the instantaneous rate of change of cost with respect to quantity:

$$\text{Marginal cost} = MC = C'(q).$$

Marginal cost is represented by the slope of the cost curve.

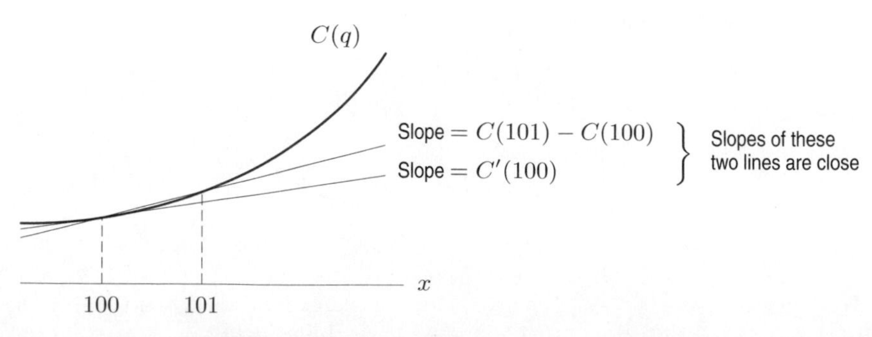

Figure 2.53: Marginal cost: Slope of one of these lines

Similarly, if the revenue generated by q flights is $R(q)$, then the additional revenue generated by increasing the number of flights from 100 to 101 is

$$\text{Marginal revenue} = R(101) - R(100).$$

Now $R(101) - R(100)$ is the average rate of change of revenue between 100 and 101 flights. As before, the average rate of change is approximately equal to the instantaneous rate of change, so economists often define

$$\text{Marginal revenue} = MR = R'(q).$$

Marginal revenue is represented by the slope of the revenue curve.

Example 2 If $C(q)$ and $R(q)$ for the airline are given in Figure 2.54, should the company add the 101st flight?

Solution The marginal revenue is the slope of the revenue curve at $q = 100$. The marginal cost is the slope of the graph of C at $q = 100$. Figure 2.54 suggests that the slope at the point A is smaller than the slope at B, so $MC < MR$ for $q = 100$. This means that the airline will make more in extra revenue than it will spend in extra costs if it runs another flight, so it should go ahead and run the 101st flight.

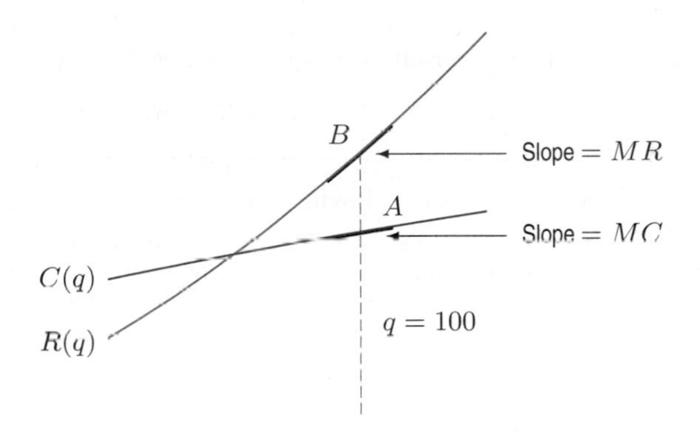

Figure 2.54: Cost and revenue for Example 2

Example 3 The graph of a cost function is given in Figure 2.55. Does it cost more to produce the 500th item or the 2000th? Does it cost more to produce the 3000th item or the 4000th? At approximately what production level is marginal cost smallest? What is the total cost at this production level?

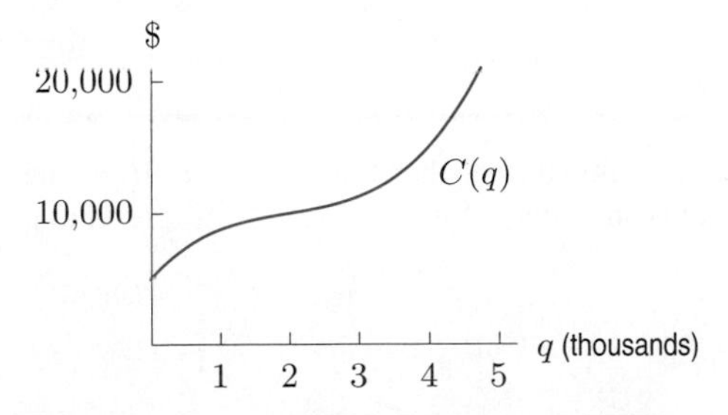

Figure 2.55: Estimating marginal cost: Where is marginal cost smallest?

Solution The cost to produce an additional item is the marginal cost, which is represented by the slope of the cost curve. Since the slope of the cost function in Figure 2.55 is greater at $q = 0.5$ (when the quantity produced is 0.5 thousand, or 500) than at $q = 2$, it costs more to produce the 500th item than the 2000th item. Since the slope is greater at $q = 4$ than $q = 3$, it costs more to produce the 4000th item than the 3000th item.

The slope of the cost function is close to zero at $q = 2$, and is positive everywhere else, so the slope is smallest at $q = 2$. The marginal cost is smallest at a production level of 2000 units. Since $C(2) \approx 10{,}000$, the total cost to produce 2000 units is about \$10,000.

Example 4 If the revenue and cost functions, R and C, are given by the graphs in Figure 2.56, sketch graphs of the marginal revenue and marginal cost functions, MR and MC.

Figure 2.56: Total revenue and total cost for Example 4

Solution The revenue graph is a line through the origin, with equation

$$R = pq$$

where p represents the constant price, so the slope is p and

$$MR = R'(q) = p.$$

The total cost is increasing, so the marginal cost is always positive. For small q values, the graph of the cost function is concave down, so the marginal cost is decreasing. For larger q, say $q > 100$, the graph of the cost function is concave up and the marginal cost is increasing. Thus, the marginal cost has a minimum at about $q = 100$. (See Figure 2.57.)

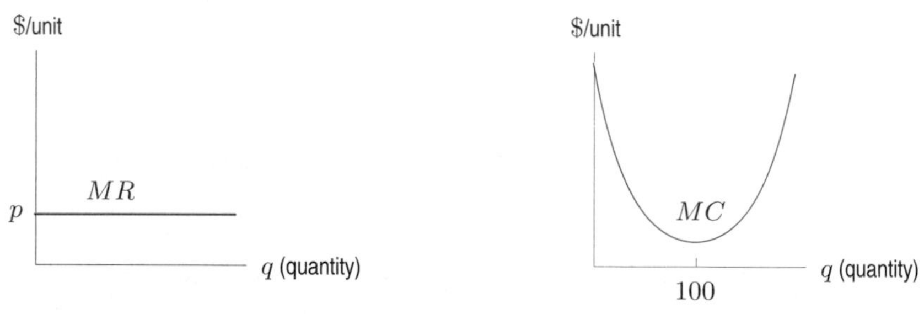

Figure 2.57: Marginal revenue and costs for Example 4

Problems for Section 2.5

1. In Figure 2.58, estimate the marginal cost when the level of production is 10,000 units and interpret it.

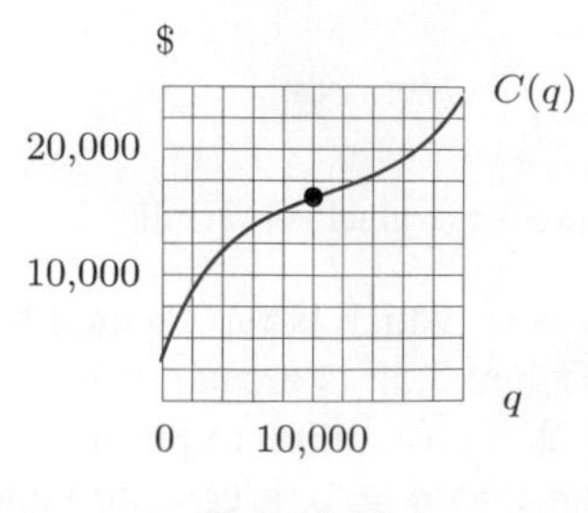

Figure 2.58

2. In Figure 2.59, estimate the marginal revenue when the level of production is 600 units and interpret it.

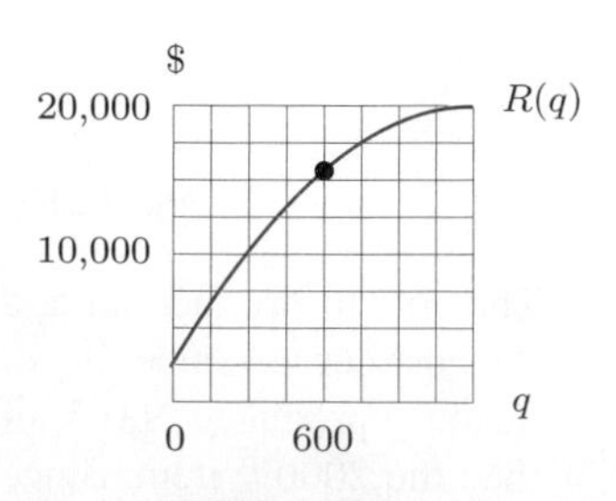

Figure 2.59

3. The function $C(q)$ gives the cost in dollars to produce q barrels of olive oil.

(a) What are the units of marginal cost?
(b) What is the practical meaning of the statement $MC = 3$ for $q = 100$?

4. It costs \$4800 to produce 1295 items and it costs \$4830 to produce 1305 items. What is the approximate marginal cost at a production level of 1300 items?

5. In Figure 2.60, is marginal cost greater at $q = 5$ or at $q = 30$? At $q = 20$ or at $q = 40$? Explain.

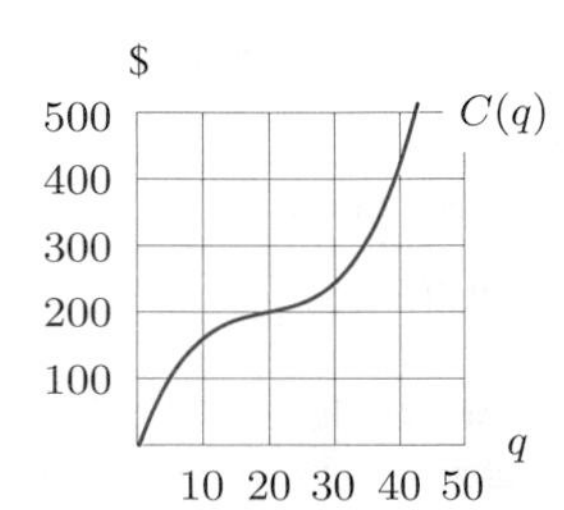

Figure 2.60

6. Figure 2.61 shows part of the graph of cost and revenue for a car manufacturer. Which is greater, marginal cost or marginal revenue, at

(a) q_1? **(b)** q_2?

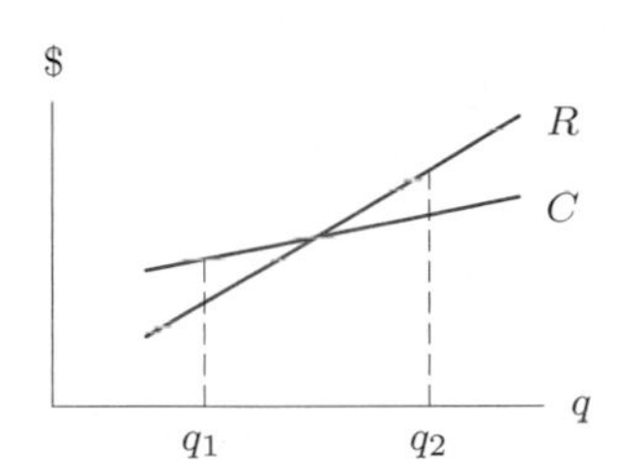

Figure 2.61

7. Let $C(q)$ represent the total cost of producing q items. Suppose $C(15) = 2300$ and $C'(15) = 108$. Estimate the total cost of producing: **(a)** 16 items **(b)** 14 items.

8. To produce 1000 items, the total cost is \$5000 and the marginal cost is \$25 per item. Estimate the costs of producing 1001 items, 999 items, and 1100 items.

9. Let $C(q)$ represent the cost and $R(q)$ represent the revenue, in dollars, of producing q items.

(a) If $C(50) = 4300$ and $C'(50) = 24$, estimate $C(52)$.
(b) If $C'(50) = 24$ and $R'(50) = 35$, approximately how much profit is earned by the 51st item?
(c) If $C'(100) = 38$ and $R'(100) = 35$, should the company produce the 101st item? Why or why not?

10. Cost and revenue functions for a charter bus company are shown in Figure 2.62. Should the company add a 50th bus? How about a 90th? Explain your answers using marginal revenue and marginal cost.

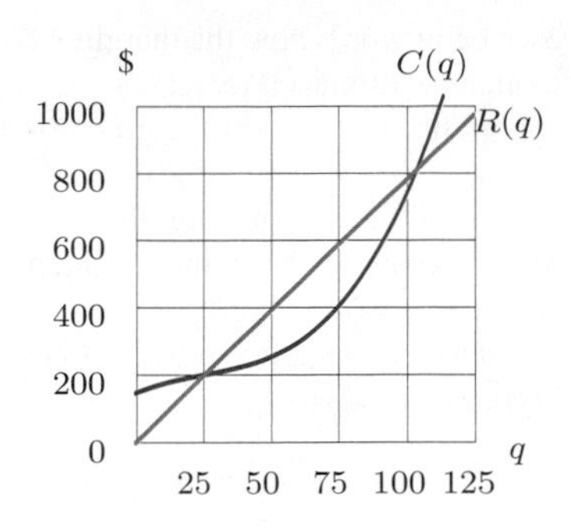

Figure 2.62

11. For q units of a product, a manufacturer's cost is $C(q)$ dollars and revenue is $R(q)$ dollars, with $C(500) = 7200$, $R(500) = 9400$, $MC(500) = 15$, and $MR(500) = 20$.

(a) What is the profit or loss at $q = 500$?
(b) If production is increased from 500 to 501 units, by approximately how much does profit change?

12. A company's cost of producing q liters of a chemical is $C(q)$ dollars; this quantity can be sold for $R(q)$ dollars. Suppose $C(2000) = 5930$ and $R(2000) = 7780$.

(a) What is the profit at a production level of 2000?
(b) If $MC(2000) = 2.1$ and $MR(2000) = 2.5$, what is the approximate change in profit if q is increased from 2000 to 2001? Should the company increase or decrease production from $q = 2000$?
(c) If $MC(2000) = 4.77$ and $MR(2000) = 4.32$, should the company increase or decrease production from $q = 2000$?

13. An industrial production process costs $C(q)$ million dollars to produce q million units; these units then sell for $R(q)$ million dollars. If $C(2.1) = 5.1$, $R(2.1) = 6.9$, $MC(2.1) = 0.6$, and $MR(2.1) = 0.7$, calculate

(a) The profit earned by producing 2.1 million units
(b) The change in revenue if production increases from 2.1 to 2.14 million units.
(c) The change in revenue if production decreases from 2.1 to 2.05 million units.
(d) The change in profit in parts (b) and (c).

14. The cost of recycling q tons of paper is given in the following table. Estimate the marginal cost at $q = 2000$. Give units and interpret your answer in terms of cost. At approximately what production level does marginal cost appear smallest?

q (tons)	1000	1500	2000	2500	3000	3500
$C(q)$ (dollars)	2500	3200	3640	3825	3900	4400

15. Let $C(q)$ be the total cost of producing a quantity q of a certain product. See Figure 2.63.

(a) What is the meaning of $C(0)$?

(b) Describe in words how the marginal cost changes as the quantity produced increases.
(c) Explain the concavity of the graph (in terms of economics).
(d) Explain the economic significance (in terms of marginal cost) of the point at which the concavity changes.
(e) Do you expect the graph of $C(q)$ to look like this for all types of products?

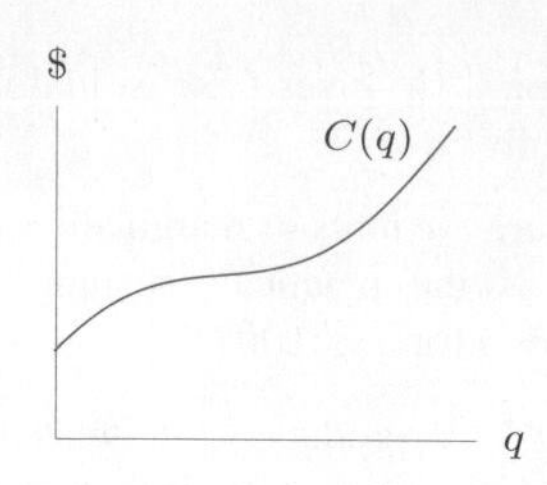

Figure 2.63

CHAPTER SUMMARY

- **Rate of change**
  Average, instantaneous
- **Estimating derivatives**
  Estimate derivatives from a graph, table of values, or formula.
- **Interpretation of derivatives**
  Rate of change, slope, using units, instantaneous velocity.
- **Marginality**
  Marginal cost and marginal revenue
- **Second derivative**
  Concavity
- **Derivatives and graphs**
  Understand relation between sign of f' and whether f is increasing or decreasing. Sketch graph of f' from graph of f. Marginal analysis.

REVIEW PROBLEMS FOR CHAPTER TWO

1. For the function shown in Figure 2.64, at what labeled points is the slope of the graph positive? Negative? At which labeled point does the graph have the greatest (i.e., most positive) slope? The least slope (i.e., negative and with the largest magnitude)?

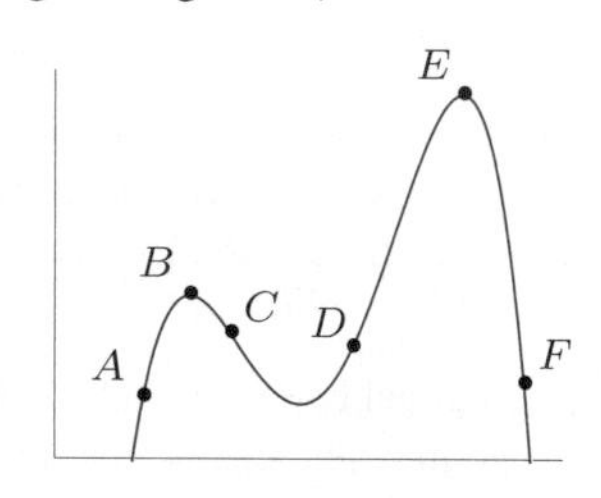

Figure 2.64

2. The function in Figure 2.65 has $f(4) = 25$ and $f'(4) = 1.5$. Find the coordinates of the points A, B, C.

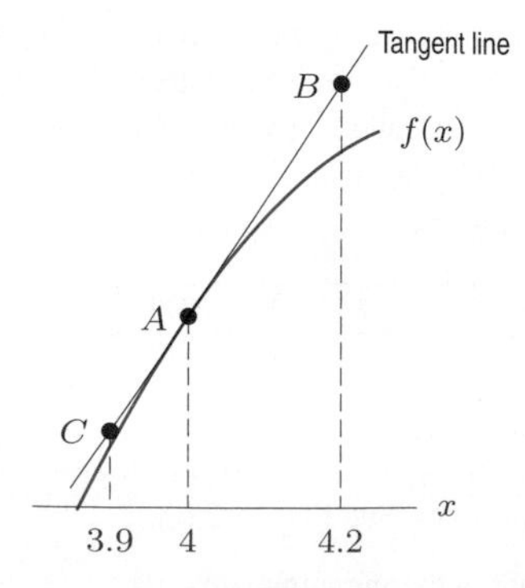

Figure 2.65

3. Estimate $f'(2)$ for $f(x) = 3^x$. Explain your reasoning.

4. In a time of t seconds, a particle moves a distance of s meters from its starting point, where $s = 3t^2$.

(a) Find the average velocity between $t = 1$ and $t = 1 + h$ if:

(i) $h = 0.1$, (ii) $h = 0.01$, (iii) $h = 0.001$.

(b) Use your answers to part (a) to estimate the instantaneous velocity of the particle at time $t = 1$.

5. The population of the world reached 1 billion in 1804, 2 billion in 1927, 3 billion in 1960, 4 billion in 1974, 5 billion in 1987 and 6 billion in 1999. Find the average rate of change of the population of the world, in people per minute, during each of these intervals. [That is, from 1804 to 1927, 1927 to 1960, etc.]

6. In a time of t seconds, a particle moves a distance of s meters from its starting point, where $s = \sin(2t)$.

(a) Find the average velocity between $t = 1$ and $t = 1 + h$ if:

(i) $h = 0.1$, (ii) $h = 0.01$, (iii) $h = 0.001$.

(b) Use your answers to part (a) to estimate the instantaneous velocity of the particle at time $t = 1$.

7. Given the numerical values shown, find approximate values for the derivative of $f(x)$ at each of the x-values given. Where is the rate of change of $f(x)$ positive? Where is it negative? Where does the rate of change of $f(x)$ seem to be greatest?

x	0	1	2	3	4	5	6	7	8
$f(x)$	18	13	10	9	9	11	15	21	30

Sketch the graphs of the derivatives of the functions shown in Problems 8–13. Be sure your sketches are consistent with the important features of the graphs of the original functions.

8.

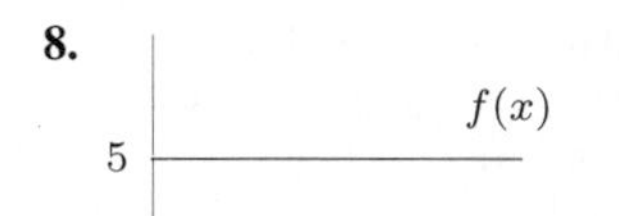

9.

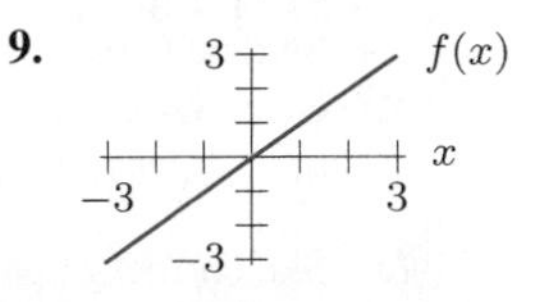

10.

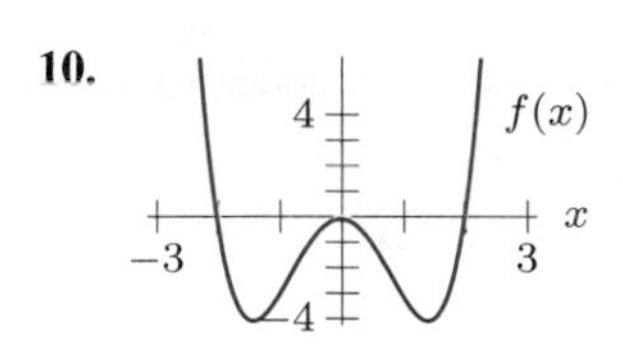

11.

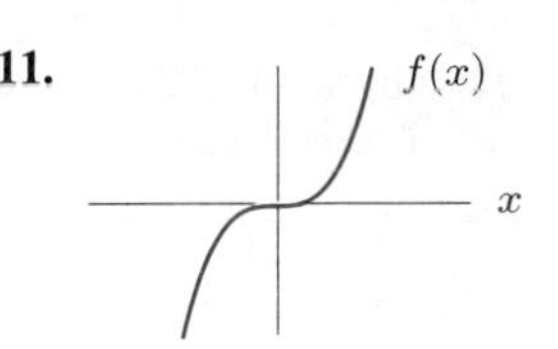

12.

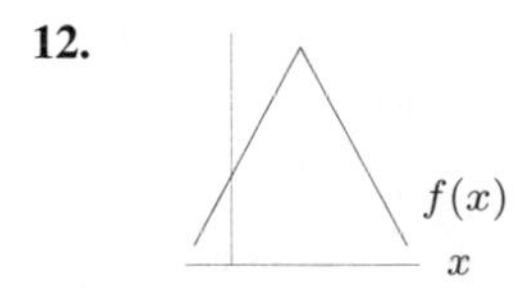

13.

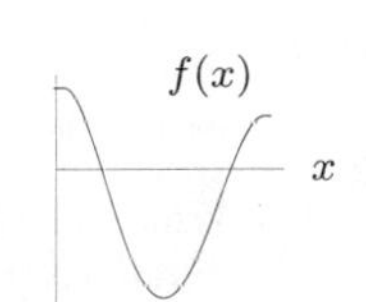

14. A vehicle moving along a straight road has distance $f(t)$ from its starting point at time t. Which of the graphs in Figure 2.66 could be $f'(t)$ for the following scenarios? (Assume the scales on the vertical axes are all the same.)

(a) A bus on a popular route, with no traffic
(b) A car with no traffic and all green lights
(c) A car in heavy traffic conditions

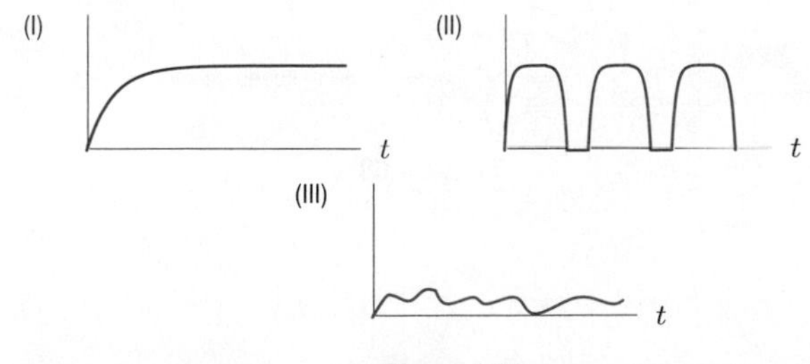

Figure 2.66

15. The temperature, H, in degrees Celsius, of a cup of coffee placed on the kitchen counter is given by $H = f(t)$, where t is in minutes since the coffee was put on the counter.

(a) Is $f'(t)$ positive or negative? Give a reason for your answer.
(b) What are the units of $f'(20)$? What is its practical meaning in terms of the temperature of the coffee?

16. The temperature, T, in degrees Fahrenheit, of a cold yam placed in a hot oven is given by $T = f(t)$, where t is the time in minutes since the yam was put in the oven.

(a) What is the sign of $f'(t)$? Why?
(b) What are the units of $f'(20)$? What is the practical meaning of the statement $f'(20) = 2$?

17. Suppose that $f(x)$ is a function with $f(100) = 35$ and $f'(100) = 3$. Estimate $f(102)$.

18. Suppose that $f(t)$ is a function with $f(25) = 3.6$ and $f'(25) = -0.2$. Estimate $f(26)$ and $f(30)$.

19. A mutual fund is currently valued at \$80 per share and its value per share is increasing at a rate of \$0.50 a day. Let $V = f(t)$ be the value of the share t days from now.

(a) Express the information given about the mutual fund in term of f and f'.
(b) Assuming that the rate of growth stays constant, estimate and interpret $f(10)$.

20. The average weight, W, in pounds, of an adult is a function, $W = f(c)$, of the average number of Calories per day, c, consumed.

(a) Interpret the statements $f(1800) = 155$ and $f'(2000) = 0$ in terms of diet and weight.
(b) What are the units of $f'(c) = dW/dc$?

21. Investing \$1000 at an annual interest rate of $r\%$, compounded continuously, for 10 years gives you a balance of \$$B$, where $B = g(r)$. Give a financial interpretation of the statements:

(a) $g(5) \approx 1649$.
(b) $g'(5) \approx 165$. What are the units of $g'(5)$?

22. Suppose $P(t)$ is the monthly payment, in dollars, on a mortgage which will take t years to pay off. What are the units of $P'(t)$? What is the practical meaning of $P'(t)$? What is its sign?

23. Let $f(x)$ be the elevation in feet of the Mississippi river x miles from its source. What are the units of $f'(x)$? What can you say about the sign of $f'(x)$?

24. An economist is interested in how the price of a certain item affects its sales. At a price of \$$p$, a quantity, q, of the item is sold. If $q = f(p)$, explain the meaning of each of the following statements:

(a) $f(150) = 2000$ **(b)** $f'(150) = -25$

25. A company's revenue from car sales, C (in thousands of dollars), is a function of advertising expenditure, a, in thousands of dollars, so $C = f(a)$.

(a) What does the company hope is true about the sign of f'?
(b) What does the statement $f'(100) = 2$ mean in practical terms? How about $f'(100) = 0.5$?
(c) Suppose the company plans to spend about \$100,000 on advertising. If $f'(100) = 2$, should the company spend more or less than \$100,000 on advertising? What if $f'(100) = 0.5$?

26. At one of the labeled points on the graph in Figure 2.67 both dy/dx and d^2y/dx^2 are positive. Which is it?

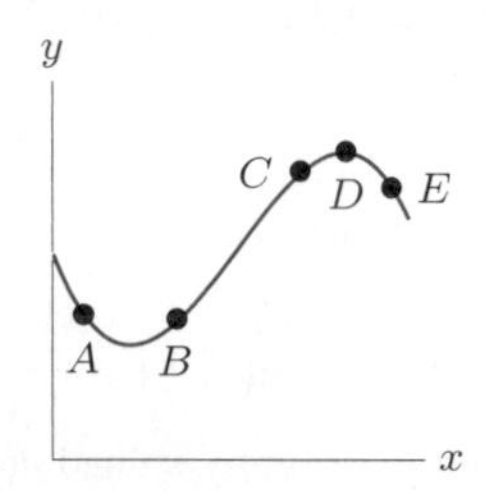

Figure 2.67

27. Sketch the graph of a function whose first and second derivatives are everywhere positive.

28. Sketch the graph of a function whose first derivative is everywhere negative and whose second derivative is positive for some x-values and negative for other x-values.

29. IBM-Peru uses second derivatives to assess the relative success of various advertising campaigns. They assume that all campaigns produce some increase in sales. If a graph of sales against time shows a positive second derivative during a new advertising campaign, what does this suggest to IBM management? Why? What does a negative second derivative suggest?

30. A high school principal is concerned about the drop in the percentage of students who graduate from her school, shown in the following table.

Year entered school, t	1992	1995	1998	2001	2004
Percent graduating, P	62.4	54.1	48.0	43.5	41.8

(a) Calculate the average rate of change of P for each of the three-year intervals between 1992 and 2004.
(b) Does d^2P/dt^2 appear to be positive or negative between 1992 and 2004?
(c) Explain why the values of P and dP/dt are troublesome to the principal.
(d) Explain why the sign of d^2P/dt^2 and the magnitude of dP/dt in the year 2001 may give the principal some cause for optimism.

31. Students were asked to evaluate $f'(4)$ from the following table which shows values of the function f:

x	1	2	3	4	5	6
$f(x)$	4.2	4.1	4.2	4.5	5.0	5.7

- Student A estimated the derivative as $f'(4) \approx \dfrac{f(5) - f(4)}{5 - 4} = 0.5$.
- Student B estimated the derivative as $f'(4) \approx \dfrac{f(4) - f(3)}{4 - 3} = 0.3$.
- Student C suggested that they should split the difference and estimate the average of these two results, that is, $f'(4) \approx \frac{1}{2}(0.5 + 0.3) = 0.4$.

(a) Sketch the graph of f, and indicate how the three estimates are represented on the graph.
(b) Explain which answer is likely to be best.

32. Given all of the following information about a function f, sketch its graph.

- $f(x) = 0$ at $x = -5, x = 0$, and $x = 5$
- $f(x) \to \infty$ as $x \to -\infty$
- $f(x) \to -3$ as $x \to \infty$
- $f'(x) = 0$ at $x = -3, x = 2.5$, and $x = 7$

33. Figure 2.68 shows the rate at which energy, $f(v)$, is consumed by a bird flying at speed v meters/sec.

(a) What rate of energy consumption is needed by the bird to keep aloft, without moving forward?
(b) What does the shape of the graph tell you about how birds fly?
(c) Sketch $f'(v)$.

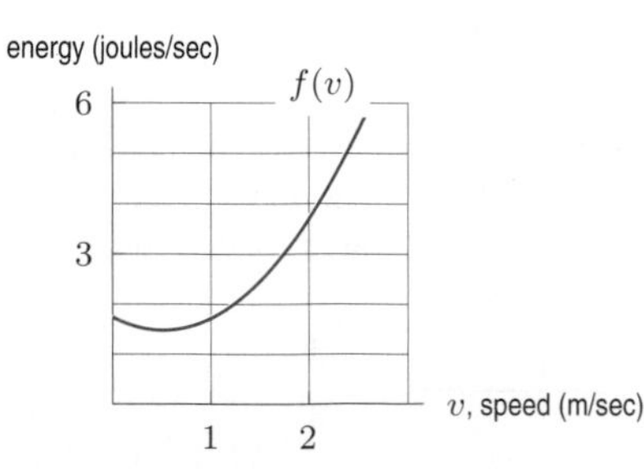

Figure 2.68

PROJECTS FOR CHAPTER TWO

1. Estimating the Temperature of a Yam

Suppose you put a yam in a hot oven, maintained at a constant temperature of 200°C. As the yam picks up heat from the oven, its temperature rises.[16]

(a) Draw a possible graph of the temperature T of the yam against time t (minutes) since it is put into the oven. Explain any interesting features of the graph, and in particular explain its concavity.

(b) Suppose that, at $t = 30$, the temperature T of the yam is $120°$ and increasing at the (instantaneous) rate of 2°/min. Using this information, plus what you know about the shape of the T graph, estimate the temperature at time $t = 40$.

(c) Suppose in addition you are told that at $t = 60$, the temperature of the yam is $165°$. Can you improve your estimate of the temperature at $t = 40$?

(d) Assuming all the data given so far, estimate the time at which the temperature of the yam is $150°$.

2. Temperature and Illumination

Alone in your dim, unheated room, you light a single candle rather than curse the darkness. Depressed with the situation, you walk directly away from the candle, sighing. The temperature (in degrees Fahrenheit) and illumination (in % of one candle power) decrease as your distance (in feet) from the candle increases. In fact, you have tables showing this information.

Distance (feet)	Temperature (°F)
0	55
1	54.5
2	53.5
3	52
4	50
5	47
6	43.5

Distance (feet)	Illumination (%)
0	100
1	85
2	75
3	67
4	60
5	56
6	53

You are cold when the temperature is below $40°$. You are in the dark when the illumination is at most 50% of one candle power.

(a) Two graphs are shown in Figures 2.69 and 2.70. One is temperature as a function of distance and one is illumination as a function of distance. Which is which? Explain.

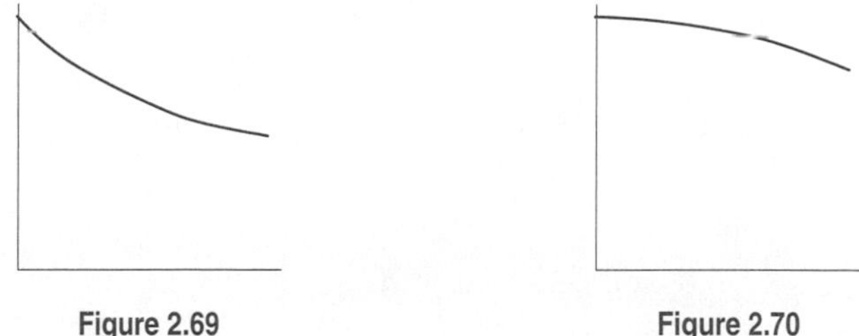

Figure 2.69 Figure 2.70

(b) What is the average rate at which the temperature is changing when the illumination drops from 75% to 56%?

(c) You can still read your watch when the illumination is about 65%. Can you still read your watch at 3.5 feet? Explain.

[16]From Peter D. Taylor, *Calculus: The Analysis of Functions* (Toronto: Wall & Emerson, Inc., 1992).

(d) Suppose you know that at 6 feet the instantaneous rate of change of the temperature is $-4.5°$F/ft and the instantaneous rate of change of illumination is -3% candle power/ft. Estimate the temperature and the illumination at 7 feet.

(e) Are you in the dark before you are cold, or vice-versa?

FOCUS ON THEORY

LIMITS, CONTINUITY, AND THE DEFINITION OF THE DERIVATIVE

The velocity at a single instant in time is surprisingly difficult to define precisely. Consider the statement "At the instant it crossed the finish line, the horse was traveling at 42 mph." How can such a claim be substantiated? A photograph taken at that instant will show the horse motionless—it is no help at all. There is some paradox in trying to quantify the property of motion at a particular instant in time, since by focusing on a single instant we stop the motion!

A similar difficulty arises whenever we attempt to measure the rate of change of anything—for example, oil leaking out of a damaged tanker. The statement "One hour after the ship's hull ruptured, oil was leaking at a rate of 200 barrels per second" seems not to make sense. We could argue that at any given instant *no* oil is leaking.

Problems of motion were of central concern to Zeno and other philosophers as early as the fifth century BC. The approach that we took, made famous by Newton's calculus, is to stop looking for a simple notion of speed at an instant, and instead to look at speed over small intervals containing the instant. This method sidesteps the philosophical problems mentioned earlier but brings new ones of its own.

Definition of the Derivative Using Average Rates

On page 100 of Section 2.1, we defined the derivative as the instantaneous rate of change of a function. We can estimate a derivative by computing average rates of change over smaller and smaller intervals. We use this idea to give a symbolic definition of the derivative. Letting h represent the size of the interval, we have

$$\begin{array}{c}\text{Average rate of change}\\ \text{between } x \text{ and } x+h\end{array} = \frac{f(x+h)-f(x)}{(x+h)-x} = \frac{f(x+h)-f(x)}{h}.$$

To find the derivative, or instantaneous rate of change at the point x, we use smaller and smaller intervals. To find the derivative exactly, we take the limit as h, the size of the interval, shrinks to zero, so we say

$$\text{Derivative} = \text{Limit, as } h \text{ approaches zero, of } \frac{f(x+h)-f(x)}{h}.$$

Finally, instead of writing the phrase "limit, as h approaches 0," we use the notation $\lim\limits_{h\to 0}$. This leads to the following symbolic definition:

For any function f, we define the **derivative function**, f', by

$$f'(x) = \lim_{h\to 0} \frac{f(x+h)-f(x)}{h},$$

provided the limit exists. The function f is said to be **differentiable** at any point x at which the derivative function is defined.

Notice that we have replaced the original difficulty of computing velocity at a point by an argument that the average rates of change approach a number as the time intervals shrink in size. In a sense, we have traded one hard question for another, since we don't yet have any idea how to be certain what number the average velocities are approaching.

The Idea of a Limit

We used a limit to define the derivative. Now we look a bit more at the idea of the limit of a function at the point c. Provided the limit exists:

> We write $\lim_{x \to c} f(x)$ to represent the number approached by $f(x)$ as x approaches c.

Example 1 Investigate $\lim_{x \to 2} x^2$.

Solution Notice that we can make x^2 as close to 4 as we like by taking x sufficiently close to 2. (Look at the values of 1.9^2, 1.99^2, 1.999^2, and 2.1^2, 2.01^2, 2.001^2 in Table 2.11; they seem to be approaching 4.) We write

$$\lim_{x \to 2} x^2 = 4,$$

which is read "the limit, as x approaches 2, of x^2 is 4." Notice that the limit does not ask what happens *at* $x = 2$, so it is not sufficient to substitute 2 to find the answer. The limit describes behavior of a function *near* a point, not *at* the point.

Table 2.11 *Values of x^2 near $x = 2$*

x	1.9	1.99	1.999	2.001	2.01	2.1
x^2	3.61	3.96	3.996	4.004	4.04	4.41

Example 2 Use a graph to estimate $\lim_{x \to 0} \frac{2^x - 1}{x}$.

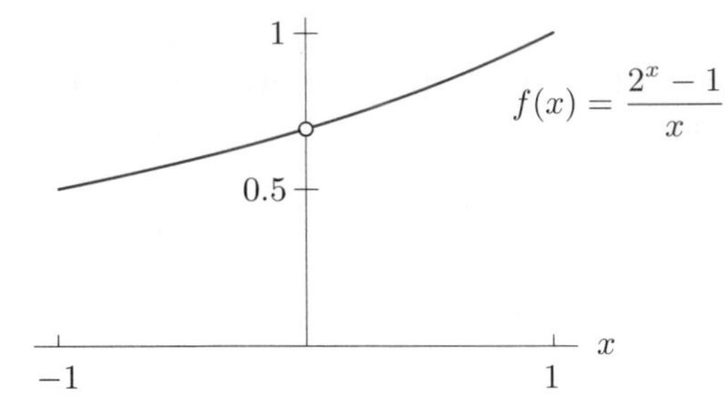

Figure 2.71: Find the limit as $x \to 0$ of $\frac{2^x - 1}{x}$

Solution Notice that the expression $\frac{2^x - 1}{x}$ is undefined at $x = 0$. To find out what happens to this expression as x approaches 0, look at a graph of $f(x) = \frac{2^x - 1}{x}$. Figure 2.71 shows that as x approaches 0 from either side, the value of $\frac{2^x - 1}{x}$ appears to approach 0.7. If we zoom in on the graph near $x = 0$, we can estimate the limit with greater accuracy, giving

$$\lim_{x \to 0} \frac{2^x - 1}{x} \approx 0.693.$$

Example 3 Estimate $\lim_{h\to 0} \dfrac{(3+h)^2-9}{h}$ numerically.

Solution The limit is the value approached by this expression as h approaches 0. The values in Table 2.12 seem to be approaching 6 as $h \to 0$. So it is a reasonable guess that

$$\lim_{h\to 0} \frac{(3+h)^2-9}{h} = 6.$$

However, we cannot be sure that the limit is *exactly* 6 by looking at the table. To calculate the limit exactly requires algebra.

Table 2.12 *Values of* $\left((3+h)^2-9\right)/h$

h	-0.1	-0.01	-0.001	0.001	0.01	0.1
$\left((3+h)^2-9\right)/h$	5.9	5.99	5.999	6.001	6.01	6.1

Example 4 Use algebra to find $\lim_{h\to 0} \dfrac{(3+h)^2-9}{h}$.

Solution Expanding the numerator gives

$$\frac{(3+h)^2-9}{h} = \frac{9+6h+h^2-9}{h} = \frac{6h+h^2}{h}.$$

Since taking the limit as $h \to 0$ means looking at values of h near, but not equal, to 0, we can cancel a common factor of h, giving

$$\lim_{h\to 0} \frac{(3+h)^2-9}{h} = \lim_{h\to 0} \frac{6h+h^2}{h} = \lim_{h\to 0}(6+h).$$

As h approaches 0, the values of $(6+h)$ approach 6, so

$$\lim_{h\to 0} \frac{(3+h)^2-9}{h} = \lim_{h\to 0}(6+h) = 6.$$

Continuity

Roughly speaking, a function is said to be *continuous* on an interval if its graph has no breaks, jumps, or holes in that interval. A continuous function has a graph that can be drawn without lifting the pencil from the paper.

Example: The function $f(x) = 3x^2 - x^2 + 2x + 1$ is continuous on any interval. (See Figure 2.72.)

Example: The function $f(x) = 1/x$ is not defined at $x = 0$. It is continuous on any interval not containing the origin. (See Figure 2.73.)

Example: Suppose $p(x)$ is the price of mailing a first-class letter weighing x ounces. It costs 34¢ for one ounce or less, 57¢ between the first and second ounces, and so on. So the graph (in Figure 2.74) is a series of steps. This function is not continuous on intervals such as $(0, 2)$ because the graph jumps at $x = 1$.

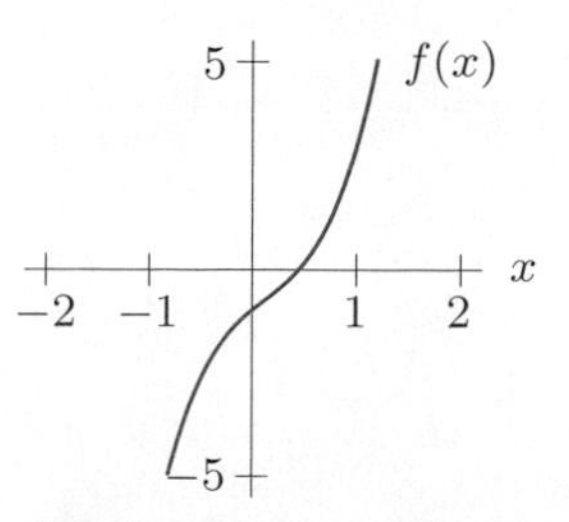

Figure 2.72: The graph of $f(x) = 3x^3 - x^2 + 2x - 1$

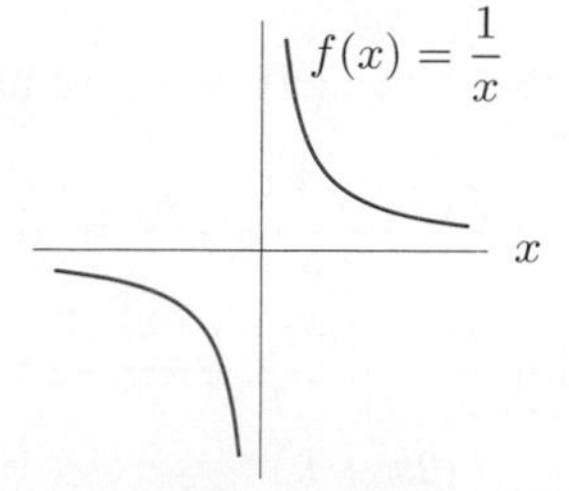

Figure 2.73: Graph of $f(x) = 1/x$: Not defined at 0

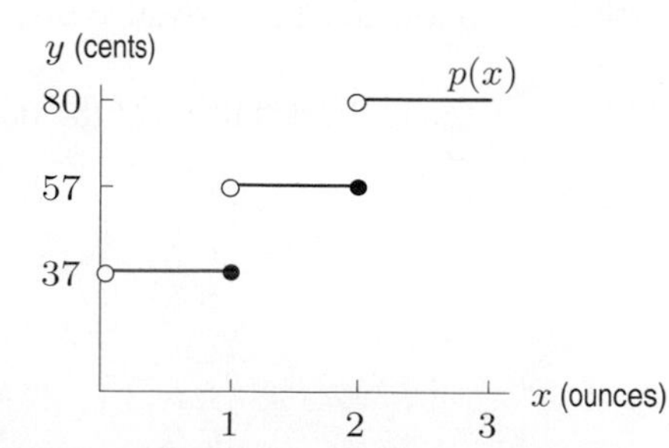

Figure 2.74: Cost of mailing a letter

What Does Continuity Mean Numerically?

Continuity is important in practical work because it means that small errors in the independent variable lead to small errors in the value of the function.

Example: Suppose that $f(x) = x^2$ and that we want to compute $f(\pi)$. Knowing f is continuous tells us that taking $x = 3.14$ should give a good approximation to $f(\pi)$, and that we can get a better approximation to $f(\pi)$ by using more decimals of π.

Example: If $p(x)$ is the cost of mailing a letter weighing x ounces, then $p(0.99) = p(1) = 34¢$, whereas $p(1.01) = 57¢$, because as soon as we get over 1 ounce, the price jumps up to $57¢$. So a small difference in the weight of a letter can lead to a significant difference in its mailing cost. Hence p is not continuous at $x = 1$.

Definition of Continuity

We now define continuity using limits. The idea of continuity rules out breaks, jumps, or holes by demanding that the behavior of a function *near* a point be consistent with its behavior *at* the point:

> The function f is **continuous** at $x = c$ if f is defined at $x = c$ and
>
> $$\lim_{x \to c} f(x) = f(c).$$
>
> The function is **continuous on an interval** (a, b) if it is continuous at every point in the interval.

Which Functions are Continuous?

Requiring a function to be continuous on an interval is not asking very much, as any function whose graph is an unbroken curve over the interval is continuous. For example, exponential functions, polynomials, and sine and cosine are continuous on every interval. Functions created by adding, multiplying, or composing continuous functions are also continuous.

Using the Definition to Calculate Derivatives

By estimating the derivative of the function $f(x) = x^2$ at several points, we guessed in Example 5 of Section 2.2 that the derivative of x^2 is $f'(x) = 2x$. In order to show that this formula is correct, we have to use the symbolic definition of the derivative given on page 135.

In evaluating the expression

$$\lim_{h \to 0} \frac{f(x+h) - f(x)}{h},$$

we simplify the difference quotient first, and then take the limit as h approaches zero.

Example 5 Show that the derivative of $f(x) = x^2$ is $f'(x) = 2x$.

Solution Using the definition of the derivative with $f(x) = x^2$, we have

$$\begin{aligned} f'(x) &= \lim_{h \to 0} \frac{f(x+h) - f(x)}{h} = \lim_{h \to 0} \frac{(x+h)^2 - x^2}{h} \\ &= \lim_{h \to 0} \frac{x^2 + 2xh + h^2 - x^2}{h} = \lim_{h \to 0} \frac{2xh + h^2}{h} \\ &= \lim_{h \to 0} \frac{h(2x+h)}{h}. \end{aligned}$$

To take the limit, look at what happens when h is close to 0, but do not let $h = 0$. Since $h \neq 0$, we cancel the common factor of h, giving

$$f'(x) = \lim_{h\to 0} \frac{h(2x+h)}{h} = \lim_{h\to 0} (2x+h) = 2x,$$

because as h gets close to zero, $2x + h$ gets close to $2x$. So

$$f'(x) = \frac{d}{dx}(x^2) = 2x.$$

Example 6 Show that if $f(x) = 3x - 2$, then $f'(x) = 3$.

Solution Since the slope of the linear function $f(x) = 3x - 2$ is 3 and the derivative is the slope, we see that $f'(x) = 3$. We can also use the definition to get this result:

$$f'(x) = \lim_{h\to 0} \frac{f(x+h) - f(x)}{h} = \lim_{h\to 0} \frac{(3(x+h)-2) - (3x-2)}{h}$$
$$= \lim_{h\to 0} \frac{3x + 3h - 2 - 3x + 2}{h} = \lim_{h\to 0} \frac{3h}{h}.$$

To find the limit, look at what happens when h is close to, but not equal to, 0. Simplifying, we get

$$f'(x) = \lim_{h\to 0} \frac{3h}{h} = \lim_{h\to 0} 3 = 3.$$

Problems on Limits and the Definition of the Derivative

1. On Figure 2.75, mark lengths that represent the quantities in parts (a) – (e). (Pick any h, with $h > 0$.)

(a) $a + h$ **(b)** h **(c)** $f(a)$
(d) $f(a+h)$ **(e)** $f(a+h) - f(a)$

(f) Using your answers to parts (a)–(e), show how the quantity $\dfrac{f(a+h) - f(a)}{h}$ can be represented as the slope of a line on the graph.

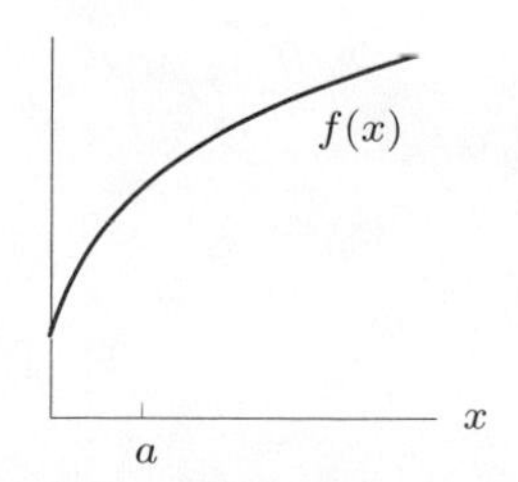

Figure 2.75

2. On Figure 2.76, mark lengths that represent the quantities in parts (a)–(e). (Pick any h, with $h > 0$.)

(a) $a + h$ **(b)** h **(c)** $f(a)$
(d) $f(a+h)$ **(e)** $f(a+h) - f(a)$

(f) Using your answers to parts (a)–(e), represent the quantity $\dfrac{f(a+h) - f(a)}{h}$ as the slope of a line on the graph.

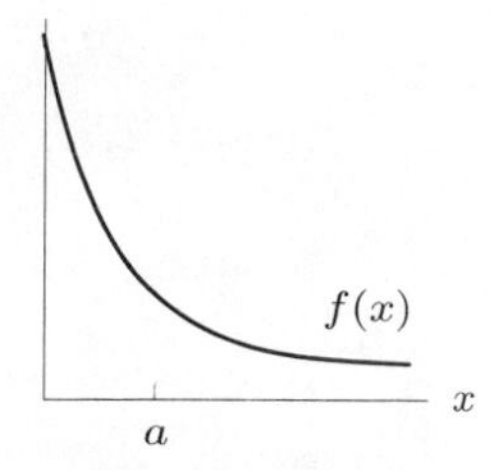

Figure 2.76

Use a graph to estimate the limits in Problems 3–4.

3. $\lim_{x\to 0} \frac{\sin x}{x}$ (with x in radians)

4. $\lim_{x\to 0} \frac{5^x - 1}{x}$

Estimate the limits in Problems 5–8 by substituting smaller and smaller values of h. For trigonometric functions, use radians. Give answers to one decimal place.

5. $\lim_{h\to 0} \frac{(3+h)^3 - 27}{h}$

6. $\lim_{h\to 0} \frac{7^h - 1}{h}$

7. $\lim_{h\to 0} \frac{e^{1+h} - e}{h}$

8. $\lim_{h\to 0} \frac{\cos h - 1}{h}$

In Problems 9–12, does the function $f(x)$ appear to be continuous on the interval $0 \le x \le 2$? If not, what about on the interval $0 \le x \le 0.5$?

9. 

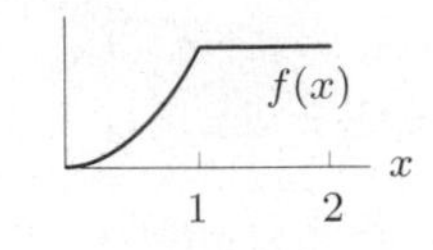

10.

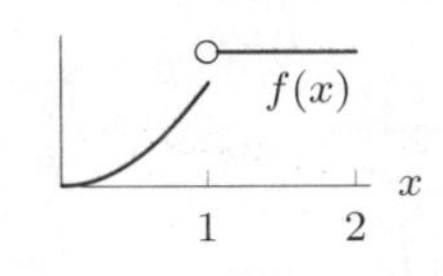

11.

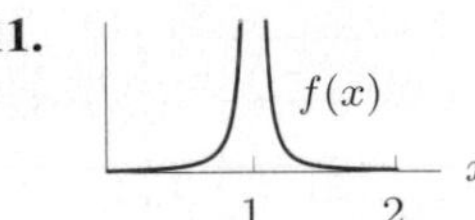

12.

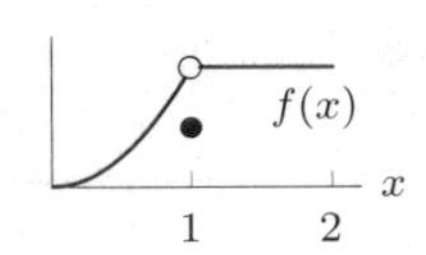

Are the functions in Problems 13–18 continuous on the given intervals?

13. $f(x) = x + 2$ on $-3 \le x \le 3$

14. $f(x) = 2^x$ on $0 \le x \le 10$

15. $f(x) = x^2 + 2$ on $0 \le x \le 5$

16. $f(x) = \frac{1}{x-1}$ on $2 \le x \le 3$

17. $f(x) = \frac{1}{x-1}$ on $0 \le x \le 2$

18. $f(x) = \frac{1}{x^2+1}$ on $0 \le x \le 2$

Which of the functions described in Problems 19–23 are continuous?

19. The number of people in a village as a function of time.

20. The weight of a baby as a function of time during the second month of the baby's life.

21. The number of pairs of pants as a function of the number of yards of cloth from which they are made. Each pair requires 3 yards.

22. The distance traveled by a car in stop-and-go traffic as a function of time.

23. You start in North Carolina and go westward on Interstate 40 toward California. Consider the function giving the local time of day as a function of your distance from your starting point.

Use the definition of the derivative to show how the formulas in Problems 24–33 are obtained.

24. If $f(x) = 5x$, then $f'(x) = 5$.

25. If $f(x) = 3x - 2$, then $f'(x) = 3$.

26. If $f(x) = x^2 + 4$, then $f'(x) = 2x$.

27. If $f(x) = 3x^2$, then $f'(x) = 6x$.

28. If $f(x) = -2x^3$, then $f'(x) = -6x^2$.

29. If $f(x) = x - x^2$, then $f'(x) = 1 - 2x$.

30. If $f(x) = 1 - x^3$, then $f'(x) = -3x^2$.

31. If $f(x) = 5x^2 + 1$, then $f'(x) = 10x$.

32. If $f(x) = 2x^2 + x$, then $f'(x) = 4x + 1$.

33. If $f(x) = 1/x$, then $f'(x) = -1/x^2$.

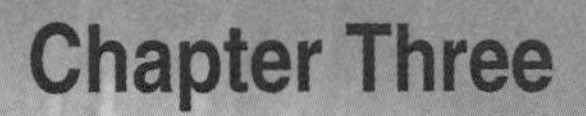

Chapter Three

SHORT-CUTS TO DIFFERENTIATION

In this chapter we calculate the derivatives of functions given by formulas. These functions include power, polynomial, exponential, logarithmic, and periodic functions. The chapter also contains general rules, such as the product, quotient, and chain rules, which allow us to differentiate combinations of functions.

3.1 DERIVATIVE FORMULAS FOR POWERS AND POLYNOMIALS

The derivative of a function at a point represents a slope and a rate of change. In Chapter 2, we learned how to estimate values of the derivative of a function given by a graph or by a table. Now, we learn how to find a formula for the derivative of a function given by a formula.

Derivative of a Constant Function

The graph of a constant function $f(x) = k$ is a horizontal line, with a slope of 0 everywhere. Therefore, its derivative is 0 everywhere. (See Figure 3.1.)

$$\text{If } f(x) = k, \text{ then } f'(x) = 0.$$

For example, $\frac{d}{dx}(5) = 0$.

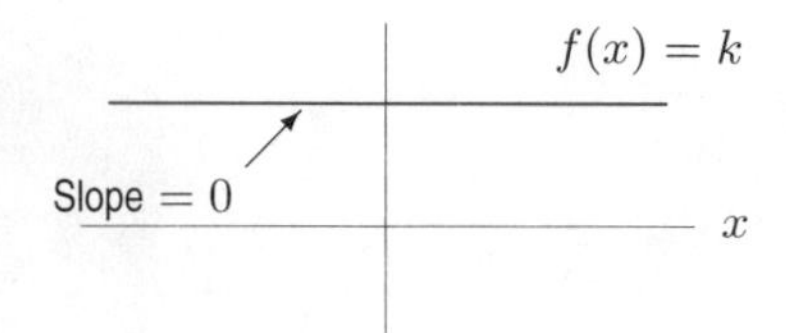

Figure 3.1: A constant function

Derivative of a Linear Function

We already know that the slope of a line is constant. This tells us that the derivative of a linear function is constant.

$$\text{If } f(x) = b + mx, \text{ then } f'(x) = \text{Slope} = m.$$

For example, $\frac{d}{dx}(5 - \frac{3}{2}x) = -\frac{3}{2}$.

Derivative of a Constant Times a Function

Figure 3.2 shows the graph of $y = f(x)$ and of three multiples: $y = 3f(x)$, $y = \frac{1}{2}f(x)$, and $y = -2f(x)$. What is the relationship between the derivatives of these functions? In other words, for a particular x-value, how are the slopes of these graphs related?

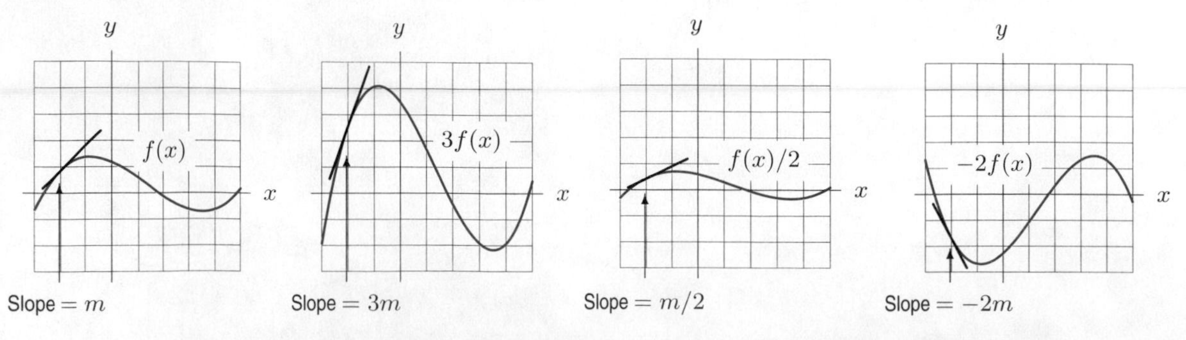

Figure 3.2: A function and its multiples: Derivative of multiple is multiple of derivative

Multiplying by a constant stretches or shrinks the graph (and reflects it about the x-axis if the constant is negative). This changes the slope of the curve at each point. If the graph has been stretched, the "rises" have all been increased by the same factor, whereas the "runs" remain the same. Thus, the slopes are all steeper by the same factor. If the graph has been shrunk, the slopes are all smaller by the same factor. If the graph has been reflected about the x-axis, the slopes will all have their signs reversed. Thus, if a function is multiplied by a constant, c, so is its derivative:

Derivative of a Constant Multiple

If c is a constant,

$$\frac{d}{dx}[cf(x)] = cf'(x).$$

Derivatives of Sums and Differences

Values of two functions, $f(x)$ and $g(x)$, and their sum $f(x) + g(x)$, are listed in Table 3.1.

Table 3.1 *Sum of functions*

x	$f(x)$	$g(x)$	$f(x) + g(x)$
0	100	0	100
1	110	0.2	110.2
2	130	0.4	130.4
3	160	0.6	160.6
4	200	0.8	200.8

We see that adding the increments of $f(x)$ and the increments of $g(x)$ gives the increments of $f(x) + g(x)$. For example, as x increases from 0 to 1, $f(x)$ increases by 10 and $g(x)$ increases by 0.2, while $f(x) + g(x)$ increases by $110.2 - 100 = 10.2$. Similarly, as x increases from 3 to 4, $f(x)$ increases by 40 and $g(x)$ by 0.2, while $f(x) + g(x)$ increases by $200.8 - 160.6 = 40.2$.

From this example, we see that the rate at which $f(x)+g(x)$ is increasing is the sum of the rates at which $f(x)$ and $g(x)$ are increasing. Similar reasoning applies to the difference, $f(x) - g(x)$. In terms of derivatives:

Derivative of Sum and Difference

$$\frac{d}{dx}[f(x) + g(x)] = f'(x) + g'(x) \quad \text{and} \quad \frac{d}{dx}[f(x) - g(x)] = f'(x) - g'(x).$$

Powers of x

We start by looking at $f(x) = x^2$ and $g(x) = x^3$. We show in the Focus on Theory section at the end of this chapter that

$$f'(x) = \frac{d}{dx}(x^2) = 2x \quad \text{and} \quad g'(x) = \frac{d}{dx}(x^3) = 3x^2.$$

The graphs of $f(x) = x^2$ and $g(x) = x^3$ and their derivatives are shown in Figures 3.3 and 3.4. Notice $f'(x) = 2x$ has the behavior we expect. It is negative for $x < 0$ (when f is decreasing), zero for $x = 0$, and positive for $x > 0$ (when f is increasing). Similarly, $g'(x) = 3x^2$ is zero when $x = 0$, but positive everywhere else, as g is increasing everywhere else. These examples are special cases of the power rule.

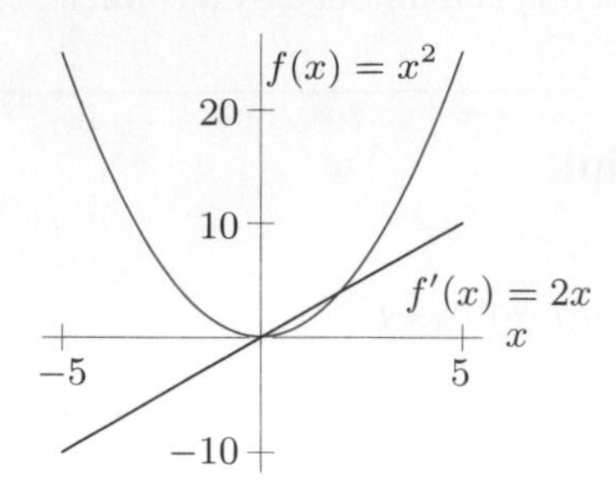

Figure 3.3: Graphs of $f(x) = x^2$ and its derivative $f'(x) = 2x$

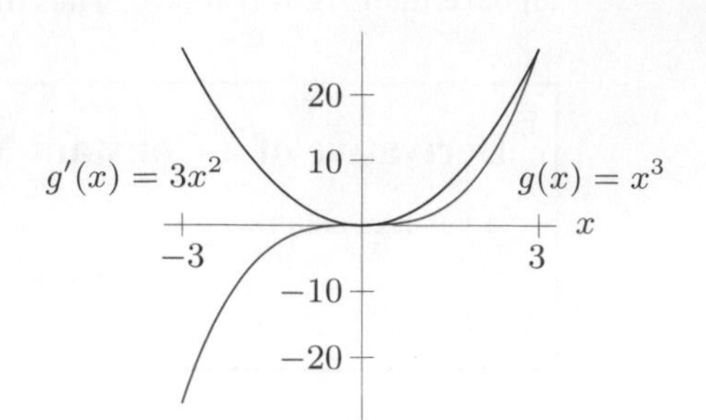

Figure 3.4: Graphs of $g(x) = x^3$ and its derivative $g'(x) = 3x^2$

The Power Rule

For any constant real number n,

$$\frac{d}{dx}(x^n) = nx^{n-1}.$$

Example 1 Find the derivative of (a) $h(x) = x^8$ (b) $P(t) = t^7$.

Solution (a) $h'(x) = 8x^7$. (b) $P'(t) = 7t^6$.

Derivatives of Polynomials

Using the derivatives of powers, constant multiples, and sums, we can differentiate any polynomial.

Example 2 Differentiate (a) $A(t) = 3t^5$ (b) $r(p) = p^5 + p^3$ (c) $f(x) = 5x^2 - 7x^3$.

Solution (a) Using the constant multiple rule: $A'(t) = \frac{d}{dt}(3t^5) = 3\frac{d}{dt}(t^5) = 3 \cdot 5t^4 = 15t^4$.

(b) Using the sum rule: $r'(p) = \frac{d}{dp}(p^5 + p^3) = \frac{d}{dp}(p^5) + \frac{d}{dp}(p^3) = 5p^4 + 3p^2$.

(c) Using both rules together:

$$\begin{aligned} \frac{d}{dx}(5x^2 - 7x^3) &= \frac{d}{dx}(5x^2) - \frac{d}{dx}(7x^3) && \text{Derivative of difference} \\ &= 5\frac{d}{dx}(x^2) - 7\frac{d}{dx}(x^3) && \text{Derivative of multiple} \\ &= 5(2x) - 7(3x^2) = 10x - 21x^2. \end{aligned}$$

Example 3 Find the derivatives of (a) $5x^2 + 3x + 2$ (b) $\sqrt{3}x^7 - \frac{x^5}{5} + \pi$.

Solution

(a)
$$\begin{aligned}\frac{d}{dx}(5x^2 + 3x + 2) &= 5\frac{d}{dx}(x^2) + 3\frac{d}{dx}(x) + \frac{d}{dx}(2)\\ &= 5 \cdot 2x + 3 \cdot 1 + 0 \quad \text{(Since the derivative of a constant, } \tfrac{d}{dx}(2)\text{, is zero.)}\\ &= 10x + 3.\end{aligned}$$

(b)
$$\begin{aligned}\frac{d}{dx}\left(\sqrt{3}x^7 - \frac{x^5}{5} + \pi\right) &= \sqrt{3}\frac{d}{dx}(x^7) - \frac{1}{5}\frac{d}{dx}(x^5) + \frac{d}{dx}(\pi)\\ &= \sqrt{3} \cdot 7x^6 - \frac{1}{5} \cdot 5x^4 + 0 \quad \text{(Since } \pi \text{ is a constant, } \tfrac{d}{dx}(\pi) = 0.\text{)}\\ &= 7\sqrt{3}x^6 - x^4.\end{aligned}$$

We can use the power rule to differentiate negative and fractional powers.

Example 4 Use the power rule to differentiate (a) $\frac{1}{x^3}$ (b) $\sqrt{x}$ (c) $2t^{4.5}$.

Solution

(a) For $n = -3$: $\quad \frac{d}{dx}\left(\frac{1}{x^3}\right) = \frac{d}{dx}(x^{-3}) = -3x^{-3-1} = -3x^{-4} = -\frac{3}{x^4}.$

(b) For $n = 1/2$: $\quad \frac{d}{dx}(\sqrt{x}) = \frac{d}{dx}\left(x^{1/2}\right) = \frac{1}{2}x^{(1/2)-1} = \frac{1}{2}x^{-1/2} = \frac{1}{2\sqrt{x}}.$

(c) For $n = 4.5$: $\quad \frac{d}{dt}\left(2t^{4.5}\right) = 2\left(4.5t^{4.5-1}\right) = 9t^{3.5}.$

Using the Derivative Formulas

Since the slope of the tangent line to a curve is given by the derivative, we use differentiation to find the equation of the tangent line.

Example 5 Find an equation for the tangent line at $x = 1$ to the graph of

$$y = x^3 + 2x^2 - 5x + 7.$$

Sketch the graph of the curve and its tangent line on the same axes.

Solution Differentiating gives

$$\frac{dy}{dx} = 3x^2 + 2(2x) - 5(1) + 0 = 3x^2 + 4x - 5,$$

so the slope of the tangent line at $x = 1$ is

$$m = \left.\frac{dy}{dx}\right|_{x=1} = 3(1)^2 + 4(1) - 5 = 2.$$

When $x = 1$, we have $y = 1^3 + 2(1^2) - 5(1) + 7 = 5$, so the point $(1, 5)$ lies on the tangent line. Using the formula $y - y_0 = m(x - x_0)$ gives

$$\begin{aligned}y - 5 &= 2(x - 1)\\ y &= 3 + 2x.\end{aligned}$$

The equation of the tangent line is $y = 3 + 2x$. See Figure 3.5.

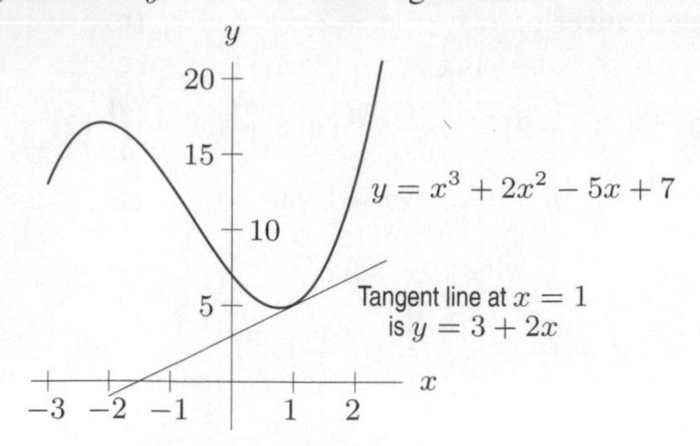

Figure 3.5: Find the equation for this tangent line

Example 6 Find and interpret the second derivatives of (a) $f(x) = x^2$ (b) $g(x) = x^3$.

Solution

(a) Differentiating $f(x) = x^2$ gives $f'(x) = 2x$, so $f''(x) = \frac{d}{dx}(2x) = 2$. Since f'' is always positive, the graph of f is concave up, as expected for a parabola opening upward. (See Figure 3.6.)

(b) Differentiating $g(x) = x^3$ gives $g'(x) = 3x^2$, so $g''(x) = \frac{d}{dx}(3x^2) = 3\frac{d}{dx}(x^2) = 3 \cdot 2x = 6x$. This is positive for $x > 0$ and negative for $x < 0$, which means that the graph of $g(x) = x^3$ is concave up for $x > 0$ and concave down for $x < 0$. (See Figure 3.7.)

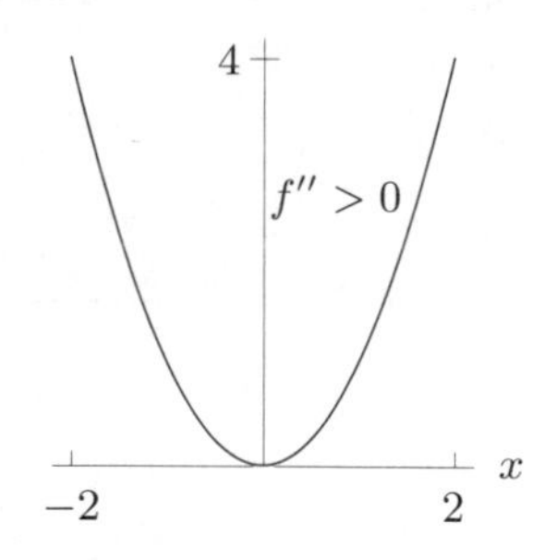

Figure 3.6: Graph of $f(x) = x^2$ with $f''(x) = 2$

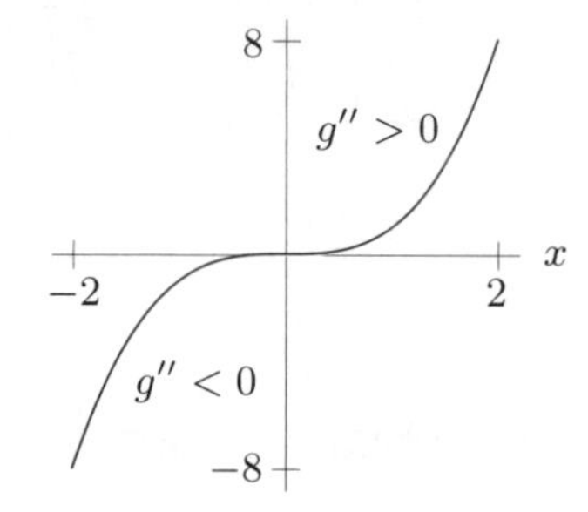

Figure 3.7: Graph of $g(x) = x^3$ with $g''(x) = 6x$

Example 7 Find the velocity of a body at time t if its position, in meters, is given as a function of time t, in seconds, by

$$s = -4.9t^2 + 5t + 6.$$

Solution The velocity, v, is the derivative of the position:

$$v = \frac{ds}{dt} = \frac{d}{dt}(-4.9t^2 + 5t + 6) = -9.8t + 5 \text{ meters/sec.}$$

Example 8 Figure 3.8 shows the graph of a cubic polynomial. Both graphically and algebraically, describe the behavior of the derivative of this cubic.

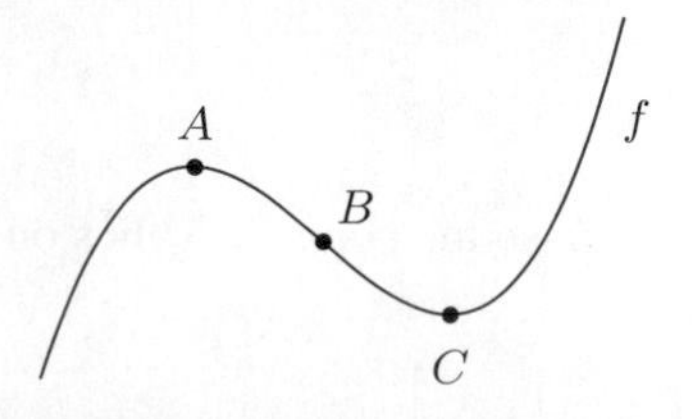

Figure 3.8: The cubic of Example 8

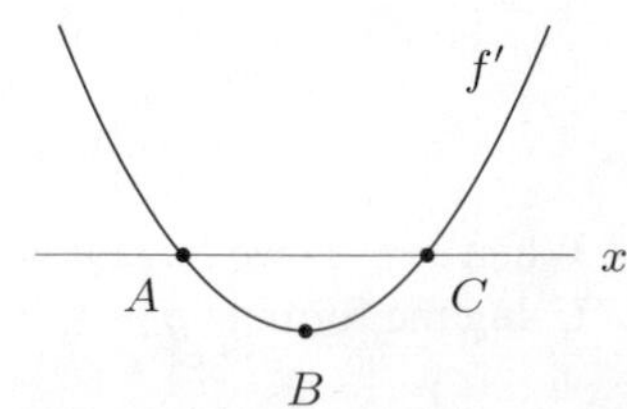

Figure 3.9: Derivative of the cubic of Example 8

Solution Graphical approach: Suppose we move along the curve from left to right. To the left of A, the slope is positive; it starts very positive and decreases until the curve reaches A, where the slope is 0. Between A and C the slope is negative. Between A and B the slope is decreasing (getting more negative); it is most negative at B. Between B and C the slope is negative but increasing; at C the slope is zero. From C to the right, the slope is positive and increasing. The graph of the derivative function is shown in Figure 3.9.

Algebraic approach: f is a cubic that goes to $+\infty$ as $x \to +\infty$ so

$$f(x) = ax^3 + bx^2 + cx + d$$

with $a > 0$. Hence,

$$f'(x) = 3ax^2 + 2bx + c,$$

whose graph is a parabola opening upward, as in Figure 3.9.

Problems for Section 3.1

For Problems 1–36, find the derivative. Assume a, b, c, k are constants.

1. $y = 5$

2. $y = 3x$

3. $y = x^{12}$

4. $y = x^{-12}$

5. $y = x^{4/3}$

6. $y = 8t^3$

7. $y = 3t^4 - 2t^2$

8. $y = 5x + 13$

9. $f(x) = \dfrac{1}{x^4}$

10. $f(q) = q^3 + 10$

11. $y = x^2 + 5x + 9$

12. $y = 6x^3 + 4x^2 - 2x$

13. $y = 3x^2 + 7x - 9$

14. $y = 8t^3 - 4t^2 + 12t - 3$

15. $y = 4.2q^2 - 0.5q + 11.27$

16. $y = -3x^4 - 4x^3 - 6x + 2$

17. $g(t) = \dfrac{1}{t^5}$

18. $f(z) = -\dfrac{1}{z^{6.1}}$

19. $y = \dfrac{1}{r^{7/2}}$

20. $y = \sqrt{x}$

21. $h(\theta) = \dfrac{1}{\sqrt[3]{\theta}}$

22. $f(x) = \sqrt{\dfrac{1}{x^3}}$

23. $y = 3t^5 - 5\sqrt{t} + \dfrac{7}{t}$

24. $y = z^2 + \dfrac{1}{2z}$

25. $y = 3t^2 + \dfrac{12}{\sqrt{t}} - \dfrac{1}{t^2}$

26. $h(t) = \dfrac{3}{t} + \dfrac{4}{t^2}$

27. $y = \sqrt{x}(x + 1)$

28. $h(\theta) = \theta(\theta^{-1/2} - \theta^{-2})$

29. $f(x) = kx^2$

30. $y = ax^2 + bx + c$

31. $Q = aP^2 + bP^3$

32. $v = at^2 + \dfrac{b}{t^2}$

33. $P = a + b\sqrt{t}$

34. $V = \frac{4}{3}\pi r^2 b$

35. $w = 3ab^2 q$

36. $h(x) = \dfrac{ax + b}{c}$

37. Let $f(t) = t^2 - 4t + 5$.
- **(a)** Find $f'(t)$.
- **(b)** Find $f'(1)$ and $f'(2)$.
- **(c)** Use a graph of $f(t)$ to check that your answers to part (b) are reasonable. Explain.

38. Let $f(x) = x^2 + 1$. Compute the derivatives $f'(0)$, $f'(1)$, $f'(2)$, and $f'(-1)$. Check your answers graphically.

39. Let $f(x) = x^2 + 3x - 5$. Find $f'(0)$, $f'(3)$, $f'(-2)$.

40. The height of a sand dune (in centimeters) is represented by $f(t) = 700 - 3t^2$, where t is measured in years since 2005. Find $f(5)$ and $f'(5)$. Using units, explain what each means in terms of the sand dune.

41. Find the rate of change of a population of size $P(t) = t^3 + 4t + 1$ at time $t = 2$.

42. If $f(t) = 2t^3 - 4t^2 + 3t - 1$, find $f'(t)$ and $f''(t)$.

43. If $f(t) = t^4 - 3t^2 + 5t$, find $f'(t)$ and $f''(t)$.

44. Zebra mussels are freshwater shellfish that first appeared in the St. Lawrence River in the early 1980s and have spread throughout the Great Lakes. Suppose that t months after they appeared in a small bay, the number of zebra mussels is given by $Z(t) = 300t^2$. How many zebra mussels are in the bay after four months? At what rate is the population growing at that time? Give units.

45.
- **(a)** Find the equation of the tangent line to $f(x) = x^3$ at the point where $x = 2$.
- **(b)** Graph the tangent line and the function on the same axes. If the tangent line is used to estimate values of the function, will the estimates be overestimates or underestimates?

46. Find the equation of the line tangent to the graph of $f(t) = 6t - t^2$ at $t = 4$. Sketch the graph of $f(t)$ and the tangent line on the same axes.

47. Find the equation of the line tangent to the graph of $f(x) = 2x^3 - 5x^2 + 3x - 5$ at $x = 1$.

48. Suppose W is proportional to r^3. The derivative dW/dr is proportional to what power of r?

49. The cost to produce q items is $C(q) = 1000 + 2q^2$ dollars. Find the marginal cost of producing the 25th item. Interpret your answer in terms of costs.

50. The demand curve for a product is given by $q = 300 - 3p$, where p is the price of the product and q is the quantity consumers will buy at that price.

(a) Write the revenue as a function of price.
(b) Find the marginal revenue when the price is \$10, and interpret your answer in terms of revenue.
(c) For what prices is the marginal revenue positive? For what prices is it negative?

51. The yield, Y, of an apple orchard (measured in bushels of apples per acre) is a function of the amount x of fertilizer in pounds used per acre. Suppose

$$Y = f(x) = 320 + 140x - 10x^2.$$

(a) What is the yield if 5 pounds of fertilizer is used per acre?
(b) Find $f'(5)$. Give units with your answer and interpret it in terms of apples and fertilizer.
(c) Given your answer to part (b), should more or less fertilizer be used? Explain.

52. The demand for a product is given, for $p, q \geq 0$, by

$$p = f(q) = 50 - 0.03q^2.$$

(a) Find the p- and q-intercepts for this function and interpret them in terms of demand for this product.
(b) Find $f(20)$ and give units with your answer. Explain what it tells you in terms of demand.
(c) Find $f'(20)$ and give units with your answer. Explain what it tells you in terms of demand.

53. The cost (in dollars) of producing q items is given by $C(q) = 0.08q^3 + 75q + 1000$.

(a) Find the marginal cost function.
(b) Find $C(50)$ and $C'(50)$. Give units with your answers and explain what each is telling you about costs of production.

54. Let $f(x) = x^2 - 4x + 8$. For what x-values is $f'(x) = 0$?

55. Let $f(x) = x^3 - 6x^2 - 15x + 20$. Find $f'(x)$ and all values of x for which $f'(x) = 0$. Explain the relationship between these values of x and the graph of $f(x)$.

56. The graph of $y = x^3 - 9x^2 - 16x + 1$ has a slope of 5 at two points. Find the coordinates of the points.

57. Show that for any power function $f(x) = x^n$, we have $f'(1) = n$.

58. If the demand curve is a line, we can write $p = b + mq$, where p is the price of the product, q is the quantity sold at that price, and b and m are constants.

(a) Write the revenue as a function of quantity sold.
(b) Find the marginal revenue function.

59. A ball is dropped from the top of the Empire State building. The height, y, of the ball above the ground (in feet) is given as a function of time, t, (in seconds) by

$$y = 1250 - 16t^2.$$

(a) Find the velocity of the ball at time t. What is the sign of the velocity? Why is this to be expected?
(b) When does the ball hit the ground, and how fast is it going at that time? Give your answer in feet per second and in miles per hour (1 ft/sec = 15/22 mph).

3.2 EXPONENTIAL AND LOGARITHMIC FUNCTIONS

The Exponential Function

What do we expect the graph of the derivative of the exponential function $f(x) = a^x$ to look like? The graph of an exponential function with $a > 1$ is shown in Figure 3.10. The function increases slowly for $x < 0$ and more rapidly for $x > 0$, so the values of f' are small for $x < 0$ and larger for $x > 0$. Since the function is increasing for all values of x, the graph of the derivative must lie above the x-axis. In fact, the graph of f' resembles the graph of f itself. We will see how this observation holds for $f(x) = 2^x$ and $g(x) = 3^x$.

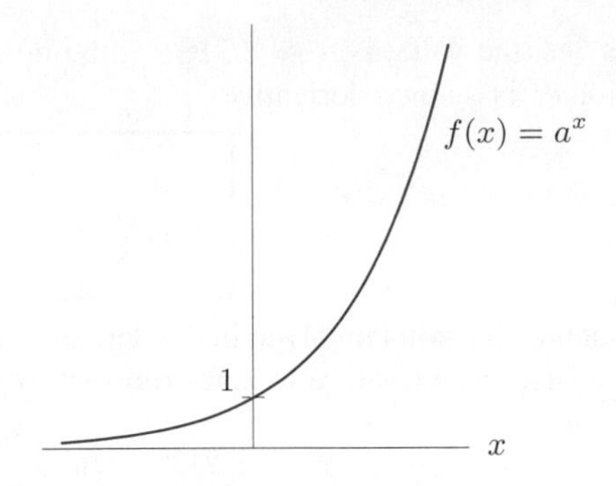

Figure 3.10: $f(x) = a^x$, with $a > 1$

The Derivatives of 2^x and 3^x

On page 101, we estimated the derivative of $f(x) = 2^x$ at $x = 0$:

$$f'(0) \approx 0.693.$$

By estimating the derivative at other values of x, we obtain the graph in Figure 3.11. Since the graph of f' looks like the graph of f stretched vertically, we assume that f' is a multiple of f. Since $f'(0) \approx 0.693 = 0.693 \cdot 1 = 0.693f(0)$, the multiplier is approximately 0.693, which suggests that

$$\frac{d}{dx}(2^x) = f'(x) \approx (0.693)2^x.$$

Similarly, in Figure 3.12, the derivative of $g(x) = 3^x$ is a multiple of g, with multiplier $g'(0) \approx 1.0986$. So

$$\frac{d}{dx}(3^x) = g'(x) \approx (1.0986)3^x.$$

The Derivative of a^x and the Number e

The calculation of the derivative of $f(x) = a^x$, for $a > 0$, is similar to that of 2^x and 3^x. The derivative is again proportional to the original function. When $a = 2$, the constant of proportionality (0.6931) is less than 1, and the derivative is smaller than the original function. When $a = 3$, the constant of proportionality (1.0986) is more than 1, and the derivative is greater than the original function. Is there an in-between case, when derivative and function are exactly equal? In other words:

$$\text{Is there a value of } a \text{ that makes } \frac{d}{dx}(a^x) = a^x?$$

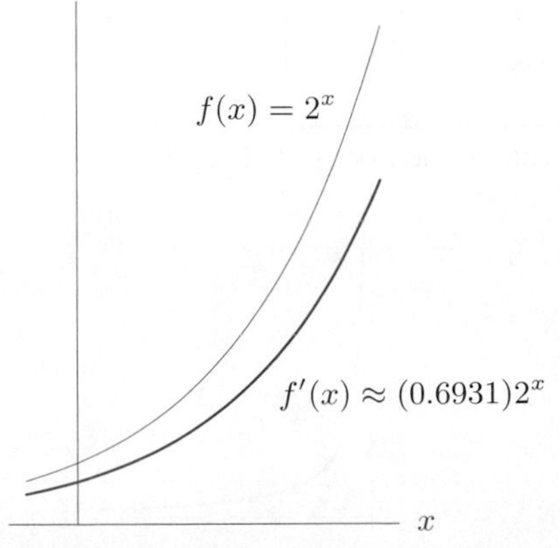

Figure 3.11: Graph of $f(x) = 2^x$ and its derivative

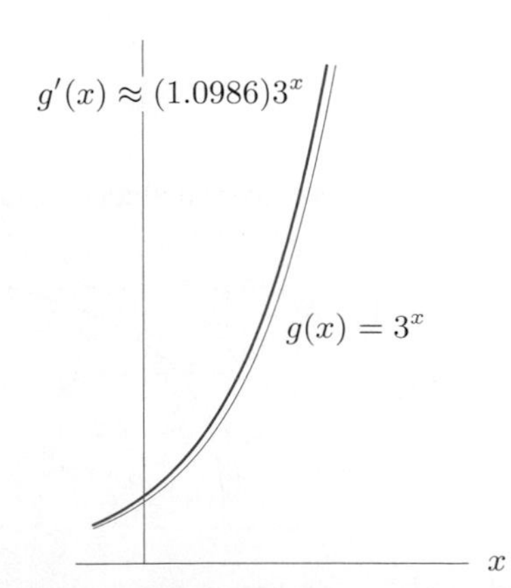

Figure 3.12: Graph of $g(x) = 3^x$ and its derivative

The answer is yes: the value is $a \approx 2.718\ldots$, the number e introduced in Chapter 1. This means that the function e^x is its own derivative:

$$\frac{d}{dx}(e^x) = e^x.$$

It turns out that the constants involved in the derivatives of 2^x and 3^x are natural logarithms. In fact, since $0.6931 \approx \ln 2$ and $1.0986 \approx \ln 3$, we (correctly) guess that

$$\frac{d}{dx}(2^x) = (\ln 2)2^x \quad \text{and} \quad \frac{d}{dx}(3^x) = (\ln 3)3^x.$$

In the Focus on Theory section at the end of this chapter, we show that, in general:

The Exponential Rule

For any positive constant a,

$$\frac{d}{dx}(a^x) = (\ln a)a^x.$$

Since $\ln a$ is a constant, the derivative of a^x is proportional to a^x. Many quantities have rates of change that are proportional to themselves; for example, the simplest model of population growth has this property. The fact that the constant of proportionality is 1 when $a = e$ makes e a particularly useful base for exponential functions.

Example 1 Differentiate $2 \cdot 3^x + 5e^x$.

Solution We have $\dfrac{d}{dx}(2 \cdot 3^x + 5e^x) = 2\dfrac{d}{dx}(3^x) + 5\dfrac{d}{dx}(e^x) = 2 \ln 3 \cdot 3^x + 5e^x.$

The Derivative of ln x

What does the graph of the derivative of the logarithmic function $f(x) = \ln x$ look like? Figure 3.13 shows that $\ln x$ is increasing, so its derivative is positive. The graph of $f(x) = \ln x$ is concave down, so the derivative is decreasing. Furthermore, the slope of $f(x) = \ln x$ is very large near $x = 0$ and very small for large x, so the derivative tends to $+\infty$ for x near 0 and tends to 0 for very large x. See Figure 3.14. It turns out that

$$\frac{d}{dx}(\ln x) = \frac{1}{x}.$$

We give an algebraic justification for this rule in the Focus on Theory section on page 171.

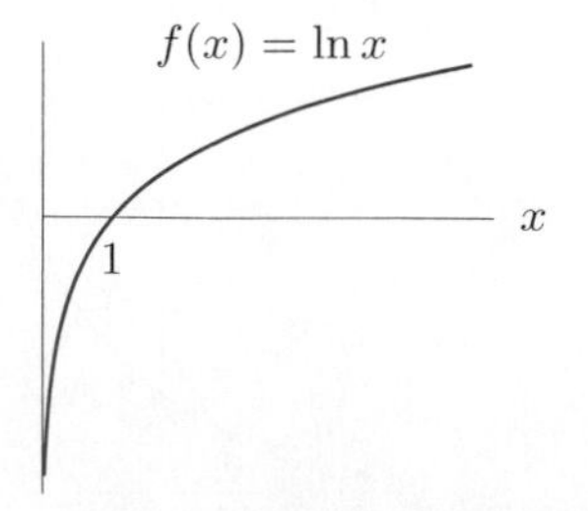

Figure 3.13: Graph of $f(x) = \ln x$

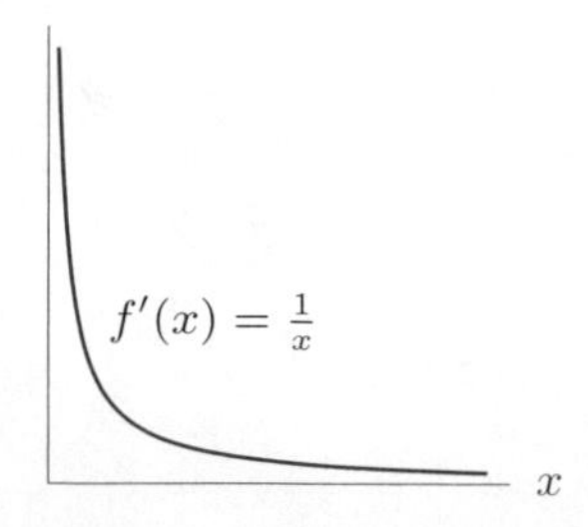

Figure 3.14: Graph of the derivative of $f(x) = \ln x$

Example 2 Differentiate $y = 5\ln t + 7e^t - 4t^2 + 12$.

Solution We have

$$\begin{aligned}\frac{d}{dt}(5\ln t + 7e^t - 4t^2 + 12) &= 5\frac{d}{dt}(\ln t) + 7\frac{d}{dt}(e^t) - 4\frac{d}{dt}(t^2) + \frac{d}{dt}(12)\\ &= 5\left(\frac{1}{t}\right) + 7(e^t) - 4(2t) + 0\\ &= \frac{5}{t} + 7e^t - 8t.\end{aligned}$$

Using the Derivative Formulas

Example 3 In Chapter 1, we saw that the population of McAllen, Texas, P, can be modeled by

$$P = 570(1.037)^t,$$

where P is thousands of people, and t is years since the start of 2000. At what rate was the population growing at the beginning of 2005? Give units with your answer.

Solution The instantaneous rate of growth is the derivative, so we want dP/dt when $t = 5$. We have:

$$\frac{dP}{dt} = \frac{d}{dt}(570(1.037)^t) = 570(\ln 1.037)(1.037)^t = 20.709(1.037)^t.$$

Substituting $t = 5$ gives

$$20.709(1.037)^5 = 24.835.$$

The population of McAllen was growing at a rate of about 24.835 thousand, or 24,835, people per year at the start of 2005.

Example 4 Find the equation of the tangent line to the graph of $f(x) = \ln x$ at the point where $x = 2$. Draw a graph with $f(x)$ and the tangent line on the same axes.

Solution Since $f'(x) = 1/x$, the slope of the tangent line at $x = 2$ is $f'(2) = 1/2 = 0.5$. When $x = 2$, $y = \ln 2 = 0.693$, so a point on the tangent line is $(2, 0.693)$. Substituting into the equation for a line, we have:

$$\begin{aligned}y - 0.693 &= 0.5(x - 2)\\ y &= -0.307 + 0.5x.\end{aligned}$$

The equation of the tangent line is $y = -0.307 + 0.5x$. See Figure 3.15.

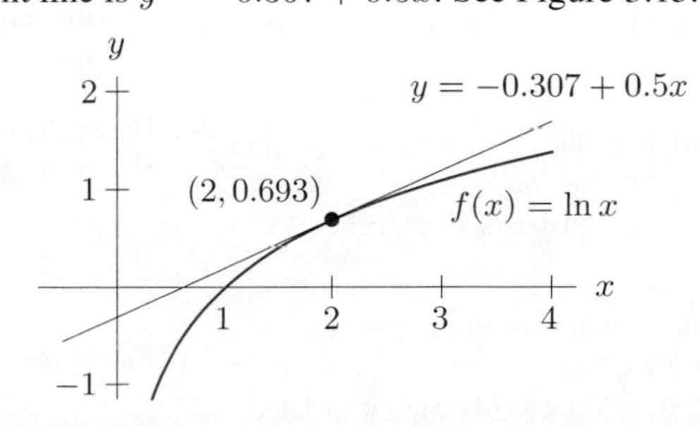

Figure 3.15: Graph of $f(x) = \ln x$ and a tangent line

Problems for Section 3.2

Differentiate the functions in Problems 1–22. Assume that A, B, and C are constants.

1. $f(x) = 2e^x + x^2$
2. $P = 3t^3 + 2e^t$
3. $y = 5t^2 + 4e^t$
4. $f(x) = x^3 + 3^x$
5. $y = 2^x + \dfrac{2}{x^3}$
6. $y = 5 \cdot 5^t + 6 \cdot 6^t$
7. $f(x) = 2^x + 2 \cdot 3^x$
8. $y = 4 \cdot 10^x - x^3$
9. $y = 3x - 2 \cdot 4^x$
10. $y = 5 \cdot 2^x - 5x + 4$
11. $P(t) = 3000(1.02)^t$
12. $P(t) = 12.41(0.94)^t$
13. $P(t) = Ce^t$.
14. $y = B + Ae^t$
15. $f(x) = Ae^x - Bx^2 + C$
16. $y = 10^x + \dfrac{10}{x}$
17. $R = 3 \ln q$
18. $D = 10 - \ln p$
19. $y = t^2 + 5 \ln t$
20. $R(q) = q^2 - 2 \ln q$
21. $y = x^2 + 4x - 3 \ln x$
22. $f(t) = Ae^t + B \ln t$

23. For $f(t) = 4 - 2e^t$, find $f'(-1)$, $f'(0)$, and $f'(1)$. Graph $f(t)$, and draw tangent lines at $t = -1$, $t = 0$, and $t = 1$. Do the slopes of the lines match the derivatives you found?

24. Find the equation of the tangent line to the graph of $y = 3^x$ at $x = 1$. Check your work by sketching a graph of the function and the tangent line on the same axes.

25. **(a)** Find the slope of the graph of $f(x) = 1 - e^x$ at the point where it crosses the x-axis.
    (b) Find the equation of the tangent line to the curve at this point.

26. Worldwide production of solar power, in megawatts, can be modeled by $f(t) = 1040(1.3)^t$, where t is years[1] since 2000. Find $f(0)$, $f'(0)$, $f(15)$, and $f'(15)$. Give units and interpret your answers in terms of solar power.

27. During the 1990s, the population of Hungary was approximated by

$$P = 10.8(0.994)^t,$$

where P is in millions and t is in years since 1990. Assume the trend continues.

(a) What does this model predict for the population of Hungary in the year 2010?
(b) How fast (in people/year) does this model predict Hungary's population will be decreasing in 2010?

28. Certain pieces of antique furniture increased very rapidly in value in the 1990s and 2000s. For example, the value of a particular rocking chair is well approximated by

$$V = 75(1.35)^t,$$

where V is in dollars and t is the number of years since 1995. Find the rate, in dollars per year, at which the value is increasing.

29. With a yearly inflation rate of 5%, prices are given by

$$P = P_0(1.05)^t,$$

where P_0 is the price in dollars when $t = 0$ and t is time in years. Suppose $P_0 = 1$. How fast (in cents/year) are prices rising when $t = 10$?

30. For the cost function $C = 1000 + 300 \ln q$ (in dollars), find the cost and the marginal cost at a production level of 500. Interpret your answers in economic terms.

31. The *Global 2000 Report* gave the world's population, P, as 4.1 billion in 1975 and growing at 2% annually.
    (a) Give a formula for P in terms of time, t, measured in years since 1975.
    (b) Find $\dfrac{dP}{dt}$, $\left.\dfrac{dP}{dt}\right|_{t=0}$, and $\left.\dfrac{dP}{dt}\right|_{t=25}$. What do each of these represent in practical terms?

32. In 1990, the population of Mexico was about 84 million and growing at 2.6% annually, while the population of the US was about 250 million and growing at 0.7% annually. Which population was growing faster, if we measure growth rates in people/year? Explain your answer.

33. **(a)** Find the equation of the tangent line to $y = \ln x$ at $x = 1$.
    (b) Use it to calculate approximate values for $\ln(1.1)$ and $\ln(2)$.
    (c) Using a graph, explain whether the approximate values are smaller or larger than the true values. Would the same result have held if you had used the tangent line to estimate $\ln(0.9)$ and $\ln(0.5)$? Why?

34. Using the equation of the tangent line to the graph of e^x at $x = 0$, show that

$$e^x \geq 1 + x$$

for all values of x. A sketch may be helpful.

35. Find the value of c in Figure 3.16, where the line l tangent to the graph of $y = 2^x$ at $(0, 1)$ intersects the x-axis.

[1] www.bp.com, accessed May 29, 2005.

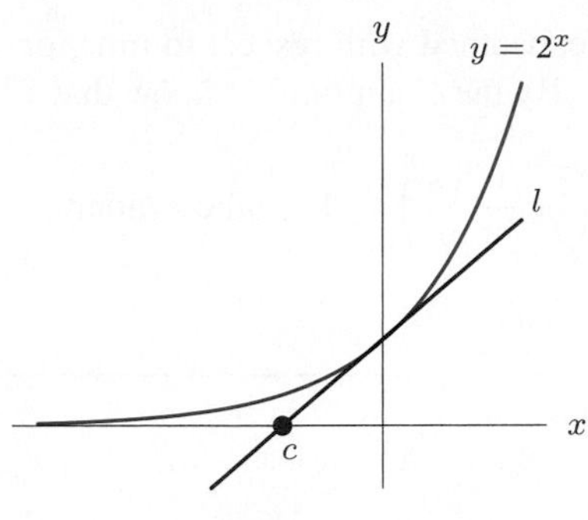

Figure 3.16

36. Find the quadratic polynomial $g(x) = ax^2 + bx + c$ which best fits the function $f(x) = e^x$ at $x = 0$, in the sense that

$$g(0) = f(0), \text{ and } g'(0) = f'(0), \text{ and } g''(0) = f''(0).$$

Using a computer or calculator, sketch graphs of f and g on the same axes. What do you notice?

3.3 THE CHAIN RULE

We now see how to differentiate composite functions such as $f(t) = \ln(3t)$ and $g(x) = e^{-x^2}$.

The Derivative of a Composition of Functions

Suppose $y = f(z)$ with $z = g(t)$ for some inside function g and outside function f, where f and g are differentiable. A small change in t, called Δt, generates a small change in z, called Δz. In turn, Δz generates a small change in y, called Δy. Provided Δt and Δz are not zero, we can say

$$\frac{\Delta y}{\Delta t} = \frac{\Delta y}{\Delta z} \cdot \frac{\Delta z}{\Delta t}.$$

Since the derivative $\frac{dy}{dt}$ is the limit of the quotient $\frac{\Delta y}{\Delta t}$ as Δt gets smaller and smaller, this suggests

The Chain Rule

If $y = f(z)$ and $z = g(t)$ are differentiable, then the derivative of $y = f(g(t))$ is given by

$$\frac{dy}{dt} = \frac{dy}{dz} \cdot \frac{dz}{dt}.$$

In words, the derivative of a composite function is the derivative of the outside function times the derivative of the inside function:

$$\frac{d}{dt}(f(g(t)) = f'(g(t)) \cdot g'(t).$$

The following example shows us how to interpret the chain rule in practical terms.

Example 1 The amount of gas, G, in gallons, consumed by a car depends on the distance traveled, s, in miles, and s depends on the time, t, in hours. If 0.05 gallons of gas is consumed for each mile traveled, and the car is traveling at 30 miles/hr, how fast is gas being consumed? Give units.

Solution We expect the rate of gas consumption to be in gallons/hr. We are told that

$$\text{Rate gas is consumed with respect to distance} = \frac{dG}{ds} = 0.05 \text{ gallons/mile}$$

$$\text{Rate distance is increasing with respect to time} = \frac{ds}{dt} = 30 \text{ miles/hr}.$$

We want to calculate the rate at which gas is being consumed with respect to time, or dG/dt. We think of G as a function of s, and s as a function of t. By the chain rule we know that

$$\frac{dG}{dt} = \frac{dG}{ds} \cdot \frac{ds}{dt} = \left(0.05 \frac{\text{gallons}}{\text{mile}}\right) \cdot \left(30 \frac{\text{miles}}{\text{hour}}\right) = 1.5 \text{ gallons/hour.}$$

Thus, gas is being consumed at a rate of 1.5 gallons/hour.

The Chain Rule for Functions Given by Formulas

In order to use the chain rule to differentiate a composite function, we first rewrite the function using a new variable z to represent the inside function:

$$y = (t+1)^4 \quad \text{is the same as} \quad y = z^4 \quad \text{where} \quad z = t+1.$$

Example 2 Use a new variable z for the inside function to express each of the following as a composite function:
(a) $y = \ln(3t)$ (b) $P = e^{-0.03t}$ (c) $w = 5(2r+3)^2$.

Solution
(a) The inside function is $3t$, so we have $y = \ln z$ with $z = 3t$.
(b) The inside function is $-0.03t$, so we have $P = e^z$ with $z = -0.03t$.
(c) The inside function is $2r+3$, so we have $w = 5z^2$ with $z = 2r+3$.

Example 3 Find the derivative of the following functions: (a) $y = (4t^2+1)^7$ (b) $P = e^{3t}$.

Solution
(a) Here $z = 4t^2+1$ is the inside function; $y = z^7$ is the outside function. Since $dy/dz = 7z^6$ and $dz/dt = 8t$, we have

$$\frac{dy}{dt} = \frac{dy}{dz} \cdot \frac{dz}{dt} = (7z^6)(8t) = 7(4t^2+1)^6(8t) = 56t(4t^2+1)^6.$$

(b) Let $z = 3t$ and $P = e^z$. Then $dP/dz = e^z$ and $dz/dt = 3$, so

$$\frac{dP}{dt} = \frac{dP}{dz} \cdot \frac{dz}{dt} = e^z \cdot 3 = e^{3t} \cdot 3 = 3e^{3t}.$$

The derivative rules give us

$$\frac{d}{dt}(t^n) = nt^{n-1} \qquad \frac{d}{dt}(e^t) = e^t \qquad \frac{d}{dt}(\ln t) = \frac{1}{t}.$$

Using the chain rule in addition, we have the following results.

> If z is a differentiable function of t, then
>
> $$\frac{d}{dt}(z^n) = nz^{n-1}\frac{dz}{dt}, \qquad \frac{d}{dt}(e^z) = e^z\frac{dz}{dt}, \qquad \frac{d}{dt}(\ln z) = \frac{1}{z}\frac{dz}{dt}$$

Example 4 Differentiate (a) $(3t^3 - t)^5$ (b) $\ln(q^2 + 1)$ (c) e^{-x^2}.

Solution (a) Let $z = 3t^3 - t$, giving

$$\frac{d}{dt}(3t^3 - t)^5 = \frac{d}{dt}(z^5) = 5z^4\frac{dz}{dt} = 5(3t^3 - t)^4(9t^2 - 1).$$

(b) We have $z = q^2 + 1$, so

$$\frac{d}{dq}(\ln(q^2 + 1)) = \frac{d}{dq}(\ln z) = \frac{1}{z}\frac{dz}{dq} = \frac{1}{q^2 + 1}(2q).$$

(c) Since $z = -x^2$, the derivative is

$$\frac{d}{dx}(e^{-x^2}) = \frac{d}{dx}(e^z) = e^z\frac{dz}{dx} = e^{-x^2}(-2x) = -2xe^{-x^2}.$$

As we see in the following example, it is often faster to use the chain rule without introducing the new variable, z.

Example 5 Differentiate
(a) $(x^2 + 4)^3$ (b) $5\ln(2t^2 + 3)$ (c) $\sqrt{1 + 2e^{5t}}$

Solution (a) We have

$$\begin{aligned}\frac{d}{dx}\left((x^2 + 4)^3\right) &= 3(x^2 + 4)^2 \cdot \frac{d}{dx}(x^2 + 4)\\ &= 3(x^2 + 4)^2 \cdot 2x\\ &= 6x(x^2 + 4)^2.\end{aligned}$$

(b) We have

$$\begin{aligned}\frac{d}{dt}\left(5\ln(2t^2 + 3)\right) &= 5 \cdot \frac{1}{2t^2 + 3} \cdot \frac{d}{dt}(2t^2 + 3)\\ &= 5 \cdot \frac{1}{2t^2 + 3} \cdot 4t\\ &= \frac{20t}{2t^2 + 3}.\end{aligned}$$

(c) Here we use the chain rule twice, giving

$$\begin{aligned}\frac{d}{dt}\left((1 + 2e^{5t})^{1/2}\right) &= \frac{1}{2}(1 + 2e^{5t})^{-1/2} \cdot \frac{d}{dl}(1 + 2e^{5t})\\ &= \frac{1}{2}(1 + 2e^{5t})^{-1/2} \cdot 2e^{5t} \cdot \frac{d}{dt}(5t)\\ &= \frac{1}{2}(1 + 2e^{5t})^{-1/2} \cdot 2e^{5t} \cdot 5\\ &= \frac{5e^{5t}}{\sqrt{1 + 2e^{5t}}}.\end{aligned}$$

Example 6 Let $h(x) = f(g(x))$ and $k(x) = g(f(x))$. Use Figure 3.17 to estimate: (a) $h'(1)$ (b)$k'(2)$

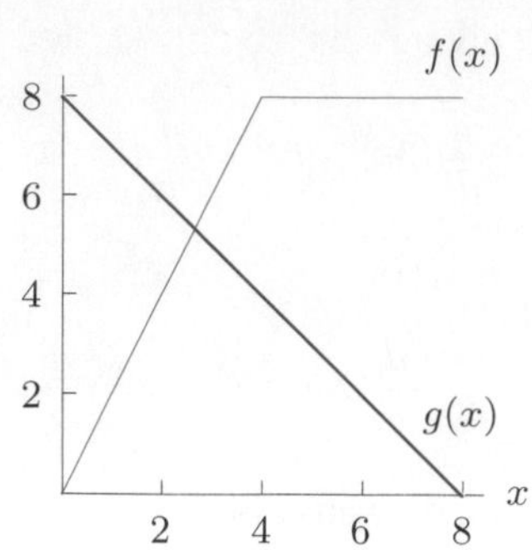

Figure 3.17: Graphs of f and g for Example 6

Solution (a) The chain rule tells us that $h'(x) = f'(g(x)) \cdot g'(x)$, so

$$\begin{aligned} h'(1) &= f'(g(1)) \cdot g'(1) \\ &= f'(7) \cdot g'(1) \\ &= 0 \cdot (-1) \\ &= 0. \end{aligned}$$

We use the slopes of the lines in Figure 3.17 to find the derivatives $f'(7) = 0$ and $g'(1) = -1$.

(b) The chain rule tells us that $k'(x) = g'(f(x)) \cdot f'(x)$, so

$$\begin{aligned} k'(2) &= g'(f(2)) \cdot f'(2) \\ &= g'(4) \cdot f'(2) \\ &= (-1) \cdot 2 \\ &= -2. \end{aligned}$$

We use slopes to compute the derivatives $g'(4) = -1$ and $f'(2) = 2$.

Since functions of the form e^{kt} where k is a constant are often useful, we calculate the derivative of e^{kt}. We have $z = kt$, and so $dz/dt = k$. Thus, if k is a constant,

$$\boxed{\frac{d}{dt}(e^{kt}) = ke^{kt}.}$$

Example 7 Find the derivative of $P = 5 + 3x^2 - 7e^{-0.2x}$.

Solution The derivative is

$$\frac{dP}{dx} = 0 + 3(2x) - 7(-0.2e^{-0.2x}) = 6x + 1.4e^{-0.2x}.$$

Example 8 Suppose $1000 is deposited into a bank account that pays 8% annual interest, compounded continuously.

(a) Find a formula $f(t)$ for the balance t years after the initial deposit.
(b) Find $f(10)$ and $f'(10)$ and explain what your answers mean in terms of money.

Solution (a) The balance is $f(t) = 1000e^{0.08t}$.
(b) Substituting $t = 10$ gives

$$f(10) = 1000e^{(0.08)(10)} = 2225.54.$$

This means that the balance is $2225.54 after 10 years.

To find $f'(10)$, we compute $f'(t) = 1000(0.08e^{0.08t}) = 80e^{0.08t}$. Therefore,

$$f'(10) = 80e^{(0.08)(10)} = 178.04.$$

This means that after 10 years, the balance is growing at the rate of about $178 per year.

Problems for Section 3.3

Find the derivative of the functions in Problems 1–34.

1. $(4x^2 + 1)^7$
2. $f(x) = (x + 1)^{99}$
3. $R = (q^2 + 1)^4$
4. $w = (t^2 + 1)^{100}$
5. $w = (t^3 + 1)^{100}$
6. $w = (5r - 6)^3$
7. $y = \sqrt{s^3 + 1}$
8. $f(t) = e^{3t}$
9. $y = e^{0.7t}$
10. $y = e^{-4t}$
11. $P = e^{-0.2t}$
12. $P = 50e^{-0.6t}$
13. $P = 200e^{0.12t}$
14. $y = 12 - 3x^2 + 2e^{3x}$
15. $C = 12(3q^2 - 5)^3$
16. $f(x) = 6e^{5x} + e^{-x^2}$
17. $y = 5e^{5t+1}$
18. $w = e^{-3t^2}$
19. $w = e^{\sqrt{s}}$
20. $y = \ln(5t + 1)$
21. $f(x) = \ln(1 - x)$
22. $f(t) = \ln(t^2 + 1)$
23. $f(x) = \ln(1 - e^{-x})$
24. $f(x) = \ln(e^x + 1)$
25. $f(t) = 5\ln(5t + 1)$
26. $g(t) = \ln(4t + 9)$
27. $y = 5 + \ln(3t + 2)$
28. $Q = 100(t^2 + 5)^{0.5}$
29. $y = 5x + \ln(x + 2)$
30. $y = (5 + e^x)^2$
31. $P = (1 + \ln x)^{0.5}$
32. $\sqrt{e^x + 1}$
33. $f(x) = \sqrt{1 - x^2}$
34. $f(\theta) = (e^\theta + e^{-\theta})^{-1}$
35. Find the equation of the tangent line to $f(x) = (x - 1)^3$ at the point where $x = 2$.
36. Find the equation of the tangent line to $y = e^{-2t}$ at $t = 0$. Check by sketching the graphs of $y = e^{-2t}$ and the tangent line on the same axes.
37. Find the equation of the tangent line to $f(x) = 10e^{-0.2x}$ at $x = 4$.
38. A firm estimates that the total revenue, R, received from the sale of q goods is given by
$$R = \ln(1 + 1000q^2).$$
Calculate the marginal revenue when $q = 10$.
39. According to the US Census, the world population P, in billions, is approximately
$$P = 6.342e^{0.011t},$$
where t is in years since January 1, 2004. At what rate was the world's population increasing on that date? Give your answer in millions of people per year.
40. The cost of producing a quantity, q, of a product is given by
$$C(q) = 1000 + 30e^{0.05q} \quad \text{dollars.}$$
Find the cost and the marginal cost when $q = 50$. Interpret these answers in economic terms.
41. The demand curve for a product is given by
$$q = f(p) = 10{,}000e^{-0.25p},$$
where q is the quantity sold and p is the price of the product, in dollars. Find $f(2)$ and $f'(2)$. Explain in economic terms what information each of these answers gives you.
42. At a time t hours after it was administered, the concentration of a drug in the body is $f(t) = 27e^{-0.14t}$ ng/ml. What is the concentration 4 hours after it was administered? At what rate is the concentration changing at that time?

43. The balance, $\$B$, in a bank account t years after a deposit of \$5000 is given by $B = 5000e^{0.08t}$. At what rate is the balance in the account changing at $t = 5$ years? Use units to interpret your answer in financial terms.

44. With time, t, in minutes, the temperature, H, in degrees Celsius, of a bottle of water put in the refrigerator at $t = 0$ is given by

$$H = 4 + 16e^{-0.02t}.$$

How fast is the water cooling initially? After 10 minutes? Give units.

45. A fish population is approximated by $P(t) = 10e^{0.6t}$, where t is in months. Calculate and use units to explain what each of the following tells us about the population:

(a) $P(12)$ **(b)** $P'(12)$

46. The distance, s, of a moving body from a fixed point is given as a function of time by $s = 20e^{t/2}$. Find the velocity, v, of the body as a function of t.

47. If you invest P dollars in a bank account at an annual interest rate of $r\%$, then after t years you will have B dollars, where

$$B = P\left(1 + \frac{r}{100}\right)^t.$$

(a) Find dB/dt, assuming P and r are constant. In terms of money, what does dB/dt represent?

(b) Find dB/dr, assuming P and t are constant. In terms of money, what does dB/dr represent?

48. Some economists suggest that an extra year of education increases a person's wages, on average, by about 14%. Assume you could make \$10 per hour with your current level of education and that inflation increases wages at a continuous rate of 3.5% per year.

(a) How much would you make per hour with four additional years of education?

(b) What is the difference between your wages in 20 years time with and without the additional four years of education?

(c) Is the difference you found in part (b) increasing with time? If so, at what rate? (Assume the number of additional years of education stays fixed at four.)

In Problems 49–52, use Figure 3.18 to evaluate the derivative.

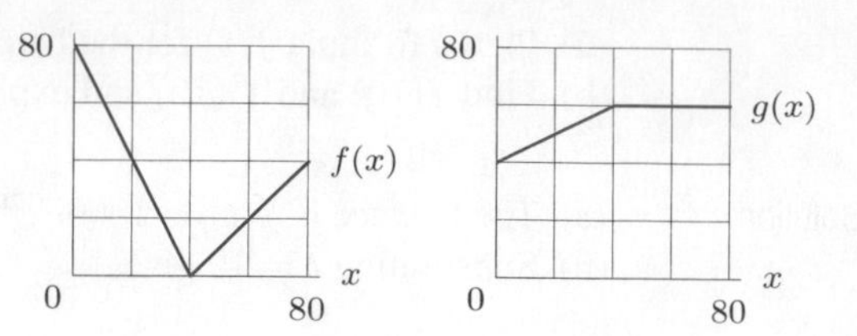

Figure 3.18

49. $\frac{d}{dx}f(g(x))|_{x=30}$

50. $\frac{d}{dx}f(g(x))|_{x=70}$

51. $\frac{d}{dx}g(f(x))|_{x=30}$

52. $\frac{d}{dx}g(f(x))|_{x=70}$

For Problems 53–56, let $h(x) = f(g(x))$ and $k(x) = g(f(x))$. Use Figure 3.19 to estimate the derivatives.

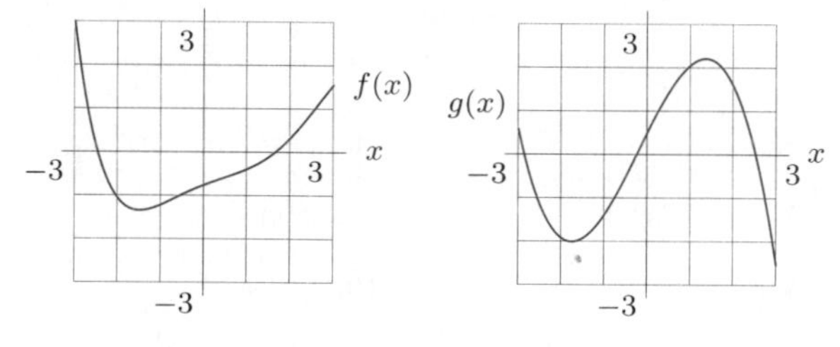

Figure 3.19

53. $h'(1)$

54. $k'(1)$

55. $h'(2)$

56. $k'(2)$

3.4 THE PRODUCT AND QUOTIENT RULES

This section shows how to find the derivatives of products and quotients of functions.

The Product Rule

Suppose we know the derivatives of $f(x)$ and $g(x)$ and want to calculate the derivative of the product, $f(x)g(x)$. We start by looking at an example. Let $f(x) = x$ and $g(x) = x^2$. Then

$$f(x)g(x) = x \cdot x^2 = x^3,$$

so the derivative of the product is $3x^2$. Notice that the derivative of the product is *not* equal to the product of the derivatives, since $f'(x) = 1$ and $g'(x) = 2x$, so $f'(x)g'(x) = (1)(2x) = 2x$. In general, we have the following rule, which is justified on page 172 in the Focus on Theory section.

The Product Rule

If $u = f(x)$ and $v = g(x)$ are differentiable functions, then

$$(fg)' = f'g + fg'.$$

The product rule can also be written

$$\frac{d(uv)}{dx} = \frac{du}{dx} \cdot v + u \cdot \frac{dv}{dx}.$$

In words:

The derivative of a product is the derivative of the first times the second, plus the first times the derivative of the second.

We check that this rule gives the correct answers for $f(x) = x$ and $g(x) = x^2$. The derivative of $f(x)g(x)$ is

$$f'(x)g(x) + f(x)g'(x) = 1(x^2) + x(2x) = x^2 + 2x^2 = 3x^2.$$

This is the answer we expect for the derivative of $f(x)g(x) = x \cdot x^2 = x^3$.

Example 1 Differentiate (a) x^2e^{2x} (b) $t^3 \ln(t+1)$ (c) $(3x^2 + 5x)e^x$.

Solution (a) Using the product rule, we have

$$\begin{aligned}\frac{d}{dx}(x^2e^{2x}) &= \frac{d}{dx}(x^2) \cdot e^{2x} + x^2\frac{d}{dx}(e^{2x}) \\ &= (2x)e^{2x} + x^2(2e^{2x}) \\ &= 2xe^{2x} + 2x^2e^{2x}.\end{aligned}$$

(b) Differentiating using the product rule gives

$$\begin{aligned}\frac{d}{dt}(t^3\ln(t+1)) &= \frac{d}{dt}(t^3) \cdot \ln(t+1) + t^3\frac{d}{dt}(\ln(t+1)) \\ &= (3t^2)\ln(t+1) + t^3\left(\frac{1}{t+1}\right) \\ &= 3t^2\ln(t+1) + \frac{t^3}{t+1}.\end{aligned}$$

(c) The product rule gives

$$\begin{aligned}\frac{d}{dx}((3x^2+5x)e^x) &= \left(\frac{d}{dx}(3x^2+5x)\right)e^x + (3x^2+5x)\frac{d}{dx}(e^x) \\ &= (6x+5)e^x + (3x^2+5x)e^x \\ &= (3x^2+11x+5)e^x.\end{aligned}$$

Example 2 Find the derivative of $C = \dfrac{e^{2t}}{t}$.

Solution We write $C = e^{2t}t^{-1}$ and use the product rule:

$$\begin{aligned}\frac{d}{dt}(e^{2t}t^{-1}) &= \frac{d}{dt}(e^{2t}) \cdot t^{-1} + e^{2t}\frac{d}{dt}(t^{-1}) \\ &= (2e^{2t}) \cdot t^{-1} + e^{2t}(-1)t^{-2} \\ &= \frac{2e^{2t}}{t} - \frac{e^{2t}}{t^2}.\end{aligned}$$

Example 3 A demand curve for a product has the equation $p = 80e^{-0.003q}$, where p is price and q is quantity.

(a) Find the revenue as a function of quantity sold.
(b) Find the marginal revenue function.

Solution

(a) Since Revenue = Price × Quantity, we have $R = pq = (80e^{-0.003q})q = 80qe^{-0.003q}$.
(b) The marginal revenue function is the derivative of revenue with respect to quantity. The product rule gives

$$\begin{aligned}\text{Marginal Revenue} &= \frac{d}{dq}(80qe^{-0.003q}) \\ &= \left(\frac{d}{dq}(80q)\right)e^{-0.003q} + 80q\left(\frac{d}{dq}(e^{-0.003q})\right) \\ &= (80)e^{-0.003q} + 80q(-0.003e^{-0.003q}) \\ &= (80 - 0.24q)e^{-0.003q}.\end{aligned}$$

The Quotient Rule

Suppose we want to differentiate a function of the form $Q(x) = f(x)/g(x)$. (Of course, we have to avoid points where $g(x) = 0$.) We want a formula for Q' in terms of f' and g'. We have the following rule, which is justified on page 172 of the Focus on Theory section.

The Quotient Rule

If $u = f(x)$ and $v = g(x)$ are differentiable functions, then

$$\left(\frac{f}{g}\right)' = \frac{f'g - fg'}{g^2},$$

or equivalently,

$$\frac{d}{dx}\left(\frac{u}{v}\right) = \frac{\frac{du}{dx}\cdot v - u\cdot\frac{dv}{dx}}{v^2}.$$

In words:

The derivative of a quotient is the derivative of the numerator times the denominator minus the numerator times the derivative of the denominator, all over the denominator squared.

Example 4 Differentiate (a) $\dfrac{5x^2}{x^3+1}$ (b) $\dfrac{1}{1+e^x}$ (c) $\dfrac{e^x}{x^2}$.

Solution (a) Using the quotient rule

$$\frac{d}{dx}\left(\frac{5x^2}{x^3+1}\right) = \frac{\left(\frac{d}{dx}(5x^2)\right)(x^3+1) - 5x^2\frac{d}{dx}(x^3+1)}{(x^3+1)^2} = \frac{10x(x^3+1) - 5x^2(3x^2)}{(x^3+1)^2}$$

$$= \frac{-5x^4 + 10x}{(x^3+1)^2}.$$

(b) Differentiating using the quotient rule yields

$$\frac{d}{dx}\left(\frac{1}{1+e^x}\right) = \frac{\left(\frac{d}{dx}(1)\right)(1+e^x) - 1\frac{d}{dx}(1+e^x)}{(1+e^x)^2} = \frac{0(1+e^x) - 1(0+e^x)}{(1+e^x)^2}$$
$$= \frac{-e^x}{(1+e^x)^2}.$$

(c) The quotient rule gives

$$\frac{d}{dx}\left(\frac{e^x}{x^2}\right) = \frac{\left(\frac{d}{dx}(e^x)\right)x^2 - e^x\left(\frac{d}{dx}(x^2)\right)}{(x^2)^2} = \frac{e^x x^2 - e^x(2x)}{x^4}$$
$$= e^x\left(\frac{x^2-2x}{x^4}\right) = e^x\left(\frac{x-2}{x^3}\right).$$

Problems for Section 3.4

1. If $f(x) = (2x+1)(3x-2)$, find $f'(x)$ two ways: by using the product rule and by multiplying out. Do you get the same result?

2. If $f(x) = x^2(x^3+5)$, find $f'(x)$ two ways: by using the product rule and by multiplying out before taking the derivative. Do you get the same result? Should you?

For Problems 3–33, find the derivative. Assume that a, b, c, and k are constants.

3. $f(x) = xe^x$

4. $f(t) = te^{-2t}$

5. $y = 5xe^{x^2}$

6. $y = t^2(3t+1)^3$

7. $y = x\ln x$

8. $y = (t^2+3)e^t$

9. $z = (3t+1)(5t+2)$

10. $y = (t^3 - 7t^2 + 1)e^t$

11. $P = t^2 \ln t$

12. $f(t) = \dfrac{5}{t} + \dfrac{6}{t^2}$

13. $f(x) = \dfrac{x^2+3}{x}$

14. $R = 3qe^{-q}$

15. $y = te^{-t^2}$

16. $f(z) = \sqrt{z}e^{-z}$

17. $g(p) = p\ln(2p+1)$

18. $f(t) = te^{5-2t}$

19. $f(w) = (5w^2+3)e^{w^2}$

20. $y = x \cdot 2^x$

21. $w = (t^3+5t)(t^2-7t+2)$

22. $z = (te^{3t} + e^{5t})^9$

23. $f(x) = \dfrac{x}{e^x}$

24. $w = \dfrac{3z}{1+2z}$

25. $z = \dfrac{1-t}{1+t}$

26. $y = \dfrac{e^x}{1+e^x}$

27. $w = \dfrac{3y+y^2}{5+y}$

28. $y = \dfrac{1+z}{\ln z}$

29. $f(t) = ae^{bt}$

30. $f(x) = (ax^2+b)^3$

31. $f(x) = axe^{-bx}$

32. $f(x) = \dfrac{ax+b}{cx+k}$

33. $g(\alpha) = e^{\alpha e^{-2\alpha}}$

34. If $f(x) = (3x+8)(2x-5)$, find $f'(x)$ and $f''(x)$.

35. Find the equation of the tangent line to the graph of $f(x) = x^2e^{-x}$ at $x = 0$. Check by graphing this function and the tangent line on the same axes.

36. Find the equation of the tangent line to the graph of $f(x) = \dfrac{2x-5}{x+1}$ at the point at which $x = 0$.

37. If p is price in dollars and q is quantity, demand for a product is given by

$$q = 5000e^{-0.08p}.$$

(a) What quantity is sold at a price of \$10?

(b) Find the derivative of demand with respect to price when the price is \$10 and interpret your answer in terms of demand.

38. The demand for a product is given in Problem 37. Find the revenue and the derivative of revenue with respect to price at a price of \$10. Interpret your answers in economic terms.

39. The quantity of a drug, Q mg, present in the body t hours after an injection of the drug is given is

$$Q = f(t) = 100te^{-0.5t}.$$

Find $f(1)$, $f'(1)$, $f(5)$, and $f'(5)$. Give units and interpret the answers.

40. The quantity demanded of a certain product, q, is given in terms of p, the price, by

$$q = 1000e^{-0.02p}$$

(a) Write revenue, R, as a function of price.
(b) Find the rate of change of revenue with respect to price.
(c) Find the revenue and rate of change of revenue with respect to price when the price is \$10. Interpret your answers in economic terms.

41. A drug concentration curve is given by $C = f(t) = 20te^{-0.04t}$, with C in mg/ml and t in minutes.

(a) Graph C against t. Is $f'(15)$ positive or negative? Is $f'(45)$ positive or negative? Explain.
(b) Find $f(30)$ and $f'(30)$ analytically. Interpret them in terms of the concentration of the drug in the body.

42. If $\frac{d}{dt}(tf(t)) = 1 + f(t)$, what is $f'(t)$?

43. Let $h(x) = t(x)s(x)$ and $p(x) = t(x)/s(x)$, where $t(x)$ and $s(x)$ are shown in Figure 3.20. Estimate:

(a) $h'(1)$ **(b)** $h'(0)$ **(c)** $p'(0)$

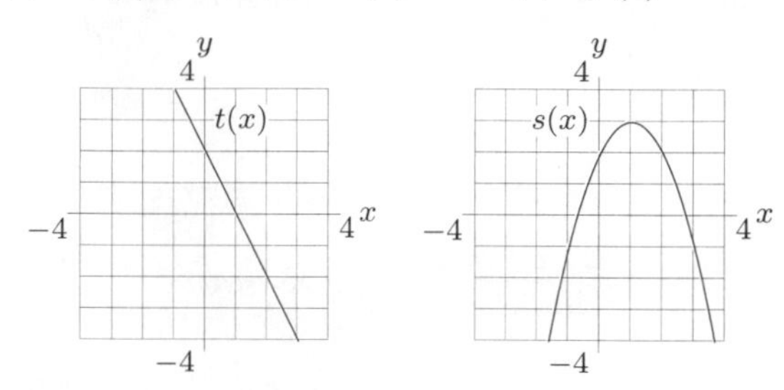

Figure 3.20

44. The quantity, q, of a certain skateboard sold depends on the selling price, p, in dollars, so we write $q = f(p)$. You are given that $f(140) = 15{,}000$ and $f'(140) = -100$.

(a) What do $f(140) = 15{,}000$ and $f'(140) = -100$ tell you about the sales of skateboards?
(b) The total revenue, R, earned by the sale of skateboards is given by $R = pq$. Find $\left.\frac{dR}{dp}\right|_{p=140}$.
(c) What is the sign of $\left.\frac{dR}{dp}\right|_{p=140}$? If the skateboards are currently selling for \$140, what happens to revenue if the price is increased to \$141?

45. The derivative f' gives the (absolute) rate of change of a quantity f, and f'/f gives the relative rate of change of the quantity. In this problem, we show that the product rule is equivalent to an additive rule for relative rates of change. Assume $h = f \cdot g$ with $f \neq 0$ and $g \neq 0$.

(a) Show that the additive rule

$$\frac{f'}{f} + \frac{g'}{g} = \frac{h'}{h}$$

implies the product rule, by multiplying through by h and using the fact that $h = f \cdot g$.
(b) Show that the product rule implies the additive rule in part (a), by starting with the product rule and dividing through by $h = f \cdot g$.

3.5 DERIVATIVES OF PERIODIC FUNCTIONS

Since the sine and cosine functions are periodic, their derivatives must be periodic also. (Why?) Let's look at the graph of $f(x) = \sin x$ in Figure 3.21 and estimate the derivative function graphically.

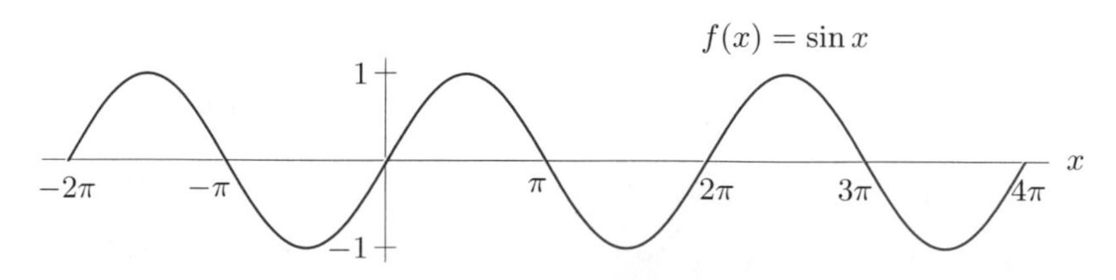

Figure 3.21: The sine function

First we might ask ourselves where the derivative is zero. (At $x = \pm\pi/2, \pm 3\pi/2, \pm 5\pi/2$, etc.) Then ask where the derivative is positive and where it is negative. (Positive for $-\pi/2 < x < \pi/2$; negative for $\pi/2 < x < 3\pi/2$, etc.) Since the largest positive slopes are at $x = 0, 2\pi$, and so on, and the largest negative slopes are at $x = \pi, 3\pi$, and so on, we get something like the graph in Figure 3.22.

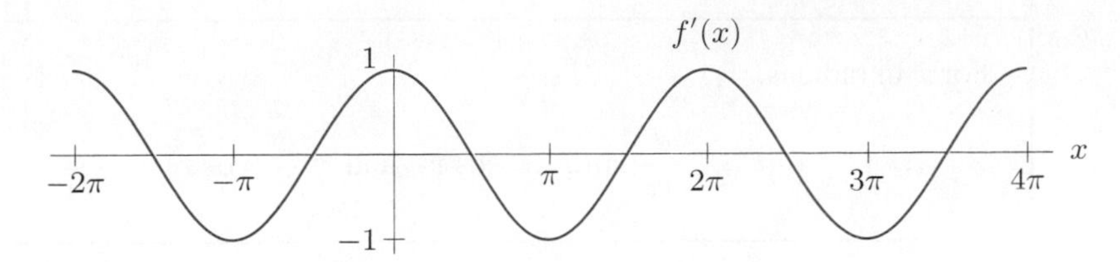

Figure 3.22: Derivative of $f(x) = \sin x$

The graph of the derivative in Figure 3.22 looks suspiciously like the graph of the cosine function. This might lead us to conjecture, quite correctly, that the derivative of the sine is the cosine.

Of course, we cannot be sure, just from the graphs, that the derivative of the sine really is the cosine. However, this is in fact true.

One thing we can do is to check that the derivative function in Figure 3.22 has amplitude 1 (as it must if it is the cosine). That means we have to convince ourselves that the derivative of $f(x) = \sin x$ is 1 when $x = 0$. The next example suggests that this is true when x is in radians.

Example 1 Using a calculator, estimate the derivative of $f(x) = \sin x$ at $x = 0$. Make sure your calculator is set in radians.

Solution We use the average rate of change of $\sin x$ on the small interval $0 \le x \le 0.01$ to compute

$$f'(0) \approx \frac{\sin(0.01) - \sin(0)}{0.01 - 0} = \frac{0.0099998 - 0}{0.01} = 0.99998 \approx 1.0.$$

The derivative of $f(x) = \sin x$ at $x = 0$ is approximately 1.0.

Warning: It is important to notice that in the previous example x was in *radians*; any conclusions we have drawn about the derivative of $\sin x$ are valid *only* when x is in radians.

Example 2 Starting with the graph of the cosine function, sketch a graph of its derivative.

Solution The graph of $g(x) = \cos x$ is in Figure 3.23(a). Its derivative is 0 at $x = 0, \pm\pi, \pm 2\pi$, and so on; it is positive for $-\pi < x < 0$, $\pi < x < 2\pi$, and so on, and it is negative for $0 < x < \pi$, $2\pi < x < 3\pi$, and so on. The derivative is in Figure 3.23(b).

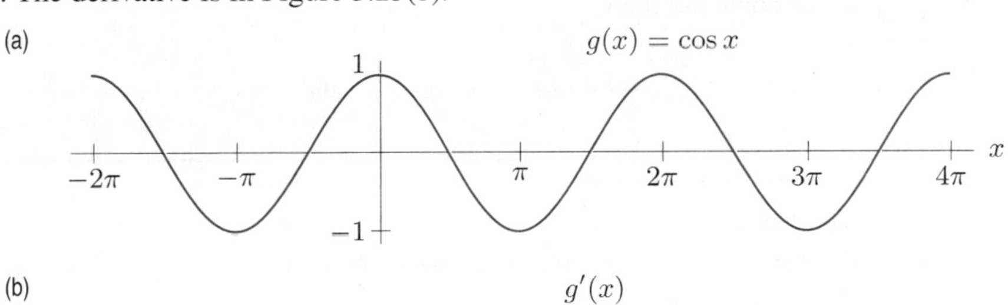

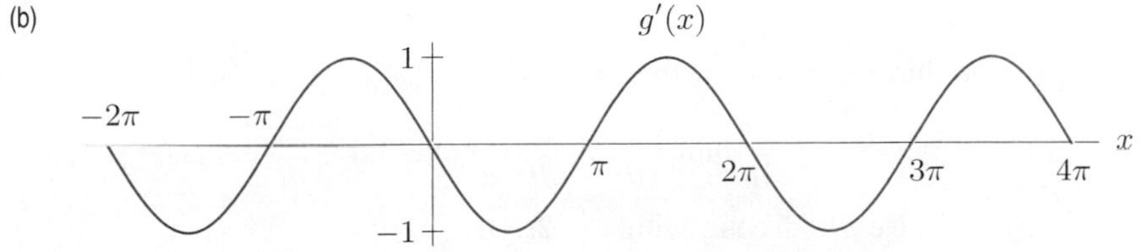

Figure 3.23: $g(x) = \cos x$ and its derivative, $g'(x)$

As we did with the sine, we'll use the graphs to make a conjecture. The derivative of the cosine in Figure 3.23(b) looks exactly like the graph of sine, except reflected about the x-axis. It turns out that the derivative of $\cos x$ is $-\sin x$.

For x in radians,

$$\frac{d}{dx}(\sin x) = \cos x \quad \text{and} \quad \frac{d}{dx}(\cos x) = -\sin x.$$

Example 3 Differentiate (a) $5\sin t - 8\cos t$ (b) $5 - 3\sin x + x^3$

Solution (a) Differentiating gives

$$\frac{d}{dt}(5\sin t - 8\cos t) = 5\frac{d}{dt}(\sin t) - 8\frac{d}{dt}(\cos t) = 5(\cos t) - 8(-\sin t) = 5\cos t + 8\sin t$$

(b) We have

$$\frac{d}{dx}(5 - 3\sin x + x^3) = \frac{d}{dx}(5) - 3\frac{d}{dx}(\sin x) + \frac{d}{dx}(x^3) = 0 - 3(\cos x) + 3x^2 = -3\cos x + 3x^2.$$

The chain rule tells us how to differentiate composite functions involving the sine and cosine. Suppose $y = \sin(3t)$, so $y = \sin z$ and $z = 3t$, so

$$\frac{dy}{dt} = \frac{dy}{dz} \cdot \frac{dz}{dt} = \cos z \frac{dz}{dt} = \cos(3t) \cdot 3 = 3\cos(3t).$$

In general,

If z is a differentiable function of t, then

$$\frac{d}{dt}(\sin z) = \cos z \frac{dz}{dt} \quad \text{and} \quad \frac{d}{dt}(\cos z) = -\sin z \frac{dz}{dt}$$

In many applications, $z = kt$ for some constant k. Then we have:

If k is a constant, then

$$\frac{d}{dt}(\sin kt) = k\cos kt \quad \text{and} \quad \frac{d}{dt}(\cos kt) = -k\sin kt$$

Example 4 Differentiate:

(a) $\sin(t^2)$ (b) $5\cos(2t)$ (c) $t\sin t$.

Solution (a) We have $y = \sin z$ with $z = t^2$, so

$$\frac{d}{dt}(\sin(t^2)) = \frac{d}{dt}(\sin z) = \cos z \frac{dz}{dt} = \cos(t^2) \cdot 2t = 2t\cos(t^2).$$

(b) We have $y = 5\cos z$ with $z = 2t$, so

$$\frac{d}{dt}(5\cos(2t)) = 5\frac{d}{dt}(\cos(2t)) = 5(-2\sin(2t)) = -10\sin(2t).$$

(c) We use the product rule:

$$\frac{d}{dt}(t\sin t) = \frac{d}{dt}(t) \cdot \sin t + t\frac{d}{dt}(\sin t) = 1 \cdot \sin t + t(\cos t) = \sin t + t\cos t.$$

Problems for Section 3.5

Differentiate the functions in Problems 1–20. Assume that A and B are constants.

1. $y = 5\sin x$

2. $P = 3 + \cos t$

3. $y = t^2 + 5\cos t$

4. $y = B + A\sin t$

5. $R(q) = q^2 - 2\cos q$

6. $y = 5\sin x - 5x + 4$

7. $f(x) = \sin(3x)$

8. $R = \sin(5t)$

9. $W = 4\cos(t^2)$

10. $y = 2\cos(5t)$

11. $y = \sin(x^2)$

12. $y = A\sin(Bt)$

13. $z = \cos(4\theta)$

14. $y = 6\sin(2t) + \cos(4t)$

15. $f(x) = x^2\cos x$

16. $f(x) = 2x\sin(3x)$

17. $f(\theta) = \theta^3\cos\theta$

18. $z = \dfrac{e^{t^2} + t}{\sin(2t)}$

19. $f(t) = \dfrac{t^2}{\cos t}$

20. $f(\theta) = \dfrac{\sin\theta}{\theta}$

21. Find the equation of the tangent line to the graph of $y = \sin x$ at $x = \pi$. Graph the function and the tangent line on the same axes.

22. The depth of the water, y, in meters, in the Bay of Fundy, Canada, is given as a function of time, t, in hours after midnight, by the function

$$y = 10 + 7.5\cos(0.507t).$$

How quickly is the tide rising or falling (in meters/hour) at each of the following times?

(a) 6:00 am
(b) 9:00 am
(c) Noon
(d) 6:00 pm

23. If t is the number of months since June, the number of bird species, N, found in an Ohio forest oscillates approximately according to the formula

$$N = f(t) = 19 + 9\cos\left(\frac{\pi}{6}t\right).$$

(a) Graph $f(t)$ for $0 \le t \le 24$ and describe what it shows. Use the graph to decide whether $f'(1)$ and $f'(10)$ are positive or negative.
(b) Find $f'(t)$.
(c) Find and interpret $f(1)$, $f'(1)$, $f(10)$, and $f'(10)$.

24. Is the graph of $y = \sin(x^4)$ increasing or decreasing when $x = 10$? Is it concave up or concave down?

25. A company's monthly sales, $S(t)$, are seasonal and given as a function of time, t, in months, by

$$S(t) = 2000 + 600\sin\left(\frac{\pi}{6}t\right).$$

(a) Graph $S(t)$ for $t = 0$ to $t = 12$. What is the maximum monthly sales? What is the minimum monthly sales? If $t = 0$ is January 1, when during the year are sales highest?
(b) Find $S(2)$ and $S'(2)$. Interpret in terms of sales.

26. On page 68 of Section 1.10 the depth, y, in feet, of water in Portland, Maine is given in terms of t, the number of hours since midnight, by

$$y = 4.9 + 4.4\cos\left(\frac{\pi}{6}t\right).$$

(a) Find dy/dt. What does dy/dt represent, in terms of water level?
(b) For $0 \le t \le 24$, when is dy/dt zero? (Figure 1.101 on page 68 may be helpful.) Explain what it means (in terms of water level) for dy/dt to be zero.

CHAPTER SUMMARY

- **Derivatives of elementary functions**
  Powers, polynomials, exponential functions, logarithms, periodic functions
- **Derivatives of sums, differences, constant multiples**
- **Chain rule**
- **Product and quotient rules**
- **Tangent line approximation**

REVIEW PROBLEMS FOR CHAPTER THREE

Find the derivatives for the functions in Problems 1–40. Assume k is a constant.

1. $f(t) = 6t^4$

2. $f(x) = x^3 - 3x^2 + 5x$

3. $P(t) = e^{2t}$

4. $W = r^3 + 5r - 12$

5. $C = e^{0.08q}$

6. $y = 5e^{-0.2t}$

7. $y = xe^{3x}$

8. $s(t) = (t^2 + 4)(5t - 1)$

9. $g(t) = e^{(1+3t)^2}$

10. $f(x) = x^2 + 3\ln x$

11. $Q(t) = 5t + 3e^{1.2t}$

12. $g(z) = (z^2 + 5)^3$

13. $f(x) = 6(5x - 1)^3$

14. $f(z) = \ln(z^2 + 1)$

15. $h(x) = (1 + e^x)^{10}$

16. $q = 100e^{-0.05p}$

17. $y = x^2 \ln x$

18. $s(t) = t^2 + 2\ln t$

19. $P = 4t^2 + 7\sin t$

20. $R(t) = (\sin t)^5$

21. $h(t) = \ln\left(e^{-t} - t\right)$

22. $f(x) = \sin(2x)$

23. $g(x) = \dfrac{25x^2}{e^x}$

24. $h(t) = \dfrac{t+4}{t-4}$

25. $y = x^2 \cos x$

26. $h(x) = \ln(1 + e^x)$

27. $h(w) = e^{\ln w + 1}$

28. $h(x) = \ln(x^3 + x)$

29. $q(x) = \dfrac{1 + e^x}{1 - e^{-x}}$

30. $q(x) = \dfrac{x}{1+x}$

31. $f(x) = xe^x$

32. $h(x) = \sin(e^x)$

33. $h(x) = \cos(x^3)$

34. $z = \dfrac{3t+1}{5t+2}$

35. $z = \dfrac{t^2 + 5t + 2}{t+3}$

36. $h(p) = \dfrac{1+p^2}{3+2p^2}$

37. $y = \dfrac{3^x}{3} + \dfrac{33}{\sqrt{x}}$

38. $f(t) = \sin\sqrt{e^t + 1}$

39. $g(y) = e^{2e^{(y^3)}}$

40. $g(t) = \dfrac{\ln(kt) + t}{\ln(kt) - t}$

41. Let $f(x) = x^3 - 4x^2 + 7x - 11$. Find $f'(0)$, $f'(2)$, $f'(-1)$.

42. **(a)** Use a graph of $P(q) = 6q - q^2$ to determine whether each of the following derivatives is positive, negative, or zero: $P'(1)$, $P'(3)$, $P'(4)$. Explain.
(b) Find $P'(q)$ and the three derivatives in part (a).

43. Find the equation of the line tangent to the graph of f at $(1, 1)$, where f is given by $f(x) = 2x^3 - 2x^2 + 1$.

44. **(a)** Use the formula for the area of a circle of radius r, $A = \pi r^2$, to find dA/dr.
(b) The result from part (a) should look familiar. What does dA/dr represent geometrically?
(c) Use the difference quotient to explain the observation you made in part (b).

45. Since January 1, 1960, the population of Slim Chance has been described by the formula

$$P = 35{,}000(0.98)^t,$$

where P is the population of the city t years after the start of 1960. At what rate was the population changing on January 1, 2005?

46. Find the equation of the tangent line to the graph of $P(t) = t\ln t$ at $t = 2$. Graph the function $P(t)$ and the tangent line $Q(t)$ on the same axes.

47. The world's population is about $f(t) = 6e^{0.013t}$ billion, where t is time in years since 1999. Find $f(0)$, $f'(0)$, $f(10)$, and $f'(10)$. Using units, interpret your answers in terms of population.

48. With length, l, in meters, the period T, in seconds, of a pendulum is given by

$$T = 2\pi\sqrt{\frac{l}{9.8}}.$$

(a) How fast does the period increase as l increases?
(b) Does this rate of change increase or decrease as l increases?

49. Figure 3.24 shows the tangent line approximation to $f(x)$ near $x = a$.

(a) Find a, $f(a)$, $f'(a)$.
(b) Estimate $f(2.1)$ and $f(1.98)$. Are these under or overestimates? Which estimate would you expect to be more accurate and why?

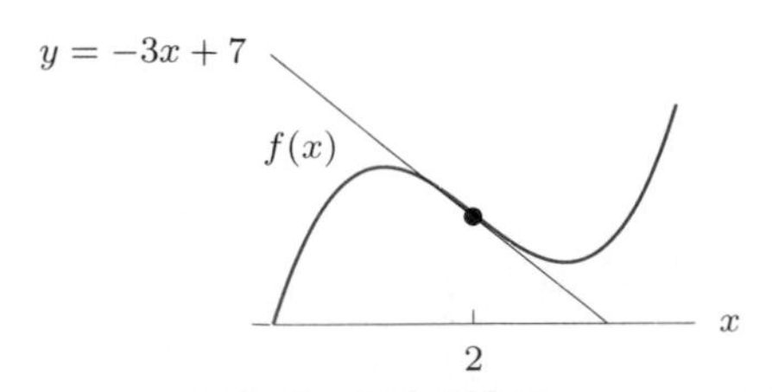

Figure 3.24

50. One gram of radioactive carbon-14 decays according to the formula

$$Q = e^{-0.000121t},$$

where Q is the number of grams of carbon-14 remaining after t years.

(a) Find the rate at which carbon-14 is decaying (in grams/year).
(b) Sketch the rate you found in part (a) against time.

51. The temperature, H, in degrees Fahrenheit (°F), of a can of soda that is put into a refrigerator to cool is given as a function of time, t, in hours, by

$$H = 40 + 30e^{-2t}.$$

(a) Find the rate at which the temperature of the soda is changing (in °F/hour).
(b) What is the sign of dH/dt? Explain.
(c) When, for $t \geq 0$, is the magnitude of dH/dt largest? In terms of the can of soda, why is this?

52. Use the fact that, for $a > 0$,

$$\frac{d}{dx}(a^x) = (\ln a)a^x$$

to explain for which values of a the function a^x is increasing and for which values it is decreasing.

53. The value of a certain automobile purchased in 2005 can be approximated by the function $V(t) = 25(0.85)^t$, where t is the time, in years, from the date of purchase, and V is the value, in thousands of dollars.

(a) Evaluate and interpret $V(4)$.
(b) Find an expression for $V'(t)$, including units.
(c) Evaluate and interpret $V'(4)$.
(d) Use $V(t)$, $V'(t)$, and any other considerations you think are relevant to write a paragraph in support of or in opposition to the following statement: "From a monetary point of view, it is best to keep this vehicle as long as possible."

54. A boat at anchor is bobbing up and down in the sea. The vertical distance, y, in feet, between the sea floor and the boat is given as a function of time, t, in minutes, by

$$y = 15 + \sin(2\pi t).$$

(a) Find the vertical velocity, v, of the boat at time t.
(b) Make rough sketches of y and v against t.

55. Given $r(2) = 4$, $s(2) = 1$, $s(4) = 2$, $r'(2) = -1$, $s'(2) = 3$, and $s'(4) = 3$, compute the following derivatives, or state what additional information you would need to be able to compute the derivative.

(a) $H'(2)$ if $H(x) = r(x) + s(x)$
(b) $H'(2)$ if $H(x) = 5s(x)$
(c) $H'(2)$ if $H(x) = r(x) \cdot s(x)$
(d) $H'(2)$ if $H(x) = \sqrt{r(x)}$

56. Given $F(2) = 1, F'(2) = 5, F(4) = 3, F'(4) = 7$ and $G(4) = 2, G'(4) = 6, G(3) = 4, G'(3) = 8$, find:

(a) $H(4)$ if $H(x) = F(G(x))$
(b) $H'(4)$ if $H(x) = F(G(x))$
(c) $H(4)$ if $H(x) = G(F(x))$
(d) $H'(4)$ if $H(x) = G(F(x))$
(e) $H'(4)$ if $H(x) = F(x)/G(x)$

57. Find the equations of the tangent lines to the graph of $f(x) = \sin x$ at $x = 0$ and at $x = \pi/3$. Use each tangent line to approximate $\sin(\pi/6)$. Would you expect these results to be equally accurate, since they are taken equally far away from $x = \pi/6$ but on opposite sides? If the accuracy is different, can you account for the difference?

For Problems 58–63, let $h(x) = f(x) \cdot g(x)$, and $k(x) = f(x)/g(x)$, and $l(x) = g(x)/f(x)$. Use Figure 3.25 to estimate the derivatives.

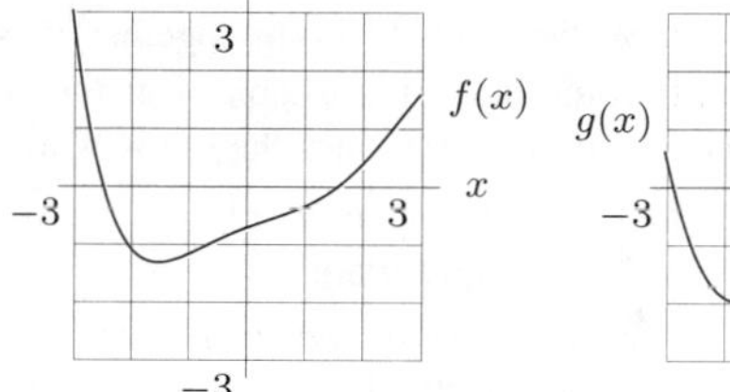

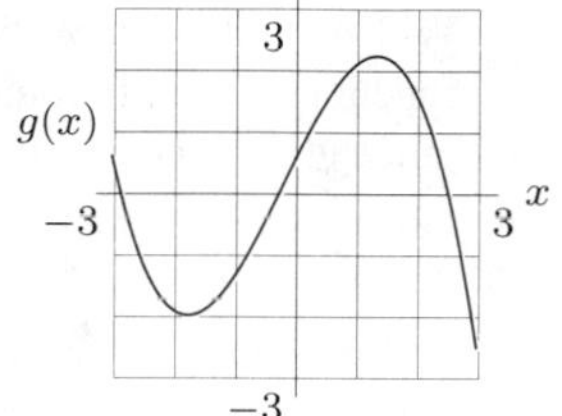

Figure 3.25

58. $h'(1)$ **59.** $k'(1)$
60. $h'(2)$ **61.** $k'(2)$
62. $l'(1)$ **63.** $l'(2)$

64. On what intervals is the function $f(x) = x^4 - 4x^3$ both decreasing and concave up?

65. Given $p(x) = x^n - x$, find the intervals over which p is a decreasing function when:

(a) $n = 2$ **(b)** $n = \frac{1}{2}$ **(c)** $n = -1$

66. Using a graph to help you, find the equations of all lines through the origin tangent to the parabola

$$y = x^2 - 2x + 4.$$

Sketch the lines on the graph.

67. A museum has decided to sell one of its paintings and to invest the proceeds. If the picture is sold between the years 2000 and 2020 and the money from the sale is invested in a bank account earning 5% interest per year compounded annually, then $B(t)$, the balance in the year 2020, depends on the year, t, in which the painting is sold and the sale price $P(t)$. If t is measured from the year 2000 so that $0 < t < 20$ then

$$B(t) = P(t)(1.05)^{20-t}.$$

(a) Explain why $B(t)$ is given by this formula.

(b) Show that the formula for $B(t)$ is equivalent to
$$B(t) = (1.05)^{20}\frac{P(t)}{(1.05)^t}.$$
(c) Find $B'(10)$, given that $P(10) = 150{,}000$ and $P'(10) = 5000$.

68. Find the mean and variance of the normal distribution of statistics using the information in parts (a) and (b) with
$$m(t) = e^{\mu t + \sigma^2 t^2/2}.$$
(a) Mean $= m'(0)$
(b) Variance $= m''(0) - (m'(0))^2$

69. Imagine you are zooming in on the graph of each of the following functions near the origin:

$y = x$ $\quad y = \sqrt{x}$ $\quad y = x^2$
$y = x^3 + \frac{1}{2}x^2$ $\quad y = x^3$ $\quad y = \ln(x+1)$
$y = \frac{1}{2}\ln(x^2+1)$ $\quad y = \sqrt{2x - x^2}$

Which of them look the same? Group together those functions which become indistinguishable near the origin, and give the equations of the lines they look like.

70. Given a power function of the form $f(x) = ax^n$, with $f'(2) = 3$ and $f'(4) = 24$, find n and a.

71. A yam is put in a hot oven, maintained at a constant temperature 200°C. At time $t = 30$ minutes, the temperature T of the yam is 120° and is increasing at an (instantaneous) rate of 2°/min. Newton's law of cooling (or, in our case, warming) implies that the temperature at time t is given by
$$T(t) = 200 - ae^{-bt}.$$
Find a and b.

72. Given a number $a > 1$, the equation
$$a^x = 1 + x$$
has the solution $x = 0$. Are there any other solutions? How does your answer depend on the value of a? [Hint: Graph the functions on both sides of the equation.]

73. The temperature Y in degrees Fahrenheit of a yam in a hot oven t minutes after it is placed there is given by
$$Y(t) = 350(1 - 0.7e^{-0.008t}).$$
(a) What was the temperature of the yam when it was placed in the oven?
(b) What is the temperature of the oven?
(c) When does the yam reach 175° F?
(d) Estimate the rate at which the temperature of the yam is increasing when $t = 20$.

74. For positive constants c and k, the *Monod growth curve*,
$$P = \frac{cr}{k+r},$$
describes the growth of a population, P, as a function of the available quantity of a resource, r. If the resource varies periodically with time, $r = 10\sin(\pi t/6) + 10$, find dP/dt.

75. A dose, D, of a drug causes a temperature change, T, in a patient. For C a positive constant, T is given by
$$T = \left(\frac{C}{2} - \frac{D}{3}\right)D^3.$$
(a) What is the rate of change of temperature change with respect to dose?
(b) For what doses does the temperature change increase as the dose increases?

76. Figure 3.26 shows the number of gallons, G, of gasoline used on a trip of M miles.

(a) The function f is linear on each of the intervals $0 < M < 70$ and $70 < M < 100$. What is the slope of these lines? What are the units of these slopes?
(b) What is gas consumption (in miles per gallon) during the first 70 miles of this trip? During the next 30 miles?
(c) Figure 3.27 shows distance traveled, M (in miles), as a function of time t, in hours since the start of the trip. Describe this trip in words. Give a possible explanation for what happens one hour into the trip. What do your answers to part (b) tell you about the trip?
(d) If we let $G = k(t) = f(h(t))$, estimate $k(0.5)$ and interpret your answer in terms of the trip.
(e) Find $k'(0.5)$ and $k'(1.5)$. Give units and interpret your answers.

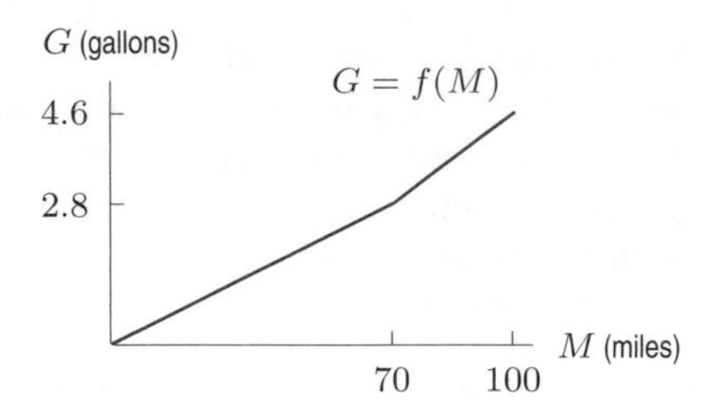

Figure 3.26

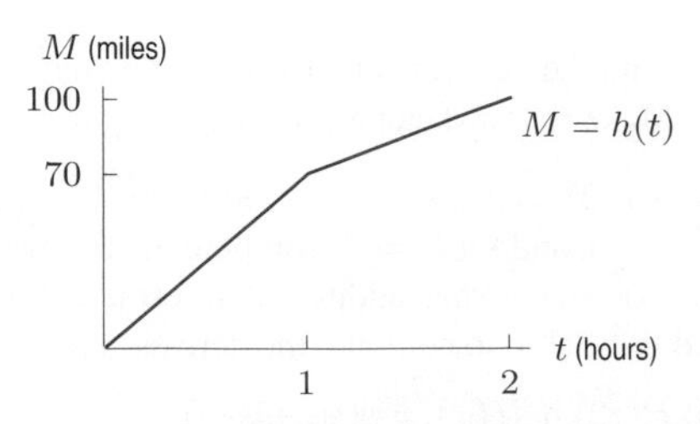

Figure 3.27:

PROJECTS FOR CHAPTER THREE

1. Coroner's Rule of Thumb

Coroners estimate time of death using the rule of thumb that a body cools about $2°$F during the first hour after death and about $1°$F for each additional hour. Assuming an air temperature of $68°$F and a living body temperature of $98.6°$F , the temperature $T(t)$ in °F of a body at a time t hours since death is given by

$$T(t) = 68 + 30.6e^{-kt}.$$

(a) For what value of k will the body cool by $2°$F in the first hour?

(b) Using the value of k found in part (a), after how many hours will the temperature of the body be decreasing at a rate of $1°$F per hour?

(c) Using the value of k found in part (a), show that, 24 hours after death, the coroner's rule of thumb gives approximately the same temperature as the formula.

2. Air Pressure and Altitude

Air pressure at sea level is 30 inches of mercury. At an altitude of h feet above sea level, the air pressure, P, in inches of mercury, is given by

$$P = 30e^{-3.23\times10^{-5}h}$$

(a) Sketch a graph of P against h.

(b) Find the equation of the tangent line at $h = 0$.

(c) A rule of thumb used by travelers is that air pressure drops about 1 inch for every 1000-foot increase in height above sea level. Write a formula for the air pressure given by this rule of thumb.

(d) What is the relation between your answers to parts (b) and (c)? Explain why the rule of thumb works.

(e) Are the predictions made by the rule of thumb too large or too small? Why?

FOCUS ON THEORY

ESTABLISHING THE DERIVATIVE FORMULAS

The graph of $f(x) = x^2$ suggests that the derivative of x^2 is $f'(x) = 2x$. However, as we saw in the Focus on Theory section in Chapter 2, to be sure that this formula is correct, we have to use the definition:

$$f'(x) = \lim_{h\to 0} \frac{f(x+h) - f(x)}{h}.$$

As in Chapter 2, we simplify the difference quotient and then take the limit as h approaches zero.

Example 1 Confirm that the derivative of $g(x) = x^3$ is $g'(x) = 3x^2$.

Solution Using the definition, we calculate $g'(x)$:

$$\begin{aligned} g'(x) &= \lim_{h\to 0} \frac{g(x+h) - g(x)}{h} = \lim_{h\to 0} \frac{(x+h)^3 - x^3}{h} \\ \text{Multiplying out} \longrightarrow &= \lim_{h\to 0} \frac{x^3 + 3x^2h + 3xh^2 + h^3 - x^3}{h} \\ &= \lim_{h\to 0} \frac{3x^2h + 3xh^2 + h^3}{h} \\ \text{Simplifying} \longrightarrow &= \lim_{h\to 0} (3x^2 + 3xh + h^2) = 3x^2. \end{aligned}$$

Looking at what happens as $h \to 0$

So $g'(x) = \dfrac{d}{dx}(x^3) = 3x^2$.

Example 2 Give an informal justification that the derivative of $f(x) = e^x$ is $f'(x) = e^x$.

Solution Using $f(x) = e^x$, we have

$$\begin{aligned} f'(x) &= \lim_{h\to 0} \frac{f(x+h) - f(x)}{h} = \lim_{h\to 0} \frac{e^{x+h} - e^x}{h} \\ &= \lim_{h\to 0} \frac{e^x e^h - e^x}{h} = \lim_{h\to 0} e^x \left(\frac{e^h - 1}{h}\right). \end{aligned}$$

What is the limit of $\dfrac{e^h - 1}{h}$ as $h \to 0$? The graph of $\dfrac{e^h - 1}{h}$ in Figure 3.28 suggests that $\dfrac{e^h - 1}{h}$ approaches 1 as $h \to 0$. In fact, it can be proved that the limit equals 1, so

$$f'(x) = \lim_{h\to 0} e^x \left(\frac{e^h - 1}{h}\right) = e^x \cdot 1 = e^x.$$

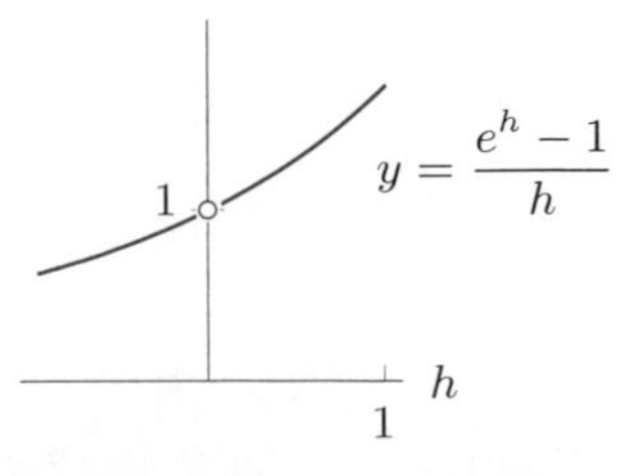

Figure 3.28: What is $\lim_{h\to 0} \frac{e^h - 1}{h}$?

Example 3 Show that if $f(x) = 2x^2 + 1$, then $f'(x) = 4x$.

Solution We use the definition of the derivative with $f(x) = 2x^2 + 1$:

$$
\begin{aligned}
f'(x) &= \lim_{h\to 0} \frac{f(x+h) - f(x)}{h} = \lim_{h\to 0} \frac{(2(x+h)^2 + 1) - (2x^2 + 1)}{h} \\
&= \lim_{h\to 0} \frac{2(x^2 + 2xh + h^2) + 1 - 2x^2 - 1}{h} = \lim_{h\to 0} \frac{2x^2 + 4xh + 2h^2 + 1 - 2x^2 - 1}{h} \\
&= \lim_{h\to 0} \frac{4xh + 2h^2}{h} = \lim_{h\to 0} \frac{h(4x + 2h)}{h}
\end{aligned}
$$

To find the limit, look at what happens when h is close to 0, but $h \neq 0$. Simplifying, we have

$$f'(x) = \lim_{h\to 0} \frac{h(4x + 2h)}{h} = \lim_{h\to 0} (4x + 2h) = 4x$$

because as h gets close to 0, we know that $4x + 2h$ gets close to $4x$.

Using the Chain Rule to Establish Derivative Formulas

We use the chain rule to justify the formulas for derivatives of $\ln x$ and of a^x.

Derivative of $\ln x$

We'll differentiate an identity that involves $\ln x$. On page 40, we have $e^{\ln x} = x$. Differentiating gives

$$\frac{d}{dx}(e^{\ln x}) = \frac{d}{dx}(x) = 1.$$

On the left side, since e^x is the outside function and $\ln x$ is the inside function, the chain rule gives

$$\frac{d}{dx}(e^{\ln x}) = e^{\ln x} \cdot \frac{d}{dx}(\ln x).$$

Thus, as we said on page 150,

$$\frac{d}{dx}(\ln x) = \frac{1}{e^{\ln x}} = \frac{1}{x}.$$

Derivative of a^x

Graphical arguments suggest that the derivative of a^x is proportional to a^x. Now we show that the constant of proportionality is $\ln a$. For $a > 0$, we use the identity from page 40

$$\ln(a^x) = x \ln a.$$

On the left side, using $\frac{d}{dx}(\ln x) = \frac{1}{x}$ and the chain rule gives

$$\frac{d}{dx}(\ln a^x) = \frac{1}{a^x} \cdot \frac{d}{dx}(a^x).$$

Since $\ln a$ is a constant, differentiating the right side gives

$$\frac{d}{dx}(x \ln a) = \ln a.$$

Since the two sides are equal, we have

$$\frac{1}{a^x}\frac{d}{dx}(a^x) = \ln a.$$

Solving for $\frac{d}{dx}(a^x)$ gives the result of Section 3.2. For $a > 0$,

$$\frac{d}{dx}(a^x) = (\ln a)a^x.$$

The Product Rule

Suppose we want to calculate the derivative of the product of differentiable functions, $f(x)g(x)$, using the definition of the derivative. Notice that in the second step below, we are adding and subtracting the same quantity: $f(x)g(x+h)$.

$$\begin{aligned}\frac{d[f(x)g(x)]}{dx} &= \lim_{h\to 0}\frac{f(x+h)g(x+h)-f(x)g(x)}{h}\\ &= \lim_{h\to 0}\frac{f(x+h)g(x+h)-f(x)g(x+h)+f(x)g(x+h)-f(x)g(x)}{h}\\ &= \lim_{h\to 0}\left[\frac{f(x+h)-f(x)}{h}\cdot g(x+h)+f(x)\cdot\frac{g(x+h)-g(x)}{h}\right]\end{aligned}$$

Taking the limit as $h \to 0$ gives the product rule:

$$(f(x)g(x))' = f'(x)\cdot g(x) + f(x)\cdot g'(x).$$

The Quotient Rule

Let $Q(x) = f(x)/g(x)$ be the quotient of differentiable functions. Assuming that $Q(x)$ is differentiable, we can use the product rule on $f(x) = Q(x)g(x)$:

$$f'(x) = Q'(x)g(x) + Q(x)g'(x).$$

Substituting for $Q(x)$ gives

$$f'(x) = Q'(x)g(x) + \frac{f(x)}{g(x)}g'(x).$$

Solving for $Q'(x)$ gives

$$Q'(x) = \frac{f'(x) - \frac{f(x)}{g(x)}g'(x)}{g(x)}.$$

Multiplying the top and bottom by $g(x)$ to simplify gives the quotient rule:

$$\left(\frac{f(x)}{g(x)}\right)' = \frac{f'(x)g(x) - f(x)g'(x)}{(g(x))^2}.$$

Problems on Establishing the Derivative Formulas

For Problems 1–7, use the definition of the derivative to obtain the following results.

1. If $f(x) = 2x+1$, then $f'(x) = 2$.
2. If $f(x) = 5x^2$, then $f'(x) = 10x$.
3. If $f(x) = 2x^2+3$, then $f'(x) = 4x$.
4. If $f(x) = x^2+x$, then $f'(x) = 2x+1$.
5. If $f(x) = 4x^2+1$, then $f'(x) = 8x$.
6. If $f(x) = x^4$, then $f'(x) = 4x^3$. [Hint: $(x+h)^4 = x^4+4x^3h+6x^2h^2+4xh^3+h^4$.]
7. If $f(x) = x^5$, then $f'(x) = 5x^4$. [Hint: $(x+h)^5 = x^5+5x^4h+10x^3h^2+10x^2h^3+5xh^4+h^5$.]
8. **(a)** Use a graph of $g(h) = \dfrac{2^h-1}{h}$ to explain why we believe that $\lim\limits_{h\to 0}\dfrac{2^h-1}{h} \approx 0.6931$.

   (b) Use the definition of the derivative and the result from part (a) to explain why, if $f(x) = 2^x$, we believe that $f'(x) \approx (0.6931)2^x$.
9. Use the definition of the derivative to show that if $f(x) = C$, where C is a constant, then $f'(x) = 0$.
10. Use the definition of the derivative to show that if $f(x) = b+mx$, for constants m and b, then $f'(x) = m$.
11. Use the definition of the derivative to show that if $f(x) = k\cdot u(x)$, where k is a constant and $u(x)$ is a function, then $f'(x) = k\cdot u'(x)$.
12. Use the definition of the derivative to show that if $f(x) = u(x)+v(x)$, for functions $u(x)$ and $v(x)$, then $f'(x) = u'(x)+v'(x)$.

FOCUS ON PRACTICE

Find derivatives for the functions in Problems 1–63. Assume a, b, c, and k are constants.

1. $f(t) = t^2 + t^4$
2. $g(x) = 5x^4$
3. $y = 5x^3 + 7x^2 - 3x + 1$
4. $s(t) = 6t^{-2} + 3t^3 - 4t^{1/2}$
5. $f(x) = \dfrac{1}{x^2} + 5\sqrt{x} - 7$
6. $P(t) = 100e^{0.05t}$
7. $f(x) = 5e^{2x} - 2 \cdot 3^x$
8. $P(t) = 1{,}000(1.07)^t$
9. $D(p) = e^{p^2} + 5p^2$
10. $y = t^2 e^{5t}$
11. $y = x^2\sqrt{x^2+1}$
12. $f(x) = \ln\left(x^2 + 1\right)$
13. $s(t) = 8\ln(2t+1)$
14. $g(w) = w^2 \ln(w)$
15. $f(x) = 2^x + x^2 + 1$
16. $P(t) = \sqrt{t^2+4}$
17. $C(q) = (2q+1)^3$
18. $g(x) = 5x(x+3)^2$
19. $P(t) = be^{kt}$
20. $f(x) = ax^2 + bx + c$
21. $y = x^2\ln(2x+1)$
22. $f(t) = \left(e^t + 4\right)^3$
23. $f(x) = 5\sin(2x)$
24. $W(r) = r^2 \cos r$
25. $g(t) = 3\sin(5t) + 4$
26. $y = e^{3t}\sin(2t)$
27. $y = 2e^x + 3\sin x + 5$
28. $f(t) = 3t^2 - 4t + 1$
29. $y = 17x + 24x^{1/2}$
30. $g(x) = -\frac{1}{2}(x^5 + 2x - 9)$
31. $f(x) = 5x^4 + \dfrac{1}{x^2}$
32. $y = \dfrac{e^{2x}}{x^2+1}$
33. $f(x) = \dfrac{x^2+3x+2}{x+1}$
34. $y = \left(\dfrac{x^2+2}{3}\right)^2$
35. $g(x) = \sin(2-3x)$
36. $f(z) = \dfrac{z^2+1}{3z}$
37. $q(r) = \dfrac{3r}{5r+2}$
38. $y = x\ln x - x + 2$
39. $j(x) = \ln(e^{ax} + b)$
40. $g(t) = \dfrac{t-4}{t+4}$
41. $h(w) = (w^4 - 2w)^5$
42. $h(w) = w^3\ln(10w)$
43. $f(x) = \ln(\sin x + \cos x)$
44. $w(r) = \sqrt{r^4+1}$
45. $h(w) = -2w^{-3} + 3\sqrt{w}$
46. $h(x) = \sqrt{\dfrac{x^2+9}{x+3}}$
47. $v(t) = t^2 e^{-ct}$
48. $f(x) = \dfrac{x}{1+\ln x}$
49. $g(\theta) = e^{\sin\theta}$
50. $p(t) = e^{4t+2}$
51. $j(x) = \dfrac{x^3}{a} + \dfrac{a}{b}x^2 - cx$
52. $f(z) = \dfrac{z^2+1}{\sqrt{z}}$
53. $h(r) = \dfrac{r^2}{2r+1}$
54. $g(x) = 2x - \dfrac{1}{\sqrt[3]{x}} + 3^x - e$
55. $f(t) = 2te^t - \dfrac{1}{\sqrt{t}}$
56. $w = \dfrac{5-3z}{5+3z}$
57. $f(x) = \dfrac{x^3}{9}(3\ln x - 1)$
58. $g(x) = \dfrac{x^2+\sqrt{x}+1}{x^{3/2}}$
59. $y = \left(x^2+5\right)^3\left(3x^3-2\right)^2$
60. $f(x) = \dfrac{a^2-x^2}{a^2+x^2}$
61. $w(r) = \dfrac{ar^2}{b+r^3}$
62. $H(t) = (at^2+b)e^{-ct}$
63. $g(w) = \dfrac{5}{(a^2-w^2)^2}$

Chapter Four

USING THE DERIVATIVE

In this chapter, the derivative is used to understand the behavior of a function. We see how to locate its maximum and minimum values and its points of inflection, and we see how to analyze the relationship between average and marginal costs.

As we saw in Chapter 2, the derivatives of a function and the function itself are connected in the following way:

- If $f' > 0$ on an interval, then f is increasing on that interval.
- If $f' < 0$ on an interval, then f is decreasing on that interval.
- If $f'' > 0$ on an interval, then the graph of is concave up on that interval.
- If $f'' < 0$ on an interval, then the graph of f is concave down on that interval.

We can do more with these principles now than we could in Chapter 2 because we now have formulas for the derivatives of the elementary functions.

4.1 LOCAL MAXIMA AND MINIMA

What Derivatives Tell Us About a Function and its Graph

When we graph a function on a computer or calculator, we often see only part of the picture. Information given by the first and second derivatives can help identify regions with interesting behavior.

Example 1 Use a computer or calculator to sketch a useful graph of the function

$$f(x) = x^3 - 9x^2 - 48x + 52.$$

Solution Since f is a cubic polynomial, we expect a graph that is roughly S-shaped. Graphing this function with $-10 \le x \le 10$, $-10 \le y \le 10$, gives the two nearly vertical lines in Figure 4.1. We know that there is more going on than this, but how do we know where to look?

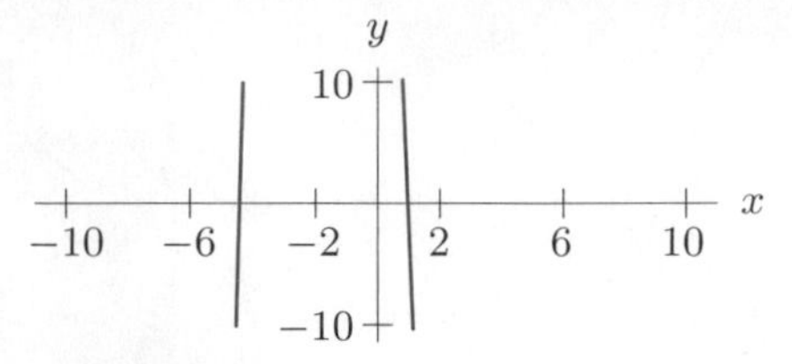

Figure 4.1: Unhelpful graph of $f(x) = x^3 - 9x^2 - 48x + 52$

We use the derivative to determine where the function is increasing and where it is decreasing. The derivative of f is

$$f'(x) = 3x^2 - 18x - 48.$$

To find where $f' > 0$ or $f' < 0$, we first find where $f' = 0$, that is, where $3x^2 - 18x - 48 = 0$. Factoring gives $3(x-8)(x+2) = 0$, so $x = -2$ or $x = 8$. Since $f' = 0$ *only* at $x = -2$ and $x = 8$, and since f' is continuous, f' cannot change sign on any of the three intervals $x < -2$, or $-2 < x < 8$, or $8 < x$. How can we tell the sign of f' on each of these intervals? The easiest way is to pick a point and substitute into f'. For example, since $f'(-3) = 33 > 0$, we know f' is positive for $x < -2$, so f is increasing for $x < -2$. Similarly, since $f'(0) = -48$ and $f'(10) = 72$, we know that f decreases between $x = -2$ and $x = 8$ and increases for $x > 8$. Summarizing:

f increasing ↗	$x = -2$	f decreasing ↘	$x = 8$	f increasing ↗	x
$f' > 0$	$f' = 0$	$f' < 0$	$f' = 0$	$f' > 0$	

We find that $f(-2) = 104$ and $f(8) = -396$. Hence, on the interval $-2 < x < 8$ the function decreases from a high of 104 to a low of -396. (Now we see why not much showed up in our first calculator graph.) One more point on the graph is easy to get: the y-intercept, $f(0) = 52$. With just these three points we can get a much more helpful graph. By setting the plotting window to $-10 \le x \le 20$ and $-400 \le y \le 400$, we get Figure 4.2, which gives much more insight into the behavior of $f(x)$ than the graph in Figure 4.1.

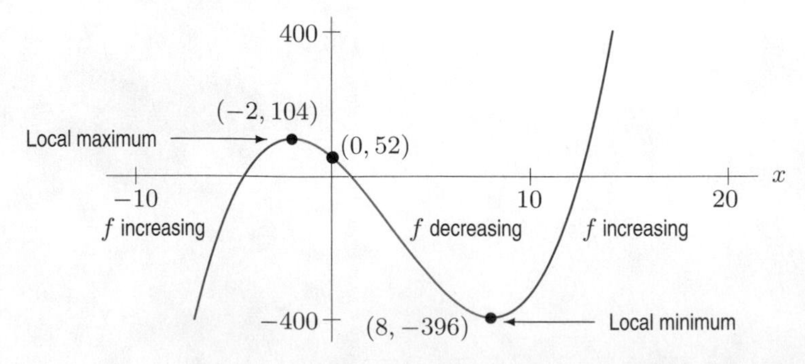

Figure 4.2: Useful graph of $f(x) = x^3 - 9x^2 - 48x + 52$. Notice that the scales on the x-and y-axes are different

Local Maxima and Minima

We are often interested in points such as those marked local maximum and local minimum in Figure 4.2. We have the following definition:

> Suppose p is a point in the domain of f:
> - f has a **local minimum** at p if $f(p)$ is less than or equal to the values of f for points near p.
> - f has a **local maximum** at p if $f(p)$ is greater than or equal to the values of f for points near p.

We use the adjective "local" because we are describing only what happens near p.

How Do We Detect a Local Maximum or Minimum?

In the preceding example, the points $x = -2$ and $x = 8$, where $f'(x) = 0$, played a key role in leading us to local maxima and minima. We give a name to such points:

> For any function f, a point p in the domain of f where $f'(p) = 0$ or $f'(p)$ is undefined is called a **critical point** of the function. In addition, the point $(p, f(p))$ on the graph of f is also called a critical point. A **critical value** of f is the value, $f(p)$, of the function at a critical point, p.

Notice that "critical point of f" can refer either to points in the domain of f or to points on the graph of f. You will know which meaning is intended from the context.

Geometrically, at a critical point where $f'(p) = 0$, the line tangent to the graph of f at p is horizontal. At a critical point where $f'(p)$ is undefined, there is no horizontal tangent to the graph—there is either a vertical tangent or no tangent at all. (For example, $x = 0$ is a critical point for the absolute value function $f(x) = |x|$.) However, most of the functions we will work with will be differentiable everywhere, and therefore most of our critical points will be of the $f'(p) = 0$ variety.

The critical points divide the domain of f into intervals on which the sign of the derivative remains the same, either positive or negative. Therefore, if f is defined on the interval between two successive critical points, its graph cannot change direction on that interval; it is either going up or it is going down. We have the following result:

> If a function, continuous on an interval (its domain), has a local maximum or minimum at p, then p is a critical point or an endpoint of the interval.

A function may have any number of critical points or none at all. (See Figures 4.3–4.5.)

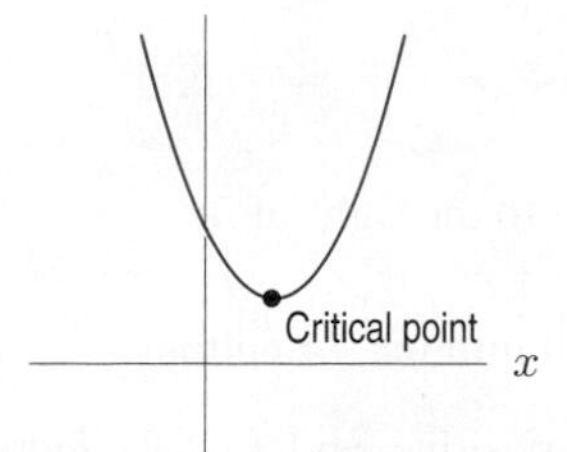

Figure 4.3: A quadratic: One critical point

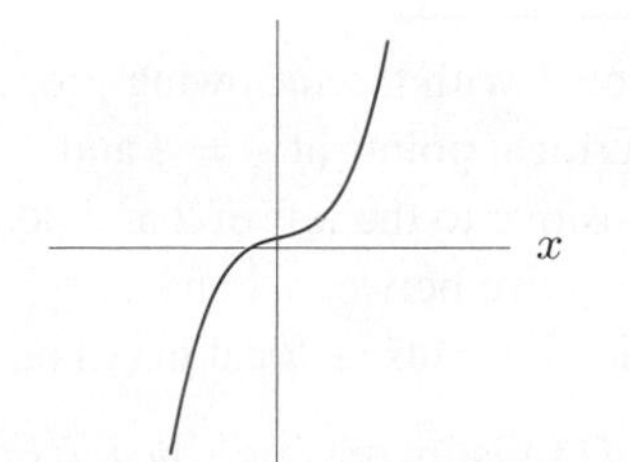

Figure 4.4: $f(x) = x^3 + x + 1$: No critical points

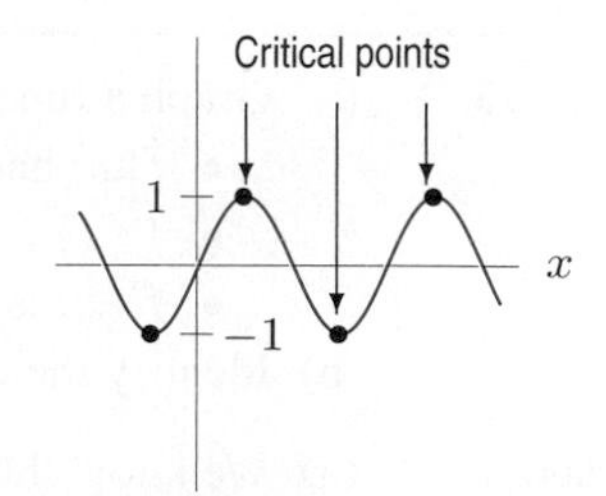

Figure 4.5: Many critical points

Testing For Local Maxima and Minima

If f' has different signs on either side of a critical point p with $f'(p) = 0$, then the graph changes direction at p and looks like one of those in Figure 4.6. We have the following criteria:

First Derivative Test for Local Maxima and Minima

Suppose p is a critical point of a continuous function f.

- If f changes from decreasing to increasing at p, then f has a local minimum at p.
- If f changes from increasing to decreasing at p, then f has a local maximum at p.

Alternatively, the concavity of the graph of f gives another way of distinguishing between local maxima and minima:

Second Derivative Test for Local Maxima and Minima

Suppose p is a critical point of a continuous function f, and $f'(p) = 0$.

- If f is concave up at p, then f has a local minimum at p.
- If f is concave down at p, then f has a local maximum at p.

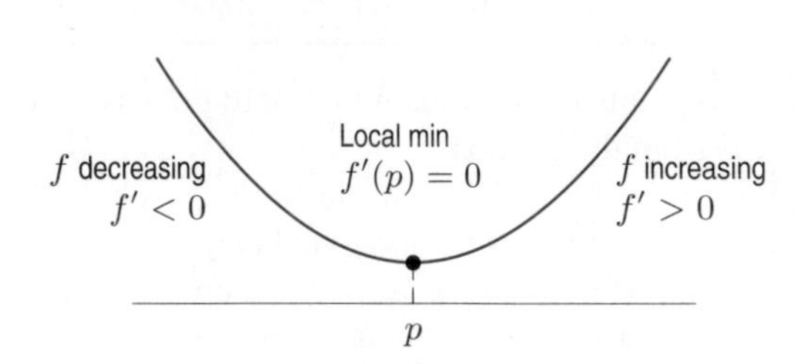

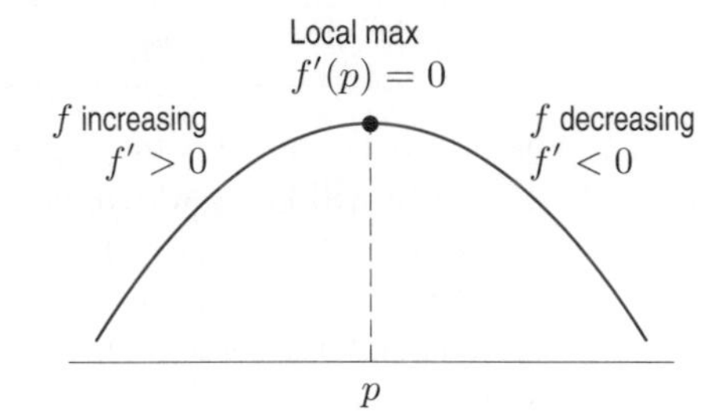

Figure 4.6: Changes in direction at a critical point, p: Local maxima and minima

Example 2 Use the second derivative test to confirm that $f(x) = x^3 - 9x^2 - 48x + 52$ has a local maximum at $x = -2$ and a local minimum at $x = 8$.

Solution In Example 1, we calculated $f'(x) = 3x^2 - 18x - 48 = 3(x - 8)(x + 2)$, so $f'(8) = f'(-2) = 0$. Differentiating again gives $f''(x) = 6x - 18$. Since $f''(8) = 6 \cdot 8 - 18 = 30$ and $f''(-2) = 6(-2) - 18 = -30$, the second derivative test confirms that $x = 8$ is a local minimum and $x = -2$ is a local maximum.

Example 3

(a) Graph a function f with the following properties:
- $f(x)$ has critical points at $x = 2$ and $x = 5$;
- $f'(x)$ is positive to the left of 2 and positive to the right of 5;
- $f'(x)$ is negative between 2 and 5.

(b) Identify the critical points as local maxima, local minima, or neither.

Solution

(a) We know that $f(x)$ is increasing when $f'(x)$ is positive, and $f(x)$ is decreasing when $f'(x)$ is negative. The function is increasing to the left of 2 and increasing to the right of 5, and it is decreasing between 2 and 5. A possible sketch is given in Figure 4.7.

(b) We see that the function has a local maximum at $x = 2$ and a local minimum at $x = 5$.

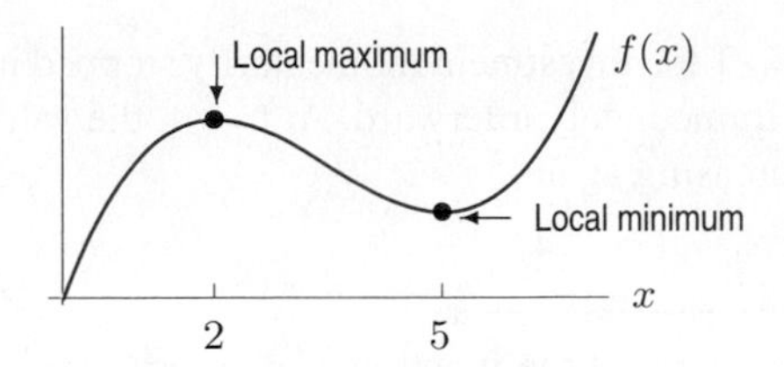

Figure 4.7: A function with critical points at $x = 2$ and $x = 5$

Warning!

Not every critical point of a function is a local maximum or minimum. For instance, consider $f(x) = x^3$, graphed in Figure 4.8. The derivative is $f'(x) = 3x^2$ so $x = 0$ is a critical point. But $f'(x) = 3x^2$ is positive on both sides of $x = 0$, so f increases on both sides of $x = 0$. There is neither a local maximum nor a local minimum for $f(x)$ at $x = 0$.

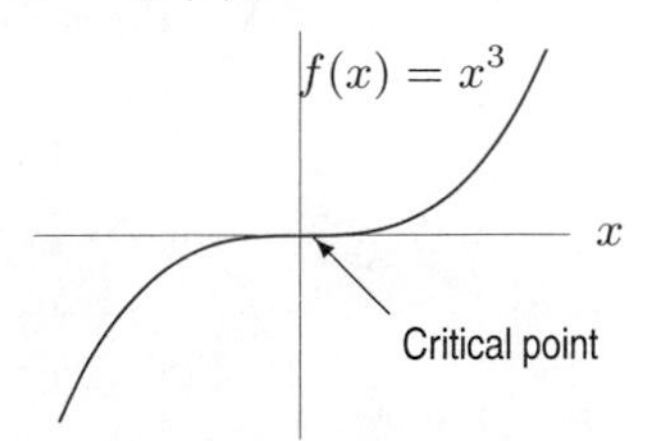

Figure 4.8: A critical point which is neither a local maximum nor minimum.

Example 4 The value of an investment at time t is given by $S(t)$. The rate of change, $S'(t)$, of the value of the investment is shown in Figure 4.9.

(a) What are the critical points of the function $S(t)$?
(b) Identify each critical point as either a local maximum, a local minimum, or neither.
(c) Explain the financial significance of each of the critical points.

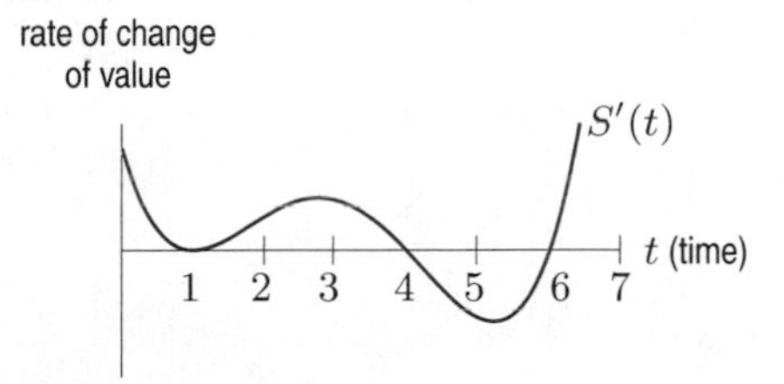

Figure 4.9: Graph of $S'(t)$, the rate of change of the value of the investment

Solution

(a) The critical points of S occur at times t when $S'(t) = 0$. We see in Figure 4.9 that $S'(t) = 0$ at $t = 1$, 4, and 6, so the critical points occur at $t = 1$, 4, and 6.
(b) In Figure 4.9, we see that $S'(t)$ is positive to the left of 1 and between 1 and 4, that $S'(t)$ is negative between 4 and 6, and that $S'(t)$ is positive to the right of 6. Therefore $S(t)$ is increasing to the left of 1 and between 1 and 4 (with a slope of zero at 1), decreasing between 4 and 6, and increasing again to the right of 6. A possible sketch of $S(t)$ is given in Figure 4.10. We see that S has neither a local maximum nor a local minimum at the critical point $t = 1$, but that it has a local maximum at $t = 4$ and a local minimum at $t = 6$.

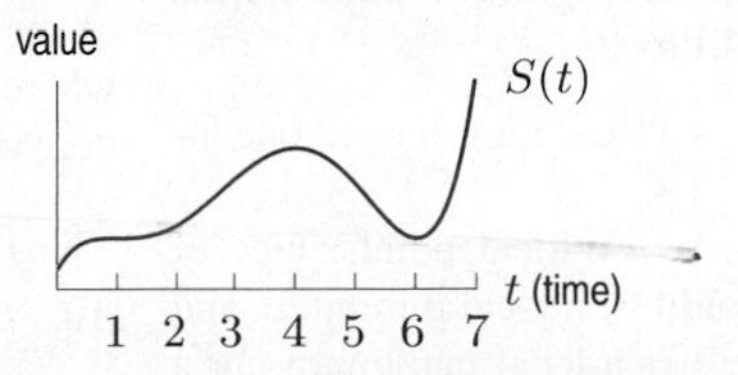

Figure 4.10: Possible graph of the function representing the value of the investment at time t

(c) At time $t = 1$ the investment momentarily stopped increasing in value, though it started increasing again immediately afterward. At $t = 4$, the value peaked and began to decline. At $t = 6$, it started increasing again.

Example 5 Find the critical point of the function $f(x) = x^2 + bx + c$. What is its graphical significance?

Solution Since $f'(x) = 2x + b$, the critical point x satisfies the equation: $2x + b = 0$. Thus, the critical point is at $x = -b/2$. The graph of f is a parabola and the critical point is its vertex. See Figure 4.11.

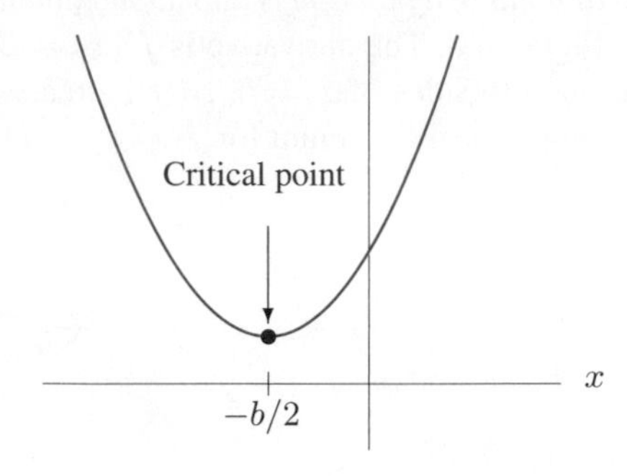

Figure 4.11: Critical point of the parabola $f(x) = x^2 + bx + c$. (Sketched with $b, c > 0$)

Problems for Section 4.1

In Problems 1–4, indicate all critical points of the function f. How many critical points are there? Identify each critical point as a local maximum, a local minimum, or neither.

1.
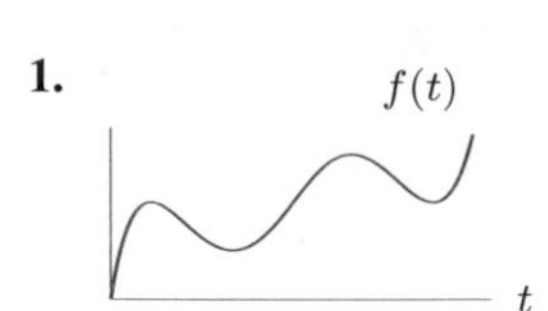

2.
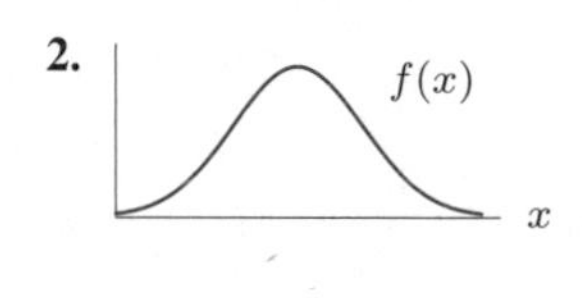

3.
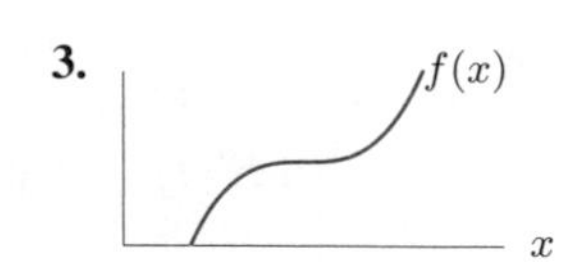

4.
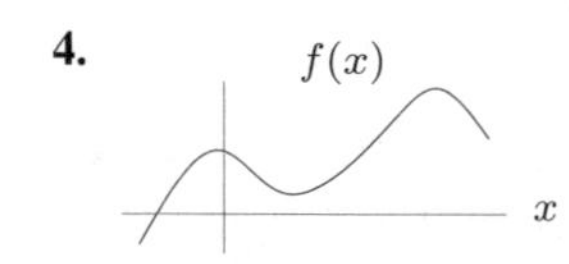

5. During an illness a person ran a fever. His temperature rose steadily for eighteen hours, then went steadily down for twenty hours. When was there a critical point for his temperature as a function of time?

6. **(a)** Graph a function with two local minima and one local maximum.

(b) Graph a function with two critical points. One of these critical points should be a local minimum, and the other should be neither a local maximum nor a local minimum.

7. Graph two continuous functions f and g, each of which has exactly five critical points, the points A–E in Figure 4.12, and which satisfy the following conditions:

(a) $f(x) \to \infty$ as $x \to -\infty$ and $f(x) \to \infty$ as $x \to \infty$

(b) $g(x) \to -\infty$ as $x \to -\infty$ and $g(x) \to 0$ as $x \to \infty$

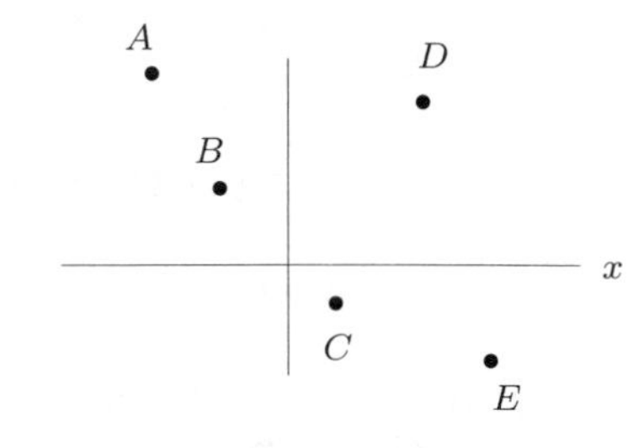

Figure 4.12

Using a calculator or computer, graph the functions in Problems 8–13. Describe briefly in words the interesting features of the graph including the location of the critical points and where the function is increasing/decreasing. Then use the derivative and algebra to explain the shape of the graph.

8. $f(x) = x^3 - 6x + 1$

9. $f(x) = x^3 + 6x + 1$

10. $f(x) = 3x^5 - 5x^3$

11. $f(x) = e^x - 10x$

12. $f(x) = x \ln x, \quad x > 0$

13. $f(x) = x + 2 \sin x$

Problems 14–15 show the graph of a derivative function f'. Indicate on a sketch the x-values that are critical points of the function f itself. Identify each critical point as a local maximum, a local minimum, or neither.

14. **15.**

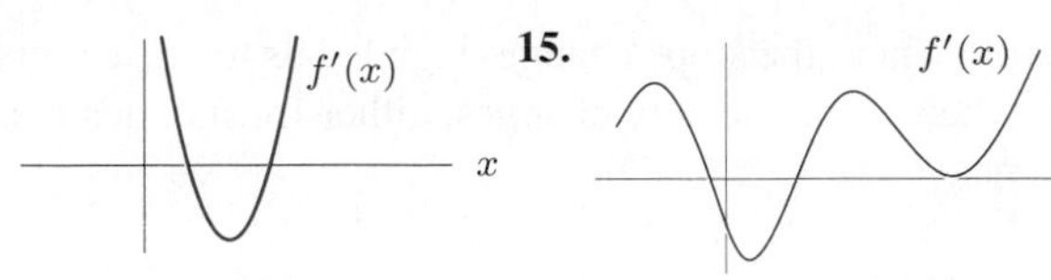

16. Figure 4.13 is a graph of f'. For what values of x does f have a local maximum? A local minimum?

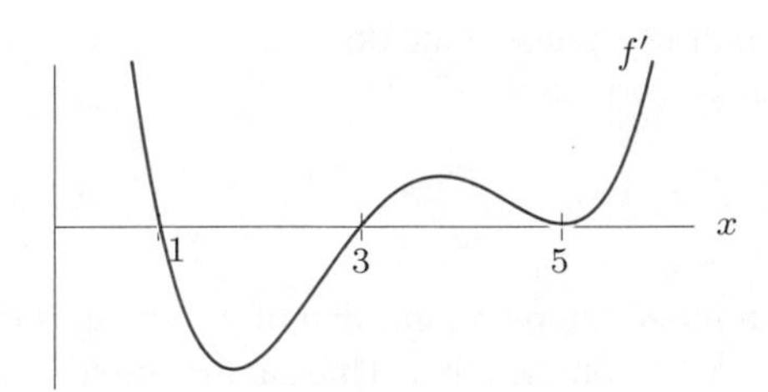

Figure 4.13: Graph of f' (not f)

17. On the graph of f' in Figure 4.14, indicate the x-values that are critical points of the function f itself. Are they local maxima, local minima, or neither?

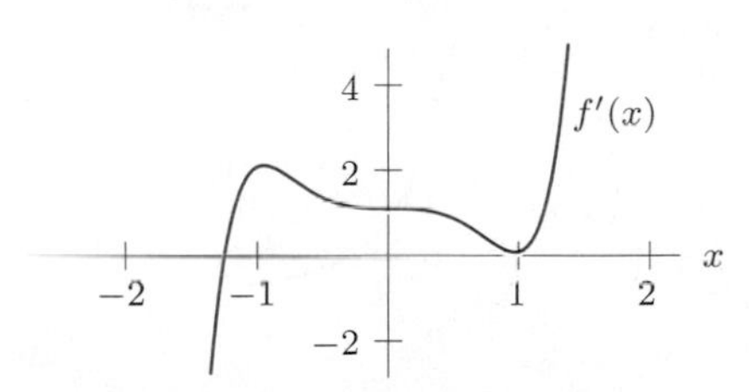

Figure 4.14: Graph of f' (not f)

18. The derivative of $f(t)$ is given by $f'(t) = t^3 - 6t^2 + 8t$ for $0 \le t \le 5$. Graph $f'(t)$, and describe how the function $f(t)$ changes over the interval $t = 0$ to $t = 5$. When is $f(t)$ increasing and when is it decreasing? Where does $f(t)$ have a local maximum and where does it have a local minimum?

19. Consumer demand for a certain product is changing over time, and the rate of change of this demand, $f'(t)$, in units/week, is given, in week t, in the following table.

t	0	1	2	3	4	5	6	7	8	9	10
$f'(t)$	12	10	4	−2	−3	−1	3	7	11	15	10

(a) When is the demand for this product increasing? When is it decreasing?

(b) Approximately when is demand at a local maximum? A local minimum?

20. Suppose f has a continuous derivative whose values are given in the following table.

(a) Estimate the x-coordinates of critical points of f for $0 \le x \le 10$.

(b) For each critical point, indicate if it is a local maximum of f, local minimum, or neither.

x	0	1	2	3	4	5	6	7	8	9	10
$f'(x)$	5	2	1	−2	−5	−3	−1	2	3	1	−1

21. The function $f(x) = x^4 - 4x^3 + 8x$ has a critical point at $x = 1$. Use the second derivative test to identify it as a local maximum or local minimum.

22. Find and classify the critical points of $f(x) = x^3(1-x)^4$ as local maxima and minima.

In Problems 23–24, find constants a and b so that the minimum for the parabola $f(x) = x^2 + ax + b$ is at the given point. [Hint: Begin by finding the critical point in terms of a.]

23. $(3, 5)$

24. $(-2, -3)$

25. Sketch several members of the family $y = x^3 - ax^2$ on the same axes. Discuss the effect of the parameter a on the graph. Find all critical points for this function.

26. For what values of a and b does $f(x) = a(x - b\ln x)$ have a local minimum at the point $(2, 5)$? Figure 4.15 shows a graph of $f(x)$ with $a = 1$ and $b = 1$.

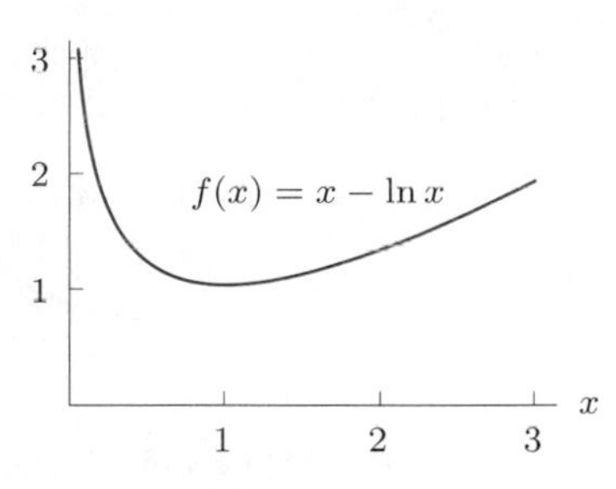

Figure 4.15

27. Find the value of a so that the function $f(x) = xe^{ax}$ has a critical point at $x = 3$.

28. Assume f has a derivative everywhere and has just one critical point, at $x = 3$. In parts (a)–(d), you are given additional conditions. In each case decide whether $x = 3$ is a local maximum, a local minimum, or neither. Explain your reasoning. Sketch possible graphs for all four cases.

(a) $f'(1) = 3$ and $f'(5) = -1$

(b) $f(x) \to \infty$ as $x \to \infty$ and as $x \to -\infty$

(c) $f(1) = 1$, $f(2) = 2$, $f(4) = 4$, $f(5) = 5$

(d) $f'(2) = -1$, $f(3) = 1$, $f(x) \to 3$ as $x \to \infty$

29. **(a)** On a computer or calculator, graph $f(\theta) = \theta - \sin\theta$. Can you tell whether the function has any zeros in the interval $0 \le \theta \le 1$?

(b) Find f'. What does the sign of f' tell you about the zeros of f in the interval $0 \le \theta \le 1$?

4.2 INFLECTION POINTS

Concavity and Inflection Points

A study of the points on the graph of a function where the slope changes sign led us to critical points. Now we will study the points on the graph where the concavity changes, either from concave up to concave down, or from concave down to concave up.

> A point at which the graph of a function f changes concavity is called an **inflection point** of f.

The words "inflection point of f" can refer either to a point in the domain of f or to a point on the graph of f. The context of the problem will tell you which is meant.

How Do You Locate an Inflection Point?

Since the concavity of the graph of f changes at an inflection point, the sign of f'' changes there: it is positive on one side of the inflection point and negative on the other. Thus, at the inflection point, f'' is zero or undefined. (See Figure 4.16.)

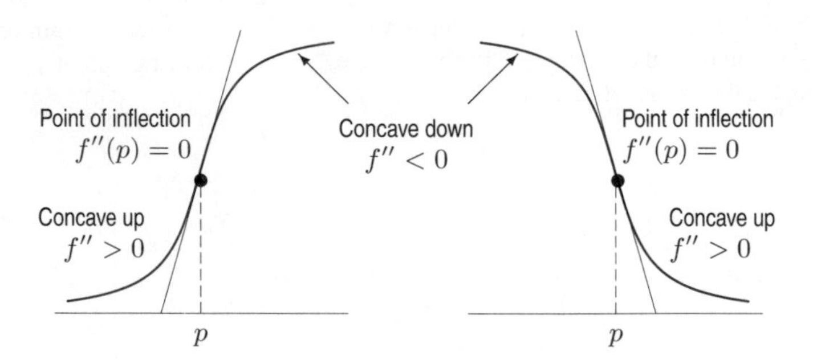

Figure 4.16: Change in concavity (from positive to negative or vice versa) at point p

---

Example 1 Find the inflection points of $f(x) = x^3 - 9x^2 - 48x + 52$.

Solution In Figure 4.17, part of the graph of f is concave up and part is concave down, so the function must have an inflection point. However, it is difficult to locate the inflection point accurately by examining the graph. To find the inflection point exactly, calculate where the second derivative is zero.[1] Since $f'(x) = 3x^2 - 18x - 48$,

$$f''(x) = 6x - 18 \qquad \text{so} \qquad f''(x) = 0 \quad \text{when} \quad x = 3.$$

The graph of $f(x)$ changes concavity at $x = 3$, so $x = 3$ is an inflection point.

[1] For a polynomial, the second derivative cannot be undefined.

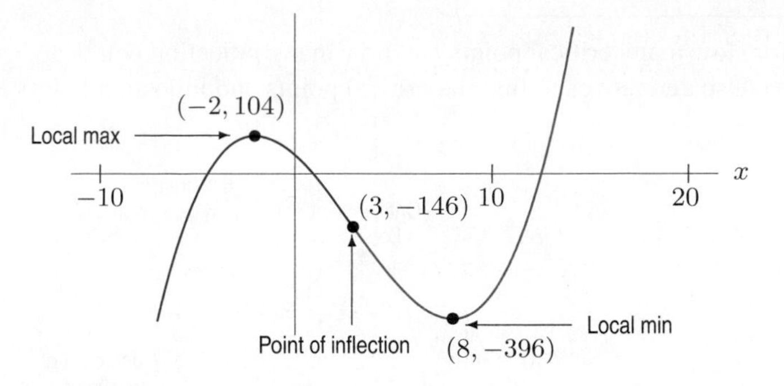

Figure 4.17: Graph of $f(x) = x^3 - 9x^2 - 48x + 52$ showing the inflection point at $x = 3$

Example 2 Graph a function f with the following properties: f has a critical point at $x = 4$ and an inflection point at $x = 8$; the value of f' is negative to the left of 4 and positive to the right of 4; the value of f'' is positive to the left of 8 and negative to the right of 8.

Solution Since f' is negative to the left of 4 and positive to the right of 4, the value of $f(x)$ is decreasing to the left of 4 and increasing to the right of 4. The values of f'' tell us that the graph of $f(x)$ is concave up to the left of 8 and concave down to the right of 8. A possible sketch is given in Figure 4.18.

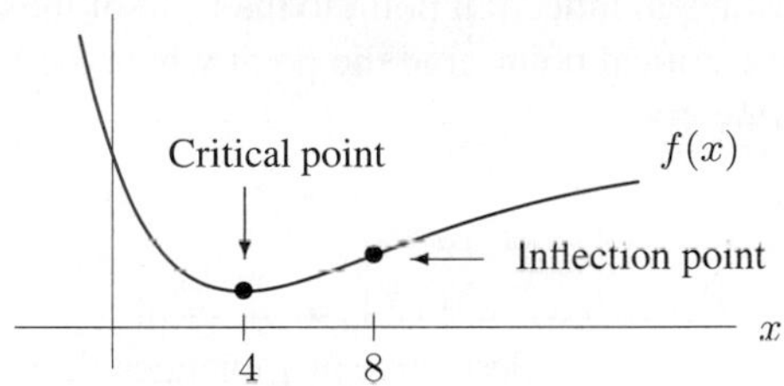

Figure 4.18: A function with a critical point at $x = 4$ and an inflection point at $x = 8$

Example 3 Figure 4.19 shows a population growing toward a limiting population, L. There is an inflection point on the graph at the point where the population reaches $L/2$. What is the significance of the inflection point to the population?

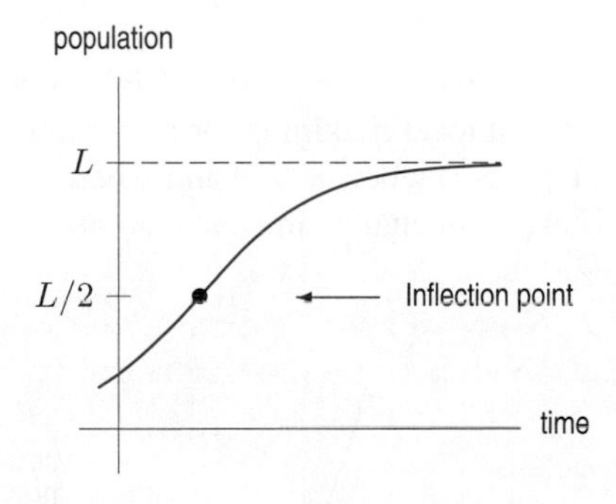

Figure 4.19: Inflection point on graph of a population growing toward a limiting population, L

Solution At times before the inflection point, the population is increasing faster every year. At times after the inflection point the population is increasing slower every year. At the inflection point, the population is growing fastest.

Example 4 (a) How many critical points and how many inflection points does the function $f(x) = xe^{-x}$ have?
(b) Use derivatives to find the critical points and inflection points exactly.

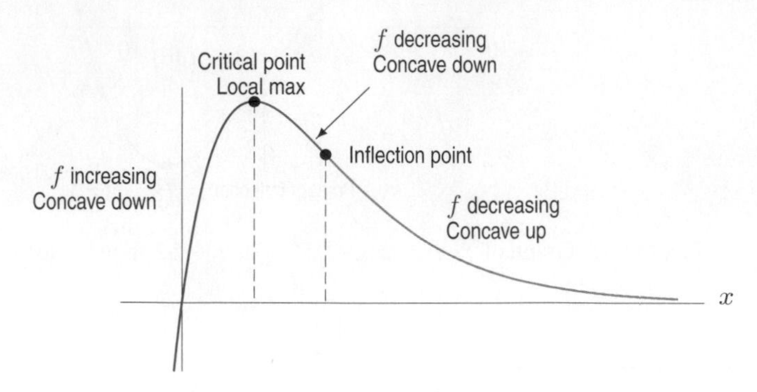

Figure 4.20: Graph of $f(x) = xe^{-x}$

Solution

(a) Figure 4.20 shows the graph of $f(x) = xe^{-x}$. It appears to have one critical point, which is a local maximum. Are there any inflection points? Since the graph of the function is concave down at the critical point and concave up for large x, the graph of the function changes concavity, so there must be an inflection point to the right of the critical point.

(b) To find the critical point, find the point where the first derivative of f is zero or undefined. The product rule gives

$$f'(x) = x(-e^{-x}) + (1)(e^{-x}) = (1-x)e^{-x}.$$

We have $f'(x) = 0$ when $x = 1$, so the critical point is at $x = 1$. To find the inflection point, we find where the second derivative of f changes sign. Using the product rule on the first derivative, we have

$$f''(x) = (1-x)(-e^{-x}) + (-1)(e^{-x}) = (x-2)e^{-x}.$$

We have $f''(x) = 0$ when $x = 2$. Since $f''(x) > 0$ for $x > 2$ and $f''(x) < 0$ for $x < 2$, the concavity changes sign at $x = 2$. So the inflection point is at $x = 2$.

Warning!

Not every point x where $f''(x) = 0$ (or f'' is undefined) is an inflection point (just as not every point where $f' = 0$ is a local maximum or minimum). For instance, $f(x) = x^4$ has $f''(x) = 12x^2$ so $f''(0) = 0$, but $f'' > 0$ when $x > 0$ and when $x < 0$, so the graph of f is concave up on both sides of $x = 0$. There is *no* change in concavity at $x = 0$. (See Figure 4.21.)

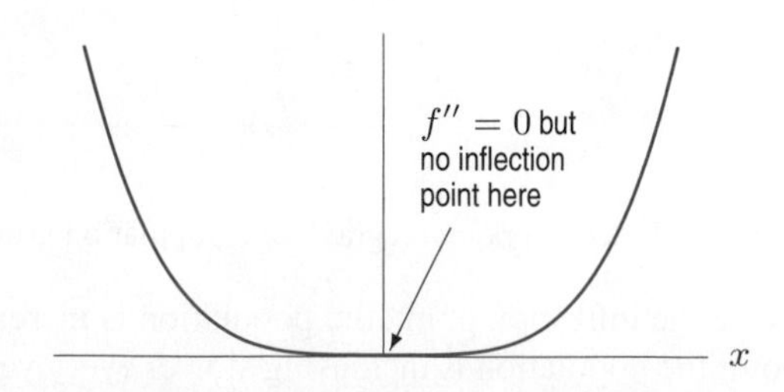

Figure 4.21: Graph of $f(x) = x^4$

Example 5 Suppose that water is being poured into the vase in Figure 4.22 at a constant rate measured in liters per minute. Graph $y = f(t)$, the depth of the water against time, t. Explain the concavity, and indicate the inflection points.

Solution Notice that the volume of water in the vase increases at a constant rate.

At first the water level, y, rises quite slowly because the base of the vase is wide, and so it takes a lot of water to make the depth increase. However, as the vase narrows, the rate at which the water level rises increases. This means that initially y is increasing at an increasing rate, and the graph is concave up. The water level is rising fastest, so the rate of change of the depth y is at a maximum, when the water reaches the middle of the vase, where the diameter is smallest; this is an inflection point. (See Figure 4.23.) After that, the rate at which the water level changes starts to decrease, and so the graph is concave down.

Figure 4.22: A vase

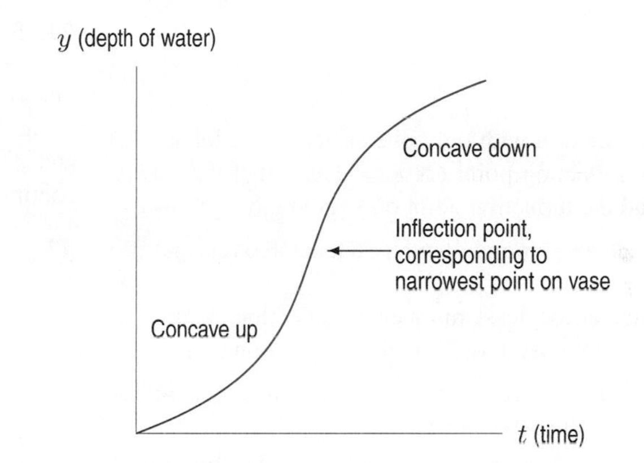

Figure 4.23: Graph of depth of water in the vase, y, against time, t

Example 6 What is the concavity of the graph of $f(x) = ax^2 + bx + c$?

Solution We have $f'(x) = 2ax + b$ and $f''(x) = 2a$. The second derivative of f has the same sign as a. If $a > 0$, the graph is concave up everywhere, an upward opening parabola. If $a < 0$, the graph is concave down everywhere, a downward opening parabola. (See Figure 4.24.)

Figure 4.24: Concavity of $f(x) = ax^2 + bx + c$

Problems for Section 4.2

In Problems 1–4, indicate the approximate locations of all inflection points. How many inflection points are there?

1.

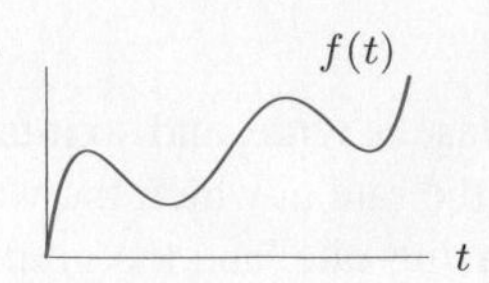

2.

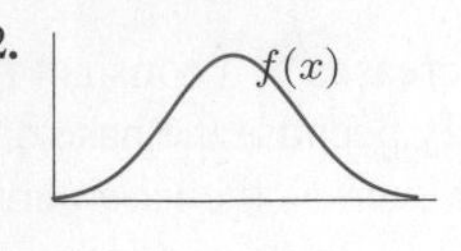

3.

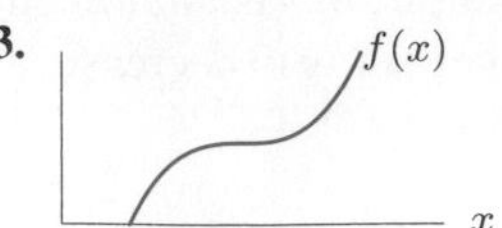

4.

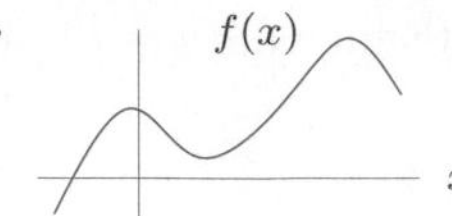

5. Graph a function with only one critical point (at $x = 5$) and one inflection point (at $x = 10$). Label the critical point and the inflection point on your graph.

6. **(a)** Graph a polynomial with two local maxima and two local minima.
(b) What is the least number of inflection points this function must have? Label the inflection points.

7. Graph a function which has a critical point and an inflection point at the same place.

8. When I got up in the morning I put on only a light jacket because, although the temperature was dropping, it seemed that the temperature would not go much lower. But I was wrong. Around noon a northerly wind blew up and the temperature began to drop faster and faster. The worst was around 6 pm when, fortunately, the temperature started going back up.

(a) When was there a critical point in the graph of temperature as a function of time?
(b) When was there an inflection point in the graph of temperature as a function of time?

9. During a flood, the water level in a river first rose faster and faster, then rose more and more slowly until it reached its highest point, then went back down to its pre-flood level. Consider water depth as a function of time.

(a) Is the time of highest water level a critical point or an inflection point of this function?
(b) Is the time when the water first began to rise more slowly a critical point or an inflection point?

10. For $f(x) = x^3 - 18x^2 - 10x + 6$, find the inflection point algebraically. Graph the function with a calculator or computer and confirm your answer.

In each of Problems 11–20, use the first derivative to find all critical points and use the second derivative to find all inflection points. Use a graph to identify each critical point as a local maximum, a local minimum, or neither.

11. $f(x) = x^2 - 5x + 3$

12. $f(x) = x^3 - 3x + 10$

13. $f(x) = 2x^3 + 3x^2 - 36x + 5$

14. $f(x) = \dfrac{x^3}{6} + \dfrac{x^2}{4} - x + 2$

15. $f(x) = x^4 - 2x^2$

16. $f(x) = 3x^4 - 4x^3 + 6$

17. $f(x) = x^4 - 8x^2 + 5$

18. $f(x) = x^4 - 4x^3 + 10$

19. $f(x) = x^5 - 5x^4 + 35$

20. $f(x) = 3x^5 - 5x^3$

21. Find the inflection points of $f(x) = x^4 + x^3 - 3x^2 + 2$.

For Problems 22–25, sketch a possible graph of $y = f(x)$, using the given information about the derivatives $y' = f'(x)$ and $y'' = f''(x)$. Assume that the function is defined and continuous for all real x.

22.

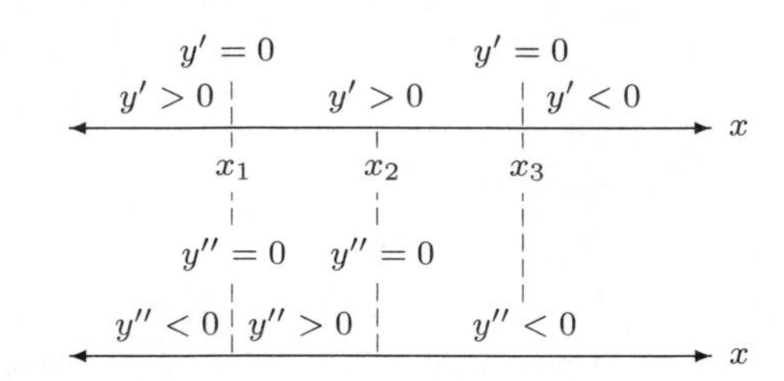

23.

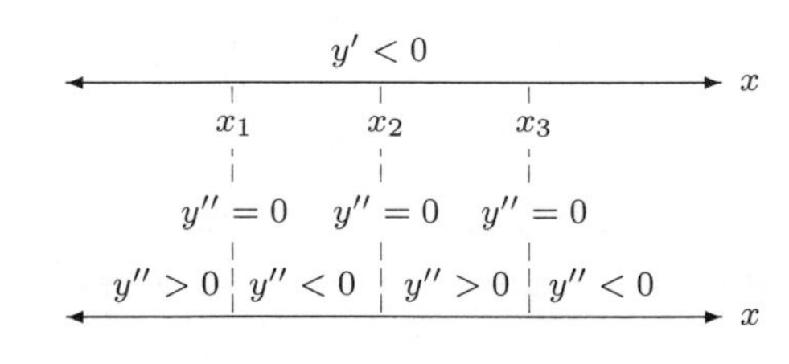

24.

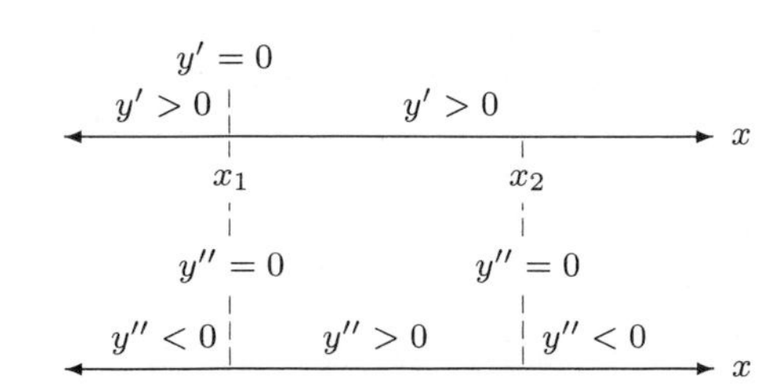

25.

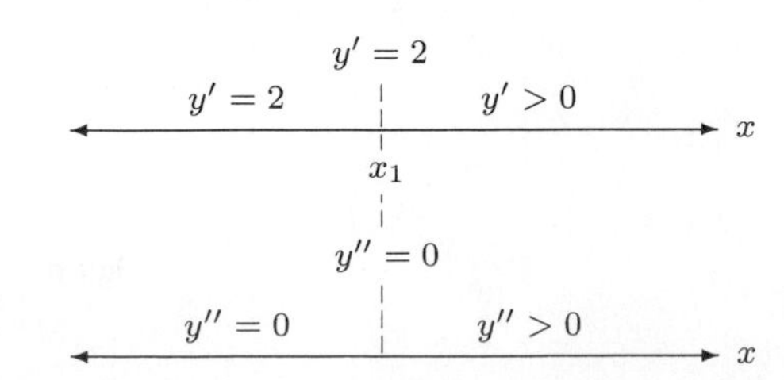

26. In 1774, Captain James Cook left 10 rabbits on a small Pacific island. The rabbit population is approximated by

$$P(t) = \frac{2000}{1 + e^{5.3-0.4t}}$$

with t measured in years since 1774. Using a calculator or computer:

(a) Graph P. Does the population level off?
(b) Estimate when the rabbit population grew most rapidly. How large was the population at that time?
(c) Find the inflection point on the graph and explain its significance for the rabbit population.
(d) What natural causes could lead to the shape of the graph of P?

27. **(a)** Water is flowing at a constant rate (i.e., constant volume per unit time) into a cylindrical container standing vertically. Sketch a graph showing the depth of water against time.
(b) Water is flowing at a constant rate into a cone-shaped container standing on its point. Sketch a graph showing the depth of the water against time.

28. If water is flowing at a constant rate (i.e., constant volume per unit time) into the Grecian urn in Figure 4.25, sketch a graph of the depth of the water against time. Mark on the graph the time at which the water reaches the widest point of the urn.

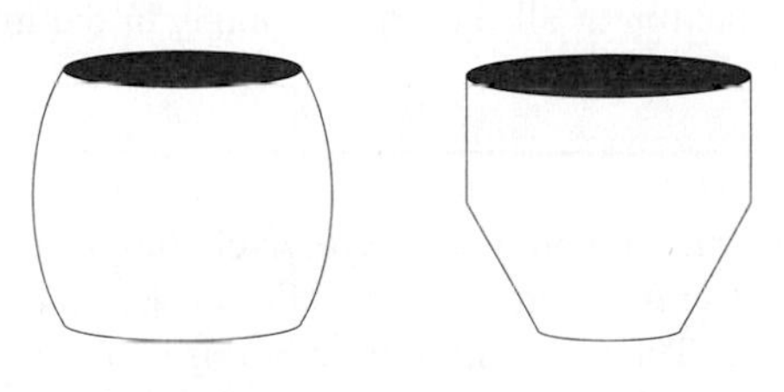

Figure 4.25 Figure 4.26

29. If water is flowing at a constant rate (i.e., constant volume per unit time) into the vase in Figure 4.26, sketch a graph of the depth of the water against time. Mark on the graph the time at which the water reaches the corner of the vase.

30. Water flows at a constant rate into the left side of the W-shaped container in Figure 4.27. Sketch a graph of the height, H, of the water in the left side of the container as a function of time, t. The container starts empty.

Figure 4.27

31. The vase in Figure 4.28 is filled with water at a constant rate (i.e., constant volume per unit time).

(a) Graph $y = f(t)$, the depth of the water, against time, t. Show on your graph the points at which the concavity changes.
(b) At what depth is $y = f(t)$ growing most quickly? Most slowly? Estimate the ratio between the growth rates at these two depths.

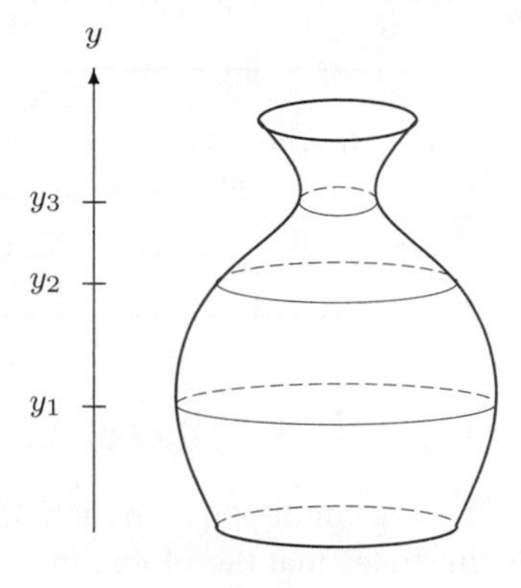

Figure 4.28

32. Assume that the polynomial f has exactly two local maxima and one local minimum, and that these are the only critical points of f.

(a) Sketch a possible graph of f.
(b) What is the largest number of zeros f could have?
(c) What is the least number of zeros f could have?
(d) What is the least number of inflection points f could have?
(e) What is the smallest degree f could have?
(f) Find a possible formula for $f(x)$.

33. Indicate on Figure 4.29 approximately where the inflection points of $f(x)$ are if the graph shows

(a) The function $f(x)$ **(b)** The derivative $f'(x)$
(c) The second derivative $f''(x)$

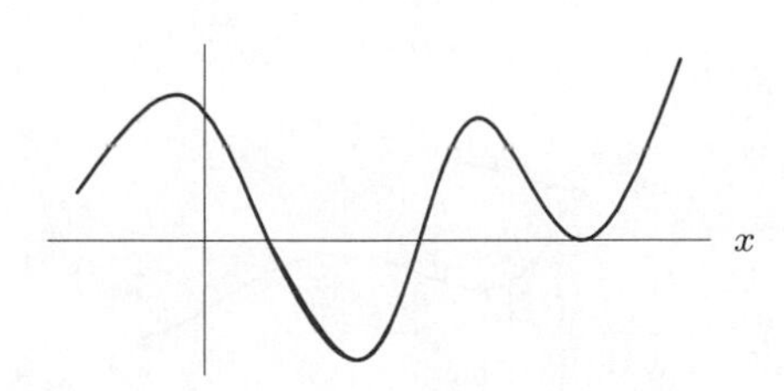

Figure 4.29

4.3 GLOBAL MAXIMA AND MINIMA

Global Maxima and Minima

The techniques for finding maximum and minimum values make up the field called *optimization*. Local maxima and minima occur where a function takes larger or smaller values than at nearby points. However, we are often interested in where a function is larger or smaller than at all other points. For example, a firm trying to maximize its profit may do so by minimizing its costs. We make the following definition:

For any function f:
- f has a **global minimum** at p if $f(p)$ is less than or equal to all values of f.
- f has a **global maximum** at p if $f(p)$ is greater than or equal to all values of f.

How Do We Find Global Maxima and Minima?

If f is a continuous function defined on an interval $a \leq x \leq b$ (including its endpoints), Figure 4.30 illustrates that the global maximum or minimum of f occurs either at a local maximum or a local minimum, respectively, or at one of the endpoints, $x = a$ or $x = b$, of the interval.

To find the global maximum and minimum of a continuous function on an interval including endpoints: Compare values of the function at all the critical points in the interval and at the endpoints.

What if the continuous function is defined on an interval $a < x < b$ (excluding its endpoints), or on the entire real line which has no endpoints? The function graphed in Figure 4.31 has no global maximum because the function has no largest value. The global minimum of this function coincides with one of the local minima and is marked. A function defined on the entire real line or on an interval excluding endpoints may or may not have a global maximum or a global minimum.

To find the global maximum and minimum of a continuous function on an interval excluding endpoints or on the entire real line: Find the values of the function at all the critical points and sketch a graph.

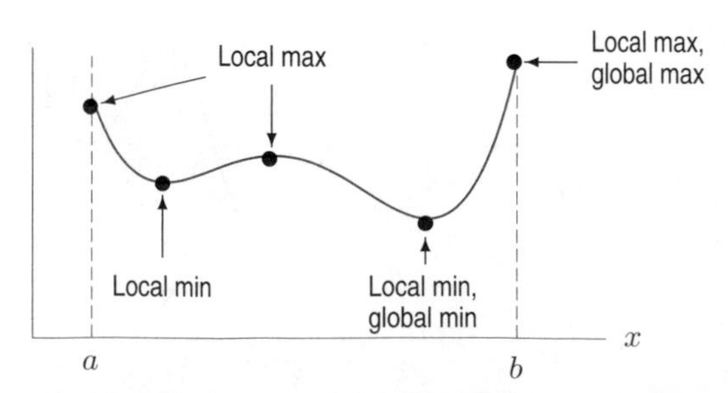

Figure 4.30: Global maximum and minimum on an interval domain, $a \leq x \leq b$

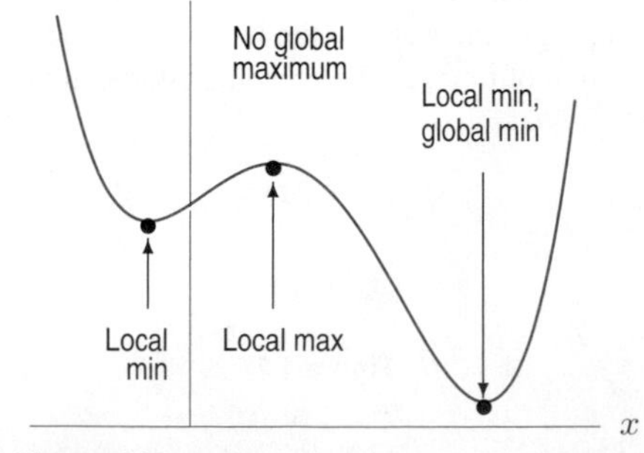

Figure 4.31: Global maximum and minimum on the entire real line

Example 1 Find the global maximum and minimum of $f(x) = x^3 - 9x^2 - 48x + 52$ on the interval $-5 \le x \le 14$.

Solution We have calculated the critical points of this function previously using

$$f'(x) = 3x^2 - 18x - 48 = 3(x+2)(x-8),$$

so $x = -2$ and $x = 8$ are critical points. Since the global maxima and minima occur at a critical point or at an endpoint of the interval, we evaluate f at these four points:

$$f(-5) = -58, \quad f(-2) = 104, \quad f(8) = -396, \quad f(14) = 360.$$

Comparing these four values, we see that the global maximum is 360 and occurs at $x = 14$, and that the global minimum is -396 and occurs at $x = 8$. See Figure 4.32.

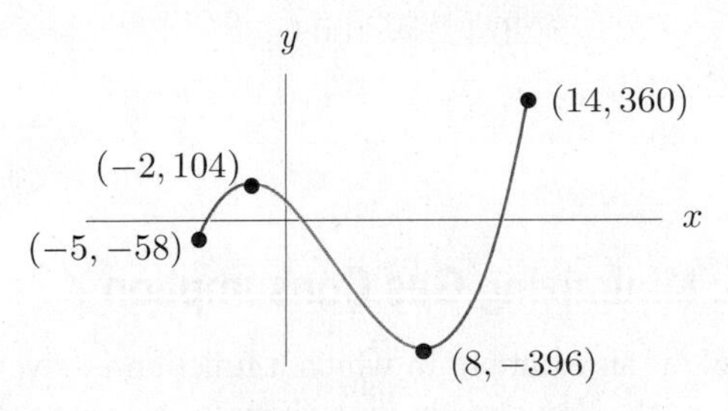

Figure 4.32: Global maximum and minimum on the interval $-5 \le x \le 14$

Example 2 For time, $t \ge 0$, in days, the rate at which photosynthesis takes place in the leaf of a plant, represented by the rate at which oxygen is produced, is approximated by[2]

$$p(t) = 100(e^{-0.02t} - e^{-0.1t}).$$

When is photosynthesis occurring fastest? What is that rate?

Solution To find the global maximum value of $p(t)$, we first find critical points. We differentiate, set equal to zero, and solve for t:

$$\begin{aligned} p'(t) = 100(-0.02e^{-0.02t} &+ 0.1e^{-0.1t}) = 0 \\ -0.02e^{-0.02t} &= -0.1e^{-0.1t} \\ \frac{e^{-0.02t}}{e^{-0.1t}} &= \frac{0.1}{0.02} \\ e^{-0.02t+0.1t} &= 5 \\ e^{0.08t} &= 5 \\ 0.08t &= \ln 5 \\ t &= \frac{\ln 5}{0.08} = 20.12 \text{ days.} \end{aligned}$$

Differentiating again gives

$$p''(t) = 100(0.0004e^{-0.02t} - 0.01e^{-0.1t})$$

and substituting $t = 20.12$ gives $p''(20.12) = -0.107$, so $t = 20.12$ is a local maximum. However, there is only one critical point, so this local maximum is the global maximum. See Figure 4.33.

[2]Examples adapted from Rodney Gentry, *Introduction to Calculus for the Biological and Health Sciences*, (Reading: Addison Wesley, 1978).

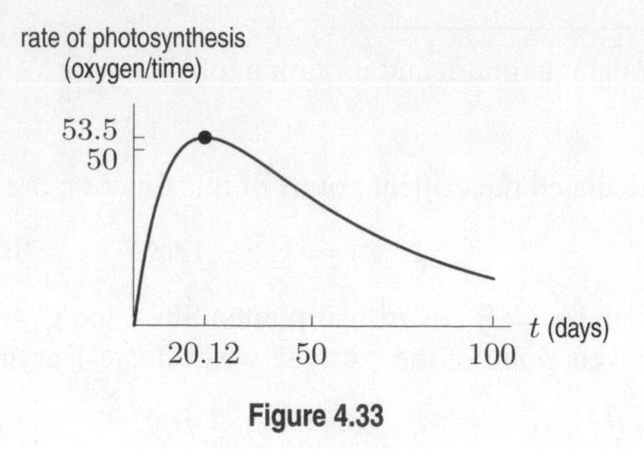

Figure 4.33

When $t = 20.12$ days, the rate, in units of oxygen per unit time, is

$$p(20.12) = 100\left(e^{-0.02(20.12)} - e^{-0.1(20.12)}\right) = 53.50.$$

A Graphical Example: Minimizing Gas Consumption

Next we look at an example in which a function is given graphically and the optimum values are read from a graph. You already know how to estimate the optimum values of $f(x)$ from a graph of $f(x)$—read off the highest and lowest values. In this example, we see how to estimate the optimum value of the quantity $f(x)/x$ from a graph of $f(x)$ against x.

The question we investigate is how to set driving speeds to maximize fuel efficiency.[3] We assume that gas consumption, g (in gallons/hour), as a function of velocity, v (in mph) is as shown in Figure 4.34. We want to minimize the gas consumption per *mile*, not the gas consumption per hour. Let $G = g/v$ represent the average gas consumption per mile. (The units of G are gallons/mile.)

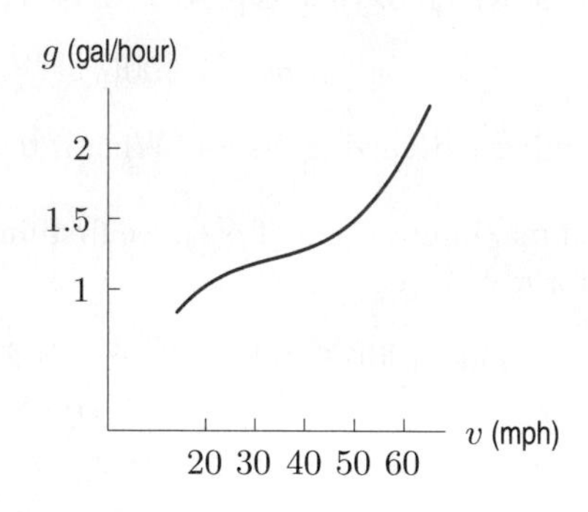

Figure 4.34: Gas consumption versus velocity

Example 3 Using Figure 4.34, estimate the velocity which minimizes $G = g/v$.

Solution We want to find the minimum value of $G = g/v$ when g and v are related by the graph in Figure 4.34. We could use Figure 4.34 to sketch a graph of G against v and estimate a critical point. But there is an easier way. Figure 4.35 shows that g/v is the slope of the line from the origin to the point P. Where on the curve should P be to make the slope a minimum? From the possible positions of the line shown in Figure 4.35, we see that the slope of the line is both a local and global minimum when the line is tangent to the curve. From Figure 4.36, we can see that the velocity at this point is about 50 mph. Thus to minimize gas consumption per mile, we should drive about 50 mph.

[3] Adapted from Peter D. Taylor, *Calculus: The Analysis of Functions* (Toronto: Wall & Emerson, 1992).

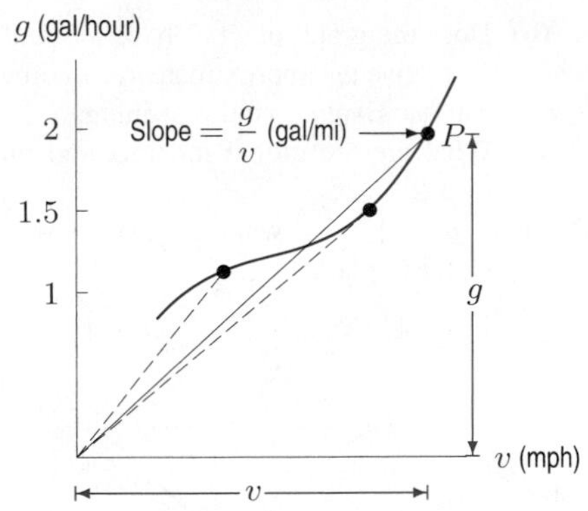

Figure 4.35: Graphical representation of gas consumption per mile, $G = g/v$

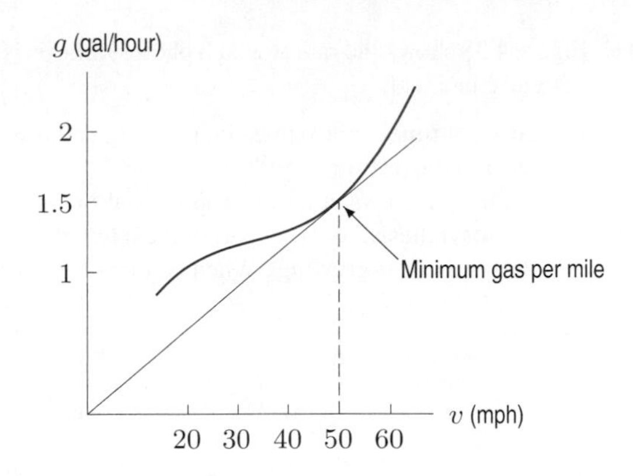

Figure 4.36: Velocity for maximum fuel efficiency

Problems for Section 4.3

For Problems 1–2, indicate all critical points on the given graphs. Which correspond to local minima, local maxima, global maxima, global minima, or none of these? (Note that the graphs are on closed intervals.)

1.

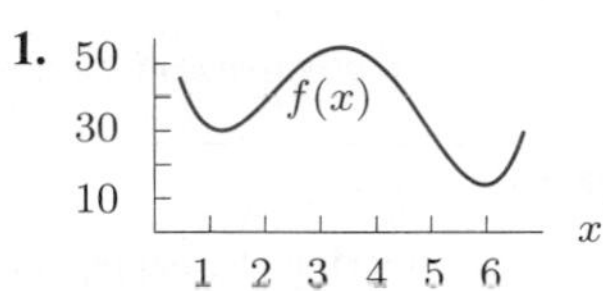

2.

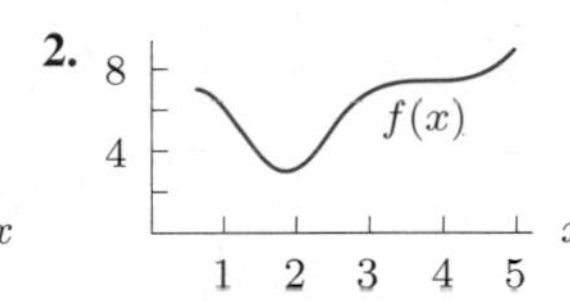

3. A grapefruit is tossed straight up with an initial velocity of 50 ft/sec. The grapefruit is 5 feet above the ground when it is released. Its height at time t is given by

$$y = -16t^2 + 50t + 5.$$

How high does it go before returning to the ground?

4. For each interval, use Figure 4.37 to choose the statement that gives the location of the global maximum and global minimum of f on the interval.

(a) $4 \le x \le 12$ **(b)** $11 \le x \le 16$
(c) $4 \le x \le 9$ **(d)** $8 \le x \le 18$

(I) Maximum at right endpoint, minimum at left endpoint.
(II) Maximum at right endpoint, minimum at critical point.
(III) Maximum at left endpoint, minimum at right endpoint.
(IV) Maximum at left endpoint, minimum at critical point.

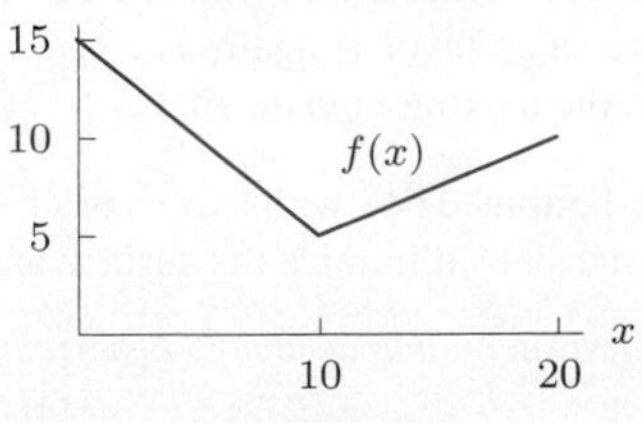

Figure 4.37

In Problems 5–8, graph a function with the given properties.

5. Has local minimum and global minimum at $x = 3$ but no local or global maximum.

6. Has local minimum at $x = 3$, local maximum at $x = 8$, but no global maximum or minimum.

7. Has local and global minimum at $x = 3$, local and global maximum at $x = 8$.

8. Has no local or global maxima or minima.

9. True or false? Give an explanation for your answer. The global maximum of $f(x) = x^2$ on every closed interval is at one of the endpoints of the interval.

In Problems 10–13, sketch the graph of a function on the interval $0 \le x \le 10$ with the given properties.

10. Has local minimum at $x = 3$, local maximum at $x = 8$, but global maximum and global minimum at the endpoints of the interval.

11. Has local and global maximum at $x = 3$, local and global minimum at $x = 10$.

12. Has local and global minimum at $x = 3$, local and global maximum at $x = 8$.

13. Has global maximum at $x = 0$, global minimum at $x = 10$, and no other local maxima or minima.

14. Plot the graph of $f(x) = x^3 - e^x$ using a graphing calculator or computer to find all local and global maxima and minima for: **(a)** $-1 \le x \le 4$ **(b)** $-3 \le x \le 2$

15. For some positive constant C, a patient's temperature change, T, due to a dose, D, of a drug is given by

$$T = \left(\frac{C}{2} - \frac{D}{3}\right) D^2.$$

(a) What dosage maximizes the temperature change?
(b) The sensitivity of the body to the drug is defined as dT/dD. What dosage maximizes sensitivity?

16. Figure 4.38 shows the rate at which photosynthesis is taking place in a leaf.

 (a) At what time, approximately, is photosynthesis proceeding fastest for $t \geq 0$?
 (b) If the leaf grows at a rate proportional to the rate of photosynthesis, for what part of the interval $0 \leq t \leq 200$ is the leaf growing? When is it growing fastest?

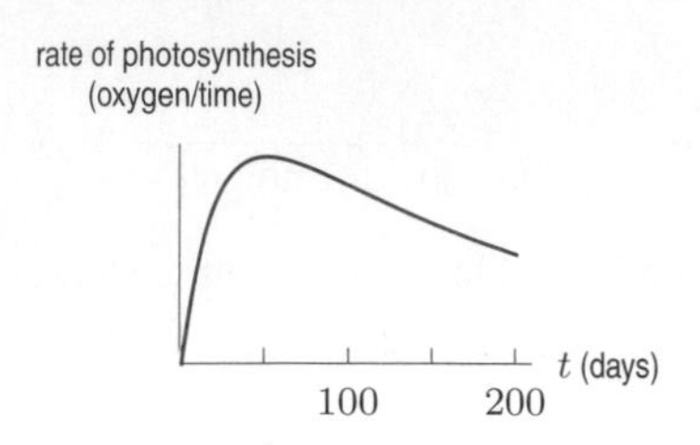

Figure 4.38

For each of the functions in Problems 17–21, do the following:

 (a) Find f' and f''.
 (b) Find the critical points of f.
 (c) Find any inflection points of f.
 (d) Evaluate f at its critical points and at the endpoints of the given interval. Identify local and global maxima and minima of f in the interval.
 (e) Graph f.

17. $f(x) = x^3 - 3x^2 \quad (-1 \leq x \leq 3)$
18. $f(x) = 2x^3 - 9x^2 + 12x + 1 \, (-0.5 \leq x \leq 3)$
19. $f(x) = x^3 - 3x^2 - 9x + 15 \quad (-5 \leq x \leq 4)$
20. $f(x) = x + \sin x \quad (0 \leq x \leq 2\pi)$
21. $f(x) = e^{-x} \sin x \quad (0 \leq x \leq 2\pi)$

In Problems 22–27, find the exact global maximum and minimum values of the function. The domain is all real numbers unless otherwise specified.

22. $g(x) = 4x - x^2 - 5$
23. $f(x) = x + 1/x$ for $x > 0$
24. $g(t) = te^{-t}$ for $t > 0$
25. $f(x) = x - \ln x$ for $x > 0$
26. $f(t) = \dfrac{t}{1+t^2}$
27. $f(t) = (\sin^2 t + 2) \cos t$
28. Figure 4.39 gives the derivative of $g(x)$ on $-2 \leq x \leq 2$.

 (a) Write a few sentences describing the behavior of $g(x)$ on this interval.
 (b) Does the graph of $g(x)$ have any inflection points? If so, give the approximate x-coordinates of their locations. Explain your reasoning.
 (c) What are the global maxima and minima of g on $[-2, 2]$?
 (d) If $g(-2) = 5$, what do you know about $g(0)$ and $g(2)$? Explain.

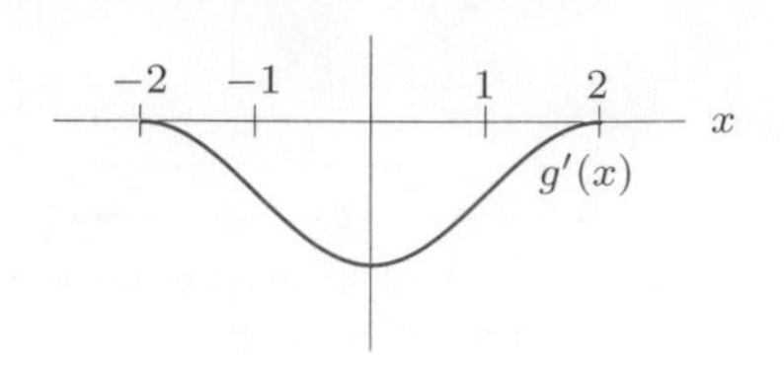

Figure 4.39

29. When you cough, your windpipe contracts. The speed, v, with which air comes out depends on the radius, r, of your windpipe. If R is the normal (rest) radius of your windpipe, then for $r \leq R$, the speed is given by:

$$v = a(R - r)r^2 \quad \text{where } a \text{ is a positive constant.}$$

What value of r maximizes the speed?

30. The energy expended by a bird per day, E, depends on the time spent foraging for food per day, F hours. Foraging for a shorter time requires better territory, which then requires more energy for its defense.[4] Find the foraging time that minimizes energy expenditure if

$$E = 0.25F + \frac{1.7}{F^2}.$$

31. Find the dimensions of the rectangle with perimeter 200 meters that has the largest area.

32. If you have 100 feet of fencing and want to enclose a rectangular area up against a long, straight wall, what is the largest area you can enclose?

33. A landscape architect plans to enclose a 3000 square foot rectangular region in a botanical garden. She will use shrubs costing \$25 per foot along three sides and fencing costing \$10 per foot along the fourth side. Find the minimum total cost.

34. A closed box has a fixed surface area A and a square base with side x.

 (a) Find a formula for its volume, V, as a function of x.
 (b) Sketch a graph of V against x.
 (c) Find the maximum value of V.

35. A square-bottomed box with a top has a fixed volume, V. What dimensions minimize the surface area?

[4] Adapted from Graham Pyke, reported by J. R. Krebs and N. B. Davis in *An Introduction to Behavioural Ecology* (Oxford: Blackwell, 1987).

36. On the west coast of Canada, crows eat whelks (a shellfish). To open the whelks, the crows drop them from the air onto a rock. If the shell does not smash the first time, the whelk is dropped again.[5] The average number of drops, n, needed when the whelk is dropped from a height of x meters is approximated by

$$n(x) = 1 + \frac{27}{x^2}.$$

(a) Give the total vertical distance the crow travels upward to open a whelk as a function of drop height, x.
(b) Crows are observed to drop whelks from the height that minimizes the total vertical upward distance traveled per whelk. What is this height?

37. During a flu outbreak in a school of 763 children, the number of infected children, I, was expressed in terms of the number of susceptible (but still healthy) children, S, by the expression[6]

$$I = 192 \ln\left(\frac{S}{762}\right) - S + 763.$$

What is the maximum possible number of infected children?

38. **(a)** Find the critical points of $p(1-p)^4$.
(b) Classify the critical points as local maxima, local minima, or neither.
(c) What are the maximum and minimum values of $p(1-p)^4$ on $0 \le x \le 1$?

39. An apple tree produces, on average, 400 kg of fruit each season. However, if more than 200 trees are planted per km^2, crowding reduces the yield by 1 kg for each tree over 200.

(a) Express the total yield from one square kilometer as a function of the number of trees on it. Graph this function.
(b) How many trees should a farmer plant on each square kilometer to maximize yield?

40. The number of offspring in a population may not be a linear function of the number of adults. The Ricker curve, used to model fish populations, claims that $y = axe^{-bx}$, where x is the number of adults, y is the number of offspring, and a and b are positive constants.

(a) Find and classify all critical points of the Ricker curve.
(b) Is there a global maximum? What does this imply about populations?

41. A chemical reaction converts substance A to substance Y; the presence of Y catalyzes the reaction. At the start of the reaction, the quantity of A present is a grams. At time t seconds later, the quantity of Y present is y grams. The rate of the reaction, in grams/sec, is given by

$$\text{Rate} = ky(a-y), \quad k \text{ is a positive constant.}$$

(a) For what values of y is the rate nonnegative? Graph the rate against y.
(b) For what values of y is the rate a maximum?

42. In a chemical reaction, substance A combines with substance B to form substance Y. At the start of the reaction, the quantity of A present is a grams, and the quantity of B present is b grams. Assume $a < b$. At time t seconds after the start of the reaction, the quantity of Y present is y grams. For certain types of reactions, the rate of the reaction, in grams/sec, is given by

$$\text{Rate} = k(a-y)(b-y), \quad k \text{ is a positive constant.}$$

(a) For what values of y is the rate nonnegative? Graph the rate against y.
(b) Use your graph to find the value of y at which the rate of the reaction is fastest.

43. The oxygen supply, S, in the blood depends on the hematocrit, H, the percentage of red blood cells in the blood:

$$S = aHe^{-bH} \quad \text{for positive constants } a, b.$$

(a) What value of H maximizes the oxygen supply? What is the maximum oxygen supply?
(b) How does increasing the value of the constants a and b change the maximum value of S?

44. The quantity of a drug in the bloodstream t hours after a tablet is swallowed is given, in mg, by

$$q(t) = 20(e^{-t} - e^{-2t}).$$

(a) How much of the drug is in the bloodstream at time $t = 0$?
(b) When is the maximum quantity of drug in the bloodstream? What is that maximum?
(c) In the long run, what happens to the quantity?

45. When birds lay eggs, they do so in clutches of several at a time. When the eggs hatch, each clutch gives rise to a brood of baby birds. We want to determine the clutch size which maximizes the number of birds surviving to adulthood per brood. If the clutch is small, there are few baby birds in the brood; if the clutch is large, there are so many baby birds to feed that most die of starvation.

[5] Adapted from Reto Zach, reported by J. R. Krebs and N. B. Davis in *An Introduction to Behavioural Ecology* (Oxford: Blackwell, 1987).

[6] Data from Communicable Disease Surveillance Centre (UK), reported in "Influenza in a Boarding School", *British Medical Journal*; March 4, 1978.

[7] Data from C. M. Perrins and D. Lack reported by J. R. Krebs and N. B. Davies in *An Introduction to Behavioural Ecology* (Oxford: Blackwell, 1987).

The number of surviving birds per brood as a function of clutch size is shown by the benefit curve in Figure 4.40.[7]

(a) Estimate the clutch size which maximizes the number of survivors per brood.

(b) Suppose also that there is a biological cost to having a larger clutch: the female survival rate is reduced by large clutches. This cost is represented by the dotted line in Figure 4.40. If we take cost into account by assuming that the optimal clutch size in fact maximizes the vertical distance between the curves, what is the new optimal clutch size?

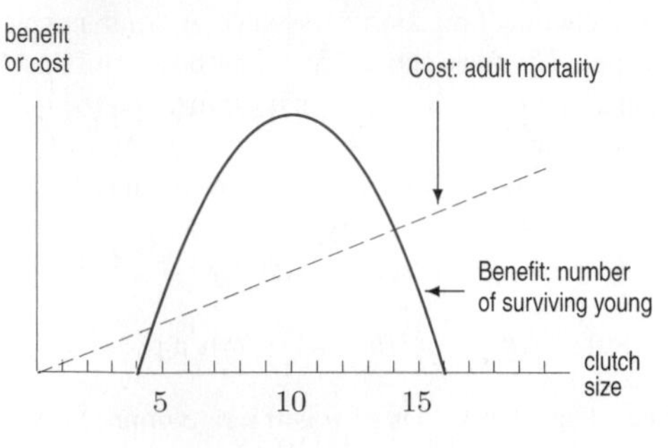

Figure 4.40

46. Let $f(v)$ be the amount of energy consumed by a flying bird, measured in joules per second (a joule is a unit of energy), as a function of its speed v (in meters/sec). Let $a(v)$ be the amount of energy consumed by the same bird, measured in joules per meter.

(a) Suggest a reason (in terms of the way birds fly) for the shape of the graph of $f(v)$ in Figure 4.41.

(b) What is the relationship between $f(v)$ and $a(v)$?

(c) Where is $a(v)$ a minimum?

(d) Should the bird try to minimize $f(v)$ or $a(v)$ when it is flying? Why?

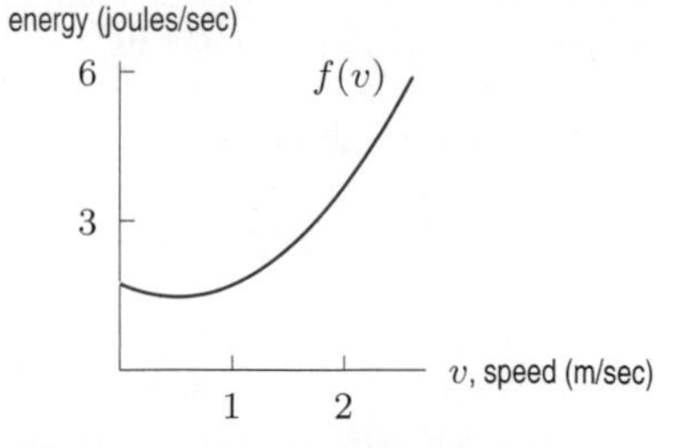

Figure 4.41

47. As an epidemic spreads through a population, the number of infected people, I, is expressed as a function of the number of susceptible people, S, by

$$I = k\ln\left(\frac{S}{S_0}\right) - S + S_0 + I_0, \quad \text{for } k, S_0, I_0 > 0.$$

(a) Find the maximum number of infected people.

(b) The constant k is a characteristic of the particular disease; the constants S_0 and I_0 are the values of S and I when the disease starts. Which of the following affects the maximum possible value of I? Explain.

- The particular disease, but not how it starts.
- How the disease starts, but not the particular disease.
- Both the particular disease and how it starts.

48. The hypotenuse of a right triangle has one end at the origin and one end on the curve $y = x^2e^{-3x}$, with $x \geq 0$. One of the other two sides is on the x-axis, the other side is parallel to the y-axis. Find the maximum area of such a triangle. At what x-value does it occur?

49. A right triangle has one vertex at the origin and one vertex on the curve $y = e^{-x/3}$ for $1 \leq x \leq 5$. One of the two perpendicular sides is along the x-axis; the other is parallel to the y-axis. Find the maximum and minimum areas for such a triangle.

50. A person's blood pressure, p, in millimeters of mercury (mm Hg) is given, for t in seconds, by

$$p = 100 + 20\sin(2.5\pi t).$$

(a) What are the maximum and minimum values of blood pressure?

(b) What is the interval between successive maxima?

(c) Show your answers on a graph of blood pressure against time.

51. A pigeon is released from a boat (point B in Figure 4.42) floating on a lake. Because of falling air over the cool water, the energy required to fly one meter over the lake is twice the corresponding energy e required for flying over the bank ($e = 3$ joule/meter). To minimize the energy required to fly from B to the loft, L, the pigeon heads to a point P on the bank and then flies along the bank to L. The distance $\overline{AL}$ is 2000 m, and $\overline{AB}$ is 500 m. The angle at A is a right angle.

(a) Express the energy required to fly from B to L via P as a function of the angle θ (the angle BPA).

(b) What is the optimal angle θ?

(c) Does your answer change if $\overline{AL}$, $\overline{AB}$, and e have different numerical values?

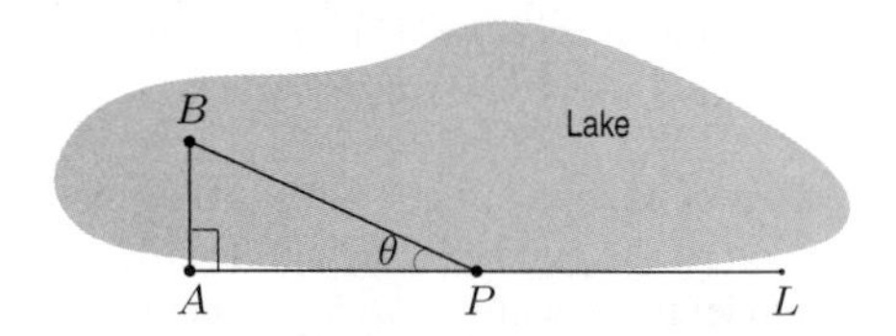

Figure 4.42

52. The bell-shaped curve of statistics has formula

$$p(x) = \frac{1}{\sigma\sqrt{2\pi}}e^{-(x-\mu)^2/(2\sigma^2)}$$

where μ is the mean and σ is the standard deviation.

(a) Where does $p(x)$ have a maximum?

(b) Does $p(x)$ have a point of inflection? If so, where?

4.4 PROFIT, COST, AND REVENUE

Maximizing Profit

A fundamental issue for a producer of goods is how to maximize profit. For a quantity, q, the profit $\pi(q)$ is the difference between the revenue, $R(q)$, and the cost, $C(q)$, of supplying that quantity. Thus, $\pi(q) = R(q) - C(q)$. The marginal cost, $MC = C'$, is the derivative of C; marginal revenue is $MR = R'$.

Now we look at how to maximize total profit, given functions for revenue and cost. The next example suggests a criterion for identifying the optimal production level.

Example 1 Estimate the maximum profit if the revenue and cost are given by the curves R and C, respectively, in Figure 4.43.

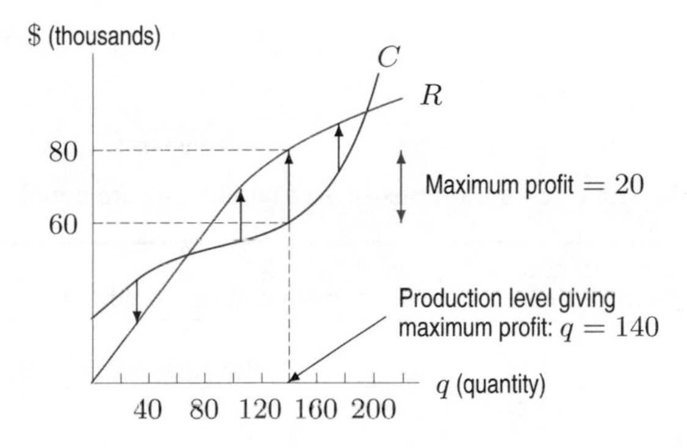

Figure 4.43: Maximum profit at $q = 140$

Solution Since profit is revenue minus cost, the profit is represented by the vertical distance between the cost and revenue curves, marked by the vertical arrows in Figure 4.43. When revenue is below cost, the company is taking a loss; when revenue is above cost, the company is making a profit. The maximum profit must occur between about $q = 70$ and $q = 200$, which is the interval in which the company is making a profit. Profit is maximized when the vertical distance between the curves is largest (and revenue is above cost). This occurs at approximately $q = 140$.

The profit accrued at $q = 140$ is the vertical distance between the curves, so the maximum profit = \$80,000 − \$60,000 = \$20,000.

Maximum Profit Can Occur Where $MR = MC$

We now analyze the marginal costs and marginal revenues near the optimal point. Zooming in on Figure 4.43 around $q = 140$ gives Figure 4.44.

At a production level q_1 to the left of 140 in Figure 4.44, marginal cost is less than marginal revenue. The company would make more money by producing more units, so production should be increased (toward a production level of 140). At any production level q_2 to the right of 140, marginal cost is greater than marginal revenue. The company would lose money by producing more units and would make more money by producing fewer units. Production should be adjusted down toward 140.

What about the marginal revenue and marginal cost at $q = 140$? Since $MC < MR$ to the left of 140, and $MC > MR$ to the right of 140, we expect $MC = MR$ at 140. In this example, profit is maximized at the point where the slopes of the cost and revenue graphs are equal.

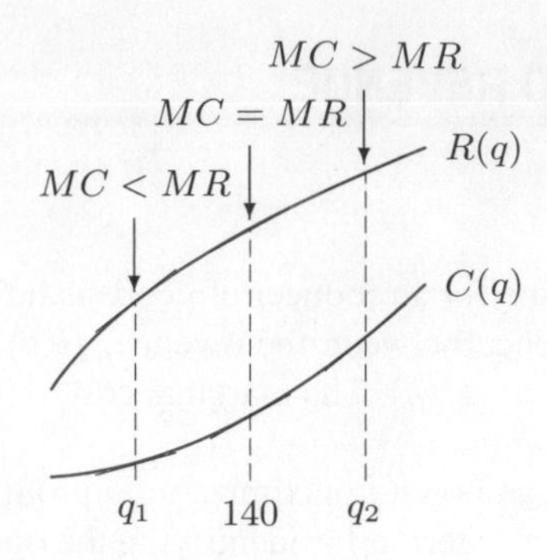

Figure 4.44: Example 1: Maximum profit occurs where $MC = MR$

We can get the same result analytically. Global maxima and minima of a function can only occur at critical points of the function or at the endpoints of the interval. To find critical points of π, look for zeros of the derivative:

$$\pi'(q) = R'(q) - C'(q) = 0.$$

So

$$R'(q) = C'(q),$$

that is, the slopes of the graphs of $R(q)$ and $C(q)$ are equal at q. In economic language,

> The maximum (or minimum) profit can occur where
>
> $$\text{Marginal profit} = 0,$$
>
> that is, where
>
> $$\text{Marginal revenue} = \text{Marginal cost}.$$

Of course, maximum or minimum profit does not *have* to occur where $MR = MC$; either one could occur at an endpoint. Example 2 shows how to visualize maxima and minima of the profit on a graph of marginal revenue and marginal cost.

Example 2 The total revenue and total cost curves for a product are given in Figure 4.45.

(a) Sketch the marginal revenue and marginal cost, MR and MC, on the same axes. Mark the two quantities where marginal revenue equals marginal cost. What is the significance of these two quantities? At which quantity is profit maximized?

(b) Graph the profit function $\pi(q)$.

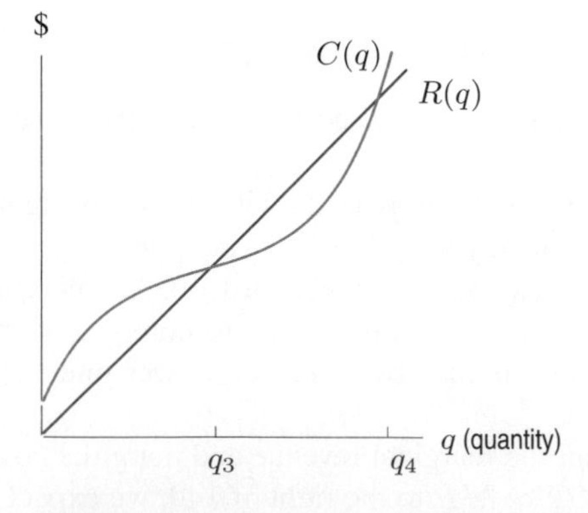

Figure 4.45: Total revenue and total cost

Solution (a) Since $R(q)$ is a straight line with positive slope, the graph of its derivative, MR, is a horizontal line. (See Figure 4.46.) Since $C(q)$ is always increasing, its derivative, MC, is always positive. As q increases, the cost curve changes from concave down to concave up, so the derivative of the cost function, MC, changes from decreasing to increasing. (See Figure 4.46.) The local minimum on the marginal cost curve corresponds to the inflection point of $C(q)$.

Where is profit maximized? We know that the maximum profit can occur when Marginal revenue = Marginal cost, that is where the curves in Figure 4.46 cross at q_1 and q_2. Do these points give the maximum profit?

We first consider q_1. To the left of q_1, we have $MR < MC$, so $\pi' = MR - MC$ is negative and the profit function is decreasing there. To the right of q_1, we have $MR > MC$, so π' is positive and the profit function is increasing. This behavior, decreasing and then increasing, means that the profit function has a local minimum at q_1. This is certainly not the production level we want.

What happens at q_2? To the left of q_2, we have $MR > MC$, so π' is positive and the profit function is increasing. To the right of q_2, we have $MR < MC$, so π' is negative and the profit function is decreasing. This behavior, increasing and then decreasing, means that the profit function has a local maximum at q_2. The global maximum profit occurs either at the production level q_2 or at an endpoint (the largest and smallest possible production levels). Since the profit is negative at the endpoints (see Figure 4.45), the global maximum occurs at q_2.

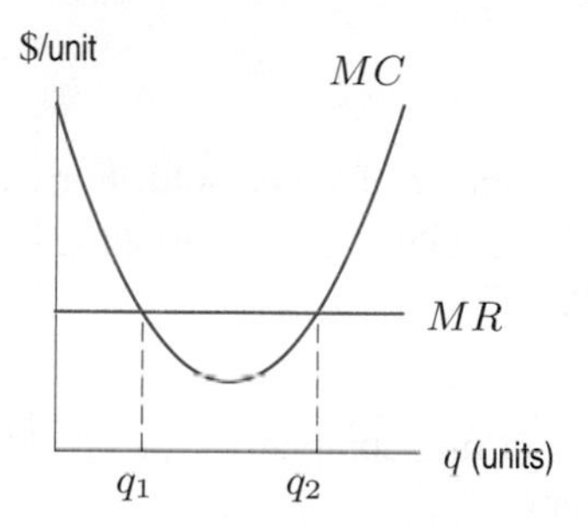

Figure 4.46: Marginal revenue and marginal cost

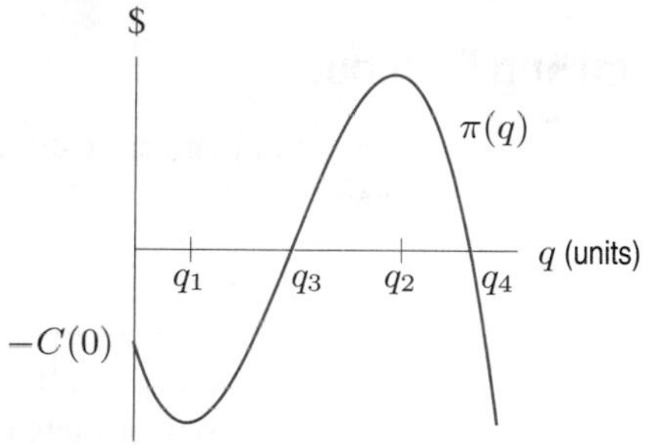

Figure 4.47: Profit function

(b) The graph of the profit function is in Figure 4.47. At the maximum and minimum, the slope of the profit curve is zero:

$$\pi'(q_1) = \pi'(q_2) = 0.$$

Note that since $R(0) = 0$ and $C(0)$ represents the fixed costs of production, we have

$$\pi(0) = R(0) - C(0) = -C(0).$$

Therefore the vertical intercept of the profit function is a negative number, equal in magnitude to the size of the fixed cost.

Example 3 Find the quantity which maximizes profit if the total revenue and total cost (in dollars) are given by

$$R(q) = 5q - 0.003q^2$$
$$C(q) = 300 + 1.1q$$

where q is quantity and $0 \leq q \leq 1000$ units. What production level gives the minimum profit?

Solution We begin by looking for production levels that give Marginal revenue = Marginal cost. Since

$$MR = R'(q) = 5 - 0.006q$$
$$MC = C'(q) = 1.1,$$

$MR = MC$ leads to

$$5 - 0.006q = 1.1$$
$$q = \frac{3.9}{0.006} = 650 \text{ units.}$$

Does this represent a local maximum or minimum of the profit π? To decide, look to the left and right of 650 units.

When $q = 649$, we have $MR = \$1.106$ per unit, which is greater than $MC = \$1.10$ per unit.

Thus, producing one more unit (the 650th) brings in more revenue than it costs, so profit increases.

When $q = 651$, we have $MR = \$1.094$ per unit, which is less than $MC = \$1.10$ per unit.

It is not profitable to produce the 651st unit. We conclude that $q = 650$ gives a local maximum for the profit function π.

To check whether $q = 650$ gives a global maximum, we compare the profit at the endpoints, $q = 0$ and $q = 1000$, with the profit at $q = 650$.

At $q = 0$, the only cost is \$300 (the fixed costs) and there is no revenue, so $\pi(0) = -\$300$.
At $q = 1000$, we have $R(1000) = \$2000$ and $C(1000) = \$1400$, so $\pi(1000) = \$600$.
At $q = 650$, we have $R(650) = \$1982.50$ and $C(650) = \$1015$, so $\pi(650) = \$967.50$.

Therefore, the maximum profit is obtained at a production level of $q = 650$ units. The minimum profit (a loss) occurs when $q = 0$ and there is no production at all.

Maximizing Revenue

For some companies, costs do not depend on the number of items sold. For example, a city bus company with a fixed schedule has the same costs no matter how many people ride the buses. In such a situation, profit is maximized by maximizing revenue.

Example 4 At a price of \$80 for a half-day trip, a white-water rafting company attracts 300 customers. Every \$5 decrease in price attracts an additional 30 customers.

(a) Find the demand equation.
(b) Express revenue as a function of price.
(c) What price should the company charge per trip to maximize revenue?

Solution (a) We first find the equation relating price to demand. If price, p, is 80, the number of trips sold, q, is 300. If p is 75, then q is 330, and so on. See Table 4.1. Because demand changes by a constant (30 people) for every \$5 drop in price, q is a linear function of p. Then

$$\text{Slope} = \frac{300 - 330}{80 - 75} = -\frac{30}{5} = -6 \text{ people/dollar},$$

so the demand equation is $q = -6p + b$. Since $p = 80$ when $q = 300$, we have

$$\begin{aligned} 300 &= -6 \cdot 80 + b \\ b &= 300 + 6 \cdot 80 = 780. \end{aligned}$$

The demand equation is $q = -6p + 780$.

(b) Since revenue $R = p \cdot q$, revenue as a function of price is

$$R(p) = p(-6p + 780) = -6p^2 + 780p.$$

(c) Figure 4.48 shows this revenue function has a maximum. To find it, we differentiate:

$$\begin{aligned} R'(q) &= -12p + 780 = 0 \\ p &= \frac{780}{12} = 65. \end{aligned}$$

The maximum revenue is achieved when the price is $65.

Table 4.1 *Demand for rafting trips*

Price, p	Number of trips sold, q
80	300
75	330
70	360
65	390
...	...

Figure 4.48: Revenue for a rafting company as a function of price

Problems for Section 4.4

1. Table 4.2 shows cost, $C(q)$, and revenue, $R(q)$.

(a) At approximately what production level, q, is profit maximized? Explain your reasoning.
(b) What is the price of the product?
(c) What are the fixed costs?

Table 4.2

q	0	500	1000	1500	2000	2500	3000
$R(q)$	0	1500	3000	4500	6000	7500	9000
$C(q)$	3000	3800	4200	4500	4800	5500	7400

2. Figure 4.49 shows cost and revenue. For what production levels is the profit function positive? Negative? Estimate the production at which profit is maximized.

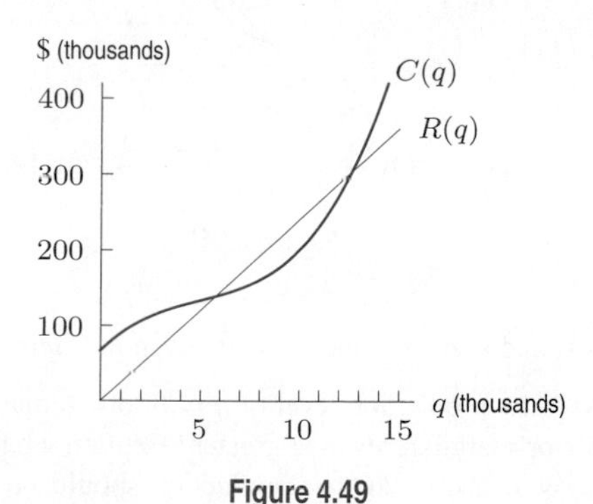

Figure 4.49

3. Using the cost and revenue graphs in Figure 4.50, sketch the following functions. Label the points q_1 and q_2.

(a) Total profit **(b)** Marginal cost
(c) Marginal revenue

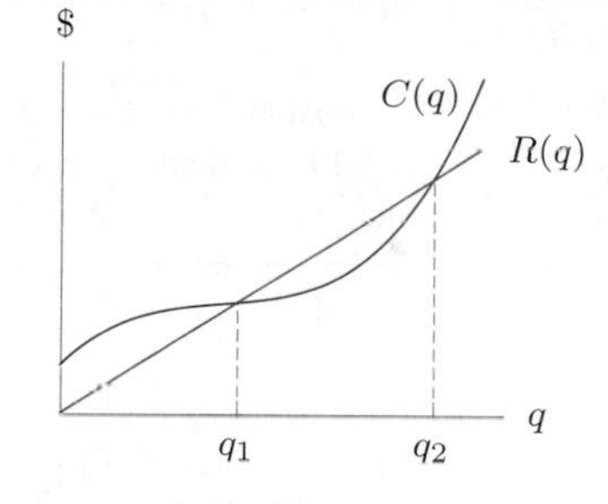

Figure 4.50

4. A demand function is $p = 400 - 2q$, where q is the quantity of the good sold for price $\$p$.

(a) Find an expression for the total revenue, R, in terms of q.
(b) Differentiate R with respect to q to find the marginal revenue, MR, in terms of q. Calculate the marginal revenue when $q = 10$.
(c) Calculate the change in total revenue when production increases from $q = 10$ to $q = 11$ units. Confirm that a one unit increase in q gives a reasonable approximation to the exact value of MR obtained in part (b).

5. Let $C(q)$ represent the cost, $R(q)$ the revenue, and $\pi(q)$ the total profit, in dollars, of producing q items.

(a) If $C'(50) = 75$ and $R'(50) = 84$, approximately how much profit is earned by the 51st item?
(b) If $C'(90) = 71$ and $R'(90) = 68$, approximately how much profit is earned by the 91st item?
(c) If $\pi(q)$ is a maximum when $q = 78$, how do you think $C'(78)$ and $R'(78)$ compare? Explain.

6. Figure 4.46 on page 197 shows the points, q_1 and q_2, where marginal revenue equals marginal cost.

(a) On the graph of the corresponding total cost and total revenue functions in Figure 4.51, label the points q_1 and q_2. Using slopes, explain the significance of these points.

(b) Explain in terms of profit why one is a local minimum and one is a local maximum.

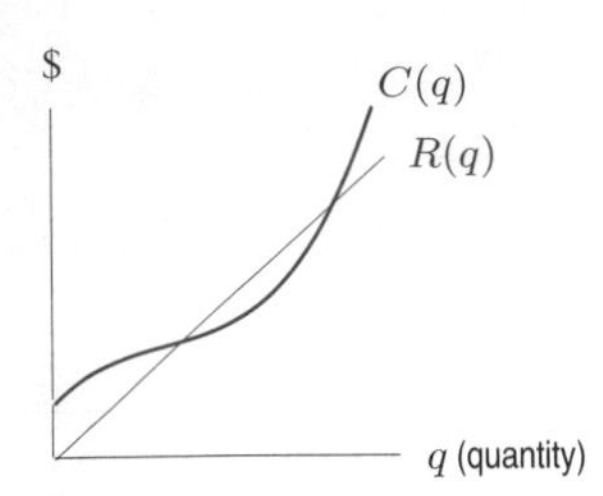

Figure 4.51

7. Table 4.3 shows marginal cost, MC, and marginal revenue, MR.

(a) Use the marginal cost and marginal revenue at a production of $q = 5000$ to determine whether production should be increased or decreased from 5000.

(b) Estimate the production level that maximizes profit.

Table 4.3

q	5000	6000	7000	8000	9000	10000
MR	60	58	56	55	54	53
MC	48	52	54	55	58	63

8. Marginal revenue and marginal cost are given in the following table. Estimate the production levels that could maximize profit. Explain.

q	1000	2000	3000	4000	5000	6000
MR	78	76	74	72	70	68
MC	100	80	70	65	75	90

9. Figure 4.52 shows cost and revenue for a product.

(a) Estimate the production level that maximizes profit.

(b) Graph marginal revenue and marginal cost for this product on the same coordinate system. Label on this graph the production level that maximizes profit.

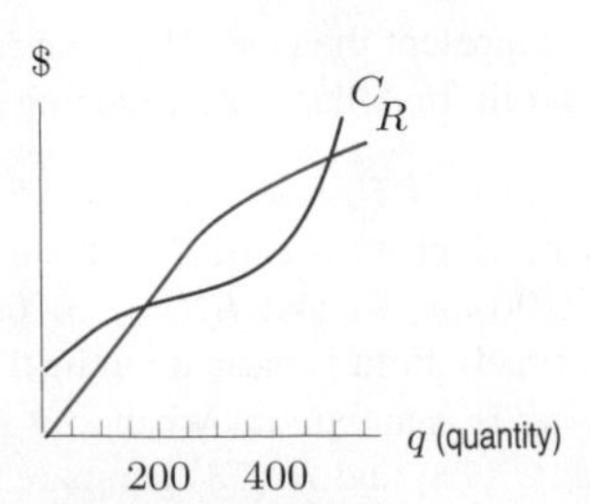

Figure 4.52

10. Figure 4.53 shows graphs of marginal cost and marginal revenue. Estimate the production levels that could maximize profit. Explain your reasoning.

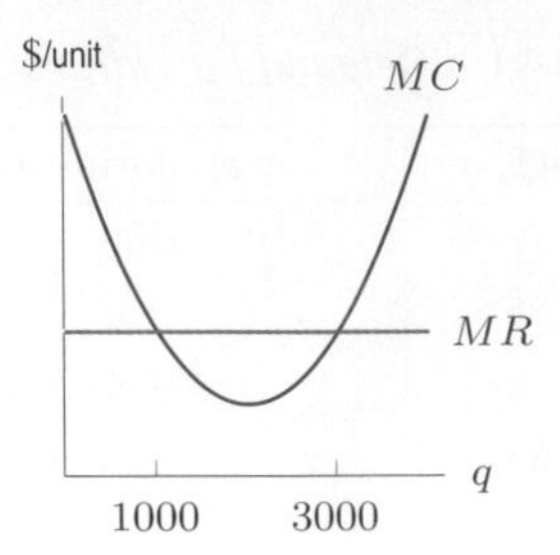

Figure 4.53

11. The marginal cost and marginal revenue of a company are $MC(q) = 0.03q^2 - 1.4q + 34$ and $MR(q) = 30$, where q is the number of items manufactured. To increase profits, should the company increase or decrease production from each of the following levels?

(a) 25 items **(b)** 50 items **(c)** 80 items

12. A manufacturing process has marginal costs given in the table; the item sells for $30 per unit. At how many quantities, q, does the profit appear to be a maximum? In what intervals do these quantities appear to lie?

q	0	10	20	30	40	50	60
MC ($/unit)	34	23	18	19	26	39	58

13. Cost and revenue functions are given in Figure 4.54. Approximately what quantity maximizes profits?

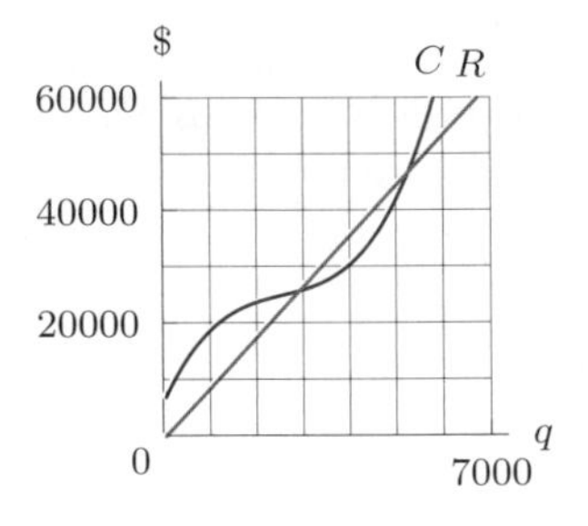

Figure 4.54

14. Cost and revenue functions are given in Figure 4.54.

(a) At a production level of $q = 3000$, is marginal cost or marginal revenue greater? Explain what this tells you about whether production should be increased or decreased.

(b) Answer the same questions for $q = 5000$.

15. When production is 2000, marginal revenue is $4 per unit and marginal cost is $3.25 per unit. Do you expect maximum profit to occur at a production level above or below 2000? Explain.

16. Revenue is given by $R(q) = 450q$ and cost is given by $C(q) = 10{,}000 + 3q^2$. At what quantity is profit maximized? What is the total profit at this production level?

17. The demand equation for a product is $p = 45 - 0.01q$. Write the revenue as a function of q and find the quantity that maximizes revenue. What price corresponds to this quantity? What is the total revenue at this price?

18. Revenue and cost functions for a company are given in Figure 4.55.

(a) Estimate the marginal cost at $q = 400$.
(b) Should the company produce the 500th item? Why?
(c) Estimate the quantity which maximizes profit.

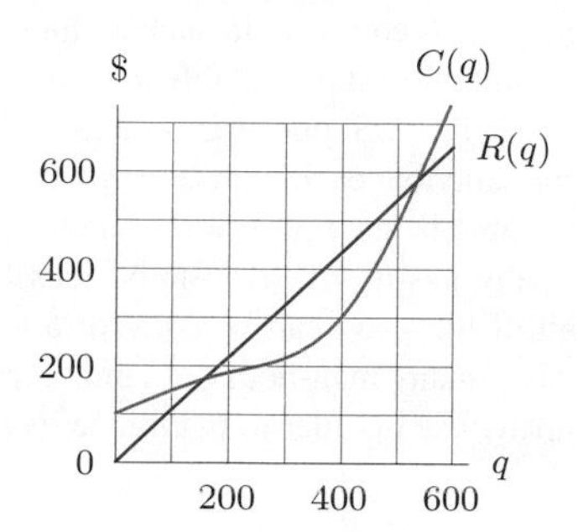

Figure 4.55

19. The following table gives the cost and revenue, in dollars, for different production levels, q.

(a) At approximately what production level is profit maximized?
(b) What price is charged per unit for this product?
(c) What are the fixed costs of production?

q	0	100	200	300	400	500
$R(q)$	0	500	1000	1500	2000	2500
$C(q)$	700	900	1000	1100	1300	1900

20. An ice cream company finds that at a price of \$4.00, demand is 4000 units. For every \$0.25 decrease in price, demand increases by 200 units. Find the price and quantity sold that maximize revenue.

21. At a price of \$8 per ticket, a musical theater group can fill every seat in the theater, which has a capacity of 1500. For every additional dollar charged, the number of people buying tickets decreases by 75. What ticket price maximizes revenue?

22. The demand for tickets to an amusement park is given by $p = 70 - 0.02q$, where p is the price of a ticket in dollars and q is the number of people attending at that price.

(a) What price generates an attendance of 3000 people? What is the total revenue at that price? What is the total revenue if the price is \$20?
(b) Write the revenue function as a function of attendance, q, at the amusement park.
(c) What attendance maximizes revenue?
(d) What price should be charged to maximize revenue?
(e) What is the maximum revenue? Can we determine the corresponding profit?

23. The demand equation for a quantity q of a product at price p, in dollars, is $p = -5q + 4000$. Companies producing the product report the cost, C, in dollars, to produce a quantity q is $C = 6q + 5$ dollars.

(a) Express a company's profit, in dollars, as a function of q.
(b) What production level earns the company the largest profit?
(c) What is the largest profit possible?

24. **(a)** Production of an item has fixed costs of \$10,000 and variable costs of \$2 per item. Express the cost, C, of producing q items.
(b) The relationship between price, p, and quantity, q, demanded is linear. Market research shows that 10,100 items are sold when the price is \$5 and 12,872 items are sold when the price is \$4.50. Express q as a function of price p.
(c) Express the profit earned as a function of q.
(d) How many items should the company produce to maximize profit? (Give your answer to the nearest integer.) What is the profit at that production level?

25. You run a small furniture business. You sign a deal with a customer to deliver up to 400 chairs, the exact number to be determined by the customer later. The price will be \$90 per chair up to 300 chairs, and above 300, the price will be reduced by \$0.25 per chair (on the whole order) for every additional chair over 300 ordered. What are the largest and smallest revenues your company can make under this deal?

26. A warehouse selling cement has to decide how often and in what quantities to reorder. It is cheaper, on average, to place large orders, because this reduces the ordering cost per unit. On the other hand, larger orders mean higher storage costs. The warehouse always reorders cement in the same quantity, q. The total weekly cost, C, of ordering and storage is given by

$$C = \frac{a}{q} + bq, \quad \text{where } a, b \text{ are positive constants.}$$

(a) Which of the terms, a/q and bq, represents the ordering cost and which represents the storage cost?
(b) What value of q gives the minimum total cost?

27. A business sells an item at a constant rate of r units per month. It reorders in batches of q units, at a cost of $a + bq$ dollars per order. Storage costs are k dollars per item per month, and, on average, $q/2$ items are in storage, waiting to be sold. [Assume r, a, b, k are positive constants.]

(a) How often does the business reorder?

(b) What is the average monthly cost of reordering?

(c) What is the total monthly cost, C of ordering and storage?

(d) Obtain Wilson's lot size formula, the optimal batch size which minimizes cost.

28. The marginal revenue and marginal cost for a certain item are graphed in Figure 4.56. Do the following quantities maximize profit for the company? Explain your answer.

(a) $q = a$ **(b)** $q = b$

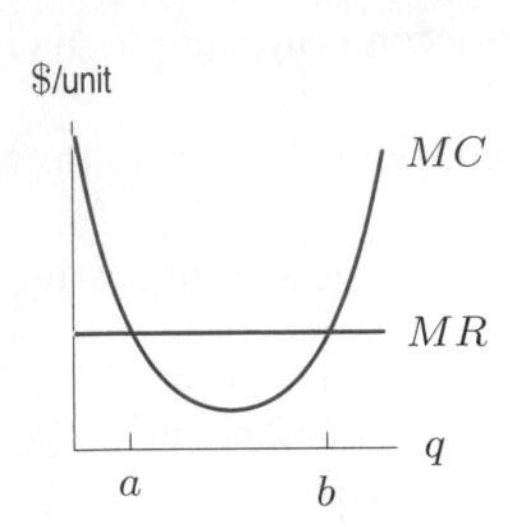

Figure 4.56

29. A company manufactures only one product. The quantity, q, of this product produced per month depends on the amount of capital, K, invested (i.e., the number of machines the company owns, the size of its building, and so on) and the amount of labor, L, available each month. We assume that q can be expressed as a *Cobb-Douglas production function*:

$$q = cK^{\alpha}L^{\beta}$$

where c, α, β are positive constants, with $0 < \alpha < 1$ and $0 < \beta < 1$. In this problem we will see how the Russian government could use a Cobb-Douglas function to estimate how many people a newly privatized industry might employ. A company in such an industry has only a small amount of capital available to it and needs to use all of it, so K is fixed. Suppose L is measured in man-hours per month, and that each man-hour costs the company w rubles (a ruble is the unit of Russian currency). Suppose the company has no other costs besides labor, and that each unit of the good can be sold for a fixed price of p rubles. How many man-hours of labor per month should the company use in order to maximize its profit?

4.5 AVERAGE COST

To maximize profit, a company arranges production to equalize marginal cost and marginal revenue. But how do we know if the company makes money? It turns out that whether the maximum profit is positive or negative is determined by the company's average cost of production. Average cost also tells us about the behavior of similar companies in an industry. If average costs are low, more companies will enter the market; if average costs are high, companies will leave the market.

In this section, we see how average cost can be calculated and visualized, and the relationship between average and marginal cost.

What Is Average Cost?

The average cost is the cost per unit of producing a certain quantity; it is the total cost divided by the number of units produced.

> If the cost of producing a quantity q is $C(q)$, then the **average cost**, $a(q)$, of producing a quantity q is given by
>
> $$a(q) = \frac{C(q)}{q}.$$

Although both are measured in the same units, for example, dollars per item, be careful not to confuse the average cost with the marginal cost (the cost of producing the next item).

Example 1 A salsa company has cost function $C(q) = 0.01q^3 - 0.6q^2 + 13q + 1000$ (in dollars), where q is the number of cases of salsa produced. If 100 cases are produced, find the average cost per case.

Solution The total cost of producing the 100 cases is given by

$$C(100) = 0.01(100^3) - 0.6(100^2) + 13(100) + 1000 = \$6300.$$

We find the average cost per case by dividing by 100, the number of cases produced.

$$\text{Average cost} = \frac{6300}{100} = 63 \text{ dollars/case.}$$

If 100 cases of salsa are produced, the average cost is $63 per case.

Visualizing Average Cost on the Total Cost Curve

We know that average cost is $a(q) = C(q)/q$. Since we can subtract zero from any number without changing it, we can write

$$a(q) = \frac{C(q)}{q} = \frac{C(q) - 0}{q - 0}.$$

This expression gives the slope of the line joining the points $(0, 0)$ and $(q, C(q))$ on the cost curve. See Figure 4.57.

$$\begin{array}{c}\text{Average cost}\\ \text{to produce } q \text{ items}\end{array} = \frac{C(q)}{q} = \begin{array}{c}\text{Slope of the line from the origin}\\ \text{to point } (q, C(q)) \text{ on cost curve.}\end{array}$$

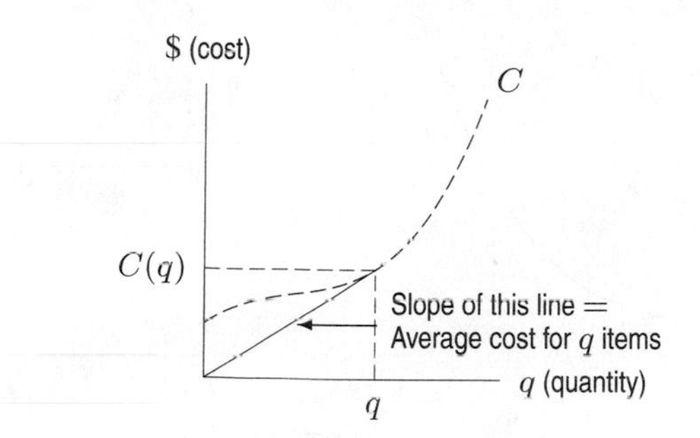

Figure 4.57: Average cost is the slope of the line from the origin to a point on the cost curve

Minimizing Average Cost

We use the graphical representation of average cost to investigate the relationship between average and marginal cost, and to identify the production level which minimizes average cost.

Example 2 A cost function, in dollars, is $C(q) = 1000 + 20q$, where q is the number of units produced. Find and compare the marginal cost to produce the 100th unit and the average cost of producing 100 units. Illustrate your answer on a graph.

Solution The cost function is linear with fixed costs of $1000 and variable costs of $20 per unit. Thus,

$$\text{Marginal cost} = C'(q) = 20 \text{ dollars per unit.}$$

This means that after 99 units have been produced, it costs an additional $20 to produce the next unit. In contrast,

$$\text{Average cost of producing 100 units} = a(100) = \frac{C(100)}{100} = \frac{3000}{100} = 30 \text{ dollars/unit.}$$

Notice that the average cost includes the fixed costs of $1000 spread over the entire production, whereas marginal cost does not. Thus, the average cost is greater than the marginal cost in this example. See Figure 4.58.

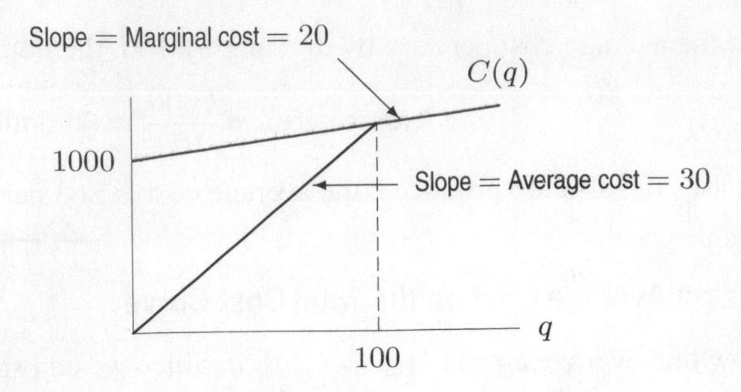

Figure 4.58: Average cost > Marginal cost

Example 3 Mark on the cost graph in Figure 4.59 the quantity at which the average cost is minimized.

Solution In Figure 4.60, the average costs at q_1, q_2, q_3, and q_4 are given by the slopes of the lines from the origin to the curve. These slopes are steep for small q, become less steep as q increases, and then get steeper again. Thus, as q increases, the average cost decreases and then increases, so there is a minimum value. In Figure 4.60 the minimum occurs at the point q_0 where the line from the origin is tangent to the cost curve.

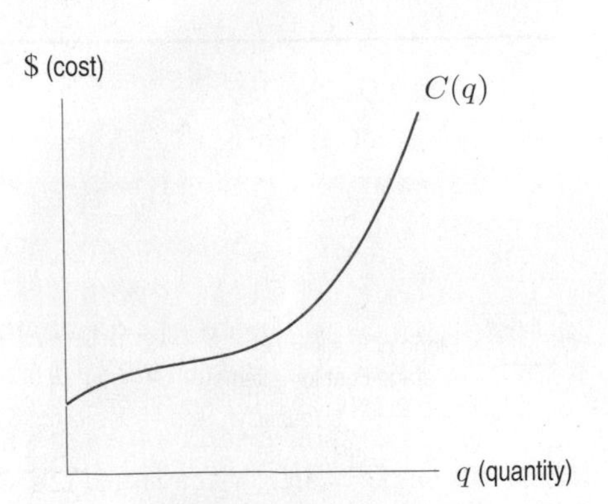

Figure 4.59: A cost function

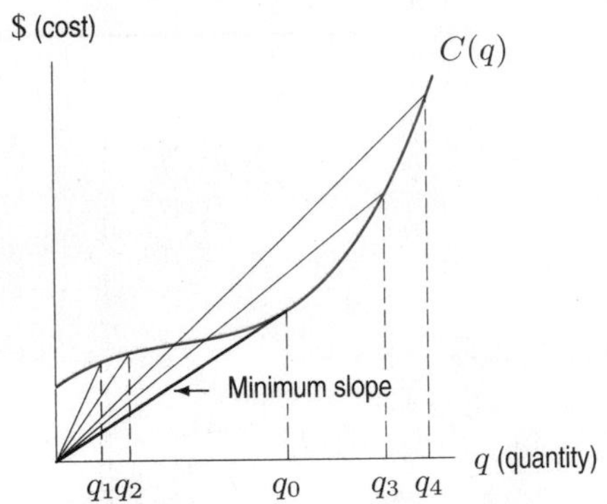

Figure 4.60: Minimum average cost occurs at q_0 where line is tangent to cost curve

---

In Figure 4.60, notice that average cost is a minimum (at q_0) when average cost equals marginal cost. The next example shows what happens when marginal cost and average cost are not equal.

Example 4 Suppose 100 items are produced at an average cost of \$2 per item. Find the average cost of producing 101 items if the marginal cost to produce the 101st item is: (a) \$1 (b) \$3.

Solution If 100 items are produced at an average cost of \$2 per item, the total cost of producing the items is $100 \cdot \$2 = \$200$.

(a) Since the marginal, or additional, cost to produce the 101st item is \$1, the total cost of producing 101 items is $\$200 + \$1 = \$201$. The average cost to produce these items is $201/101$, or \$1.99 per item. The average cost has gone down.

(b) In this case, the marginal cost to produce the 101st item is \$3. The total cost to produce 101 items is \$203 and the average cost is $203/101$, or \$2.01 per item. The average cost has gone up.

Notice that in Example 4 (a), where it costs less than the average to produce an additional item, average cost decreases as production increases. In Example 4 (b), where it costs more than the average to produce an additional item, average cost increases with production. We summarize:

Relationship Between Average Cost and Marginal Cost

- If marginal cost is less than average cost, then increasing production decreases average cost.
- If marginal cost is greater than average cost, then increasing production increases average cost.
- Marginal cost equals average cost at critical points of average cost.

Example 5 Show analytically that critical points of average cost occur when marginal cost equals average cost.

Solution Since $a(q) = C(q)/q = C(q)q^{-1}$, we use the product rule to find $a'(q)$:

$$a'(q) = C'(q)(q^{-1}) + C(q)(-q^{-2}) = \frac{C'(q)}{q} + \frac{-C(q)}{q^2} = \frac{qC'(q) - C(q)}{q^2}.$$

At critical points we have $a'(q) = 0$, so

$$\frac{qC'(q) - C(q)}{q^2} = 0$$

Therefore, we have

$$\begin{aligned} qC'(q) - C(q) &= 0 \\ qC'(q) &= C(q) \\ C'(q) &= \frac{C(q)}{q}. \end{aligned}$$

In other words, at a critical point:

$$\text{Marginal cost} = \text{Average cost.}$$

Example 6 A total cost function, in thousands of dollars, is given by $C(q) = q^3 - 6q^2 + 15q$, where q is in thousands and $0 \leq q \leq 5$.

(a) Graph $C(q)$. Estimate visually the quantity at which average cost is minimized.
(b) Graph the average cost function. Use it to estimate the minimum average cost.
(c) Determine analytically the exact value of q at which average cost is minimized.
(d) Graph the marginal cost function on the same axes as the average cost.
(e) Show that at the minimum average cost, Marginal cost $=$ Average cost. Explain how you can see this result on your graph of average and marginal costs.

Solution

(a) A graph of $C(q)$ is in Figure 4.61. Average cost is minimized at the point where a line from the origin to the point on the curve has minimum slope. This occurs where the line is tangent to the curve, which is at approximately $q = 3$, corresponding to a production of 3000 units.

(b) Since average cost is total cost divided by quantity, we have

$$a(q) = \frac{C(q)}{q} = \frac{q^3 - 6q^2 + 15q}{q} = q^2 - 6q + 15.$$

Figure 4.62 suggests that the minimum average cost occurs at $q = 3$.

(c) Average cost is minimized at a critical point of $a(q) = q^2 - 6q + 15$. Differentiating gives

$$\begin{aligned} a'(q) = 2q - 6 &= 0 \\ q &= 3. \end{aligned}$$

The minimum occurs at $q = 3$.

(d) See Figure 4.62. Marginal cost is the derivative of $C(q) = q^3 - 6q^2 + 15q$,

$$MC(q) = 3q^2 - 12q + 15.$$

(e) At $q = 3$, we have

$$\text{Marginal cost } = 3 \cdot 3^2 - 12 \cdot 3 + 15 = 6.$$
$$\text{Average cost } = 3^2 - 6 \cdot 3 + 15 = 6.$$

Thus, marginal and average cost are equal at $q = 3$. This result can be seen in Figure 4.62 since the marginal cost curve cuts the average cost curve at the minimum average cost.

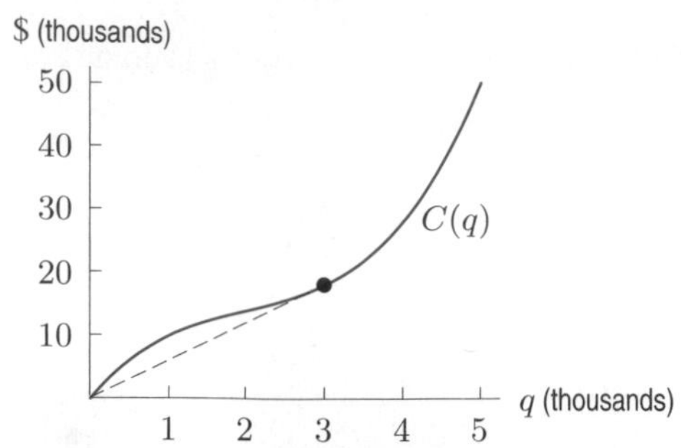

Figure 4.61: Cost function, showing the minimum average cost

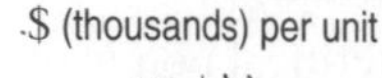

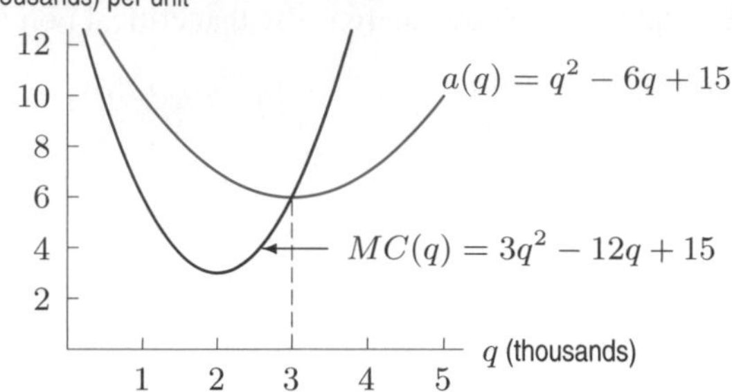

Figure 4.62: Average and marginal cost functions, showing minimum average cost

Problems for Section 4.5

1. Figure 4.63 shows cost with $q = 10{,}000$ marked.
   - **(a)** Find the average cost when the production level is 10,000 units and interpret it.
   - **(b)** Represent your answer to part (a) graphically.
   - **(c)** At approximately what production level is average cost minimized?

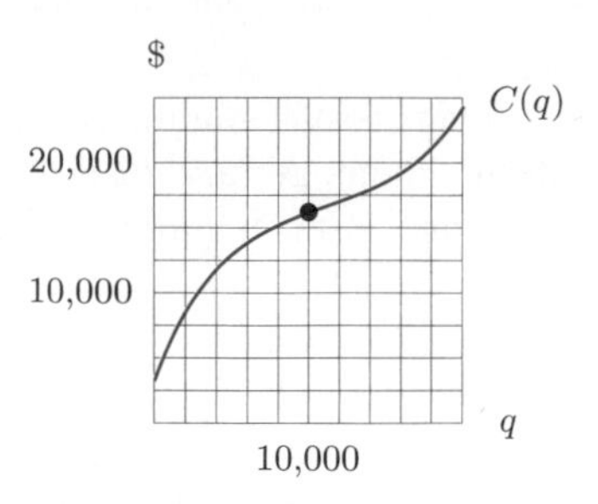

Figure 4.63

2. The cost of producing q items is $C(q) = 2500 + 12q$ dollars.
   - **(a)** What is the marginal cost of producing the 100th item? the 1000th item?
   - **(b)** What is the average cost of producing 100 items? 1000 items?

3. The cost function is $C(q) = 1000 + 20q$. Find the marginal cost to produce the 200th unit and the average cost of producing 200 units.

4. The graph of a cost function is given in Figure 4.64.
   - **(a)** At $q = 25$, estimate the following quantities and represent your answers graphically.
     (i) Average cost (ii) Marginal cost
   - **(b)** At approximately what value of q is average cost minimized?

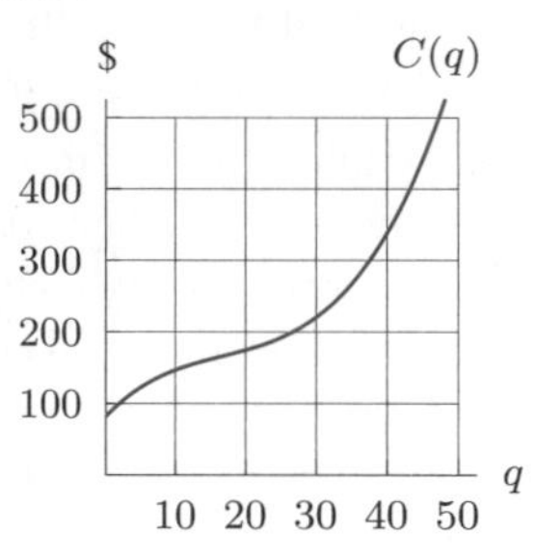

Figure 4.64

5. Graph the average cost function corresponding to the total cost function shown in Figure 4.65.

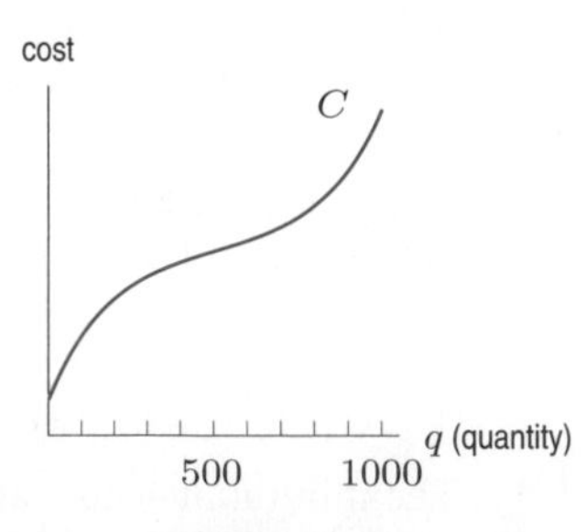

Figure 4.65

6. For each cost function in Figure 4.66, is there a value of q at which average cost is minimized? If so, approximately where? Explain your answer.

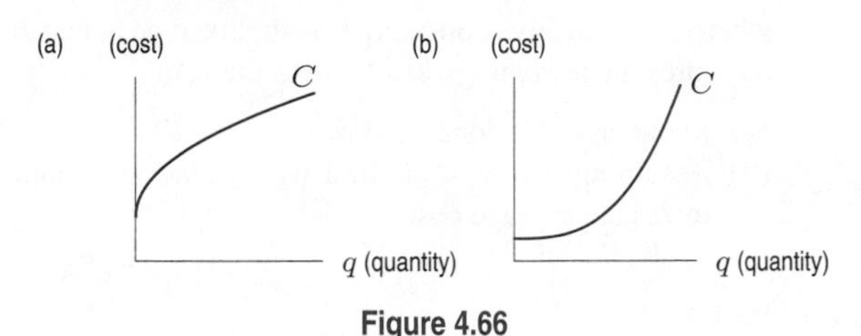

Figure 4.66

7. You are the manager of a firm that produces slippers that sell for \$20. You are producing 1200 slippers each month, at an average cost of \$2 per slipper. The marginal cost at a production level of 1200 is \$3 per slipper.

(a) Are you making or losing money?
(b) Will increasing production increase or decrease your average cost? Your profit?
(c) Would you recommend that production be increased or decreased?

8. The total cost of production, in thousands of dollars, is $C(q) = q^3 - 12q^2 + 60q$, where q is in thousands and $0 \leq q \leq 8$.

(a) Graph $C(q)$. Estimate visually the quantity at which average cost is minimized.
(b) Determine analytically the exact value of q at which average cost is minimized.

9. The average cost per item to produce q items is given by

$$a(q) = 0.01q^2 - 0.6q + 13, \quad \text{for} \quad q > 0.$$

(a) What is the total cost, $C(q)$, of producing q goods?
(b) What is the minimum marginal cost? What is the practical interpretation of this result?
(c) At what production level is the average cost a minimum? What is the lowest average cost?
(d) Compute the marginal cost at $q = 30$. How does this relate to your answer to part (c)? Explain this relationship both analytically and in words.

10. The marginal cost at a production level of 2000 units of an item is \$10 per unit and the average cost of producing 2000 units is \$15 per unit. If the production level were increased slightly above 2000, would the following quantities increase or decrease, or is it impossible to tell?

(a) Average cost **(b)** Profit

11. An agricultural worker in Uganda is planting clover to increase the number of bees making their home in the region. There are 100 bees in the region naturally, and for every acre put under clover, 20 more bees are found in the region.

(a) Draw a graph of the total number, $N(x)$, of bees as a function of x, the number of acres devoted to clover.
(b) Explain, both geometrically and algebraically, the shape of the graph of:
(i) The marginal rate of increase of the number of bees with acres of clover, $N'(x)$.
(ii) The average number of bees per acre of clover, $N(x)/x$.

12. A developer has recently purchased a laundromat and an adjacent factory. For years, the laundromat has taken pains to keep the smoke from the factory from soiling the air used by its clothes dryers. Now that the developer owns both the laundromat and the factory, she could install filters in the factory's smokestacks to reduce the emission of smoke, instead of merely protecting the laundromat from it. The cost of filters for the factory and the cost of protecting the laundromat against smoke depend on the number of filters used, as shown in the table.

Number of filters	Total cost of filters	Total cost of protecting laundromat from smoke
0	\$0	\$127
1	\$5	\$63
2	\$11	\$31
3	\$18	\$15
4	\$26	\$6
5	\$35	\$3
6	\$45	\$0
7	\$56	\$0

(a) Make a table which shows, for each possible number of filters (0 through 7), the marginal cost of the filter, the average cost of the filters, and the marginal savings in protecting the laundromat from smoke.
(b) Since the developer wishes to minimize the total costs to both her businesses, what should she do? Use the table from part (a) to explain your answer.
(c) What should the developer do if, in addition to the cost of the filters, the filters must be mounted on a rack which costs \$100?
(d) What should the developer do if the rack costs \$50?

13. Figure 4.67 shows the average cost, $a(q) = b + mq$.

(a) Show that $C'(q) = b + 2mq$.
(b) Graph the marginal cost $C'(q)$.

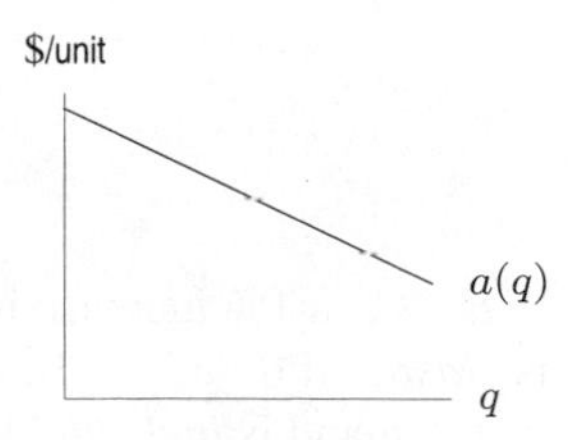

Figure 4.67

14. Show analytically that if marginal cost is less than average cost, then the derivative of average cost with respect to quantity satisfies $a'(q) < 0$.

15. Show analytically that if marginal cost is greater than average cost, then the derivative of average cost with respect to quantity satisfies $a'(q) > 0$.

16. A reasonably realistic model of a firm's costs is given by the *short-run Cobb-Douglas cost curve*

$$C(q) = Kq^{1/a} + F,$$

where a is a positive constant, F is the fixed cost, and K measures the technology available to the firm.

(a) Show that C is concave down if $a > 1$.
(b) Assuming that $a < 1$, find what value of q minimizes the average cost.

4.6 ELASTICITY OF DEMAND

The sensitivity of demand to changes in price varies with the product. For example, a change in the price of light bulbs may not affect the demand for light bulbs much, because people need light bulbs no matter what their price. However, a change in the price of a particular make of car may have a significant effect on the demand for that car, because people can switch to another make.

Elasticity of Demand

We want to find a way to measure this sensitivity of demand to price changes. Our measure should work for products as diverse as light bulbs and cars. The prices of these two items are so different that it makes little sense to talk about absolute changes in price: Changing the price of light bulbs by \$1 is a substantial change, whereas changing the price of a car by \$1 is not. Instead, we use the percent change in price. How, for example, does a 1% increase in price affect the demand for the product?

Let Δp denote the change in the price p of a product and Δq denote the corresponding change in quantity q demanded. The percent change in price is $\Delta p/p$ and the percent change in demand is $\Delta q/q$. Notice that Δp and Δq usually have opposite signs (because increasing the price decreases demand). Then the effect of a price change on demand is measured by the absolute value of the ratio

$$\left|\frac{\text{Percent change in demand}}{\text{Percent change in price}}\right| = \left|\frac{\Delta q/q}{\Delta p/p}\right| = \left|\frac{\Delta q}{q}\cdot\frac{p}{\Delta p}\right| = \left|\frac{p}{q}\cdot\frac{\Delta q}{\Delta p}\right|$$

For small changes in p, we approximate $\Delta q/\Delta p$ by the derivative dq/dp. We define:

> The **elasticity of demand**[8] for a product, E, is given approximately by
>
> $$E \approx \left|\frac{\Delta q/q}{\Delta p/p}\right|, \quad \text{or exactly by} \quad E = \left|\frac{p}{q}\cdot\frac{dq}{dp}\right|.$$

Increasing the price of an item by 1% causes a drop of approximately $E\%$ in the quantity of goods demanded. For small changes, Δp, in price,

$$\frac{\Delta q}{q} \approx -E\frac{\Delta p}{p}.$$

If $E > 1$, a 1% increase in price causes demand to drop by more than 1%, and we say that demand is *elastic*. If $0 \le E < 1$, a 1% increase in price causes demand to drop by less than 1%, and we say that demand is *inelastic*. In general, a larger elasticity causes a larger percent change in demand for a given percent change in price.

[8]When it is necessary to distinguish it from other elasticities, this quantity is called the elasticity of demand with respect to price, or the price elasticity of demand.

Example 1 Raising the price of hotel rooms from \$75 to \$80 per night reduces weekly sales from 100 rooms to 90 rooms.

(a) Approximate the elasticity of demand for rooms at a price of \$75.
(b) Should the owner raise the price?

Solution (a) The percent change in the price is

$$\frac{\Delta p}{p} = \frac{5}{75} = 0.067 = 6.7\%$$

and the percent change in demand is

$$\frac{\Delta q}{q} = \frac{-10}{100} = -0.1 = -10\%.$$

The elasticity of demand is approximated by the ratio

$$E \approx \left|\frac{\Delta q/q}{\Delta p/p}\right| = \frac{0.10}{0.067} = 1.5.$$

The elasticity is greater than 1 because the percent change in the demand is greater than the percent change in the price.

(b) At a price of \$75 per room,

$$\text{Revenue} = (100 \text{ rooms})(\$75 \text{ per room}) = \$7500 \text{ per week.}$$

At a price of \$80 per room,

$$\text{Revenue} = (90 \text{ rooms})(\$80 \text{ per room}) = \$7200 \text{ per week.}$$

A price increase results in loss of revenue, so the price should not be raised.

Example 2 The demand curve for a product is given by $q = 1000 - 2p^2$, where p is the price. Find the elasticity at $p = 10$ and at $p = 15$. Interpret your answers.

Solution We first find the derivative $dq/dp = -4p$. At a price of $p = 10$, we have $dq/dp = -4 \cdot 10 = -40$, and the quantity demanded is $q = 1000 - 2 \cdot 10^2 = 800$. At this price, the elasticity is

$$E = \left|\frac{p}{q} \cdot \frac{dq}{dp}\right| = \left|\frac{10}{800}(-40)\right| = 0.5.$$

The demand is inelastic at a price of $p = 10$: a 1% increase in price results in approximately a 0.5% decrease in demand.

At a price of \$15, we have $q = 550$ and $dq/dp = -60$. The elasticity is

$$E = \left|\frac{p}{q} \cdot \frac{dq}{dp}\right| = \left|\frac{15}{550}(-60)\right| = 1.64.$$

The demand is elastic: a 1% increase in price results in approximately a 1.64% decrease in demand.

Revenue and Elasticity of Demand

Elasticity enables us to analyze the effect of a price change on revenue. An increase in price usually leads to a fall in demand. However, the revenue may increase or decrease. The revenue $R = pq$ is the product of two quantities, and as one increases, the other decreases. Elasticity measures the relative significance of these two competing changes.

Example 3 Three hundred units of an item are sold when the price of the item is $10. When the price of the item is raised by $1, what is the effect on revenue if the quantity sold drops by

(a) 10 units? (b) 100 units?

Solution Since Revenue = Price · Quantity, when the price is $10, we have

$$\text{Revenue} = 10 \cdot 300 = \$3000.$$

(a) At a price of $11, the quantity sold is $300 - 10 = 290$, so

$$\text{Revenue} = 11 \cdot 290 = \$3190.$$

Thus, raising the price has increased revenue.

(b) At a price of $11, the quantity sold is $300 - 100 = 200$, so

$$\text{Revenue} = 11 \cdot 200 = \$2200.$$

Thus, raising the price has decreased revenue.

Elasticity allows us to predict whether revenue increases or decreases with a price increase.

Example 4 The item in Example 3 (a) is wool whose demand equation is $q = 400 - 10p$. The item in Example 3 (b) is houseplants, whose demand equation is $q = 1300 - 100p$. Find the elasticity of wool and houseplants.

Solution For wool, $q = 400 - 10p$, so $dq/dp = -10$. Thus,

$$E_{\text{Wool}} = \left|\frac{p}{q}\frac{dq}{dp}\right| = \left|\frac{10}{300}(-10)\right| = \frac{1}{3}.$$

For houseplants, $q = 1300 - 100p$, so $dq/dp = -100$. Thus,

$$E_{\text{Houseplants}} = \left|\frac{p}{q}\frac{dq}{dp}\right| = \left|\frac{10}{300}(-100)\right| = \frac{10}{3}.$$

Notice that $E_{\text{Wool}} < 1$ and revenue increases with an increase in price; $E_{\text{Houseplants}} > 1$ and revenue decreases with an increase in price. In the next example we see the relationship between elasticity and maximum revenue.

Example 5 Table 4.4 shows the demand, q, revenue, R, and elasticity, E, for the product in Example 2 at several prices. What price brings in the greatest revenue? What is the elasticity at that price?

Solution Table 4.4 suggests that maximum revenue is achieved at a price of about $13, and at that price, E is about 1. At prices below $13, we have $E < 1$, so the reduction in demand caused by a price increase is small; thus, raising the price increases revenue. At prices above $13, we have $E > 1$, so the increase in demand caused by a price decrease is relatively large; thus lowering the price increases revenue.

Table 4.4 *Revenue and elasticity at different points*

Price p	10	11	12	13	14	15
Demand q	800	758	712	662	608	550
Revenue R	8000	8338	8544	8606	8512	8250
Elasticity E	0.5	0.64	0.81	1.02	1.29	1.64
	Inelastic	Inelastic	Inelastic	Elastic	Elastic	Elastic

Example 6 shows that revenue does have a local maximum when $E = 1$. We summarize as follows:

Relationship Between Elasticity and Revenue

- If $E < 1$, demand is inelastic and revenue is increased by raising the price.
- If $E > 1$, demand is elastic and revenue is increased by lowering the price.
- $E = 1$ occurs at critical points of the revenue function.

Example 6 Show analytically that critical points of the revenue function occur when $E = 1$.

Solution We think of revenue as a function of price. Using the product rule to differentiate $R = pq$, we have

$$\frac{dR}{dp} = \frac{d}{dp}(pq) = p\frac{dq}{dp} + \frac{dp}{dp}q = p\frac{dq}{dp} + q.$$

At a critical point the derivative dR/dp equals zero, so we have

$$\begin{aligned} p\frac{dq}{dp} + q &= 0 \\ p\frac{dq}{dp} &= -q \\ \frac{p}{q}\frac{dq}{dp} &= -1 \\ E &= 1. \end{aligned}$$

Elasticity of Demand for Different Products

Different products generally have different elasticities. See Table 4.5. If there are close substitutes for a product, or if the product is a luxury rather than a necessity, a change in price generally has a large effect on demand, and the demand for the product is elastic. On the other hand, if there are no close substitutes or if the product is a necessity, changes in price have a relatively small effect on demand, and the demand is inelastic. For example, demand for salt, penicillin, eyeglasses, and lightbulbs are inelastic over the usual range of prices for these products.

Table 4.5 *Elasticity of demand (with respect to price) for selected farm products*[9]

Product	Elasticity	Product	Elasticity
Cabbage	0.25	Oranges	0.62
Potatoes	0.27	Cream	0.69
Wool	0.33	Apples	1.27
Peanuts	0.38	Peaches	1.49
Eggs	0.43	Fresh tomatoes	2.22
Milk	0.49	Lettuce	2.58
Butter	0.62	Fresh peas	2.83

[9]Estimated by the US Department of Agriculture and reported in W. Adams & J. Brock, *The Structure of American Industry*, 10th ed (Engelwood Cliffs: Prentice Hall, 2000).

Problems for Section 4.6

1. The elasticity of a good is $E = 0.5$. What is the effect on demand of:

(a) A 3% price increase? **(b)** A 3% price decrease?

2. The elasticity of a good is $E = 2$. What is the effect on demand of:

(a) A 3% price increase? **(b)** A 3% price decrease?

3. What is the elasticity for peaches in Table 4.5? Explain what this number tells you about the effect of price increases on the demand for peaches. Is the demand for peaches elastic or inelastic? Is this what you expect? Explain.

4. What is the elasticity for potatoes in Table 4.5? Explain what this number tells you about the effect of price increases on the demand for potatoes. Is the demand for potatoes elastic or inelastic? Is this what you expect? Explain.

5. Would you expect the demand for high definition television sets to be elastic or inelastic? Explain.

6. There are many brands of laundry detergent. Would you expect elasticity of demand for any particular brand to be high or low? Explain.

7. There is only one company offering local telephone service in a town. Would you expect elasticity of demand for telephone service to be high or low? Explain.

8. The demand curve for a product is given by $q = 200 - 2p^2$. Find the elasticity of demand when the price is \$5. Is the demand inelastic or elastic, or neither?

9. The demand curve for a product is given by $p = 90 - 10q$. Find the elasticity of demand when $p = 50$. If this price rises by 2% calculate the corresponding percentage change in demand.

10. School organizations raise money by selling candy door to door. The table shows p, the price of the candy, and q, the quantity sold at that price.

p	\$1.00	\$1.25	\$1.50	\$1.75	\$2.00	\$2.25	\$2.50
q	2765	2440	1980	1660	1175	800	430

(a) Estimate the elasticity of demand at a price of \$1.00. At this price, is the demand elastic or inelastic?
(b) Estimate the elasticity at each of the prices shown. What do you notice? Give an explanation for why this might be so.
(c) At approximately what price is elasticity equal to 1?
(d) Find the total revenue at each of the prices shown. Confirm that the total revenue appears to be maximized at approximately the price where $E = 1$.

11. What are the units of elasticity if:

(a) Price p is in dollars and quantity q is in tons?
(b) Price p is in yen and quantity q is in liters?
(c) What can you conclude in general?

12. The demand for yams is given by $q = 5000 - 10p^2$, where q is in pounds of yams and p is the price of a pound of yams.

(a) If the current price of yams is \$2 per pound, how many pounds will be sold?
(b) Is the demand at \$2 elastic or inelastic? Is it more accurate to say "People want yams and will buy them no matter what the price" or "Yams are a luxury item and people will stop buying them if the price gets too high"?

13. The demand for yams is given in Problem 12.

(a) At a price of \$2 per pound, what is the total revenue for the yam farmer?
(b) Write revenue as a function of price, and then find the price that maximizes revenue.
(c) What quantity is sold at the price you found in part (b), and what is the total revenue?
(d) Show that $E = 1$ at the price you found in part (b).

14. It has been estimated that the elasticity of demand for slaves in the American South before the civil war was equal to 0.86 (fairly high) in the cities and equal to 0.05 (very low) in the countryside.[10]

(a) Why might this be?
(b) Where do you think the staunchest defenders of slavery were from, the cities or the countryside?

15. Find the exact price that maximizes revenue for sales of the product in Example 2.

16. If $E = 2$ for all prices p, how can you maximize revenue?

17. If $E = 0.5$ for all prices p, how can you maximize revenue?

18. **(a)** If the demand equation is $pq = k$ for a positive constant k, compute the elasticity of demand.
(b) Explain the answer to part (a) in terms of the revenue function.

19. Show that a demand equation $q = k/p^r$, where r is a positive constant, gives constant elasticity $E = r$.

20. A linear demand function is given in Figure 4.68. Economists compute elasticity of demand E for any quantity q_0 using the formula

$$E = d_1/d_2,$$

where d_1 and d_2 are the vertical distances shown in Figure 4.68.

(a) Explain why this formula works.

[10] Donald McCloskey, *The Applied Theory of Price*, p. 134, (New York: Macmillan, 1982).

(b) Determine the prices, p, at which (i) $E > 1$ (ii) $E < 1$ (iii) $E = 1$

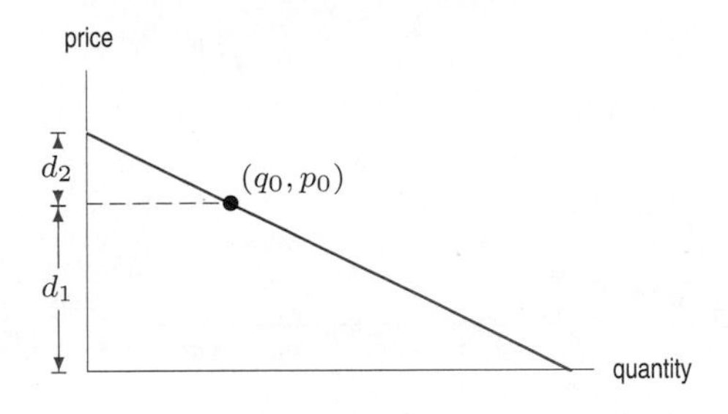

Figure 4.68

21. If p is price and E is the elasticity of demand for a good, show analytically that

$$\text{Marginal revenue} = p(1 - 1/E).$$

22. Suppose cost is proportional to quantity, $C(q) = kq$. Show that a firm earns maximum profit when

$$\frac{\text{Profit}}{\text{Revenue}} = \frac{1}{E}$$

[Hint: Combine the result of Problem 21 with the fact that profit is maximized when $MR = MC$.]

23. Elasticity of cost with respect to quantity is defined as $E_{C,q} = q/C \cdot dC/dq$.

(a) What does this elasticity tell you about sensitivity of cost to quantity produced?
(b) Show that $E_{C,q} =$ Marginal cost/Average cost.

24. If q is the quantity of chicken demanded as a function of the price p of beef, the *cross-price* elasticity of demand for chicken with respect to the price of beef is defined as $E_{\text{cross}} = |p/q \cdot dq/dp|$. What does E_{cross} tell you about the sensitivity of the quantity of chicken bought to changes in the price of beef?

25. The *income* elasticity of demand for a product is defined as $E_{\text{income}} = |I/q \cdot dq/dI|$ where q is the quantity demanded as a function of the income I of the consumer. What does E_{income} tell you about the sensitivity of the quantity of the product purchased to changes in the income of the consumer?

4.7 LOGISTIC GROWTH

In 1923, eighteen koalas were introduced to Kangaroo Island, off the coast of Australia.[11] The koalas thrived on the island and their population grew to about 5000 in 1997. Is it reasonable to expect the population to continue growing exponentially? Since there is only a finite amount of space on the island, the population cannot grow without bound forever. Instead we expect that there is a maximum population that the island can sustain. Population growth with an upper bound can be modeled with a *logistic* or *inhibited growth model.*

Modeling the US Population

Population projections first became important to political philosophers in the late eighteenth century. As concern for scarce resources has grown, so has the interest in accurate population projections. In the US, the population is recorded every ten years by a census. The first such census was in 1790. Table 4.6 contains the census data from 1790 to 2000.

Table 4.6 *US Population,[12] in millions, 1790–2000*

Year	Population	Year	Population	Year	Population	Year	Population
1790	3.9	1850	23.1	1910	92.0	1960	179.3
1800	5.3	1860	31.4	1920	105.7	1970	203.3
1810	7.2	1870	38.6	1930	122.8	1980	226.5
1820	9.6	1880	50.2	1940	131.7	1990	248.7
1830	12.9	1890	62.9	1950	150.7	2000	281.4
1840	17.1	1900	76.0				

[11] *Watertown Daily Times*, April 18, 1997.
[12] *The World Almanac and Book of Facts 2005*, pp. 622–623 (New York).

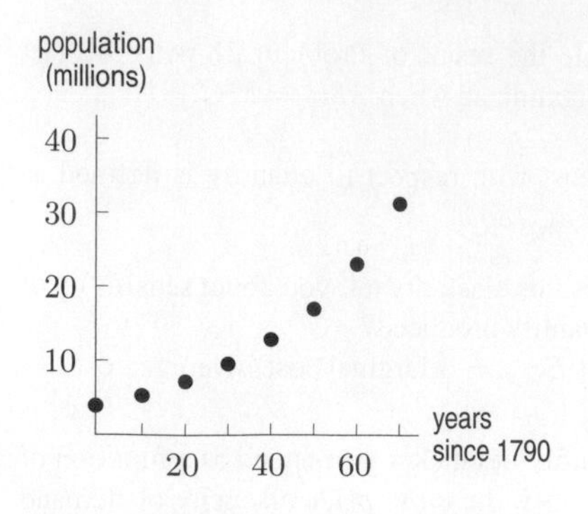

Figure 4.69: US Population, 1790–1860

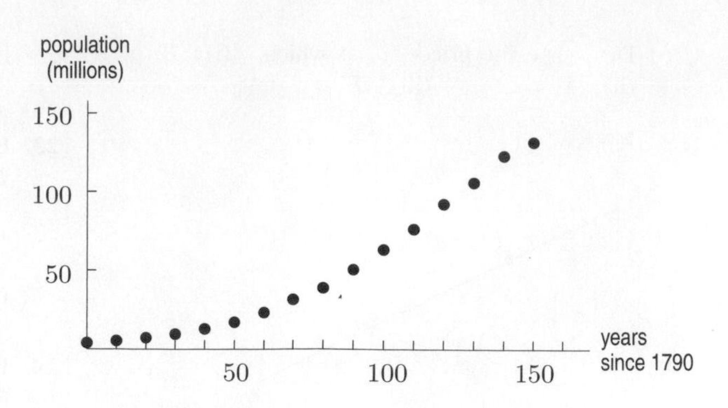

Figure 4.70: US Population, 1790–1940

Figure 4.69 suggests that the population grew exponentially during the years 1790–1860. However, after 1860 the rate of growth began to decrease. See Figure 4.70.

The Years 1790–1860: An Exponential Model

We begin by modeling the US population for the years 1790–1860 using an exponential function. If t is the number of years since 1790 and P is the population in millions, regression gives the exponential function that fits the data as approximately[13]

$$P = 3.9(1.03)^t.$$

Thus, between 1790 and 1860, the US population was growing at an annual rate of about 3%.

The function $P = 3.9(1.03)^t$ is plotted in Figure 4.71 with the data; it fits the data remarkably well. Of course, since we used the data from throughout the 70-year period, we should expect good agreement throughout that period. What is surprising is that if we had used only the populations in 1790 and 1800 to create our exponential function, the predictions would still be very accurate. It is amazing that a person in 1800 could predict the population 60 years later so accurately, especially when one considers all the wars, recessions, epidemics, additions of new territory, and immigration that took place from 1800 to 1860.

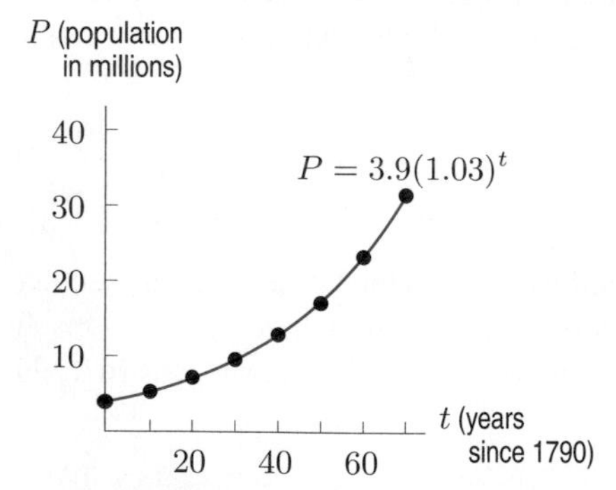

Figure 4.71: An exponential model for the US population, 1790–1860

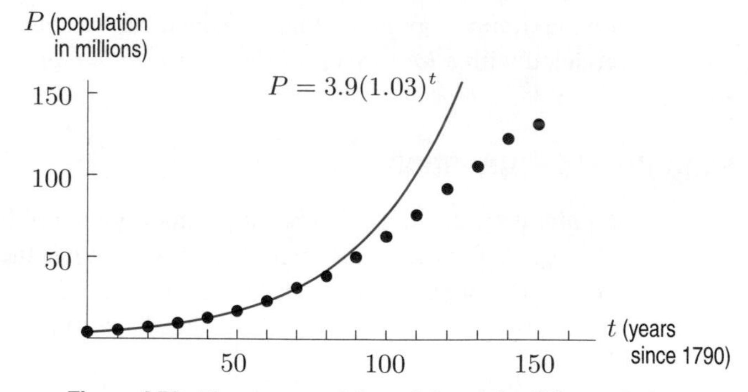

Figure 4.72: The exponential model and the US population, 1790-1940. Not a good fit beyond 1860

The Years 1790–1940: A Logistic Model

How well does the exponential function fit the US population beyond 1860? Figure 4.72 shows a graph of the US population from 1790 until 1940 with the exponential function $P = 3.9(1.03)^t$. The exponential function which fit the data so well for the years 1790–1860 does not fit very well beyond 1860. We must look for another way to model this data.

The graph of the function given by the data in Figure 4.70 is concave up for small values of t, but then appears to become concave down and to be leveling off. This kind of growth is modeled with a *logistic function.* If t is in years since 1790, the function

$$P = \frac{187}{1 + 47e^{-0.0318t}},$$

[13] See the section on Fitting Formulas to Data on page 78. Different algorithms may give different formulas.

which is graphed in Figure 4.73, fits the data well up to 1940. Such a formula is found by logistic regression on a calculator or computer.[14]

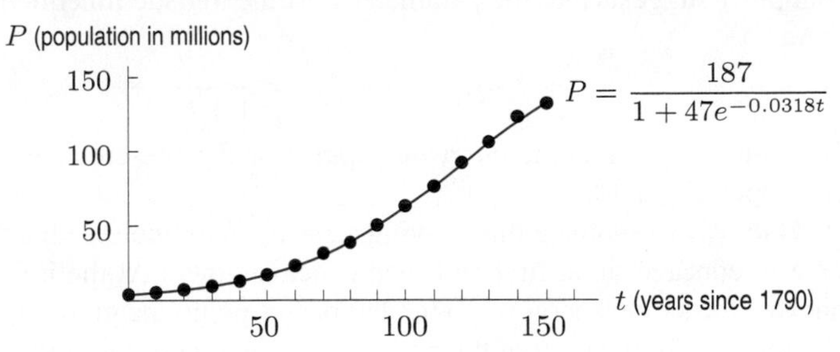

Figure 4.73: A logistic model for US population, 1790–1940

The Logistic Function

A logistic function, such as that used to model the US population, is everywhere increasing. Its graph is concave up at first, then becomes concave down, and levels off at a horizontal asymptote. As we saw in the US population model, a logistic function is approximately exponential for small[15] values of t. A logistic function can be used to model the sales of a new product and the spread of a virus.

> For positive constants L, C, and k, a **logistic function** has the form
>
> $$P = f(t) = \frac{L}{1+Ce^{-kt}}.$$

The general logistic function has three parameters: L, C, and k. In Example 1, we investigate the effect of two of these parameters on the graph; Problem 6 on page 219 considers the third.

Example 1 Consider the logistic function $P = \dfrac{L}{1+100e^{-kt}}$.

(a) Let $k = 1$. Graph P for several values for L. Explain the effect of the parameter L.
(b) Now let $L = 1$. Graph P for several values for k. Explain the effect of the parameter k.

Solution

(a) See Figure 4.74. Notice that the graph levels off at the value L. The parameter L determines the horizontal asymptote and the upper bound for P.

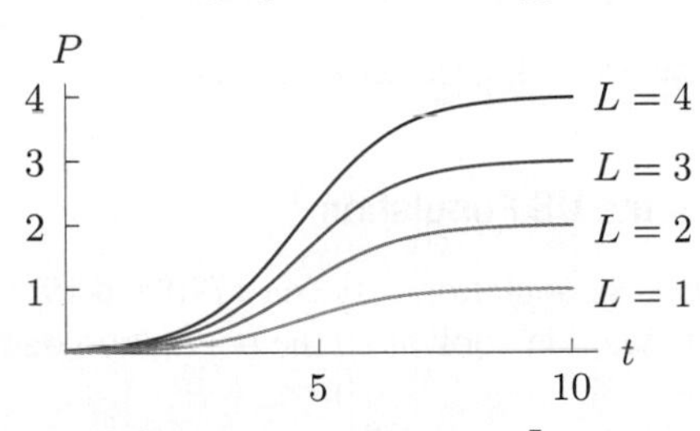

Figure 4.74: Graph of $P = \dfrac{L}{1+100e^{-t}}$ for various values of L

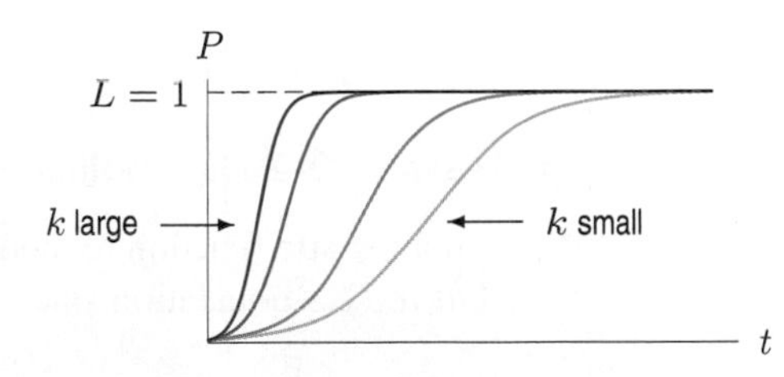

Figure 4.75: Graph of $P = \dfrac{1}{1+100e^{-kt}}$ with various values of k

(b) See Figure 4.75. Notice that as k increases, the curve approaches the asymptote more rapidly. The parameter k affects the steepness of the curve.

[14] See the section on Fitting Formulas to Data on page 78.

[15] Just how small is small enough depends on the values of the parameters C and k.

The Carrying Capacity and the Point of Diminishing Returns

Example 1 suggests that the parameter L of the logistic function is the value at which P levels off, where

$$P = \frac{L}{1 + Ce^{-kt}}.$$

This value L is called the *carrying capacity* and represents the largest population an environment can support.

One way to estimate the carrying capacity is to find the inflection point. The graph of a logistic curve is concave up at first and then concave down. At the inflection point, where the concavity changes, the slope is largest. To the left of this point, the graph is concave up and the rate of growth is increasing. To the right of this point, the graph is concave down and the rate of growth is diminishing. The inflection point is called the *point of diminishing returns*. Problem 39 on page 230 shows that this point is at $P = L/2$. See Figure 4.76. Companies sometimes watch for this concavity change in the sales of a new product and use it to estimate the maximum potential sales.

Properties of the logistic function $P = \dfrac{L}{1 + Ce^{-kt}}$

- The limiting value L represents the carrying capacity for P.
- The point of diminishing returns is the inflection point where P is growing the fastest. It occurs where $P = L/2$.
- The logistic function is approximately exponential for small values of t, with growth rate k.

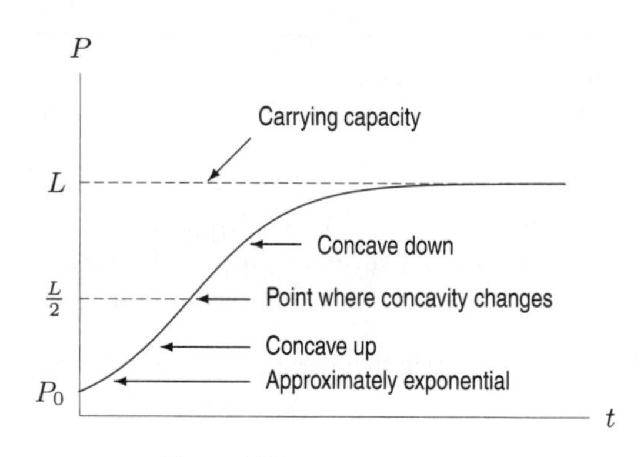

Figure 4.76: Logistic growth

The Years 1790–2000: Another look at the US Population

We used a logistic function to model the US population between 1790 and 1940. How well does this model fit the US population since 1940? We now look at all the population data from 1790 to 2000.

Example 2 If t is in years since 1790 and P is in millions, we used the following logistic function to model the US population between 1790 and 1940:

$$P = \frac{187}{1 + 47e^{-0.0318t}}.$$

According to this function, what is the maximum US population? Is this prediction accurate? How well does this logistic model fit the growth of the US population since 1940?

Table 4.7 *Predicted versus actual US population, in millions, 1940–2000 (logistic model)*

Year	1940	1950	1960	1970	1980	1990	2000
Actual	131.7	150.7	179.3	203.3	226.5	248.7	281.4
Predicted	133.7	145.0	154.4	162.1	168.2	172.9	176.6

Solution Table 4.7 shows the actual US population between 1940 and 2000 and the predicted values using this logistic model. According to the formula for the logistic function, the upper bound for the population is $L = 187$ million. However, Table 4.7 shows that the actual US population was above this figure by 1970. The fit between the logistic function and the actual population is not a good one beyond 1940. See Figure 4.77.

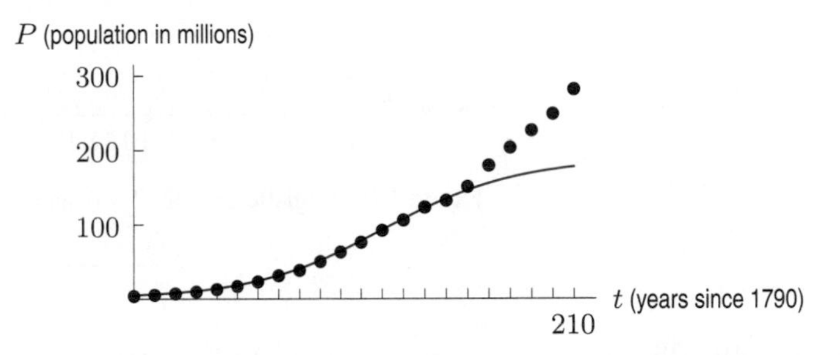

Figure 4.77: The logistic model and the US population, 1790–2000

Despite World War II, which depressed population growth between 1942 and 1945, in the last half of the 1940s the US population surged. The 1950s saw a population growth of 28 million, leaving our logistic model in the dust. This surge in population is referred to as the baby boom.

Once again we have reached a point where our model is no longer useful. This should not lead you to believe that a reasonable mathematical model cannot be found; rather it points out that no model is perfect and that when one model fails, we seek a better one. Just as we abandoned the exponential model in favor of the logistic model for the US population, we could look further.

Sales Predictions

Total sales of a new product often follow a logistic model. For example, when a new compact disc (CD) appears on the market, sales first increase rapidly as word of the CD spreads. Eventually, most of the people who want the CD have already bought it and sales slow down. The graph of total sales against time is concave up at first and then concave down, with the upper bound L equal to the maximum potential sales.

Example 3 Table 4.8 shows the total sales (in thousands) of a new CD since it was introduced.

(a) Find the point where concavity changes in this function. Use it to estimate the maximum potential sales, L.

(b) Using logistic regression, fit a logistic function to this data. What maximum potential sales does this function predict?

Table 4.8 *Total sales of a new CD since its introduction*

t (months)	0	1	2	3	4	5	6	7
P (total sales in 1000s)	0.5	2	8	33	95	258	403	496

Solution (a) The rate of change of total sales increases until $t = 5$ and decreases after $t = 5$, so the inflection point is at approximately $t = 5$, when $P = 258$. So $L/2 = 258$ and $L = 516$. The maximum potential sales for this CD are estimated to be 516,000.

(b) Logistic regression gives the following function:

$$P = \frac{532}{1 + 869e^{-1.33t}}.$$

Maximum potential sales predicted by this function are $L = 532$, or about 532,000 CDs. See Figure 4.78.

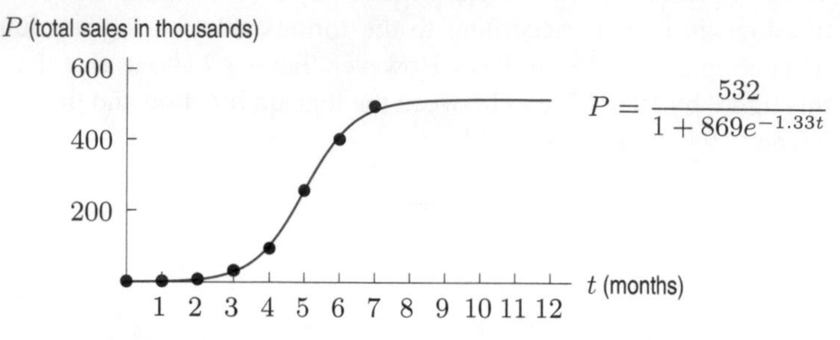

Figure 4.78: Logistic growth: Total sales of a CD

Dose-Response Curves

A *dose-response curve* plots the intensity of physiological response to a drug as a function of the dose administered. As the dose increases, the intensity of the response increases, so a dose-response function is increasing. The intensity of the response is generally scaled as a percentage of the maximum response. The curve cannot go above the maximum response (or 100%), so the curve levels off at a horizontal asymptote. Dose-response curves are generally concave up for low doses and concave down for high doses. A dose-response curve can be modeled by a logistic function with the independent variable being the dose of the drug, not time.

A dose-response curve shows the amount of drug needed to produce the desired effect, as well as the maximum effect attainable and the dose required to obtain it. The slope of the dose-response curve gives information about the therapeutic safety margin of the drug.

Drugs need to be administered in a dose which is large enough to be effective but not so large as to be dangerous. Figure 4.79 shows two different dose-response curves: one with a small slope and one with a large slope. In Figure 4.79(a), there is a broad range of dosages at which the drug is both safe and effective. In Figure 4.79(b), where the slope of the curve is steep, the range of dosages at which the drug is both safe and effective is small. If the slope of the dose-response curve is steep, a small mistake in the dosage can have dangerous results. Administration of such a drug is difficult.

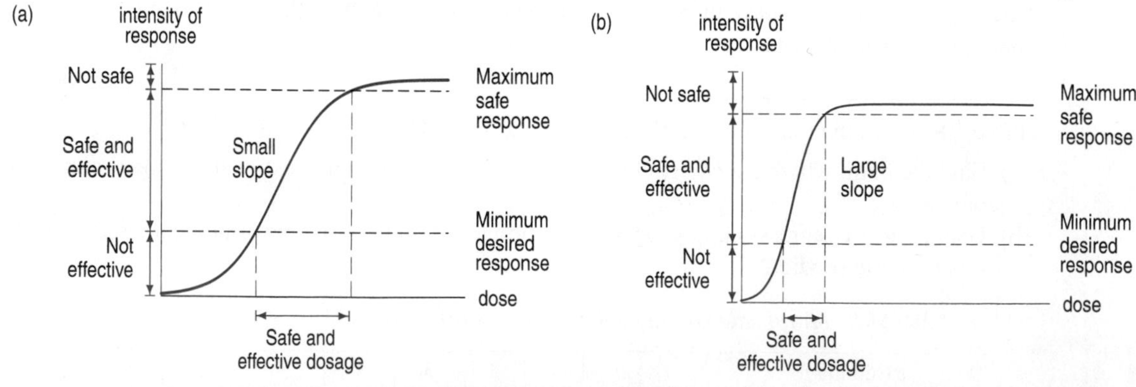

Figure 4.79: What does the slope of the dose-response curve tell us?

Example 4 Figure 4.80 shows dose-response curves for three different drugs used for the same purpose. Discuss the advantages and disadvantages of the three drugs.

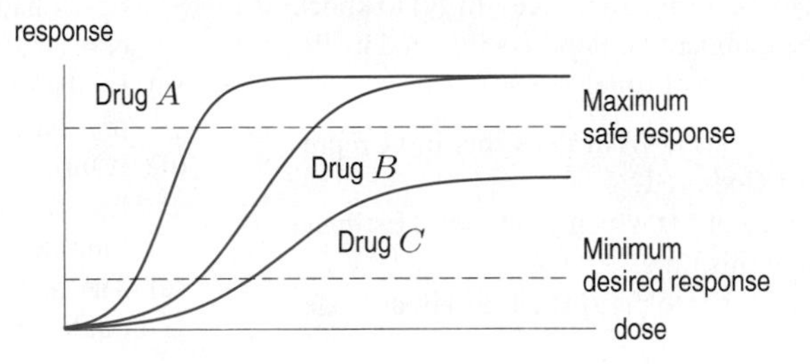

Figure 4.80: What are the advantages and disadvantages of each of these drugs?

Solution Drugs A and B exhibit the same maximum response, while the maximum response of Drug C is significantly less; however all three drugs reach the minimum desired response. The potency of Drugs B and C (the dose required to reach desired effect) is significantly less than the potency of Drug A. (Potency, however, is a relatively unimportant characteristic of a drug, since a less potent drug can simply be given in larger doses.) Drug A has a steeper slope than either of the other two. Both Drugs A and B can exceed the maximum safe response. Thus, Drug C may be the preferred drug despite its lower maximum effect because it is the safest to administer.

Problems for Section 4.7

1. If t is in years since 1990, one model for the population of the world, P, in billions, is

$$P = \frac{40}{1 + 11e^{-0.08t}}.$$

(a) What does this model predict for the maximum sustainable population of the world?
(b) Graph P against t.
(c) According to this model, when will the earth's population reach 20 billion? 39.9 billion?

2. A rumor spreads among a group of 400 people. The number of people, $N(t)$, who have heard the rumor by time t in hours since the rumor started to spread can be approximated by a function of the form

$$N(t) = \frac{400}{1 + 399e^{-0.4t}}.$$

(a) Find $N(0)$ and interpret it.
(b) How many people will have heard the rumor after 2 hours? After 10 hours?
(c) Graph $N(t)$.
(d) Approximately how long will it take until half the people have heard the rumor? Virtually everyone?
(e) Approximately when is the rumor spreading fastest?

3. The rate of sales of an automobile anti-theft device are given in the following table.

(a) When is the point of diminishing returns reached?
(b) What are the total sales at this point?
(c) Assuming logistic growth in sales, use your answer to part (b) to estimate total potential sales of the device.

Months	1	2	3	4	5	6
Sales per month	140	520	680	750	700	550

4. The following table shows the total sales, in thousands, since a new game was brought to market.

(a) Plot this data and mark on your plot the point of diminishing returns.
(b) Predict total possible sales of this game, using the point of diminishing returns.

Month	0	2	4	6	8	10	12	14
Sales	0	2.3	5.5	9.6	18.2	31.8	42.0	50.8

5. Write a paragraph explaining why sales of a new product often follow a logistic curve. Explain the benefit to the company of watching for the point of diminishing returns.

6. Investigate the effect of the parameter C on the logistic curve $P = \frac{10}{1 + Ce^{-t}}$. Substitute several values for C and explain, with a graph and with words, the effect of C on the graph.

7. Figure 4.81 shows the spread of the Code-red computer virus during July 2001. Most of the growth took place starting at midnight on July 19; on July 20, the virus attacked the White House, trying (unsuccessfully) to knock its site off-line. The number of computers infected by the virus is a logistic function of time.

(a) Estimate $\lim_{t\to\infty} f(t)$. What does this limit represent in terms of Code-red?
(b) Estimate the value of t at which $f''(t) = 0$. Estimate the value of n at this time.
(c) What does the answer to part (b) tell us about Code-red?
(d) How are the answers to parts (a) and (b) related?

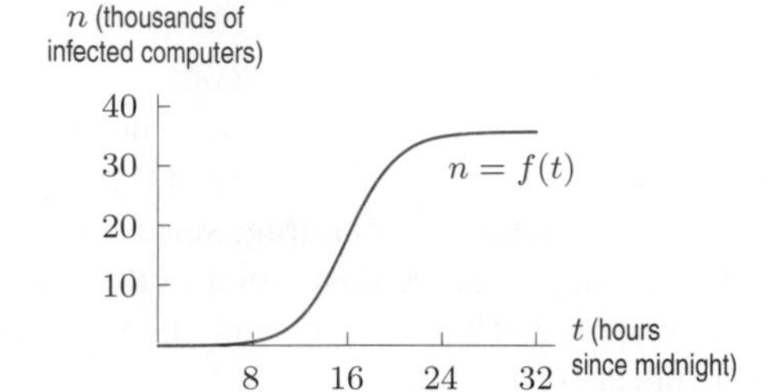

Figure 4.81

8. (a) Draw a logistic curve. Label the carrying capacity L and the point of diminishing returns t_0.
(b) Draw the derivative of the logistic curve. Mark the point t_0 on the horizontal axis.
(c) A company keeps track of the rate of sales (for example, sales per week) rather than total sales. Explain how the company can tell on a graph of rate of sales when the point of diminishing returns is reached.

9. The Tojolobal Mayan Indian community in Southern Mexico has available a fixed amount of land.[16] The proportion, P, of land in use for farming t years after 1935 is modeled with the logistic function

$$P = \frac{1}{1 + 3e^{-0.0275t}}.$$

(a) What proportion of the land was in use for farming in 1935?
(b) What is the long run prediction of this model?
(c) When was half the land in use for farming?
(d) When is the proportion of land used for farming increasing most rapidly?

10. In the spring of 2003, SARS (Severe Acute Respiratory Syndrome) spread rapidly in several Asian countries and Canada. Table 4.9 gives the total number, P, of SARS cases reported in Hong Kong[17] by day t, where $t = 0$ is March 17, 2003.

(a) Find the average rate of change of P for each interval in Table 4.9.
(b) In early April 2003, there was fear that the disease would spread at an ever-increasing rate for a long time. What is the earliest date by which epidemiologists had evidence to indicate that the rate of new cases had begun to slow?
(c) Explain why an exponential model for P is not appropriate.
(d) It turns out that a logistic model fits the data well. Estimate the value of t at the inflection point. What limiting value of P does this point predict?
(e) The best-fitting logistic function for this data turns out to be

$$P = \frac{1760}{1 + 17.53e^{-0.1408t}}.$$

What limiting value of P does this function predict?

Table 4.9 *Total number of SARS cases in Hong Kong by day t (where t = 0 is March 17, 2003)*

t	P	t	P	t	P	t	P
0	95	26	1108	54	1674	75	1739
5	222	33	1358	61	1710	81	1750
12	470	40	1527	68	1724	87	1755
19	800	47	1621				

11. Substitute $t = 0, 10, 20, \ldots, 70$ into the exponential function used in this section to model the US population 1790–1860. Compare the predicted values of the population with the actual values.

12. In this section, a logistic function was used to model the US population. Use this function to predict the US population in each of the census years from 1790–1940. Compare the predicted and actual values.

13. A curve representing the total number of people, P, infected with a virus often has the shape of a logistic curve of the form $P = \dfrac{L}{1 + Ce^{-kt}}$, with time t in weeks. Suppose that 10 people originally have the virus and that in the early stages the number of people infected is increasing approximately exponentially, with a continuous growth rate of 1.78. It is estimated that, in the long run, approximately 5000 people will become infected.

(a) What should we use for the parameters k and L?
(b) Use the fact that when $t = 0$, we have $P = 10$, to find C.
(c) Now that you have estimated L, k, and C, what is the logistic function you are using to model the data? Graph this function.
(d) Estimate the length of time until the rate at which people are becoming infected starts to decrease. What is the value of P at this point?

[16] Adapted from J. S. Thomas and M. C. Robbins, "The Limits to Growth in a Tojolobal Maya Ejido," *Geoscience and Man 26*, pp. 9–16, (Baton Rouge: Geoscience Publications, 1988).
[17] www.who.int/csr/country/en, accessed July 13, 2003.

14. If R is percent of maximum response and x is dose in mg, the dose-response curve for a drug is given by

$$R = \frac{100}{1 + 100e^{-0.1x}}.$$

(a) Graph this function.
(b) What dose corresponds to a response of 50% of the maximum? This is the inflection point, at which the response is increasing the fastest.
(c) For this drug, the minimum desired response is 20% and the maximum safe response is 70%. What range of doses is both safe and effective for this drug?

15. Dose-response curves for three different products are given in Figure 4.82.

(a) For the desired response, which drug requires the largest dose? The smallest dose?
(b) Which drug has the largest maximum response? The smallest?
(c) Which drug is the safest to administer? Explain.

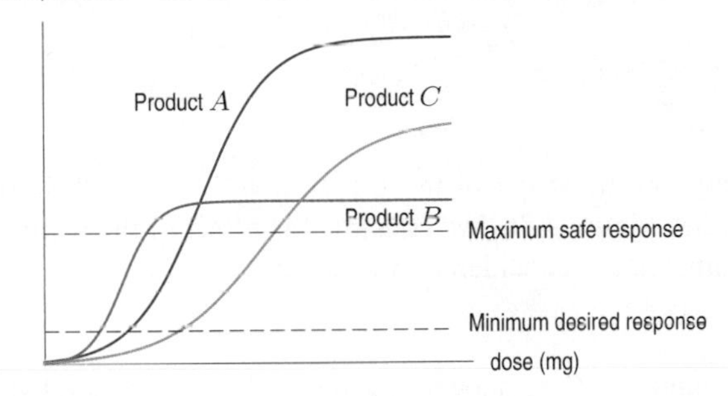

Figure 4.82

16. A dose-response curve is given by $R = f(x)$, where R is percent of maximum response and x is the dose of the drug in mg. The curve has the shape shown in Figure 4.79 on page 218. The inflection point is at $(15, 50)$ and $f'(15) = 11$.

(a) Explain what $f'(15)$ tells you in terms of dose and response for this drug.
(b) Is $f'(10)$ greater than or less than 11? Is $f'(20)$ greater than or less than 11? Explain.

17. Explain why it is safer to use a drug for which the derivative of the dose-response curve is smaller.

There are two kinds of dose-response curves. One type, discussed in this section, plots the intensity of response against the dose of the drug. We now consider a dose-response curve in which the percentage of subjects showing a specific response is plotted against the dose of the drug. In Problems 18–19, the curve on the left shows the percentage of subjects exhibiting the desired response at the given dose, and the curve on the right shows the percentage of subjects for which the given dose is lethal.

18. In Figure 4.83, what range of doses appears to be both safe and effective for 99% of all patients?

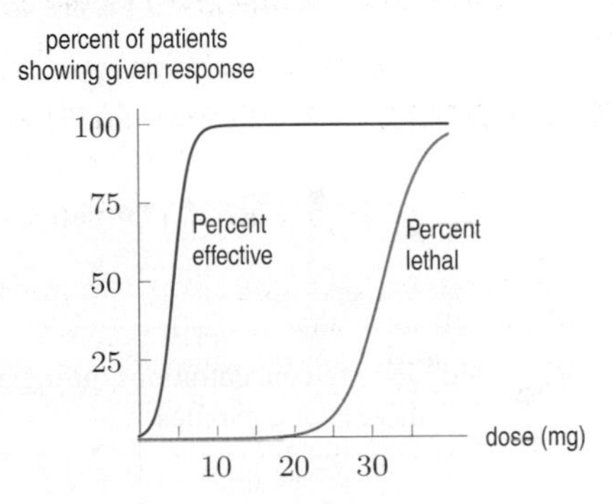

Figure 4.83

19. In Figure 4.84, discuss the possible outcomes and what percent of patients fall in each outcome when 50 mg of the drug is administered.

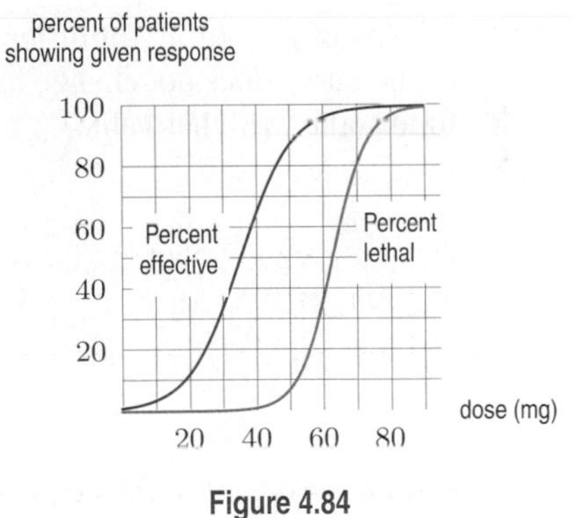

Figure 4.84

4.8 THE SURGE FUNCTION AND DRUG CONCENTRATION

Nicotine in the Blood

When a person smokes a cigarette, the nicotine from the cigarette enters the person's body through the lungs, is absorbed into the blood, and spreads throughout the body. Most cigarettes contain between 0.5 and 2.0 mg of nicotine; approximately 20% (between 0.1 and 0.4 mg) is actually inhaled and absorbed into the person's bloodstream. As the nicotine leaves the blood, the smoker feels the need for another cigarette. The half-life of nicotine in the bloodstream is about two hours. The lethal dose is considered to be about 60 mg.

The nicotine level in the blood rises as a person smokes, and tapers off when smoking ceases. Table 4.10 shows blood nicotine concentration (in ng/ml) during and after the use of cigarettes. (Smoking occurred during the first ten minutes and the experimental data shown represent average values for ten people.)[18]

The points in Table 4.10 are plotted in Figure 4.85. Functions with this behavior are called *surge functions*. They have equations of the form $y = ate^{-bt}$, where a and b are positive constants.

Table 4.10 *Blood nicotine concentrations during and after the use of cigarettes*

t (minutes)	0	5	10	15	20	25	30	45	60	75	90	105	120
C (ng/ml)	4	12	17	14	13	12	11	9	8	7.5	7	6.5	6

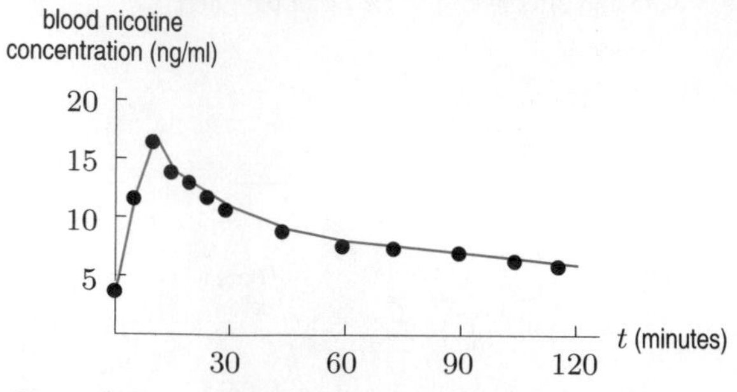

Figure 4.85: Blood nicotine concentrations during and after the use of cigarettes

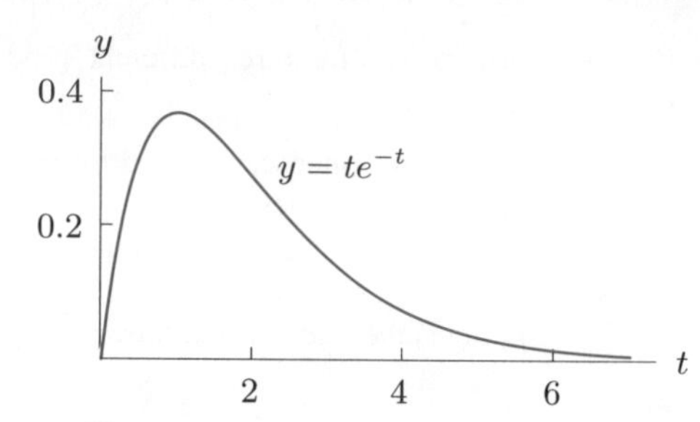

Figure 4.86: One member of the family $y = ate^{-bt}$, with $a = 1$ and $b = 1$

The Family of Functions $y = ate^{-bt}$

What effect do the parameters a and b have on the shape of the graph of $y = ate^{-bt}$? Start by looking at the graph with $a = 1$ and $b = 1$. See Figure 4.86. We consider the effect of the parameter b on the graph of $y = ate^{-bt}$ now; the parameter a is considered in Problem 12 on page 227.

The Effect of the Parameter b on $y = te^{-bt}$

Graphs of $y = te^{-bt}$ for different positive values of b are shown in Figure 4.87. The general shape of the curve does not change as b changes, but as b decreases, the curve rises for a longer period of time and to a higher value.

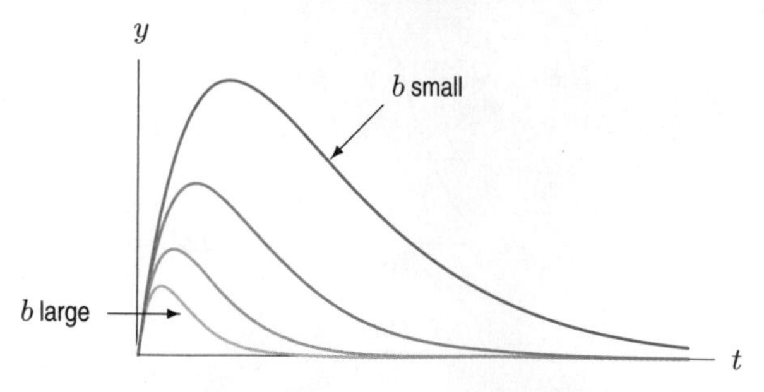

Figure 4.87: Graph of $y = te^{-bt}$, with b varying

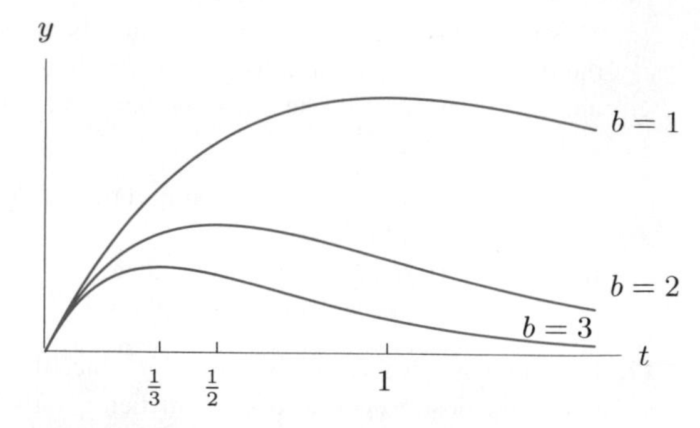

Figure 4.88: How does the maximum depend on b?

We see in Figure 4.88 that, when $b = 1$, the maximum occurs at about $t = 1$. When $b = 2$, it occurs at about $t = \frac{1}{2}$, and when $b = 3$, it occurs at about $t = \frac{1}{3}$. The next example shows that the maximum of the function $y = te^{-bt}$ occurs at $t = 1/b$.

Example 1 For $b > 0$, show that the maximum value of $y = te^{-bt}$ occurs at $t = 1/b$ and increases as b decreases.

[18]Benowitz, Porched, Skeiner, Jacog, "Nicotine Absorption and Cardiovascular Effects with Smokeless Tobacco Use: Comparison with Cigarettes and Nicotine Gum," *Clinical Pharmacology and Therapeutics* 44 (1988): 24.

Solution The maximum occurs at a critical point where $dy/dt = 0$. Differentiating gives

$$\frac{dy}{dt} = 1 \cdot e^{-bt} + t\left(-be^{-bt}\right) = e^{-bt} - bte^{-bt} = e^{-bt}(1 - bt).$$

So $dy/dt = 0$ where

$$1 - bt = 0$$
$$t = \frac{1}{b}$$

Substituting $t = 1/b$ shows that at the maximum,

$$y = \frac{1}{b}e^{-b(1/b)} = \frac{e^{-1}}{b}.$$

So, for $b > 0$, as b increases, the maximum value of y decreases and vice versa.

> The **surge function** $y = ate^{-bt}$ increases rapidly and then decreases toward zero with a maximum at $t = 1/b$.

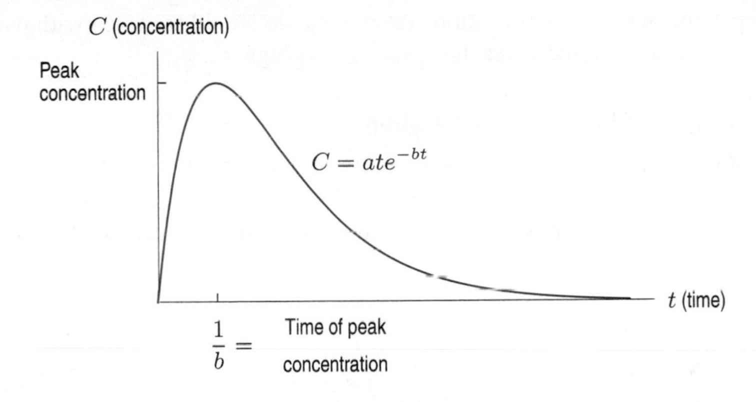

Figure 4.09: Curve showing drug concentration as a function of time

Drug Concentration Curves

When the concentration, C, of a drug in the body is plotted against the time, t, since the drug was administered, the curve generally has the shape shown in Figure 4.89. This is called a *drug concentration curve*, and is modeled using a function of the form $C = ate^{-bt}$. Figure 4.89 shows the peak concentration (the maximum concentration of the drug in the body) and the length of time until peak concentration is reached.

Factors Affecting Drug Absorption

Drug interactions and the age of the patient can affect the drug concentration curve. In Problems 9 and 10, we see that food intake can also affect the rate of absorption of a drug, and (perhaps most surprising) that drug concentration curves can vary markedly between different commercial versions of the same drug.

Example 2 Figure 4.90 shows the drug concentration curves for paracetamol (acetaminophen) alone and for paracetamol taken in conjunction with propantheline. Figure 4.91 shows drug concentration curves for patients known to be slow absorbers of the drug, for paracetamol alone and for paracetamol in conjunction with metoclopramide. Discuss the effects of the additional drugs on peak concentration and the time to reach peak concentration.[19]

[19]Graeme. S. Avery, ed. *Drug Treatment: Principle and Practice of Clinical Pharmacology and Therapeutics*, (Sydney: Adis Press, 1976).

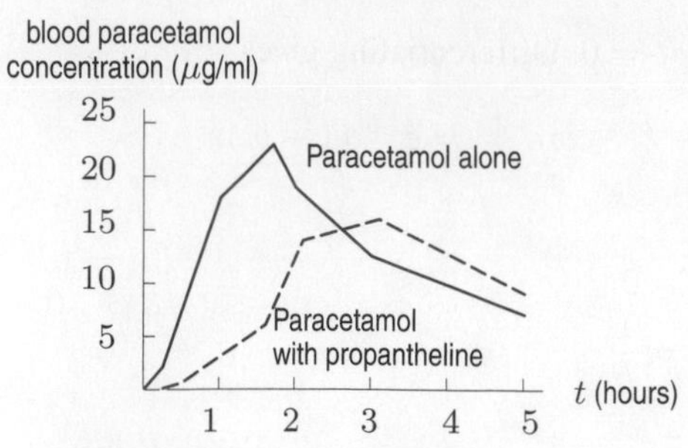

Figure 4.90: Drug concentration curves for paracetamol, normal patients

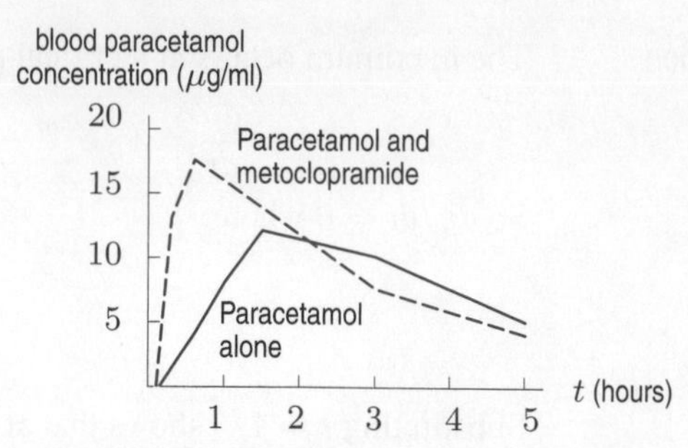

Figure 4.91: Drug concentration curves for paracetamol, patients with slow absorption

Solution Figure 4.90 shows it takes about 1.5 hours for the paracetamol to reach its peak concentration, and that the maximum concentration reached is about 23 μg of paracetamol per ml of blood. However, if propantheline is administered with the paracetamol, it takes much longer to reach the peak concentration (about three hours, or approximately double the time), and the peak concentration is much lower, at about 16 μg/ml.

Comparing the curves for paracetamol alone in Figures 4.90 and 4.91 shows that the time to reach peak concentration is the same (about 1.5 hours), but the maximum concentration is lower for patients with slow absorption. When metoclopramide is given with paracetamol in Figure 4.91, the peak concentration is reached faster and is higher.

Minimum Effective Concentration

The minimum effective concentration of a drug is the blood concentration necessary to achieve a pharmacological response. The time at which this concentration is reached is referred to as onset; termination occurs when the drug concentration falls below this level. See Figure 4.92.

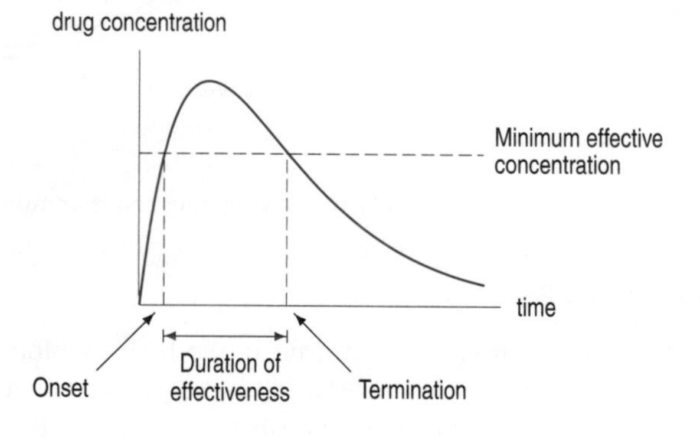

Figure 4.92: When is the drug effective?

Example 3 Depo-Provera was approved for use in the US in 1992 as a contraceptive. Figure 4.93 shows the drug concentration curve for a dose of 150 mg given intramuscularly.[20] The minimum effective concentration is about 4 ng/ml. How often should the drug be administered?

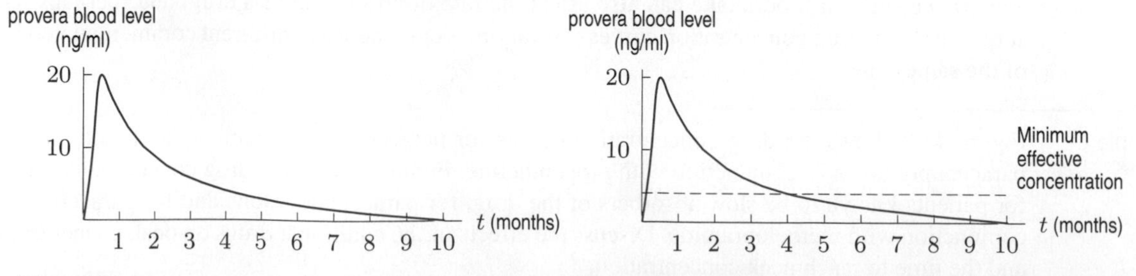

Figure 4.93: Drug concentration curve for Depo-Provera **Figure 4.94**: When should the next dose be administered?

[20]Robert M. Julien, *A Primer of Drug Action*, (W. H. Freeman and Co, 1995).

Solution The minimum effective concentration on the drug concentration curve is plotted as a dotted horizontal line at 4 ng/ml. See Figure 4.94. We see that the drug becomes effective almost immediately and ceases to be effective after about four months. Doses should be given about every four months.

Although the dosage interval is four months, notice that it takes ten months after injections are discontinued for Depo-Provera to be entirely eliminated from the body. Fertility during that period is unpredictable.

Problems for Section 4.8

1. If time, t, is in hours and concentration, C, is in ng/ml, the drug concentration curve for a drug is given by

$$C = 12.4te^{-0.2t}.$$

(a) Graph this curve.
(b) How many hours does it take for the drug to reach its peak concentration? What is the concentration at that time?
(c) If the minimum effective concentration is 10 ng/ml, during what time period is the drug effective?
(d) Complications can arise whenever the level of the drug is above 4 ng/ml. How long must a patient wait before being safe from complications?

2. Figure 4.95 shows drug concentration curves for anhydrous ampicillin for newborn babies and adults.[21] Discuss the differences between newborns and adults in the absorption of this drug.

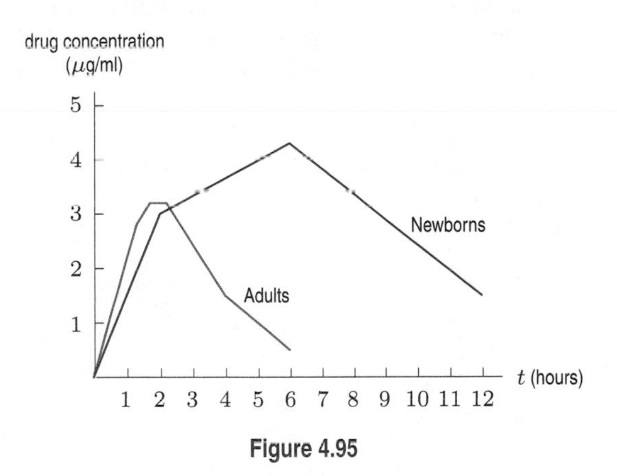

Figure 4.95

3. If t is in hours, the drug concentration curve for a drug is given by $C = 17.2te^{-0.4t}$ ng/ml. The minimum effective concentration is 10 ng/ml.

(a) If the second dose of the drug is to be administered when the first dose becomes ineffective, when should the second dose be given?
(b) If you want the onset of effectiveness of the second dose to coincide with termination of effectiveness of the first dose, when should the second dose be given?

4. Absorption of different forms of the antibiotic erythromycin may be increased, decreased, delayed or not affected by food. Figure 4.96 shows the drug concentration levels of erythromycin in healthy, fasting human volunteers who received single oral doses of 500 mg erythromycin tablets, together with either large (250 ml) or small (20 ml) accompanying volumes of water.[22] Discuss the effect of the water on the concentration of erythromycin in the blood. How are the peak concentration and the time to reach peak concentration affected? When does the effect of the volume of water wear off?

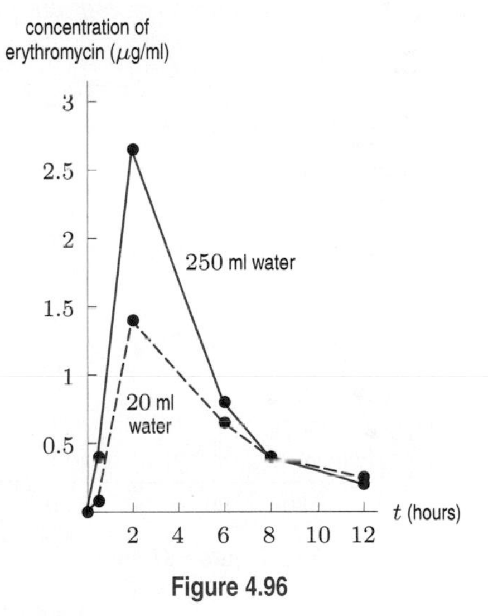

Figure 4.96

5. Hydrocodone bitartrate is a cough suppressant usually administered in a 10 mg oral dose. The peak concentration of the drug in the blood occurs 1.3 hours after consumption and the peak concentration is 23.6 ng/ml. Draw the drug concentration curve for hydrocodone bitartrate.

6. If t is in minutes since the drug was administered, the concentration, $C(t)$ in ng/ml, of a drug in a patient's bloodstream is given by

$$C(t) = 20te^{-0.03t}.$$

(a) How long does it take for the drug to reach peak concentration? What is the peak concentration?
(b) What is the concentration of the drug in the body after 15 minutes? After an hour?
(c) If the minimum effective concentration is 10 ng/ml, when should the next dose be administered?

[21] *Pediatrics*, 1973, 51, 578.
[22] J. W. Bridges and L.F. Chasseaud, *Progress in Drug Metabolism*, (New York: John Wiley and Sons, 1980).

7. Figure 4.85 on page 222 shows the concentration of nicotine in the blood during and after smoking a cigarette. Figure 4.97 shows the concentration of nicotine in the blood during and after using chewing tobacco or nicotine gum. (The chewing occurred during the first 30 minutes and the experimental data shown represent the average values for ten patients.)[23] Compare the three nicotine concentration curves (for cigarettes, chewing tobacco and nicotine gum) in terms of peak concentration, the time until peak concentration, and the rate at which the nicotine is eliminated from the bloodstream.

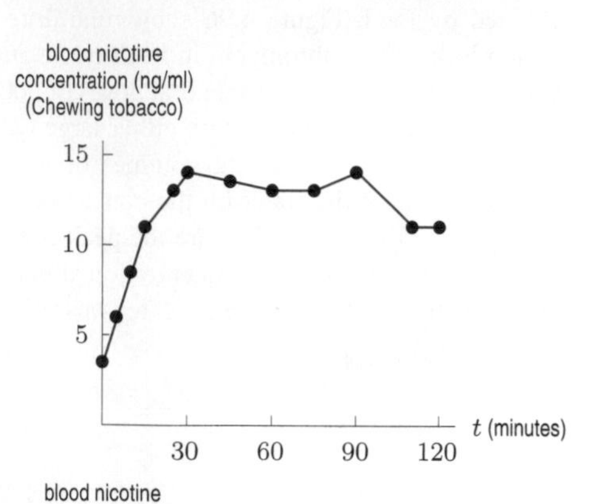

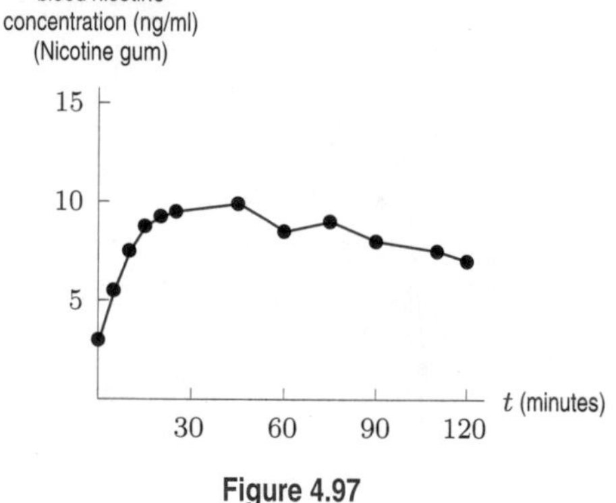

Figure 4.97

8. The method of administering a drug can have a strong influence on the drug concentration curve. Figure 4.98 shows drug concentration curves for penicillin following various routes of administration. Three milligrams per kilogram of body weight were dissolved in water and administered intravenously (IV), intramuscularly (IM), subcutaneously (SC), and orally (PO). The same quantity of penicillin dissolved in oil was administered intramuscularly (P-IM). The minimum effective concentration (MEC) is labeled on the graph.[24]

(a) Which method reaches peak concentration the fastest? The slowest?

(b) Which method has the largest peak concentration? The smallest?

(c) Which method wears off the fastest? The slowest?

(d) Which method has the longest effective duration? The shortest?

(e) When penicillin is administered orally, for approximately what time interval is it effective?

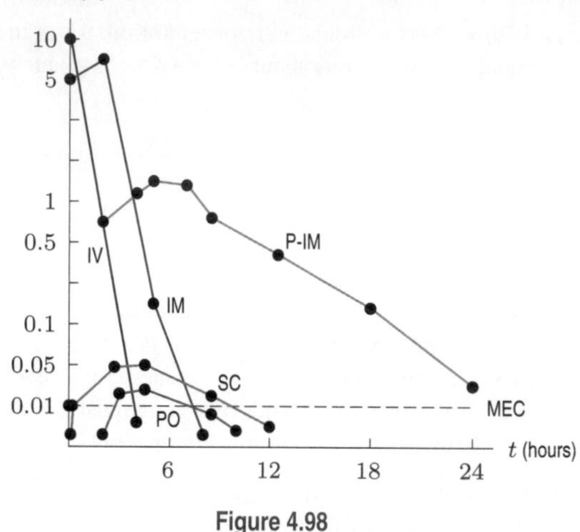

Figure 4.98

9. Figure 4.99 shows the plasma levels of canrenone in a healthy volunteer after a single oral dose of spironolactone given on a fasting stomach and together with a standardized breakfast. (Spironolactone is a diuretic agent that is partially converted into canrenone in the body.)[25] Discuss the effect of food on peak concentration and time to reach peak concentration. Is the effect of the food strongest during the first 8 hours, or after 8 hours?

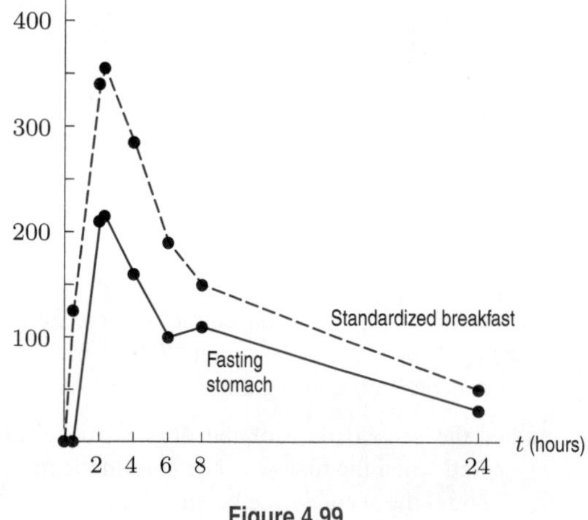

Figure 4.99

[23] Benowitz, Porchet, Skeiner, Jacob, "Nicotine Absorption and Cardiovascular Effects with Smokeless Tobacco Use: Comparison with Cigarettes and Nicotine Gum," *Clinical Pharmacology and Therapeutics* 44 (1988): 24.

[24] J. W. Bridges and L. F. Chasseaud, *Progress in Drug Metabolism*, (New York: John Wiley and Sons, 1980).

[25] Welling & Tse, *Pharmacokinetics of Cardiovascular, Central Nervous System, and Antimicrobial Drugs*, (The Royal Society of Chemistry, 1985).

10. Figure 4.100 shows drug concentration curves after oral administration of 0.5 mg of four digoxin products. All the tablets met current USP standards of potency, disintegration time, and dissolution rate.[26]

(a) Discuss differences and similarities in the peak concentration and the time to reach peak concentration.
(b) Give possible values for minimum effective concentration and maximum safe concentration that would make Product C or Product D the preferred drug.
(c) Give possible values for minimum effective concentration and maximum safe concentration that would make Product A the preferred drug.

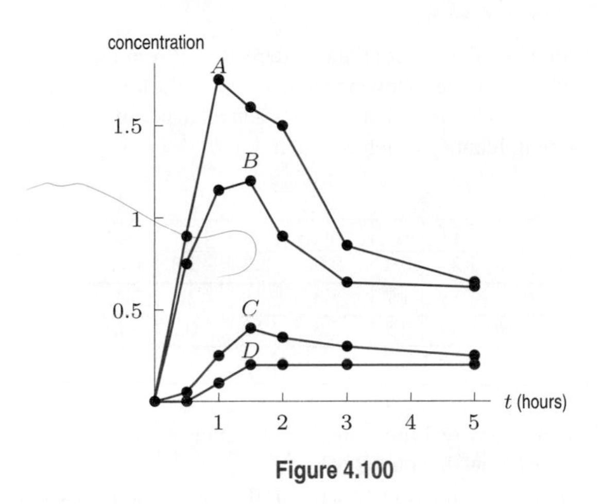

Figure 4.100

11. Figure 4.101 shows a graph of the percentage of drug dissolved against time for four tetracycline products A, B, C, and D. Figure 4.102 shows the drug concentration curves for the same four tetracycline products.[27] Discuss the effect of dissolution rate on peak concentration and time to reach peak concentration.

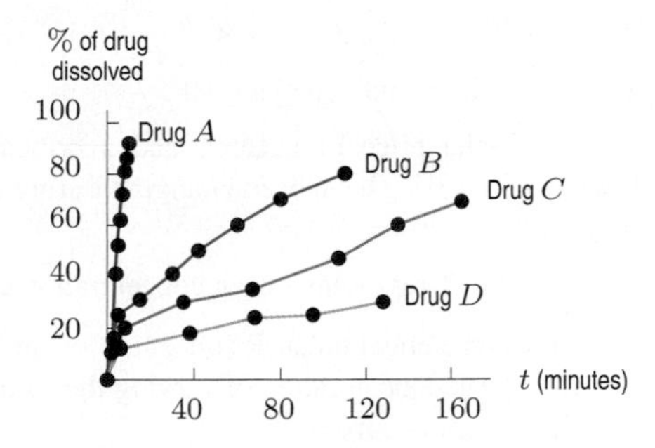

Figure 4.101

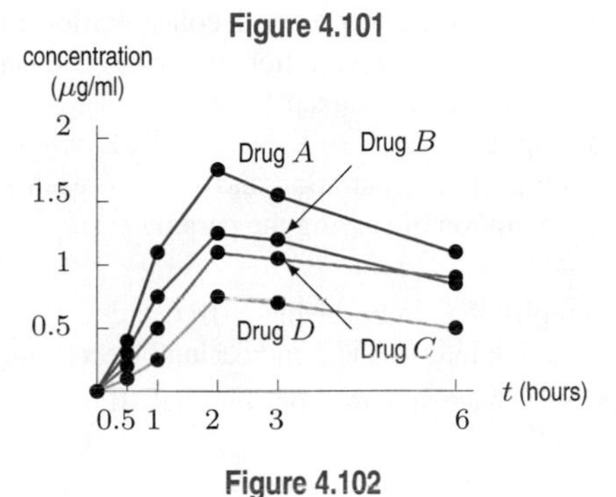

Figure 4.102

12. Let $b = 1$, and graph $C = ate^{-bt}$ using different values for a. Explain the effect of the parameter a.

CHAPTER SUMMARY

- **Using the first derivative**
  Critical points, local maxima and minima
- **Using the second derivative**
  Inflection points, concavity
- **Optimization**
  Global maxima and minima
- **Maximizing profit and revenue**
- **Average cost**
  Minimizing average cost
- **Elasticity**
- **Families of functions**
  Parameters. The surge function, drug concentration curves. The logistic function, carrying capacity, point of diminishing returns.

REVIEW PROBLEMS FOR CHAPTER FOUR

For Problems 1–2, indicate all critical points on the given graphs. Determine which correspond to local minima, local maxima, global minima, global maxima, or none of these. (Note that the graphs are on closed intervals.)

1.

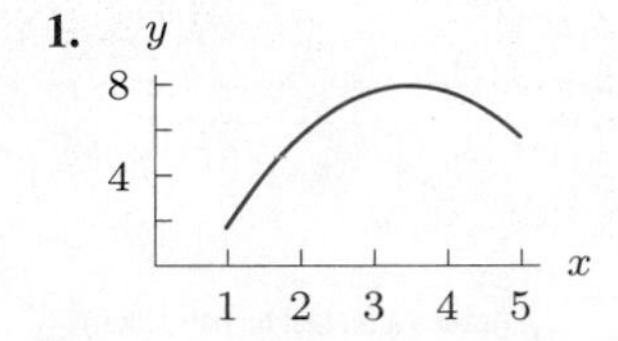

2.

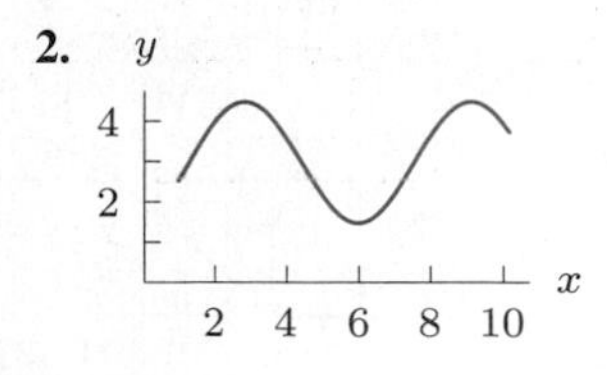

[26]Graeme S. Avery, ed. *Drug Treatment: Principles and Practice of Clinical Pharmacology and Therapeutics*, (Sydney: Adis Press, 1976).

[27]J. W. Bridges and L.F. Chasseaud, *Progress in Drug Metabolism*, (New York: John Wiley and Sons, 1980).

In Problems 3–4, find the value(s) of x for which:

(a) $f(x)$ has a local maximum or local minimum. Indicate which ones are maxima and which are minima.

(b) $f(x)$ has a global maximum or global minimum.

3. $f(x) = x^{10} - 10x$, and $0 \le x \le 2$

4. $f(x) = x - \ln x$, and $0.1 \le x \le 2$

5. On July 1, the price of a stock had a critical point. How could the price have been changing during the time around July 1?

6. Let $C = ate^{-bt}$ represent a drug concentration curve.

(a) Discuss the effect on peak concentration and time to reach peak concentration of varying the parameter a while keeping b fixed.

(b) Discuss the effect on peak concentration and time to reach peak concentration of varying the parameter b while keeping a fixed.

(c) Suppose $a = b$, so $C = ate^{-at}$. Discuss the effect on peak concentration and time to reach peak concentration of varying the parameter a.

For the graphs of f' in Problems 7–10 decide:

(a) Over what intervals is f increasing? Decreasing?

(b) Does f have maxima or minima? If so, which, and where?

7.

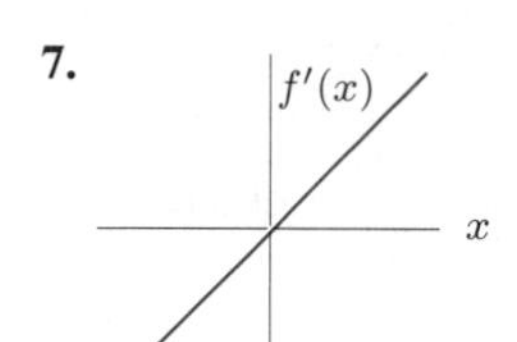

8.

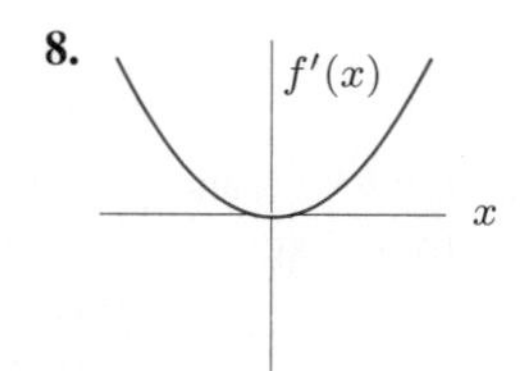

9.

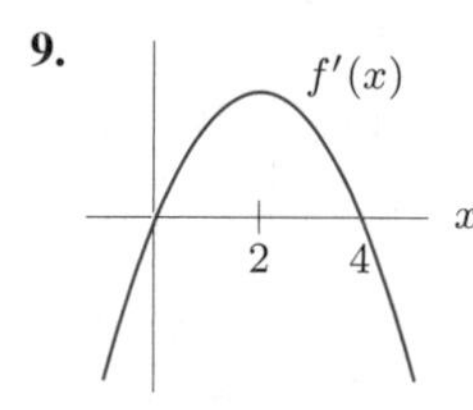

10.

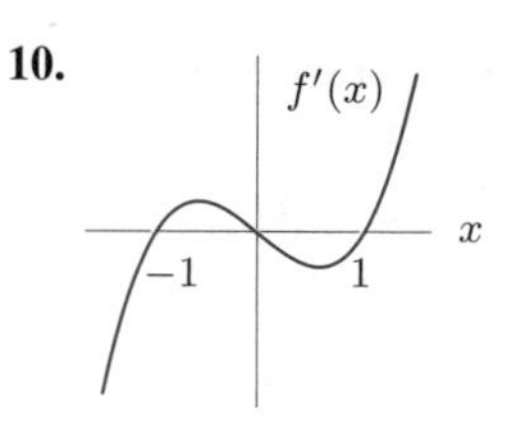

Problems 11–14 concern $f(t)$ in Figure 4.103, which gives the length of a human fetus as a function of its age.

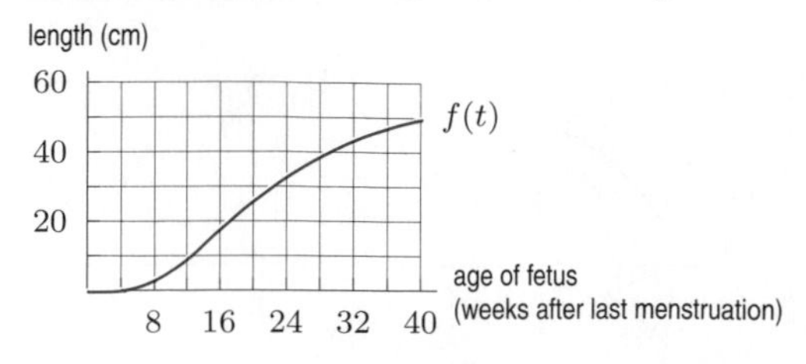

Figure 4.103

11. **(a)** What are the units of $f'(24)$?

(b) What is the biological meaning of $f'(24) = 1.6$?

12. **(a)** Which is greater, $f'(20)$ or $f'(36)$?

(b) What does your answer say about fetal growth?

13. **(a)** At what time does the inflection point occur?

(b) What is the biological significance of this point?

14. Estimate

(a) $f'(20)$ **(b)** $f'(36)$

(c) The average rate of change of length over the 40 weeks shown.

15. Find constants a and b in the function $f(x) = axe^{bx}$ such that $f(\frac{1}{3}) = 1$ and the function has a local maximum at $x = \frac{1}{3}$.

16. Suppose f has a continuous derivative. From the values of $f'(\theta)$ in the following table, estimate the θ values with $1 < \theta < 2.1$ at which $f(\theta)$ has a local maximum or minimum. Identify which is which.

θ	1.0	1.1	1.2	1.3	1.4	1.5
$f'(\theta)$	2.4	0.3	−2.0	−3.5	−3.3	−1.7
θ	1.6	1.7	1.8	1.9	2.0	2.1
$f'(\theta)$	0.8	2.8	3.6	2.8	0.7	−1.6

17. How many real roots does the equation $x^5 + x + 7 = 0$ have? How do you know?
[Hint: How many critical points does this function have?]

18. For the function, f, graphed in Figure 4.104:

(a) Sketch $f'(x)$.

(b) Where does $f'(x)$ change its sign?

(c) Where does $f'(x)$ have local maxima or minima?

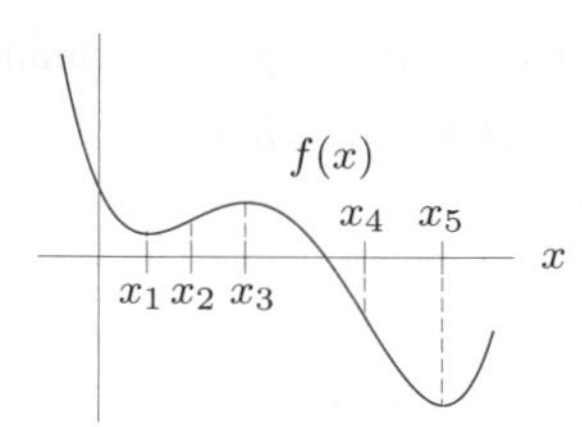

Figure 4.104

19. Using your answer to Problem 18 as a guide, write a short paragraph (using complete sentences) which describes the relationships between the following features of a function f:

- The local maxima and minima of f.
- The points at which the graph of f changes concavity.
- The sign changes of f'.
- The local maxima and minima of f'.

20. The function $y = t(x)$ is positive and continuous with a global maximum at the point $(3, 3)$. Graph $t(x)$ if $t'(x)$ and $t''(x)$ have the same sign for $x < 3$, but opposite signs for $x > 3$.

21. Each of the graphs in Figure 4.105 belongs to one of the following families of functions. In each case, identify which family is most likely:

an exponential function,

a logarithmic function,

a polynomial (What is the degree? Is the leading coefficient positive or negative?),

a periodic function,

a logistic function,

a surge function.

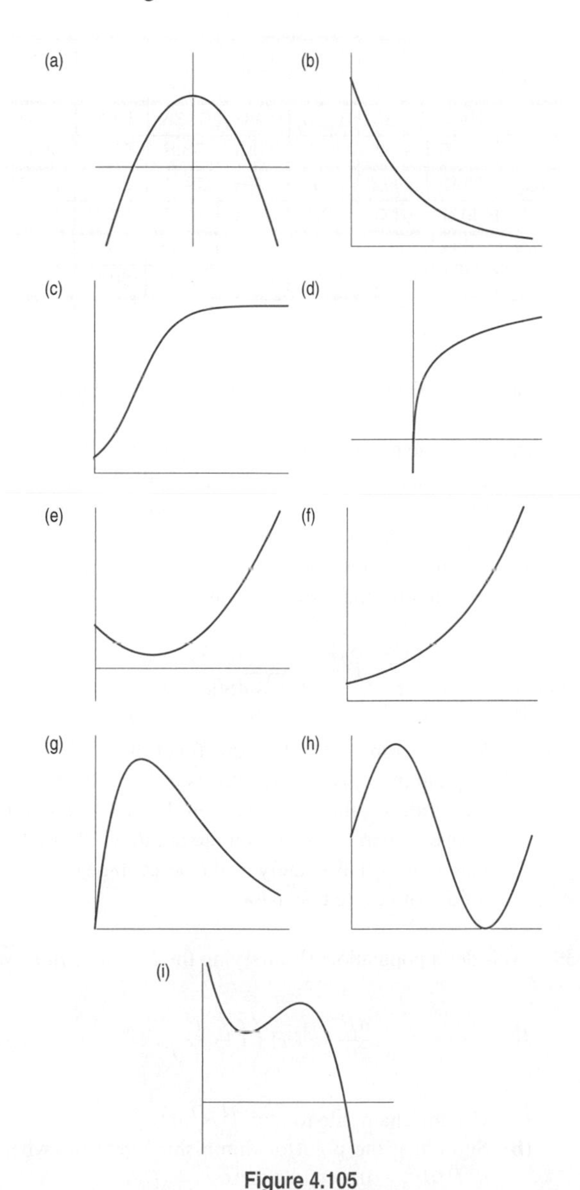

Figure 4.105

22. The total cost of producing q units of a product is given by $C(q) = q^3 - 60q^2 + 1400q + 1000$ for $0 \le q \le 50$; the product sells for \$788 per unit. What production level maximizes profit? Find the total cost, total revenue, and total profit at this production level. Graph the cost and revenue functions on the same axes, and label the production level at which profit is maximized, and the corresponding cost, revenue, and profit. [Hint: Costs can go as high as \$46,000.]

23. The total cost $C(q)$ of producing q goods is given by:

$$C(q) = 0.01q^3 - 0.6q^2 + 13q.$$

(a) What is the fixed cost?

(b) What is the maximum profit if each item is sold for \$7? (Assume you sell everything you produce.)

(c) Suppose exactly 34 goods are produced. They all sell when the price is \$7 each, but for each \$1 increase in price, 2 fewer goods are sold. Should the price be raised, and if so by how much?

24. The demand equation for a product is $p = b_1 - a_1q$ and the cost function is $C(q) = b_2 + a_2q$, where p is the price of the product and q is the quantity sold. Find the value of q, in terms of the positive constants b_1, a_1, b_2, a_2, that maximizes profit.

25. A manufacturer's cost of producing a product is given in Figure 4.106. The manufacturer can sell the product for a price p each (regardless of the quantity sold), so that the total revenue from selling a quantity q is $R(q) = pq$.

(a) The difference $\pi(q) = R(q) - C(q)$ is the total profit. For which quantity q_0 is the profit a maximum? Mark your answer on a sketch of the graph.

(b) What is the relationship between p and $C'(q_0)$? Explain your result both graphically and analytically. What does this mean in terms of economics? (Note that p is the slope of the line $R(q) = pq$. Note also that $\pi(q)$ has a maximum at $q = q_0$, so $\pi'(q_0) = 0$.)

(c) Graph $C'(q)$ and p (as a horizontal line) on the same axes. Mark q_0 on the q-axis.

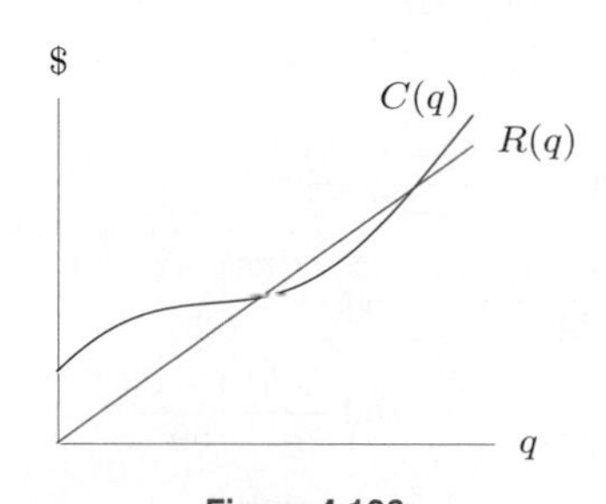

Figure 4.106

26. Let $C(q) = 0.04q^3 - 3q^2 + 75q + 96$ be the total cost of producing q items.

(a) Find the average cost per item as a function of q.
(b) Use a graphing calculator or computer to graph average cost against q.
(c) For what values of q is the average cost per item decreasing? Increasing?
(d) For what value of q is the average cost per item smallest? What is the smallest average cost per item at that point?

27. Graph a cost function where the minimum average cost of $25 per unit is achieved by producing 15,000 units.

28. Figure 4.107 shows the cost, $C(q)$, and the revenue, $R(q)$, for a quantity q. Label the following points on the graph:

(a) The point F representing the fixed costs.
(b) The point B representing the break-even level of production.
(c) The point M representing the level of production at which marginal cost is a minimum.
(d) The point A representing the level of production at which average cost $a(q) = C(q)/q$ is a minimum.
(e) The point P representing the level of production at which profit is a maximum.

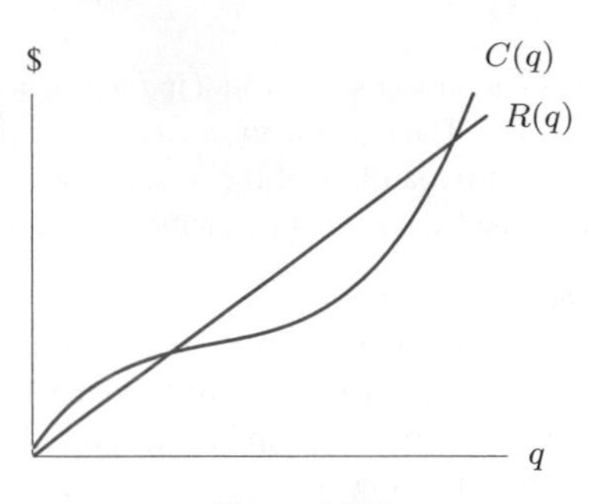

Figure 4.107

In Problems 29–34, cost, $C(q)$, is a positive, increasing, concave down function of quantity produced, q. Which one of the two numbers is the larger?

29. $C'(2)$ and $C'(3)$

30. $C'(5)$ and $\dfrac{C(5) - C(3)}{5 - 3}$

31. $\dfrac{C(100) - C(50)}{50}$ and $\dfrac{C(75) - C(50)}{25}$

32. $\dfrac{C(100) - C(50)}{50}$ and $\dfrac{C(75) - C(25)}{50}$

33. $C'(3)$ and $C(3)/3$

34. $C(10)/10$ and $C(25)/25$

35. The demand for a product is $q = 2000 - 5p$ where q is units sold at a price of p dollars. Find the elasticity if the price is $20, and interpret your answer in terms of demand.

36. Show analytically that if elasticity of demand satisfies $E > 1$, then the derivative of revenue with respect to price satisfies $dR/dp < 0$.

37. Show analytically that if elasticity of demand satisfies $E < 1$, then the derivative of revenue with respect to price satisfies $dR/dp > 0$.

38. The following table gives the percentage, P, of households with cable television.[28]

Year	1977	1978	1979	1980	1981	1982	1983
P	16.6	17.9	19.4	22.6	28.3	35.0	40.5
Year	1984	1985	1986	1987	1988	1989	1990
P	43.7	46.2	48.1	50.5	53.8	57.1	59.0
Year	1991	1992	1993	1994	1995	1996	1997
P	60.6	61.5	62.5	63.4	65.7	66.7	67.3
Year	1998	1999	2000	2001	2002	2003	
P	67.4	68.0	67.8	69.2	68.9	68.0	

(a) Explain why a logistic model is reasonable for this data.
(b) Estimate the point of diminishing returns. What limiting value L does this point predict? Does this limiting value appear to be accurate, given the percentages for 2002 and 2003?
(c) If t is in years since 1977, the best fitting logistic function for this data turns out to be

$$P = \frac{68.8}{1 + 3.486e^{-0.237t}},$$

What limiting value does this function predict?
(d) Explain in terms of percentages of households what the limiting value is telling you. Do you think your answer to part (c) is an accurate prediction? What do you think will ultimately be the percentage of households with cable television?

39. Consider a population P satisfying the *logistic equation*

$$\frac{dP}{dt} = kP\left(1 - \frac{P}{L}\right).$$

(a) Use the chain rule to find d^2P/dt^2.
(b) Show that the point of diminishing returns, where $d^2P/dt^2 = 0$, occurs where $P = L/2$.

[28] *The World Almanac and Book of Facts 2005*, p. 310 (New York).

40. Figure 4.108 shows the concentration of bemetizide (a diuretic) in the blood after single oral doses of 25 mg alone, or 25 mg bemetizide and 50 mg triamterene in combination.[29] If the minimum effective concentration in a patient is 40 ng/ml, compare the effect of combining bemetizide with triamterene on peak concentration, time to reach peak concentration, time until onset of effectiveness, and duration of effectiveness. Under what circumstances might it be wise to use the triamterene with the bemetizide?

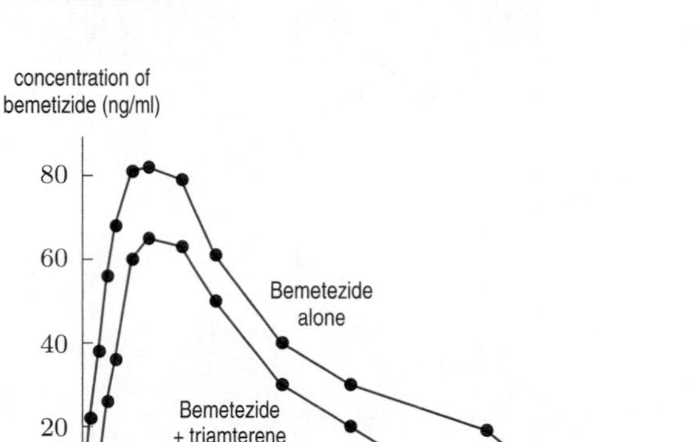

Figure 4.108

In Problems 41–43, find the exact global maximum and minimum values of the function.

41. $h(z) = \dfrac{1}{z} + 4z^2$ for $z > 0$

42. $g(t) = \dfrac{1}{t^3 + 1}$ for $t \geq 0$

43. $f(x) = \dfrac{1}{(x-1)^2 + 2}$

44. For $f(x) = \sin(x^2)$ between $x = 0$ and $x = 3$, find the coordinates of all intercepts, critical points, and inflection points to two decimal points.

45. **(a)** Graph $f(x) = x + a \sin x$ for $a = 0.5$ and $a = 3$.
(b) For what values of a is $f(x)$ increasing for all x?

46. **(a)** Graph $f(x) = x^2 + a \sin x$ for $a = 1$ and $a = 20$.
(b) For what values of a is $f(x)$ concave up for all x?

47. The distance, s, traveled by a cyclist, who starts at 1 pm, is given in Figure 4.109. Time, t, is in hours since noon.

(a) Explain why the quantity, s/t, is represented by the slope of a line from the origin to the point (t, s) on the graph.
(b) Estimate the time at which the quantity s/t is a maximum.
(c) What is the relationship between the quantity s/t and the instantaneous speed of the cyclist at the time you found in part (b)?

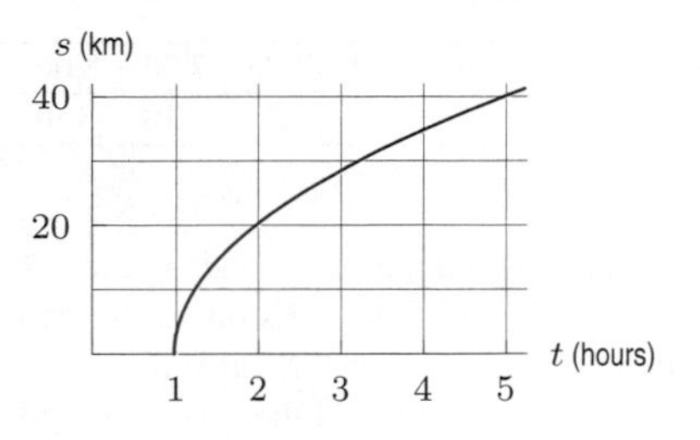

Figure 4.109

48. A bird such as a starling feeds worms to its young. To collect worms, the bird flies to a site where worms are to be found, picks up several in its beak, and flies back to its nest. The *loading curve* in Figure 4.110 shows how the number of worms (the load) a starling collects depends on the time it has been searching for them.[30] The curve is concave down because the bird can pick up worms more efficiently when its beak is empty; when its beak is partly full, the bird becomes much less efficient. The traveling time (from nest to site and back) is represented by the distance PO in Figure 4.110. The bird wants to maximize the rate at which it brings worms to the nest, where

$$\text{Rate worms arrive} = \frac{\text{Load}}{\text{Traveling time + Searching time}}$$

(a) Draw a line in Figure 4.110 whose slope is this rate.
(b) Using the graph, estimate the load which maximizes this rate.
(c) If the traveling time is increased, does the optimal load increase or decrease? Why?

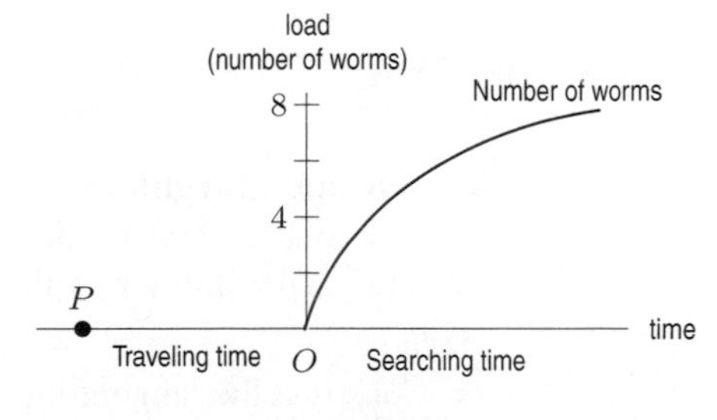

Figure 4.110

[29] Welling & Tse, *Pharmacokinetics of Cardiovascular, Central Nervous System, and Antimicrobial Drugs*, (The Royal Society of Chemistry, 1985).

[30] Alex Kacelnick (1984). Reported by J. R. Krebs and N. B. Davis, *An Introduction to Behavioural Ecology* (Oxford: Blackwell, 1987).

49. Table 4.11 shows the total number of cars in a University of Arizona parking lot at 30-minute intervals.[31]

Table 4.11 *Total number of cars, C, at time t*

t	5:00	5:30	6:00	6:30	7:00	7:30	8:00	8:30	9:00
C	4	5	8	18	50	110	170	200	200

(a) Graph the total number of cars in the UA parking lot as a function of time. Estimate the capacity of the parking lot, and when it was full.

(b) Construct a table, and then plot a graph, of the rate of arrival of cars as a function of time.

(c) From your graph in part (b), estimate when rush hour occurred at UA.

(d) Explain the relationship between the points on the graphs in parts (a) and (b) where rush hour occurred.

50. A line goes through the origin and a point on the curve $y = x^2e^{-3x}$, for $x \geq 0$. Find the maximum slope of such a line. At what x-value does it occur?

51. A rectangle has one side on the x-axis, one side on the y-axis, one vertex at the origin and one on the curve $y = e^{-2x}$ for $x \geq 0$. Find the

(a) Maximum area **(b)** Minimum perimeter

52. A single cell of a bee's honey comb has the shape shown in Figure 4.111. The surface area of this cell is given by

$$A = 6hs + \frac{3}{2}s^2\left(\frac{-\cos\theta}{\sin\theta} + \frac{\sqrt{3}}{\sin\theta}\right)$$

where h, s, θ are as shown in the picture.

(a) Keeping h and s fixed, for what angle, θ, is the surface area a minimum?

(b) Measurements on bee's cells have shown that the angle actually used by bees is about $\theta = 55°$. Comment.

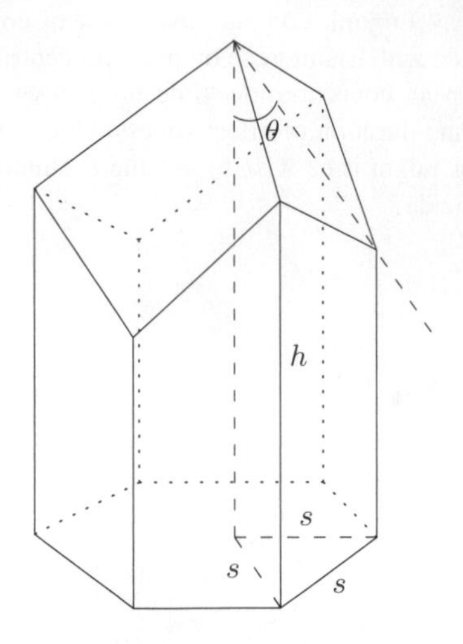

Figure 4.111

53. An organism has size W at time t. For positive constants A, b, and c, the Gompertz growth function gives

$$W = Ae^{-e^{b-ct}}, \quad t \geq 0.$$

(a) Find the intercepts and asymptotes.

(b) Find the critical points and inflection points.

(c) Graph W for various values of A, b, and c.

(d) A certain organism grows fastest when it is about 1/3 of its final size. Would the Gompertz growth function be useful in modeling its growth? Explain.

PROJECTS FOR CHAPTER FOUR

1. Average and Marginal Costs

The total cost of producing a quantity q is $C(q)$. The average cost $a(q)$ is given in Figure 4.112. The following rule is used by economists to determine the marginal cost $C'(q_0)$, for any q_0:

- Construct the tangent line t_1 to $a(q)$ at q_0.
- Let t_2 be the line with the same vertical intercept as t_1 but with twice the slope of t_1.

Then $C'(q_0)$ is the vertical distance shown in Figure 4.112. Explain why this rule works.

[31] Adapted from Nancy Roberts et al. *Introduction to Computer Simulation*, p. 93 (Reading: Addison-Wesley 1983).

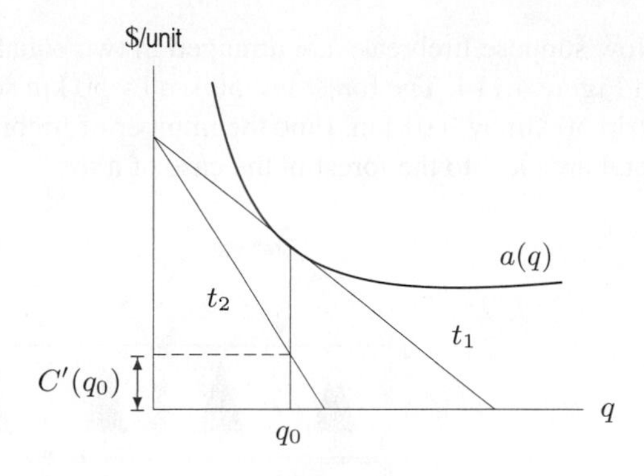

Figure 4.112

2. Firebreaks

The summer of 2000 was devastating for forests in the western US: over 3.5 million acres of trees were lost to fires, making this the worst fire season in 30 years. This project studies a fire management technique called *firebreaks*, which reduce the damage done by forest fires. A firebreak is a strip where trees have been removed in a forest so that a fire started on one side of the strip will not spread to the other side. Having many firebreaks helps confine a fire to a small area. On the other hand, having too many firebreaks involves removing large swaths of trees.[32]

(a) A forest in the shape of a 50 km by 50 km square has firebreaks in rectangular strips 50 km by 0.01 km. The trees between two firebreaks are called a stand of trees. All firebreaks in this forest are parallel to each other and to one edge of the forest, with the first firebreak at the edge of the forest. The firebreaks are evenly spaced throughout the forest. (For example, Figure 4.113 shows four firebreaks.) The total area lost in the case of a fire is the area of the stand of trees in which the fire started plus the area of all the firebreaks.

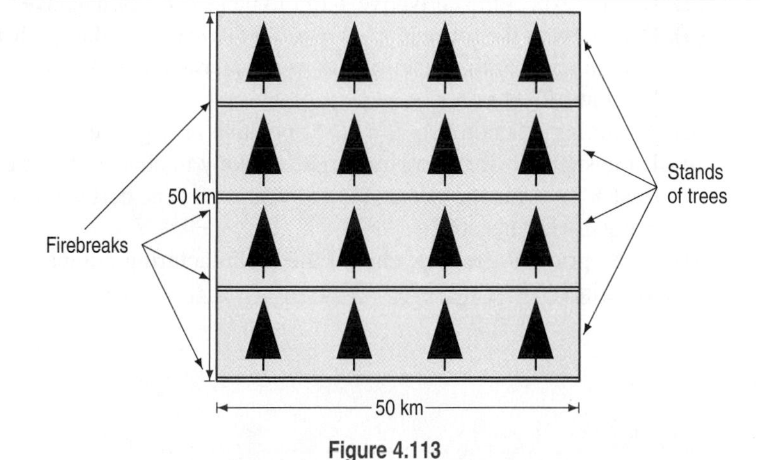

Figure 4.113

(i) Find the number of firebreaks that minimizes the total area lost to the forest in the case of a fire.

(ii) If a firebreak is 50 km by b km, find the optimal number of firebreaks as a function of b. If the width, b, of a firebreak is quadrupled, how does the optimal number of firebreaks change?

[32] Adapted from D. Quinney and R. Harding, *Calculus Connections* (New York: John Wiley & Sons, 1996).

(b) Now suppose firebreaks are arranged in two equally spaced sets of parallel lines, as shown in Figure 4.114. The forest is a 50 km by 50 km square, and each firebreak is a rectangular strip 50 km by 0.01 km. Find the number of firebreaks in each direction that minimizes the total area lost to the forest in the case of a fire.

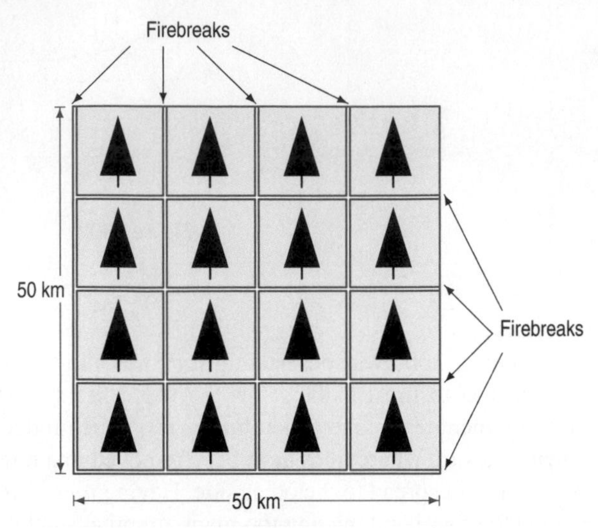

Figure 4.114

3. Production and the Price of Raw Materials

The production function $f(x)$ gives the number of units of an item that a manufacturing company can produce from x units of raw material. The company buys the raw material at price w dollars per unit and sells all it produces at a price of p dollars per unit. The quantity of raw material that maximizes profit is denoted by x^*.

(a) Do you expect the derivative $f'(x)$ to be positive or negative? Justify your answer.

(b) Explain why the formula $\pi(x) = pf(x) - wx$ gives the profit $\pi(x)$ that the company earns as a function of the quantity x of raw materials that it uses.

(c) Evaluate $f'(x^*)$.

(d) Assuming it is nonzero, is $f''(x^*)$ positive or negative?

(e) If the supplier of the raw materials is likely to change the price w, then it is appropriate to treat x^* as a function of w. Find a formula for the derivative dx^*/dw and decide whether it is positive or negative.

(f) If the price w goes up, should the manufacturing company buy more or less of the raw material?

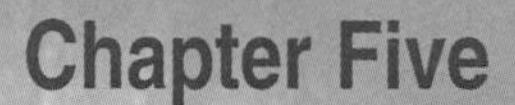

Chapter Five

ACCUMULATED CHANGE: THE DEFINITE INTEGRAL

Chapter 2 discussed the rate of change of a function, which led us to the derivative. Now we consider the reverse process: Getting information about the original function from the rate of change. This leads us to the *definite integral*, which can also be used to calculate the area under a curve.

The connection between the derivative and the definite integral is given by the Fundamental Theorem of Calculus. This shows us that calculating derivatives and calculating definite integrals are, in a sense, reverse processes.

5.1 DISTANCE AND ACCUMULATED CHANGE

In Chapter 2, we used the derivative to find the rate of change of a function. Here we see how to go in the other direction. If we know the rate of change, can we find the original function? We start by finding the distance traveled from the velocity.

How Do We Measure Distance Traveled?

The rate of change of distance with respect to time is velocity. If we are given the velocity, can we find the distance traveled? Suppose the velocity was 50 miles per hour throughout a four hour trip. What is the total distance traveled? Since

$$\text{Distance} = \text{Velocity} \times \text{Time},$$

we have

$$\begin{aligned}\text{Distance traveled} &= (50 \text{ miles/hour}) \times (4 \text{ hours}) \\ &= 200 \text{ miles}.\end{aligned}$$

The graph of velocity against time is the horizontal line in Figure 5.1. Notice that the distance traveled is represented by the shaded area under the graph.

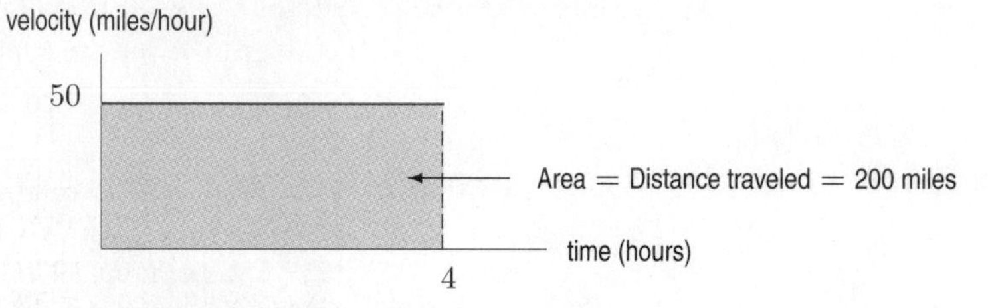

Figure 5.1: Area shaded represents distance traveled in 4 hours at 50 mph

Now let's see what happens if the velocity is not constant.

Example 1 Suppose that you travel 30 miles/hour for 2 hours, then 40 miles/hour for $1/2$ hour, then 20 miles/hour for 4 hours. What is the total distance you traveled?

Solution We compute the distances traveled for each of the three legs of the trip and add them to find the total distance traveled:

$$\begin{aligned}\text{Distance} &= (30 \text{ miles/hour})(2 \text{ hours}) + (40 \text{ miles/hour})(1/2 \text{ hour}) + (20 \text{ miles/hour})(4 \text{ hours}) \\ &= 60 \text{ miles} + 20 \text{ miles} + 80 \text{ miles} \\ &= 160 \text{ miles}.\end{aligned}$$

You travel 160 miles on this trip.

A Thought Experiment: How Far Did the Car Go?

In Example 1, the velocity was constant over intervals. Of course, this is not always the case; we now look at an example where the velocity is continually changing.

Velocity Data Every Two Seconds

Suppose a car is moving with increasing velocity and suppose we measure the car's velocity every two seconds, obtaining the data in Table 5.1:

Table 5.1 *Velocity of car every two seconds*

Time (sec)	0	2	4	6	8	10
Velocity (ft/sec)	20	30	38	44	48	50

How far has the car traveled? Since we don't know how fast the car is moving at every moment, we can't calculate the distance exactly, but we can make an estimate. The velocity is increasing, so the car is going at least 20 ft/sec for the first two seconds. Since Distance $=$ Velocity $\times$ Time, the car goes at least $20 \cdot 2 = 40$ feet during the first two seconds. Likewise, it goes at least $30 \cdot 2 = 60$ feet during the next two seconds, and so on. During the ten-second period it goes at least

$$20 \cdot 2 + 30 \cdot 2 + 38 \cdot 2 + 44 \cdot 2 + 48 \cdot 2 = 360 \text{ feet.}$$

Thus, 360 feet is an underestimate of the total distance traveled during the ten seconds.

To get an overestimate, we can reason in a similar way: During the first two seconds, the car's velocity is at most 30 ft/sec, so it moves at most $30 \cdot 2 = 60$ feet. In the next two seconds it moves at most $38 \cdot 2 = 76$ feet, and so on. Therefore, over the ten-second period it moves at most

$$30 \cdot 2 + 38 \cdot 2 + 44 \cdot 2 + 48 \cdot 2 + 50 \cdot 2 = 420 \text{ feet.}$$

Therefore,

$$360 \text{ feet} \leq \text{Total distance traveled} \leq 420 \text{ feet.}$$

There is a difference of 60 feet between the upper and lower estimates.

Velocity Data Every One Second

What if we want a more accurate estimate? We could make more frequent velocity measurements, say every second. The data is in Table 5.2.

As before, we get a lower estimate for each second by using the velocity at the beginning of that second. During the first second the velocity is at least 20 ft/sec, so the car travels at least $20 \cdot 1 = 20$ feet. During the next second the car moves at least 26 feet, and so on. So we say

$$\begin{aligned} \text{New lower estimate} &= 20 \cdot 1 + 26 \cdot 1 + 30 \cdot 1 + 35 \cdot 1 + 38 \cdot 1 \\ &\quad + 42 \cdot 1 + 44 \cdot 1 + 46 \cdot 1 + 48 \cdot 1 + 49 \cdot 1 \\ &= 378 \text{ feet.} \end{aligned}$$

Notice that this is greater than the old lower estimate of 360 feet.

We get a new upper estimate by considering the velocity at the end of each second. During the first second the velocity is at most 26 ft/sec, and so the car moves at most $26 \cdot 1 = 26$ feet; in the next second it moves at most 30 feet, and so on.

$$\begin{aligned} \text{New upper estimate} &= 26 \cdot 1 + 30 \cdot 1 + 35 \cdot 1 + 38 \cdot 1 + 42 \cdot 1 \\ &\quad + 44 \cdot 1 + 46 \cdot 1 + 48 \cdot 1 + 49 \cdot 1 + 50 \cdot 1 \\ &= 408 \text{ feet.} \end{aligned}$$

This is less than the old upper estimate of 420 feet. Now we know that

$$378 \text{ feet} \leq \text{Total distance traveled} \leq 408 \text{ feet.}$$

Notice that the difference between the new upper and lower estimates is now 30 feet, half of what it was before. By halving the interval of measurement, we have halved the difference between the upper and lower estimates.

Table 5.2 *Velocity of car every second*

Time (sec)	0	1	2	3	4	5	6	7	8	9	10
Velocity (ft/sec)	20	26	30	35	38	42	44	46	48	49	50

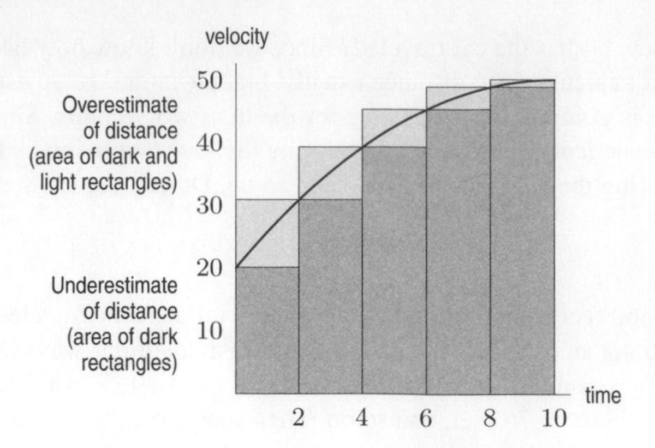

Figure 5.2: Shaded area estimates distance traveled. Velocity measured every 2 seconds

Visualizing Distance on the Velocity Graph

Consider the two-second data in Table 5.1. We can represent both upper and lower estimates on a graph of the velocity against time. The velocity can be graphed by plotting these data and drawing a smooth curve through the data points. (See Figure 5.2.)

We use the fact that for a rectangle, Area $=$ Height $\times$ Width. The area of the first dark rectangle is $20 \cdot 2 = 40$, the lower estimate of the distance moved during the first two seconds. The area of the second dark rectangle is $30 \cdot 2 = 60$, the lower estimate for the distance moved in the next two seconds. The total area of the dark rectangles represents the lower estimate for the total distance moved during the ten seconds.

If the dark and light rectangles are considered together, the first area is $30 \cdot 2 = 60$, the upper estimate for the distance moved in the first two seconds. The second area is $38 \cdot 2 = 76$, the upper estimate for the next two seconds. Continuing this calculation suggests that the upper estimate for the total distance is represented by the sum of the areas of the dark and light rectangles. Therefore, the area of the light rectangles alone represents the difference between the two estimates.

Figure 5.3 shows a graph of the one-second data. The area of the dark rectangles again represents the lower estimate, and the area of the dark and light rectangles together represents the upper estimate. The total area of the light rectangles is smaller in Figure 5.3 than in Figure 5.2, so the underestimate and overestimate are closer for the one-second data than for the two-second data.

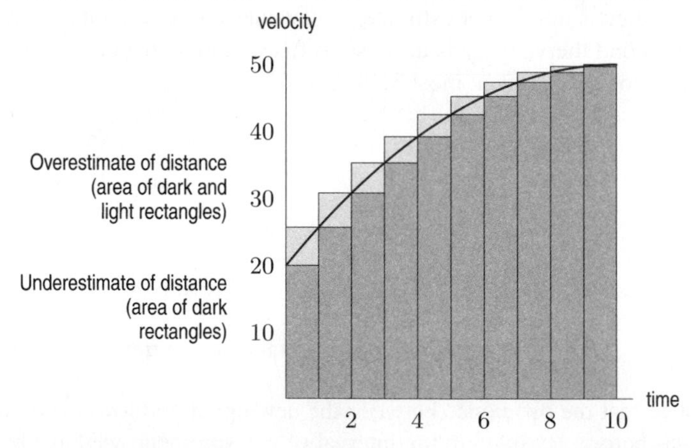

Figure 5.3: Shaded area estimates distance traveled. Velocity measured every second

Visualizing Distance on the Velocity Graph: Area Under Curve

As we make more frequent velocity measurements, the rectangles used to estimate the distance traveled fit the curve more closely. See Figures 5.4 and 5.5. In the limit, as the number of subdivisions increases, we see that the distance traveled is given by the area between the velocity curve and the horizontal axis. See Figure 5.6 on the next page. In general:

> If the velocity is positive, the total distance traveled is the area under the velocity curve.

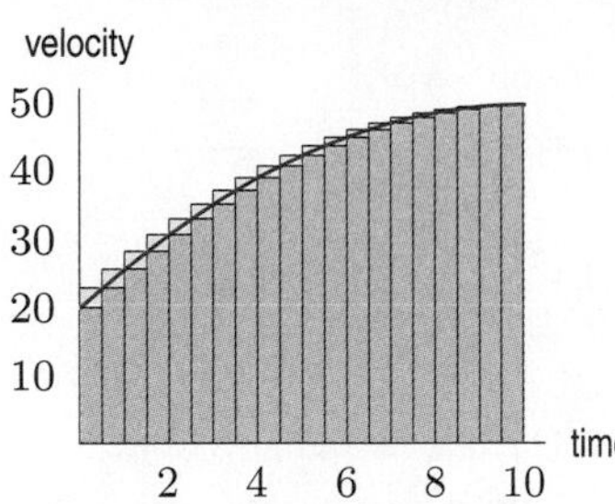

Figure 5.4: Velocity measured every 1/2 second

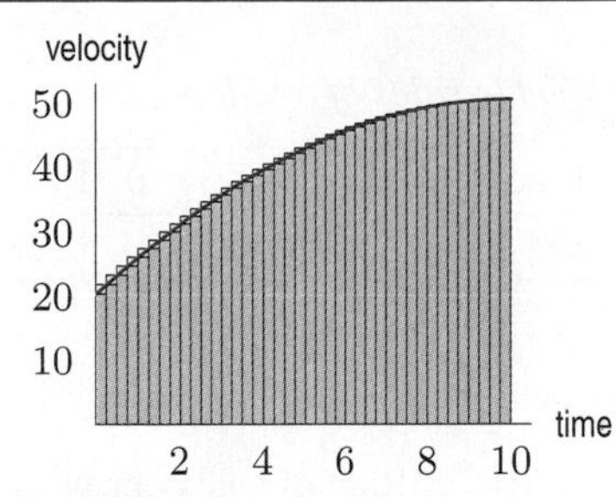

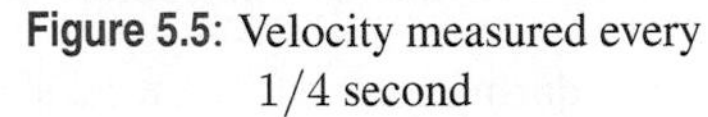

Figure 5.5: Velocity measured every 1/4 second

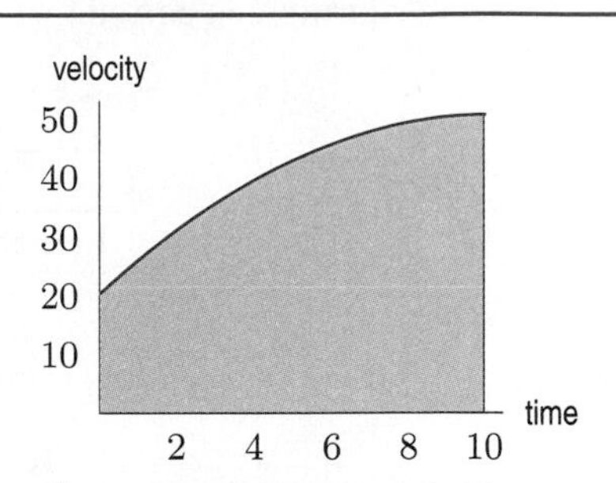

Figure 5.6: Distance traveled is area under curve

Example 2 With time t in seconds, the velocity of a bicycle, in feet per second, is given by $v(t) = 5t$. How far does the bicycle travel in 3 seconds?

Solution The velocity is linear. See Figure 5.7. The distance traveled is the area between the line $v(t) = 5t$ and the t-axis. Since this region is a triangle of height 15 and base 3

$$\text{Distance traveled} = \text{Area of triangle} = \frac{1}{2} \cdot 15 \cdot 3 = 22.5 \text{ feet.}$$

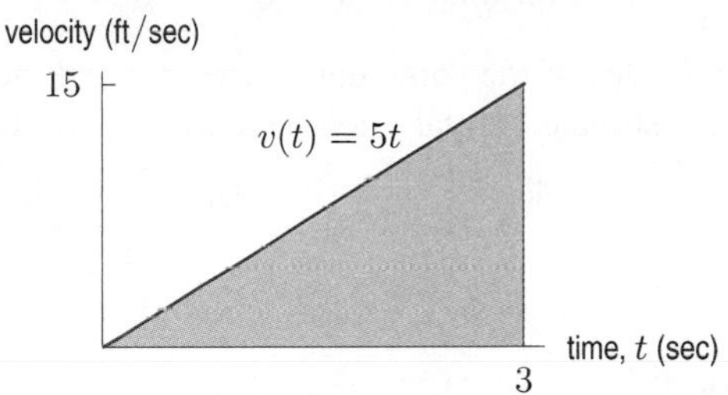

Figure 5.7: Shaded area represents distance traveled

Approximating Total Change from Rate of Change

We have seen how to use the rate of change of distance (the velocity) to calculate the total distance traveled. We can use the same method to find total change from rate of change of other quantities.

Example 3 A city's population grows at the rate of 5000 people/year for 3 years and then grows at the rate of 3000 people/year for the next 4 years. What is the total change in the population of the city during this 7-year period?

Solution The units of people/year remind us that we are given a rate of change of the population (people) with respect to time (years). If the rate of change is constant, we know that

$$\text{Total change in population} = \text{Rate of change per year} \times \text{Number of years.}$$

Thus, the total change in this population is

$$\begin{aligned}\text{Total change} &= (5000 \text{ people/year})(3 \text{ years}) + (3000 \text{ people/year})(4 \text{ years}) \\ &= 15{,}000 \text{ people} + 12{,}000 \text{ people} \\ &= 27{,}000 \text{ people.}\end{aligned}$$

The change in population is 27,000 people. Notice that this does not tell us the population of the city at the end of 7 years; it tells us the change in the population. For example, if the population were initially 100,000, it would be 127,000 at the end of the 7-year period.

Example 4 The rate of sales (in games per week) of a new video game is shown in Table 5.3. Assuming that the rate of sales increased throughout the 20-week period, estimate the total number of games sold during this period.

Table 5.3 *Weekly sales of a video game*

Time (weeks)	0	5	10	15	20
Rate of sales (games per week)	0	585	892	1875	2350

Solution If the rate of sales is constant, we have

$$\text{Total sales} = \text{Rate of sales per week} \times \text{Number of weeks}.$$

How many games were sold during the first five weeks? During this time, sales went from 0 to 585 games per week. If we assume that 585 games were sold every week, we get an overestimate for the sales in the first five weeks of $(585 \text{ games/week})(5 \text{ weeks}) = 2925$ games. Similar overestimates for each of the five-week periods gives an overestimate for the entire 20-week period:

$$\text{Overestimate for total sales} = 585 \cdot 5 + 892 \cdot 5 + 1875 \cdot 5 + 2350 \cdot 5 = 28{,}510 \text{ games.}$$

We underestimate the total sales by taking the lower value for rate of sales during each of the five-week periods:

$$\text{Underestimate for total sales} = 0 \cdot 5 + 585 \cdot 5 + 892 \cdot 5 + 1875 \cdot 5 = 16{,}760 \text{ games.}$$

Thus, the total sales of the game during the 20-week period is between $16{,}760$ and $28{,}510$ games. A good single estimate of total sales is the average of these two numbers:

$$\text{Total sales} \approx \frac{16{,}760 + 28{,}510}{2} = 22{,}635 \text{ games.}$$

Problems for Section 5.1

1. A car comes to a stop six seconds after the driver applies the brakes. While the brakes are on, the following velocities are recorded:

Time since brakes applied (sec)	0	2	4	6
Velocity (ft/sec)	88	45	16	0

(a) Give lower and upper estimates for the distance the car traveled after the brakes were applied.
(b) On a sketch of velocity against time, show the lower and upper estimates of part (a).

2. A car starts moving at time $t = 0$ and goes faster and faster. Its velocity is shown in the following table. Estimate how far the car travels during the 12 seconds.

t (seconds)	0	3	6	9	12
Velocity (ft/sec)	0	10	25	45	75

3. The velocity of a car is $f(t) = 5t$ meters/sec. Use a graph of $f(t)$ to find the exact distance traveled by the car, in meters, from $t = 0$ to $t = 10$ seconds.

4. Two cars start at the same time and travel in the same direction along a straight road. Figure 5.8 gives the velocity, v, of each car as a function of time, t. Which car:

(a) Attains the larger maximum velocity?
(b) Stops first?
(c) Travels farther?

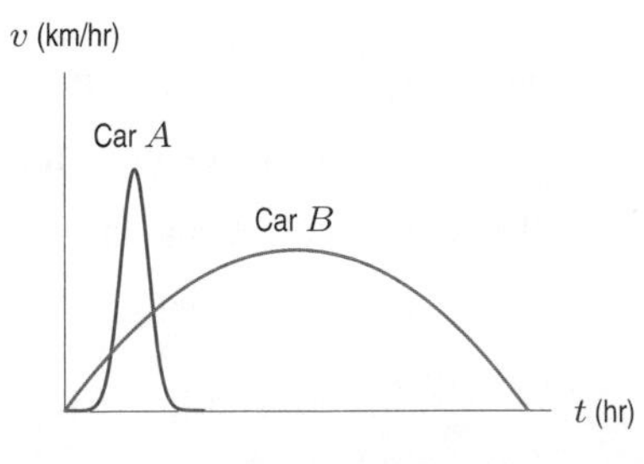

Figure 5.8

5. Two cars travel in the same direction along a straight road. Figure 5.9 shows the velocity, v, of each car at time t. Car B starts 2 hours after car A and car B reaches a maximum velocity of 50 km/hr.

(a) For approximately how long does each car travel?
(b) Estimate car A's maximum velocity.
(c) Approximately how far does each car travel?

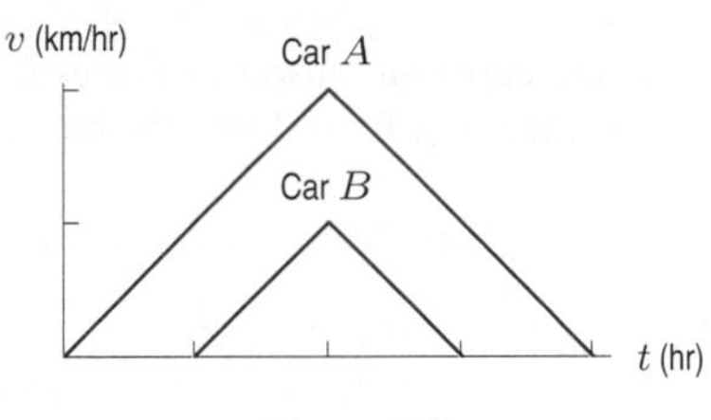

Figure 5.9

6. Figure 5.10 shows the velocity, v, of an object (in meters/sec). Estimate the total distance the object traveled between $t = 0$ and $t = 6$.

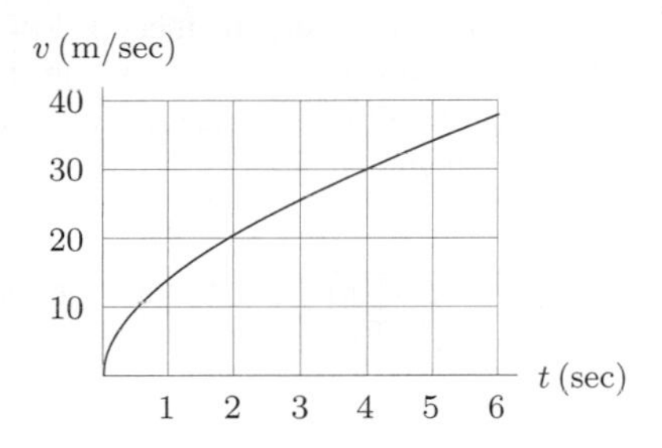

Figure 5.10

7. A car accelerates smoothly from 0 to 60 mph in 10 seconds with the velocity given in Figure 5.11. Estimate how far the car travels during the 10-second period.

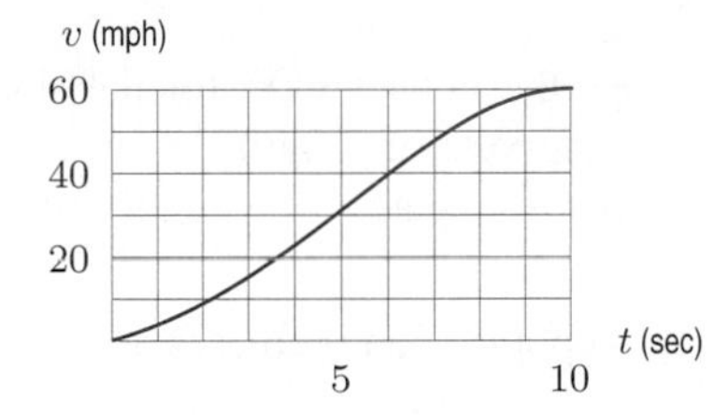

Figure 5.11

8. The following table gives world oil consumption, in billions of barrels per year.[1] Estimate total oil consumption during this 20-year period.

Year	1980	1985	1990	1995	2000
Oil (bn barrels/yr)	22.3	21.3	23.9	24.9	27.0

9. Filters at a water treatment plant become less effective over time. The rate at which pollution passes through the filters into a nearby lake is given in the following table.

(a) Estimate the total quantity of pollution entering the lake during the 30-day period.
(b) Your answer to part (a) is only an estimate. Give bounds (lower and upper estimates) between which the true quantity of pollution must lie. (Assume the rate of pollution is continually increasing.)

Day	0	6	12	18	24	30
Rate (kg/day)	7	8	10	13	18	35

10. The rate of change of the world's population, in millions of people per year, is given in the following table.

(a) Use this data to estimate the total change in the world's population between 1950 and 2000.
(b) The world population was 2555 million people in 1950 and 6085 million people in 2000. Calculate the true value of the total change in the population. How does this compare with your estimate in part (a)?

Year	1950	1960	1970	1980	1990	2000
Rate of change	37	41	78	77	86	79

11. A village wishes to measure the quantity of water that is piped to a factory during a typical morning. A gauge on the water line gives the flowrate (in cubic meters per hour) at any instant. The flowrate is about 100 m^3/hr at 6 am and increases steadily to about 280 m^3/hr at 9 am. Using only this information, give your best estimate of the total volume of water used by the factory between 6 am and 9 am.

12. **(a)** Sketch a graph of the velocity function for the trip described in Example 1 on page 236.
(b) Represent the total distance traveled on this graph.

13. Graph the rate of sales against time for the video game data in Example 4. Represent graphically the overestimate and the underestimate calculated in that example.

14. Roger runs a marathon. His friend Jeff rides behind him on a bicycle and clocks his speed every 15 minutes. Roger starts out strong, but after an hour and a half he is so exhausted that he has to stop. Jeff's data follow:

Time since start (min)	0	15	30	45	60	75	90
Speed (mph)	12	11	10	10	8	7	0

(a) Assuming that Roger's speed is never increasing, give upper and lower estimates for the distance Roger ran during the first half hour.
(b) Give upper and lower estimates for the distance Roger ran in total during the entire hour and a half.

[1] www.bp.com/centres/energy/world_stat_rev/oil/reserves.asp.

15. A car initially going 50 ft/sec brakes at a constant rate (constant negative acceleration), coming to a stop in 5 seconds.

(a) Graph the velocity from $t = 0$ to $t = 5$.
(b) How far does the car travel?
(c) How far does the car travel if its initial velocity is doubled, but it brakes at the same constant rate?

16. Figure 5.12 shows the rate of change of a fish population. Estimate the total change in the population during this 12-month period.

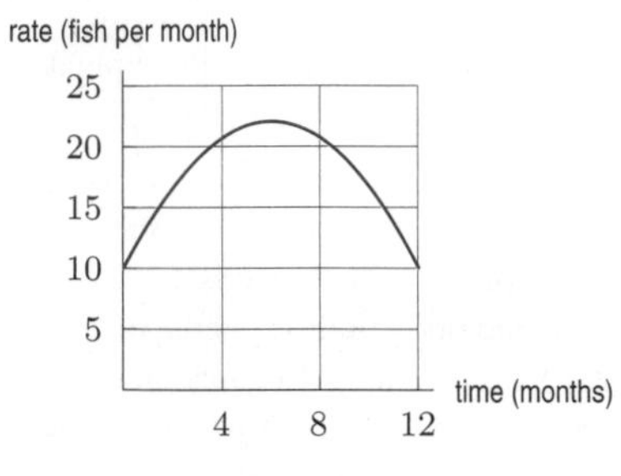

Figure 5.12

17. Your velocity is given by $v(t) = t^2+1$ in m/sec, with t in seconds. Estimate the distance, s, traveled between $t = 0$ and $t = 5$. Explain how you arrived at your estimate.

18. An old rowboat has sprung a leak. Water is flowing into the boat at a rate, $r(t)$, given in the following table.

t minutes	0	5	10	15
$r(t)$ liters/min	12	20	24	16

(a) Compute upper and lower estimates for the volume of water that has flowed into the boat during the 15 minutes.
(b) Draw a graph to illustrate the lower estimate.

19. The value of a mutual fund increases at a rate of $R = 500e^{0.04t}$ dollars per year, where t is years since 2005.

(a) Using $t = 0, 2, 4, 6, 8, 10$, make a table of values for R.
(b) Use the table to estimate the total change in the value of the mutual fund between 2005 and 2015.

20. A car speeds up at a constant rate from 10 to 70 mph over a period of half an hour. Its fuel efficiency (in miles per gallon) at various speeds is shown in the table. Make lower and upper estimates of the quantity of fuel used during the half hour.

Speed (mph)	10	20	30	40	50	60	70
Fuel efficiency (mpg)	15	18	21	23	24	25	26

5.2 THE DEFINITE INTEGRAL

In Section 5.1 we saw how to approximate total change given the rate of change. We now see how to make the approximation more accurate.

Improving the Approximation: The Role of n and Δt

To approximate total change, we construct a sum. We use the notation Δt for the size of the t-intervals used. We use n to represent the number of subintervals of length Δt. In the following example, we see how decreasing Δt (and increasing n) improves the accuracy of the approximation.

Example 1 If t is in hours since the start of a 20-hour period, a bacteria population increases at a rate given by

$$f(t) = 3 + 0.1t^2 \text{ millions of bacteria per hour.}$$

Make an underestimate of the total change in the number of bacteria over this period using

(a) $\Delta t = 4$ hours (b) $\Delta t = 2$ hours (c) $\Delta t = 1$ hour

Solution (a) The rate of change is $f(t) = 3 + 0.1t^2$. If we use $\Delta t = 4$, we measure the rate every 4 hours and $n = 20/4 = 5$. See Table 5.4. An underestimate for the population change during the first 4 hours is $(3.0 \text{ million/hour})(4 \text{ hours}) = 12$ million. Combining the contributions from all the subintervals gives the underestimate:

$$\text{Total change} \approx 3.0 \cdot 4 + 4.6 \cdot 4 + 9.4 \cdot 4 + 17.4 \cdot 4 + 28.6 \cdot 4 = 252.0 \text{ million bacteria.}$$

The rate of change is graphed in Figure 5.13 (a); the area of the shaded rectangles represents this underestimate. Notice that $n = 5$ is the number of rectangles in the graph.

Table 5.4 *Rate of change with $\Delta t = 4$ using $f(t) = 3 + 0.1t^2$ million bacteria/hour*

t (hours)	0	4	8	12	16	20
$f(t)$	3.0	4.6	9.4	17.4	28.6	43.0

(b) If we use $\Delta t = 2$, we measure $f(t)$ every 2 hours and $n = 20/2 = 10$. See Table 5.5. The underestimate is

$$\text{Total change} \approx 3.0 \cdot 2 + 3.4 \cdot 2 + 4.6 \cdot 2 + \cdots + 35.4 \cdot 2 = 288.0 \text{ million bacteria.}$$

Figure 5.13 (b) suggests that this estimate is more accurate than the estimate made in part (a).

Table 5.5 *Rate of change with $\Delta t = 2$ using $f(t) = 3 + 0.1t^2$ million bacteria/hour*

t (hours)	0	2	4	6	8	10	12	14	16	18	20
$f(t)$	3.0	3.4	4.6	6.6	9.4	13.0	17.4	22.6	28.6	35.4	43.0

(c) If we use $\Delta t = 1$, then $n = 20$ and a similar calculation shows that we have

$$\text{Total change} \approx 307.0 \text{ million bacteria.}$$

The shaded area in Figure 5.13 (c) represents this estimate; it is the most accurate of the three.

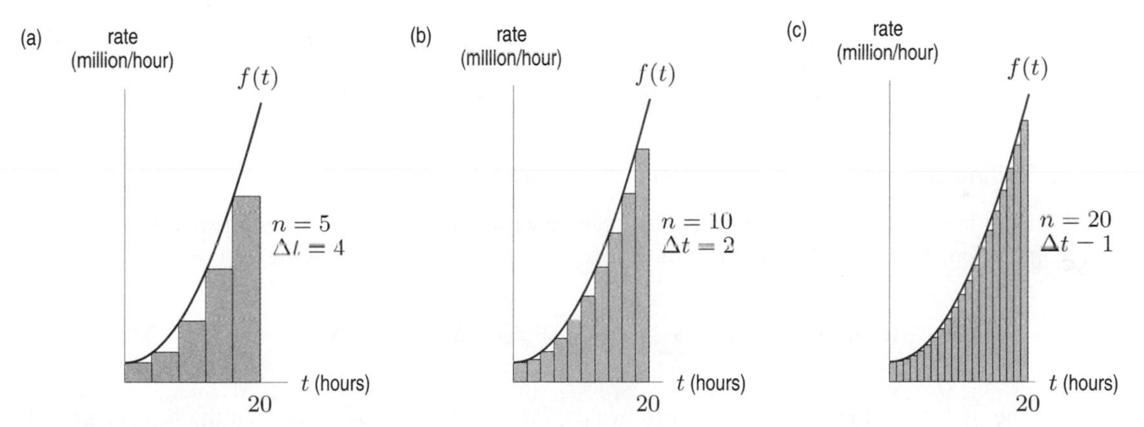

Figure 5.13: More and more accurate estimates of total change from rate of change. In each case, $f(t)$ is the rate of change, and the shaded area approximates total change. Largest n and smallest Δt give the best estimate.

Notice that as n gets larger, the estimate improves and the area of the shaded rectangles approaches the area under the curve.

Left- and Right-Hand Sums

Suppose we have a function $f(t)$ that is continuous for $a \leq t \leq b$. We divide the interval from a to b into n equal subdivisions, each of width Δt, so

$$\Delta t = \frac{b - a}{n}.$$

We let $t_0, t_1, t_2, \ldots, t_n$ be endpoints of the subdivisions, as in Figures 5.14 and 5.15. We construct two sums, similar to the overestimates and underestimates in Section 3.1. For a *left-hand sum*, we use the values of the function from the left end of the interval. For a *right-hand sum*, we use the values of the function from the right end of the interval. We have:

$$\text{Left-hand sum} = f(t_0)\Delta t + f(t_1)\Delta t + \cdots + f(t_{n-1})\Delta t$$

and

$$\text{Right-hand sum} = f(t_1)\Delta t + f(t_2)\Delta t + \cdots + f(t_n)\Delta t.$$

These sums represent the shaded areas in Figures 5.14 and 5.15, provided $f(t) \geq 0$. In Figure 5.14, the first rectangle has width Δt and height $f(t_0)$, since the top of its left edge just touches the curve, and hence it has area $f(t_0)\Delta t$. The second rectangle has width Δt and height $f(t_1)$, and hence has area $f(t_1)\Delta t$, and so on. The sum of all these areas is the left-hand sum. The right-hand sum, shown in Figure 5.15, is constructed in the same way, except that each rectangle touches the curve on its right edge instead of its left.

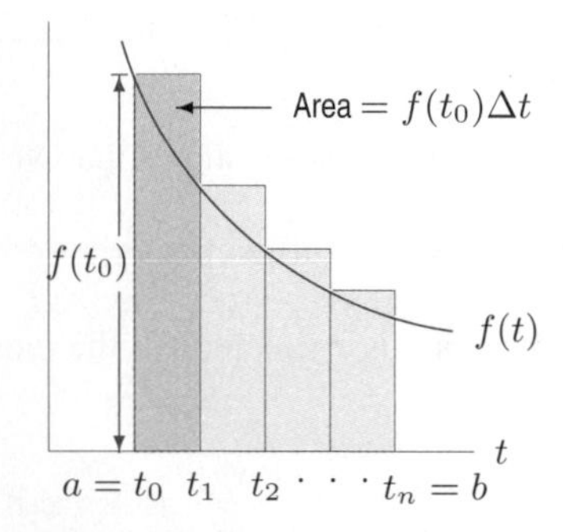

Figure 5.14: Left-hand sum: Area of rectangles

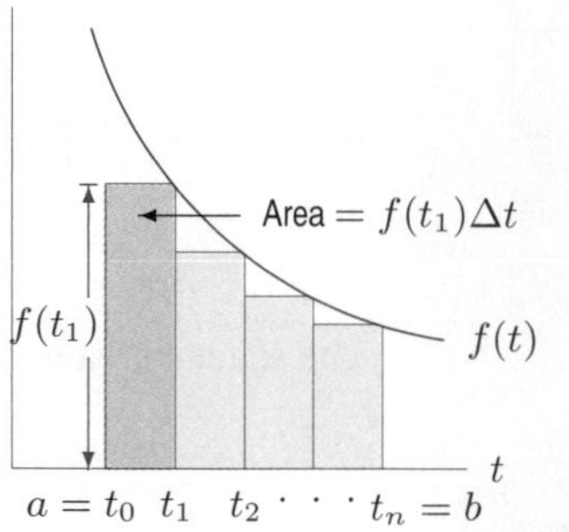

Figure 5.15: Right-hand sum: Area of rectangles

Writing Left- and Right-Hand Sums Using Sigma Notation

Both the left-hand and right-hand sums can be written more compactly using *sigma*, or summation, notation. The symbol $\sum$ is a capital sigma, or Greek letter "S." We write

$$\text{Right-hand sum} = \sum_{i=1}^{n} f(t_i)\Delta t = f(t_1)\Delta t + f(t_2)\Delta t + \cdots + f(t_n)\Delta t.$$

The $\sum$ tells us to add terms of the form $f(t_i)\Delta t$. The "$i = 1$" at the base of the sigma sign tells us to start at $i = 1$, and the "n" at the top tells us to stop at $i = n$.

In the left-hand sum we start at $i = 0$ and stop at $i = n - 1$, so we write

$$\text{Left-hand sum} = \sum_{i=0}^{n-1} f(t_i)\Delta t = f(t_0)\Delta t + f(t_1)\Delta t + \cdots + f(t_{n-1})\Delta t.$$

Taking the Limit to Obtain the Definite Integral

If f is a rate of change of some quantity, then the left-hand sum and the right-hand sum approximate the total change in the quantity. For most functions f, the approximation is improved by increasing the value of n. To find the total change exactly, we take larger and larger values of n and look at the values approached by the left and right sums. This is called taking the *limit* of these sums as n goes to infinity and is written $\lim\limits_{n\to\infty}$. If f is continuous for $a \leq t \leq b$, the limits of the left- and right-hand sums exist and are equal. The *definite integral* is the common limit of these sums.

Suppose f is continuous for $a \leq t \leq b$. The **definite integral** of f from a to b, written

$$\int_a^b f(t)\,dt,$$

is the limit of the left-hand or right-hand sums with n subdivisions of $[a, b]$ as n gets arbitrarily large. In other words, if $t_0, t_1, \ldots t_n$ are the endpoints of the subdivisions,

$$\int_a^b f(t)\,dt = \lim_{n\to\infty} (\text{Left-hand sum}) = \lim_{n\to\infty} \left(\sum_{i=0}^{n-1} f(t_i)\Delta t \right)$$

and

$$\int_a^b f(t)\,dt = \lim_{n\to\infty} (\text{Right-hand sum}) = \lim_{n\to\infty} \left(\sum_{i=1}^{n} f(t_i)\Delta t \right).$$

Each of these sums is called a *Riemann sum*, f is called the *integrand*, and a and b are called the *limits of integration.*

The "$\int$" notation comes from an old-fashioned "S," which stands for "sum" in the same way that $\sum$ does. The "dt" in the integral comes from the factor Δt. Notice that the limits on the $\sum$ symbol are 0 and $n-1$ for the left-hand sum, and 1 and n for the right-hand sum, whereas the limits on the $\int$ sign are a and b.

When $f(t)$ is positive, the left- and right-hand sums are represented by the sums of areas of rectangles, so the definite integral is represented graphically by an area.

Computing a Definite Integral

In practice, we often approximate definite integrals numerically using a calculator or computer. They compute sums for larger and larger values of n, and eventually give a value for the integral. Different calculators and computers may give slightly different estimates, owing to round-off error and the fact that they may use different approximation methods.

Example 2 Compute $\int_1^3 t^2\,dt$ and represent this integral as an area.

Solution Using a calculator, we find

$$\int_1^3 t^2\,dt = 8.667.$$

The integral represents the area between $t = 1$ and $t = 3$ under the curve $f(t) = t^2$. See Figure 5.16.

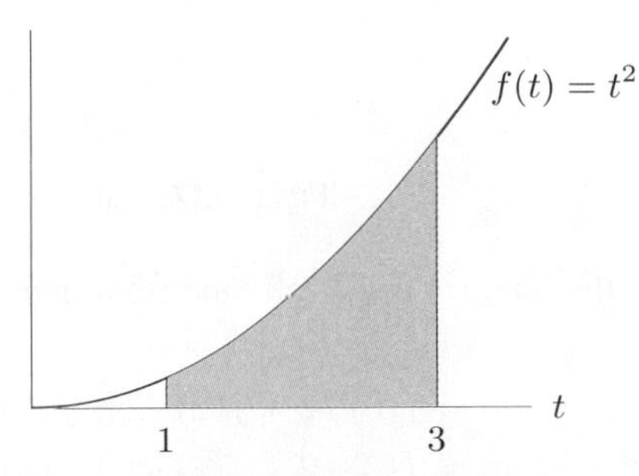

Figure 5.16: Shaded area $= \int_1^3 t^2\,dt$

Estimating a Definite Integral from a Table or Graph

If we have a formula for the integrand, $f(x)$, we can calculate the integral $\int_a^b f(x)\,dx$ using a calculator or computer. If, however, we have only a table of values or a graph of $f(x)$, we can still estimate the integral.

Example 3 Values for a function $f(t)$ are in the following table. Estimate $\displaystyle\int_{20}^{30} f(t)dt$.

t	20	22	24	26	28	30
$f(t)$	5	7	11	18	29	45

Solution Since we have only a table of values, we use left- and right-hand sums to approximate the integral. The values of $f(t)$ are spaced 2 units apart, so $\Delta t = 2$ and $n = (30 - 20)/2 = 5$. Calculating the left-hand and right-hand sums gives

$$\begin{aligned}\text{Left-hand sum} &= f(20)\cdot 2 + f(22)\cdot 2 + f(24)\cdot 2 + f(26)\cdot 2 + f(28)\cdot 2\\ &= 5\cdot 2 + 7\cdot 2 + 11\cdot 2 + 18\cdot 2 + 29\cdot 2\\ &= 10 + 14 + 22 + 36 + 58\\ &= 140.\end{aligned}$$

$$\begin{aligned}\text{Right-hand sum} &= f(22)\cdot 2 + f(24)\cdot 2 + f(26)\cdot 2 + f(28)\cdot 2 + f(30)\cdot 2\\ &= 7\cdot 2 + 11\cdot 2 + 18\cdot 2 + 29\cdot 2 + 45\cdot 2\\ &= 14 + 22 + 36 + 58 + 90\\ &= 220.\end{aligned}$$

Both left- and right-hand sums approximate the integral. We generally get a better estimate by averaging the two:

$$\int_{20}^{30} f(t)dt \approx \frac{140 + 220}{2} = 180.$$

Example 4 The function $f(x)$ is graphed in Figure 5.17. Estimate $\displaystyle\int_0^6 f(x)\,dx$.

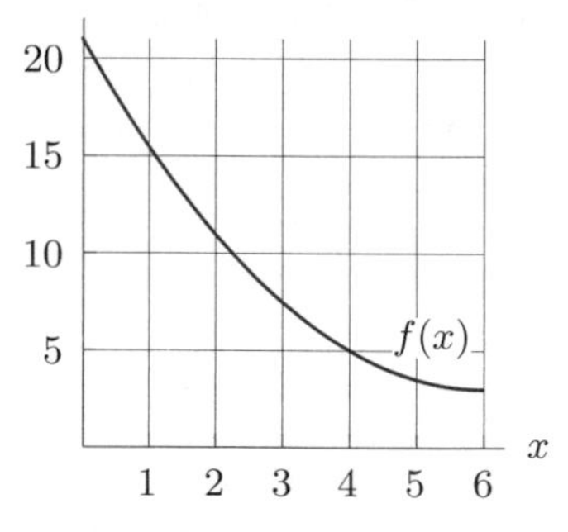

Figure 5.17: Estimate $\int_0^6 f(x)\,dx$

Solution We approximate the integral using left- and right-hand sums with $n = 3$, so $\Delta x = 2$. Figures 5.18 and 5.19 give

$$\begin{aligned}\text{Left-hand sum} &= f(0)\cdot 2 + f(2)\cdot 2 + f(4)\cdot 2 = 21\cdot 2 + 11\cdot 2 + 5\cdot 2 = 74,\\ \text{Right-hand sum} &= f(2)\cdot 2 + f(4)\cdot 2 + f(6)\cdot 2 = 11\cdot 2 + 5\cdot 2 + 3\cdot 2 = 38.\end{aligned}$$

We estimate the integral by taking the average:

$$\int_0^6 f(x)\,dx \approx \frac{74 + 38}{2} = 56.$$

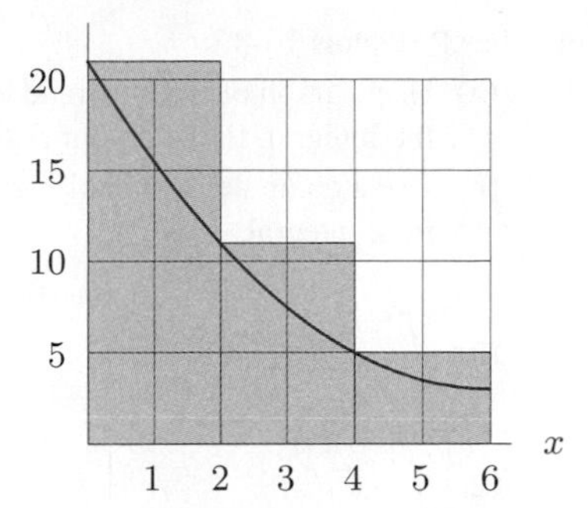

Figure 5.18: Area of shaded region is left-hand sum with $n = 3$

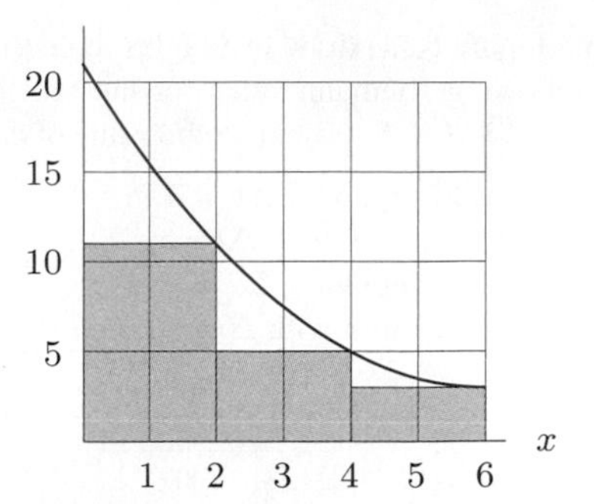

Figure 5.19: Area of shaded region is right-hand sum with $n = 3$

Alternatively, since the integral equals the area under the curve between $x = 0$ and $x = 6$, we can estimate it by counting grid squares. Each grid square has area $5 \cdot 1 = 5$, and the region under $f(x)$ includes about 10.5 grid squares, so the area is about $10.5 \cdot 5 = 52.5$.

Rough Estimates of a Definite Integral

When calculating an integral using a calculator or computer, it is useful to have a rough idea of the value you expect. This helps detect errors in entering the integral.

Example 5 Three people calculated $\displaystyle\int_1^3 \frac{1}{t}\,dt$ on a calculator and got values 0.023, 11.984, and 1.526. Explain how you can be sure that none of these values is correct.

Solution Figure 5.20 shows left- and right-hand approximations to $\int_1^3 \frac{1}{t}\,dt$ with $n = 1$. We see that the right-hand sum $2(1/3)$ is an underestimate of $\int_1^3 \frac{1}{t}\,dt$. Since 0.023 is less than $2/3$, this value must be wrong. Similarly, we see that the left-hand sum $2(1)$ is an overestimate of the integral. Since 11.984 is larger than 2, this value is wrong. Since the graph is concave up, the value of the integral is closer to the smaller value of the two sums, $2/3$, than to the larger value, 2. Thus, the integral is less than the average of the two sums, $4/3$. Since $1.526 > 4/3$, this value is wrong too.

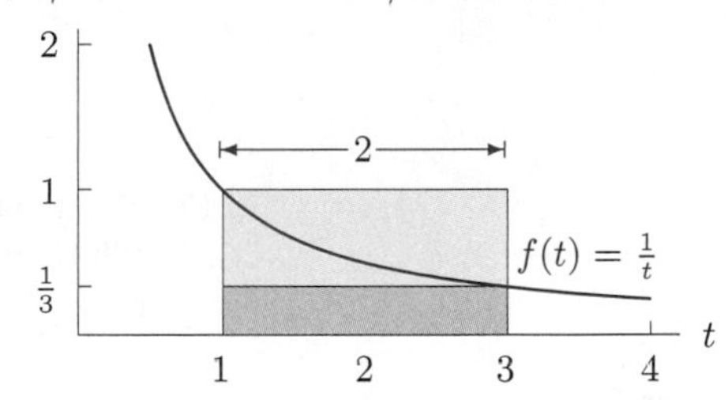

Figure 5.20: Left- and right-hand sums with $n = 1$

Problems for Section 5.2

1. Use the following table to estimate $\int_0^{25} f(x)dx$.

x	0	5	10	15	20	25
$f(x)$	100	82	69	60	53	49

2. Use the following table to estimate $\int_3^4 W(t)\,dt$. What are n and Δt?

t	3.0	3.2	3.4	3.6	3.8	4.0
$W(t)$	25	23	20	15	9	2

3. Use the following table to estimate $\int_0^{15} f(x)\,dx$.

x	0	3	6	9	12	15
$f(x)$	50	48	44	36	24	8

4. Use the table to estimate $\int_0^{40} f(x)dx$. What values of n and Δx did you use?

x	0	10	20	30	40
$f(x)$	350	410	435	450	460

5. Using Figure 5.21, draw rectangles representing each of the following Riemann sums for the function f on the interval $0 \leq t \leq 8$. Calculate the value of each sum.

(a) Left-hand sum with $\Delta t = 4$
(b) Right-hand sum with $\Delta t = 4$
(c) Left-hand sum with $\Delta t = 2$
(d) Right-hand sum with $\Delta t = 2$

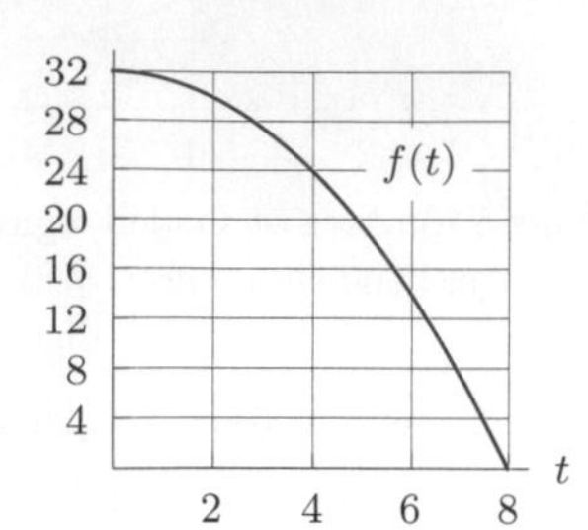

Figure 5.21

6. Use Figure 5.22 to estimate $\int_0^{20} f(x)\,dx$.

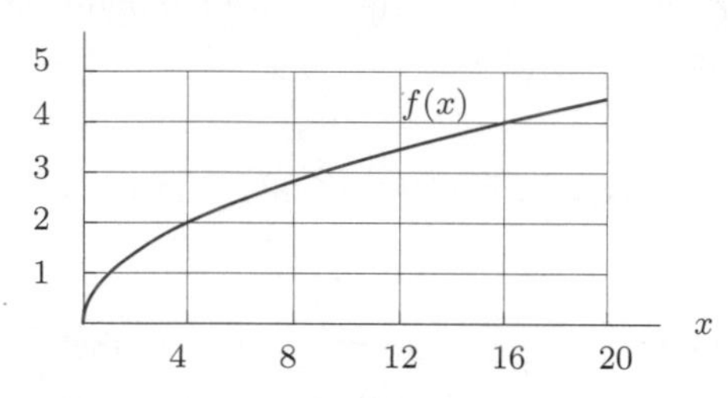

Figure 5.22

7. Use Figure 5.23 to estimate $\int_{-10}^{15} f(x)dx$.

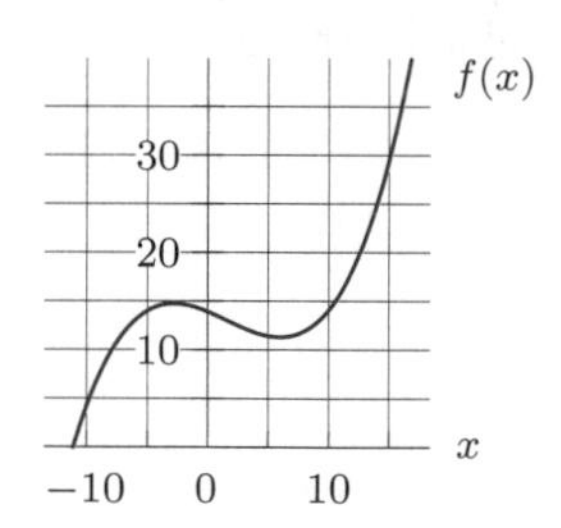

Figure 5.23

Use the graphs in Problems 8–9 to estimate $\int_0^3 f(x)\,dx$

8.

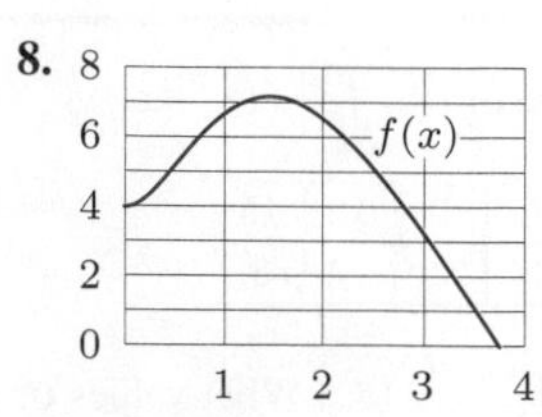

9.

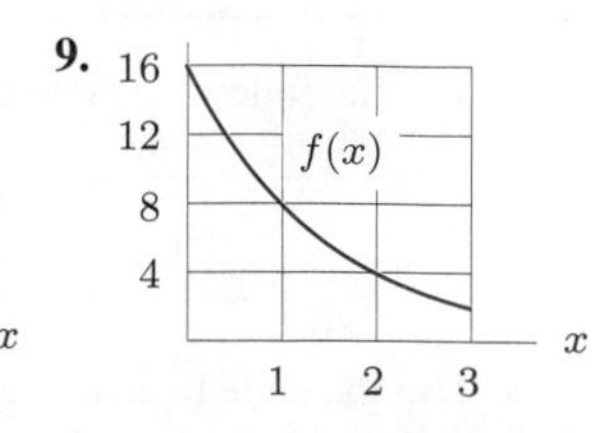

For Problems 10–13:

(a) Use a graph of the integrand to make a rough estimate of the integral. Explain your reasoning.
(b) Use a computer or calculator to find the value of the definite integral.

10. $\displaystyle\int_0^1 x^3\,dx$

11. $\displaystyle\int_0^3 \sqrt{x}\,dx$

12. $\displaystyle\int_0^1 3^t\,dt$

13. $\displaystyle\int_1^2 x^x\,dx$

14. The rate of change of a quantity is given by $f(t) = t^2+1$. Make an underestimate and an overestimate of the total change in the quantity between $t = 0$ and $t = 8$ using

(a) $\Delta t = 4$ **(b)** $\Delta t = 2$ **(c)** $\Delta t = 1$

What is n in each case? Graph $f(t)$ and shade rectangles to represent each of your six answers.

15. **(a)** Use a calculator or computer to find $\int_0^6 (x^2+1)\,dx$. Represent this value as the area under a curve.

(b) Estimate $\int_0^6 (x^2+1)\,dx$ using a left-hand sum with $n = 3$. Represent this sum graphically on a sketch of $f(x) = x^2 + 1$. Is this sum an overestimate or underestimate of the true value found in part (a)?

(c) Estimate $\int_0^6 (x^2+1)\,dx$ using a right-hand sum with $n = 3$. Represent this sum on your sketch. Is this sum an overestimate or underestimate?

16. Using Figure 5.24, find the value of $\int_1^6 f(x)\,dx$.

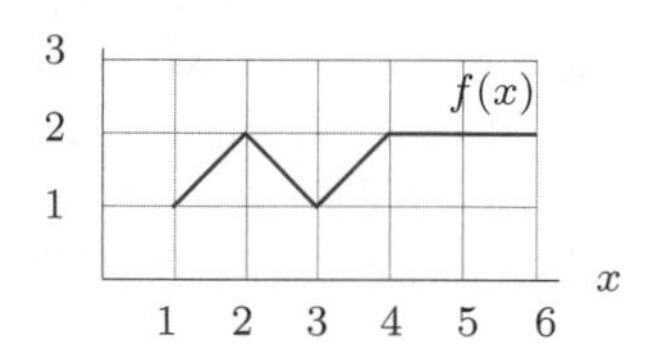

Figure 5.24

17. The graph of a function $f(t)$ is given in Figure 5.25. Which of the following four numbers could be an estimate of $\int_0^1 f(t)dt$ accurate to two decimal places? Explain how you chose your answer.

(a) -98.35 **(b)** 71.84
(c) 100.12 **(d)** 93.47

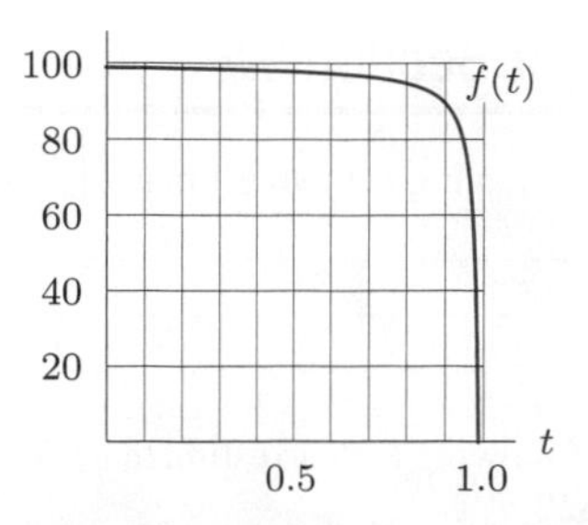

Figure 5.25

In Problems 18–27, use a calculator or computer to evaluate the integral.

18. $\int_0^5 x^2\,dx$

19. $\int_1^5 (3x+1)^2\,dx$

20. $\int_1^4 \frac{1}{\sqrt{1+x^2}}\,dx$

21. $\int_{-1}^1 \frac{1}{e^t}\,dt$

22. $\int_{1.1}^{1.7} 10(0.85)^t\,dt$

23. $\int_1^2 2^x\,dx$

24. $\int_1^2 (1.03)^t\,dt$

25. $\int_1^3 \ln x\,dx$

26. $\int_{1.1}^{1.7} e^t \ln t\,dt$

27. $\int_{-3}^3 e^{-t^2}\,dt$

28. Use the expressions for left and right sums on page 244 and Table 5.6.

(a) If $n = 4$, what is Δt? What are t_0, t_1, t_2, t_3, t_4? What are $f(t_0), f(t_1), f(t_2), f(t_3), f(t_4)$?
(b) Find the left and right sums using $n = 4$.
(c) If $n = 2$, what is Δt? What are t_0, t_1, t_2? What are $f(t_0), f(t_1), f(t_2)$?
(d) Find the left and right sums using $n = 2$.

Table 5.6

t	15	17	19	21	23
$f(t)$	10	13	18	20	30

29. Use the expressions for left and right sums on page 244 and Table 5.7.

(a) If $n = 4$, what is Δt? What are t_0, t_1, t_2, t_3, t_4? What are $f(t_0), f(t_1), f(t_2), f(t_3), f(t_4)$?
(b) Find the left and right sums using $n = 4$.
(c) If $n = 2$, what is Δt? What are t_0, t_1, t_2? What are $f(t_0), f(t_1), f(t_2)$?
(d) Find the left and right sums using $n = 2$.

Table 5.7

t	0	4	8	12	16
$f(t)$	25	23	22	20	17

5.3 THE DEFINITE INTEGRAL AS AREA

The Definite Integral as an Area: When $f(x)$ is Positive

If $f(x)$ is continuous and positive, each term $f(x_0)\Delta x, f(x_1)\Delta x, \ldots$ in a left- or right-hand Riemann sum represents the area of a rectangle. See Figure 5.26. As the width Δx of the rectangles approaches zero, the rectangles fit the curve of the graph more exactly, and the sum of their areas gets closer to the area under the curve shaded in Figure 5.27. In other words:

When $f(x)$ is positive and $a < b$:

$$\text{Area under graph of } f \text{ between } a \text{ and } b = \int_a^b f(x)\,dx.$$

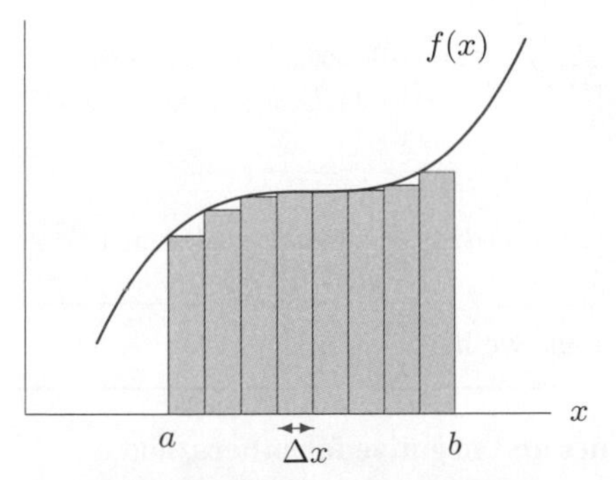

Figure 5.26: Area of rectangles approximating the area under the curve

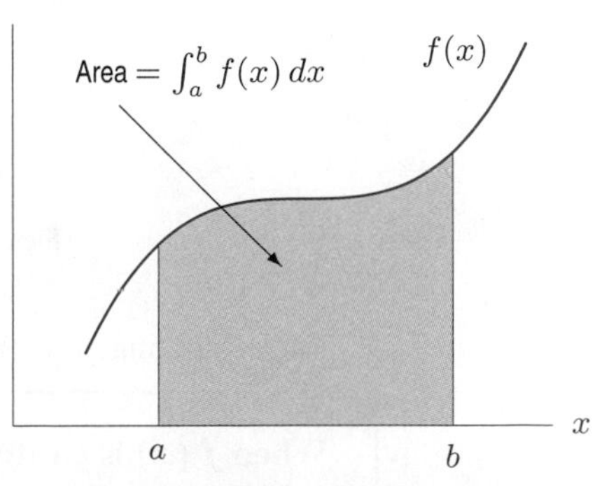

Figure 5.27: Shaded area is the definite integral $\int_a^b f(x)\,dx$

Example 1 Find the area under the graph of $y = 10x(3^{-x})$ between $x = 0$ and $x = 3$.

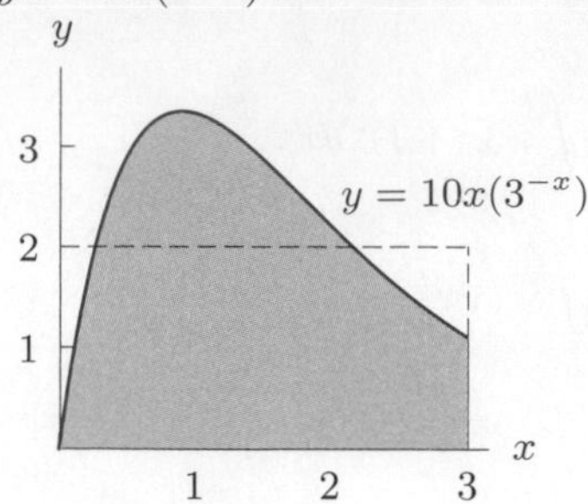

Figure 5.28: Area shaded $= \int_0^3 10x(3^{-x})\,dx$

Solution The area we want is shaded in Figure 5.28. A rough estimate of this area is 6, since it has about the same area as a rectangle of width 3 and height 2. To find the area more accurately, we say

$$\text{Area shaded} = \int_0^3 10x(3^{-x})dx.$$

Using a calculator or computer to evaluate the integral, we obtain

$$\text{Area shaded} = \int_0^3 10x(3^{-x})\,dx = 6.967 \approx 7 \text{ square units.}$$

Relationship Between Definite Integral and Area: When $f(x)$ is Not Positive

We assumed in drawing Figure 5.27 that the graph of $f(x)$ lies above the x-axis. If the graph lies below the x-axis, then each value of $f(x)$ is negative, so each $f(x)\Delta x$ is negative, and the area gets counted negatively. In that case, the definite integral is the negative of the area between the graph of f and the horizontal axis.

Example 2 What is the relation between the definite integral $\int_{-1}^{1}(x^2-1)\,dx$ and the area between the parabola $y = x^2 - 1$ and the x-axis?

Solution The parabola lies below the x-axis between $x = -1$ and $x = 1$. (See Figure 5.29.) So,

$$\int_{-1}^{1}(x^2-1)\,dx = -\text{Area} = -1.33.$$

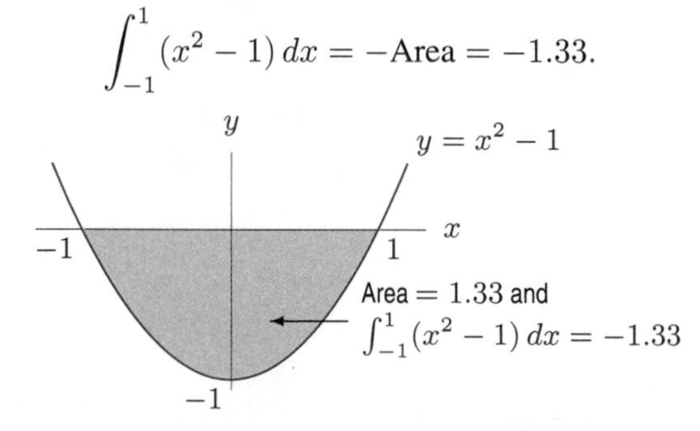

Figure 5.29: Integral $\int_{-1}^{1}(x^2-1)\,dx$ is negative of shaded area

Summarizing, assuming $f(x)$ is continuous, we have:

> **When $f(x)$ is positive for some x-values and negative for others**, and $a < b$:
> $\int_a^b f(x)\,dx$ is the sum of the areas above the x-axis, counted positively, and the areas below the x-axis, counted negatively.

In the following example, we break up the integral. The properties that allow us to do this are listed on page 272.

Example 3 Interpret the definite integral $\int_0^4 (x^3 - 7x^2 + 11x)\,dx$ in terms of areas.

Solution Figure 5.30 shows the graph of $f(x) = x^3 - 7x^2 + 11x$ crossing below the x-axis at about $x = 2.38$. The integral is the area above the x-axis, A_1, minus the area below the x-axis, A_2. Computing the integral with a calculator or computer shows

$$\int_0^4 (x^3 - 7x^2 + 11x)\,dx = 2.67.$$

Breaking the integral into two parts and calculating each one separately gives

$$\int_0^{2.38} (x^3 - 7x^2 + 11x)\,dx = 7.72 \quad \text{and} \quad \int_{2.38}^4 (x^3 - 7x^2 + 11x)\,dx = -5.05,$$

so $A_1 = 7.72$ and $A_2 = 5.05$. Then, as we would expect,

$$\int_0^4 (x^3 - 7x^2 + 11x)\,dx = A_1 - A_2 = 7.72 - 5.05 = 2.67.$$

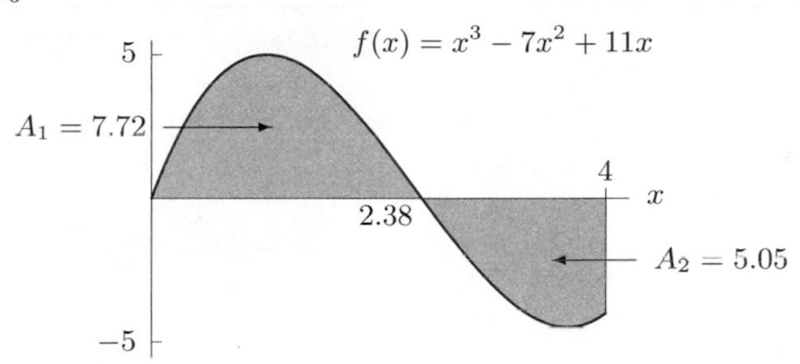

Figure 5.30: Integral $\int_0^4 (x^3 - 7x^2 + 11x)\,dx = A_1 - A_2$

Example 4 Find the total area of the shaded regions in Figure 5.30.

Solution We saw in Example 3 that $A_1 = 7.72$ and $A_2 = 5.05$. Thus we have

$$\text{Total shaded area} = A_1 + A_2 = 7.72 + 5.05 = 12.77.$$

Example 5 For each of the functions graphed in Figure 5.31, decide whether $\int_0^5 f(x)\,dx$ is positive, negative or approximately zero.

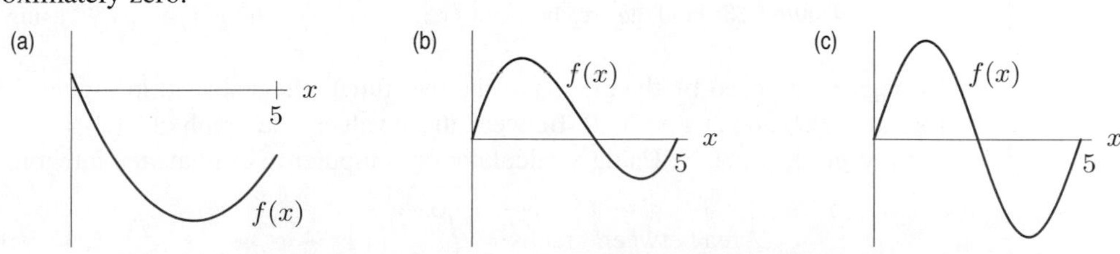

Figure 5.31: Is $\int_0^5 f(x)dx$ positive, negative or zero?

Solution

(a) The graph lies almost entirely below the x-axis, so the integral is negative.
(b) The graph lies partly below the x-axis and partly above the x-axis. However, the area above the x-axis is larger than the area below the x-axis, so the integral is positive.
(c) The graph lies partly below the x-axis and partly above the x-axis. Since the areas above and below the x-axis appear to be approximately equal in size, the integral is approximately zero.

Area Between Two Curves

We can use rectangles to approximate the area between two curves. If $g(x) \le f(x)$, as in Figure 5.32, the height of a rectangle is $f(x) - g(x)$. The area of the rectangle is $(f(x) - g(x))\Delta x$, and we have the following result:

> If $g(x) \le f(x)$ for $a \le x \le b$:
>
> $$\text{Area between graphs of } f(x) \text{ and } g(x) \text{ for } a \le x \le b = \int_a^b (f(x) - g(x))\,dx.$$

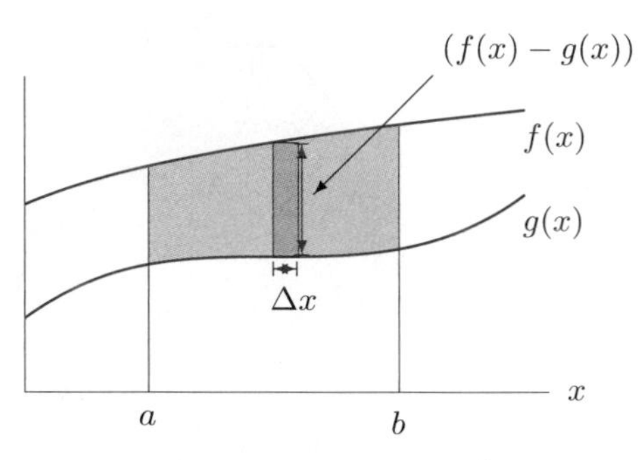

Figure 5.32: Area between two curves $= \int_a^b (f(x) - g(x))\,dx$

Example 6 Graphs of $f(x) = 4x - x^2$ and $g(x) = \frac{1}{2}x^{3/2}$ for $x \ge 0$ are shown in Figure 5.33. Use a definite integral to estimate the area enclosed by the graphs of these two functions.

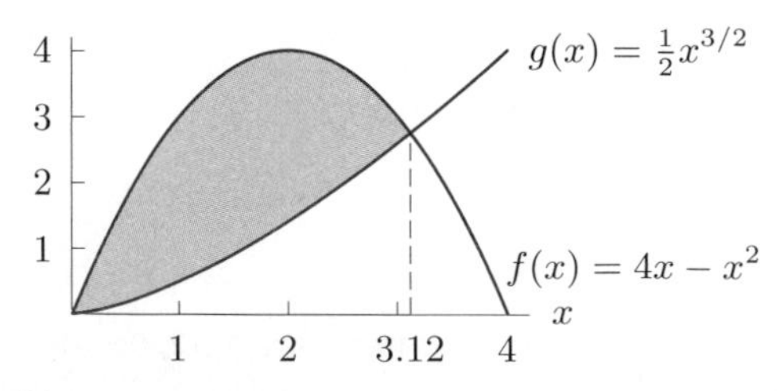

Figure 5.33: Find the area between $f(x) = 4x - x^2$ and $g(x) = \frac{1}{2}x^{3/2}$ using an integral

Solution

The region enclosed by the graphs of the two functions is shaded in Figure 5.33. The two graphs cross at $x = 0$ and at $x \approx 3.12$. Between these values, the graph of $f(x) = 4x - x^2$ lies above the graph of $g(x) = \frac{1}{2}x^{3/2}$. Using a calculator or computer to evaluate the integral, we get

$$\text{Area between graphs} = \int_0^{3.12} \left((4x - x^2) - \tfrac{1}{2}x^{3/2} \right) dx = 5.906.$$

Problems for Section 5.3

1. Find the area under $P = 100(0.6)^t$ between $t = 0$ and $t = 8$.

2. Find the area under $y = x^3 + 2$ between $x = 0$ and $x = 2$. Sketch this area.

3. **(a)** What is the area between the graph of $f(x)$ in Figure 5.34 and the x-axis, between $x = 0$ and $x = 5$?
(b) What is $\int_0^5 f(x)\,dx$?

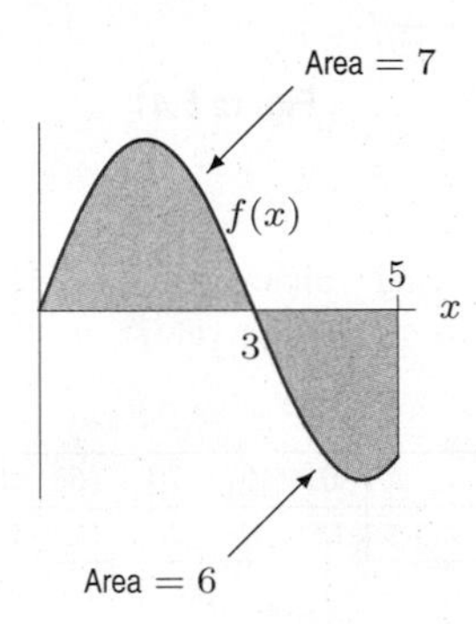

Figure 5.34

For the functions in Problems 4–7, decide whether $\int_{-3}^{3} f(x)dx$ is positive, negative, or approximately zero.

4.

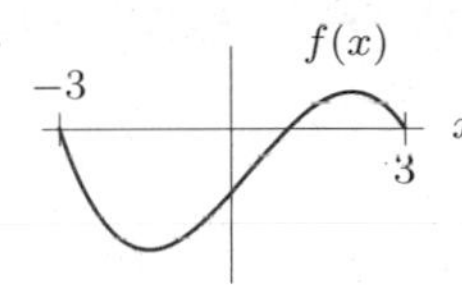

5.

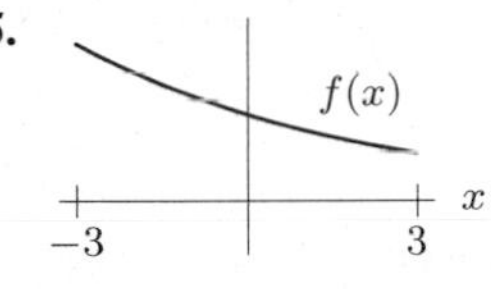

6.

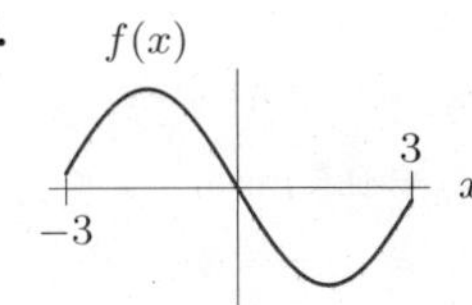

7.

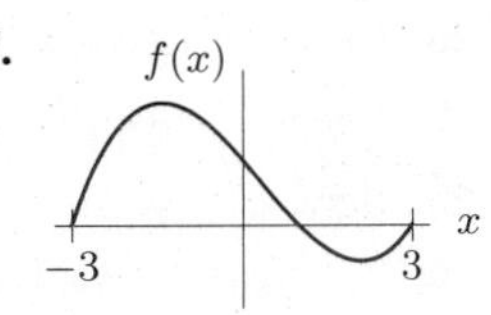

8. **(a)** Estimate (by counting the squares) the total area shaded in Figure 5.35.
(b) Using Figure 5.35, estimate $\int_0^8 f(x)\,dx$.
(c) Why are your answers to parts (a) and (b) different?

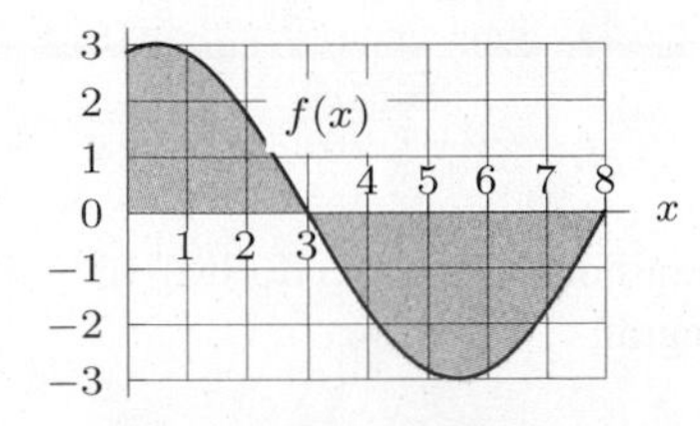

Figure 5.35

9. Using Figure 5.36, estimate $\int_{-3}^{5} f(x)dx$.

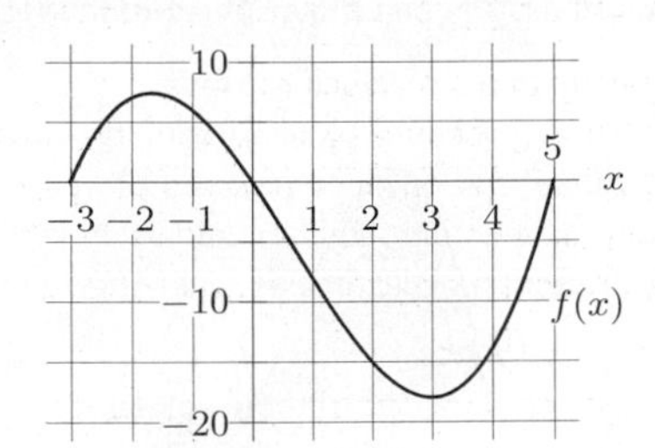

Figure 5.36

10. Given $\int_{-1}^{0} f(x)\,dx = 0.25$ and Figure 5.37, estimate:

(a) $\int_0^1 f(x)\,dx$ **(b)** $\int_{-1}^{1} f(x)\,dx$
(c) The total shaded area.

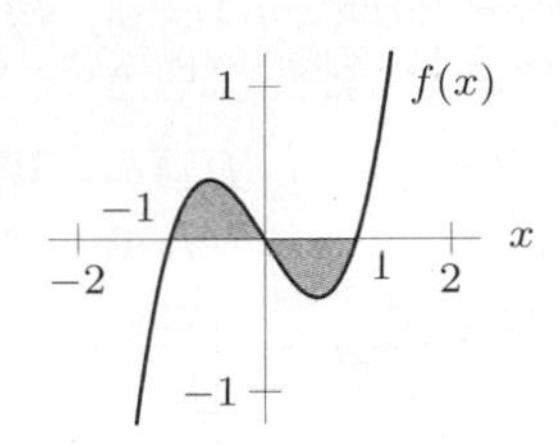

Figure 5.37

11. Given $\int_{-2}^{0} f(x)dx = 4$ and Figure 5.38, estimate:

(a) $\int_0^2 f(x)dx$ **(b)** $\int_{-2}^{2} f(x)dx$
(c) The total shaded area.

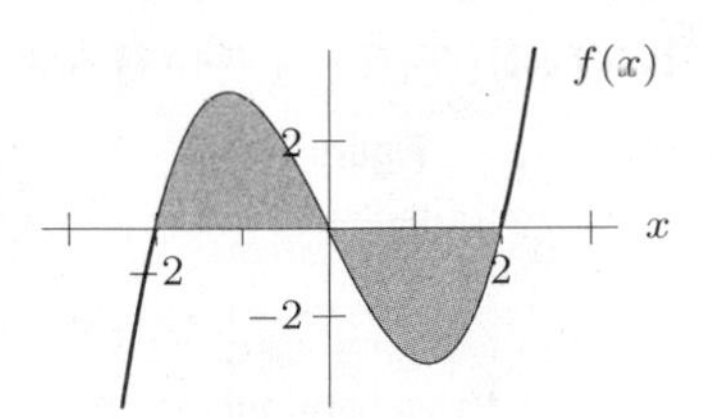

Figure 5.38

In Problems 12–15, match the graph with one of the following possible values for the integral $\int_0^5 f(x)\,dx$:

I. − 10.4 II. − 2.1 III. 5.2 IV. 10.4

12.

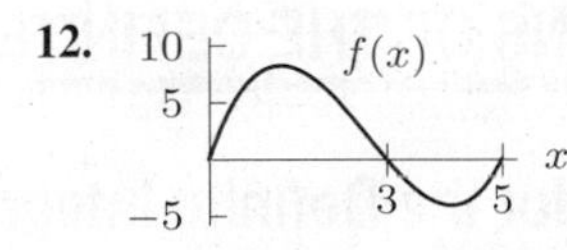

13.

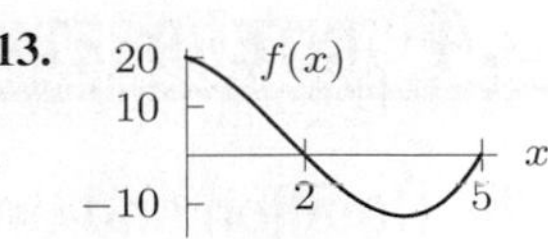

14.

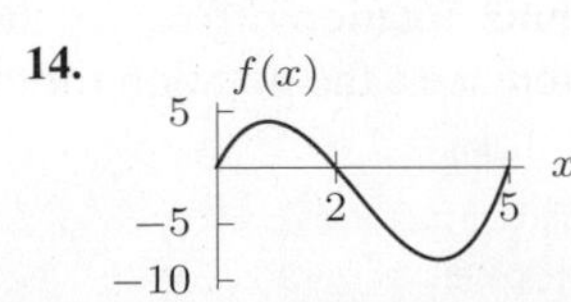

15.

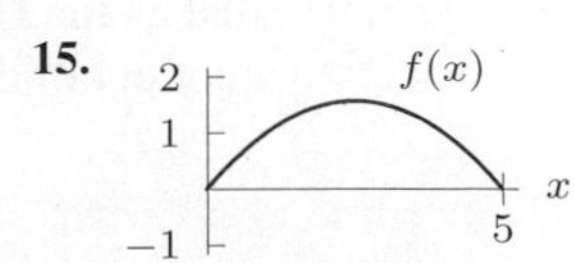

16. Use Figure 5.39 to find the values of

(a) $\int_a^b f(x)\,dx$ **(b)** $\int_b^c f(x)\,dx$
(c) $\int_a^c f(x)\,dx$ **(d)** $\int_a^c |f(x)|\,dx$

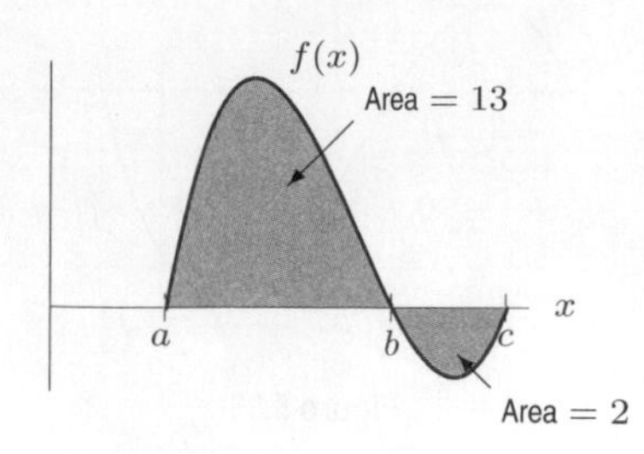

Figure 5.39

17. Using Figure 5.40, list the following integrals in increasing order (from smallest to largest). Which integrals are negative, which are positive? Give reasons.

I. $\int_a^b f(x)\,dx$ II. $\int_a^c f(x)\,dx$ III. $\int_a^e f(x)\,dx$
IV. $\int_b^e f(x)\,dx$ V. $\int_b^c f(x)\,dx$

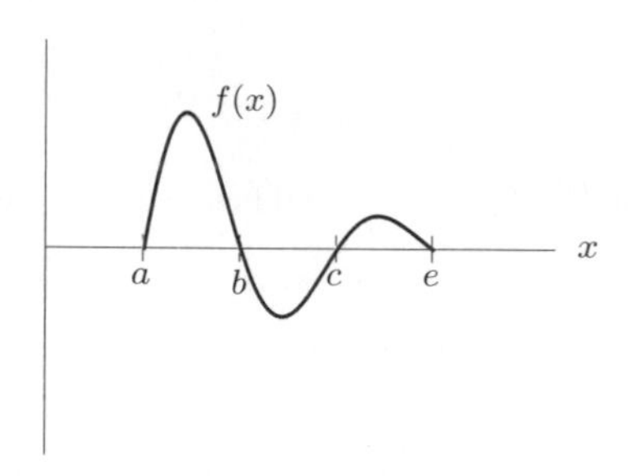

Figure 5.40

18. **(a)** Graph $f(x) = x(x+2)(x-1)$.
(b) Find the total area between the graph and the x-axis between $x = -2$ and $x = 1$.
(c) Find $\int_{-2}^{1} f(x)\,dx$ and interpret it in terms of areas.

19. Find the area between the graph of $y = x^2 - 2$ and the x-axis, between $x = 0$ and $x = 3$.

20. Compute the definite integral $\int_0^4 \cos\sqrt{x}\,dx$ and interpret the result in terms of areas.

21. **(a)** Using Figure 5.41, find $\int_{-3}^{0} f(x)\,dx$.
(b) If the area of the shaded region is A, estimate $\int_{-3}^{4} f(x)\,dx$.

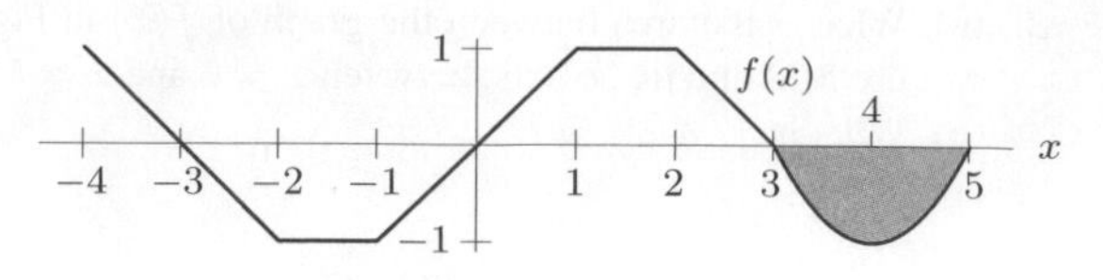

Figure 5.41

22. Use the following table to estimate the area between $f(x)$ and the x-axis on the interval $0 \le x \le 20$.

x	0	5	10	15	20
$f(x)$	15	18	20	16	12

For Problems 23–24, compute the definite integral and interpret the result in terms of areas.

23. $\displaystyle\int_1^4 \frac{x^2-3}{x}\,dx.$ **24.** $\displaystyle\int_1^4 (x - 3\ln x)\,dx.$

In Problems 25–32, use an integral to find the specified area.

25. Under $y = 6x^3 - 2$ for $5 \le x \le 10$.

26. Under $y = 2\cos(t/10)$ for $1 \le t \le 2$.

27. Under $y = 5\ln(2x)$ and above $y = 3$ for $3 \le x \le 5$.

28. Between $y = \sin x + 2$ and $y = 0.5$ for $6 \le x \le 10$.

29. Between $y = \cos x + 7$ and $y = \ln(x-3)$, $5 \le x \le 7$.

30. Above the curve $y = x^4 - 8$ and below the x-axis.

31. Above the curve $y = -e^x + e^{2(x-1)}$ and below the x-axis, for $x \ge 0$.

32. Under $y = \cos t$ and above $y = \sin t$ for $0 \le t \le \pi$.

5.4 INTERPRETATIONS OF THE DEFINITE INTEGRAL

The Notation and Units for the Definite Integral

Just as the Leibniz notation dy/dx for the derivative reminds us that the derivative is the limit of a quotient of differences, the notation for the definite integral,

$$\int_a^b f(x)\,dx,$$

reminds us that an integral is a limit of a sum. The integral sign is a misshapen S. Since the terms being added are products of the form "$f(x)$ times a difference in x," we have the following result:

> The unit of measurement for $\int_a^b f(x)\,dx$ is the product of the units for $f(x)$ and the units for x.

For example, if x and $f(x)$ have the same units, then the integral $\int_a^b f(x)\,dx$ is measured in square units, say cm $\times$ cm $=$ cm^2. This is what we would expect, since the integral represents an area.

Similarly, if $f(t)$ is velocity in meters/second and t is time in seconds, then the integral

$$\int_a^b f(t)\,dt$$

has units of (meters/sec) $\times$ (sec) $=$ meters, which is what we expect since the integral represents change in position. We saw in Section 5.1 that total change could be approximated by a Riemann sum formed using the rate of change. In the limit, we have:

> If $f(t)$ is a rate of change of a quantity, then
>
> $$\text{Total change in quantity between } t = a \text{ and } t = b = \int_a^b f(t)\,dt$$

The units correspond: If $f(t)$ is a rate of change, with units of quantity/time, then $f(t)\Delta t$ and the definite integral have units of (quantity/time) $\times$ (time) $=$ quantity.

Example 1 A bacteria colony initially has a population of 14 million bacteria. Suppose that t hours later the population is growing at a rate of $f(t) = 2^t$ million bacteria per hour.

(a) Give a definite integral that represents the total change in the bacteria population during the time from $t - 0$ to $t - 2$.

(b) Find the population at time $t = 2$.

Solution (a) Since $f(t) = 2^t$ gives the rate of change of population, we have

$$\text{Change in population between } t = 0 \text{ and } t = 2 = \int_0^2 2^t\,dt.$$

(b) Using a calculator, we find $\int_0^2 2^t\,dt = 4.328$. The bacteria population was 14 million at time $t = 0$ and increased 4.328 million between $t = 0$ and $t = 2$. Therefore, at time $t = 2$,

$$\text{Population} = 14 + 4.328 = 18.328 \text{ million bacteria.}$$

Example 2 Suppose that $C(t)$ represents the cost per day to heat your home in dollars per day, where t is time measured in days and $t = 0$ corresponds to January 1, 2005. Interpret $\displaystyle\int_0^{90} C(t)\,dt$.

Solution The units for the integral $\int_0^{90} C(t)\,dt$ are (dollars/day) $\times$ (days) $=$ dollars. The integral represents the cost in dollars to heat your house for the first 90 days of 2005, namely the months of January, February, and March.

Example 3 A man starts 50 miles away from his home and takes a trip in his car. He moves on a straight line, and his home lies on this line. His velocity is given in Figure 5.42.

(a) When is the man closest to his home? Approximately how far away is he then?
(b) When is the man farthest from his home? How far away is he then?

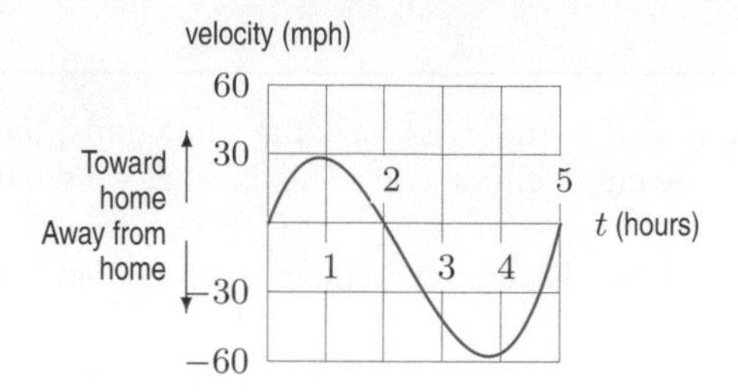

Figure 5.42: Velocity of trip starting 50 miles from home

Solution What happens on this trip? The velocity function is positive the first two hours and negative between $t = 2$ and $t = 5$. So the man moves toward his home during the first two hours, then turns around at $t = 2$ and moves away from his home. The distance he travels is represented by the area between the graph of velocity and the t-axis; since the area below the axis is greater than the area above the axis, we see that he ends up farther away from home than when he started. Thus he is closest to home at $t = 2$ and farthest from home at $t = 5$. We can estimate how far he went in each direction by estimating areas.

(a) The man starts out 50 miles from home. The distance the man travels during the first two hours is the area under the curve between $t = 0$ and $t = 2$. This area corresponds to about one grid square. Since each grid square has area (30 miles/hour)(1 hour) = 30 miles, the man travels about 30 miles toward home. He is closest to home after 2 hours, and he is about 20 miles away at that time.
(b) Between $t = 2$ and $t = 5$, the man moves away from his home. Since this area is equal to about 3.5 grid squares, which is $(3.5)(30)$ =105 miles, he has moved 105 miles farther from home. He was already 20 miles from home, so at $t = 5$ he is about 125 miles from home. He is farthest from home at $t = 5$.

Notice that the man has covered a total distance of $30 + 105 = 135$ miles. However, he went toward his home for 30 miles and away from his home for 105 miles. His *net* change in position is 75 miles.

Example 4 The rates of growth of the populations of two species of plants (measured in new plants per year) are shown in Figure 5.43. Assume that the populations of the two species are equal at time $t = 0$.

(a) Which population is larger after one year? After two years?
(b) How much does the population of species 1 increase during the first two years?

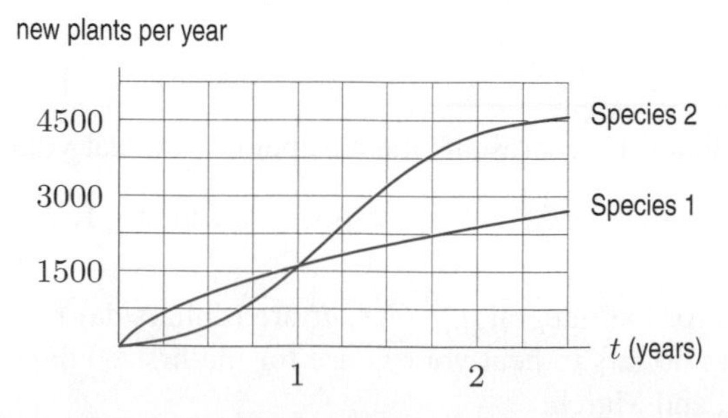

Figure 5.43: Population growth rates for two species of plants

Solution (a) The rate of growth of the population of species 1 is higher than that of species 2 throughout the first year, so the population of species 1 is larger after one year. After two years, the situation is less clear, since the population of species 1 increased faster for the first year and that of species 2 for the second. However, if $r(t)$ is the rate of growth of a population, we have

$$\text{Total change in population during first two years} = \int_0^2 r(t)dt.$$

This integral is the area under the graph of $r(t)$. For $t = 0$ to $t = 2$, the area under the species 1 graph in Figure 5.43 is smaller than the area under the species 2 graph, so the population of species 2 is larger after two years.

(b) The population change for species 1 is the area of the region under the graph of $r(t)$ between $t = 0$ and $t = 2$ in Figure 5.43. The region consists of about 16.5 grid squares, each of area (750 plants/year)(0.25 year) = 187.5 plants, giving a total of (16.5)(187.5) = 3093.75 plants. The population of species 1 increases by about 3100 plants during the two years.

Bioavailability of Drugs

In pharmacology, the definite integral is used to measure *bioavailability*; that is, the overall presence of a drug in the bloodstream during the course of a treatment. Unit bioavailability represents 1 unit concentration of the drug in the bloodstream for 1 hour. For example, a concentration of 3 μg/cm^3 in the blood for 2 hours has bioavailability of $3 \cdot 2 = 6$ (μg/cm^3)-hours.

Ordinarily the concentration of a drug in the blood is not constant. Typically, the concentration in the blood increases as the drug is absorbed into the bloodstream, and then decreases as the drug is broken down and excreted.[2] (See Figure 5.44.)

Suppose that we want to calculate the bioavailability of a drug that is in the blood with concentration $C(t)\mu$g/cm^3 at time t for the time period $0 \le t \le T$. Over a small interval Δt, we estimate

$$\text{Bioavailability} \approx \text{Concentration} \times \text{Time} = C(t)\Delta t.$$

Summing over all subintervals gives

$$\text{Total bioavailability} \approx \sum C(t)\Delta t.$$

In the limit as $n \to \infty$, where n is the number of intervals of width Δt, the sum becomes an integral. So for $0 \le t \le T$, we have

$$\text{Bioavailability} = \int_0^T C(t)dt.$$

That is, the total bioavailability of a drug is equal to the area under the drug concentration curve.

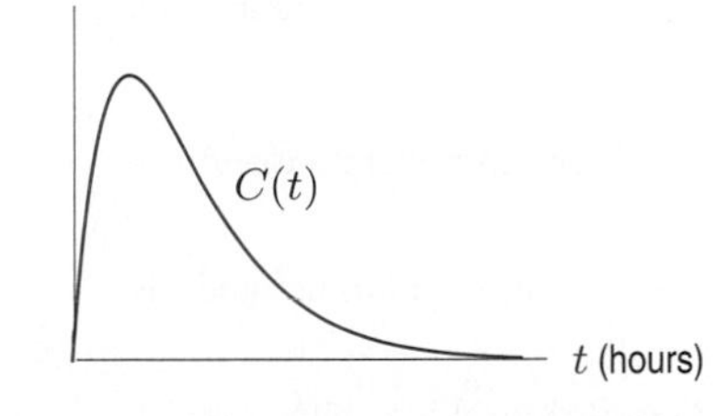

Figure 5.44: Curve showing drug concentration as a function of time

Example 5 Blood concentration curves[3] of two drugs are shown in Figure 5.45. Describe the differences and similarities between the two drugs in terms of peak concentration, speed of absorption into the bloodstream, and total bioavailability.

[2] *Drug Treatment*, Graeme S. Avery (Ed.), (Sydney: Adis Press, 1976).
[3] *Drug Treatment*, Graeme S. Avery (Ed.), (Sydney: Adis Press, 1976).

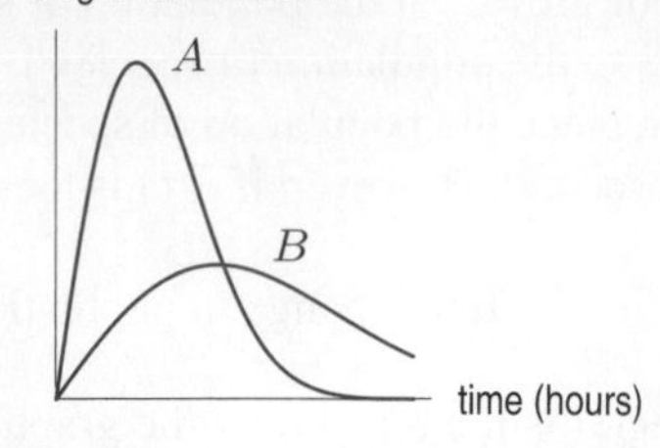

Figure 5.45: Concentration curves of two drugs

Solution Drug A has a peak concentration more than twice as high as that of drug B. Because drug A achieves peak concentration sooner than drug B, drug A appears to be absorbed more rapidly into the blood stream than drug B. Finally, drug A has greater total bioavailability, since the area under the graph of the concentration function for drug A is greater than the area under the graph for drug B.

Problems for Section 5.4

1. The following table gives the emissions, E, of nitrogen oxides in millions of metric tons per year in the US.[4] Let t be the number of years since 1970 and $E = f(t)$.

(a) What are the units and meaning of $\int_0^{30} f(t)dt$?

(b) Estimate $\int_0^{30} f(t)dt$.

Year	1970	1975	1980	1985	1990	1995	2000
E	26.9	26.4	27.1	25.8	25.5	25.0	22.6

2. Annual coal production in the US (in quadrillion BTU per year) is given in the table.[5] Estimate the total amount of coal produced in the US between 1960 and 1990. If $r = f(t)$ is the rate of coal production t years since 1960, write an integral to represent the 1960–1990 coal production.

Year	1960	1965	1970	1975	1980	1985	1990
Rate	10.82	13.06	14.61	14.99	18.60	19.33	22.46

In Problems 3–6, explain in words what the integral represents and give units.

3. $\int_1^3 v(t)\,dt$, where $v(t)$ is velocity in meters/sec and t is time in seconds.

4. $\int_0^6 a(t)\,dt$, where $a(t)$ is acceleration in km/hr^2 and t is time in hours.

5. $\int_{2000}^{2004} f(t)\,dt$, where $f(t)$ is the rate at which the world's population is growing in year t, in billion people per year.

6. $\int_0^5 s(x)\,dx$, where $s(x)$ is rate of change of salinity (salt concentration) in gm/liter per cm in sea water, where x is depth below the surface of the water in cm.

7. Oil leaks out of a tanker at a rate of $r = f(t)$ gallons per minute, where t is in minutes. Write a definite integral expressing the total quantity of oil which leaks out of the tanker in the first hour.

8. A cup of coffee at 90°C is put into a 20°C room when $t = 0$. The coffee's temperature is changing at a rate of $r(t) = -7(0.9^t)$ °C per minute, with t in minutes. Estimate the coffee's temperature when $t = 10$.

9. After a foreign substance is introduced into the blood, the rate at which antibodies are made is given by

$$r(t) = \frac{t}{t^2+1} \text{ thousands of antibodies per minute,}$$

where time, t, is in minutes. Assuming there are no antibodies present at time $t = 0$, find the total quantity of antibodies in the blood at the end of 4 minutes.

10. Figure 5.46 shows the rate of change of the quantity of water in a water tower, in liters per day, during the month of April. If the tower had 12,000 liters of water in it on April 1, estimate the quantity of water in the tower on April 30.

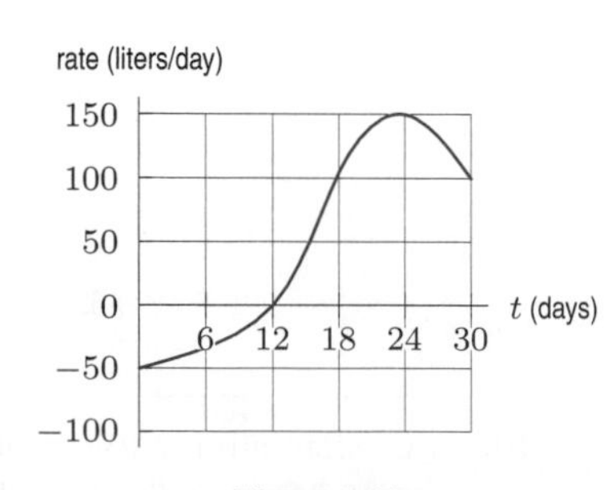

Figure 5.46

[4] *The World Almanac and Book of Facts 2005*, p. 177 (New York).
[5] *World Almanac*, 1995.

11. Figure 5.47 shows the weight growth rate of a human fetus.

(a) What property of a graph of weight as a function of age corresponds to the fact that the function in Figure 5.47 is increasing?
(b) Estimate the weight of a baby born in week 40.

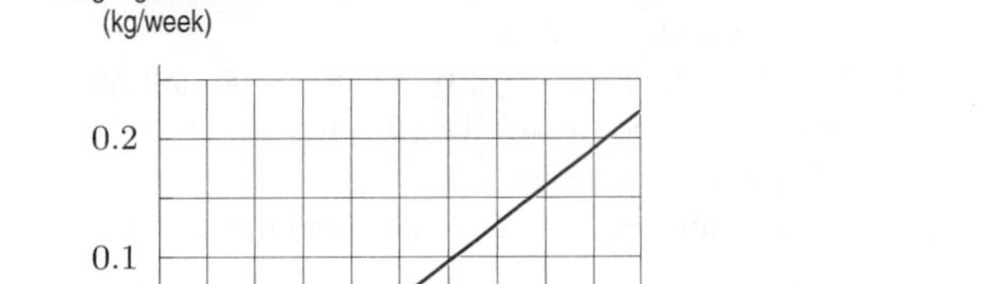

Figure 5.47: Rate of increase of fetal weight

12. A forest fire covers 2000 acres at time $t = 0$. The fire is growing at a rate of $8\sqrt{t}$ acres per hour, where t is in hours. How many acres are covered 24 hours later?

13. Water is pumped out of a holding tank at a rate of $5 - 5e^{-0.12t}$ liters/minute, where t is in minutes since the pump is started. If the holding tank contains 1000 liters of water when the pump is started, how much water does it hold one hour later?

14. Figure 5.48 shows the rate of growth of two trees. If the two trees are the same height at time $t = 0$, which tree is taller after 5 years? After 10 years?

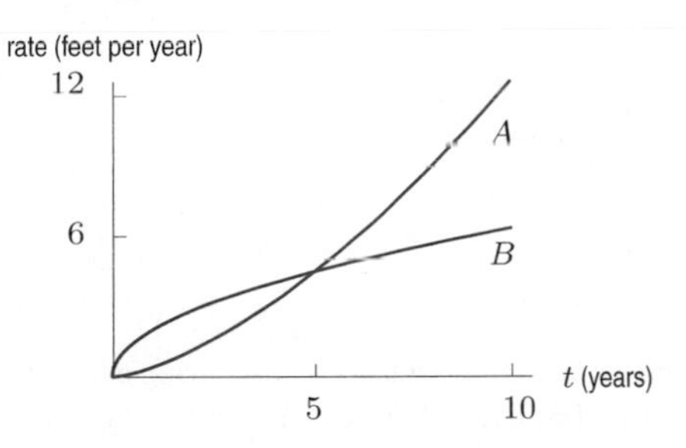

Figure 5.48

15. Figure 5.49 shows the number of sales per month made by two salespeople. Which person has the most total sales after 6 months? After the first year? At approximately what times (if any) have they sold roughly equal total amounts? Approximately how many total sales has each person made at the end of the first year?

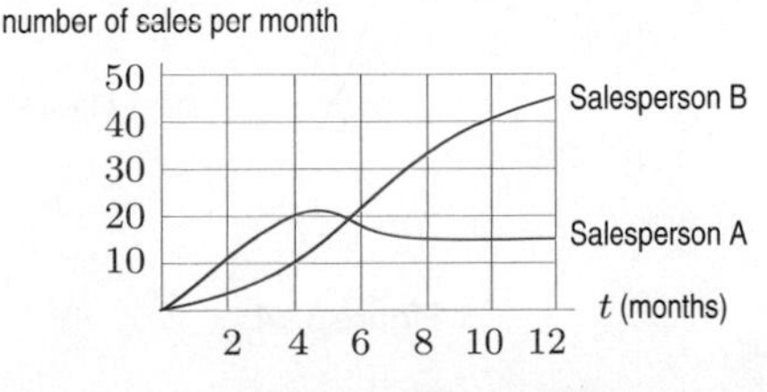

Figure 5.49

16. Height velocity graphs are used by endocrinologists to follow the progress of children with growth deficiencies. Figure 5.50 shows the height velocity curves of an average boy and an average girl between ages 3 and 18.

(a) Which curve is for girls and which is for boys? Explain how you can tell.
(b) About how much does the average boy grow between ages 3 and 10?
(c) The growth spurt associated with adolescence and the onset of puberty occurs between ages 12 and 15 for the average boy and between ages 10 and 12.5 for the average girl. Estimate the height gained by each average child during this growth spurt.
(d) When fully grown, about how much taller is the average man than the average woman? (The average boy and girl are about the same height at age 3.)

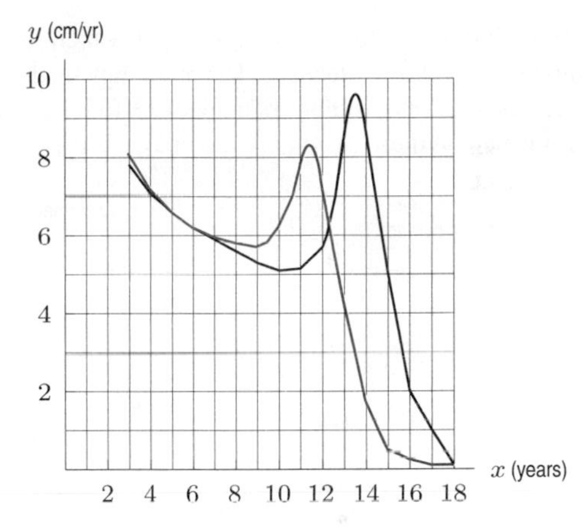

Figure 5.50

17. The birth rate, B, in births per hour, of a bacteria population is given in Figure 5.51. The curve marked D gives the death rate, in deaths per hour, of the same population.

(a) Explain what the shape of each of these graphs tells you about the population.
(b) Use the graphs to find the time at which the net rate of increase of the population is at a maximum.
(c) At time $t = 0$ the population has size N. Sketch the graph of the total number born by time t. Also sketch the graph of the number alive at time t. Estimate the time at which the population is a maximum.

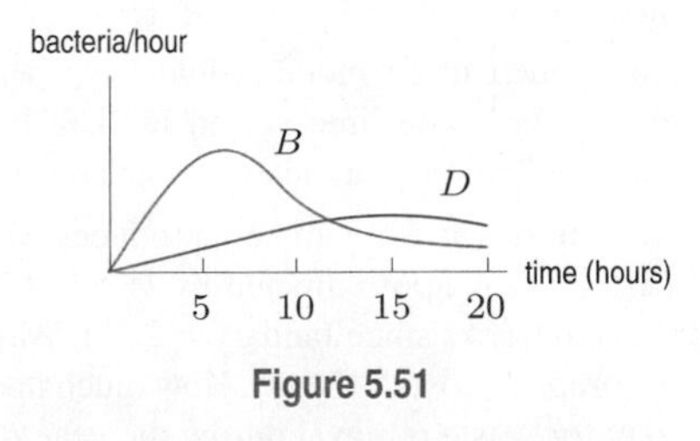

Figure 5.51

18. The rates of consumption of stores of protein and fat in the human body during 8 weeks of starvation are shown in Figure 5.52. Does the body burn more fat or more protein during this period?

Figure 5.52

A healthy human heart pumps about 5 liters of blood per minute. Problems 19–20 refer to Figure 5.53, which shows the response of the heart to bleeding. The pumping rate drops and then returns to normal if the person recovers fully, or drops to zero if the person dies.

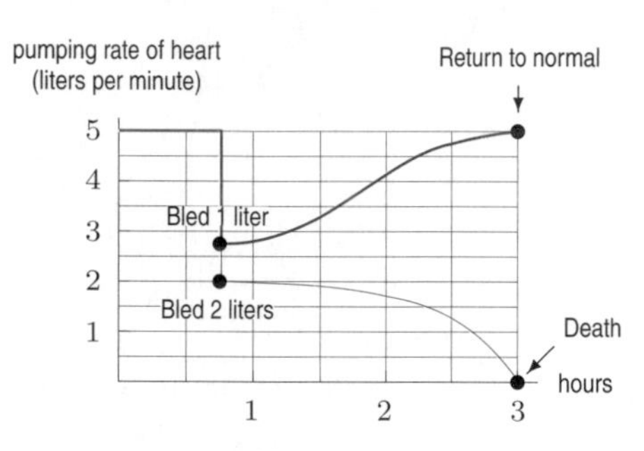

Figure 5.53

19. **(a)** If the body is bled 2 liters, how much blood is pumped during the three hours leading to death?

(b) If $f(t)$ is the pumping rate in liters per minute at time t hours, express your answer to part (a) as a definite integral.

(c) How much more blood would have been pumped during the same time period if there had been no bleeding? Illustrate your answer on the graph.

20. **(a)** If the body is bled 1 liter, how much blood is pumped during the three hours leading to full recovery?

(b) If $g(t)$ is the pumping rate in liters per minute at time t hours, express your answer to part (a) as a definite integral.

(c) How much more blood would have been pumped during the same time period if there had been no bleeding? Show your answer as an area on the graph.

21. The amount of waste a company produces, W, in metric tons per week, is approximated by $W = 3.75e^{-0.008t}$, where t is in weeks since January 1, 2005. Waste removal for the company costs \$15/ton. How much does the company pay for waste removal during the year 2005?

22. Since 1987, when it was \$26,000, average per capita income in the US has been increasing at a rate of

$$r(t) = 480(1.024)^t \quad \text{dollars per year.}$$

Estimate the average per capita income in 2010.

23. The velocity of a car (in miles per hour) is given by $v(t) = 40t - 10t^2$, where t is in hours.

(a) Write a definite integral for the distance the car travels during the first three hours.

(b) Sketch a graph of velocity against time and represent the distance traveled during the first three hours as an area on your graph.

(c) Use a computer or calculator to find this distance.

24. Your velocity is $v(t) = \ln(t^2 + 1)$ ft/sec for t in seconds, $0 \le t \le 3$. Estimate the distance traveled during this time.

Problems 25–28 show the velocity, in cm/sec, of a particle moving along the x-axis. Compute the particle's change in position, left (negative) or right (positive), between times $t = 0$ and $t = 5$ seconds.

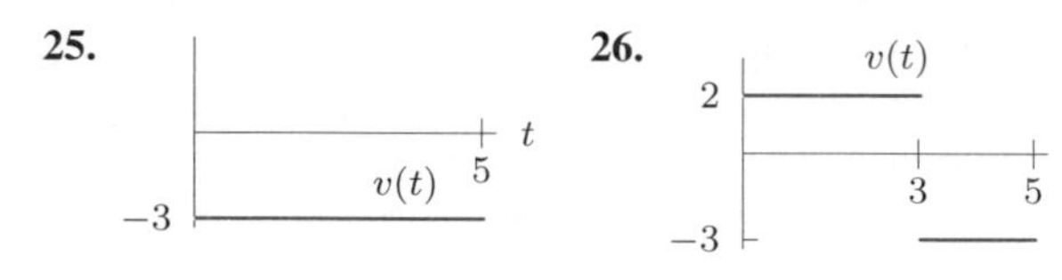

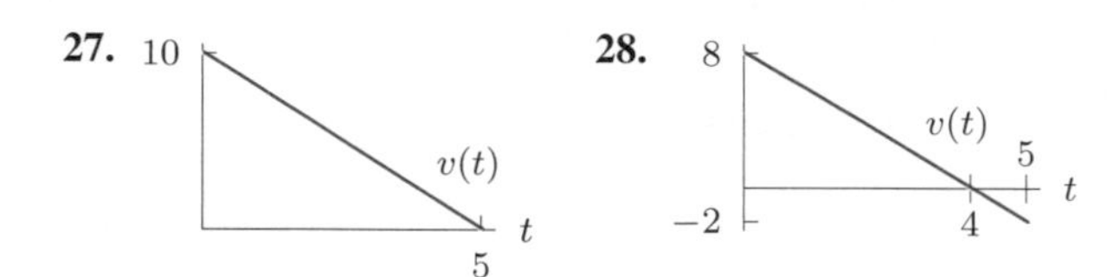

29. Figure 5.54 gives your velocity during a trip starting from home. Positive velocities take you away from home and negative velocities take you toward home. Where are you at the end of the 5 hours? When are you farthest from home? How far away are you at that time?

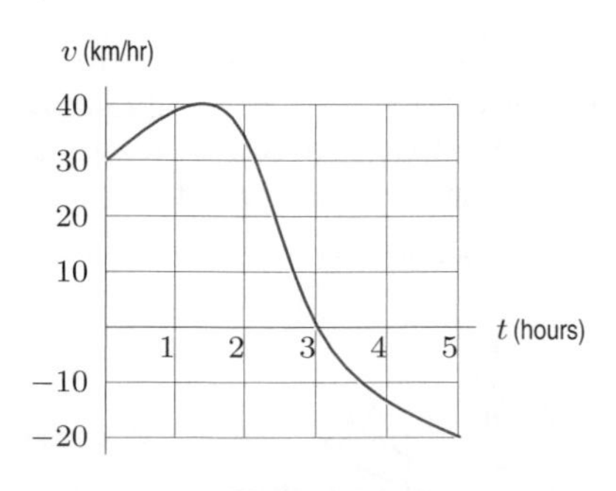

Figure 5.54

30. A bicyclist is pedaling along a straight road for one hour with a velocity v shown in Figure 5.55. She starts out five kilometers from the lake and positive velocities take her toward the lake. [Note: The vertical lines on the graph are at 10 minute ($1/6$ hour) intervals.]

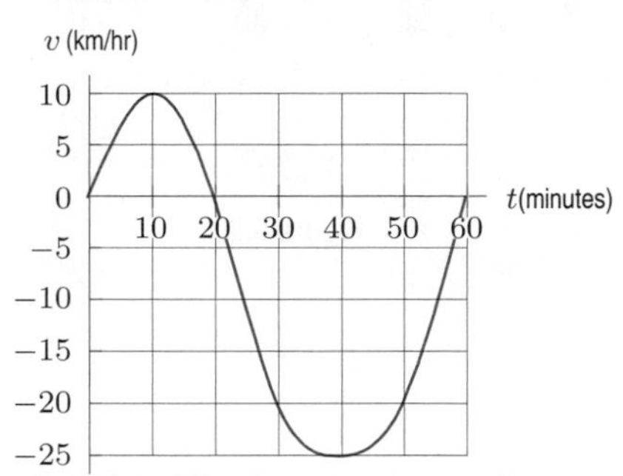

Figure 5.55

(a) Does the cyclist ever turn around? If so, at what time(s)?
(b) When is she going the fastest? How fast is she going then? Toward the lake or away?
(c) When is she closest to the lake? Approximately how close to the lake does she get?
(d) When is she farthest from the lake? Approximately how far from the lake is she then?

31. Figure 5.56 shows plasma concentration curves for two drugs used to slow a rapid heart rate. Compare the two products in terms of level of peak concentration, time until peak concentration, and overall bioavailability.

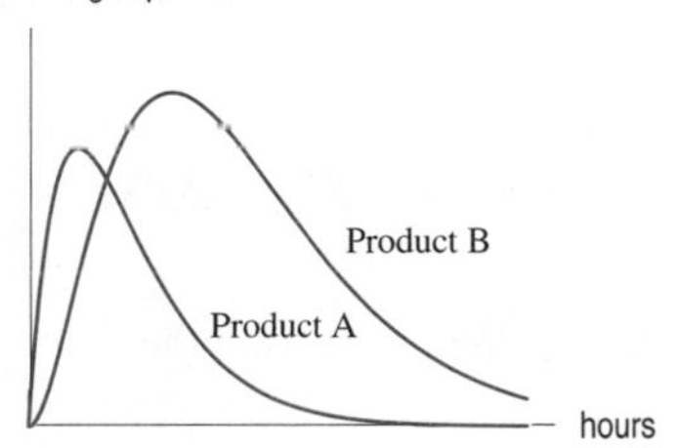

Figure 5.56

32. Figure 5.57 compares the concentration in blood plasma for two pain relievers. Compare the two products in terms of level of peak concentration, time until peak concentration, and overall bioavailability.

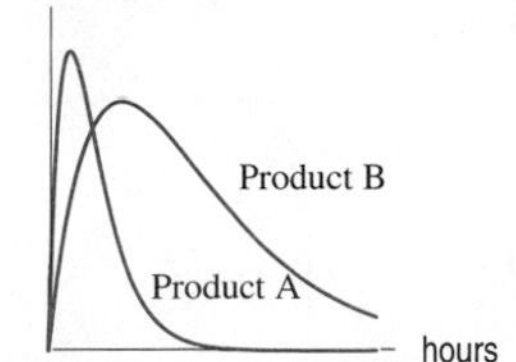

Figure 5.57

33. Draw plasma concentration curves for two drugs A and B if product A has the highest peak concentration, but product B is absorbed more quickly and has greater overall bioavailability.

34. A two-day environmental clean up started at 9 am on the first day. The number of workers fluctuated as shown in Figure 5.58. If the workers were paid \$10 per hour, how much was the total personnel cost of the clean up?

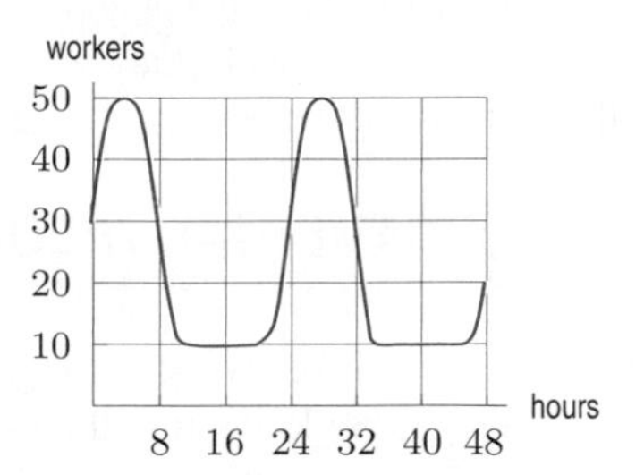

Figure 5.58

35. Suppose in Problem 34 that the workers were paid \$10 per hour for work during the time period 9 am to 5 pm and were paid \$15 per hour for work during the rest of the day. What would the total personnel costs of the clean up have been under these conditions?

36. At the site of a spill of radioactive iodine, radiation levels were four times the maximum acceptable limit, so an evacuation was ordered. If R_0 is the initial radiation level (at $t = 0$) and t is the time in hours, the radiation level $R(t)$, in millirems/hour, is given by

$$R(t) = R_0(0.996)^t.$$

(a) How long does it take for the site to reach the acceptable level of radiation of 0.6 millirems/hour?
(b) How much total radiation (in millirems) has been emitted by that time?

37. If you jump out of an airplane and your parachute fails to open, your downward velocity (in meters per second) t seconds after the jump is approximated by

$$v(t) = 49(1 - (0.8187)^t).$$

(a) Write an expression for the distance you fall in T seconds.
(b) If you jump from 5000 meters above the ground, estimate, using trial and error, how many seconds you fall before hitting the ground.

38. The Montgolfier brothers (Joseph and Etienne) were eighteenth-century pioneers in the field of hot-air ballooning. Had they had the appropriate instruments, they might have left us a record, like that shown in Figure 5.59, of one of their early experiments. The graph shows their vertical velocity, v, with upward as positive.

(a) Over what intervals was the acceleration positive? Negative?
(b) What was the greatest altitude achieved, and at what time?
(c) This particular flight ended on top of a hill. How do you know that it did, and what was the height of the hill above the starting point?

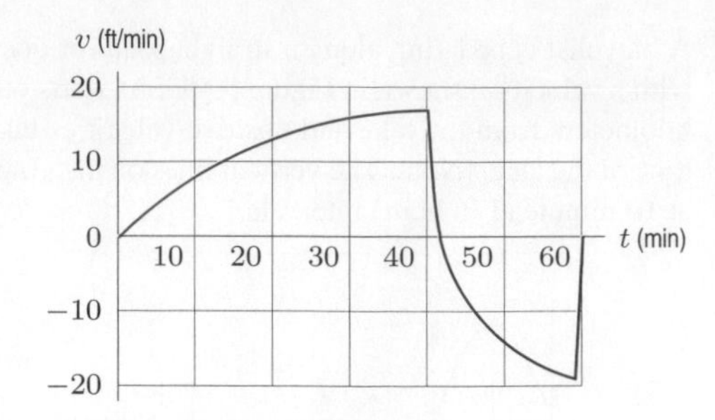

Figure 5.59

5.5 THE FUNDAMENTAL THEOREM OF CALCULUS

In Section 5.2, we saw that the definite integral gives the total change in a quantity from the rate of change. We know that the rate of change of a quantity $F(t)$ is given by the derivative $F'(t)$.

To compute the total change, we break the interval $a \leq t \leq b$ into n equal subintervals at t_0, $t_1, t_2, \cdots t_n$. We take $t_0 = a$ and $t_n = b$, and write Δt for the length of each subinterval, so

$$\Delta t = \frac{b-a}{n}.$$

On the first subinterval the rate of change of F can be approximated by $F'(t_1)$. So

$$\text{Change in } F = \text{Rate} \times \text{Time} \approx F'(t_1)\Delta t.$$

Similarly, for the second subinterval, the rate of change of F can be approximated by $F'(t_2)$, so

$$\text{Change in } F = \text{Rate} \times \text{Time} \approx F'(t_2)\Delta t.$$

Continuing in this way, we see that the total change can be approximated by the right-hand sum:

$$\begin{matrix}\text{Total change in } F \\ \text{between } a \text{ and } b\end{matrix} \approx F'(t_1)\Delta t + F'(t_2)\Delta t + \cdots + F'(t_n)\Delta t = \sum_{i=1}^{n} F'(t_i)\Delta t.$$

This approximation becomes better as n gets larger. When we take the limit, the sum becomes an integral and we have

$$\begin{matrix}\text{Total change in } F \\ \text{between } a \text{ and } b\end{matrix} = \lim_{n\to\infty} \sum_{i=1}^{n} F'(t_i)\Delta t = \int_a^b F'(t)\,dt.$$

On the other hand, the total change in F between a and b is also given by $F(b) - F(a)$, so we have the following result:

> **The Fundamental Theorem of Calculus**
>
> If $F'(t)$ is continuous for $a \leq t \leq b$, then
>
> $$\int_a^b F'(t)\,dt = F(b) - F(a).$$
>
> In words:
> The definite integral of the derivative of a function gives the total change in the function.

The Fundamental Theorem can be used when the rate, $F'(t)$, is known and we want to find the total change $F(b) - F(a)$. The Focus on Theory Section on page 271 gives another version of the Fundamental Theorem.

Example 1 Figure 5.60 shows $F'(t)$, the rate of change of the value, $F(t)$, of an investment over a 5-month period.

(a) When is the value of the investment increasing in value and when is it decreasing?
(b) Does the investment increase or decrease in value during the 5 months?

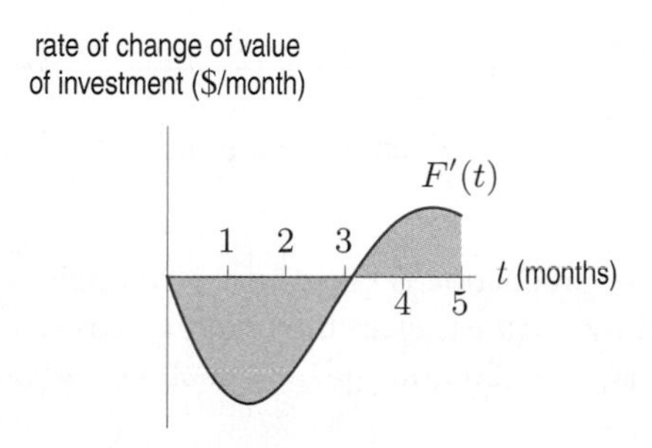

Figure 5.60: Did the investment increase or decrease in value over these 5 months?

Solution

(a) The investment decreased in value during the first 3 months, since the rate of change of value is negative then. The value rose during the last 2 months.
(b) We want to find the total change in the value of the investment between $t = 0$ and $t = 5$. Since the total change is the integral of the rate of change, $F'(t)$, we are looking for

$$\text{Total change in value} = \int_0^5 F'(t)dt.$$

The integral equals the shaded area above the t-axis minus the shaded area below the t-axis. Since in Figure 5.60 the area below the axis is greater than the area above the axis, the integral is negative. The total change in value of the investment during this time is negative, so it decreased in value.

Marginal Cost and Change in Total Cost

Suppose $C(q)$ represents the cost of producing q items. The derivative, $C'(q)$, is the marginal cost. Since marginal cost $C'(q)$ is the rate of change of the cost function with respect to quantity, by the Fundamental Theorem, the integral

$$\int_a^b C'(q)\,dq$$

represents the total change in the cost function between $q = a$ and $q = b$. In other words, the integral gives the amount it costs to increase production from a units to b units.

The cost of producing 0 units is the fixed cost $C(0)$. The area under the marginal cost curve between $q = 0$ and $q = b$ is the total increase in cost between a production of 0 and a production of b. This is called the *total variable cost*. Adding this to the fixed cost gives the total cost to produce b units. In summary,

If $C'(q)$ is a marginal cost function and $C(0)$ is the fixed cost,

$$\text{Cost to increase production from } a \text{ units to } b \text{ units} = C(b) - C(a) = \int_a^b C'(q)\,dq$$

$$\text{Total variable cost to produce } b \text{ units} = \int_0^b C'(q)\,dq$$

$$\begin{aligned}\text{Total cost of producing } b \text{ units} &= \text{Fixed cost} + \text{Total variable cost}\\ &= C(0) + \int_0^b C'(q)\,dq\end{aligned}$$

Example 2 A marginal cost curve is given in Figure 5.61. If the fixed cost is \$1000, estimate the total cost of producing 250 items.

Solution The total cost of production is Fixed cost + Variable cost. The variable cost of producing 250 items is represented by the area under the marginal cost curve. The area in Figure 5.61 between $q = 0$ and $q = 250$ is about 20 grid squares. Each grid square has area (2 dollars/item)(50 items) = 100 dollars, so

$$\text{Total variable cost} = \int_0^{250} C'(q)\,dq \approx 20(100) = 2000.$$

The total cost to produce 250 items is given by :

$$\begin{aligned}\text{Total cost} &= \text{Fixed cost} + \text{Total variable cost}\\ &\approx \$1000 + \$2000 = \$3000.\end{aligned}$$

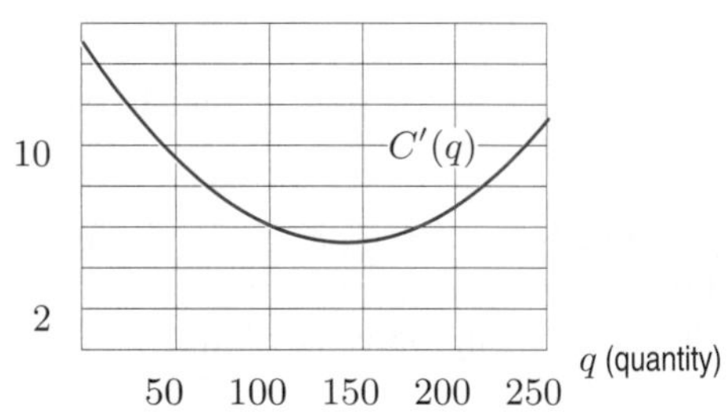

Figure 5.61: A marginal cost curve

Problems for Section 5.5

1. If the marginal cost function $C'(q)$ is measured in dollars per ton, and q gives the quantity in tons, what are the units of measurement for $\int_{800}^{900} C'(q)\,dq$? What does this integral represent?

2. The marginal cost function of a product, in dollars per unit, is $C'(q) = q^2 - 50q + 700$. If fixed costs are \$500, find the total cost to produce 50 items.

3. The total cost in dollars to produce q units of a product is $C(q)$. Fixed costs are \$20,000. The marginal cost is

$$C'(q) = 0.005q^2 - q + 56.$$

(a) On a graph of $C'(q)$, illustrate graphically the total variable cost of producing 150 units.
(b) Estimate $C(150)$, the total cost to produce 150 units.
(c) Find the value of $C'(150)$ and interpret your answer in terms of costs of production.
(d) Use parts (b) and (c) to estimate $C(151)$.

4. A marginal cost function $C'(q)$ is given in Figure 5.62. If the fixed costs are \$10,000, estimate:

(a) The total cost to produce 30 units.
(b) The additional cost if the company increases production from 30 units to 40 units.
(c) The value of $C'(25)$. Interpret your answer in terms of costs of production.

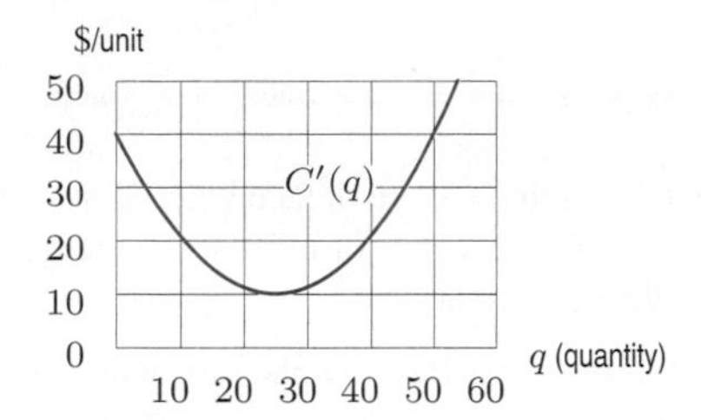

Figure 5.62

5. The population of Tokyo grew at the rate shown in Figure 5.63. Estimate the change in population between 1970 and 1990.

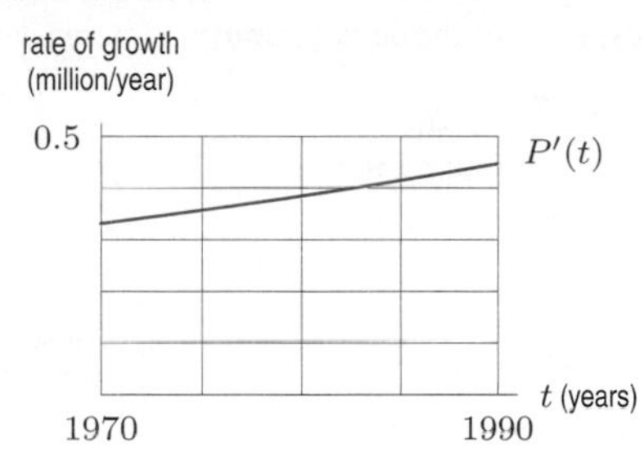

Figure 5.63

6. The marginal cost function for a company is given by

$$C'(q) = q^2 - 16q + 70 \text{ dollars/unit,}$$

where q is the quantity produced. If $C(0) = 500$, find the total cost of producing 20 units. What is the fixed cost and what is the total variable cost for this quantity?

7. The marginal cost $C'(q)$ (in dollars per unit) of producing q units is given in the following table.

(a) If fixed cost is \$10,000, estimate the total cost of producing 400 units.
(b) How much would the total cost increase if production were increased one unit, to 401 units?

q	0	100	200	300	400	500	600
$C'(q)$	25	20	18	22	28	35	45

8. The marginal cost function of producing q mountain bikes is

$$C'(q) = \frac{600}{0.3q + 5}.$$

(a) If the fixed cost in producing the bicycles is \$2000, find the total cost to produce 30 bicycles.
(b) If the bikes are sold for \$200 each, what is the profit (or loss) on the first 30 bicycles?
(c) Find the marginal profit on the 31st bicycle.

9. The marginal revenue function on sales of q units of a product is $R'(q) = 200 - 12\sqrt{q}$ dollars per unit.

(a) Graph $R'(q)$.
(b) Estimate the total revenue if sales are 100 units.
(c) What is the marginal revenue at 100 units? Use this value and your answer to part (b) to estimate the total revenue if sales are 101 units.

10. Figure 5.64 shows $P'(t)$, the rate of change of the price of stock in a certain company at time t.

(a) At what time during this five-week period was the stock at its highest value? At its lowest value?
(b) If $P(t)$ represents the price of the stock, arrange the following quantities in increasing order:

$$P(0),\ P(1),\ P(2),\ P(3),\ P(4),\ P(5).$$

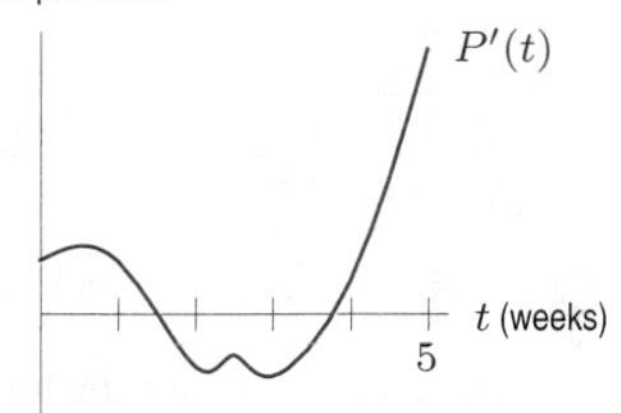

Figure 5.64

11. Ice is forming on a pond at a rate given by

$$\frac{dy}{dt} = \frac{\sqrt{t}}{2} \text{ inches per hour,}$$

where y is the thickness of the ice in inches at time t measured in hours since the ice started forming.

(a) Estimate the thickness of the ice after 8 hours.
(b) At what rate is the thickness of the ice increasing after 8 hours?

12. The net worth, $f(t)$, of a company is growing at a rate of $f'(t) = 2000 - 12t^2$ dollars per year, where t is in years since 2005. How is the net worth of the company expected to change between 2005 and 2015? If the company is worth \$40,000 in 2005, what is it worth in 2015?

13. The graph of a derivative $f'(x)$ is shown in Figure 5.65. Fill in the table of values for $f(x)$ given that $f(0) = 2$.

x	0	1	2	3	4	5	6
$f(x)$	2						

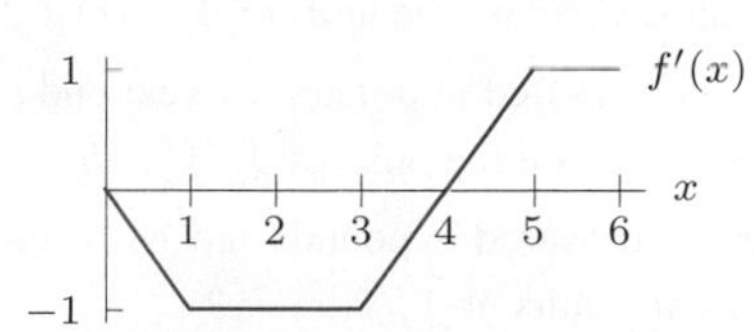

Figure 5.65: Graph of f', not f

CHAPTER SUMMARY

- **Definite integral as limit of right-hand or left-hand sums**
- **Interpretations of the definite integral**
  Total change from rate of change, change in position given velocity, area, bioavailability, total variable cost.
- **Working with the definite integral**
  Estimate definite integral from graph, table of values, or formula.
- **Fundamental Theorem of Calculus**

REVIEW PROBLEMS FOR CHAPTER FIVE

1. The velocity $v(t)$ in Table 5.8 is decreasing, $2 \le t \le 12$. Using $n = 5$ subdivisions to approximate the total distance traveled, find

(a) An upper estimate **(b)** A lower estimate

Table 5.8

t	2	4	6	8	10	12
$v(t)$	44	42	41	40	37	35

2. The velocity $v(t)$ in Table 5.9 is increasing, $0 \le t \le 12$.

(a) Find an upper estimate for the total distance traveled using

(i) $n = 4$ (ii) $n = 2$

(b) Which of the two answers in part (a) is more accurate? Why?

(c) Find a lower estimate of the total distance traveled using $n = 4$.

Table 5.9

t	0	3	6	9	12
$v(t)$	34	37	38	40	45

3. Use the following table to estimate $\int_{10}^{26} f(x)\,dx$.

x	10	14	18	22	26
$f(x)$	100	88	72	50	28

4. If $f(t)$ is measured in miles per hour and t is measured in hours, what are the units of $\int_a^b f(t)\,dt$?

5. If $f(t)$ is measured in meters/second^2 and t is measured in seconds, what are the units of $\int_a^b f(t)\,dt$?

6. If $f(t)$ is measured in dollars per year and t is measured in years, what are the units of $\int_a^b f(t)\,dt$?

7. If $f(x)$ is measured in pounds and x is measured in feet, what are the units of $\int_a^b f(x)\,dx$?

8. As coal deposits are depleted, it will become necessary to strip-mine larger and larger areas for each ton of coal. Figure 5.66 shows the number of acres of land per million tons of coal that will be defaced during strip-mining as a function of the number of million tons removed, starting from the present day.

(a) Estimate the total number of acres defaced in extracting the next 4 million tons of coal (measured from the present day). Draw four rectangles under the curve, and compute their area.

(b) Reestimate the number of acres defaced using rectangles above the curve.

(c) Use your answers to parts (a) and (b) to get a better estimate of the actual number of acres defaced.

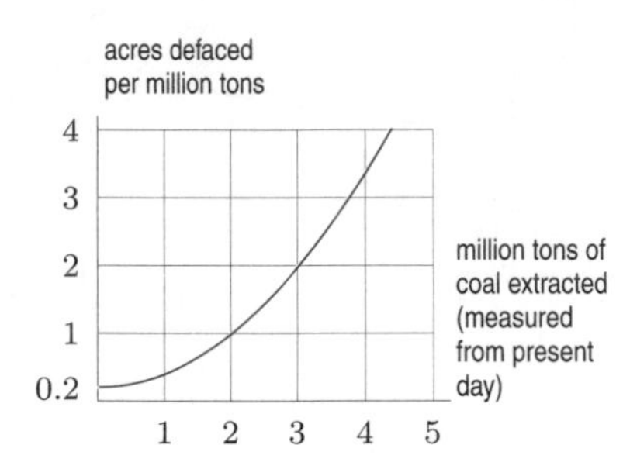

Figure 5.66

For Problems 9–14, use a calculator or computer to evaluate the integral.

9. $\displaystyle\int_0^{10} 2^{-x}\,dx$

10. $\displaystyle\int_1^{5} (x^2+1)\,dx$

11. $\displaystyle\int_0^{1} \sqrt{1+t^2}\,dt$

12. $\displaystyle\int_{-1}^{1} \frac{x^2+1}{x^2-4}\,dx$

13. $\displaystyle\int_2^{3} \frac{-1}{(r+1)^2}\,dr$

14. $\displaystyle\int_1^{3} \frac{z^2+1}{z}\,dz$

15. Find the area under the graph of $f(x) = x^2 + 2$ between $x = 0$ and $x = 6$.

In Problems 16–19, find the given area.

16. Between $y = x^2$ and $y = x^3$ for $0 \le x \le 1$.

17. Between $y = x^{1/2}$ and $y = x^{1/3}$ for $0 \le x \le 1$.

18. Between $y = 3x$ and $y = x^2$.

19. Between $y = x$ and $y = \sqrt{x}$.

20. Coal gas is produced at a gasworks. Pollutants in the gas are removed by scrubbers, which become less and less efficient as time goes on. The following measurements, made at the start of each month, show the rate at which pollutants are escaping (in tons/month) in the gas:

Time (months)	0	1	2	3	4	5	6
Rate pollutants escape	5	7	8	10	13	16	20

(a) Make an overestimate and an underestimate of the total quantity of pollutants that escape during the first month.
(b) Make an overestimate and an underestimate of the total quantity of pollutants that escape during the six months.

21. A student is speeding down Route 11 in his fancy red Porsche when his radar system warns him of an obstacle 400 feet ahead. He immediately applies the brakes, starts to slow down, and spots a skunk in the road directly ahead of him. The "black box" in the Porsche records the car's speed every two seconds, producing the following table. The speed decreases throughout the 10 seconds it takes to stop, although not necessarily at a uniform rate.

Time since brakes applied (sec)	0	2	4	6	8	10
Speed (ft/sec)	100	80	50	25	10	0

(a) What is your best estimate of the total distance the student's car traveled before coming to rest?
(b) Which one of the following statements can you justify from the information given?
(i) The car stopped before getting to the skunk.
(ii) The "black box" data is inconclusive. The skunk may or may not have been hit.
(iii) The skunk was hit by the car.

22. The velocity of a particle moving along the x-axis is given by $f(t) = 6 - 2t$ cm/sec. Use a graph of $f(t)$ to find the exact change in position of the particle from time $t = 0$ to $t = 4$ seconds.

23. A baseball thrown directly upward at 96 ft/sec has velocity $v(t) = 96 - 32t$ ft/sec at time t seconds.

(a) Graph the velocity from $t = 0$ to $t = 6$.
(b) When does the baseball reach the peak of its flight? How high does it go?
(c) How high is the baseball at time $t = 5$?

24. A news broadcast in early 1993 said the average American's annual income is changing at a rate of $r(t) = 40(1.002)^t$ dollars per month, where t is in months from January 1, 1993. How much did the average American's income change during 1993?

25. Two species of plants have the same populations at time $t = 0$ and the growth rates shown in Figure 5.67.

(a) Which species has a larger population at the end of 5 years? At the end of 10 years?
(b) Which species do you think has the larger population after 20 years? Explain.

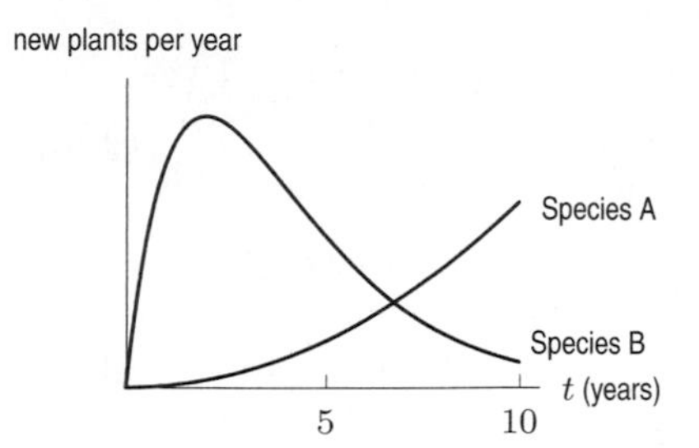

Figure 5.67

26. Figure 5.68 represents your velocity, v, on a bicycle trip along a straight road which starts 10 miles from home. Write a paragraph describing your trip: Do you start out going toward or away from home? How long do you continue in that direction and how far are you from home when you turn around? How many times do you change direction? Do you ever get home? Where are you at the end of the four-hour bike ride?

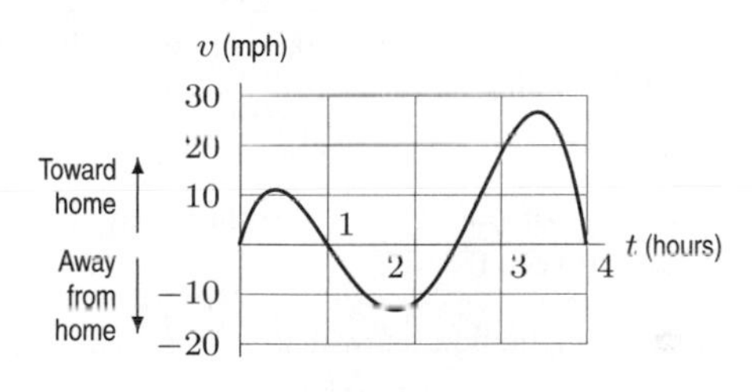

Figure 5.68

27. Figure 5.69 shows the length growth rate of a human fetus.

(a) What feature of a graph of length as a function of age corresponds to the maximum in Figure 5.69?
(b) Estimate the length of a baby born in week 40.

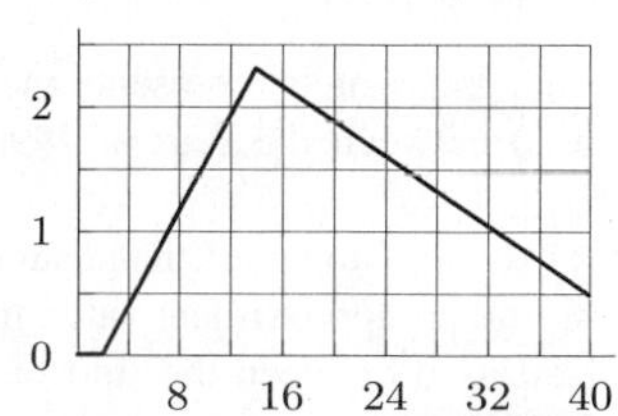

Figure 5.69

28. A bicyclist pedals along a straight road with velocity, v, given in Figure 5.70. She starts 5 miles from a lake; positive velocities take her away from the lake and negative velocities take her toward the lake. When is the cyclist farthest from the lake, and how far away is she then?

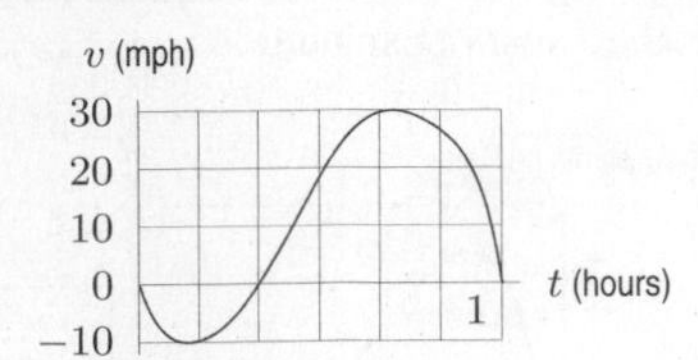

Figure 5.70

29. Using Figure 5.71, decide whether each of the following definite integrals is positive or negative.

(a) $\int_{-5}^{-4} f(x)\,dx$ **(b)** $\int_{-4}^{1} f(x)\,dx$

(c) $\int_{1}^{3} f(x)\,dx$ **(d)** $\int_{-5}^{3} f(x)\,dx$

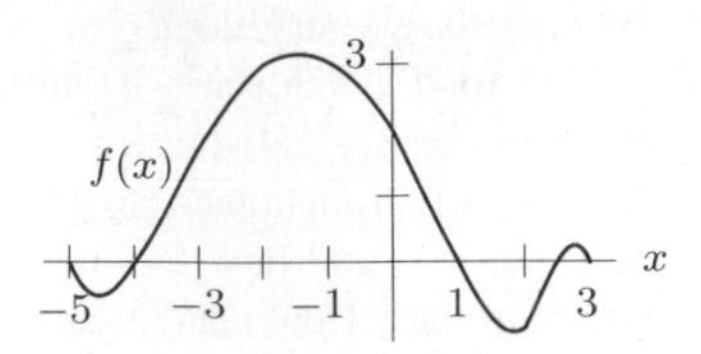

Figure 5.71

30. Using Figure 5.71, arrange the following definite integrals in ascending order:
$\int_{-5}^{-3} f(x)\,dx,\ \int_{-5}^{-1} f(x)\,dx,\ \int_{-5}^{1} f(x)\,dx,\ \int_{-5}^{3} f(x)\,dx.$

31. Use a graph of $y = 2^{-x^2}$ to explain why $\int_{-1}^{1} 2^{-x^2}\,dx$ must be between 0 and 2.

32. Without computation, show that $2 \leq \int_0^2 \sqrt{1+x^3}\,dx \leq 6$.

33. With t in seconds, the velocity of an object is $v(t) = 10 + 8t - t^2$ m/sec.

(a) Represent the distance traveled during the first 5 seconds as a definite integral and as an area.

(b) Estimate the distance traveled by the object during the first 5 seconds by estimating the area.

(c) Calculate the distance traveled.

34. The world's oil is being consumed at a continuously increasing rate, $f(t)$ (in billions of barrels per year), where t is in years since the start of 1990.

(a) Write a definite integral which represents the total quantity of oil used between the start of 1990 and the start of 2005.

(b) Suppose $r = 32(1.05)^t$. Using a left-hand sum with five subdivisions, find an approximate value for the total quantity of oil used between the start of 1990 and the start of 2005.

(c) Interpret each of the five terms in the sum from part (b) in terms of oil consumption.

35. A car moves along a straight line with velocity, in feet/second, given by

$$v(t) = 6 - 2t \quad \text{for } t \geq 0.$$

(a) Describe the car's motion in words. (When is it moving forward, backward, and so on?)

(b) The car's position is measured from its starting point. When is it farthest forward? Backward?

36. The marginal cost of drilling an oil well depends on the depth at which you are drilling; drilling becomes more expensive, per meter, as you dig deeper into the earth. The fixed costs are 1,000,000 riyals (the riyal is the unit of currency of Saudi Arabia), and, if x is the depth in meters, the marginal costs are

$$C'(x) = 4000 + 10x \quad \text{riyals/meter.}$$

Find the total cost of drilling a 500-meter well.

37. A warehouse charges its customers \$5 per day for every 10 cubic feet of space used for storage. Figure 5.72 records the storage used by one company over a month. How much will the company have to pay?

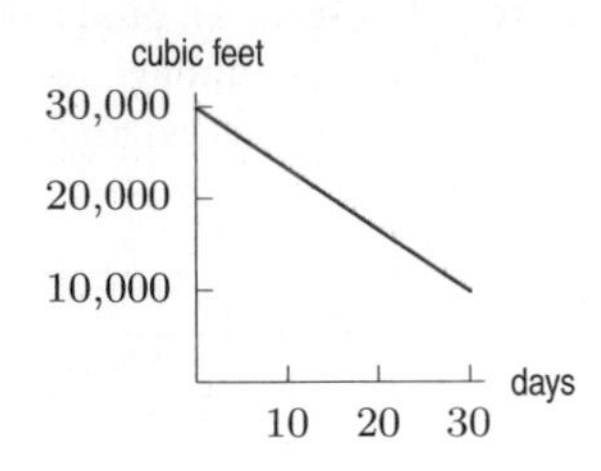

Figure 5.72

38. One of the earliest pollution problems brought to the attention of the Environmental Protection Agency (EPA) was the case of the Sioux Lake in eastern South Dakota. For years a small paper plant located nearby had been discharging waste containing carbon tetrachloride (CCl_4) into the waters of the lake. At the time the EPA learned of the situation, the chemical was entering at a rate of 16 cubic yards/year.

The agency immediately ordered the installation of filters designed to slow (and eventually stop) the flow of CCl_4 from the mill. Implementation of this program took exactly three years, during which the flow of pollutant was steady at 16 cubic yards/year. Once the filters were installed, the flow declined. From the time the filters were installed until the time the flow stopped, the rate of flow was well approximated by

$$\text{Rate (in cubic yards/year)} = t^2 - 14t + 49,$$

where t is time measured in years since the EPA learned of the situation (thus, $t \geq 3$).

(a) Draw a graph showing the rate of CCl_4 flow into the lake as a function of time, beginning at the time the EPA first learned of the situation.

(b) How many years elapsed between the time the EPA learned of the situation and the time the pollution flow stopped entirely?
(c) How much CCl_4 entered the waters during the time shown in the graph in part (a)?

39. (a) Graph $x^3 - 5x^2 + 4x$, marking $x = 1, 2, 3, 4, 5$.
(b) Use your graph and the area interpretation of the definite integral to decide which of the five numbers

$$I_n = \int_0^n (x^3 - 5x^2 + 4x)\, dx \quad \text{for } n = 1, 2, 3, 4, 5$$

is largest. Which is smallest? How many of the numbers are positive? (Don't calculate the integrals.)

40. A mouse moves back and forth in a straight tunnel, attracted to bits of cheddar cheese alternately introduced to and removed from the ends (right and left) of the tunnel. The graph of the mouse's velocity, v, is given in Figure 5.73, with positive velocity corresponding to motion toward the right end. Assuming that the mouse starts ($t = 0$) at the center of the tunnel, use the graph to estimate the time(s) at which:

(a) The mouse changes direction.
(b) The mouse is moving most rapidly to the right; to the left.
(c) The mouse is farthest to the right of center; farthest to the left.
(d) The mouse's speed (i.e., the magnitude of its velocity) is decreasing.
(e) The mouse is at the center of the tunnel.

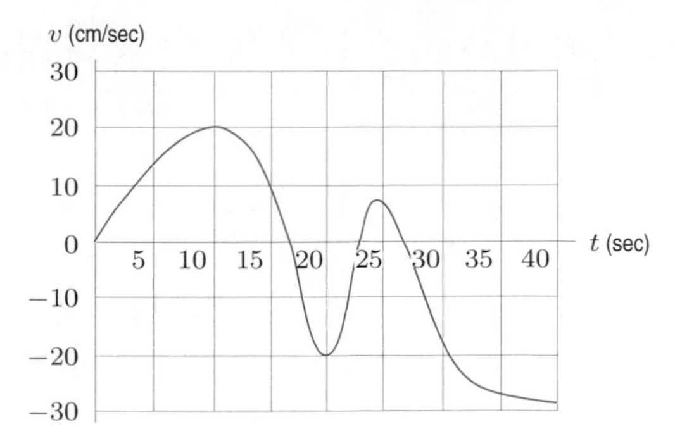

Figure 5.73

41. Pollution is being dumped into a lake at a rate which is increasing at a constant rate from 10 kg/year to 50 kg/year until a total of 270 kg has been dumped. Sketch a graph of the rate at which pollution is being dumped in the lake against time. How long does it take until 270 kg has been dumped?

For Problems 42–44, suppose $F(0) = 0$ and $F'(x) = 4 - x^2$.

42. Calculate $F(b)$ for $b = 0,\ 0.5,\ 1,\ 1.5,\ 2,\ 2.5$.

43. Using a graph of F', decide where F is increasing and where F is decreasing for $0 \le x \le 2.5$.

44. Does F have a maximum value for $0 \le x \le 2.5$? If so, what is it, and at what value of x does it occur?

PROJECTS FOR CHAPTER FIVE

1. Carbon Dioxide in Pond Water Biological activity in a pond is reflected in the rate at which carbon dioxide, CO_2, is added to or withdrawn from the water. Plants take CO_2 out of the water during the day for photosynthesis and put CO_2 into the water at night. Animals put CO_2 into the water all the time as they breathe. Biologists are interested in how the net rate at which CO_2 enters a pond varies during the day. Figure 5.74 shows this rate as a function of time of day.[6] The rate is measured in millimoles (mmol) of CO_2 per liter of water per hour; time is measured in hours past dawn. At dawn, there were 2.600 mmol of CO_2 per liter of water.

(a) What can be concluded from the fact that the rate is negative during the day and positive at night?
(b) Some scientists have suggested that plants respire (breathe) at a constant rate at night, and that they photosynthesize at a constant rate during the day. Does Figure 5.74 support this view?
(c) When was the CO_2 content of the water at its lowest? How low did it go?
(d) How much CO_2 was released into the water during the 12 hours of darkness? Compare this quantity with the amount of CO_2 withdrawn from the water during the 12 hours of daylight. How can you tell by looking at the graph whether the CO_2 in the pond is in equilibrium?
(e) Estimate the CO_2 content of the water at three hour intervals throughout the day. Use your estimates to plot a graph of CO_2 content throughout the day.

[6]Data from R. J. Beyers, *The Pattern of Photosynthesis and Respiration in Laboratory Microsystems* (Mem. 1st. Ital. Idrobiol., 1965).

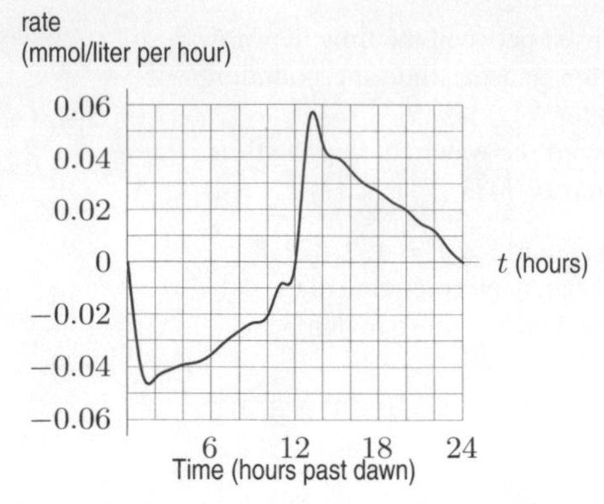

Figure 5.74: Rate at which CO_2 is entering the pond

2. Flooding in the Grand Canyon

The Glen Canyon Dam at the top of the Grand Canyon prevents natural flooding. In 1996, scientists decided an artificial flood was necessary to restore the environmental balance. Water was released through the dam at a controlled rate[7] shown in Figure 5.75. The figure also shows the rate of flow of the last natural flood in 1957.

(a) At what rate was water passing through the dam in 1996 before the artificial flood?
(b) At what rate was water passing down the river in the pre-flood season in 1957?
(c) Estimate the maximum rates of discharge for the 1996 and 1957 floods.
(d) Approximately how long did the 1996 flood last? How long did the 1957 flood last?
(e) Estimate how much additional water passed down the river in 1996 as a result of the artificial flood.
(f) Estimate how much additional water passed down the river in 1957 as a result of the flood.

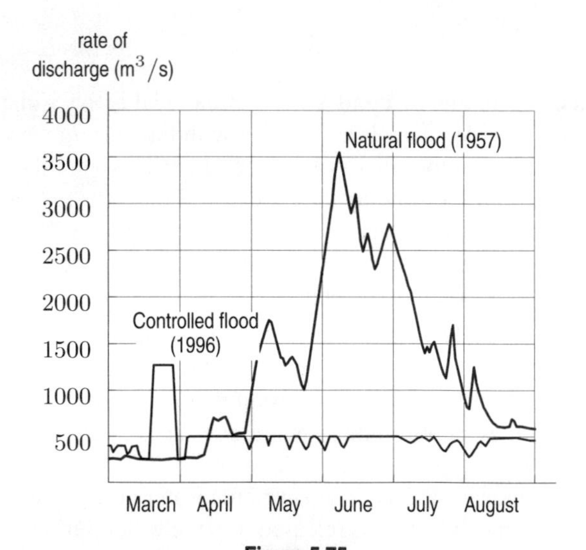

Figure 5.75

[7]Adapted from M. Collier, R. Webb, E. Andrews, "Experimental Flooding in Grand Canyon" in *Scientific American* (January 1997).

FOCUS ON THEORY

THEOREMS ABOUT DEFINITE INTEGRALS

The Second Fundamental Theorem of Calculus

The Fundamental Theorem of Calculus tells us that if we have a function F whose derivative is a continuous function f, then the definite integral of f is given by

$$\int_a^b f(t)\,dt = F(b) - F(a).$$

We now take a different point of view. If a is fixed and the upper limit is x, then the value of the integral is a function of x. We define a new function G by

$$G(x) = \int_a^x f(t)\,dt.$$

To visualize G, suppose that f is positive and $x > a$. Then $G(x)$ is the area under the graph of f in Figure 5.76. If f is continuous on an interval containing a, then it can be shown that G is defined for all x on that interval.

We now consider the derivative of G. Using the definition of the derivative,

$$G'(x) = \lim_{h\to 0} \frac{G(x+h) - G(x)}{h}.$$

Suppose f and h are positive. Then we can visualize

$$G(x) = \int_a^x f(t)\,dt$$

and

$$G(x+h) = \int_a^{x+h} f(t)\,dt$$

as areas, which leads to representing

$$G(x+h) - G(x) = \int_x^{x+h} f(t)\,dt$$

as a difference of two areas.

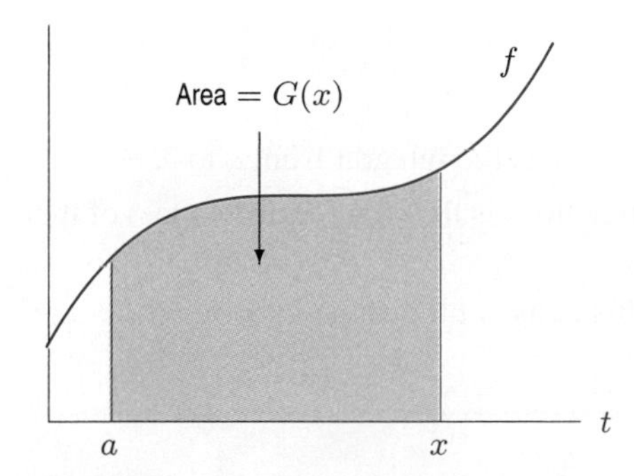

Figure 5.76: Representing $G(x)$ as an area

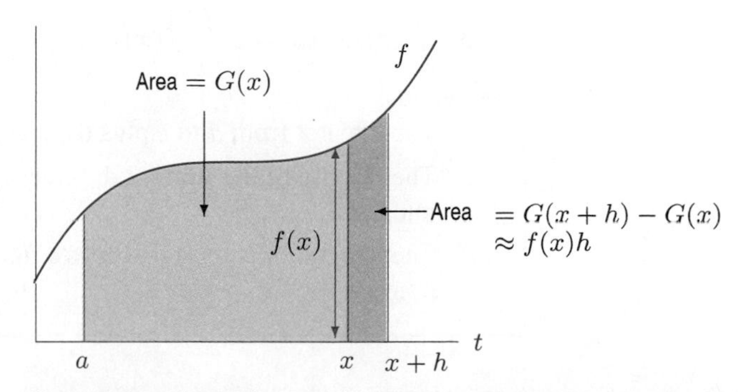

Figure 5.77: $G(x+h) - G(x)$ is the area of a roughly rectangular region

From Figure 5.77, we see that, if h is small, $G(x+h)-G(x)$ is roughly the area of a rectangle of height $f(x)$ and width h (shaded darker in Figure 5.77), so we have

$$G(x+h)-G(x)\approx f(x)h,$$

hence

$$\frac{G(x+h)-G(x)}{h}\approx f(x).$$

The same result holds when h is negative, suggesting that

$$G'(x)=\lim_{h\to 0}\frac{G(x+h)-G(x)}{h}=f(x).$$

This result is another form of the Fundamental Theorem of Calculus. It is usually stated as follows:

Second Fundamental Theorem of Calculus

If f is a continuous function on an interval, and if a is any number in that interval, then the function G defined by

$$G(x)=\int_a^x f(t)\,dt$$

has derivative f; that is, $G'(x)=f(x)$.

Properties of the Definite Integral

In this chapter, we have used the following properties to break up definite integrals.

Sums and Multiples of Definite Integrals

If a, b, and c are any numbers and f and g are continuous functions, then

1. $\displaystyle\int_a^c f(x)\,dx+\int_c^b f(x)\,dx=\int_a^b f(x)\,dx.$
2. $\displaystyle\int_a^b (f(x)\pm g(x))\,dx=\int_a^b f(x)\,dx\pm\int_a^b g(x)\,dx.$
3. $\displaystyle\int_a^b cf(x)\,dx=c\int_a^b f(x)\,dx.$

In words:

1. The integral from a to c plus the integral from c to b is the integral from a to b.
2. The integral of the sum (or difference) of two functions is the sum (or difference) of their integrals.
3. The integral of a constant times a function is that constant times the integral of the function.

These properties can best be visualized by thinking of the integrals as areas or as the limit of the sum of areas of rectangles.

Problems on the Second Fundamental Theorem of Calculus

For Problems 1–4, find $G'(x)$.

1. $G(x) = \displaystyle\int_a^x t^3\,dt$

2. $G(x) = \displaystyle\int_a^x 3^t\,dt$

3. $G(x) = \displaystyle\int_a^x te^t\,dt$

4. $G(x) = \displaystyle\int_a^x \ln y\,dy$

5. Let $F(b) = \displaystyle\int_0^b 2^x\,dx$.

(a) What is $F(0)$?

(b) Does the value of F increase or decrease as b increases? (Assume $b \geq 0$.)

(c) Estimate $F(1)$, $F(2)$, and $F(3)$.

6. For $x = 0, 0.5, 1.0, 1.5,$ and 2.0, make a table of values for $I(x) = \int_0^x \sqrt{t^4+1}\,dt$.

7. Assume that $F'(t) = \sin t \cos t$ and $F(0) = 1$. Find $F(b)$ for $b = 0,\ 0.5,\ 1,\ 1.5,\ 2,\ 2.5,$ and 3.

Let $\int_a^b f(x)\,dx = 8$, $\int_a^b (f(x))^2\,dx = 12$, $\int_a^b g(t)\,dt = 2$, and $\int_a^b (g(t))^2\,dt = 3$. Find the integrals in Problems 8–11.

8. $\int_a^b (f(x) + g(x))\,dx$

9. $\int_a^b \left((f(x))^2 - (g(x))^2\right)\,dx$

10. $\int_a^b (f(x))^2\,dx - \left(\int_a^b f(x)\,dx\right)^2$

11. $\int_a^b cf(z)\,dz$

Chapter Six

USING THE DEFINITE INTEGRAL

Chapter 5 showed how a definite integral can be used to compute an area or a total change. In this chapter, we use definite integrals to solve problems involving average value, consumer and producer surplus, present and future value, and population growth. In each case, we begin by estimating the quantity using a Riemann sum.

6.1 AVERAGE VALUE

In this section we show how to interpret the definite integral as the average value of a function.

The Definite Integral as an Average

We know how to find the average of n numbers: Add them and divide by n. But how do we find the average value of a continuously varying function? Let us consider an example. Suppose $f(t)$ is the temperature at time t, measured in hours since midnight, and that we want to calculate the average temperature over a 24-hour period. One way to start would be to average the temperatures at n equally spaced times, $t_1, t_2, \ldots, t_n$, during the day.

$$\text{Average temperature} \approx \frac{f(t_1) + f(t_2) + \cdots + f(t_n)}{n}.$$

The larger we make n, the better the approximation. We can rewrite this expression as a Riemann sum over the interval $0 \le t \le 24$ if we use the fact that $\Delta t = 24/n$, so $n = 24/\Delta t$:

$$\begin{aligned}\text{Average temperature} &\approx \frac{f(t_1) + f(t_2) + \cdots + f(t_n)}{24/\Delta t} \\ &= \frac{f(t_1)\Delta t + f(t_2)\Delta t + \cdots + f(t_n)\Delta t}{24} \\ &= \frac{1}{24}\sum_{i=1}^{n} f(t_i)\Delta t.\end{aligned}$$

As $n \to \infty$, the Riemann sum tends toward an integral, and the approximation gets better. We expect that

$$\begin{aligned}\text{Average temperature} &= \lim_{n\to\infty} \frac{1}{24}\sum_{i=1}^{n} f(t_i)\Delta t \\ &= \frac{1}{24}\int_0^{24} f(t)\,dt.\end{aligned}$$

Generalizing for any function f, if $a < b$, we have

$$\begin{array}{c}\text{Average value of } f \\ \text{on the interval from } a \text{ to } b\end{array} = \frac{1}{b-a}\int_a^b f(x)\,dx.$$

The units of $f(x)$ are the same as the units of the average value of $f(x)$.

How to Visualize the Average on a Graph

The definition of average value tells us that

$$(\text{Average value of } f)\cdot(b-a) = \int_a^b f(x)\,dx.$$

Let's interpret the integral as the area under the graph of f. If $f(x)$ is positive, then the average value of f is the height of a rectangle whose base is $(b-a)$ and whose area is the same as the area between the graph of f and the x-axis. (See Figure 6.1.)

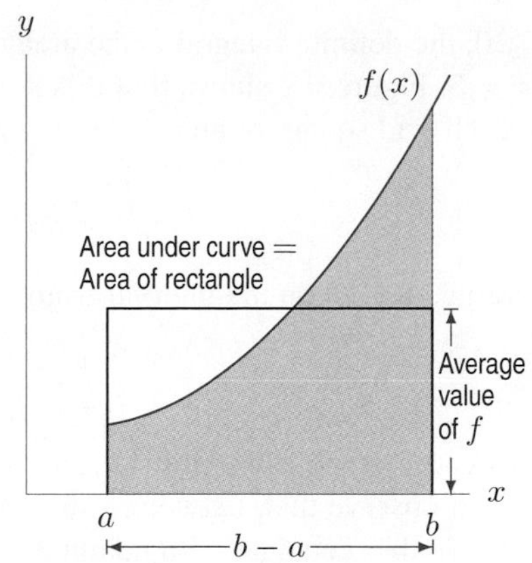

Figure 6.1: Area and average value

Example 1 Suppose that $C(t)$ represents the daily cost of heating your house, in dollars per day, where t is time in days and $t = 0$ corresponds to January 1, 2005. Interpret $\frac{1}{90-0}\int_0^{90} C(t)\,dt$.

Solution The units for the integral $\int_0^{90} C(t)\,dt$ are (dollars/day)×(days) = dollars. The integral represents the total cost in dollars to heat your house for the first 90 days of 2005, namely the months of January, February, and March. The expression $\frac{1}{90-0}\int_0^{90} C(t)\,dt$ represents the average cost per day to heat your house during the first 90 days of 2005. It is measured in (1/days) × (dollars) = dollars/day, the same units as $C(t)$.

Example 2 On page 33, we saw that the population of McAllen could be modeled by the function

$$P = f(t) = 570(1.037)^t,$$

where P is in thousands of people and t is in years since 2000. Use this function to predict the average population of McAllen between the years 2020 and 2040.

Solution We want the average value of $f(t)$ between $t = 20$ and $t = 40$. Using a calculator to evaluate the integral, we get

$$\text{Average population} = \frac{1}{40-20}\int_{20}^{40} f(t)\,dt = \frac{1}{20}(34{,}656.2) = 1732.81.$$

The average population of McAllen between 2020 and 2040 is predicted to be about 1733 thousand people.

Example 3 (a) For the function $f(x)$ graphed in Figure 6.2, evaluate $\int_0^5 f(x)\,dx$.

(b) Find the average value of $f(x)$ on the interval $x = 0$ to $x = 5$. Check your answer graphically.

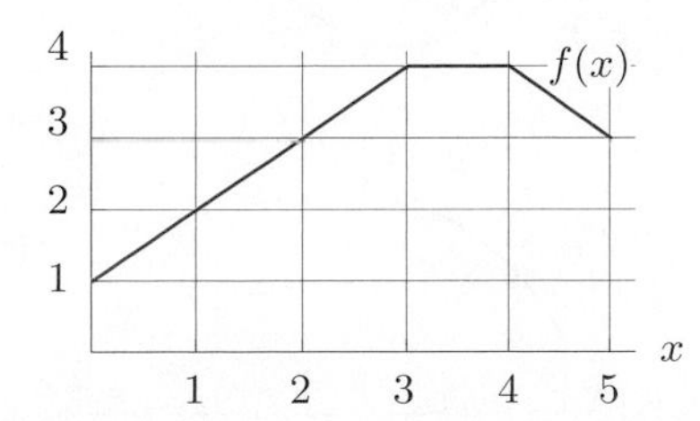

Figure 6.2: Estimate $\int_0^5 f(x)dx$

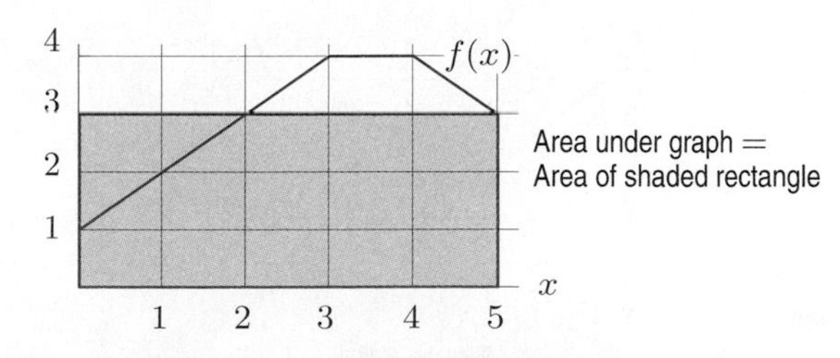

Figure 6.3: Average value of $f(x)$ is 3

Solution

(a) Since $f(x) \geq 0$, the definite integral is the area of the region under the graph of $f(x)$ between $x = 0$ and $x = 5$. Figure 6.2 shows that this region consists of 13 full grid squares and 4 half grid squares, each grid square of area 1, for a total area of 15, so

$$\int_0^5 f(x)\,dx = 15.$$

(b) The average value of $f(x)$ on the interval from 0 to 5 is given by

$$\text{Average value} = \frac{1}{5-0}\int_0^5 f(x)\,dx = \frac{1}{5}(15) = 3.$$

To check the answer graphically, draw a horizontal line at $y = 3$ on the graph of $f(x)$. (See Figure 6.3.) Then observe that, between $x = 0$ and $x = 5$, the area under the graph of $f(x)$ is equal to the area of the rectangle with height 3.

Problems for Section 6.1

1. Use Figure 6.4 to estimate the following:

(a) The integral $\int_0^5 f(x)\,dx$.
(b) The average value of f between $x = 0$ and $x = 5$ by estimating visually the average height.
(c) The average value of f between $x = 0$ and $x = 5$ by using your answer to part (a).

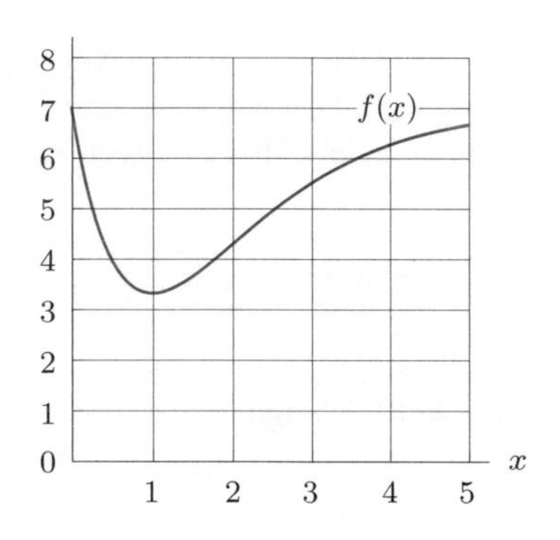

Figure 6.4

2. Find the average value of the function $f(x) = 5+4x-x^2$ between $x = 0$ and $x = 3$.

3. **(a)** Use Figure 6.5 to find $\int_0^6 f(x)\,dx$.
(b) What is the average value of f on the interval $x = 0$ to $x = 6$?

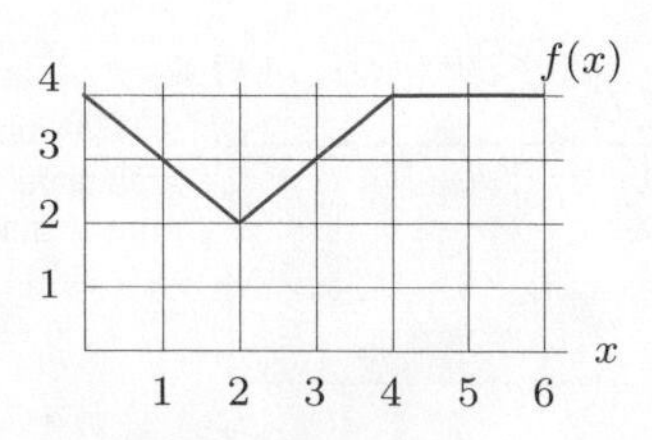

Figure 6.5

4. Find the average value of $g(t) = 1 + t$ over the interval $[0, 2]$

5. **(a)** Using Figures 6.6 and 6.7, find the average value on $0 \leq x \leq 2$ of

(i) $f(x)$ (ii) $g(x)$ (iii) $f(x)\cdot g(x)$

(b) Is the following statement true? Explain your answer.

$$\text{Average}(f)\cdot\text{Average}(g) = \text{Average}(f\cdot g)$$

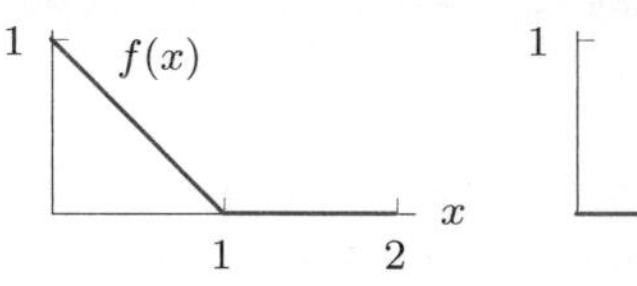

Figure 6.6

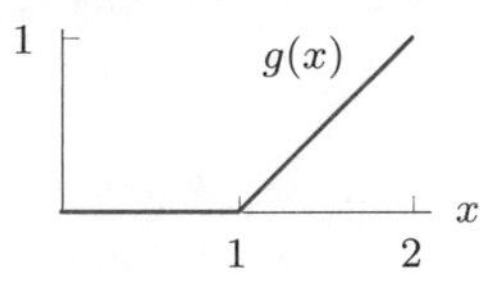

Figure 6.7

6. Find the average value of $g(t) = e^t$ over the interval $0 \leq t \leq 10$.

In Problems 7–8, estimate the average value of the function between $x = 0$ and $x = 7$.

7.

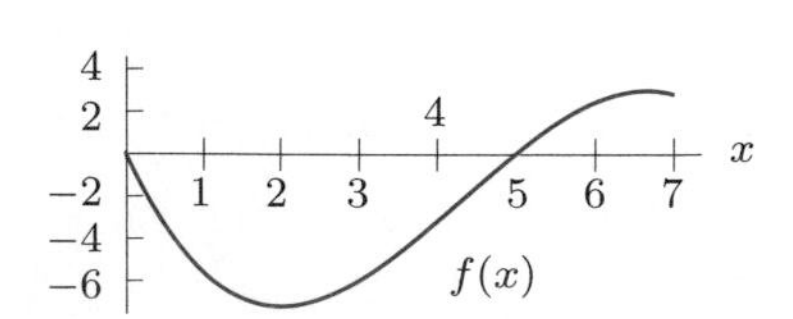

8.

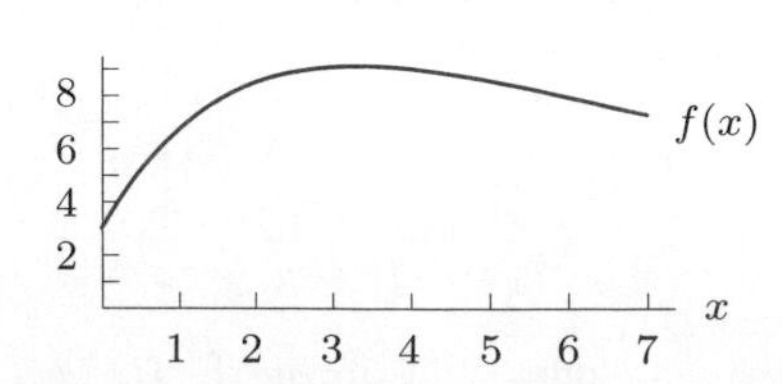

In Problems 9–10, estimate the average value of $f(x)$ from $x = a$ to $x = b$.

9.

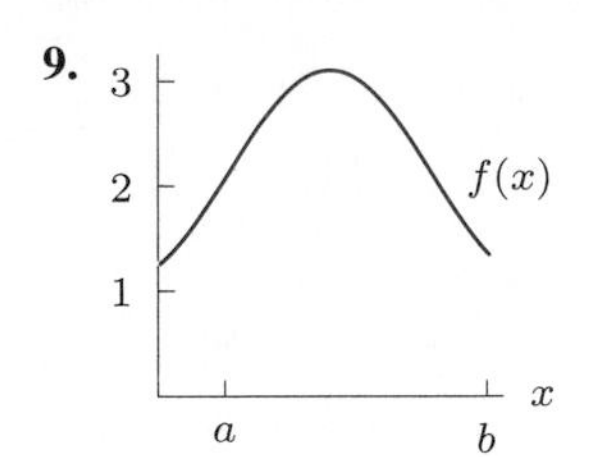

10.

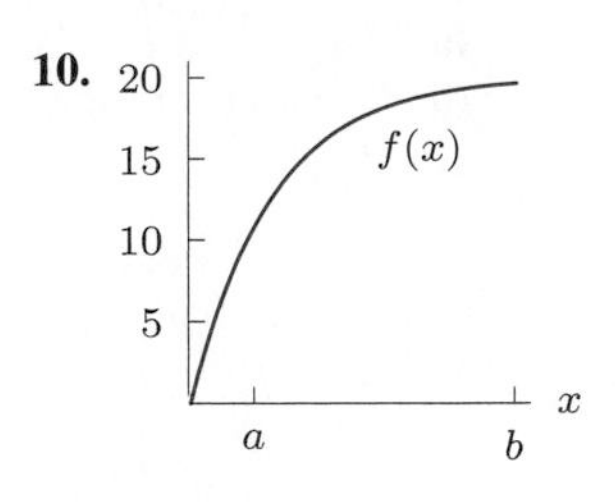

11. A service station orders 100 cases of motor oil every 6 months. The number of cases of oil remaining t months after the order arrives is modeled by

$$f(t) = 100e^{-0.5t}.$$

(a) How many cases are there at the start of the six-month period? How many cases are left at the end of the six-month period?
(b) Find the average number of cases in inventory over the six-month period.

12. If t is measured in days since June 1, the inventory $I(t)$ for an item in a warehouse is given by

$$I(t) = 5000(0.9)^t.$$

(a) Find the average inventory in the warehouse during the 90 days after June 1.
(b) Graph $I(t)$ and illustrate the average graphically.

Problems 13–14 refer to Figure 6.8, which shows human arterial blood pressure during the course of one heartbeat.

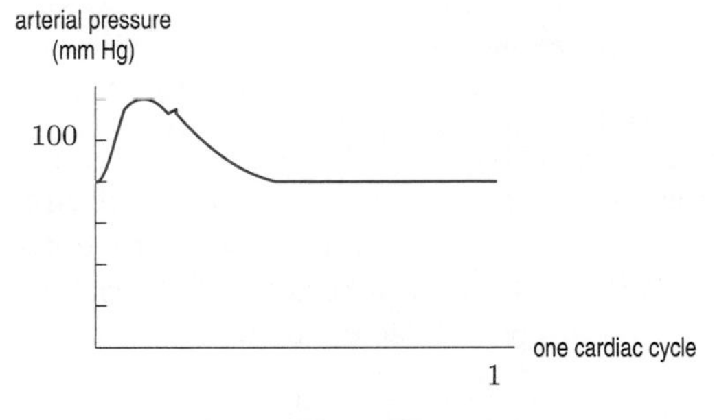

Figure 6.8

13. **(a)** Estimate the maximum blood pressure, called the systolic pressure.
(b) Estimate the minimum blood pressure, called the diastolic pressure.
(c) Calculate the average of the systolic and diastolic pressures.
(d) Is the average arterial pressure over the entire cycle greater than, less than, or equal to the answer for part (c)?

14. Estimate the average arterial blood pressure over one cardiac cycle.

15. Figure 6.9 shows the rate, $f(x)$, in thousands of algae per hour, at which a population of algae is growing, where x is in hours.

(a) Estimate the average value of the rate over the interval $x = -1$ to $x = 3$.
(b) Estimate the total change in the population over the interval $x = -3$ to $x = 3$.

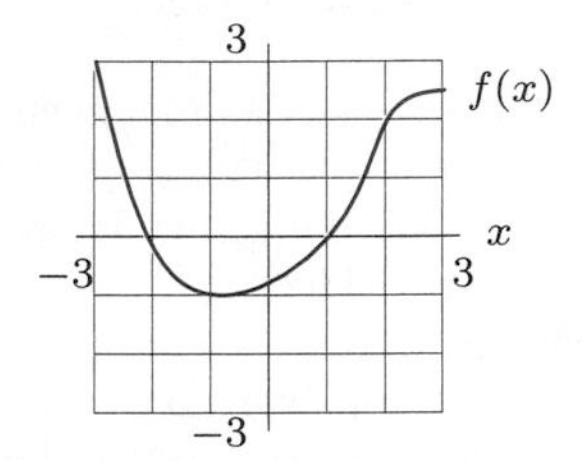

Figure 6.9

16. The population of the world t years after 2000 is predicted to be $P = 6.1e^{0.0125t}$ billion.

(a) What population is predicted in 2010?
(b) What is the predicted average population between 2000 and 2010?

17. The number of hours, H, of daylight in Madrid as a function of date is approximated by the formula

$$H = 12 + 2.4\sin[0.0172(t - 80)],$$

where t is the number of days since the start of the year. Find the average number of hours of daylight in Madrid:

(a) in January **(b)** in June **(c)** over a year

(d) Explain why the relative magnitudes of your answers to parts (a), (b), and (c) are reasonable.

18. A bar of metal is cooling from 1000°C to room temperature, 20°C. The temperature, H, of the bar t minutes after it starts cooling is given, in °C, by

$$H = 20 + 980e^{-0.1t}.$$

(a) Find the temperature of the bar at the end of one hour.
(b) Find the average value of the temperature over the first hour.
(c) Is your answer to part (b) greater or smaller than the average of the temperatures at the beginning and the end of the hour? Explain this in terms of the concavity of the graph of H.

19. For t in years since 2000, the population, P, of McAllen (in thousands), is given by

$$P = 570(1.037)^t.$$

(a) Find the average population of McAllen between 2000 and 2010.

(b) Find the average of the population of McAllen in 2000 and the population in 2010.

(c) Using the concavity of the graph of P (Figure 1.59 on page 34), explain why your answer to part (b) is larger or smaller than your answer to part (a).

20. The rate of sales (in sales per month) of a company is given, for t in months since January 1, by

$$r(t) = t^4 - 20t^3 + 118t^2 - 180t + 200.$$

(a) Graph the rate of sales per month during the first year ($t = 0$ to $t = 12$). Does it appear that more sales were made during the first half of the year, or during the second half?

(b) Estimate the total sales during the first 6 months of the year and during the last 6 months of the year.

(c) What are the total sales for the entire year?

(d) Find the average sales per month during the year.

21. Throughout much of the 20th century, the yearly consumption of electricity in the US increased exponentially at a continuous rate of 7% per year. Assume this trend continues and that the electrical energy consumed in 1900 was 1.4 million megawatt-hours.

(a) Write an expression for yearly electricity consumption as a function of time, t, in years since 1900.

(b) Find the average yearly electrical consumption throughout the 20th century.

(c) During what year was electrical consumption closest to the average for the century?

(d) Without doing the calculation for part (c), how could you have predicted which half of the century the answer would be in?

22. Using Figure 6.10, list the following numbers from least to greatest:

(a) $f'(1)$

(b) The average value of f on $0 \le x \le 4$

(c) $\int_0^1 f(x)dx$

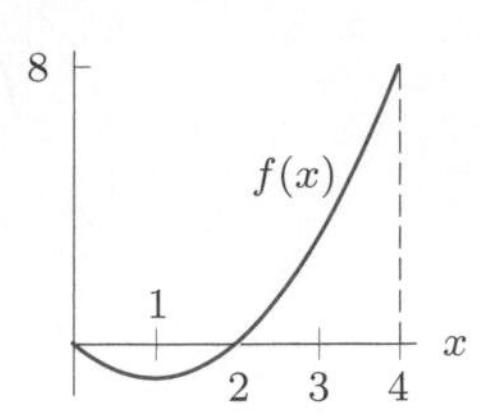

Figure 6.10

23. Using Figure 6.11, list from least to greatest,

(a) $f'(1)$.

(b) The average value of $f(x)$ on $0 \le x \le a$.

(c) The average value of the rate of change of $f(x)$, for $0 \le x \le a$.

(d) $\int_0^a f(x)\,dx$.

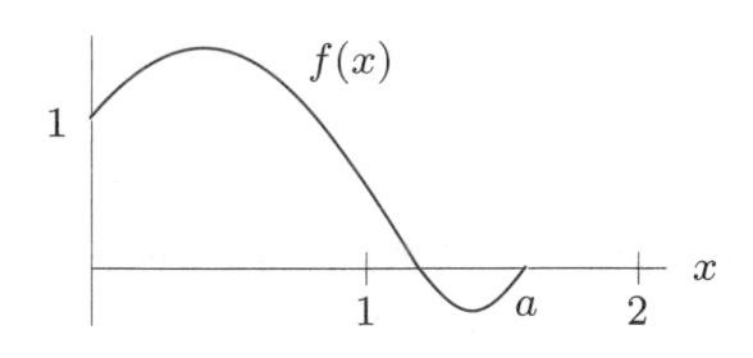

Figure 6.11

6.2 CONSUMER AND PRODUCER SURPLUS

Supply and Demand Curves

As we saw in Chapter 1, the quantity of a certain item produced and sold can be described by the supply and demand curves of the item. The *supply curve* shows what quantity, q, of the item the producers supply at different prices, p . The consumers' behavior is reflected in the *demand curve*, which shows what quantity of goods are bought at various prices. See Figure 6.12.

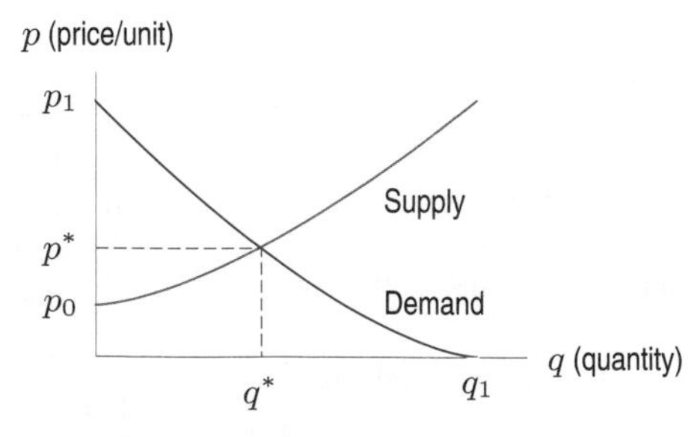

Figure 6.12: Supply and demand curves

It is assumed that the market settles at the *equilibrium price* p^* and *equilibrium quantity* q^* where the graphs cross. At equilibrium, a quantity q^* of an item is produced and sold for a price of p^* each.

Consumer and Producer Surplus

Notice that at equilibrium, a number of consumers have bought the item at a lower price than they would have been willing to pay. (For example, there are some consumers who would have been willing to pay prices up to p_1.) Similarly, there are some suppliers who would have been willing to produce the item at a lower price (down to p_0, in fact). We define the following terms:

- The **consumer surplus** measures the consumers' gain from trade. It is the total amount gained by consumers by buying the item at the current price rather than at the price they would have been willing to pay.
- The **producer surplus** measures the suppliers' gain from trade. It is the total amount gained by producers by selling at the current price, rather than at the price they would have been willing to accept.

In the absence of price controls, the current price is assumed to be the equilibrium price.

Both consumers and producers are richer for having traded. The consumer and producer surplus measure how much richer they are.

Suppose that all consumers buy the good at the maximum price they are willing to pay. Subdivide the interval from 0 to q^* into intervals of length Δq. Figure 6.13 shows that a quantity Δq of items are sold at a price of about p_1, another Δq are sold for a slightly lower price of about p_2, the next Δq for a price of about p_3, and so on. Thus, the consumers' total expenditure is about

$$p_1\Delta q + p_2\Delta q + p_3\Delta q + \cdots = \sum p_i \Delta q.$$

If the demand curve has equation[1] $p = f(q)$, and if all consumers who were willing to pay more than p^* paid as much as they were willing, then as $\Delta q \to 0$, we would have

$$\begin{array}{c}\text{Consumer}\\\text{expenditure}\end{array} = \int_0^{q^*} f(q)\,dq = \begin{array}{c}\text{Area under demand}\\\text{curve from 0 to } q^*.\end{array}$$

If all goods are sold at the equilibrium price, the consumers' actual expenditure is only p^*q^*, which is the area of the rectangle between the q-axis and the line $p = p^*$ from $q = 0$ to $q = q^*$. The consumer surplus is the difference between the total consumer expenditure if all consumers pay the maximum they are willing to pay and the actual consumer expenditure if all consumers pay the current price. The consumer surplus is represented by the area in Figure 6.14. Similarly, the producer surplus is represented by the area in Figure 6.15. (See Problems 13 and 14.) Thus:

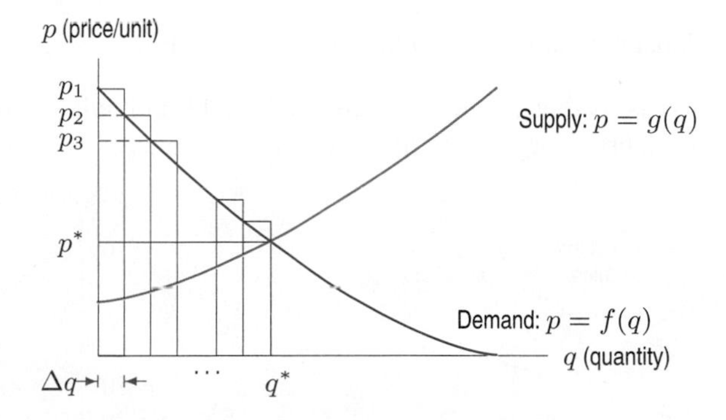

Figure 6.13: Calculation of consumer surplus

[1]Note that here p is written as a function of q.

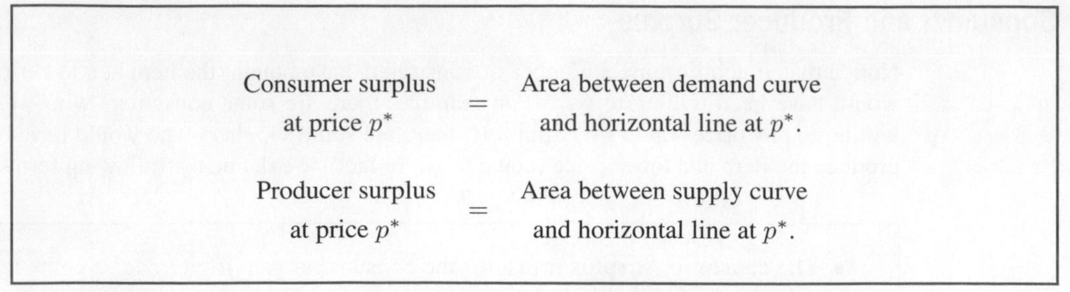

Consumer surplus at price p^*	$=$	Area between demand curve and horizontal line at p^*.
Producer surplus at price p^*	$=$	Area between supply curve and horizontal line at p^*.

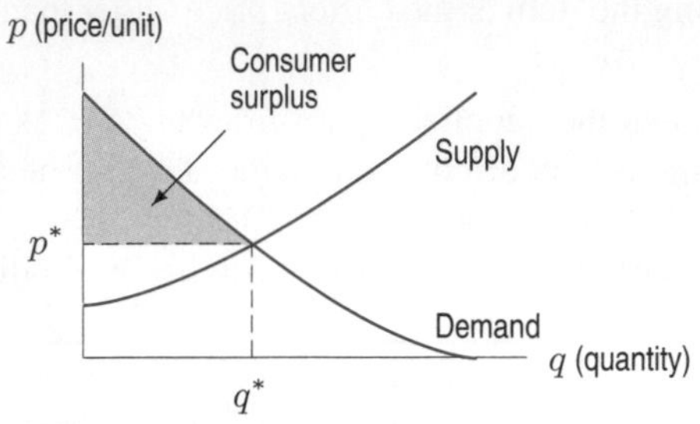

Figure 6.14: Consumer surplus

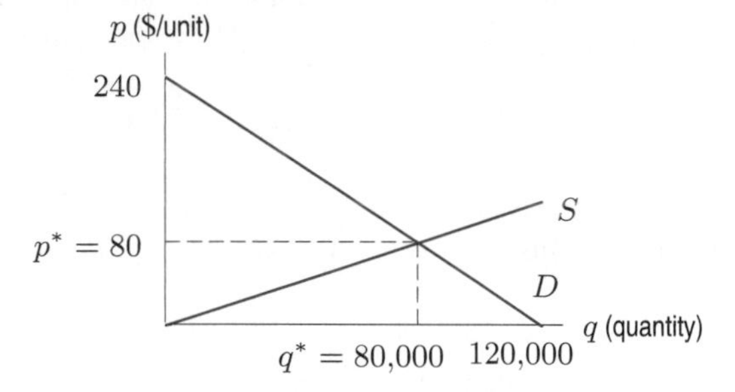

Figure 6.15: Producer surplus

Example 1 The supply and demand curves for a product are given in Figure 6.16.

Figure 6.16: Supply and demand curves for a product

(a) What are the equilibrium price and quantity?
(b) At the equilibrium price, calculate and interpret the consumer and producer surplus.

Solution (a) The equilibrium price is $p^* = \$80$ and the equilibrium quantity is $q^* = 80{,}000$ units.
(b) The consumer surplus is the area under the demand curve and above the line $p = 80$. (See Figure 6.17.) We have

$$\text{Consumer surplus} = \text{Area of triangle} = \frac{1}{2}\text{Base} \cdot \text{Height} = \frac{1}{2}80{,}000 \cdot 160 = \$6{,}400{,}000.$$

This tells us that consumers gain $6,400,000 in buying goods at the equilibrium price instead of at the price they would have been willing to pay.

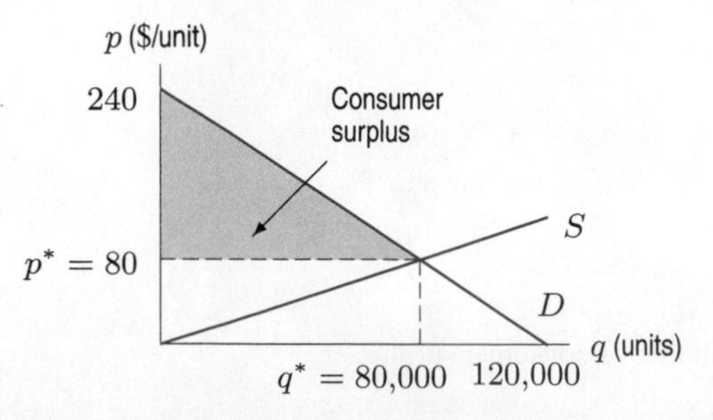

Figure 6.17: Consumer surplus

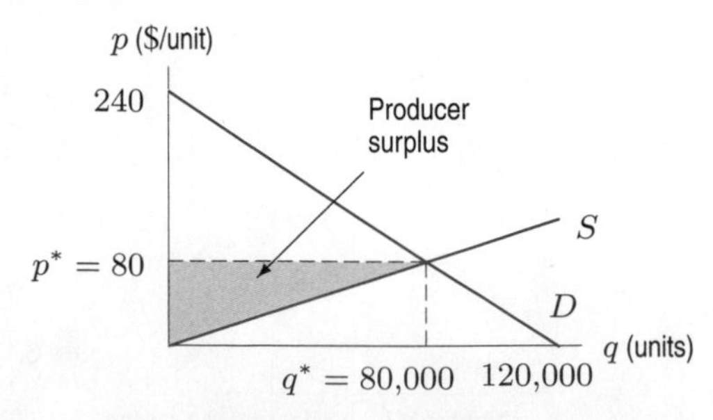

Figure 6.18: Producer surplus

The producer surplus is the area above the supply curve and below the line $p = 80$. (See Figure 6.18.) We have

$$\text{Producer surplus} = \text{Area of triangle} = \frac{1}{2}\text{Base} \cdot \text{Height} = \frac{1}{2} \cdot 80{,}000 \cdot 80 = \$3{,}200{,}000.$$

So, producers gain $3,200,000 by supplying goods at the equilibrium price instead of the price at which they would have been willing to provide the goods.

Wage and Price Controls

In a free market, the price of a product generally moves to the equilibrium price, unless outside forces keep the price artificially high or artificially low. Rent control, for example, keeps prices below market value, whereas cartel pricing or the minimum wage law raise prices above market value. What happens to consumer and producer surplus at non-equilibrium prices?

Example 2 The dairy industry has cartel pricing: the government has set milk prices artificially high. What effect does raising the price to p^+ from the equilibrium price have on:

(a) Consumer surplus?
(b) Producer surplus?
(c) Total gains from trade (that is, Consumer surplus + Producer surplus)?

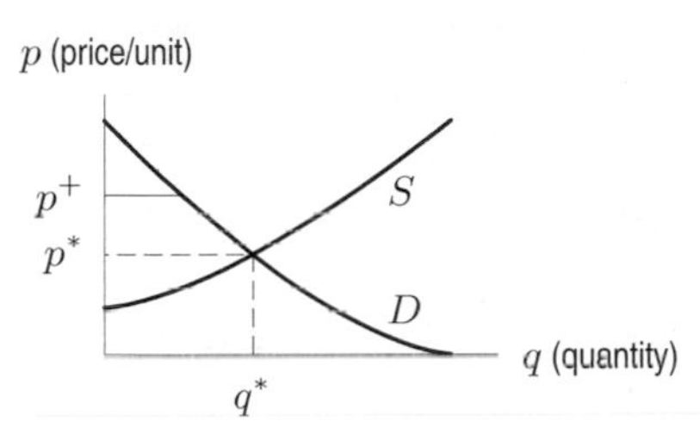

Figure 6.19: What is the effect of the artificially high price, p^+, on consumer and producer surplus? (q^* and p^* are equilibrium values)

Solution

(a) A graph of possible supply and demand curves for the milk industry is given in Figure 6.19. Suppose that the price is fixed at p^+, above the equilibrium price. Consumer surplus is the difference between the amount the consumers paid (p^+) and the amount they would have been willing to pay (given on the demand curve). This is the area shaded in Figure 6.20. This consumer surplus is less than the consumer surplus at the equilibrium price, shown in Figure 6.21.

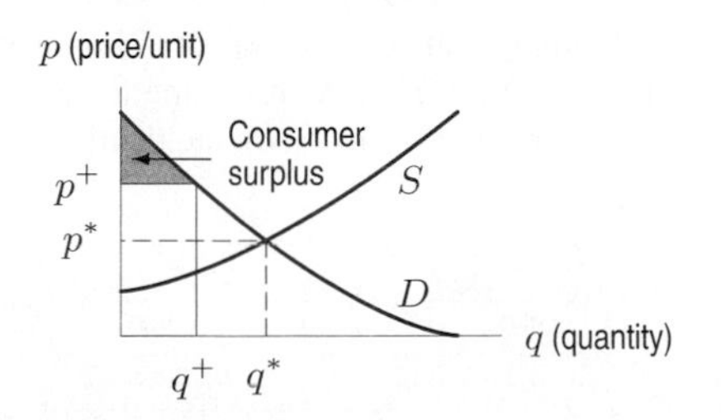

Figure 6.20: Consumer surplus: Artificial price

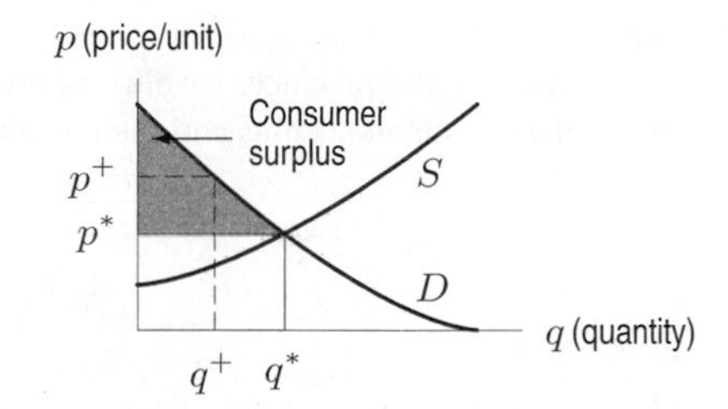

Figure 6.21: Consumer surplus: Equilibrium price

(b) At a price of p^+, the quantity sold, q^+, is less than it would have been at the equilibrium price. The producer surplus is represented by the area between p^+ and the supply curve at this reduced demand. This area is shaded in Figure 6.22. Compare this producer surplus (at the artificially high price) to the producer surplus in Figure 6.23 (at the equilibrium price). In this case, producer surplus appears to be greater at the artificial price than at the equilibrium price. (However, different supply and demand curves might lead to a different answer.)

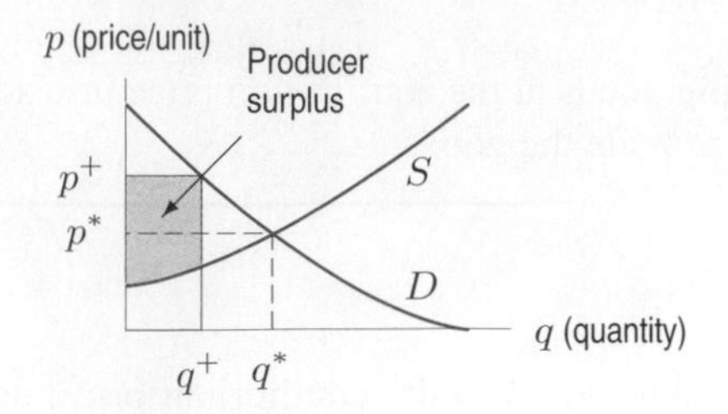

Figure 6.22: Producer surplus: Artificial price

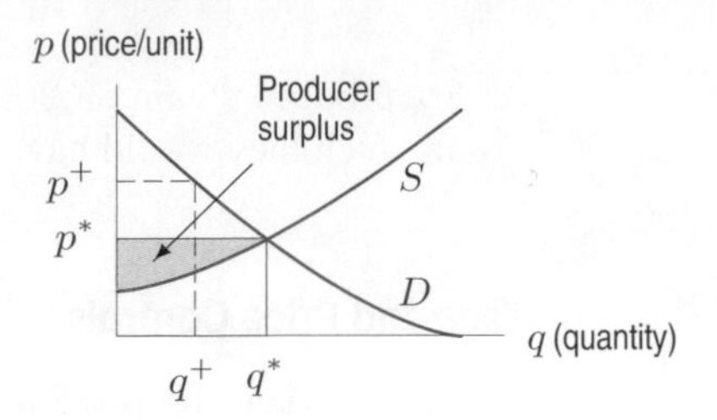

Figure 6.23: Producer surplus: Equilibrium price

(c) The total gains from trade (Consumer surplus + Producer surplus) at the price of p^+ is represented by the area shaded in Figure 6.24. The total gains from trade at the equilibrium price of p^* is represented by the area shaded in Figure 6.25. Under artificial price conditions, the total gains from trade decrease. The total financial effect of the artificially high price on all producers and consumers combined is negative.

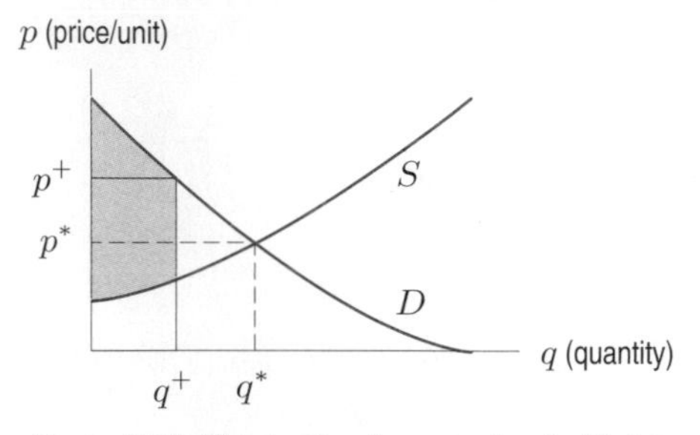

Figure 6.24: Total gains from trade: Artificial price

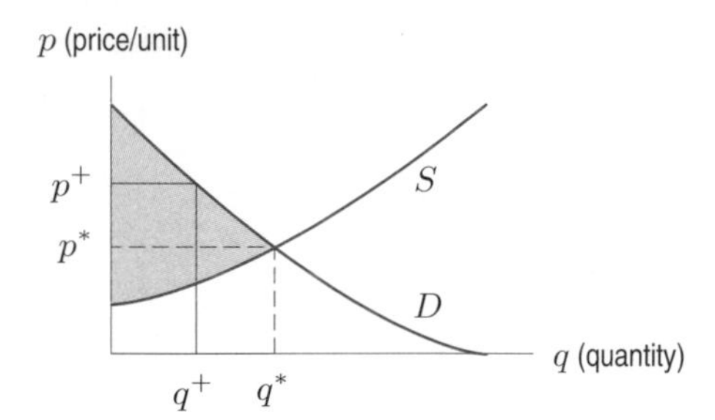

Figure 6.25: Total gains from trade: Equilibrium price

Problems for Section 6.2

1. The supply and demand curves for a product are given in Figure 6.26. Estimate the equilibrium price and quantity and the consumer and producer surplus. Shade areas representing the consumer surplus and the producer surplus.

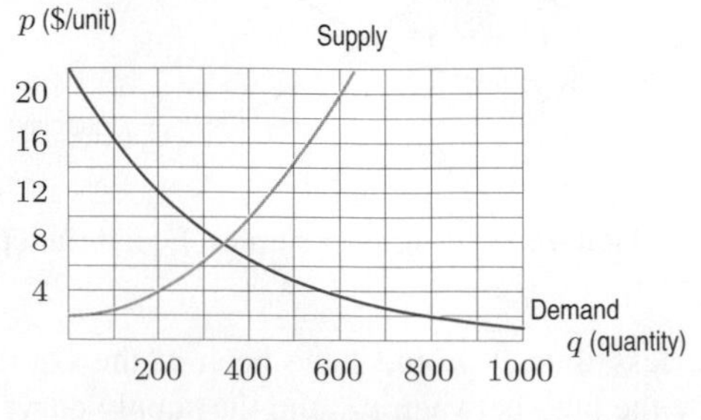

Figure 6.26

2. **(a)** What are the equilibrium price and quantity for the supply and demand curves in Figure 6.27?

(b) Shade the areas representing the consumer and producer surplus and estimate them.

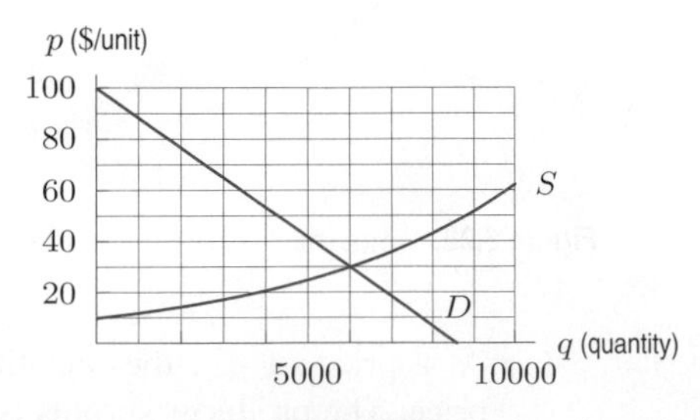

Figure 6.27

3. Given the demand curve $p = 35 - q^2$ and the supply curve $p = 3 + q^2$, find the producer surplus when the market is in equilibrium.

4. Find the consumer surplus for the demand curve $p = 100 - 3q^2$ when 5 units are sold.

5. Find the consumer surplus for the demand curve $p = 100 - 4q$ when $q = 10$.

6. Supply and demand curves for a product are in Figure 6.28.

(a) Estimate the equilibrium price and quantity.
(b) Estimate the consumer and producer surplus. Shade them.
(c) What are the total gains from trade for this product?

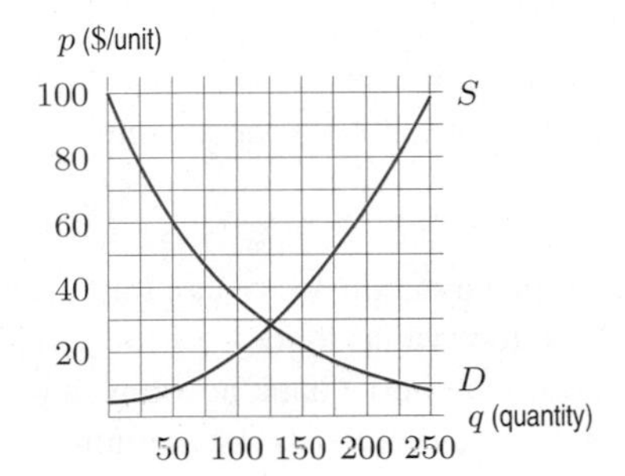

Figure 6.28

7. Supply and demand curves are in Figure 6.28. A price of $40 is artificially imposed.

(a) At the $40 price, estimate the consumer surplus, the producer surplus, and the total gains from trade.
(b) Compare your answers in this problem to your answers in Problem 6. Discuss the effect of price controls on the consumer surplus, producer surplus, and gains from trade in this case.

8. **(a)** Estimate the equilibrium price and quantity for the supply and demand curves in Figure 6.29.
(b) Estimate the consumer and producer surplus.
(c) The price is set artificially low at $p^- = 4$ dollars per unit. Estimate the consumer and producer surplus at this price. Compare your answers to the consumer and producer surplus at the equilibrium price.

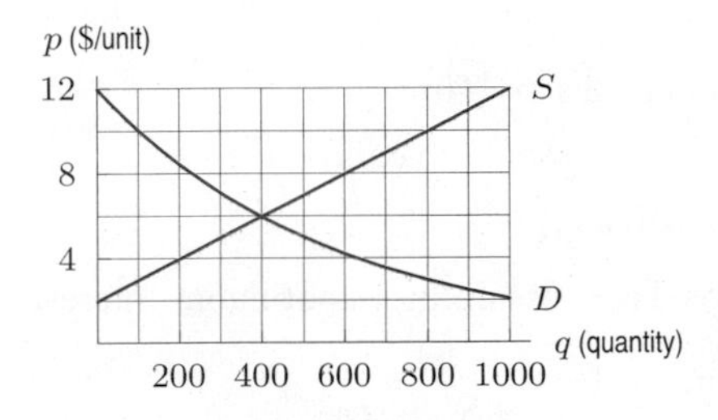

Figure 6.29

9. For a product, the supply curve is $p = 5 + 0.02q$ and the demand curve is $p = 30e^{-0.003q}$, where p is the price and q is the quantity sold at that price. Find:

(a) The equilibrium price and quantity
(b) The consumer and producer surplus

10. Sketch possible supply and demand curves where the consumer surplus at the equilibrium price is

(a) Greater than the producer surplus.
(b) Less than the producer surplus.

11. Rent controls on apartments are an example of price controls on a commodity. They keep the price artificially low (below the equilibrium price). Sketch a graph of supply and demand curves, and label on it a price p^- below the equilibrium price. What effect does forcing the price down to p^- have on:

(a) The producer surplus?
(b) The consumer surplus?
(c) The total gains from trade (Consumer surplus + Producer surplus)?

12. Supply and demand data are in Tables 6.1 and 6.2.

(a) Which table shows supply and which shows demand?
(b) Estimate the equilibrium price and quantity.
(c) Estimate the consumer and producer surplus.

Table 6.1

q (quantity)	0	100	200	300	400	500	600
p ($/unit)	60	50	41	32	25	20	17

Table 6.2

q (quantity)	0	100	200	300	400	500	600
p ($/unit)	10	14	18	22	25	28	34

In Problems 13–15, the supply and demand curves have equations $p = S(q)$ and $p = D(q)$, respectively, with equilibrium at (q^*, p^*).

13. Using Riemann sums, explain the economic significance of $\int_0^{q^*} S(q)\,dq$ to the producers.

14. Using Riemann sums, give an interpretation of producer surplus, $\int_0^{q^*} (p^* - S(q))\,dq$ analogous to the interpretation of consumer surplus.

15. Referring to Figures 6.14 and 6.15 on page 282, mark the regions representing the following quantities and explain their economic meaning:

(a) p^*q^*

(b) $\displaystyle\int_0^{q^*} D(q)\,dq$

(c) $\displaystyle\int_0^{q^*} S(q)\,dq$

(d) $\displaystyle\int_0^{q^*} D(q)\,dq - p^*q^*$

(e) $\displaystyle p^*q^* - \int_0^{q^*} S(q)\,dq$

(f) $\displaystyle\int_0^{q^*} (D(q) - S(q))\,dq$

6.3 PRESENT AND FUTURE VALUE

In Chapter 1 on page 48, we introduced the present and future value of a single payment. In this section we see how to calculate the present and future value of a continuous stream of payments.

Income Stream

When we consider payments made to or by an individual, we usually think of *discrete* payments, that is, payments made at specific moments in time. However, we may think of payments made by a company as being *continuous*. The revenues earned by a huge corporation, for example, come in essentially all the time, and therefore they can be represented by a continuous *income stream*. Since the rate at which revenue is earned may vary from time to time, the income stream is described by

$$S(t) \text{ dollars/year.}$$

Notice that $S(t)$ is a *rate* at which payments are made (its units are dollars per year, for example) and that the rate depends on the time, t, usually measured in years from the present.

Present and Future Values of an Income Stream

Just as we can find the present and future values of a single payment, so we can find the present and future values of a stream of payments. As before, the future value represents the total amount of money that you would have if you deposited an income stream into a bank account as you receive it and let it earn interest until that future date. The present value represents the amount of money you would have to deposit today (in an interest-bearing bank account) in order to match what you would get from the income stream by that future date.

When we are working with a continuous income stream, we will assume that interest is compounded continuously. If the interest rate is r, the present value, P, of a deposit, B, made t years in the future is

$$P = Be^{-rt}.$$

Suppose that we want to calculate the present value of the income stream described by a rate of $S(t)$ dollars per year, and that we are interested in the period from now until M years in the future. In order to use what we know about single deposits to calculate the present value of an income stream, we divide the stream into many small deposits, and imagine each deposited at one instant. Dividing the interval $0 \leq t \leq M$ into subintervals of length Δt:

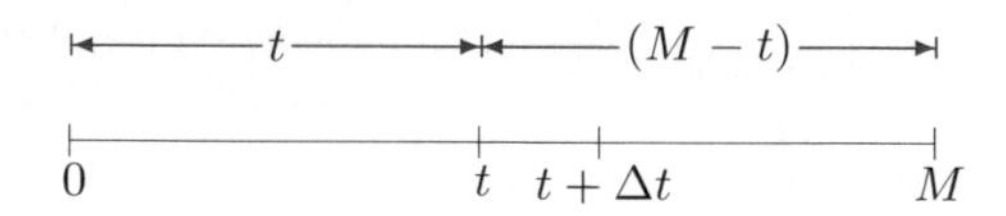

Assuming Δt is small, the rate, $S(t)$, at which deposits are being made does not vary much within one subinterval. Thus, between t and $t + \Delta t$:

$$\begin{aligned} \text{Amount paid} &\approx \text{Rate of deposits} \times \text{Time} \\ &\approx (S(t) \text{ dollars/year})(\Delta t \text{ years}) \\ &= S(t)\Delta t \text{ dollars.} \end{aligned}$$

The deposit of $S(t)\Delta t$ is made t years in the future. Thus, assuming a continuous interest rate r,

$$\begin{array}{c}\text{Present value of money} \\ \text{deposited in interval } t \text{ to } t + \Delta t\end{array} \approx S(t)\Delta t e^{-rt}.$$

Summing over all subintervals gives

$$\text{Total present value} \approx \sum S(t)e^{-rt}\Delta t \text{ dollars.}$$

In the limit as $\Delta t \to 0$, we get the following integral:

$$\text{Present value} = \int_0^M S(t)e^{-rt}dt.$$

As in Section 1.7, the value M years in the future is given by

$$\text{Future value} = \text{Present value} \cdot e^{rM}.$$

Example 1 Find the present and future values of a constant income stream of \$1000 per year over a period of 20 years, assuming an interest rate of 6% compounded continuously.

Solution Using $S(t) = 1000$ and $r = 0.06$, we have

$$\text{Present value} = \int_0^{20} 1000e^{-0.06t}dt = \$11{,}647.$$

We can get the future value, B, from the present value, P, using $B = Pe^{rt}$, so

$$\text{Future value} = 11{,}647e^{0.06(20)} = \$38{,}669.$$

Notice that since money was deposited at a rate of \$1000 a year for 20 years, the total amount deposited was \$20,000. The future value is \$38,669, so the money has almost doubled because of the interest.

Example 2 Suppose you want to have \$50,000 in 8 years time in a bank account earning 2% interest, compounded continuously.

(a) If you make one lump sum deposit now, how much should you deposit?
(b) If you deposit money continuously throughout the 8-year period, at what rate should you deposit it?

Solution

(a) If you deposit a lump sum of $\$P$, then $\$P$ is the present value of \$50,000. So, using $B = Pe^{rt}$ with $B = 50{,}000$ and $r = 0.02$ and $t = 8$:

$$50000 = Pe^{0.02(8)}$$
$$P = \frac{50000}{e^{0.02(8)}} = 42{,}607.$$

If you deposit \$42,607 into the account now, you will have \$50,000 in 8 years time.

(b) Suppose you deposit money at a constant rate of $\$S$ per year. Then

$$\text{Present value of deposits} = \int_0^8 Se^{-0.02t}\,dt$$

Since S is a constant, we can take it out in front of the integral sign:

$$\text{Present value} = S\int_0^8 e^{-0.02t}\,dt \approx S(7.3928).$$

But the present value of the continuous deposit must be the same as the present value of the lump sum deposit; that is, \$42,607. So

$$42{,}607 \approx S(7.3928)$$
$$S \approx \$5763.$$

To meet your goal of \$50,000, you need to deposit money at a continuous rate of \$5,763 per year, or about \$480 per month.

Problems for Section 6.3

1. Calculate the present value of a continuous revenue stream of $1000 per year for 5 years at an interest rate of 9% per year compounded continuously.

2. Draw a graph, with time in years on the horizontal axis, of what an income stream might look like for a company that sells sunscreen in the northeast United States.

3. Find the present and future values of an income stream of $12,000 a year for 20 years. The interest rate is 6%, compounded continuously.

4. **(a)** Find the present and future value of an income stream of $6000 per year for a period of 10 years if the interest rate, compounded continuously, is 5%.
(b) How much of the future value is from the income stream? How much is from interest?

5. A small business expects an income stream of $5000 per year for a four-year period.

(a) Find the present value of the business if the annual interest rate, compounded continuously, is

(i) 3% (ii) 10%

(b) In each case, find the value of the business at the end of the four-year period.

6. A bond is guaranteed to pay $100 + 10t$ dollars per year for 10 years, where t is in years from the present. Find the present value of this income stream, given an interest rate of 5%, compounded continuously.

7. Determine the constant income stream that needs to be invested over a period of 10 years at an interest rate of 5% per year compounded continuously to provide a present value of $5000.

8. **(a)** A bank account earns 10% interest compounded continuously. At what constant, continuous rate must a parent deposit money into such an account in order to save $100,000 in 10 years for a child's college expenses?
(b) If the parent decides instead to deposit a lump sum now in order to attain the goal of $100,000 in 10 years, how much must be deposited now?

9. A company is expected to earn $50,000 a year, at a continuous rate, for 8 years. You can invest the earnings at an interest rate of 7%, compounded continuously. You have the chance to buy the rights to the earnings of the company now for $350,000. Should you buy? Explain.

10. Your company needs $500,000 in two years time for renovations and can earn 9% interest on investments.

(a) What is the present value of the renovations?
(b) If your company deposits money continuously at a constant rate throughout the two-year period, at what rate should the money be deposited so that you have the $500,000 when you need it?

11. On April 15, 1999, Maria Grasso won the largest lottery amount awarded up to that date. She was given her choice between $197 million, paid out continuously over 26 years, or a lump sum of $104 million, paid immediately.

(a) Which option is better if the interest rate is 6%, compounded continuously? An interest rate of 5%?
(b) The winner chose the lump sum option. What assumption was she making about interest rates?

12. Intel Corporation is a leading manufacturer of integrated circuits. In 2004, Intel generated profits at a continuous rate of 7.5 billion dollars per year.[2] Assume profits continue at the same rate and that the interest rate is 8.5% per year compounded continuously.

(a) What is the present value of Intel's profits over a one-year time period?
(b) What is the value at the end of one year of Intel's profits over a one-year time period?

13. Sales of Version 6.0 of a computer software package start out high and decrease exponentially. At time t, in years, the sales are $s(t) = 50e^{-t}$ thousands of dollars per year. After two years, Version 7.0 of the software is released and replaces Version 6.0. You can invest earnings at an interest rate of 6%, compounded continuously. Calculate the total present value of sales of Version 6.0 over the two year period.

14. Hershey Foods Inc. is the largest US producer of chocolate. During 2002 and 2003, Hershey generated a net profit at a rate approximated by $54.0t + 403.58$ million dollars per year, where t is the time in years since January 1, 2002.[3] Assume this rate continues through the year 2007 and that the interest rate is 2% per year compounded continuously. Find the value, on January 1, 2007, of Hershey's net profit from January 1, 2002 to January 1, 2007.

15. McDonald's Corporation licenses and operates a chain of 31,561 fast-food restaurants throughout the world. Between 2000 and 2004, McDonald's has been generating revenue at continuous rates between 10,467 and 14,224 million dollars per year.[4] Suppose that McDonald's rate of revenue stays within this range. Use an interest rate of 9% per year compounded continuously. Fill in the blanks:

(a) The present value of McDonald's revenue over a five year time period is between ____________ and ____________ million dollars.
(b) The present value of McDonald's revenue over a twenty-five year time period is between ____________ and ____________ million dollars.

[2] www.intel.com, accessed May 28, 2005.
[3] http://media.corporate-ir.net/media_files/NYS/HSY/reports/HSY_MDA-2003.pdf, accessed May 15, 2005.
[4] http://64.26.27.40/interactive/mcd2004financialreport/md/page_01.php, accessed May 15, 2005.

16. Your company is considering buying new production machinery. You want to know how long it will take for the machinery to pay for itself; that is, you want to find the length of time over which the present value of the profit generated by the new machinery equals the cost of the machinery. The new machinery costs \$130,000 and earns profit at the continuous rate of \$80,000 per year. Use an interest rate of 8.5% per year compounded continuously.

17. An oil company discovered an oil reserve of 100 million barrels. For time $t > 0$, in years, the company's extraction plan is a linear declining function of time as follows:

$$q(t) = a - bt,$$

where $q(t)$ is the rate of extraction of oil in millions of barrels per year at time t and $b = 0.1$ and $a = 10$.

(a) How long does it take to exhaust the entire reserve?

(b) The oil price is a constant \$20 per barrel, the extraction cost per barrel is a constant \$10, and the market interest rate is 10% per year, compounded continuously. What is the present value of the company's profit?

18. In 1980 West Germany made a loan of 20 billion Deutsche Marks to the Soviet Union, to be used for the construction of a natural gas pipeline connecting Siberia to Western Russia, and continuing to West Germany (Urengoi–Uschgorod–Berlin). Assume that the deal was as follows: In 1985, upon completion of the pipeline, the Soviet Union would deliver natural gas to West Germany, at a constant rate, for all future times. Assuming a constant price of natural gas of 0.10 Deutsche Mark per cubic meter, and assuming West Germany expects 10% annual interest on its investment (compounded continuously), at what rate does the Soviet Union have to deliver the gas, in billions of cubic meters per year? Keep in mind that delivery of gas could not begin until the pipeline was completed. Thus, West Germany received no return on its investment until after five years had passed. (Note: A more complex deal of this type was actually made between the two countries.)

6.4 INTEGRATING RELATIVE GROWTH RATES

Population Growth Rates

In Chapter 5, we saw how to calculate change in a population, P, from its derivative, dP/dt, using the Fundamental Theorem of Calculus.

However, population growth rates are often not given as the derivative dP/dt. For example, we might learn that in 2004, the population, P, of Nicaragua was growing at a rate of 1.6% per year. The 1.6% per year is the *relative rate of change* (or relative growth rate) of the population, but it is not the derivative. The derivative is sometimes called the *absolute rate of change*. The relative rate of change is the absolute rate change divided by the population, so we have the following definition:

> Suppose P is a function of t,
>
> $$\text{(Absolute) rate of change of } P \text{ with respect to } t = \frac{dP}{dt} \quad \text{and} \quad \text{Relative rate of change of } P \text{ with respect to } t = \frac{1}{P}\cdot\frac{dP}{dt}$$

Notice that the absolute and relative rates of change measure different quantities. If P is population and t is years, the (absolute) rate of change is a change in population per year, whereas the relative rate of change is a percent change per year. For a quantity growing linearly, the absolute rate of change is constant. For a quantity growing exponentially, the relative rate of change is constant.

We cannot find the change in a population from the relative growth rate without more information, but we can find the percentage change in the population.

Since

$$\frac{d}{dt}(\ln P(t)) = \frac{P'(t)}{P(t)} = \text{Relative growth rate},$$

the relative growth rate of P is the rate of change of $\ln(P)$. By the Fundamental Theorem of Calculus, the integral of the relative growth rate gives the total change in $\ln(P)$:

$$\int_a^b \frac{P'(t)}{P(t)}dt = \ln(P(b)) - \ln(P(a)) = \ln\left(\frac{P(b)}{P(a)}\right).$$

Example 1 The relative rate of growth $P'(t)/P(t)$ of a population $P(t)$ over a 50-year period is given in Figure 6.30. By what factor did the population increase during the period?

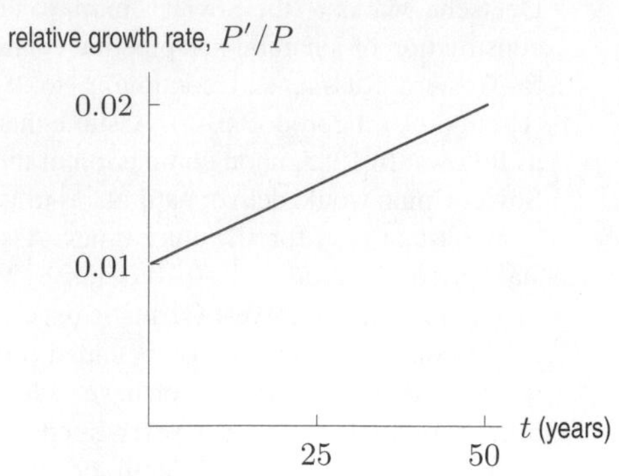

Figure 6.30: The relative growth rate of population

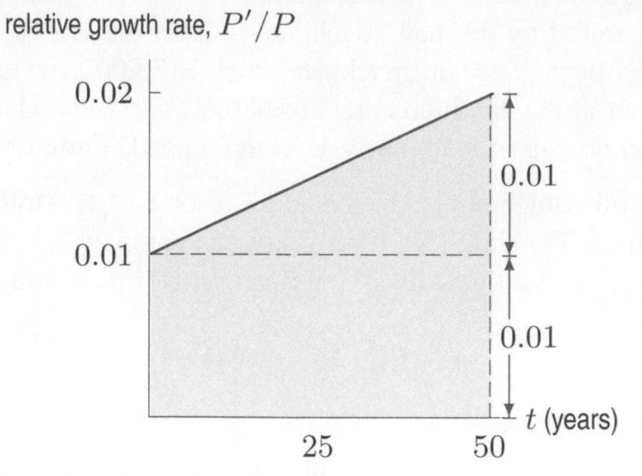

Figure 6.31: Calculating the percentage change in a population from the relative growth rate

Solution We have

$$\ln\left(\frac{P(50)}{P(0)}\right) = \int_0^{50} \frac{P'(t)}{P(t)}\,dt.$$

This integral equals the area under the graph of $P'(t)/P(t)$ between $t = 0$ and $t = 50$. See Figure 6.31. The area of a rectangle and triangle gives

$$\text{Area} = 50(0.01) + \frac{1}{2} \cdot 50(0.01) = 0.75.$$

Thus,

$$\begin{aligned}
\ln\left(\frac{P(50)}{P(0)}\right) &= 0.75 \\
\frac{P(50)}{P(0)} &= e^{0.75} = 2.1 \\
P(50) &= 2.1P(0).
\end{aligned}$$

The population more than doubled during the 50 years, increasing by a factor of about 2.1. We cannot determine the amount by which the population increased unless we know how large the population was initially.

If everything else remains constant, the relative growth rate increases if either the relative birth rate increases or the relative death rate decreases. Even if the relative birth rate decreases, we can still see an increase in the relative growth rate if the relative death rate decreases faster. This is the case with the population of the world today. The difference between the birth rate and the death rate is an important variable.

Example 2 Figure 6.32 shows relative birth rates and relative death rates for developed and developing countries.[5]

(a) Which is changing faster, the birth rate or the death rate? What does this tell you about how the populations are changing?

[5] *Food and Population: A Global Concern*, Elaine Murphy, Washington, DC: Population Reference Bureau, Inc., 1984, p. 2.

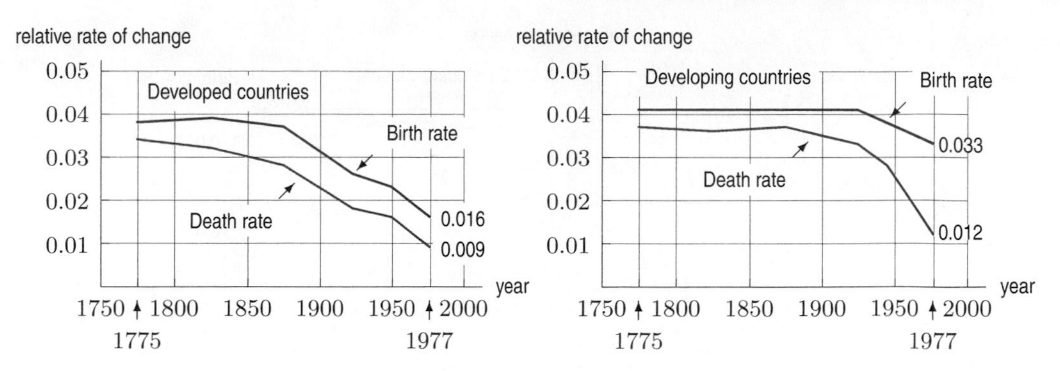

Figure 6.32: Birth and death rates in developed and developing countries, 1775-1977

(b) By what percentage did the population of developing countries increase between 1800 and 1900?

(c) By what percentage did the population of developing countries increase between 1950 and 1977?

Solution

(a) In developed countries, the birth rate is higher than the death rate, so the population is increasing. The birth rate and the death rate are both decreasing, and at approximately the same rate, so the population of developed countries is increasing at a constant relative rate.

In developing countries, the birth rate is higher than the death rate and both have been decreasing since about 1925. In recent years, the death rate has been decreasing faster than the birth rate, so the relative rate of population growth is increasing. The decline in the death rate has contributed significantly to the recent population growth in developing countries.

(b) The relative rate of growth of the population is the difference between the relative birth rate and the relative death rate, so it is represented in Figure 6.32 by the vertical distance between the birth rate curve and the death rate curve. The area of the region between the two curves from 1800 to 1900, shown in Figure 6.33, gives the change in $\ln P(t)$. The region has area approximately equal to a rectangle of height 0.005 and width 100, so has area 0.5. We have

$$\ln\left(\frac{P(1900)}{P(1800)}\right) = \int_{1800}^{1900} \frac{P'(t)}{P(t)}\,dt = 0.5$$

$$\frac{P(1900)}{P(1800)} = e^{0.5} = 1.65$$

During the nineteenth century the population of developing countries increased by about 65%, a factor of 1.65.

(c) The shaded region between the birth rate and death rate curves from 1950 to 1977 in Figure 6.33 consists of approximately 1.5 rectangles, each of which has area $(0.01)(27) = 0.27$. The area of this region is approximately (1.5) (0.27) = 0.405. We have

$$\ln\left(\frac{P(1977)}{P(1950)}\right) = \int_{1950}^{1977} \frac{P'(t)}{P(t)}\,dt = 0.405,$$

so

$$\frac{P(1977)}{P(1950)} = e^{0.405} = 1.50$$

Between 1950 and 1977, the population of developing countries increased by about 50%, a factor of 1.5.

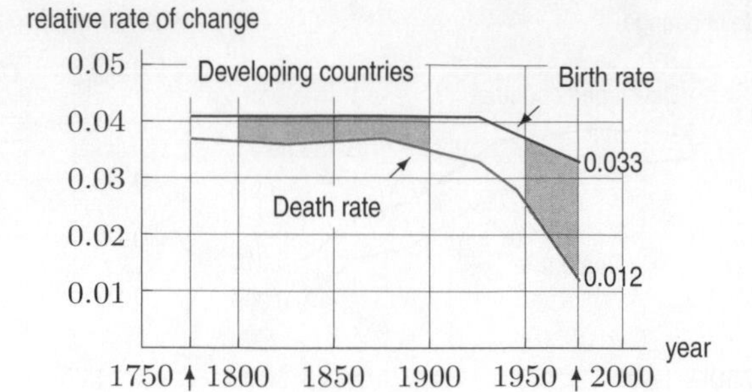

Figure 6.33: Relative growth rate of population = Relative birth rate − Relative death rate

Problems for Section 6.4

1. A population is 100 at time $t = 0$, with t in years.

(a) If the population has a constant absolute growth rate of 10 people per year, find a formula for the size of the population at time t.

(b) If the population has a constant relative growth rate of 10% per year, find a formula for the size of the population at time t.

(c) Graph both functions on the same axes.

2. Table 6.3 shows the cumulative number of AIDS deaths worldwide.[6] Find the absolute increase in AIDS deaths between 2000 and 2001 and between 2003 and 2004. Find the relative increase between 2000 and 2001 and between 2003 and 2004.

Table 6.3 *Cumulative AIDS deaths worldwide, in millions*

Year	2000	2001	2002	2003	2004
Cases	21.8	24.6	27.7	30.2	33.3

3. The size of a bacteria population is 4000. Find a formula for the size, P, of the population t hours later if the population is decreasing by

(a) 100 bacteria per hour **(b)** 5% per hour

In which case does the bacteria population reach 0 first?

In Problems 4–7, a graph of the relative rate of change of a population is given. By approximately what percentage does the population change over the 10-year period?

4. **5.**

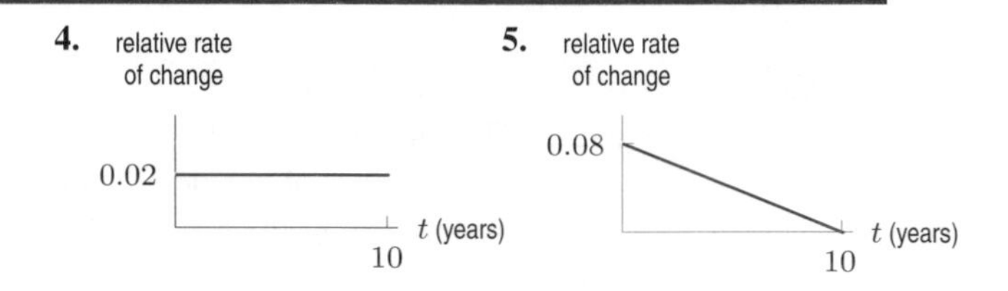

6. **7.**

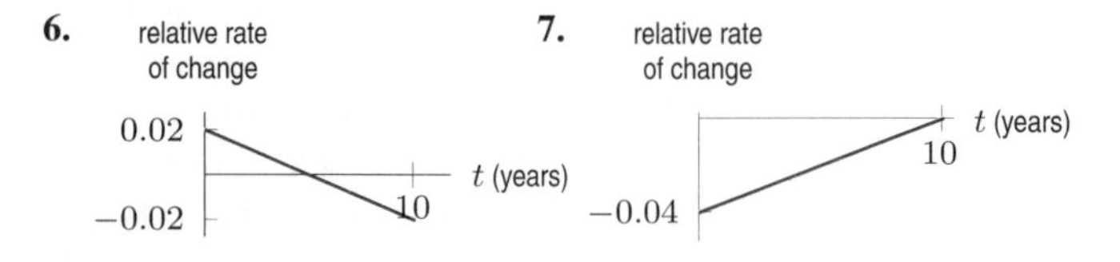

8. The relative growth rate of a population $P(t)$ is given in Figure 6.34. By what percentage does the population change over the 8-year period? If the population was 10,000 at time $t = 0$, what is the population 8 years later?

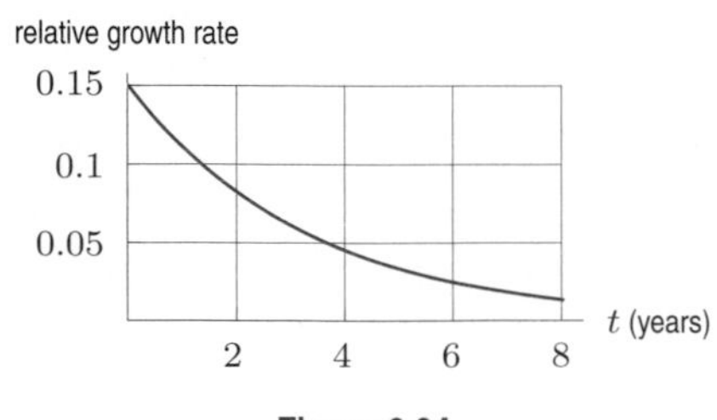

Figure 6.34

[6] www.unaids.org, May 2005.

Problems 9–12 show the relative rate of change of f for $0 \le t \le 10$. Give the intervals on which f is increasing and on which f is decreasing.

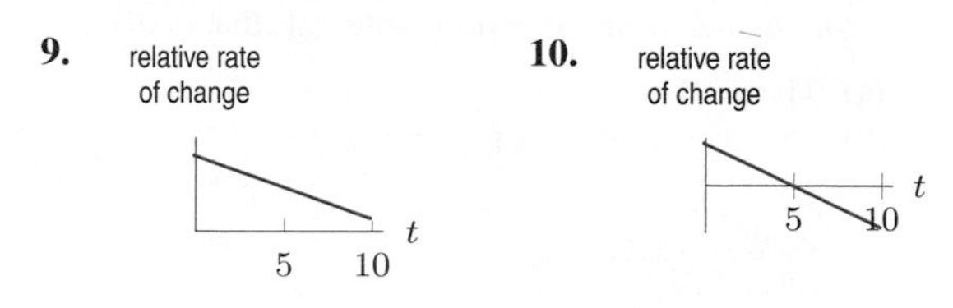

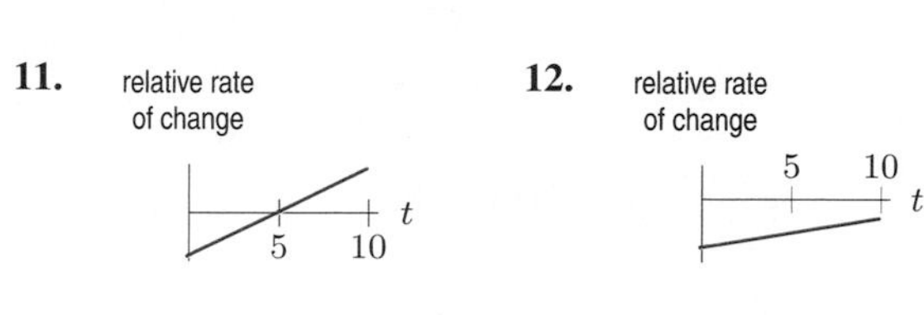

13. Figure 6.35 gives the relative growth rate of a population.

(a) On what interval is the population increasing? By what percentage does the population increase during this interval?

(b) On what interval is the population decreasing? By what percentage does the population decrease during this interval?

(c) By what percentage does the population change during the 15-year period shown?

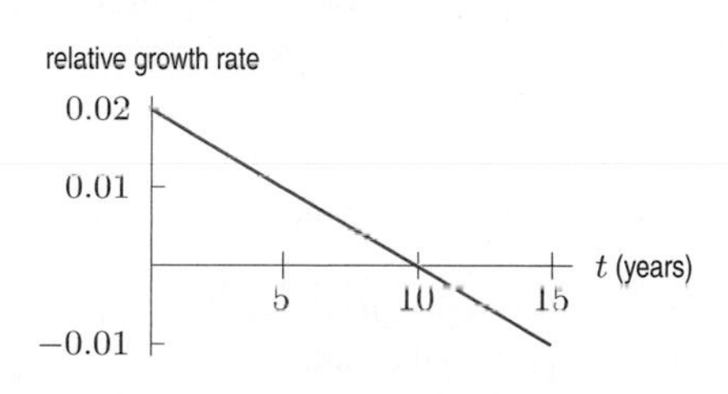

Figure 6.35

14. The population, P, in millions, of Nicaragua was 5.36 million in 2004 and growing at an annual rate of 1.62%.

(a) Write a formula for P as a function of t, where t is years since 2004.

(b) Find the projected average rate of change (or absolute growth rate) in Nicaragua between 2004 and 2005, and between 2005 and 2006. Explain why your answers are different.

(c) Use your answers to part (b) to confirm that the relative rate of change (or relative growth rate) over both time intervals was 1.62%.

In Problems 15–16, the graph shows relative birth and death rates for a population $P(t)$. Determine whether the population is increasing or decreasing, and find the percentage by which the population changes during the 10-year period.

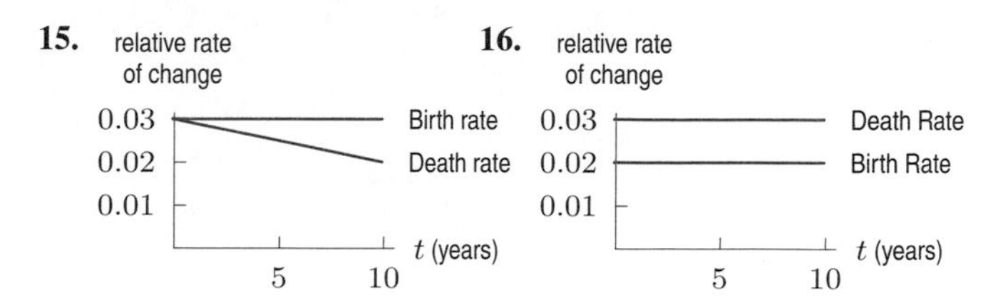

17. The number of reported burglary offenses in the US has been mostly decreasing since 1990.[7] If $P(t)$ is the number of reported burglaries as a function of year t, the relative rate of change per year, $P'(t)/P(t)$, is given in the table.

(a) Estimate the integral of the relative rate of change:
$$\int_{1990}^{2002} \frac{P'(t)}{P(t)}\,dt.$$

(b) By what percent did the number of burglaries change during the period 1990–2002?

Year	1990	1991	1992	1993	1994	1995	1996
Rel. rate	−0.03	0.03	−0.06	−0.05	−0.04	−0.04	−0.03
Year	1997	1998	1999	2000	2001	2002	
Rel. rate	−0.02	−0.05	−0.10	−0.02	0.03	0.02	

18. In 1990 humans generated $1.4 \cdot 10^{20}$ joules of energy from petroleum. At the time, it was estimated that all of the earth's petroleum would generate approximately 10^{22} joules. Assuming the use of energy generated by petroleum increases by 2% each year, how long will it be before all of our petroleum resources are used up?

CHAPTER SUMMARY

- **Average value**
- **Consumer and producer surplus**
  Wage and price controls
- **Present and future value**
  Income stream
- **Relative growth rates**

[7] The World Almanac and Book of Facts 2005, p. 162 (New York)

REVIEW PROBLEMS FOR CHAPTER SIX

1. What is the average value of the function f in Figure 6.36 over the interval $1 \le x \le 6$?

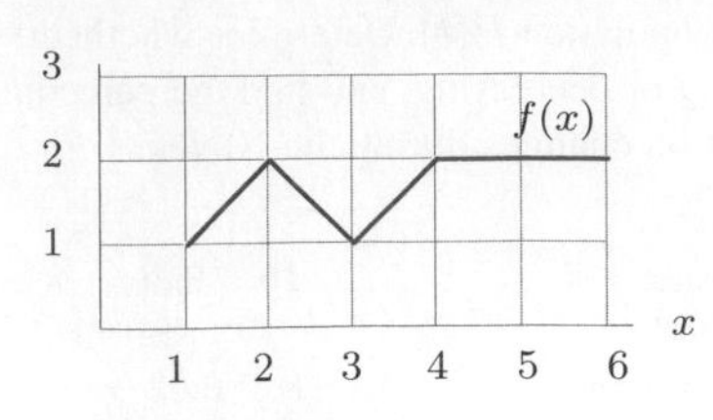

Figure 6.36

2. Find the average value of $g(t) = 2^t$ over the interval $[0, 10]$.

3. **(a)** What is the average value of $f(x) = \sqrt{1-x^2}$ over the interval $0 \le x \le 1$?
(b) How can you tell whether this average value is more or less than 0.5 without doing any calculations?

4. The value, V, of a Tiffany lamp, worth \$225 in 1965, increases at 15% per year. Its value in dollars t years after 1965 is given by

$$V = 225(1.15)^t.$$

Find the average value of the lamp over the period 1965–2000.

5. The quantity of a radioactive substance at time t is

$$Q(t) = 4(0.96)^t \text{ grams.}$$

(a) Find $Q(10)$ and $Q(20)$.
(b) Find the average of $Q(10)$ and $Q(20)$.
(c) Find the average value of $Q(t)$ over the interval $10 \le t \le 20$.
(d) Use the graph of $Q(t)$ to explain the relative sizes of your answers in parts (b) and (c).

In Problems 6–7, estimate the average value of the function $f(x)$ on the interval from $x = a$ to $x = b$.

6.

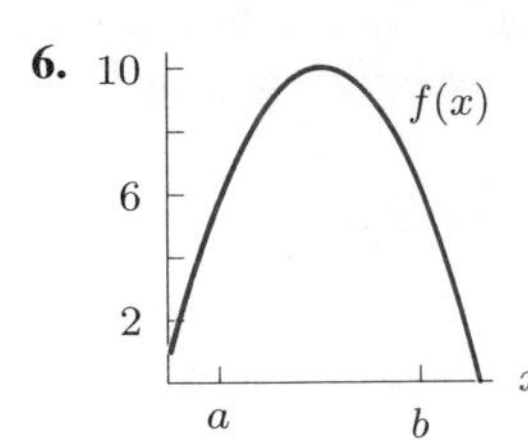

7.

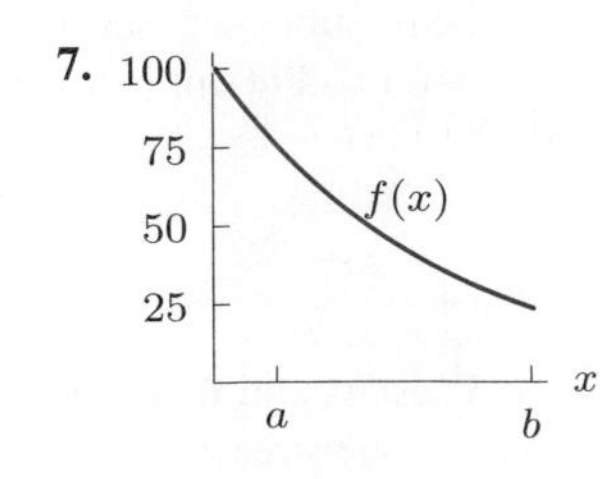

8. For the function f in Figure 6.37, write an expression involving one or more definite integrals that denotes:

(a) The average value of f for $0 \le x \le 5$.
(b) The average value of $|f|$ for $0 \le x \le 5$.

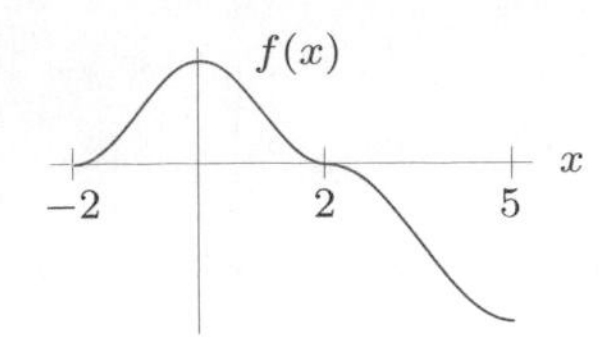

Figure 6.37

9. The function f in Figure 6.37 is symmetric about the y-axis. Consider the average value of f over the following intervals:

I. $0 \le x \le 1$ II. $0 \le x \le 2$
III. $0 \le x \le 5$ IV. $-2 \le x \le 2$

(a) For which interval is the average value of f least?
(b) For which interval is the average value of f greatest?
(c) For which pair of intervals are the average values equal?

10. The demand curve for a product has equation $p = 20e^{-0.002q}$ and the supply curve has equation $p = 0.02q + 1$ for $0 \le q \le 1000$, where q is quantity and p is price in \$/unit.

(a) Which is higher, the price at which 300 units are supplied or the price at which 300 units are demanded? Find both prices.
(b) Sketch the supply and demand curves. Find the equilibrium price and quantity.
(c) Using the equilibrium price and quantity, calculate and interpret the consumer and producer surplus.

11. Show graphically that the maximum total gains from trade occurs at the equilibrium price. Do this by showing that if outside forces keep the price artificially high or low, the total gains from trade (consumer surplus + producer surplus) are lower than at the equilibrium price.

12. In May 1991, *Car and Driver* described a Jaguar that sold for \$980,000. At that price only 50 have been sold. It is estimated that 350 could have been sold if the price had been \$560,000. Assuming that the demand curve is a straight line, and that \$560,000 and 350 are the equilibrium price and quantity, find the consumer surplus at the equilibrium price.

13. For a product, the demand curve is $p = 100e^{-0.008q}$ and the supply curve is $p = 4\sqrt{q} + 10$ for $0 \le q \le 500$, where q is quantity and p is price in dollars per unit.

(a) At a price of \$50, what quantity are consumers willing to buy and what quantity are producers willing to supply? Will the market push prices up or down?

(b) Find the equilibrium price and quantity. Does your answer to part (a) support the observation that market forces tend to push prices closer to the equilibrium price?

(c) At the equilibrium price, calculate and interpret the consumer and producer surplus.

14. The total gains from trade (consumer surplus + producer surplus) is largest at the equilibrium price. What about the consumer surplus and producer surplus separately?

(a) Suppose a price is artificially high. Can the consumer surplus at the artificial price be larger than the consumer surplus at the equilibrium price? What about the producer surplus? Sketch possible supply and demand curves to illustrate your answers.

(b) Suppose a price is artificially low. Can the consumer surplus at the artificial price be larger than the consumer surplus at the equilibrium price? What about the producer surplus? Sketch possible supply and demand curves to illustrate your answers.

15. Find the present and future values of an income stream of \$3000 per year over a 15-year period, assuming a 6% annual interest rate compounded continuously.

16. At what constant, continuous rate must money be deposited into an account if the account is to contain \$20,000 in 5 years? The account earns 6% interest compounded continuously.

17. Harley-Davidson Inc. manufactures motorcycles. During 1996 and 1997, the rate of sales was approximately $1431e^{0.134t}$ million dollars per year, where t is time in years since January 1, 1996. Assume this rate held through January 1, 2003 (the company's 100th anniversary). Using a continuous interest rate of 7.5% per year, find the value on January 1, 1996 of Harley-Davidson's sales from January 1, 1996 to January 1, 2003.

18. A recently-installed machine earns the company revenue at a continuous rate of $60{,}000t + 45{,}000$ dollars per year during the first six months of operation and at the continuous rate of 75,000 dollars per year after the first six months. The cost of the machine is \$150,000, the interest rate is 7% per year, compounded continuously, and t is time in years since the machine was installed.

(a) Find the present value of the revenue earned by the machine during the first year of operation.

(b) Find how long it will take for the machine to pay for itself; that is, how long it will take for the present value of the revenue to equal the cost of the machine?

19. The value of good wine increases with age. Thus, if you are a wine dealer, you have the problem of deciding whether to sell your wine now, at a price of $\$P$ a bottle, or to sell it later at a higher price. Suppose you know that the amount a wine-drinker is willing to pay for a bottle of this wine t years from now is $\$P(1 + 20\sqrt{t})$. Assuming continuous compounding and a prevailing interest rate of 5% per year, when is the best time to sell your wine?

20. The article "Job Scene: What's Hot, What's Not in the Next 10 Years," appearing in the *Chicago Tribune* in July 1996, contained the following quote about predicted change in jobs between 1995 and 2005:

- Employment is projected to increase to 144.7 million from 127 million – only **14** percent. The growth rate was **24** percent from 1983 to 1994.
- Jobs in services and retail trades are expected to increase by **16.2** million. Business services, health and education will account for **9.1** million new jobs.
- Manufacturing will lose **1.3** million jobs, a continuation of its decline.
- Jobs for those with master's degrees will grow the most, **28** percent.

(a) The first number in bold is 14. Is this an absolute change, a relative change, an absolute rate of change, or a relative rate of change? Find each of the other three measures of change, corresponding to 14.

(b) Identify each of the numbers in bold as an absolute change, a relative change, an absolute rate of change, or a relative rate of change.

21. A town has a population of 1000. Fill in the table assuming that the town's population grows by

(a) 50 people per year **(b)** 5% per year

Year	0	1	2	3	4	...	10
Population	1000					...	

22. Birth and death rates are often reported as births or deaths per thousand members of the population. What is the relative rate of growth of a population with a birth rate of 30 births per 1000 and a death rate of 20 deaths per 1000?

23. The relative growth rate of a population is given in Figure 6.38. By what percentage does the population change over the 10-year period?

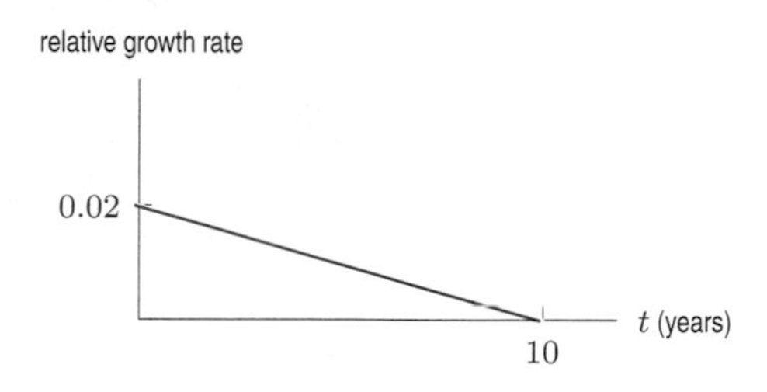

Figure 6.38

24. In everyday language, exponential growth means very fast growth. In this problem, you will see that any exponentially growing function eventually grows faster than any power function.

(a) Show that the relative growth rate of the function $f(x) = x^n$, for fixed $n > 0$ and for $x > 0$, decreases as x increases.

(b) Assume $k > 0$ is fixed. Explain why, for large x, the relative growth rate of the function $g(x) = e^{kx}$ is larger than the relative growth rate of $f(x)$.

PROJECTS FOR CHAPTER SIX

1. Distribution of Resources

Whether a resource is distributed evenly among members of a population is often an important political or economic question. How can we measure this? How can we decide if the distribution of wealth in this country is becoming more or less equitable over time? How can we measure which country has the most equitable income distribution? This problem describes a way of making such measurements. Suppose the resource is distributed evenly. Then any 20% of the population will have 20% of the resource. Similarly, any 30% will have 30% of the resource and so on. If, however, the resource is not distributed evenly, the poorest $p\%$ of the population (in terms of this resource) will not have $p\%$ of the goods. Suppose $F(x)$ represents the fraction of the resources owned by the poorest fraction x of the population. Thus $F(0.4) = 0.1$ means that the poorest 40% of the population owns 10% of the resource.

(a) What would F be if the resource were distributed evenly?

(b) What must be true of any such F? What must $F(0)$ and $F(1)$ equal? Is F increasing or decreasing? Is the graph of F concave up or concave down?

(c) Gini's index of inequality, G, is one way to measure how evenly the resource is distributed. It is defined by

$$G = 2\int_0^1 [x - F(x)]\, dx.$$

Show graphically what G represents.

(d) Graphical representations of Gini's index for two countries are given in Figures 6.39 and 6.40. Which country has the more equitable distribution of wealth? Discuss the distribution of wealth in each of the two countries.

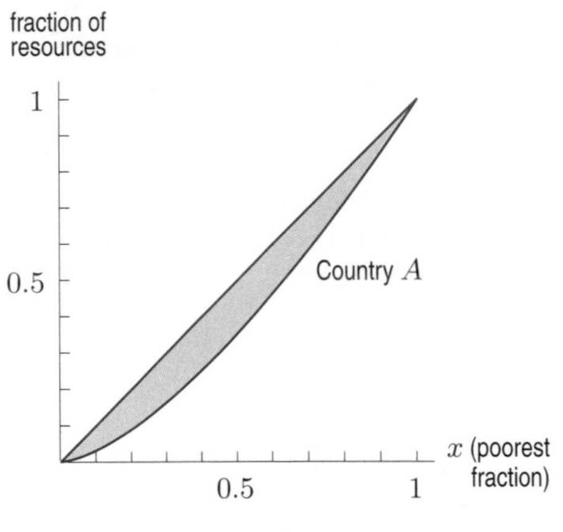

Figure 6.39: Country A

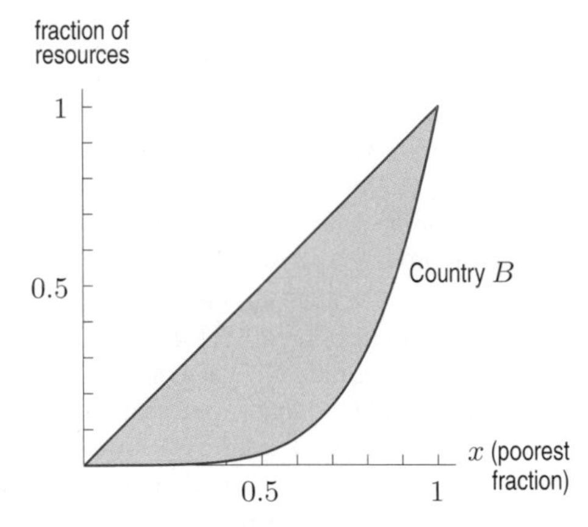

Figure 6.40: Country B

(e) What is the maximum possible value of Gini's index of inequality, G? What is the minimum possible value? Sketch graphs in each case. What is the distribution of resources in each case?

2. Yield from an Apple Orchard

Figure 6.41 is a graph of the annual yield, $y(t)$, in bushels per year, from an orchard t years after planting. The trees take about 10 years to get established, but for the next 20 years they give a substantial yield. After about 30 years, however, age and disease start to take their toll, and the annual yield falls off.[8]

(a) Represent on a sketch of Figure 6.41 the total yield, $F(M)$, up to M years, with $0 \leq M \leq 60$. Write an expression for $F(M)$ in terms of $y(t)$.

[8]From Peter D. Taylor, *Calculus: The Analysis of Functions*, (Toronto: Wall & Emerson, Inc., 1992).

(b) Sketch a graph of $F(M)$ against M for $0 \leq M \leq 60$.

(c) Write an expression for the average annual yield, $a(M)$, up to M years.

(d) When should the orchard be cut down and replanted? Assume that we want to maximize average revenue per year, and that fruit prices remain constant, so that this is achieved by maximizing average annual yield. Use the graph of $y(t)$ to estimate the time at which the average annual yield is a maximum. Explain your answer geometrically and symbolically.

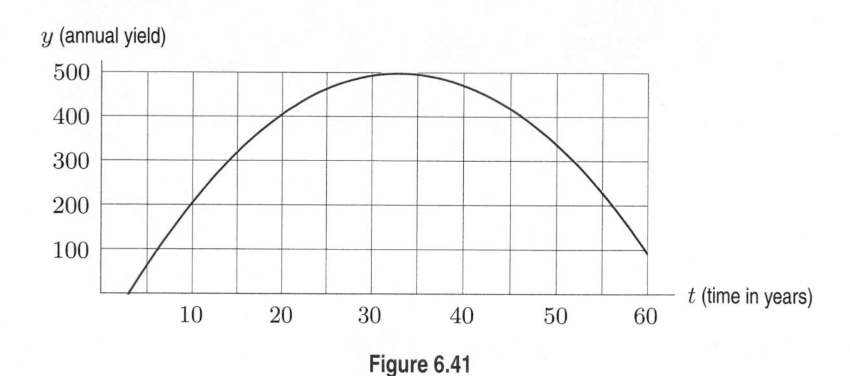

Figure 6.41

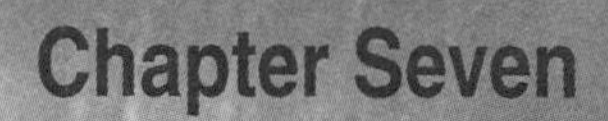

Chapter Seven

ANTIDERIVATIVES

In this chapter, we look in more detail at how to reconstruct a function from its derivative using the Fundamental Theorem of Calculus. We introduce antiderivatives and show how to construct them analytically, numerically, and graphically.

7.1 CONSTRUCTING ANTIDERIVATIVES ANALYTICALLY

What is an Antiderivative?

If the derivative of $F(x)$ is $f(x)$, that is, if $F'(x) = f(x)$, then we call $F(x)$ an *antiderivative* of $f(x)$. For example, the derivative of x^2 is $2x$, so we say that

$$x^2 \text{ is an antiderivative of } 2x.$$

Can you think of another function whose derivative is $2x$? How about $x^2 + 1$? Or $x^2 + 17$? Since, for any constant C,

$$\frac{d}{dx}(x^2 + C) = 2x + 0 = 2x,$$

any function of the form $x^2 + C$ is an antiderivative of $2x$. It can be shown that all antiderivatives of $2x$ are of this form, so we say that

$$x^2 + C \text{ is the family of antiderivatives of } 2x.$$

Once we know one antiderivative $F(x)$ for a function $f(x)$ on an interval, then all other antiderivatives of $f(x)$ are of the form $F(x) + C$.

The Indefinite Integral

We introduce a notation for the family of antiderivatives that looks like a definite integral without the limits. If all antiderivatives of $f(x)$ are of the form $F(x) + C$, we call $\int f(x)\,dx$ the *indefinite integral* of $f(x)$ and write

$$\int f(x)\,dx = F(x) + C.$$

It is important to understand the difference between

$$\int_a^b f(x)\,dx \qquad \text{and} \qquad \int f(x)\,dx.$$

The first is a number and the second is a family of functions. Because the notation is similar, the word "integration" is frequently used for the process of finding antiderivatives as well as of finding definite integrals. The context usually makes clear which is intended.

Finding Formulas for Antiderivatives

Finding antiderivatives of functions is like taking square roots of numbers: if we pick a number at random, such as 7 or 493, we may have trouble saying what its square root is without a calculator. But if we happen to pick a number such as 25 or 64, which we know is a perfect square, then we can find its square root exactly. Similarly, if we pick a function which we recognize as a derivative, then we can find its antiderivative easily.

For example, noticing that $2x$ is the derivative of x^2 tells us that x^2 is an antiderivative of $2x$. If we divide by 2, then we guess that

$$\text{An antiderivative of } x \text{ is } \frac{x^2}{2}.$$

To check this statement, take the derivative of $x^2/2$:

$$\frac{d}{dx}\left(\frac{x^2}{2}\right) = \frac{1}{2} \cdot \frac{d}{dx}x^2 = \frac{1}{2} \cdot 2x = x.$$

What about an antiderivative of x^2? The derivative of x^3 is $3x^2$, so the derivative of $x^3/3$ is $3x^2/3 = x^2$. Thus,

$$\text{An antiderivative of } x^2 \text{ is } \frac{x^3}{3}.$$

Can you see the pattern? It looks like

$$\text{An antiderivative of } x^n \text{ is } \frac{x^{n+1}}{n+1}.$$

(We assume $n \neq -1$, or we would have $x^0/0$, which does not make sense.) It is easy to check this formula by differentiation:

$$\frac{d}{dx}\left(\frac{x^{n+1}}{n+1}\right) = \frac{(n+1)x^n}{n+1} = x^n.$$

Thus, in indefinite integral notation, we see that

$$\int x^n\, dx = \frac{x^{n+1}}{n+1} + C, \quad n \neq -1.$$

Can you think of an antiderivative of the function $f(x) = 5$? We know that the derivative of $5x$ is 5, so $F(x) = 5x$ is an antiderivative of $f(x) = 5$. In general, if k is a constant, the derivative of kx is k, so we have the result:

If k is constant,

$$\int k\, dx = kx + C.$$

Example 1 Find $\int (3x + x^2)\, dx$.

Solution We know that $x^2/2$ is an antiderivative of x and that $x^3/3$ is an antiderivative of x^2, so we expect

$$\int (3x + x^2)\, dx = 3\left(\frac{x^2}{2}\right) + \frac{x^3}{3} + C.$$

Again, check your antiderivatives by differentiation it's easy to do. Here

$$\frac{d}{dx}\left(\frac{3}{2}x^2 + \frac{x^3}{3} + C\right) = \frac{3}{2} \cdot 2x + \frac{3x^2}{3} = 3x + x^2.$$

The preceding example illustrates that the sum and constant multiplication rules of differentiation work in reverse:

Properties of Antiderivatives: Sums and Constant Multiples

In indefinite integral notation,

1. $\int [f(x) \pm g(x)]\, dx = \int f(x)\, dx \pm \int g(x)\, dx$
2. $\int cf(x)\, dx = c \int f(x)\, dx.$

In words,

1. An antiderivative of the sum (or difference) of two functions is the sum (or difference) of their antiderivatives.
2. An antiderivative of a constant times a function is the constant times an antiderivative of the function.

Example 2 Find antiderivatives of each of the following: (a) x^5 (b) t^8 (c) $12x^3$ (d) $q^3 - 6q^2$

Solution

(a) $\int x^5\,dx = \frac{x^6}{6} + C.$

(b) $\int t^8\,dt = \frac{t^9}{9} + C.$

(c) $\int 12x^3\,dx = 12\left(\frac{x^4}{4}\right) + C = 3x^4 + C.$

(d) $\int q^3 - 6q^2\,dq = \frac{q^4}{4} - 6\left(\frac{q^3}{3}\right) + C = \frac{q^4}{4} - 2q^3 + C.$

To check, differentiate the antiderivative; you should get the original function.

What is the antiderivative of x^n when $n = -1$? In other words, what is an antiderivative of $1/x$? Fortunately, we know a function whose derivative is $1/x$, namely, the natural logarithm. Thus, since

$$\frac{d}{dx}(\ln x) = \frac{1}{x},$$

we know that

$$\int \frac{1}{x}\,dx = \ln x + C, \quad \text{for } x > 0.$$

If $x < 0$, then $\ln x$ is not defined, so it can't be an antiderivative of $1/x$. In this case, we can try $\ln(-x)$:

$$\frac{d}{dx}\ln(-x) = (-1)\frac{1}{-x} = \frac{1}{x}$$

so

$$\int \frac{1}{x}\,dx = \ln(-x) + C, \quad \text{for } x < 0.$$

This means $\ln x$ is an antiderivative of $1/x$ if $x > 0$, and $\ln(-x)$ is an antiderivative of $1/x$ if $x < 0$. Since $|x| = x$ when $x > 0$ and $|x| = -x$ when $x < 0$ we can collapse these two formulas into:

$$\text{An antiderivative of } \frac{1}{x} \text{ is } \ln|x|$$

on any interval that does not contain 0. Therefore

$$\boxed{\int \frac{1}{x}\,dx = \ln|x| + C.}$$

Since the exponential function e^x is its own derivative, it is also its own antiderivative; thus

$$\boxed{\int e^x\,dx = e^x + C.}$$

What about e^{kx}? We know that the derivative of e^{kx} is ke^{kx}, so, for $k \neq 0$, we have

$$\boxed{\int e^{kx}\,dx = \frac{1}{k}e^{kx} + C.}$$

Example 3 Find antiderivatives of each of the following: (a) $8x^3 + \frac{1}{x}$ (b) $12e^{0.2t}$

Solution (a) $\int 8x^3 + \frac{1}{x}\,dx = 8\left(\frac{x^4}{4}\right) + \ln|x| + C = 2x^4 + \ln|x| + C.$

(b) $\int 12e^{0.2t}\,dt = 12\left(\frac{1}{0.2}e^{0.2t}\right) + C = 60e^{0.2t} + C.$

Differentiate your answers to check your work.

Antiderivatives of Periodic Functions

The antiderivatives of the sine and cosine are easy to guess. Since

$$\frac{d}{dx}\sin x = \cos x \quad \text{and} \quad \frac{d}{dx}\cos x = -\sin x,$$

we get

$$\int \cos x\,dx = \sin x + C \quad \text{and} \quad \int \sin x\,dx = -\cos x + C.$$

Since the derivative of $\sin(kx)$ is $k\cos(kx)$ and the derivative of $\cos(kx)$ is $-k\sin(kx)$, we have, for $k \neq 0$,

$$\int \cos(kx)\,dx = \frac{1}{k}\sin(kx) + C \quad \text{and} \quad \int \sin(kx)\,dx = -\frac{1}{k}\cos(kx) + C.$$

Example 4 Find $\int (\sin x + 3\cos(5x))\,dx$.

Solution We break the antiderivative into two terms:

$$\int (\sin x + 3\cos(5x))\,dx = \int \sin x\,dx + 3\int \cos(5x)\,dx = -\cos x + \frac{3}{5}\sin(5x) + C.$$

Check by differentiating:

$$\frac{d}{dx}(-\cos x + \frac{3}{5}\sin(5x) + C) = \sin x + 3\cos(5x).$$

Problems for Section 7.1

In Problems 1–26, find an antiderivative.

1. $f(x) = 5$

2. $f(x) = 5x$

3. $f(x) = x^2$

4. $g(t) = t^2 + t$

5. $f(x) = x^4$

6. $g(t) = t^7 + t^3$

7. $f(q) = 5q^2$

8. $g(x) = 6x^3 + 4$

9. $h(y) = 3y^2 - y^3$

10. $f(t) = 2t^2 + 3t^3 + 4t^4$

11. $f(x) = 3x^2 + 5$

12. $f(x) = 6x^2 - 8x + 3$

13. $h(t) = 3t^2 + 7t + 1$

14. $g(z) = \sqrt{z}$

15. $f(x) = 5x - \sqrt{x}$

16. $p(z) = (\sqrt{z})^3$

17. $p(t) = t^3 - \frac{t^2}{2} - t$

18. $h(z) = \frac{1}{z}$

19. $r(t) = \frac{1}{t^2}$

20. $q(y) = y^4 + \frac{1}{y}$

21. $f(x) = x^6 - \frac{1}{7x^6}$

22. $p(y) = \frac{1}{y} + y + 1$

23. $g(t) = e^{-3t}$

24. $g(t) = \sin t$

25. $g(t) = 5 + \cos t$

26. $g(\theta) = \sin\theta - 2\cos\theta$

In Problems 27–32, find an antiderivative $F(x)$ with $F'(x) = f(x)$ and $F(0) = 0$. Is there only one possible solution?

27. $f(x) = 3$

28. $f(x) = -7x$

29. $f(x) = x^2$

30. $f(x) = \sqrt{x}$

31. $f(x) = 2+4x+5x^2$

32. $f(x) = e^x$

Find the indefinite integrals in Problems 33–62.

33. $\int 3x\,dx$

34. $\int (4t+7)\,dt$

35. $\int 6x^2\,dx$

36. $\int t^{12}\,dt$

37. $\int (x^3 - x)\,dx$

38. $\int (x^2+1)\,dx$

39. $\int (x^3+4x+8)\,dx$

40. $\int 5e^z\,dz$

41. $\int (q^2+5q+2)dq$

42. $\int (x^5 - 12x^3)dx$

43. $\int (6\sqrt{x})dx$

44. $\int (x^2+4x-5)dx$

45. $\int \left(\frac{5}{t^2}+\frac{6}{t^3}\right)dt$

46. $\int e^{2t}\,dt$

47. $\int \left(x+\frac{1}{\sqrt{x}}\right)dx$

48. $\int (t^2-6t+5)\,dt$

49. $\int e^{-0.05t}\,dt$

50. $\int \left(8x^3+\frac{1}{x}\right)dx$

51. $\int e^{-3t}\,dt$

52. $\int \cos\theta\,d\theta$

53. $\int \sin t\,dt$

54. $\int 30e^{-0.2t}dt$

55. $\int 100e^{4x}\,dx$

56. $\int (4x+2e^x)dx$

57. $\int (5\cos x - 3\sin x)dx$

58. $\int \sin(3x)dx$

59. $\int \cos(4x)dx$

60. $\int 6\cos(3x)dx$

61. $\int (10+8\sin(2x))dx$

62. $\int (12\sin(2x)+15\cos(5x))dx$

For Problems 63–66, find an antiderivative $F(x)$ with $F'(x) = f(x)$ and $F(0) = 5$.

63. $f(x) = x^2+1$

64. $f(x) = 6x-5$

65. $f(x) = 6e^{3x}$

66. $f(x) = 8\sin(2x)$

67. The marginal revenue function of a monopolistic producer is $MR = 20 - 4q$.

(a) Find the total revenue function.
(b) Find the corresponding demand curve.

68. A firm's marginal cost function is $MC = 3q^2+4q+6$. Find the total cost function if the fixed costs are 200.

7.2 INTEGRATION BY SUBSTITUTION

In Chapter 3, we learned rules to differentiate any function obtained by combining constants, powers of x, $\sin x$, $\cos x$, e^x, and $\ln x$, using addition, multiplication, division, or composition of functions. Such functions are called *elementary*.

In this section, we introduce integration by substitution. However, there is a great difference between looking for derivatives and looking for antiderivatives. Every elementary function has elementary derivatives, but many elementary functions do not have elementary antiderivatives. Some examples are $\sqrt{x^3+1}$, $(\sin x)/x$, and e^{-x^2}. These are ordinary functions that arise naturally, not exotic functions, yet they do not have elementary antiderivatives.

Integration by substitution reverses the chain rule. According to the chain rule,

$$\frac{d}{dx}(f(g(x))) = \underbrace{f'}_{\text{Derivative of outside}}(\overbrace{g(x)}^{\text{Inside}}) \cdot \underbrace{g'(x)}_{\text{Derivative of inside}}.$$

Thus, any function which is the result of differentiating with the chain rule is the product of two factors: the "derivative of the outside" and the "derivative of the inside." If a function has this form, its antiderivative is $f(g(x))$.

Example 1 Use the chain rule to find $f'(x)$ and then write the corresponding antidifferentiation formula.

(a) $f(x) = e^{x^2}$ (b) $f(x) = \frac{1}{6}(x^2+1)^6$ (c) $f(x) = \ln(x^2+4)$

Solution (a) Using the chain rule, we see

$$\frac{d}{dx}\left(e^{x^2}\right) = e^{x^2} \cdot 2x \quad \text{so} \quad \int e^{x^2} \cdot 2x\,dx = e^{x^2} + C.$$

(b) Using the chain rule, we see

$$\frac{d}{dx}\left(\frac{1}{6}(x^2+1)^6\right) = (x^2+1)^5 \cdot 2x \quad \text{so} \quad \int (x^2+1)^5 \cdot 2x\,dx = \frac{1}{6}(x^2+1)^6 + C.$$

(c) Using the chain rule, we see

$$\frac{d}{dx}(\ln(x^2+4)) = \frac{1}{x^2+4} \cdot 2x \quad \text{so} \quad \int \frac{1}{x^2+4} \cdot 2x\,dx = \ln(x^2+4) + C.$$

In Example 1, the derivative of each inside function is $2x$. Notice that the derivative of the inside function is a factor in the integrand in each antidifferentiation formula.

Finding an inside function whose derivative appears as a factor is key to the method of substitution. We formalize this method as follows:

To Make a Substitution in an Integral

Let w be the "inside function" and $dw = w'(x)\,dx = \dfrac{dw}{dx}dx$. Then express the integrand in terms of w.

Example 2 Make a substitution to find each of the following integrals:

(a) $\int e^{x^2} \cdot 2x\,dx$ (b) $\int (x^2+1)^5 \cdot 2x\,dx$ (c) $\int \frac{1}{x^2+4} \cdot 2x\,dx$

Solution (a) We look for an inside function whose derivative appears as a factor. In this case, the inside function is x^2, with derivative $2x$. We let $w = x^2$. Then $dw = w'(x)\,dx = 2x\,dx$. The original integrand can now be rewritten in terms of w:

$$\int e^{x^2} \cdot 2x\,dx = \int e^w\,dw = e^w + C = e^{x^2} + C.$$

By changing the variable to w, we simplified the integrand. The final step, after antidifferentiating, is to convert back to the original variable, x.

(b) Here, the inside function is x^2+1, with derivative $2x$. We let $w = x^2+1$. Then $dw = w'(x)\,dx = 2x\,dx$. Rewriting the original integral in terms of w, we have

$$\int (x^2+1)^5 \cdot 2x\,dx = \int w^5\,dw = \frac{1}{6}w^6 + C = \frac{1}{6}(x^2+1)^6 + C.$$

Again, by changing the variable to w, we simplified the integrand.

(c) The inside function is x^2+4, so we let $w = x^2+4$. Then $dw = w'(x)\,dx = 2x\,dx$. Substituting, we have

$$\int \frac{1}{x^2+4} \cdot 2x\,dx = \int \frac{1}{w}\,dw = \ln(w) + C = \ln(x^2+4) + C.$$

Notice that the derivative of the inside function must be present in the integral for this method to work. The method works, however, even when the derivative is missing a constant factor, as in the next two examples.

Example 3 Find $\int te^{(t^2+1)}\,dt$.

Solution Here the inside function is t^2+1, with derivative $2t$. Since there is a factor of t in the integrand, we try $w = t^2+1$. Then $dw = w'(t)\,dt = 2t\,dt$. Notice, however, that the original integrand has only $t\,dt$, not $2t\,dt$. We therefore write

$$\frac{1}{2}\,dw = t\,dt$$

and then substitute:

$$\int te^{(t^2+1)}\,dt = \int e^{\overbrace{(t^2+1)}^{w}} \cdot \underbrace{t\,dt}_{\frac{1}{2}dw} = \int e^w\,\frac{1}{2}\,dw = \frac{1}{2}\int e^w\,dw = \frac{1}{2}e^w + C = \frac{1}{2}e^{(t^2+1)} + C.$$

Why didn't we put $\frac{1}{2}\int e^w\,dw = \frac{1}{2}e^w + \frac{1}{2}C$ in the preceding example? Since the constant C is arbitrary, it does not really matter whether we add C or $\frac{1}{2}C$. The convention is always to add C to whatever antiderivative we have calculated.

Example 4 Find $\int x^3\sqrt{x^4+5}\,dx$.

Solution The inside function is x^4+5, with derivative $4x^3$. The integrand has a factor of x^3, and since the only thing missing is a constant factor, we try

$$w = x^4+5.$$

Then

$$dw = w'(x)\,dx = 4x^3\,dx,$$

giving

$$\frac{1}{4}dw = x^3\,dx.$$

Thus,

$$\int x^3\sqrt{x^4+5}\,dx = \int \sqrt{w}\,\frac{1}{4}dw = \frac{1}{4}\int w^{1/2}\,dw = \frac{1}{4}\cdot\frac{w^{3/2}}{3/2} + C = \frac{1}{6}(x^4+5)^{3/2} + C.$$

Warning

We saw in the preceding examples that we can apply the substitution method when a *constant* factor is missing from the derivative of the inside function. However, we may not be able to use substitution if anything other than a constant factor is missing. For example, setting $w = x^4+5$ to find

$$\int x^2\sqrt{x^4+5}\,dx$$

does us no good because $x^2\,dx$ is not a constant multiple of $dw = 4x^3\,dx$. In order to use substitution, the integrand must contain the derivative of the inside function, *to within a constant factor*.

Example 5 Find $\int \frac{t^2}{1+t^3}\,dt$.

Solution Observing that the derivative of $1+t^3$ is $3t^2$, we take $w = 1+t^3$, $dw = 3t^2\,dt$, so $\frac{1}{3}\,dw = t^2\,dt$. Thus,

$$\int \frac{t^2}{1+t^3}\,dt = \int \frac{\frac{1}{3}\,dw}{w} = \frac{1}{3}\ln|w| + C = \frac{1}{3}\ln|1+t^3| + C.$$

Since the numerator is $t^2\,dt$, we might have tried $w = t^3$. This substitution leads to the integral $\frac{1}{3}\int 1/(1+w)\,dw$. To evaluate this integral we would have to make a second substitution $u = 1+w$. There is often more than one way to do an integral by substitution.

Using Substitution with Periodic Functions

The method of substitution works with all integral formulas. In particular, it can be used for integrals involving periodic functions.

Example 6 Find $\int 3x^2\cos(x^3)\,dx$.

Solution We look for an inside function whose derivative appears—in this case x^3. We let $w = x^3$. Then $dw = w'(x)\,dx = 3x^2\,dx$. The original integrand can now be completely rewritten in terms of the new variable w:

$$\int 3x^2\cos(x^3)\,dx = \int \cos\underbrace{(x^3)}_{w}\cdot\underbrace{3x^2\,dx}_{dw} = \int \cos w\,dw = \sin w + C = \sin(x^3) + C.$$

By changing the variable to w, we can simplify the integrand. We now have $\cos w$, which can be antidifferentiated more easily. The final step, after antidifferentiating, is to convert back to the original variable, x.

Example 7 Find $\int e^{\cos\theta}\sin\theta\,d\theta$.

Solution We let $w = \cos\theta$ since its derivative is $-\sin\theta$ and there is a factor of $\sin\theta$ in the integrand. This gives

$$dw = w'(\theta)\,d\theta = -\sin\theta\,d\theta,$$

so

$$-dw = \sin\theta\,d\theta.$$

Thus,

$$\int e^{\cos\theta}\sin\theta\,d\theta = \int e^{w}\,(-dw) = (-1)\int e^{w}\,dw = -e^{w} + C = -e^{\cos\theta} + C.$$

Problems for Section 7.2

Find the integrals in Problems 1–40. Check your answers by differentiation.

1. $\int 2x(x^2+1)^5\,dx$
2. $\int \frac{4x^3}{x^4+1}\,dx$
3. $\int (x+10)^3\,dx$
4. $\int 5e^{5t+2}\,dt$
5. $\int \frac{2x}{\sqrt{x^2+1}}\,dx$
6. $\int e^{-x}\,dx$
7. $\int xe^{-x^2}\,dx$
8. $\int y(y^2+5)^8\,dy$
9. $\int t^2(t^3-3)^{10}\,dt$
10. $\int x^2(1+2x^3)^2\,dx$
11. $\int x(x^2-4)^{7/2}\,dx$
12. $\int x(x^2+3)^2\,dx$
13. $\int \frac{1}{\sqrt{4-x}}\,dx$
14. $\int \frac{dy}{y+5}$
15. $\int t\cos(t^2)\,dt$
16. $\int (2t-7)^{73}\,dt$
17. $\int (x^2+3)^2\,dx$
18. $\int \sin(3-t)\,dt$
19. $\int y^2(1+y)^2\,dy$
20. $\int \sin\theta(\cos\theta+5)^7\,d\theta$
21. $\int \sqrt{\cos 3t}\sin 3t\,dt$
22. $\int \frac{t}{1+3t^2}\,dt$
23. $\int \sin^6\theta\cos\theta\,d\theta$
24. $\int x^2e^{x^3+1}\,dx$
25. $\int \sin^6(5\theta)\cos(5\theta)\,d\theta$
26. $\int \sin^3\alpha\cos\alpha\,d\alpha$
27. $\int x\sin(x^2)dx$
28. $\int e^{3x-4}dx$
29. $\int xe^{3x^2}\,dx$
30. $\int x\sqrt{x^2+1}dx$
31. $\int \frac{q}{5q^2+8}\,dq$
32. $\int \frac{(\ln z)^2}{z}\,dz$
33. $\int \frac{e^t+1}{e^t+t}\,dt$
34. $\int \frac{y}{y^2+4}\,dy$
35. $\int \frac{\cos\sqrt{x}}{\sqrt{x}}\,dx$
36. $\int \frac{e^{\sqrt{y}}}{\sqrt{y}}\,dy$
37. $\int \frac{1+e^x}{\sqrt{x+e^x}}\,dx$
38. $\int \frac{e^x}{2+e^x}\,dx$
39. $\int \frac{x+1}{x^2+2x+19}\,dx$
40. $\int \frac{e^x-e^{-x}}{e^x+e^{-x}}\,dx$

41. If appropriate, evaluate the following integrals by substitution. If substitution is not appropriate, say so, and do not evaluate.
    - **(a)** $\int x\sin(x^2)\,dx$
    - **(b)** $\int x^2\sin x\,dx$
    - **(c)** $\int \frac{x^2}{1+x^2}\,dx$
    - **(d)** $\int \frac{x}{(1+x^2)^2}\,dx$
    - **(e)** $\int x^3e^{x^2}\,dx$
    - **(f)** $\int \frac{\sin x}{2+\cos x}\,dx$

42. Find $\int 4x(x^2+1)\,dx$ using two methods:
    - **(a)** Do the multiplication first, and then antidifferentiate.
    - **(b)** Use the substitution $w = x^2+1$.
    - **(c)** Explain how the expressions from parts (a) and (b) are different. Are they both correct?

43. **(a)** Find $\int (x+5)^2\,dx$ in two ways:
    - (i) By multiplying out
    - (ii) By substituting $w = x+5$

    (b) Are the results the same? Explain.

7.3 USING THE FUNDAMENTAL THEOREM TO FIND DEFINITE INTEGRALS

In the previous section we calculated antiderivatives. In this section we see how antiderivatives are used to calculate definite integrals exactly. This calculation is based on the Fundamental Theorem:

> **Fundamental Theorem of Calculus** If F', the derivative of F, is continuous, then
>
> $$\int_a^b F'(x)\,dx = F(b) - F(a).$$

So far we have approximated definite integrals using a graph or left- and right-hand sums. The Fundamental Theorem gives us another method of calculating definite integrals. To find $\int_a^b F'(x)\,dx$, we first try to find F, and then calculate $F(b) - F(a)$. This method of computing definite integrals has an important advantage: it gives an exact answer. However, the method only works when we can find a formula for an antiderivative $F(x)$.

Example 1 Compute $\displaystyle\int_1^3 2x\,dx$ numerically and using the Fundamental Theorem.

Solution Using a calculator, we obtain

$$\int_1^3 2x\,dx = 8.0\ldots.$$

The Fundamental Theorem allows us to compute the integral exactly. We take $F'(x) = 2x$, so $F(x) = x^2$ and we obtain

$$\int_1^3 2x\,dx = F(3) - F(1) = 3^2 - 1^2 = 8.$$

In Example 1 we used the antiderivative $F(x) = x^2$, but $F(x) = x^2 + C$ works just as well for any constant C, because the constant cancels out when we subtract $F(a)$ from $F(b)$:

$$\int_1^3 2x\,dx = F(3) - F(1) = (3^2 + C) - (1^2 + C) = 8.$$

It is helpful to introduce a shorthand notation for $F(b) - F(a)$: we write

$$F(x)\Big|_a^b = F(b) - F(a).$$

For example:

$$\int_1^3 2x\,dx = x^2\Big|_1^3 = 3^2 - 1^2 = 8.$$

Example 2 Use the Fundamental Theorem to compute the following definite integrals:

(a) $\displaystyle\int_0^2 6x^2\,dx$ (b) $\displaystyle\int_0^2 t^3\,dt$ (c) $\displaystyle\int_1^2 (8x+5)\,dx$ (d) $\displaystyle\int_0^1 8e^{2t}\,dt$

Solution (a) Since $F'(x) = 6x^2$, we take $F(x) = 6(x^3/3) = 2x^3$. So

$$\int_0^2 6x^2 dx = 6\left(\frac{x^3}{3}\right)\Big|_0^2 = 2x^3\Big|_0^2 = 2\cdot 2^3 - 2\cdot 0^3 = 16.$$

(b) Since $F'(t) = t^3$, we take $F(t) = t^4/4$, so

$$\int_0^2 t^3\,dt = F(t)\Big|_0^2 = F(2) - F(0) = \frac{2^4}{4} - \frac{0^4}{4} = \frac{16}{4} - 0 = 4.$$

(c) Since $F'(x) = 8x + 5$, we take $F(x) = 4x^2 + 5x$, giving

$$\int_1^2 (8x+5)dx = (4x^2 + 5x)\Big|_1^2 = (4\cdot 2^2 + 5\cdot 2) - (4\cdot 1^2 + 5\cdot 1)$$
$$= 26 - 9 = 17.$$

(d) Since $F'(t) = 8e^{2t}$, we take $F(t) = (8e^{2t})/2 = 4e^{2t}$, so

$$\int_0^1 8e^{2t}\,dt = 8\left(\frac{1}{2}e^{2t}\right)\bigg|_0^1 = 4e^{2t}\bigg|_0^1 = 4e^2 - 4e^0 = 25.556.$$

Example 3 Write a definite integral to represent the area under the graph of $f(t) = e^{0.5t}$ between $t = 0$ and $t = 4$. Use the Fundamental Theorem to calculate the area.

Solution The function is graphed in Figure 7.1. We have

$$\text{Area} = \int_0^4 e^{0.5t}\,dt = 2e^{0.5t}\bigg|_0^4 = 2e^{0.5(4)} - 2e^{0.5(0)} = 2e^2 - 2 = 12.778.$$

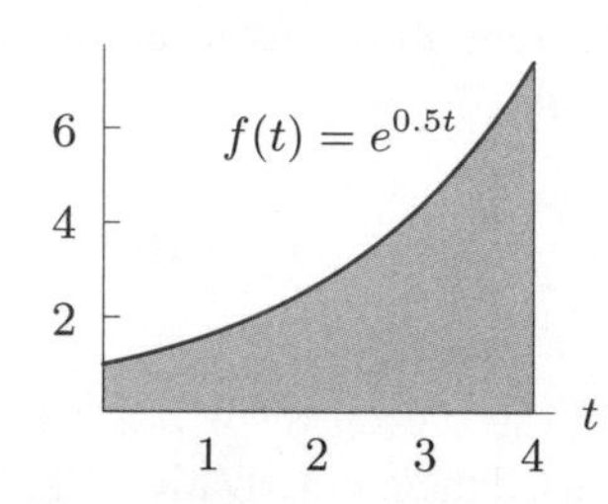

Figure 7.1: Shaded area $= \int_0^4 e^{0.5t}\,dt$

Definite Integrals by Substitution

There are two ways of computing a definite integral by substitution, as we see in the following example.

Example 4 Compute $\displaystyle\int_0^2 xe^{x^2}\,dx$.

Solution To evaluate this definite integral using the Fundamental Theorem of Calculus, we first need to find an antiderivative of $f(x) = xe^{x^2}$. The inside function is x^2, so we let $w = x^2$. Then $dw = 2x\,dx$, so $\frac{1}{2}\,dw = x\,dx$. Thus,

$$\int xe^{x^2}\,dx = \int e^w \frac{1}{2}\,dw = \frac{1}{2}e^w + C = \frac{1}{2}e^{x^2} + C.$$

Now we find the definite integral

$$\int_0^2 xe^{x^2}\,dx = \frac{1}{2}e^{x^2}\bigg|_0^2 = \frac{1}{2}(e^4 - e^0) = \frac{1}{2}(e^4 - 1).$$

There is another way to look at the same problem. After we established that

$$\int xe^{x^2}\,dx = \frac{1}{2}e^w + C,$$

our next two steps were to replace w by x^2, and then x by 2 and 0. We could have directly replaced the original limits of integration, $x = 0$ and $x = 2$, by the corresponding w limits. Since $w = x^2$, the w limits are $w = 0^2 = 0$ (when $x = 0$) and $w = 2^2 = 4$ (when $x = 2$), so we get

$$\int_{x=0}^{x=2} xe^{x^2}\,dx = \frac{1}{2}\int_{w=0}^{w=4} e^w\,dw = \frac{1}{2}e^w\bigg|_0^4 = \frac{1}{2}\left(e^4 - e^0\right) = \frac{1}{2}(e^4 - 1).$$

As we would expect, both methods give the same answer.

To Use Substitution to Find Definite Integrals

Either

- Compute the indefinite integral, expressing an antiderivative in terms of the original variable, and then evaluate the result at the original limits,

or

- Convert the original limits to new limits in terms of the new variable and do not convert the antiderivative back to the original variable.

Improper Integrals

So far, in our discussion of the definite integral $\int_a^b f(x)\,dx$, we have assumed that the interval $a \leq x \leq b$ is of finite length and the integrand f is continuous. An *improper integral* is a definite integral in which one (or both) of the limits of integration is infinite or the integrand is unbounded. An example of an improper integral is

$$\int_1^\infty \frac{1}{x^2}\,dx$$

This integral represents the area under the graph of $\frac{1}{x^2}$ from $x = 1$ infinitely far to the right. (See Figure 7.2.)

We estimate this area by letting the upper limit of integration get larger and larger. We see that

$$\int_1^{10} \frac{1}{x^2}\,dx = 0.9, \qquad \int_1^{100} \frac{1}{x^2}\,dx = 0.99, \qquad \int_1^{1000} \frac{1}{x^2}\,dx = 0.999,$$

and so on. These calculations suggest that as the upper limit of integration tends to infinity, the area tends to 1. We say that the improper integral $\int_1^\infty \frac{1}{x^2}\,dx$ *converges* to 1. To show that the integral converges to exactly 1 (and not to 1.0001, say), we need to use the Fundamental Theorem of Calculus. (See Problem 39.) It may seem strange that the region shaded in Figure 7.2 (which has infinite length) can have *finite* area. The area is finite because the values of the function $1/x^2$ shrink to zero so fast as $x \to \infty$. In other examples (where the integrand does not shrink to zero so fast), the area represented by an improper integral may not be finite. In that case, we say the improper integral *diverges*. (See Problem 43.)

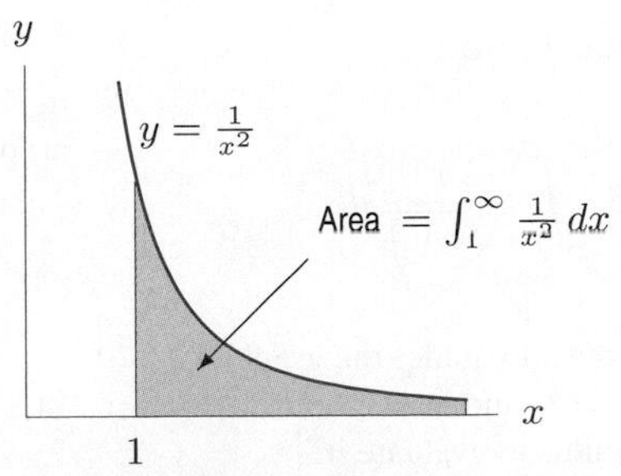

Figure 7.2: Area representation of improper integral

Problems for Section 7.3

Using the Fundamental Theorem, evaluate the definite integrals in Problems 1–20 exactly.

1. $\int_1^3 5\,dx$

2. $\int_0^4 6x\,dx$

3. $\int_1^2 (2x+3)\,dx$

4. $\int_0^2 (12x^2+1)\,dx$

5. $\int_1^2 \frac{1}{x^2}\,dx$

6. $\int_1^4 \frac{1}{\sqrt{x}}\,dx$

7. $\int_0^3 t^3\,dt$

8. $\int_0^5 3x^2\,dx$

9. $\int_1^3 6x^2\,dx$

10. $\int_1^2 5t^3\,dt$

11. $\int_0^1 (y^2+y^4)\,dy$

12. $\int_4^9 \sqrt{x}\,dx$

13. $\int_1^2 \frac{1}{x}\,dx$

14. $\int_0^2 \left(\frac{x^3}{3}+2x\right)\,dx$

15. $\int_0^1 2e^x\,dx$

16. $\int_0^1 e^{-0.2t}\,dt$

17. $\int_0^1 \sin\theta\,d\theta$

18. $\int_0^{\pi/4} (\sin t+\cos t)\,dt$

19. $\int_0^1 (6q^2+4)\,dq$

20. $\int_0^3 e^{0.05t}\,dt$

21. Use substitution to express each of the following integrals as a multiple of $\int_a^b (1/w)\,dw$ for some a and b. Then evaluate the integrals.

(a) $\int_0^1 \frac{x}{1+x^2}\,dx$ **(b)** $\int_0^{\pi/4} \frac{\sin x}{\cos x}\,dx$

Use integration by substitution and the Fundamental Theorem to evaluate the definite integrals in Problems 22–25.

22. $\int_0^2 x(x^2+1)^2\,dx$

23. $\int_0^1 2te^{-t^2}\,dt$

24. $\int_0^3 \frac{2x}{x^2+1}\,dx$

25. $\int_0^3 \frac{1}{\sqrt{t+1}}\,dt$

26. Write the definite integral for the area under the graph of $f(x) = 6x^2+1$ between $x=0$ and $x=2$. Use the Fundamental Theorem of Calculus to evaluate it.

27. Use the Fundamental Theorem to find the area under $f(x) = x^2$ between $x=1$ and $x=4$.

28. Use the Fundamental Theorem to find the average value of $f(x) = x^2+1$ on the interval $x=0$ to $x=10$. Illustrate your answer on a graph of $f(x)$.

29. Use the Fundamental Theorem of Calculus to find the average value of $f(x) = e^{0.5x}$ between $x=0$ and $x=3$. Show the average value on a graph of $f(x)$.

30. Find the exact area of the region bounded by the x-axis and the graph of $y = x^3 - x$.

31. Use the Fundamental Theorem to find the area under the graph of $f(x) = 1/(x+1)$ between $x=0$ and $x=2$.

32. Use the Fundamental Theorem to determine the value of b if the area under the graph of $f(x) = 8x$ between $x=1$ and $x=b$ is equal to 192. Assume $b>1$.

33. Use the Fundamental Theorem to determine the value of b if the area under the graph of $f(x) = x^2$ between $x=0$ and $x=b$ is equal to 100. Assume $b>0$.

34. Use the Fundamental Theorem to determine the value of b if the area under the graph of $f(x) = 4x$ between $x=1$ and $x=b$ is equal to 240. Assume $b>1$.

35. If t is in years since 1990, the population, P, of the world in billions can be modeled by $P = 5.3e^{0.014t}$.

(a) What does this model give for the world population in 1990? In 2000?
(b) Use the Fundamental Theorem to find the average population of the world during the 1990s.

36. Oil is leaking out of a ruptured tanker at the rate of $r(t) = 50e^{-0.02t}$ thousand liters per minute.

(a) At what rate, in liters per minute, is oil leaking out at $t=0$? At $t=60$?
(b) How many liters leak out during the first hour?

37. **(a)** Sketch the area represented by the improper integral $\int_0^\infty xe^{-x}\,dx$.
(b) Calculate $\int_0^b xe^{-x}\,dx$ for $b = 5, 10, 20$.
(c) The improper integral in part (a) converges. Use your answers to part (b) to estimate its value.

38. Graph $y = 1/x^2$ and $y = 1/x^3$ on the same axes. Which do you think is larger: $\int_1^\infty 1/x^2\,dx$ or $\int_1^\infty 1/x^3\,dx$? Why?

39. In this problem, you will show that the following improper integral converges to 1.

$$\int_1^\infty \frac{1}{x^2}\,dx.$$

(a) Use the Fundamental Theorem to find $\int_1^b 1/x^2\,dx$. Your answer will contain b.
(b) Now take the limit as $b \to \infty$. What does this tell you about the improper integral?

40. Decide if the improper integral $\int_0^\infty e^{-2t}\,dt$ converges, and if so, to what value, by the following method.

(a) Evaluate $\int_0^b e^{-2t}\,dt$ for $b = 3, 5, 7, 10$. What do you observe? Make a guess about the convergence of the improper integral.

(b) Find $\int_0^b e^{-2t}\,dt$ using the Fundamental Theorem. Your answer will contain b.

(c) Take a limit as $b \to \infty$. Does your answer confirm your guess?

41. **(a)** Evaluate $\int_0^b xe^{-x/10}\,dx$ for $b = 10, 50, 100, 200$.

(b) Assuming that it converges, estimate the value of $\int_0^\infty xe^{-x/10}\,dx$.

42. The rate, r, at which people get sick during an epidemic of the flu can be approximated by

$$r = 1000te^{-0.5t},$$

where r is measured in people/day and t is measured in days since the start of the epidemic.

(a) Write an improper integral representing the total number of people that get sick.

(b) Use a graph of r to represent the improper integral from part (a) as an area.

43. Consider the improper integral

$$\int_1^\infty \frac{1}{\sqrt{x}}\,dx.$$

(a) Use a calculator or computer to find $\int_1^b 1/(\sqrt{x})\,dx$ for $b = 100, 1000, 10{,}000$. What do you notice?

(b) Find $\int_1^b 1/(\sqrt{x})\,dx$ using the Fundamental Theorem of Calculus. Your answer will contain b.

(c) Now take the limit as $b \to \infty$. What does this tell you about the improper integral?

44. Find the exact area enclosed by the curve $y = x^2(1-x)^2$ and the x-axis.

45. Find the exact area below the curve $y = x^3(1-x)$ and above the x-axis.

46. An island has a carrying capacity of 1 million rabbits. (That is, no more than 1 million rabbits can be supported by the island.) The rabbit population is two at time $t = 1$ day and grows at a rate of $r(t)$ thousand rabbits/day until the carrying capacity is reached. For each of the following formulas for $r(t)$, is the carrying capacity ever reached? Explain your answer.

(a) $r(t) = 1/t^2$ **(b)** $r(t) = t$

(c) $r(t) = 1/\sqrt{t}$

47. A car moves with velocity, v, at time t in hours by

$$v(t) = \frac{60}{50^t} \text{ miles/hour.}$$

(a) Does the car ever stop?

(b) Write an integral representing the total distance traveled for $t \geq 0$.

(c) Do you think the car goes a finite distance for $t \geq 0$? If so, estimate that distance.

48. **(a)** During the 1970s, ACME Widgets sold at a continuous rate of $R = R_0e^{0.15t}$ widgets per year, where t is time in years since January 1, 1970. Suppose they were selling widgets at a rate of 1000 per year on the first day of the decade. How many widgets did they sell during the decade? How many did they sell if the rate on January 1, 1970 was 150,000,000 widgets per year?

(b) In the first case above (1000 widgets per year on January 1, 1970), how long did it take for half the widgets in the 1970s to be sold? In the second case (150,000,000 widgets per year on January 1, 1970), when had half the widgets in the 1970s been sold?

(c) In 1980 ACME began an advertising campaign claiming that half the widgets it had sold in the previous ten years were still in use. Based on your answer to part (b), about how long must a widget last in order to justify this claim?

7.4 ANALYZING ANTIDERIVATIVES GRAPHICALLY AND NUMERICALLY

In this section, we use the Fundamental Theorem to approximate values of F when the rate of change, F', and one value of the function, $F(a)$, are known.

Example 1 Suppose $F'(t) = (1.8)^t$ and $F(0) = 2$. Find the value of $F(b)$ for $b = 0, 0.1, 0.2, \ldots, 1.0$.

Solution Apply the Fundamental Theorem with $F'(t) = (1.8)^t$ and $a = 0$ to get values for $F(b)$. Since

$$F(b) - F(0) = \int_0^b F'(t)\,dt = \int_0^b (1.8)^t\,dt$$

and $F(0) = 2$, we have

$$F(b) = 2 + \int_0^b (1.8)^t\,dt.$$

Use a calculator or computer to estimate the definite integral $\int_0^b (1.8)^t\,dt$ for each value of b. For example, when $b = 0.1$, we find that $\int_0^b (1.8)^t\,dt = 0.103$. Thus $F(0.1) = 2.103$. Continuing in this way gives the values in Table 7.1.

Table 7.1 *Approximate values of F*

b	0	0.1	0.2	0.3	0.4	0.5	0.6	0.7	0.8	0.9	1.0
$F(b)$	2	2.103	2.212	2.328	2.451	2.581	2.719	2.866	3.021	3.186	3.361

Notice from the table that the function $F(b)$ is increasing between $b = 0$ and $b = 1$. This is because the derivative $F'(t) = (1.8)^t$ is positive for t between 0 and 1.

Example 2 The graph of the derivative F' of a function F is shown in Figure 7.3. Assuming that $F(20) = 150$, estimate the maximum value attained by F.

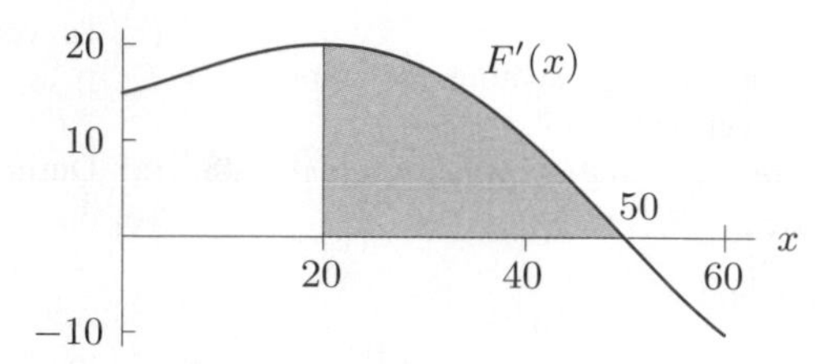

Figure 7.3: Graph of the derivative F' of some function F

Solution We know that $F(x)$ increases for $x < 50$ because the derivative of F is positive for $x < 50$. Similarly, $F(x)$ decreases for $x > 50$ because $F'(x)$ is negative for $x > 50$. Therefore, the graph of F rises until the point at which $x = 50$, and then it begins to fall. So the highest point on the graph of F is at $x = 50$ and the maximum value attained by F is $F(50)$. By the Fundamental Theorem:

$$F(50) - F(20) = \int_{20}^{50} F'(x)\,dx.$$

Since $F(20) = 150$, we have

$$F(50) = F(20) + \int_{20}^{50} F'(x)\,dx = 150 + \int_{20}^{50} F'(x)\,dx.$$

The definite integral is the area of the shaded region under the graph of F', which is roughly a triangle of base 30 and height 20. Therefore, the shaded area is about 300 and the maximum value attained by F is $F(50) \approx 150 + 300 = 450$.

Graphing a Function Given a Graph of its Derivative

Suppose we have the graph of f' and we want to sketch the graph of f. We know that when f' is positive, f is increasing, and when f' is negative, f is decreasing. If we want to know how much f increases or decreases, we compute a definite integral.

Example 3 Figure 7.4 shows the rate of change of the concentration of adrenaline, in micrograms per milliliter per minute, in a person's body. Sketch a graph of the concentration of adrenaline, in micrograms per milliliter, in the body as a function of time, in minutes.

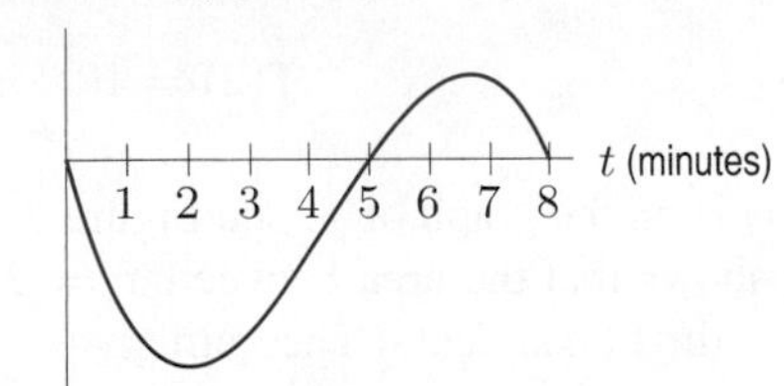

Figure 7.4: Rate of change of adrenaline concentration

Solution Since the initial concentration of the drug is not given, we can start the graph anywhere on the positive vertical axis. The rate of change is negative for $t < 5$ and positive for $t > 5$, so the concentration of adrenaline decreases until $t = 5$ and then increases. Since the area below the t-axis is greater than the area above the t-axis, the concentration of adrenaline decreases more than it increases. Thus, the concentration at $t = 8$ is less than the concentration at $t = 0$. Since the rate of change of concentration is zero at $t = 0$, 5, and 8, the graph of concentration is horizontal at these points. See Figure 7.5.

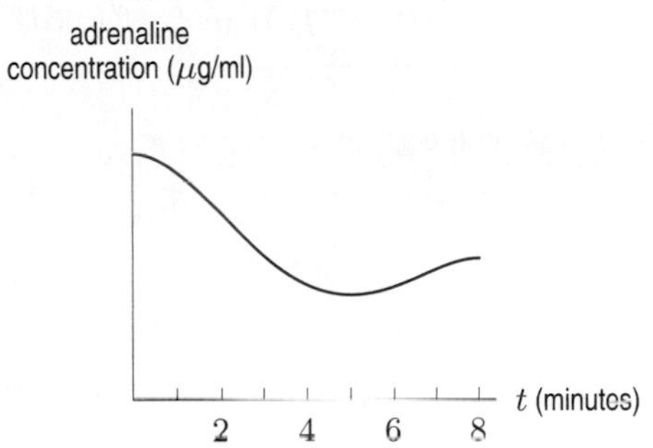

Figure 7.5: Adrenaline concentration

Example 4 Figure 7.6 shows the derivative $f'(x)$ of a function $f(x)$ and the values of some areas. If $f(0) = 10$, sketch a graph of the function $f(x)$. Give the coordinates of the local maxima and minima.

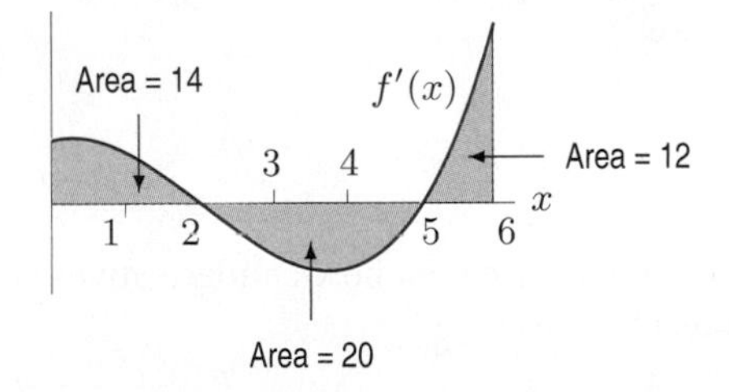

Figure 7.6: The graph of a derivative f'

Solution Figure 7.6 shows that the derivative f' is positive between 0 and 2, negative between 2 and 5, and positive between 5 and 6. Therefore, the function f is increasing between 0 and 2, decreasing between 2 and 5, and increasing between 5 and 6. (See Figure 7.7.) There is a local maximum at $x = 2$ and a local minimum at $x = 5$.

Notice that we can sketch the general shape of the graph of f without knowing any areas. The areas are used to make the graph more precise. We are told that $f(0) = 10$, so we plot the point $(0, 10)$ on the graph of f in Figure 7.8. The Fundamental Theorem and Figure 7.6 show that

$$f(2) - f(0) = \int_0^2 f'(x)\,dx = 14.$$

Therefore, the total change in f between $x = 0$ and $x = 2$ is 14. Since $f(0) = 10$, we have

$$f(2) = 10 + 14 = 24.$$

The point $(2, 24)$ is on the graph of f. See Figure 7.8.

Figure 7.6 shows that the area between $x = 2$ and $x = 5$ is 20. Since this area lies entirely below the x-axis, the Fundamental Theorem gives

$$f(5) - f(2) = \int_2^5 f'(x)\,dx = -20.$$

The total change in f is -20 between $x = 2$ and $x = 5$. Since $f(2) = 24$, we have

$$f(5) = 24 - 20 = 4.$$

Thus, the point $(5, 4)$ lies on the graph of f. Finally,

$$f(6) = f(5) + \int_5^6 f'(x)\,dx = 4 + 12 = 16,$$

so the point $(6, 16)$ is on the graph of f.

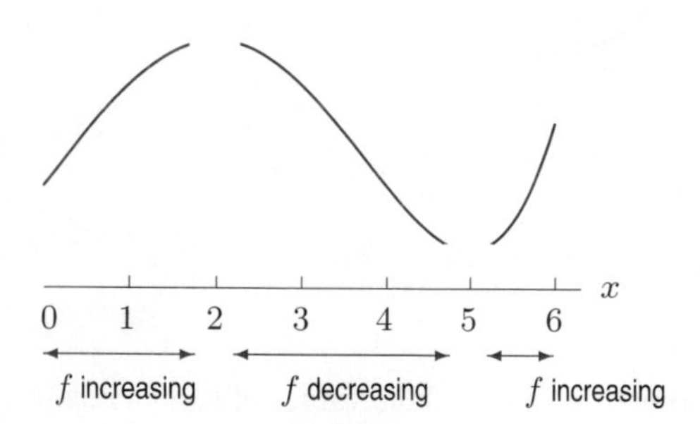

Figure 7.7: The shape of f

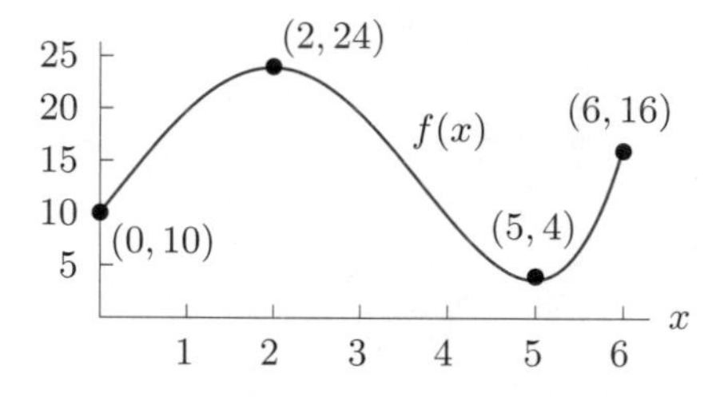

Figure 7.8: The graph of f

In this section we have seen how antiderivatives can by analyzed using the Fundamental Theorem of Calculus in the form

$$F(b) = F(0) + \int_0^b F'(t)\,dt.$$

In Section 8.2 we will see how antiderivatives can be constructed using the Second Fundamental Theorem introduced on page 272. This version of the theorem says that if f is a continuous function on an interval, and if a is any number in that interval, then the function F defined by

$$F(x) = \int_a^x f(t)\,dt$$

is an antiderivative of f; that is $F' = f$.

Problems for Section 7.4

1. Use Figure 7.9 and the fact that $P = 2$ when $t = 0$ to find values of P when $t = 1,\ 2,\ 3,\ 4$ and 5.

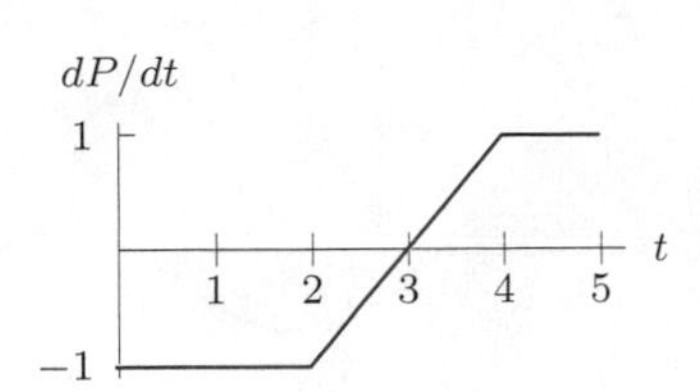

Figure 7.9

2. Figure 7.10 shows f. If $F' = f$ and $F(0) = 0$, find $F(b)$ for $b = 1, 2, 3, 4, 5, 6$.

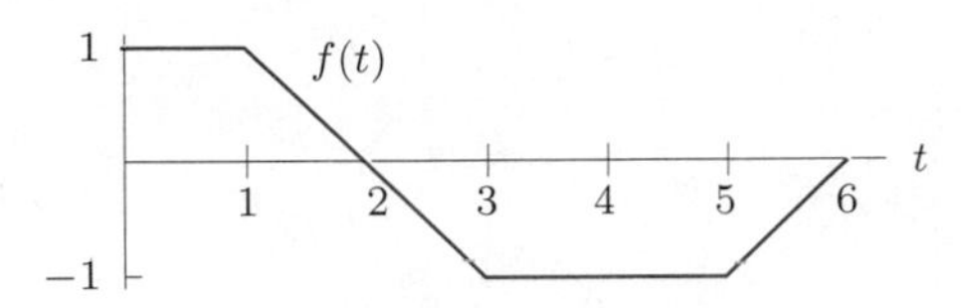

Figure 7.10

3. Figure 7.11 shows the derivative g'. If $g(0) = 0$, graph g. Give (x, y)-coordinates of all local maxima and minima.

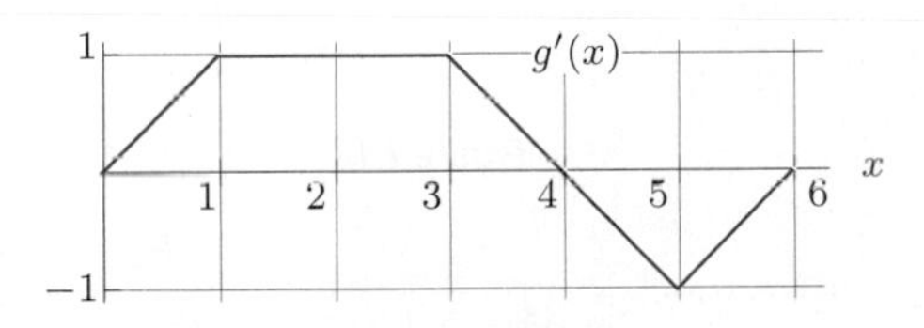

Figure 7.11

4. **(a)** Using Figure 7.12, estimate $\int_0^7 f(x)dx$.
(b) If F is an antiderivative of the same function f and $F(0) = 25$, estimate $F(7)$.

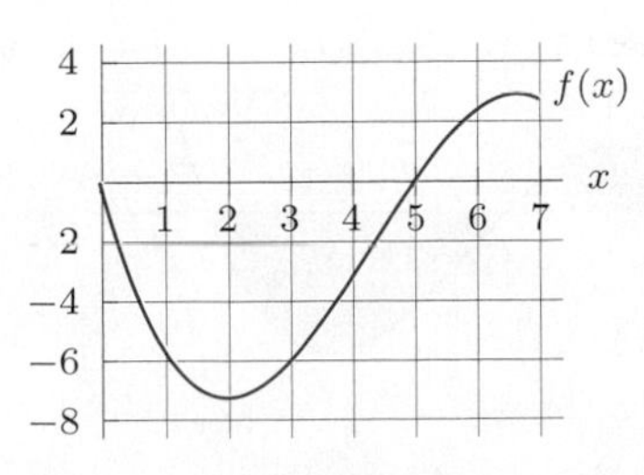

Figure 7.12

In Problems 5–8, sketch two functions F such that $F' = f$. In one case let $F(0) = 0$ and in the other, let $F(0) = 1$.

5.

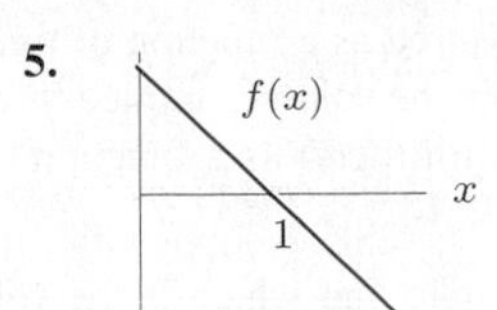

6.

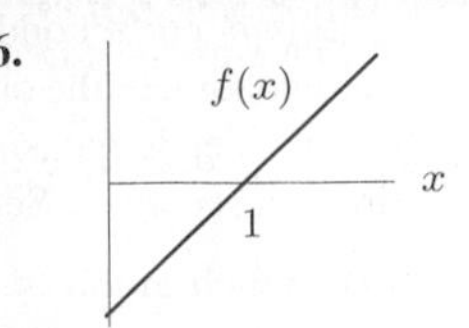

7.

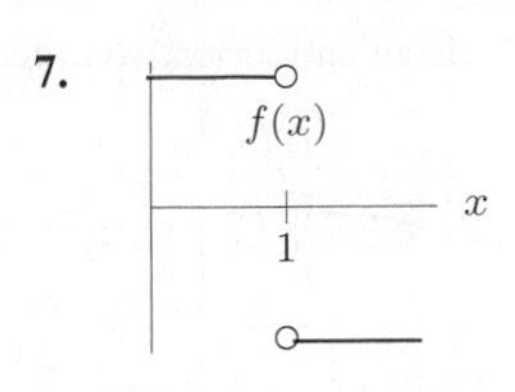

8.

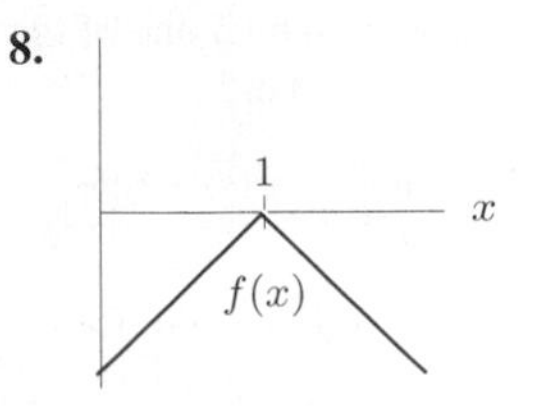

Problems 9–12 show the derivative f' of f.

(a) Where is f increasing and where is f decreasing? What are the x-coordinates of the local maxima and minima of f?

(b) Sketch a possible graph for f. (You don't need a scale on the vertical axis.)

9.

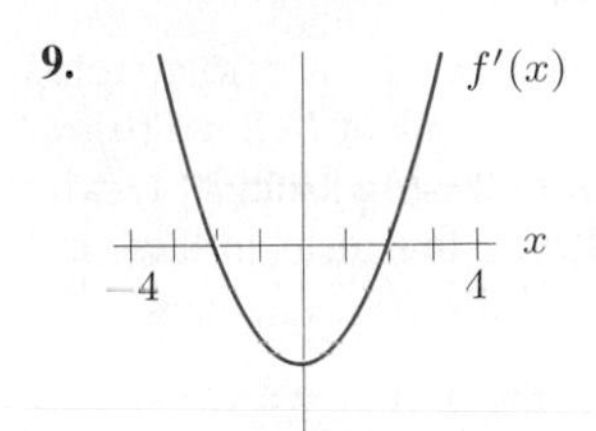

10.

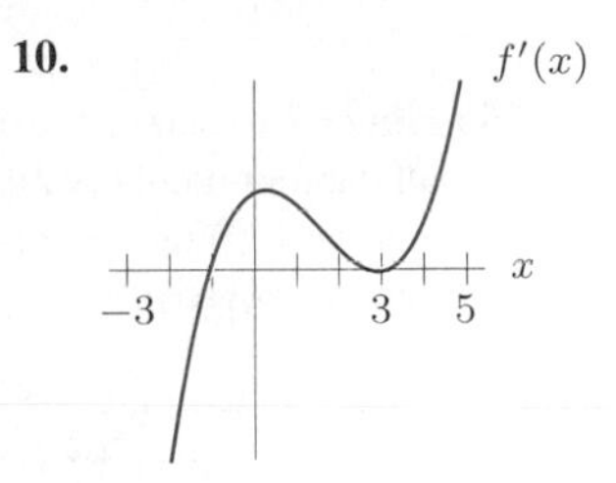

11.

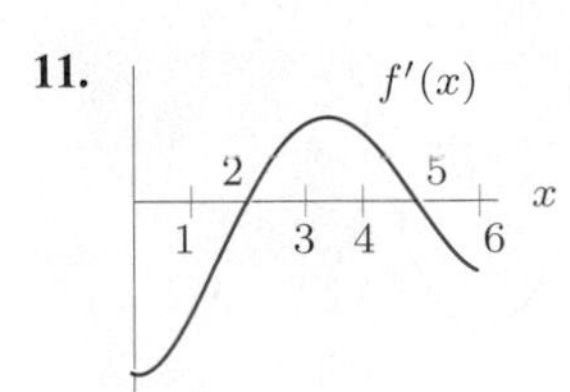

12.

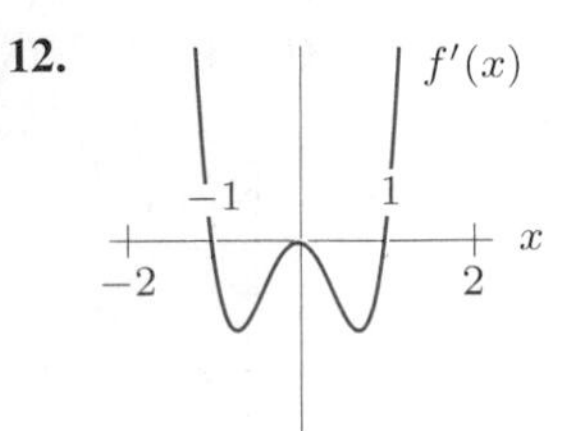

13. Figure 7.13 shows the rate at which photosynthesis is taking place in a leaf. The rate at which the leaf grows is approximately proportional to the rate of photosynthesis. Sketch the size of the leaf against time for 100 days.

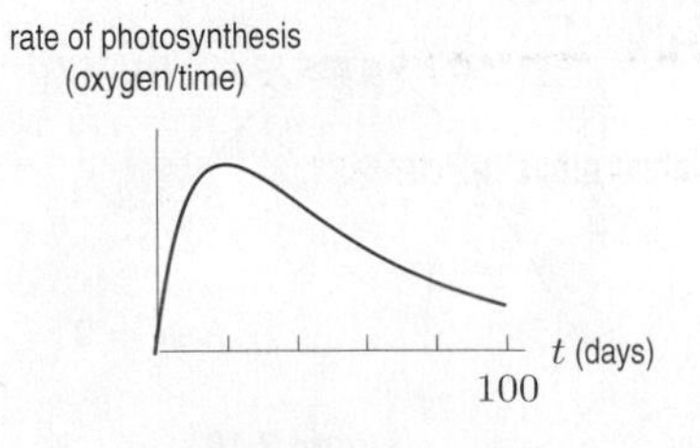

Figure 7.13

14. Urologists are physicians who specialize in the health of the bladder. In a common diagnostic test, urologists monitor the emptying of the bladder using a device that produces two graphs. In one of the graphs the flow rate (in milliliters per second) is measured as a function of time (in seconds). In the other graph, the volume emptied from the bladder is measured (in milliliters) as a function of time (in seconds). See Figure 7.14.

(a) Which graph is the flow rate and which is the volume?

(b) Which one of these graphs is an antiderivative of the other?

(I)

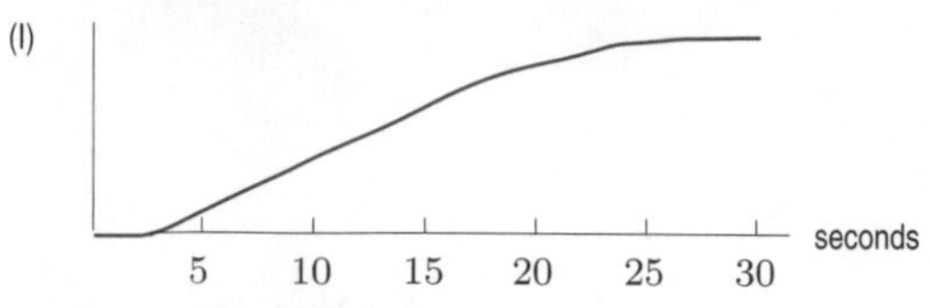

(II)

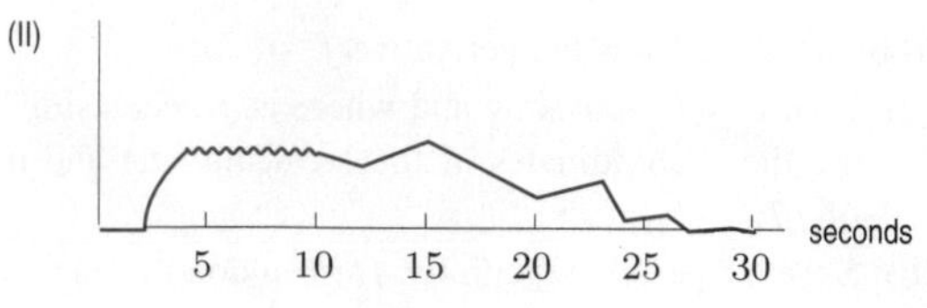

Figure 7.14

15. Figure 7.15 shows the derivative F' of F. Let $F(0) = 0$. Of the four numbers $F(1)$, $F(2)$, $F(3)$, and $F(4)$, which is largest? Which is smallest? How many of these numbers are negative?

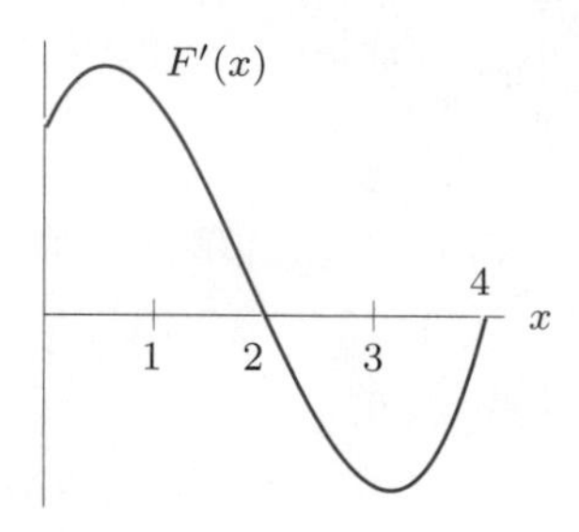

Figure 7.15

16. Figure 7.16 shows the derivative $F'(x)$. If $F(0) = 5$, find the value of $F(1)$, $F(3)$, and $F(4)$.

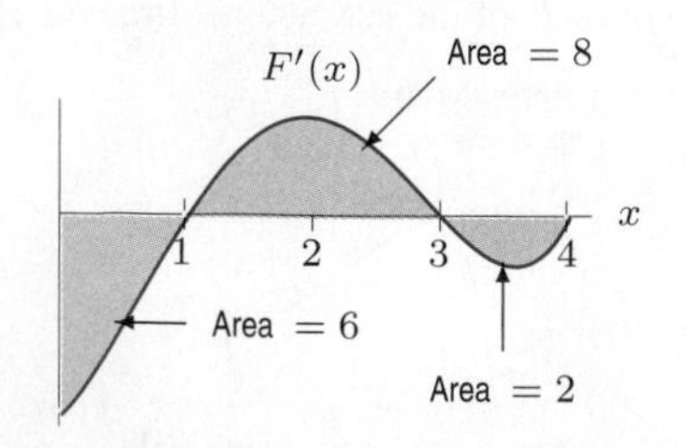

Figure 7.16

17. Using Figure 7.17, sketch a graph of an antiderivative $G(t)$ of $g(t)$ satisfying $G(0) = 5$. Label each critical point of $G(t)$ with its coordinates.

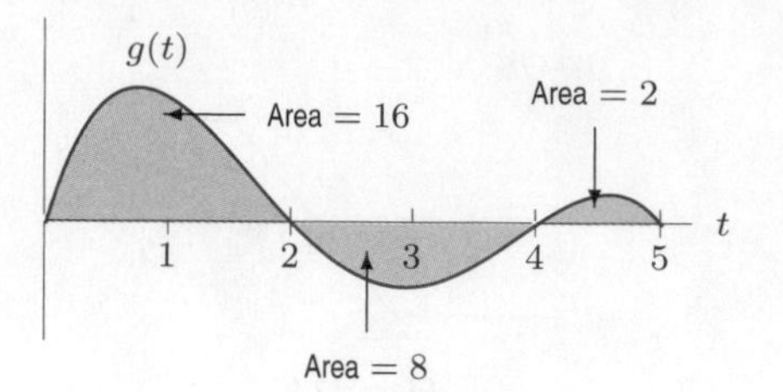

Figure 7.17

18. Figure 7.18 shows the derivative F'. If $F(0) = 14$, graph F. Give (x, y)-coordinates of all local maxima and minima.

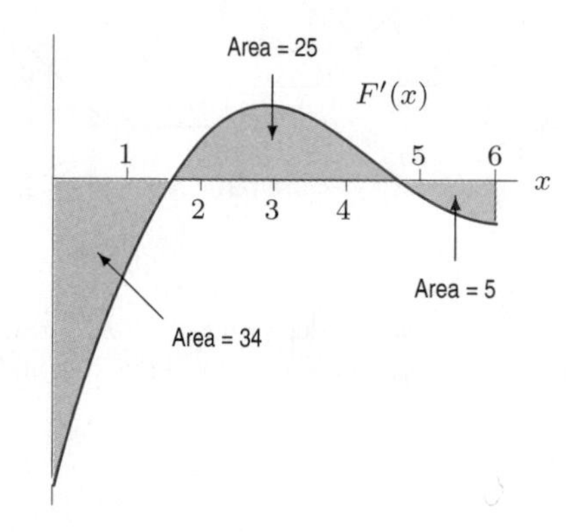

Figure 7.18

19. Figure 7.19 shows the derivative $F'(t)$. If $F(0) = 3$, find the values of $F(2)$, $F(5)$, $F(6)$. Graph $F(t)$.

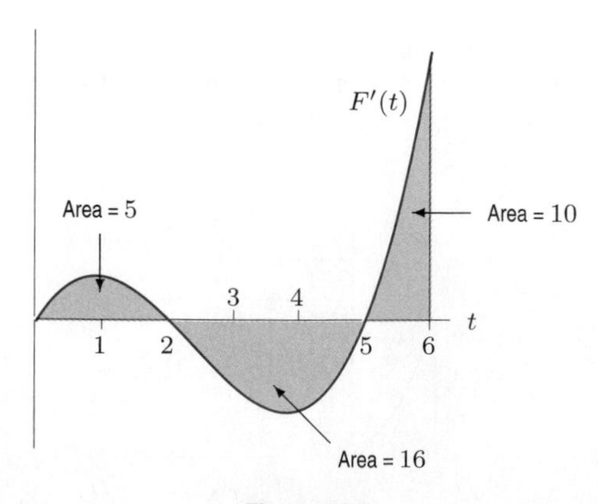

Figure 7.19

Problems 20–21 give a graph of $f'(x)$. Graph $f(x)$. Mark the points $x_1, \ldots, x_4$ on your graph and label local maxima, local minima and points of inflection.

20.

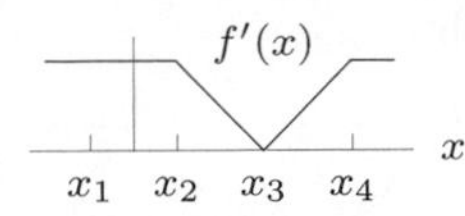

21.

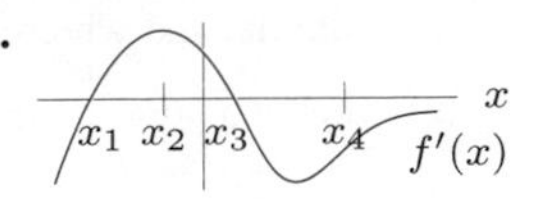

Problems 22–23 concern the graph of f' in Figure 7.20.

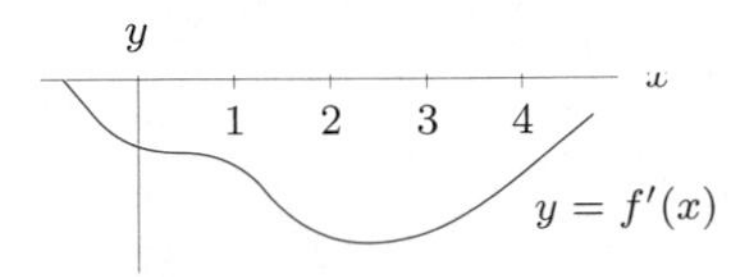

Figure 7.20: Note: Graph of f', not f

22. Which is greater, $f(0)$ or $f(1)$?

23. List the following in increasing order:

$$\frac{f(4)-f(2)}{2}, \quad f(3)-f(2), \quad f(4)-f(3).$$

For Problems 24–27, show the following quantities on Figure 7.21.

Figure 7.21

24. A length representing $f(b) - f(a)$.

25. A slope representing $\dfrac{f(b)-f(a)}{b-a}$.

26. An area representing $F(b) - F(a)$, where $F' = f$.

27. A length roughly approximating

$$\frac{F(b)-F(a)}{b-a}, \text{ where } F' = f.$$

CHAPTER SUMMARY

- **Computing antiderivatives**
  Powers and polynomials, e^{kx}, $\sin(kx)$, $\cos(kx)$, integration by substitution
- **Using antiderivatives to compute definite integrals analytically**
- **Using definite integrals to compute antiderivatives numerically and graphically**
- **Improper integrals**

REVIEW PROBLEMS FOR CHAPTER SEVEN

In Problems 1–10, find an antiderivative.

1. $k(x) = 10 + 8x^3$

2. $p(x) = x^2 - 6x + 17$

3. $f(x) = x + x^5 + x^{-5}$

4. $f(z) = e^z + 3$

5. $p(r) = 2\pi r$

6. $g(x) = \dfrac{1}{x} + \dfrac{1}{x^2} + \dfrac{1}{x^3}$

7. $g(z) = \dfrac{1}{z^3}$

8. $h(t) = \cos t$

9. $g(x) = (x+1)^3$

10. $f(x) = (2x+1)^3$

Find the indefinite integrals in Problems 11–22.

11. $\displaystyle\int 9x^2\,dx$

12. $\displaystyle\int (5x+7)\,dx$

13. $\displaystyle\int (x+1)^2\,dx$

14. $\displaystyle\int (e^x + 5)\,dx$

15. $\displaystyle\int \left(\frac{3}{t} - \frac{2}{t^2}\right)dt$

16. $\displaystyle\int \left(\frac{x+1}{x}\right)dx$

17. $\displaystyle\int (8t+3)\,dt$

18. $\displaystyle\int \left(1 + \frac{1}{p}\right)dp$

19. $\displaystyle\int \left(x^2 + \frac{1}{x}\right)dx$

20. $\displaystyle\int (3\cos x - 7\sin x)\,dx$

21. $\displaystyle\int \left(\frac{2}{x} + \pi\sin x\right)dx$

22. $\displaystyle\int (2e^x - 8\cos x)\,dx$

Find the definite integrals in Problems 23–28 using the Fundamental Theorem.

23. $\int_0^2 (3t^2 + 4t + 3)\, dt$ **24.** $\int_{-3}^{-1} \frac{2}{r^3}\, dr$

25. $\int_{-1}^{1} \cos t\, dt$ **26.** $\int_1^2 \frac{1}{2t}\, dt$

27. $\int_1^2 \frac{1}{t^2}\, dt$ **28.** $\int_2^5 (x^3 - \pi x^2)\, dx$

In Problems 29–31, find an antiderivative $F(x)$ with $F'(x) = f(x)$ and $F(0) = 0$. Is there only one possible solution?

29. $f(x) = 2x$ **30.** $f(x) = \frac{1}{4}x$ **31.** $f(x) = \sin x$

In Problems 32–41, integrate by substitution.

32. $\int 3x^2(x^3+1)^4 dx$ **33.** $\int 2qe^{q^2+1} dq$

34. $\int \frac{2x}{x^2+1} dx$ **35.** $\int \frac{1}{(3x+1)^2} dx$

36. $\int \frac{x}{\sqrt{x^2+4}} dx$ **37.** $\int (5x-7)^{10} dx$

38. $\int 100e^{-0.2t} dt$ **39.** $\int x\sqrt{3x^2+4}dx$

40. $\int x\sin(4x^2)dx$ **41.** $\int 12x^2\cos(x^3)dx$

42. Find the exact area under the graph of $f(x) = xe^{x^2}$ between $x = 0$ and $x = 2$.

43. If t is in years, and $t = 0$ is January 1, 2000, worldwide petroleum consumption, r, in quadrillion (10^{15}) BTUs per year, is modeled by

$$r = 155e^{0.015t}.$$

(a) Write a definite integral for the quantity of petroleum consumed between the start of 2000 and the start of 2005.

(b) Use the Fundamental Theorem of Calculus to evaluate the integral. Give units with your answer.

44. The derivative $F'(t)$ is graphed in Figure 7.22. Given that $F(0) = 5$, calculate $F(t)$ for $t = 1, 2, 3, 4, 5$.

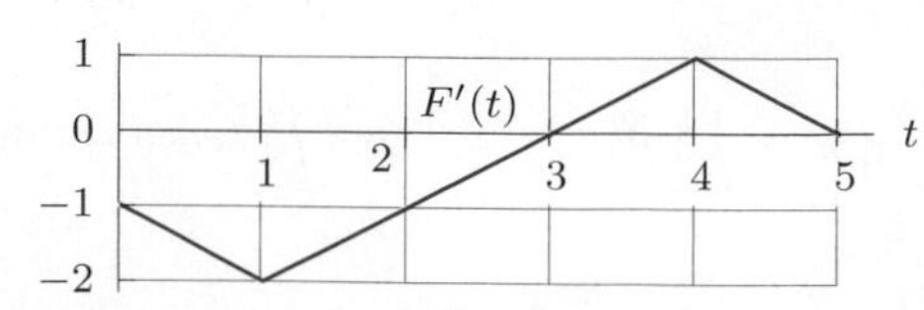

Figure 7.22

In Problems 45–46, a graph of f is given. Let $F'(x) = f(x)$.

(a) What are the critical points of $F(x)$?

(b) Which critical points are local maxima, which are local minima, and which are neither?

(c) Sketch a possible graph of $F(x)$.

45.

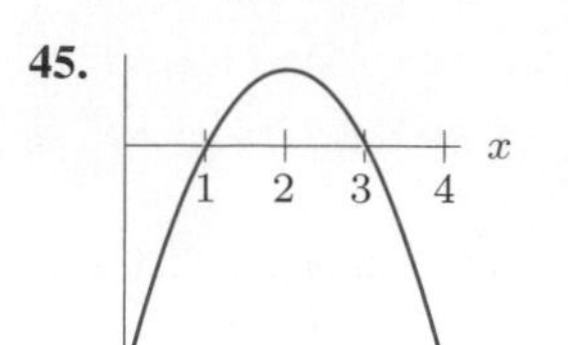

46.

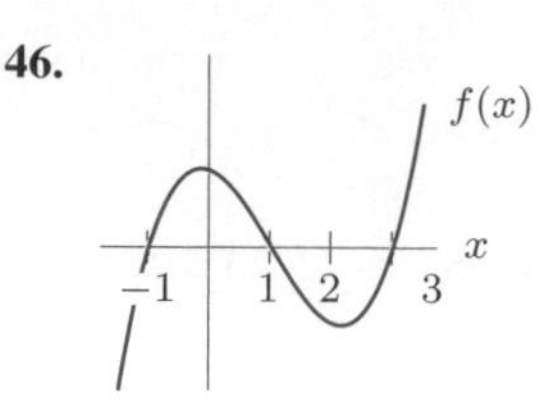

47. Use Figure 7.23 and the fact that $F(2) = 3$ to sketch the graph of $F(x)$. Label the values of at least four points.

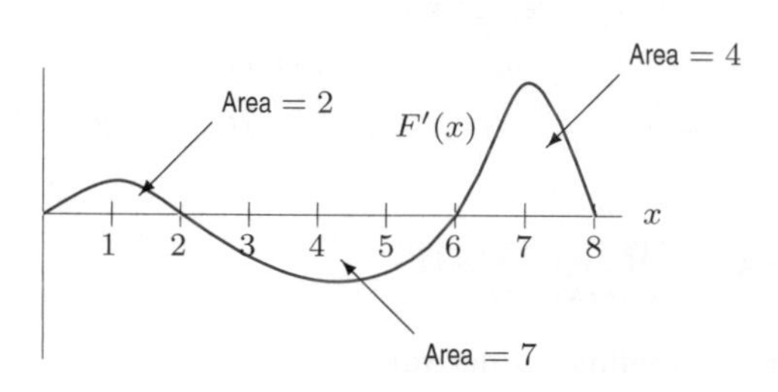

Figure 7.23

48. Suppose $\int_0^2 g(t)\, dt = 5$. Calculate the following:

(a) $\int_0^4 g(t/2)\, dt$ **(b)** $\int_0^2 g(2-t)\, dt$

49. **(a)** Graph $f(x) = e^{-x^2}$ and shade the area represented by the improper integral $\int_{-\infty}^{\infty} e^{-x^2}\, dx$.

(b) Find $\int_{-a}^{a} e^{-x^2}\, dx$ for $a = 1$, $a = 2$, $a = 3$, $a = 5$.

(c) The improper integral $\int_{-\infty}^{\infty} e^{-x^2}\, dx$ converges to a finite value. Use your answers from part (b) to estimate that value.

50. At a time t hours after taking a tablet, the rate at which a drug is being eliminated is

$$r(t) = 50\left(e^{-0.1t} - e^{-0.2t}\right) \text{ mg/hr}.$$

Assuming that all the drug is eventually eliminated, calculate the original dose.

PROJECTS FOR CHAPTER SEVEN

1. **Quabbin Reservoir** The Quabbin Reservoir in the western part of Massachusetts provides most of Boston's water. The graph in Figure 7.24 represents the flow of water in and out of the Quabbin Reservoir throughout 2004.
   (a) Sketch a possible graph for the quantity of water in the reservoir, as a function of time.
   (b) When, in the course of 2004, was the quantity of water in the reservoir largest? Smallest? Mark and label these points on the graph you drew in part (a).
   (c) When was the quantity of water increasing most rapidly? Decreasing most rapidly? Mark and label these times on both graphs.
   (d) By July 2005 the quantity of water in the reservoir was about the same as in January 2004. Draw plausible graphs for the flow into and the flow out of the reservoir for the first half of 2005. Explain your graphs.

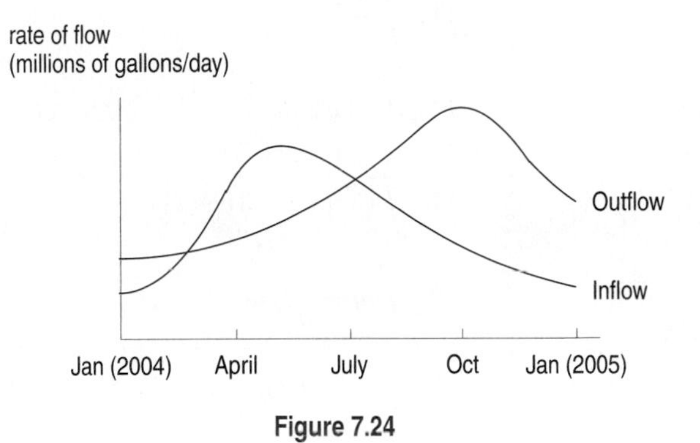

Figure 7.24

FOCUS ON PRACTICE

For Problems 1–45, evaluate the integrals. Assume a, b, A, B, P_0, h, and k are constants.

1. $\int (t^3 + 6t^2)\,dt$
2. $\int (u^4 + 5)\,du$
3. $\int (x^2 + \frac{1}{x^2})\,dx$
4. $\int e^{3r}\,dr$
5. $\int 3\sqrt{w}\,dw$
6. $\int (ax^2 + b)\,dx$
7. $\int (t^2 + 5t + 1)\,dt$
8. $\int 100e^{-0.5t}\,dt$
9. $\int (w^4 - 12w^3 + 6w^2 - 10)\,dw$
10. $\int (p^2 + \frac{5}{p})\,dp$
11. $\int \frac{dq}{\sqrt{q}}$
12. $\int 3\sin\theta\,d\theta$
13. $\int \left(\frac{4}{x} + \frac{5}{x^2}\right)\,dx$
14. $\int P_0 e^{kt}\,dt$
15. $\int (q^3 + 8q + 15)\,dq$
16. $\int 1000e^{0.075t}\,dt$
17. $\int (5\sin x + 3\cos x)\,dx$
18. $\int (10 + 5\sin x)\,dx$
19. $\int \pi r^2 h\,dr$
20. $\int (q + \frac{1}{q^3})\,dq$
21. $\int 15p^2q^4\,dp$
22. $\int 15p^2q^4\,dq$
23. $\int (3x^2 + 6e^{2x})\,dx$
24. $\int \frac{5}{w}\,dw$
25. $\int 5e^{2q}\,dq$
26. $\int \left(p^3 + \frac{1}{p}\right)\,dp$
27. $\int (Ax^3 + Bx)\,dx$
28. $\int (6\sqrt{x} + 15)\,dx$
29. $\int (x^2 + 8 + e^x)\,dx$
30. $\int 25e^{-0.04q}\,dq$
31. $\int (x^3 + 5x^2 + 6)dx$
32. $\int \left(\frac{a}{x} + \frac{b}{x^2}\right)\,dx$
33. $\int (Aq + B)dq$
34. $\int \left(\frac{6}{\sqrt{x}} + 8\sqrt{x}\right)\,dx$
35. $\int (e^{2t} + 5)dt$
36. $\int \sin(3x)dx$
37. $\int 12\cos(4x)dx$
38. $\int A\sin(Bt)dt$
39. $\int x(x^2 + 9)^6 dx$
40. $\int x\cos(x^2 + 4)dx$
41. $\int \frac{1}{y+2}dy$
42. $\int \sqrt{3x+1}dx$
43. $\int \frac{e^t}{e^t + 1}dt$
44. $\int \sin^2 x\cos x dx$
45. $\int \frac{\cos x}{\sqrt{1 + \sin x}}\,dx$

Chapter Eight

PROBABILITY

This chapter introduces the probability density and cumulative distribution functions, which show how a quantity such as income is distributed through a population. We see the relationship between these functions and how to calculate probabilities from them. The mean, median, and normal distribution are introduced.

8.1 DENSITY FUNCTIONS

Understanding the distribution of various quantities through the population can be important to decision makers. For example, the income distribution gives useful information about the economic structure of a society. In this section we look at the distribution of ages in the US. To allocate funding for education, health care, and social security, the government needs to know how many people are in each age group. We see how to represent such information by a density function.

US Age Distribution

Table 8.1 *Distribution of ages in the US in 2000*

Age group	Percentage of total population
0 – 20	29%
20 – 40	29%
40 – 60	26%
60 – 80	13%
80 – 100	3%

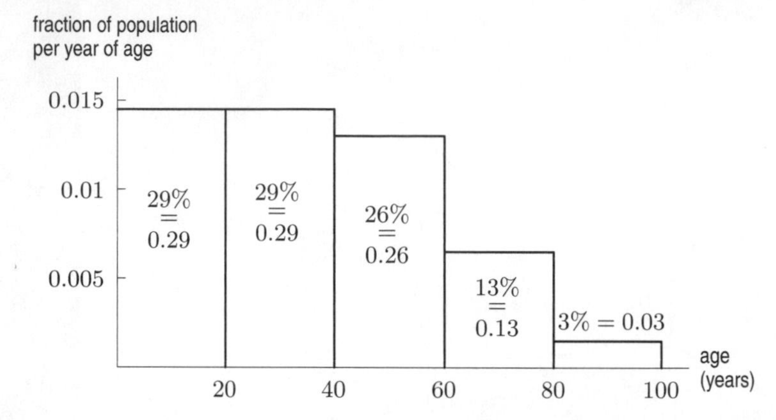

Figure 8.1: How ages were distributed in the US in 2000

Suppose we have the data in Table 8.1 showing how the ages of the US population[1] were distributed in 2000. To represent this information graphically we use a type of *histogram*[2] putting a vertical bar above each age group in such a way that the *area* of each bar represents the percentage in that age group. The total area of all the rectangles is $100\% = 1$. We only consider people who are less than 100 years old.[3] For the 0–20 age group, the base of the rectangle is 20, and we want the area to be 29%, so the height must be $29\%/20 = 1.45\%$. We treat ages as though they were continuously distributed. The category 0–20, for example, contains people who are just one day short of their twentieth birthday. Notice that the vertical axis is measured in percent/year. (See Figure 8.1.) ,

Example 1 In 2000, estimate what percentage of the US population was:

(a) Between 20 and 60 years old?
(b) Less than 10 years old?
(c) Between 75 and 80 years old?
(d) Between 80 and 85 years old?

Solution

(a) We add the percentages, so $29\% + 26\% = 55\%$.

(b) To find the percentage less than 10 years old, we could assume, for example, that the population was distributed evenly over the 0–20 group. (This means we are assuming that babies were born at a fairly constant rate over the last 20 years, which is probably reasonable.) If we make this assumption, then we can say that the population less than 10 years old was about half that in the 0–20 group, that is, 14.5%. Notice that we get the same result by computing the area of the rectangle from 0 to 10. (See Figure 8.2.)

(c) To find the population between 75 and 80 years old, since 13% of Americans in 1990 were in the 60-80 group, we might apply the same reasoning and say that $\frac{1}{4}(13\%) = 3.25\%$ of the population was in this age group. This result is represented as an area in Figure 8.2. The assumption that the population was evenly distributed is not a good one here; certainly there were more people between the ages of 60 and 65 than between 75 and 80. Thus, the estimate of 3.25% is certainly too high.

[1] www.censusscope.org/us/chart_age.html, accessed May 10, 2005
[2] There are other types of histograms which have frequency on the vertical axis.
[3] In fact, 0.02% of the population is over 100, but this is too small to be visible on the histogram.

(d) Again using the (faulty) assumption that ages in each group were distributed uniformly, we would find that the percentage between 80 and 85 was $\frac{1}{4}(3\%) = 0.75\%$. (See Figure 8.2.) This estimate is also poor—there were certainly more people in the 80–85 group than, say, the 95–100 group, and so the 0.75% estimate is too low.

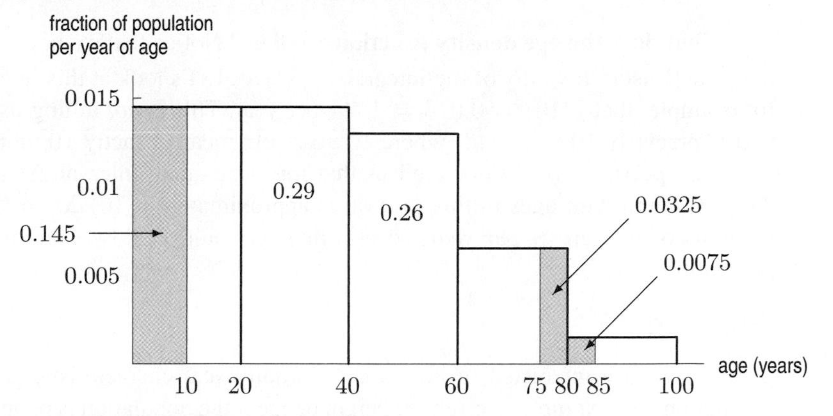

Figure 8.2: Ages in the US in 2000—various subgroups (for Example 1)

Smoothing Out the Histogram

We could get better estimates if we had smaller age groups (each age group in Figure 8.1 is 20 years, which is quite large) or if the histogram were smoother. Suppose we have the more detailed data in Table 8.2, which leads to the new histogram in Figure 8.3.

As we get more detailed information, the upper silhouette of the histogram becomes smoother, but the area of any of the bars still represents the percentage of the population in that age group. Imagine, in the limit, replacing the upper silhouette of the histogram by a smooth curve in such a way that area under the curve above one age group is the same as the area in the corresponding rectangle. The total area under the whole curve is again 100% = 1. (See Figure 8.3.)

The Age Density Function

If t is age in years, we define $p(t)$, the age *density function*, to be a function which "smooths out" the age histogram. This function has the property that

$$\text{Fraction of population between ages } a \text{ and } b = \text{Area under graph of } p \text{ between } a \text{ and } b = \int_a^b p(t)dt.$$

Table 8.2 *Ages in the US in 2000 (more detailed)*

Age group	Percentage of total population
0 – 10	14%
10 – 20	15%
20 – 30	14%
30 – 40	15%
40 – 50	15%
50 – 60	11%
60 – 70	7%
70 – 80	6%
80 – 90	2%
90 – 100	1%

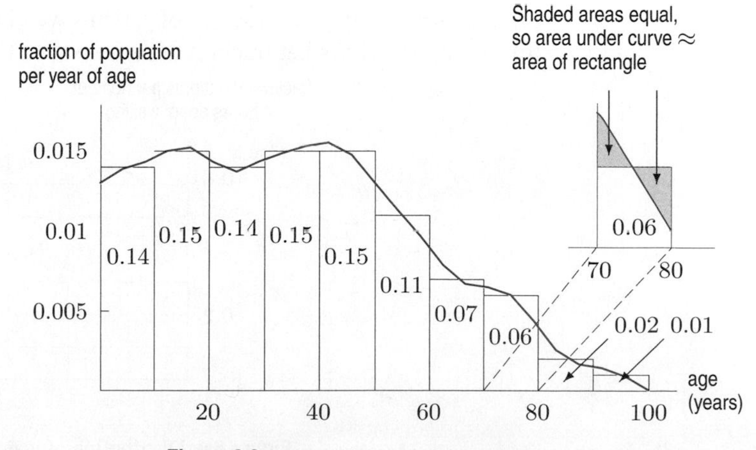

Figure 8.3: Smoothing out the age histogram

If a and b are the smallest and largest possible ages (say, $a = 0$ and $b = 100$), so that the ages of all of the population are between a and b, then

$$\int_a^b p(t)dt = \int_0^{100} p(t)dt = 1.$$

What does the age density function p tell us? Notice that we have not talked about the meaning of $p(t)$ itself, but *only* of the integral $\int_a^b p(t)\,dt$. Let's look at this in a bit more detail. Suppose, for example, that $p(10) = 0.014 = 1.4\%$ per year. This is *not* telling us that 1.4% of the population is precisely 10 years old (where 10 years old means exactly 10, not $10\frac{1}{2}$, not $10\frac{1}{4}$, not 10.1). However, $p(10) = 0.014$ does tell us that for some small interval Δt around 10, the fraction of the population with ages in this interval is approximately $p(10)\,\Delta t = 0.014\,\Delta t$. Notice also that the units of $p(t)$ are *% per year*, so $p(t)$ must be multiplied by years to give a percentage of the population.

The Density Function

Suppose we are interested in how a certain numerical characteristic, x, is distributed through a population. For example, x might be height or age if the population is people, or might be wattage for a population of light bulbs. Then we define a general density function with the following properties:

The function, $p(x)$, is a **density function** if

Fraction of population for which x is between a and b	=	Area under graph of p between a and b	=	$\int_a^b p(x)dx.$

$$\int_{-\infty}^{\infty} p(x)\,dx = 1 \quad \text{and} \quad p(x) \geq 0 \quad \text{for all } x.$$

The density function must be nonnegative if its integral always gives a fraction of the population. The fraction of the population with x between $-\infty$ and ∞ is 1 because the entire population has the characteristic x between $-\infty$ and ∞. The function p that was used to smooth out the age histogram satisfies this definition of a density function. Notice that we do not assign a meaning to the value $p(x)$ directly, but rather interpret $p(x)\,\Delta x$ as the fraction of the population with the characteristic in a short interval of length Δx around x.

Example 2 Figure 8.4 gives the density function for the amount of time spent waiting at a doctor's office.

(a) What is the longest time anyone has to wait?
(b) Approximately what fraction of patients wait between 1 and 2 hours?
(c) Approximately what fraction of patients wait less than an hour?

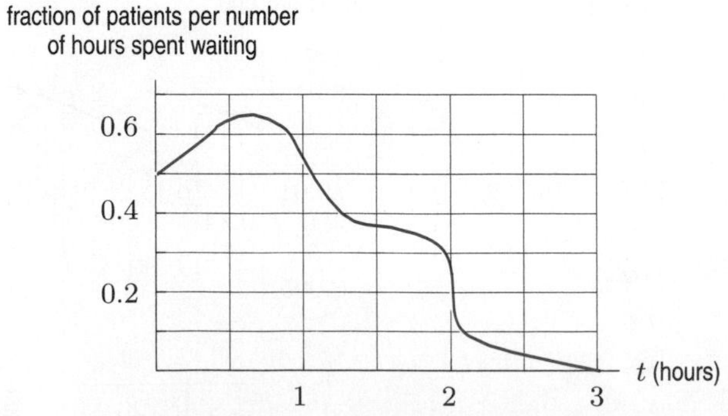

Figure 8.4: Distribution of waiting time at a doctor's office

Solution
(a) The density function is zero for all $t > 3$, so no one waits more than 3 hours. The longest time anyone has to wait is 3 hours.
(b) The fraction of patients who wait between 1 and 2 hours is equal to the area under the density curve between $t = 1$ and $t = 2$. We can estimate this area by counting squares: There are about 7.5 squares in this region, each of area $(0.5)(0.1) = 0.05$. The area is approximately $(7.5)(0.05) = 0.375$. Thus about 37.5% of patients wait between 1 and 2 hours.
(c) This fraction is equal to the area under the density function for $t < 1$. There are about 12 squares in this area, and each has area 0.05 as in part (b), so our estimate for the area is $(12)(0.05) = 0.60$. Therefore, about 60% of patients see the doctor in less than an hour.

Problems for Section 8.1

In Problems 1–2, graph a density function representing the given distribution.

1. The age at which a person dies in a society with high infant mortality and in which adults usually die between age 40 and age 60.

2. The heights of the people in an elementary school.

In Problems 3–6, given that $p(x)$ is a density function, find the value of a.

3.
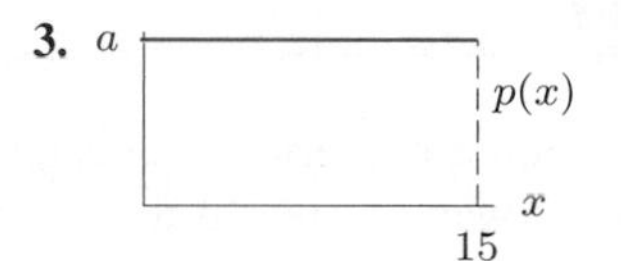

4.
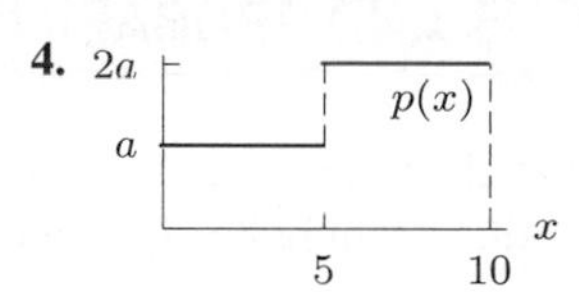

5.
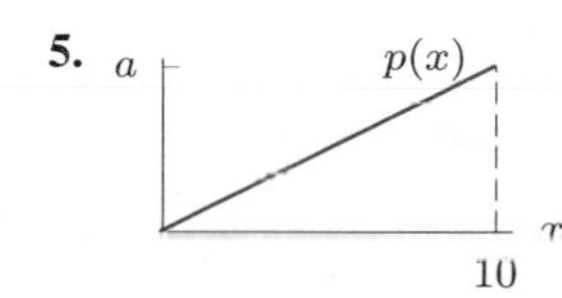

6.
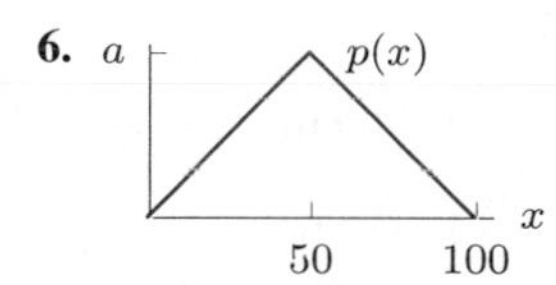

7. Figure 8.5[4] shows the distribution of elevation, in miles, across the earth's surface. Positive elevation denotes land above sea level; negative elevation shows land below sea level (i.e., the ocean floor).

(a) Describe in words the elevation of most of the earth's surface.
(b) Approximately what fraction of the earth's surface is below sea level?

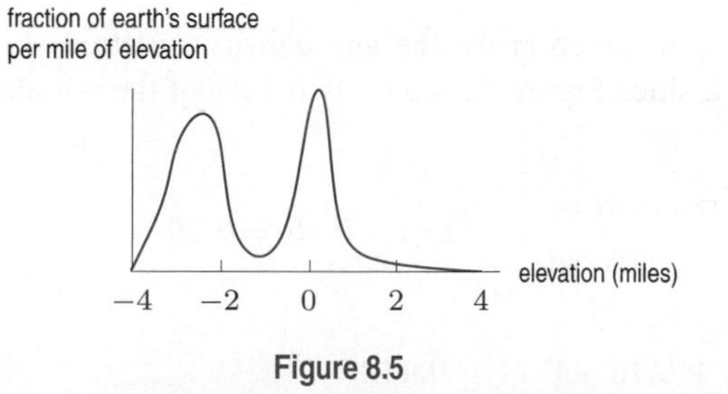

Figure 8.5

In Problems 8–11, the distribution of the heights, x, in meters, of trees is represented by the density function $p(x)$. In each case, calculate the fraction of trees which are:

(a) Less than 5 meters high
(b) More than 6 meters high
(c) Between 2 and 5 meters high

8. **9.**
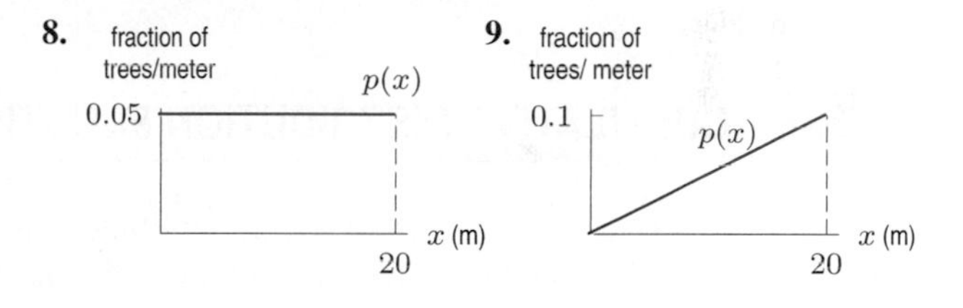

10. **11.**
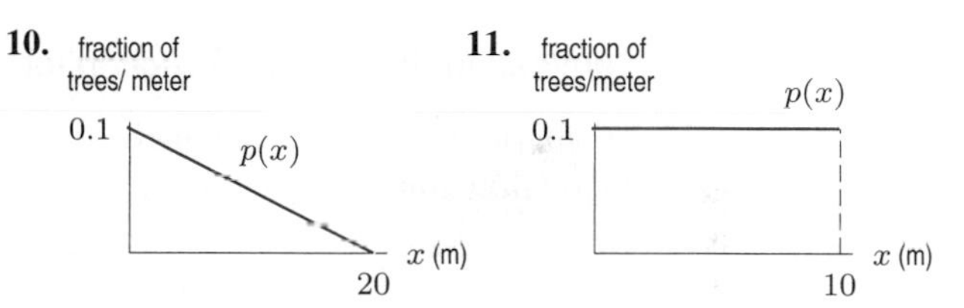

12. An insect has a life-span of no more than one year. Figure 8.6 shows the density function, $p(t)$, for the life-span.

(a) Do more insects die in the first month of their life or the twelfth month of their life?
(b) What fraction of the insects live no more than 6 months?
(c) What fraction of the insects live more than 9 months?

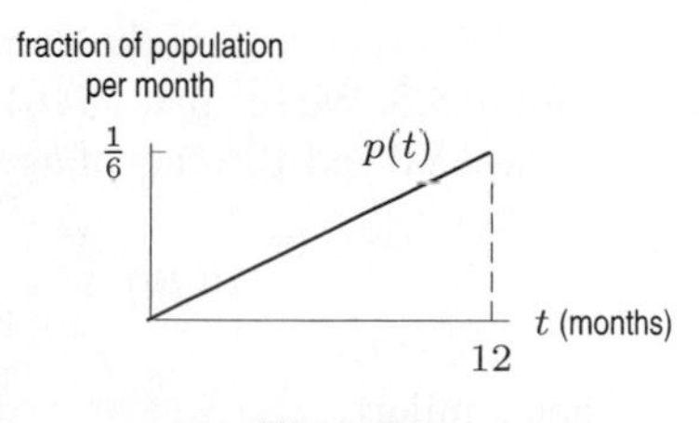

Figure 8.6

[4]Adapted from *Statistics*, by Freedman, Pisani, Purves, and Adikhari (New York: Norton).

13. The density function $p(t)$ for the length of the larval stage, in days, for a breed of insect is given in Figure 8.7. What fraction of these insects are in the larval stage for between 10 and 12 days? For less than 8 days? For more than 12 days? In which one-day interval is the length of a larval stage most likely to fall?

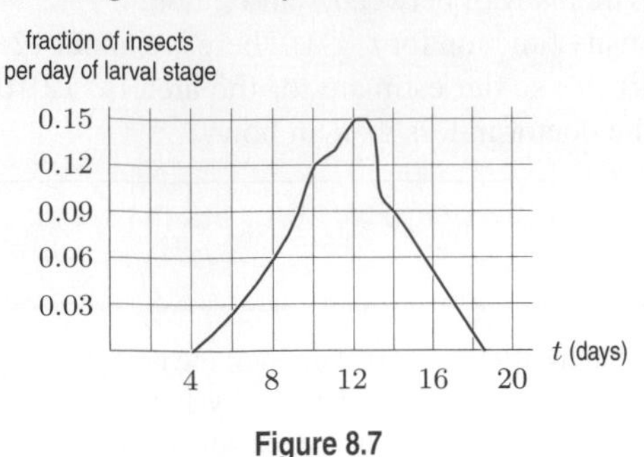

Figure 8.7

14. Figure 8.8 shows the distribution of the number of years of education completed by adults in a population. What does the shape of the graph tell you? Estimate the percentage of adults who have completed less than 10 years of education.

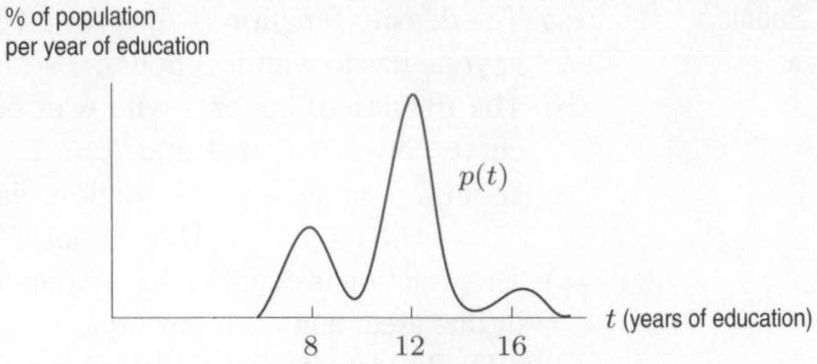

Figure 8.8

In Problems 15–17, graph a possible density function representing crop yield (in kilograms) from a field under the given circumstance.

15. All yields from 0 to 100 kg are equally likely; the field never yields more than 100 kg.

16. High yields are more likely than low. The maximum yield is 200 kg.

17. A drought makes low yields most common, and there is no yield greater than 30 kg.

8.2 CUMULATIVE DISTRIBUTION FUNCTIONS AND PROBABILITY

Section 8.1 introduced density functions which describe the way in which a numerical characteristic is distributed through a population. In this section we study another way to present the same information.

Cumulative Distribution Function for Ages

An alternative way of showing how ages are distributed in the US is by using the *cumulative distribution function* $P(t)$, defined by

$$P(t) = \begin{array}{c}\text{Fraction of population}\\ \text{of age less than } t\end{array} = \int_0^t p(x)dx.$$

Thus, P is the antiderivative of p with $P(0) = 0$, so $P(t)$ is the area under the density curve between 0 and t. See the left-hand part of Figure 8.9.

Notice that the cumulative distribution function is nonnegative and increasing (or at least nondecreasing), since the number of people younger than age t increases as t increases. Another way of seeing this is to notice that $P' = p$, and p is positive (or nonnegative). Thus the cumulative age distribution is a function which starts with $P(0) = 0$ and increases as t increases. We have $P(t) = 0$ for $t < 0$ because, when $t < 0$, there is no one whose age is less than t. The limiting value of P, as $t \to \infty$, is 1 since as t becomes very large (100 say), everyone is younger than age t, so the fraction of people with age less than t tends toward 1.

We want to find the cumulative distribution function for the age density function shown in Figure 8.3. We see that $P(10)$ is equal to 0.14, since Figure 8.3 shows that 14% of the population is between 0 and 10 years of age. Also,

$$P(20) = \begin{array}{c}\text{Fraction of the population}\\ \text{between 0 and 20 years old}\end{array} = 0.14 + 0.15 = 0.29$$

and similarly

$$P(30) = 0.14 + 0.15 + 0.14 = 0.43.$$

Continuing in this way gives the values for $P(t)$ in Table 8.3. These values were used to graph $P(t)$ in the right-hand part of Figure 8.9.

Table 8.3 *Cumulative distribution function, $P(t)$, giving fraction of US population of age less than t years*

t	0	10	20	30	40	50	60	70	80	90	100
$P(t)$	0	0.14	0.29	0.43	0.58	0.73	0.84	0.91	0.97	0.99	1.00

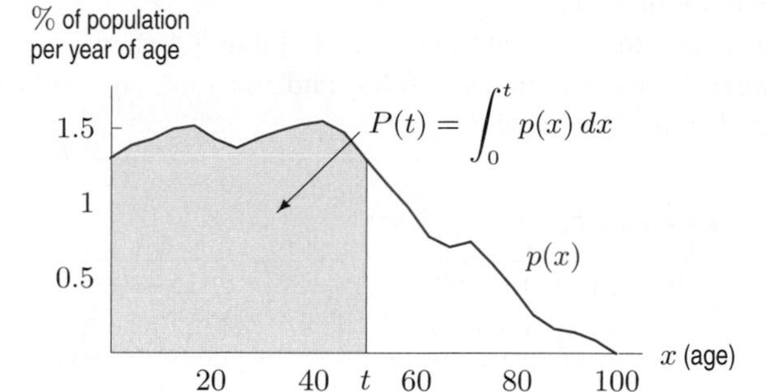

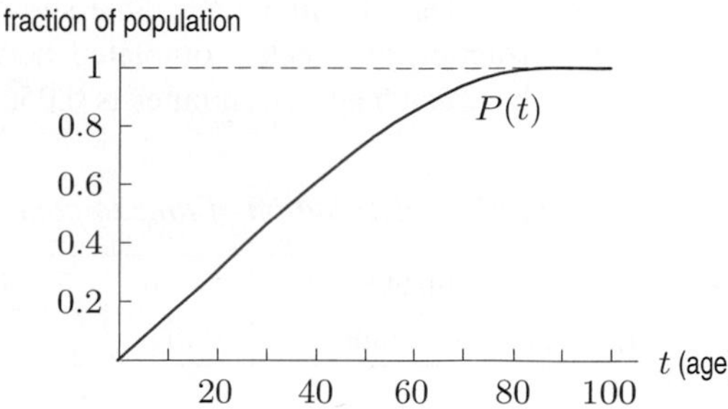

Figure 8.9: Graph of $p(x)$, the age density function, and its relation to $P(t)$, the cumulative age distribution function

Cumulative Distribution Function

A **cumulative distribution function**, $P(t)$, of a density function p, is defined by

$$P(t) = \int_{-\infty}^{t} p(x)\,dx = \text{Fraction of population having values of } x \text{ below } t.$$

Thus, P is an antiderivative of p, that is, $P' = p$.
Any cumulative distribution function has the following properties:

- P is increasing (or nondecreasing).
- $\lim_{t\to\infty} P(t) = 1$ and $\lim_{t\to-\infty} P(t) = 0$.
- Fraction of population having values of x between a and b $= \int_a^b p(x)\,dx = P(b) - P(a)$.

Example 1 The time to conduct a routine maintenance check on a machine has a cumulative distribution function $P(t)$, which gives the fraction of maintenance checks completed in time less than or equal to t minutes. Values of $P(t)$ are given in Table 8.4.

Table 8.4 *Cumulative distribution function for time to conduct maintenance checks*

t (minutes)	0	5	10	15	20	25	30
$P(t)$ (fraction completed)	0	0.03	0.08	0.21	0.38	0.80	0.98

(a) What fraction of maintenance checks are completed in 15 minutes or less?
(b) What fraction of maintenance checks take longer than 30 minutes?
(c) What fraction take between 10 and 15 minutes?
(d) Draw a histogram showing how times for maintenance checks are distributed.
(e) In which of the given 5-minute intervals is the length of a maintenance check most likely to fall?
(f) Give a rough sketch of the density function.
(g) Sketch a graph of the cumulative distribution function.

Solution

(a) The fraction of maintenance checks completed in 15 minutes is $P(15) = 0.21$, or 21%.
(b) Since $P(30) = 0.98$, we see that 98% of maintenance checks take 30 minutes or less. Therefore, only 2% take more than 30 minutes.
(c) Since 8% take 10 minutes or less and 21% take 15 minutes or less, the fraction taking between 10 and 15 minutes is $0.21 - 0.08 = 0.13$, or 13%.
(d) We begin by making a table showing how the times are distributed. Table 8.4 shows that the fraction of checks completed between 0 and 5 minutes is 0.03, and the fraction completed between 5 and 10 minutes is 0.05, and so on. See Table 8.5.

Table 8.5 *Distribution of time to conduct maintenance checks*

t (minutes)	$0-5$	$5-10$	$10-15$	$15-20$	$20-25$	$25-30$	>30
Fraction completed	0.03	0.05	0.13	0.17	0.42	0.18	0.02

The histogram in Figure 8.10 is drawn so that the area of each bar is the fraction of checks completed in the corresponding time period. For instance, the first bar has area 0.03 and width 5 minutes, so its height is $0.03/5 = 0.006$.

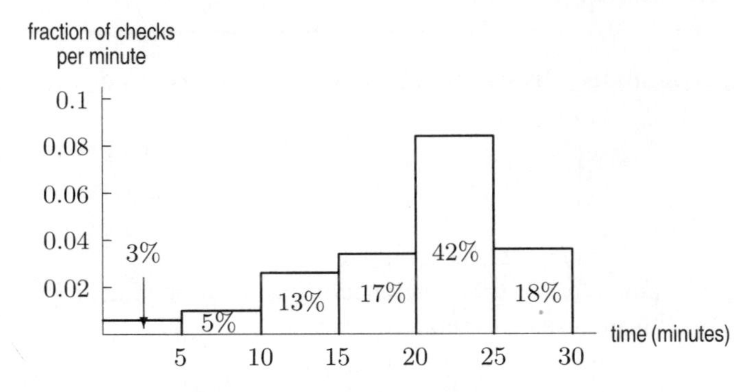

Figure 8.10: Histogram of times for maintenance checks

(e) From Figure 8.10, we see that more of the checks take between 20 and 25 minutes to complete, so this is the most likely length of time.
(f) The density function, $p(t)$, is a smoothed version of the histogram in Figure 8.10. A reasonable sketch is given in Figure 8.11.
(g) A graph of $P(t)$ is given in Figure 8.12. Since $P(t)$ is a cumulative distribution function, $P(t)$ is approaching 1 as t gets large, but is never larger than 1.

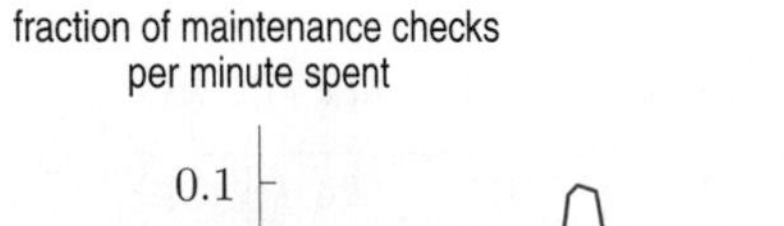

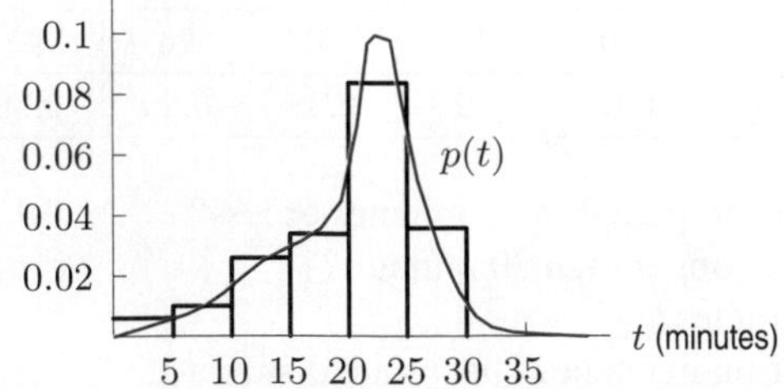

Figure 8.11: Density function for time to conduct maintenance checks

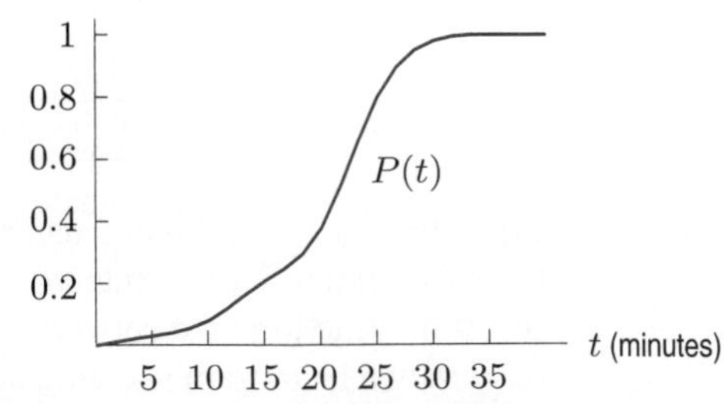

Figure 8.12: Cumulative distribution function for time to conduct maintenance checks

Probability

Suppose we pick a member of the US population at random. What is the probability that we pick a person who is between, say, the ages of 70 and 80? We saw in Table 8.2 on page 325 that 6% of the population is in this age group. We say that the probability, or chance, that the person is between 70 and 80 is 0.06. Using any age density function $p(t)$, we define probabilities as follows:

Probability that a person is between ages a and b	=	Fraction of population between ages a and b	$= \int_a^b p(t)\,dt.$

Since the cumulative distribution gives the fraction of the population younger than age t, the cumulative distribution function can also be used to calculate the probability that a randomly selected person is in a given age group.

Probability that a person is younger than age t	=	Fraction of population younger than age t	$= P(t) = \int_0^t p(x)\,dx.$

In the next example, both a density function and a cumulative distribution function are used to describe the same situation.

Example 2 Suppose you want to analyze the fishing industry in a small town. Each day, the boats bring back at least 2 tons of fish, but never more than 8 tons.

(a) Using the density function describing the daily catch in Figure 8.13, find and graph the corresponding cumulative distribution function and explain its meaning.
(b) What is the probability that the catch is between 5 and 7 tons?

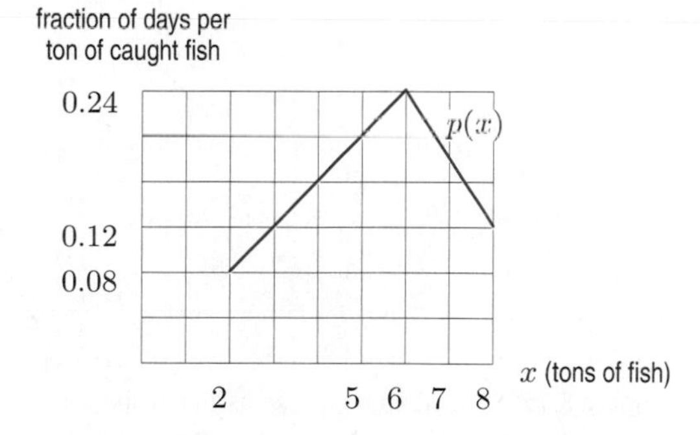

Figure 8.13: Density function of daily catch

Solution (a) The cumulative distribution function $P(t)$ is equal to the fraction of days on which the catch is less than t tons of fish. Since the catch is never less than 2 tons, we have $P(t) = 0$ for $t \leq 2$. Since the catch is always less than 8 tons, we have $P(t) = 1$ for $t \geq 8$. For t in the range $2 < t < 8$ we must evaluate the integral

$$P(t) = \int_{-\infty}^{t} p(x)dx = \int_2^t p(x)dx.$$

This integral equals the area under the graph of $p(x)$ between $x = 2$ and $x = t$. It can be computed by counting grid squares in Figure 8.13; each square has area 0.04. For example,

$$P(3) = \int_2^3 p(x)dx \approx \text{Area of 2.5 squares} = 2.5(0.04) = 0.10.$$

Table 8.6 contains values of $P(t)$; the graph is shown in Figure 8.14.

Table 8.6 *Estimates for $P(t)$ of daily catch*

t (tons of fish)	$P(t)$ (fraction of fishing days)
2	0
3	0.10
4	0.24
5	0.42
6	0.64
7	0.85
8	1

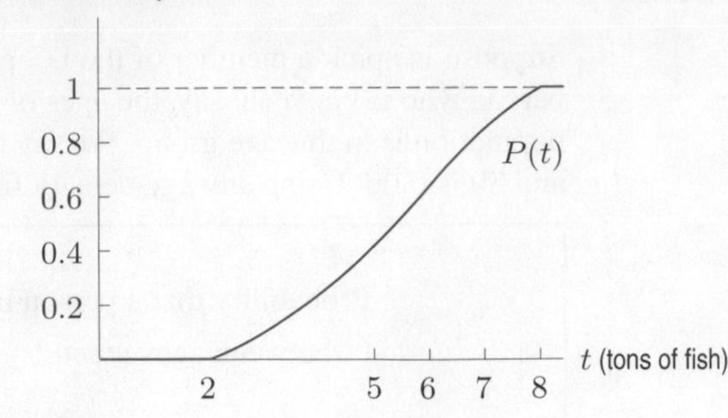

Figure 8.14: Cumulative distribution, $P(t)$, of daily catch

(b) The probability that the catch is between 5 and 7 tons can be found using either the density function p or the cumulative distribution function P. Using the density function, this probability is represented by the shaded area in Figure 8.15, which is about 10.75 squares, so

$$\begin{array}{c}\text{Probability catch is}\\ \text{between 5 and 7 tons}\end{array} = \int_5^7 p(x)\,dx \approx \text{Area of 10.75 squares} = 10.75(0.04) = 0.43.$$

The probability can be found from the cumulative distribution function as follows:

$$\begin{array}{c}\text{Probability catch is}\\ \text{between 5 and 7 tons}\end{array} = P(7) - P(5) = 0.85 - 0.42 = 0.43.$$

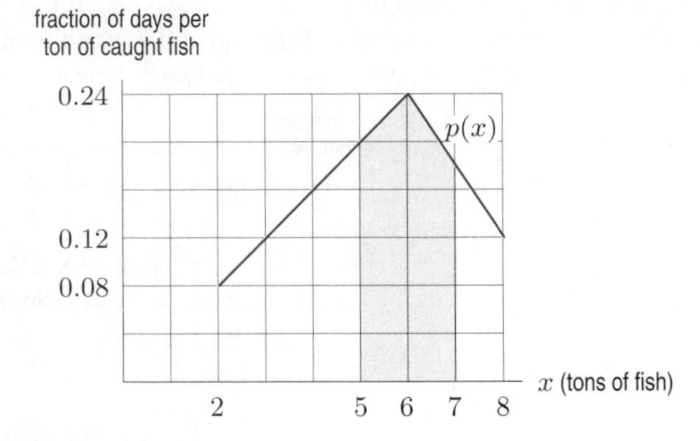

Figure 8.15: Shaded area represents the probability the catch is between 5 and 7 tons

Problems for Section 8.2

1. A congressional committee is investigating a defense contractor whose projects often incur cost overruns. The data in Table 8.7 show y, the fraction of the projects with an overrun of at most $C\%$.

(a) Plot the data with C on the horizontal axis. Is this a density function or a cumulative distribution function? Sketch a curve through these points.

(b) If you think you drew a density function in part (a), sketch the corresponding cumulative distribution function on another set of axes. If you think you drew a cumulative distribution function in part (a), sketch the corresponding density function.

(c) Based on the table, what is the probability that there will be a cost overrun of 50% or more? Between 20% and 50%? Near what percent is the cost overrun most likely to be?

Table 8.7 *Fraction, y, of overruns that are at most $C\%$*

C	−20%	−10%	0%	10%	20%	30%	40%	50%
y	0.01	0.08	0.19	0.32	0.50	0.80	0.94	0.99

2. The density function and cumulative distribution function of heights of grass plants in a meadow are in Figures 8.16 and 8.17, respectively.

(a) There are two species of grass in the meadow, a short grass and a tall grass. Explain how the graph of the density function reflects this fact.
(b) Explain how the graph of the cumulative distribution functions reflects the fact that there are two species of grass in the meadow.
(c) About what percentage of the grasses in the meadow belong to the short grass species?

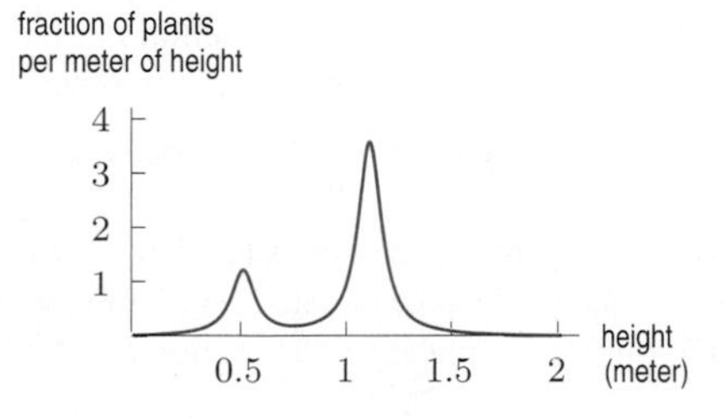

Figure 8.16

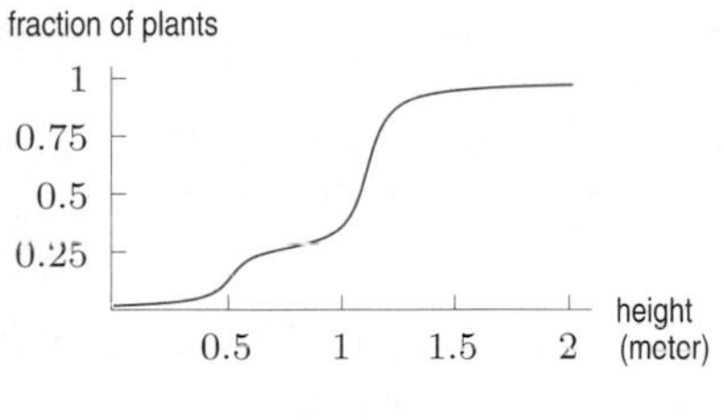

Figure 8.17

3. **(a)** Using the density function in Example 2 on page 326, fill in values for the cumulative distribution function $P(t)$ for the length of time people wait in the doctor's office.

t (hours)	0	1	2	3	4
$P(t)$ (fraction of people waiting)					

(b) Graph $P(t)$.

4. In an agricultural experiment, the quantity of grain from a given size field is measured. The yield can be anything from 0 kg to 50 kg. For each of the following situations, pick the graph that best represents the:
(i) Probability density function
(ii) Cumulative distribution function.

(a) Low yields are more likely than high yields.
(b) All yields are equally likely.
(c) High yields are more likely than low yields.

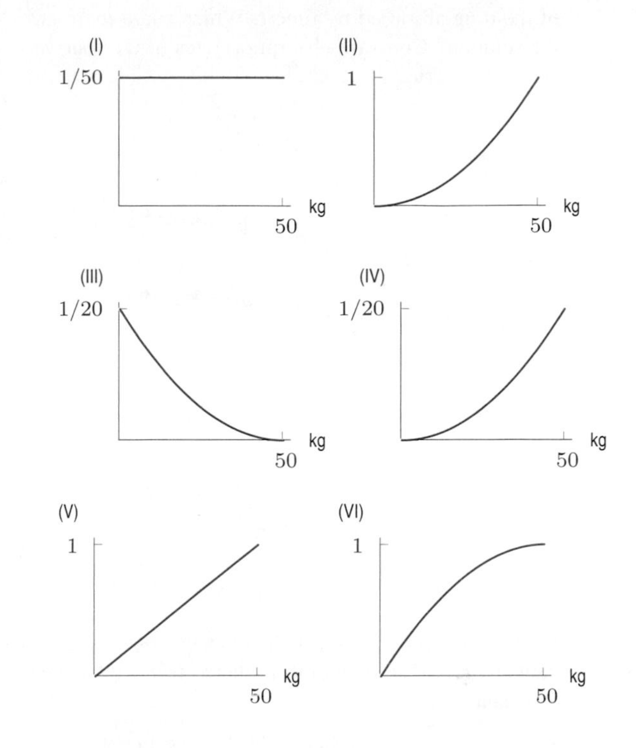

5. Show that the area under the fishing density function in Figure 8.13 on page 331 is 1. Why is this to be expected?

6. The density function for radii r (mm) of spherical raindrops during a storm is constant over the range $0 < r < 5$ and zero elsewhere.

(a) Find the density function $f(r)$ for the radii.
(b) Find the cumulative distribution function $F(r)$.

In Problems 7–9, graph a density function and a cumulative distribution function which could represent the distribution of income through a population with the given characteristics.

7. A large middle class.

8. Small middle and upper classes and many poor people.

9. Small middle class, many poor and many rich people.

10. Suppose $F(x)$ is the cumulative distribution function for heights (in meters) of trees in a forest.

(a) Explain in terms of trees the meaning of the statement $F(7) = 0.6$.
(b) Which is greater, $F(6)$ or $F(7)$? Justify your answer in terms of trees.

11. Major absorption differences have been reported for different commercial versions of the same drug. One study compared three commercial versions of timed-release theophylline capsules.[5] A theophylline solution was included for comparison. Figure 8.18 shows the cumulative distribution functions $P(t)$, which represent the fraction

[5] *Progress in Drug Metabolism*, Bridges and Chasseaud (eds.), (New York: Wiley, 1980).

of the drug absorbed by time t. Which curve represents the solution? Compare absorption rates of the four versions of the drug.

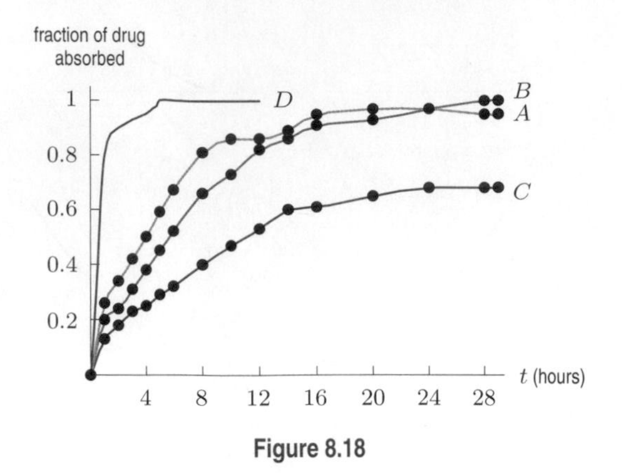

Figure 8.18

12. Figure 8.19 shows $P(t)$, the percentage of inventory of an item that has sold by time t, where t is in days and day 1 is January 1.

  (a) When did the first item sell? The last item?
  (b) On May 1 (day 121), what percentage of the inventory had been sold?
  (c) Approximately what percentage of the inventory sold during May and June (days 121 - 181)?
  (d) What percentage of the inventory remained after half of the year had passed (at day 181)?
  (e) Estimate when items went on sale and sold quickly.

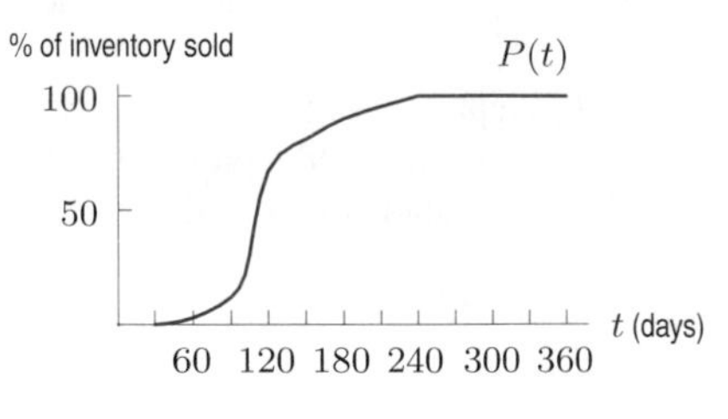

Figure 8.19

13. Sketch the density function for the cumulative distribution function in Figure 8.19.

14. An experiment is done to determine the effect of two new fertilizers A and B on the growth of a species of peas. The cumulative distribution functions of the heights of the mature peas without treatment and treated with each of A and B are graphed in Figure 8.20.

  (a) About what height are most of the unfertilized plants?
  (b) Explain in words the effect of the fertilizers A and B on the mature height of the plants.

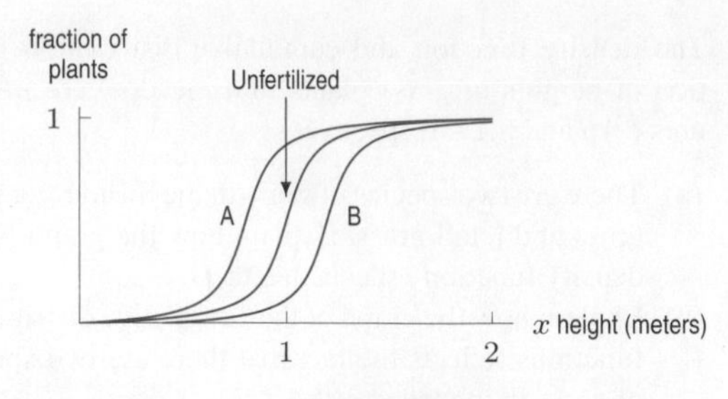

Figure 8.20

15. After measuring the duration of many telephone calls, the telephone company found their data was well-approximated by the density function $p(x) = 0.4e^{-0.4x}$, where x is the duration of a call, in minutes.

  (a) What percentage of calls last between 1 and 2 minutes?
  (b) What percentage of calls last 1 minute or less?
  (c) What percentage of calls last 3 minutes or more?
  (d) Find the cumulative distribution function.

16. A group of people have received treatment for cancer. Let t be the survival time, the number of years a person lives after the treatment. The density function giving the distribution of t is $p(t) = Ce^{-Ct}$ for some positive constant C. What is the practical meaning of the cumulative distribution function $P(t) = \int_0^t p(x)\,dx$?

For Problems 17–18, let $p(t) = -0.0375t^2 + 0.225t$ be the density function for the shelf life of a brand of banana, with t in weeks and $0 \le t \le 4$. See Figure 8.21.

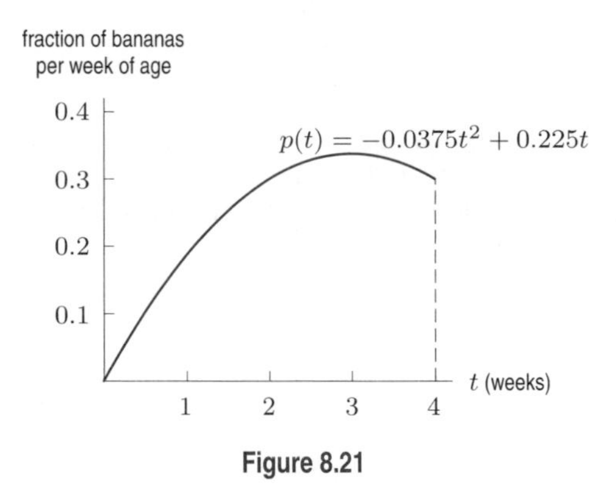

Figure 8.21

17. Find the probability that a banana will last

  (a) Between 1 and 2 weeks.
  (b) More than 3 weeks.
  (c) More than 4 weeks.

18. (a) Sketch the cumulative distribution function for the shelf life of bananas. [Note: The domain of your function should be all real numbers, including to the left of $t = 0$ and to the right of $t = 4$.]
  (b) Use the cumulative distribution function to estimate the probability that a banana lasts between 1 and 2 weeks. Check with Problem 17(a).

8.3 THE MEDIAN AND THE MEAN

It is often useful to be able to give an "average" value for a distribution. Two measures that are in common use are the *median* and the *mean*.

The Median

A **median** of a quantity x distributed through a population is a value T such that half the population has values of x less than (or equal to) T, and half the population has values of x greater than (or equal to) T. Thus, if p is the density function, a median T satisfies

$$\int_{-\infty}^{T} p(x)\,dx = 0.5.$$

In other words, half the area under the graph of p lies to the left of T. Equivalently, if P is the cumulative distribution function,

$$P(T) = 0.5.$$

Example 1 Let t days be the length of time a pair of jeans remains in a shop before it is sold. The density function of t is graphed in Figure 8.22 and given by

$$p(t) = 0.04 - 0.0008t.$$

(a) What is the longest time a pair of jeans remains unsold?
(b) Would you expect the median time till sale to be less than, equal to, or greater than 25 days?
(c) Find the median time required to sell a pair of jeans.

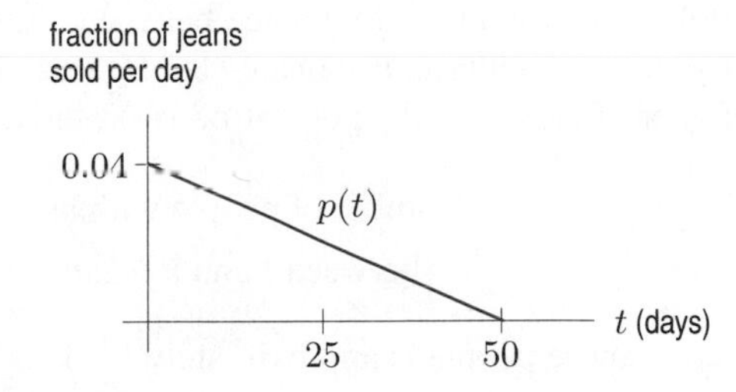

Figure 8.22: Density function for time till sale of a pair of jeans

Solution

(a) The density function is 0 for all times $t > 50$, so all jeans are sold within 50 days.
(b) The area under the graph of the density function in the interval $0 \le t \le 25$ is greater than the area under the graph in the interval $25 \le t \le 50$. So more than half the jeans are sold before their 25th day in the shop. The median time till sale is less than 25 days.
(c) Let P be the cumulative distribution function. We want to find the value of T such that

$$P(T) = \int_{-\infty}^{T} p(t)\,dt = \int_{0}^{T} p(t)\,dt = 0.5.$$

Using a calculator to evaluate the integrals, we obtain the values for P in Table 8.8.

Table 8.8 *Cumulative distribution for selling time*

T (days)	0	5	10	15	20	25
$P(T)$ (fraction of jeans sold by day T)	0	0.19	0.36	0.51	0.64	0.75

Since about half the jeans are sold within 15 days, the median time to sale is about 15 days. See Figures 8.23 and 8.24. We could also use the Fundamental Theorem of Calculus to find the median exactly. See Problem 11.

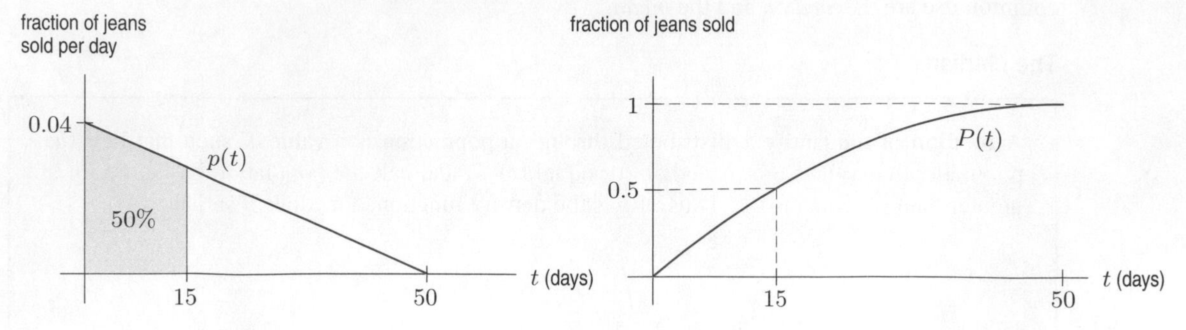

Figure 8.23: Median and density function

Figure 8.24: Median and cumulative distribution function

The Mean

Another commonly used average value is the *mean*. To find the mean of N numbers, we add the numbers and divide the sum by N. For example, the mean of the numbers 1, 2, 7, and 10 is $(1 + 2 + 7 + 10)/4 = 5$. The mean age of the entire US population is therefore defined as

$$\text{Mean age} = \frac{\sum \text{Ages of all people in the US}}{\text{Total number of people in the US}}.$$

Calculating the sum of all the ages directly would be an enormous task; we approximate the sum by an integral. We consider the people whose age is between t and $t+\Delta t$. How many are there?

The fraction of the population with age between t and $t + \Delta t$ is the area under the graph of p between these points, which is approximated by the area of the rectangle, $p(t)\Delta t$. (See Figure 8.25.) If the total number of people in the population is N, then

$$\begin{array}{c}\text{Number of people with age} \\ \text{between } t \text{ and } t + \Delta t\end{array} \approx p(t)\Delta t N.$$

The age of each of these people is approximately t, so

$$\begin{array}{c}\text{Sum of ages of people} \\ \text{between age } t \text{ and } t + \Delta t\end{array} \approx tp(t)\Delta t N.$$

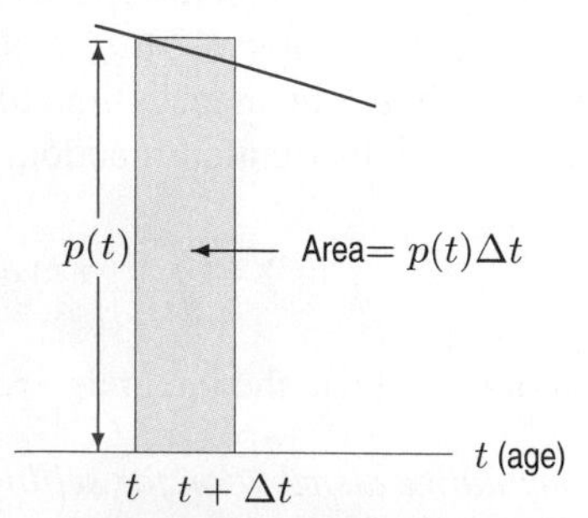

Figure 8.25: Shaded area is percentage of population with age between t and $t + \Delta t$

Therefore, adding and factoring out an N gives us

$$\text{Sum of ages of all people} \approx \left(\sum tp(t)\Delta t\right) N.$$

In the limit, as Δt shrinks to 0, the sum becomes an integral. Assuming no one is over 100 years old, we have

$$\text{Sum of ages of all people} = \left(\int_0^{100} tp(t)dt\right) N.$$

Since N is the total number of people in the US,

$$\text{Mean age} = \frac{\text{Sum of ages of all people in US}}{N} = \int_0^{100} tp(t)dt.$$

We can give the same argument for any[6] density function $p(x)$.

If a quantity has density function $p(x)$,

$$\textbf{Mean value} \text{ of the quantity} = \int_{-\infty}^{\infty} xp(x)\,dx.$$

It can be shown that the mean is the point on the horizontal axis where the region under the graph of the density function, if it were made out of cardboard, would balance.

Example 2 Find the mean time for jeans sales, using the density function of Example 1.

Solution The formula for p is $p(t) = 0.04 - 0.0008t$. We compute

$$\text{Mean time} = \int_0^{50} tp(t)\,dt = \int_0^{50} t(0.04 - 0.0008t)\,dt = 16.67 \text{ days}.$$

The mean is represented by the balance point in Figure 8.26. Notice that the mean is different from the median computed in Example 1.

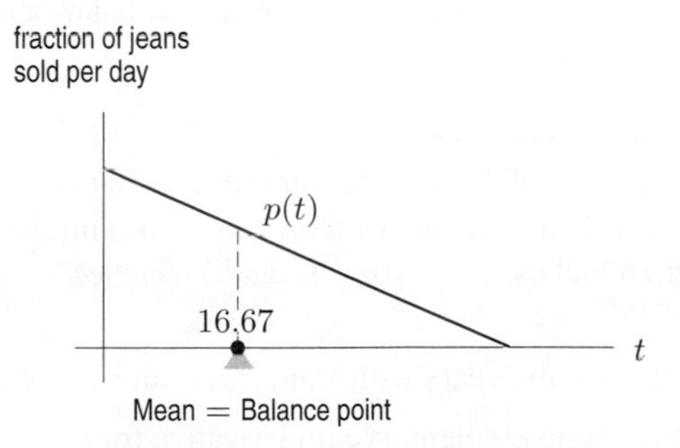

Figure 8.26: Mean sale time for jeans

[6]Provided all the relevant improper integrals converge.

Normal Distributions

How much rain do you expect to fall in your home town this year? If you live in Anchorage, Alaska, the answer is something close to 15 inches (including the snow). Of course, you don't expect exactly 15 inches. Some years there are more than 15 inches, and some years there are less. Most years, however, the amount of rainfall is close to 15 inches; only rarely is it well above or well below 15 inches. What does the density function for the rainfall look like? To answer this question, we look at rainfall data over many years. Records show that the distribution of rainfall is well-approximated by a *normal distribution*. The graph of its density function is a bell-shaped curve which peaks at 15 inches and slopes downward approximately symmetrically on either side.

Normal distributions are frequently used to model real phenomena, from grades on an exam to the number of airline passengers on a particular flight. A normal distribution is characterized by its *mean*, μ, and its *standard deviation*, σ. The mean tells us the location of the central peak. The standard deviation tells us how closely the data is clustered around the mean. A small value of σ tells us that the data is close to the mean; a large σ tells us the data is spread out. In the following formula for a normal distribution, the factor of $1/(\sigma\sqrt{2\pi})$ makes the area under the graph equal to 1.

A **normal distribution** has a density function of the form

$$p(x) = \frac{1}{\sigma\sqrt{2\pi}} e^{-(x-\mu)^2/(2\sigma^2)},$$

where μ is the mean of the distribution and σ is the standard deviation, with $\sigma > 0$.

To model the rainfall in Anchorage, we use a normal distribution with $\mu = 15$ and $\sigma = 1$. (See Figure 8.27.)

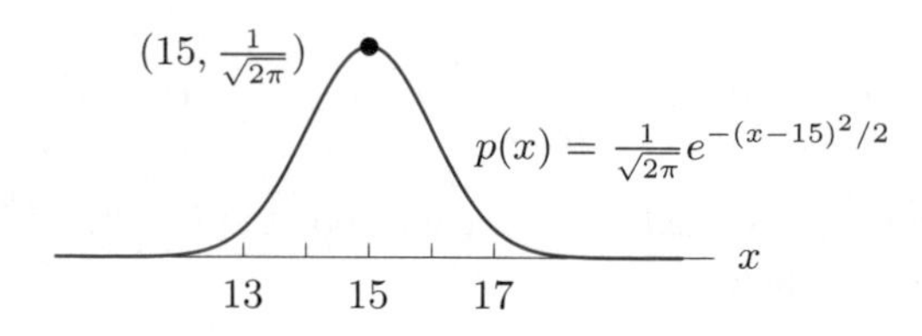

Figure 8.27: Normal distribution with $\mu = 15$ and $\sigma = 1$

Example 3 For Anchorage's rainfall, use the normal distribution with the density function with $\mu = 15$ and $\sigma = 1$ to compute the fraction of the years with rainfall between

(a) 14 and 16 inches, (b) 13 and 17 inches, (c) 12 and 18 inches.

Solution (a) The fraction of the years with annual rainfall between 14 and 16 inches is $\int_{14}^{16} \frac{1}{\sqrt{2\pi}} e^{-(x-15)^2/2}\,dx$. Since there is no elementary antiderivative for $e^{-(x-15)^2/2}$, we find the integral numerically. Its value is about 0.68.

$$\text{Fraction of years with rainfall between 14 and 16 inches} = \int_{14}^{16} \frac{1}{\sqrt{2\pi}} e^{-(x-15)^2/2}\,dx \approx 0.68.$$

(b) Finding the integral numerically again:

$$\begin{array}{c}\text{Fraction of years with rainfall}\\ \text{between 13 and 17 inches}\end{array} = \int_{13}^{17} \frac{1}{\sqrt{2\pi}} e^{-(x-15)^2/2}\, dx \approx 0.95.$$

(c)

$$\begin{array}{c}\text{Fraction of years with rainfall}\\ \text{between 12 and 18 inches}\end{array} = \int_{12}^{18} \frac{1}{\sqrt{2\pi}} e^{-(x-15)^2/2}\, dx \approx 0.997.$$

Since 0.95 is so close to 1, we expect that most of the time the rainfall will be between 13 and 17 inches a year.

Among the normal distributions, the one having $\mu = 0$, $\sigma = 1$ is called the *standard normal distribution.* Values of the corresponding cumulative distribution function are published in tables.

Problems for Section 8.3

1. Estimate the median daily catch for the fishing data given in Example 2 on page 331.

2. **(a)** Use the cumulative distribution function in Figure 8.28 to estimate the median.
(b) Describe the density function: For what values is it positive? Increasing? Decreasing? Identify all local maxima and minima values.

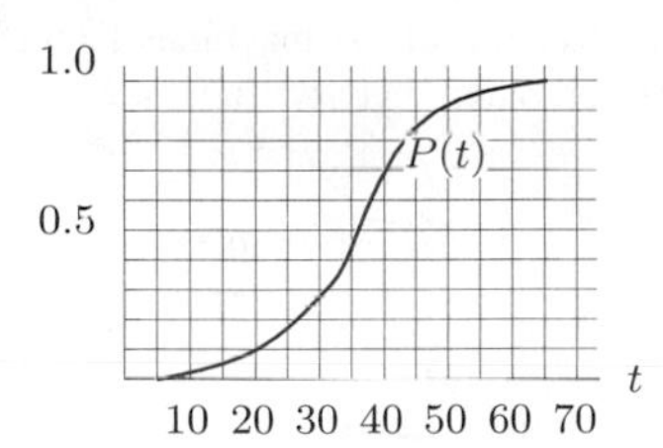

Figure 8.28

For Problems 3–4, let $p(t) = -0.0375t^2 + 0.225t$ be the density function for the shelf life of a brand of banana which lasts up to 4 weeks. Time, t, is measured in weeks and $0 \le t \le 4$.

3. Find the median shelf life of a banana using $p(t)$. Plot the median on a graph of $p(t)$. Does it look like half the area is to the right of the median and half the area is to the left?

4. Find the mean shelf life of a banana using $p(t)$. Plot the mean on a graph of $p(t)$. Does it look like the mean is the place where the density function balances?

5. Let $p(t) = 0.1e^{-0.1t}$ be the density function for the waiting time at a subway stop, with t in minutes, $0 \le t \le 60$.
(a) Graph $p(t)$. Use the graph to estimate visually the median and the mean.
(b) Calculate the median and the mean. Plot both on the graph of $p(t)$.
(c) Interpret the median and mean in terms of waiting time.

6. In 1950 an experiment was done observing the time gaps between successive cars on the Arroyo Seco Freeway. The data[7] show that, if x is time in seconds and $0 \le x \le 40$, the density function of these time gaps is

$$p(x) = 0.122e^{-0.122x}.$$

Find the median and mean time gap. Interpret them in terms of cars on the freeway.

7. Suppose that x measures the time (in hours) it takes for a student to complete an exam. All students are done within two hours and the density function for x is

$$p(x) = \begin{cases} x^3/4 & \text{if } 0 < x < 2 \\ 0 & \text{otherwise.} \end{cases}$$

(a) What proportion of students take between 1.5 and 2.0 hours to finish the exam?
(b) What is the mean time for students to complete the exam?
(c) Compute the median of this distribution.

8. Let $P(x)$ be the cumulative distribution function for the income distribution in the US in 1973 (income is measured in thousands of dollars). Some values of $P(x)$ are in the following table:

Income x (thousands)	1	4.4	7.8	12.6	20	50
$P(x)$ (%)	1	10	25	50	75	99

(a) What fraction of the population made between \$20,000 and \$50,000?
(b) What was the median income?
(c) Sketch a density function for this distribution. Where, approximately, does your density function have a maximum? What is the significance of this point, in terms of income distribution? How can you recognize this point on the graph of the density function and on the graph of the cumulative distribution?

[7]Reported by Daniel Furlough and Frank Barnes.

9. The distribution of IQ scores can be modeled by a normal distribution with mean 100 and standard deviation 15.

(a) Write the formula for the density function of IQ scores.
(b) Estimate the fraction of the population with IQ between 115 and 120.

10. The speeds of cars on a road are approximately normally distributed with a mean $\mu = 58$ km/hr and standard deviation $\sigma = 4$ km/hr.

(a) What is the probability that a randomly selected car is going between 60 and 65 km/hr?
(b) What fraction of all cars are going slower than 52 km/hr?

11. Find the median of the density function given by $p(t) = 0.04 - 0.0008t$ for $0 \leq t \leq 50$ using the Fundamental Theorem of Calculus.

CHAPTER SUMMARY

- **Density function**
- **Cumulative distribution function**
- **Probability**
- **Median**
- **Mean**
- **Normal distribution**

REVIEW PROBLEMS FOR CHAPTER EIGHT

In Problems 1–4, calculate the value of c if p is a density function.

1.

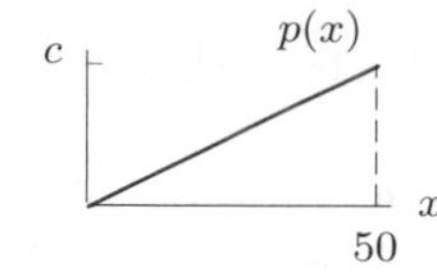

2.

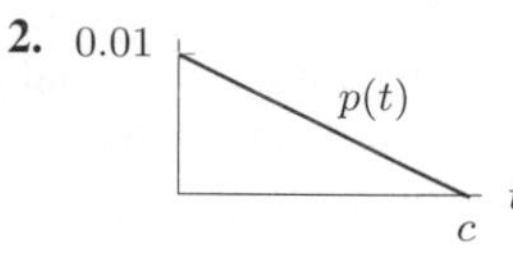

3.

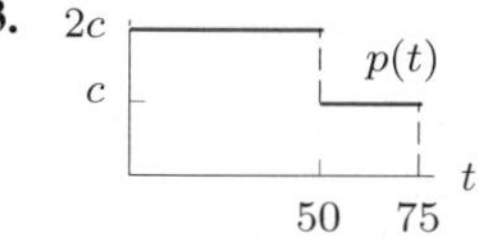

4.

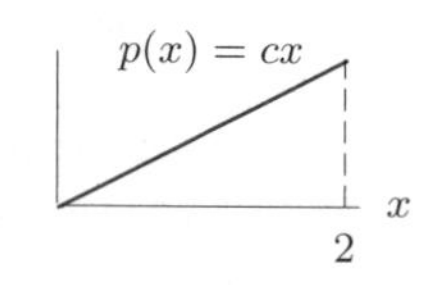

5. Suppose that $p(x)$ is the density function for heights of American men, in inches. What is the meaning of the statement $p(68) = 0.2$?

6. A large number of people take a standardized test, receiving scores described by the density function p graphed in Figure 8.29. Does the density function imply that most people receive a score near 50? Explain why or why not.

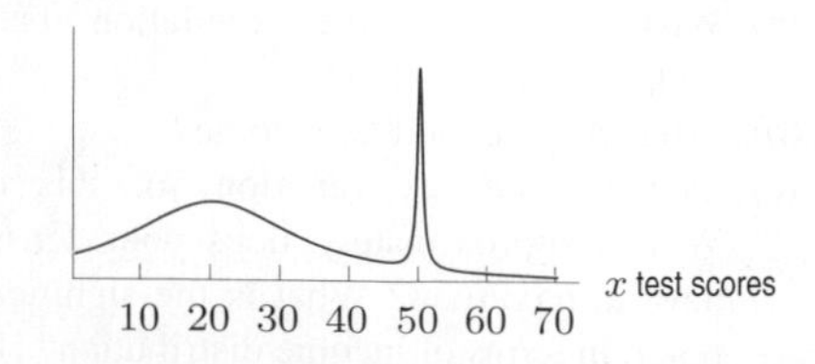

Figure 8.29

7. A machine lasts up to 10 years. Figure 8.30 shows the density function, $p(t)$, for the length of time it lasts.

(a) What is the value of C?
(b) Is a machine more likely to break in its first year or in its tenth year? In its first or second year?
(c) What fraction of the machines last 2 years or less? Between 5 and 7 years? Between 3 and 6 years?

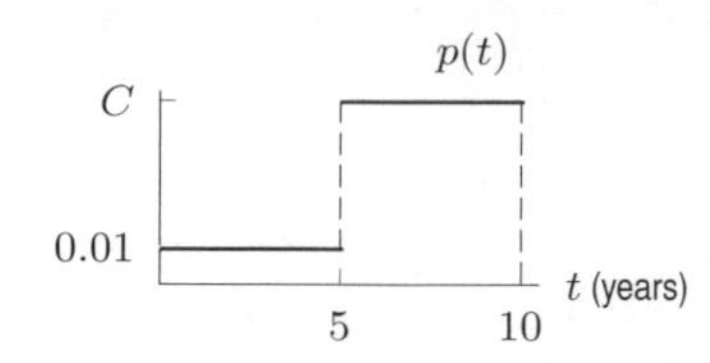

Figure 8.30

8. Which of the following functions makes the most sense as a model for the probability density representing the time (in minutes, starting from $t = 0$) that the next customer walks into a store?

(a) $p(t) = \begin{cases} \cos t & 0 \leq t \leq 2\pi \\ e^{t-2\pi} & t \geq 2\pi \end{cases}$
(b) $p(t) = 3e^{-3t}$ for $t \geq 0$
(c) $p(t) = e^{-3t}$ for $t \geq 0$
(d) $p(t) = 1/4$ for $0 \leq t \leq 4$

9. A person who travels regularly on the 9:00 am bus from Oakland to San Francisco reports that the bus is almost always a few minutes late but rarely more than five minutes late. The bus is never more than two minutes early, although it is on very rare occasions a little early.

(a) Sketch a density function, $p(t)$, where t is the number of minutes that the bus is late. Shade the region under the graph between $t = 2$ minutes and $t = 4$ minutes. Explain what this region represents.

(b) Now sketch the cumulative distribution function $P(t)$. What measurement(s) on this graph correspond to the area shaded? What do the inflection point(s) on your graph of P correspond to on the graph of p? Interpret the inflection points on the graph of P without referring to the graph of p.

10. Figure 8.31 shows a density function and the corresponding cumulative distribution function.[8]

(a) Which curve represents the density function and which represents the cumulative distribution function? Give a reason for your choice.

(b) Put reasonable values on the tick marks on each of the axes.

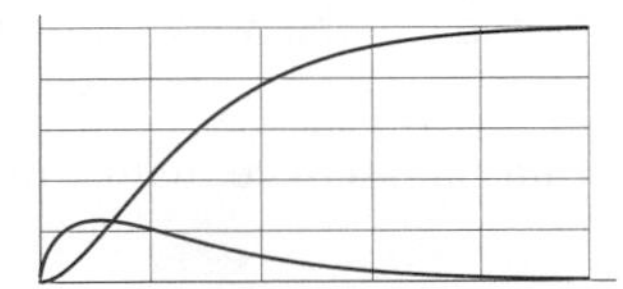

Figure 8.31

11. Students at the University of California were surveyed and asked their grade point average . (The GPA ranges from 0 to 4, where 2 is just passing.) The distribution of GPAs is shown in Figure 8.32.[9]

(a) Roughly what fraction of students are passing?

(b) Roughly what fraction of the students have honor grades (GPAs above 3)?

(c) Why do you think there is a peak around 2?

(d) Sketch the cumulative distribution function.

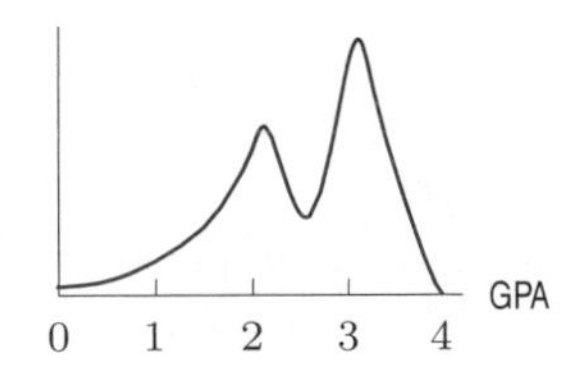

Figure 8.32

12. In southern Switzerland, most rain falls in the spring and fall; summers and winters are relatively dry. Sketch possible graphs for the density function and the cumulative distribution function of the rain distribution over the course of one year. Put the date on the horizontal axis and fraction of the year's rainfall on the vertical axis.

13. The probability of a transistor failing between $t = a$ months and $t = b$ months is given by $c\int_a^b e^{-ct}\,dt$, for some constant c.

(a) If the probability of failure within the first six months is 10%, what is c?

(b) Given the value of c in part (a), what is the probability the transistor fails within the second six months?

14. While taking a walk along the road where you live, you accidentally drop your glove, but you don't know where. The probability density $p(x)$ for having dropped the glove x kilometers from home (along the road) is

$$p(x) = 2e^{-2x} \quad \text{for } x \geq 0.$$

(a) What is the probability that you dropped it within 1 kilometer of home?

(b) At what distance y from home is the probability that you dropped it within y km of home equal to 0.95?

In Problems 15–19, a quantity x is distributed through a population with probability density function $p(x)$ and cumulative distribution function $P(x)$. Decide if the statements in Problems 15–19 are true or false. Give an explanation for your answer.

15. If $p(10) = 1/2$, then half the population has $x < 10$.

16. If $P(10) = 1/2$, then half the population has $x < 10$.

17. If $p(10) = 1/2$, then the fraction of the population lying between $x = 9.98$ and $x = 10.04$ is about 0.03.

18. If $p(10) = p(20)$, then none of the population has x values lying between 10 and 20.

19. If $P(10) = P(20)$, then none of the population has x values lying between 10 and 20.

PROJECTS FOR CHAPTER EIGHT

1. Triangular Probability Distribution

Triangular probability distributions, such as the one with density function graphed in Figure 8.33, are used in business to model uncertainty. Such a distribution can be used to model a variable where only three pieces of information are available: a lower bound ($x = a$), a most likely value ($x = c$), and an upper bound ($x = b$).

[8]Adapted from *Calculus*, by David A. Smith and Lawerence C. Moore (Lexington: D.C. Heath, 1994).
[9]Adapted from *Statistics*, by Freedman, Pisani, Purves, and Adikhari (New York: Norton, 1991).

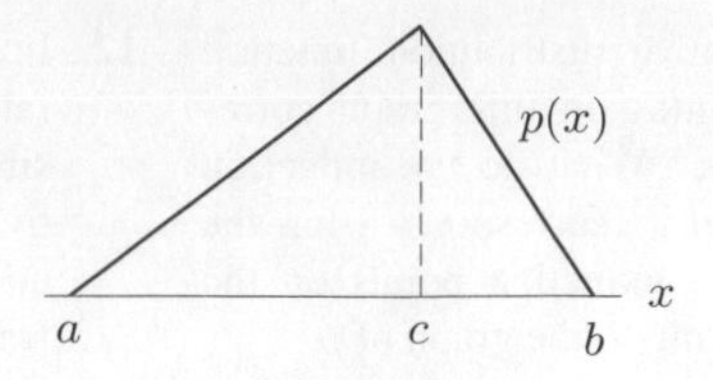

Figure 8.33

Thus, we can write the function $p(x)$ as two linear functions:

$$p(x) = \begin{cases} m_1 x + b_1 & a \le x \le c \\ m_2 x + b_2 & c < x \le b. \end{cases}$$

(a) Find the value of $p(c)$ geometrically, using the criterion that the probability that x takes on some value between a and b is 1.

Suppose a new product costs between \$6 and \$10 per unit to produce, with a most likely cost of \$9.

(b) Find $p(9)$.
(c) Use the fact that $p(6) = p(10) = 0$ and the value of $p(9)$ you found in part (b) to find m_1, m_2, b_1, and b_2.
(d) What is the probability that the production cost per unit will be less than \$8?
(e) What is the median cost?
(f) Write a formula for the cumulative probability distribution function $P(x)$ for

(i) $6 \le x \le 9$, (ii) $9 < x \le 10$.

Sketch the graph of $P(x)$.

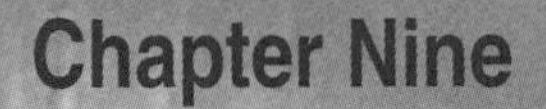

Chapter Nine

FUNCTIONS OF SEVERAL VARIABLES

Many quantities depend on more than one variable: the amount of food grown depends on the amount of rain and the amount of fertilizer used; the rate of a chemical reaction depends on the temperature and the pressure of the environment in which it proceeds; the quantity of meat purchased depends on the price of meat and the income of the buyer; the rate of fallout from a volcanic eruption depends on the distance from the volcano and the time since the eruption.

In this chapter we see how to extend the concept of the derivative to functions of two or more variables.

9.1 UNDERSTANDING FUNCTIONS OF TWO VARIABLES

To avoid flying planes with too many empty seats, airlines sell some tickets at full price and some at a discount. For a particular route, the airline's revenue, R, earned in a given time period is determined by the number of full price tickets, x, and the number of discount tickets, y, sold. We say that R is a function of x and y, and we write

$$R = f(x, y).$$

This is just like the function notation of one-variable calculus. The variable R is the dependent variable and the variables x and y are the independent variables. The letter f stands for the *function* or rule that gives the value of R corresponding to given values of x and y. The collection of all possible inputs, (x, y), is called the *domain* of f. We say a function is an *increasing* (*decreasing*) function of one of its variables if it increases (decreases) as the variable increases while the other independent variables are held constant.

A function of two variables can be represented numerically by a table of values, algebraically by a formula, or pictorially by a contour diagram. In this section we give numerical and algebraic examples; contour diagrams are introduced in Section 9.2.

Functions Given Numerically

The revenue, R, (in dollars) from a particular airline route is shown in Table 9.1 as a function of the number of full price tickets and the number of discount tickets sold.

Table 9.1 *Revenue from ticket sales as a function of x and y*

		Number of full price tickets, x			
		100	200	300	400
Number of discount tickets, y	200	75,000	110,000	145,000	180,000
	400	115,000	150,000	185,000	220,000
	600	155,000	190,000	225,000	260,000
	800	195,000	230,000	265,000	300,000
	1000	235,000	270,000	305,000	340,000

Values of x are shown across the top, values of y are down the left side, and corresponding values of $f(x, y)$ are given in the table. For example, to find the value of $f(300, 600)$, we look in the column corresponding to $x = 300$ at the row $y = 600$, where we find the number 225,000. Thus,

$$f(300, 600) = 225{,}000.$$

This means that the revenue from 300 full price tickets and 600 discount tickets is \$225,000. We see in Table 9.1 that f is an increasing function of x and an increasing function of y.

Notice how this differs from the table of values of a one-variable function, where one row or one column is enough to list the values of the function. Here many rows and columns are needed because the function has a value for every pair of values of the independent variables.

Functions Given Algebraically

The function given in Table 9.1 can be represented by a formula. Looking across the rows, we see that each additional 100 full price tickets sold raises the revenue by \$35,000, so each full price ticket must cost \$350. Similarly, looking down a column shows that an additional 200 discount tickets sold increases the revenue by \$40,000, so each discount ticket must cost \$200. Thus, the revenue function is given by the formula

$$R = 350x + 200y.$$

Example 1 Give a formula for the function $M = f(B, t)$ where M is the amount of money in a bank account t years after an initial investment of B dollars, if interest accrues at a rate of 5% per year compounded
(a) Annually (b) Continuously.

Solution (a) Annual compounding means that M increases by a factor of 1.05 every year, so

$$M = f(B, t) = B(1.05)^t.$$

(b) Continuous compounding means that M grows according to the function e^{kt}, with $k = 0.05$, so

$$M = f(B, t) = Be^{0.05t}.$$

Example 2 A car rental company charges \$40 a day and 15 cents a mile for its cars.

(a) Write a formula for the cost, C, of renting a car as a function of the number of days, d, and the number of miles driven, m.
(b) If $C = f(d, m)$, find $f(5, 300)$ and interpret it.

Solution (a) The total cost in dollars of renting a car is 40 times the numbers of days plus 0.15 times the number of miles, so

$$C = 40d + 0.15m.$$

(b) We have

$$\begin{aligned} f(5, 300) &= 40(5) + 0.15(300) \\ &= 200 + 45 \\ &= 245. \end{aligned}$$

We see that $f(5, 300) = 245$. This tells us that if we rent a car for 5 days and drive it 300 miles, it costs us \$245.

Strategy to Investigate Functions of Two Variables: Vary One Variable at a Time

We can learn a great deal about a function of two variables by letting one variable vary while holding the other fixed. This gives a function of one variable, called a *cross-section* of the original function.

Concentration of a Drug in the Blood

When a drug is injected into muscle tissue, it diffuses into the bloodstream. The concentration of the drug in the blood increases until it reaches a maximum, and then decreases. The concentration, C (in mg per liter), of the drug in the blood is a function of two variables: x, the amount (in mg) of the drug given in the injection, and t, the time (in hours) since the injection was administered. We are told that

$$C = f(x, t) = te^{-t(5-x)} \qquad \text{for } 0 \le x \le 4 \text{ and } t \ge 0.$$

Example 3 In terms of the drug concentration in the blood, explain the significance of the cross-sections:

(a) $f(4, t)$ (b) $f(x, 1)$

Solution (a) Holding x fixed at 4 means that we are considering an injection of 4 mg of the drug; letting t vary means we are watching the effect of this dose as time passes. Thus the function $f(4, t)$ describes the concentration of the drug in the blood resulting from a 4 mg injection as a function of time. Figure 9.1 shows the graph of $f(4, t) = te^{-t}$. Notice that the concentration in the blood from this dose is at a maximum at 1 hour after injection, and that the concentration in the blood eventually approaches zero.

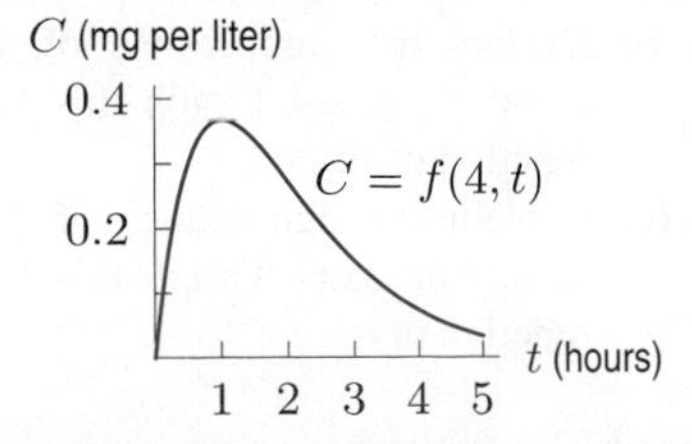

Figure 9.1: The function $f(4, t)$ shows the concentration in the blood resulting from a 4 mg injection

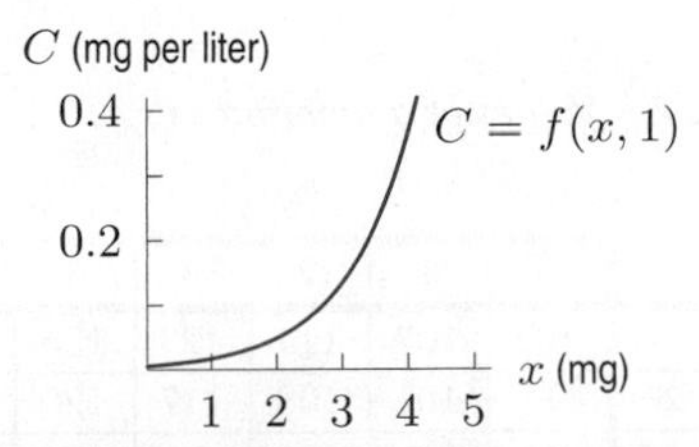

Figure 9.2: The function $f(x, 1)$ shows the concentration in the blood 1 hour after the injection

(b) Holding t fixed at 1 means that we are focusing on the blood 1 hour after the injection; letting x vary means we are considering the effect of different doses at that instant. Thus, the function $f(x,1)$ gives the concentration of the drug in the blood 1 hour after injection as a function of the amount injected. Figure 9.2 shows the graph of $f(x,1) = e^{-(5-x)} = e^{x-5}$. Notice that $f(x,1)$ is an increasing function of x. This makes sense: If we administer more of the drug, the concentration in the bloodstream is higher.

Example 4 Continue with $C = f(x,t) = te^{-t(5-x)}$. Graph the cross-sections of $f(a,t)$ for $a = 1, 2, 3$, and 4 on the same axes. Describe how the graph changes for larger values of a and explain what this means in terms of drug concentration in the blood.

Solution The one-variable function $f(a,t)$ represents the effect of an injection of a mg at time t. Figure 9.3 shows the graphs of the four functions $f(1,t) = te^{-4t}$, $f(2,t) = te^{-3t}$, $f(3,t) = te^{-2t}$, and $f(4,t) = te^{-t}$ corresponding to injections of 1, 2, 3, and 4 mg of the drug. The general shape of the graph is the same in every case: The concentration in the blood is zero at the time of injection $t = 0$, then increases to a maximum value, and then decreases toward zero again. We see that if a larger dose of the drug is administered, the peak of the graph is later and higher. This makes sense, since a larger dose will take longer to diffuse fully into the bloodstream and will produce a higher concentration when it does.

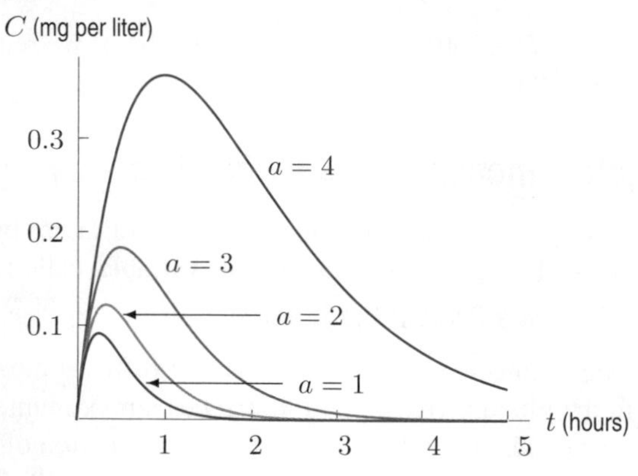

Figure 9.3: Concentration $C = f(a,t)$ of the drug resulting from an a mg injection

Problems for Section 9.1

1. The total sales of a product, S, can be expressed as a function of the price p charged for the product and the amount, a, spent on advertising, so $S = f(p,a)$. Do you expect f to be an increasing or decreasing function of p? Do you expect f to be an increasing or decreasing function of a? Why?

2. Use Table 9.2. Is f an increasing or decreasing function of x? Is f an increasing or decreasing function of y?

Table 9.2 *Values of a function $f(x,y)$*

x \ y	0	1	2	3	4	5
0	102	107	114	123	135	150
20	96	101	108	117	129	144
40	90	95	102	111	123	138
60	85	90	97	106	118	133
80	81	86	93	102	114	129

Problems 3–4 concern the cost, C, of renting a car from a company which charges $40 a day and 15 cents a mile, so $C = f(d,m) = 40d + 0.15m.$, where d is the number of days, and m is the number of miles.

3. Make a table of values for C, using $d = 1, 2, 3, 4$ and $m = 100, 200, 300, 400$. You should have 16 values in your table.

4. **(a)** Find $f(3,200)$ and interpret it.
(b) Explain the significance of $f(3,m)$ in terms of rental car costs. Graph this function, with C as a function of m.
(c) Explain the significance of $f(d,100)$ in terms of rental car costs. Graph this function, with C as a function of d.

5. The balance, B, in dollars, in a bank account depends on the amount deposited, A dollars, the annual interest rate, $r\%$, and the time, t, in months since the deposit, so $B = f(A, r, t)$.

(a) Is f an increasing or decreasing function of A? Of r? Of t?

(b) Interpret the statement $f(1250, 1, 25) \approx 1276$. Give units.

6. The monthly payments, P dollars, on a mortgage in which A dollars were borrowed at an annual interest rate of $r\%$ for t years is given by $P = f(A, r, t)$. Is f an increasing or decreasing function of A? Of r? Of t?

7. You are planning a trip whose principal cost is gasoline.

(a) Make a table showing how the daily fuel cost varies as a function of the price of gasoline (in dollars per gallon) and the number of gallons you buy each day.

(b) If your car goes 30 miles on each gallon of gasoline, make a table showing how your daily fuel cost varies as a function of your daily travel distance and the price of gas.

8. Graph the bank account function f in Example 1(a) on page 344, holding B fixed at three different values and letting t vary. Then graph f, holding t fixed at three different values and letting B vary. Explain what you see.

9. The temperature adjusted for wind-chill is a temperature which tells you how cold it feels, as a result of the combination of wind and temperature.[1] See Table 9.3.

(a) If the temperature is $0°$F and the wind speed is 15 mph, how cold does it feel?

(b) If the temperature is $35°$F, what wind speed makes it feel like $24°$F?

(c) If the temperature is $25°$F, what wind speed makes it feel like $12°$F?

(d) If the wind is blowing at 20 mph, what temperature feels like $0°$F?

Table 9.3 *Temperature adjusted for wind-chill (°F) as a function of wind speed and temperature*

	Temperature (°F)							
Wind Speed (mph)	35	30	25	20	15	10	5	0
5	31	25	19	13	7	1	−5	−11
10	27	21	15	9	3	−4	−10	−16
15	25	19	13	6	0	−7	−13	−19
20	24	17	11	4	−2	−9	−15	−22
25	23	16	9	3	−4	−11	−17	−24

10. Using Table 9.3, make tables of the temperature adjusted for wind-chill as a function of wind speed for temperatures of $20°$F and $0°$F.

11. Using Table 9.3, make tables of the temperature adjusted for wind-chill as a function of temperature for wind speeds of 5 mph and 20 mph.

12. The number, n, of new cars sold in a year is a function of the price of new cars, c, and the average price of gas, g.

(a) If c is held constant, is n an increasing or decreasing function of g? Why?

(b) If g is held constant, is n an increasing or decreasing function of c? Why?

Problems 13–17 refer to Table 9.4 which shows[2] the weekly beef consumption, C, (in lbs) of an average household as a function of p, the price of beef (in \$/lb) and I, annual household income (in \$1000s).

Table 9.4 *Quantity of beef bought (lbs/household/week)*

	p			
I	3.00	3.50	4.00	4.50
20	2.65	2.59	2.51	2.43
40	4.14	4.05	3.94	3.88
60	5.11	5.00	4.97	4.84
80	5.35	5.29	5.19	5.07
100	5.79	5.77	5.60	5.53

13. Give tables for beef consumption as a function of p, with I fixed at $I = 20$ and $I = 100$. Give tables for beef consumption as a function of I, with p fixed at $p = 3.00$ and $p = 4.00$. Comment on what you see in the tables.

14. How does beef consumption vary as a function of household income if the price of beef is held constant?

15. Make a table showing the amount of money, M, that the average household spends on beef (in dollars per household per week) as a function of the price of beef and household income.

16. Make a table of the proportion, P, of household income spent on beef per week as a function of price and income. (Note that P is the fraction of income spent on beef.)

17. Express P, the proportion of household income spent on beef per week, in terms of the original function $f(I, p)$ which gave consumption as a function of p and I.

In Problems 18–19, the fallout, V, (in kilograms per square kilometer) from a volcanic explosion depends on the distance, d, from the volcano and the time, t, since the explosion:

$$V = f(d, t) = (\sqrt{t})e^{-d}.$$

[1]Data from www.nws.noaa.gov/om/windchill, accessed on May 30, 2004.

[2]From Richard G. Lipsey, *An Introduction to Positive Economics 3rd Ed.*, (London: Weidenfeld and Nicolson, 1971).

18. On the same axes, graph cross-sections of f with $t = 1$, and $t = 2$. As distance from the volcano increases, how does the fallout change? Look at the relationship between the graphs: how does the fallout change as time passes? Explain your answers in terms of volcanoes.

19. On the same axes, graph cross-sections of f with $d = 0$, $d = 1$, and $d = 2$. As time passes since the explosion, how does the fallout change? Look at the relationship between the graphs: how does fallout change as a function of distance? Explain your answers in terms of volcanoes.

20. An airport can be cleared of fog by heating the air. The amount of heat required, $H(T, w)$ (in calories per cubic meter of fog), depends on the temperature of the air, T (in °C) and the wetness of the fog, w (in grams per cubic meter of fog). Figure 9.4 shows several graphs of H against T with w fixed.

(a) Estimate $H(20, 0.3)$ and explain what information it gives us.

(b) Make a table of values for $H(T, w)$. Use $T = 0, 10, 20, 30, 40$, and $w = 0.1, 0.2, 0.3, 0.4$.

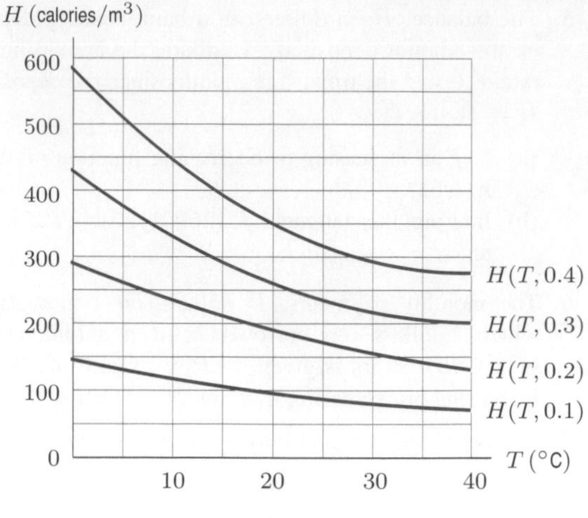

Figure 9.4

9.2 CONTOUR DIAGRAMS

How can we visualize a function of two variables? Functions of two variables are often represented by *contour diagrams*.

Weather Maps

Figure 9.5 shows a weather map from a newspaper. It shows the predicted high temperature, T, in degrees Fahrenheit (°F), throughout the US on that day. The curves on the map, called *isotherms*, separate the country into zones, according to whether T is in the 60s, 70s, 80s, 90s, or 100s. (*Iso* means same and *therm* means heat.) Notice that the isotherm separating the 80s and 90s zones connects all the points where the temperature is predicted to be exactly 90°F.

If the function $T = f(x, y)$ gives the predicted high temperature (in ° F) on this particular day as a function of latitude x and longitude y, then the isotherms are graphs of the equations

$$f(x, y) = c$$

where c is a constant. In general, such curves are called *contours*, and a graph showing selected contours of a function is called a contour diagram.

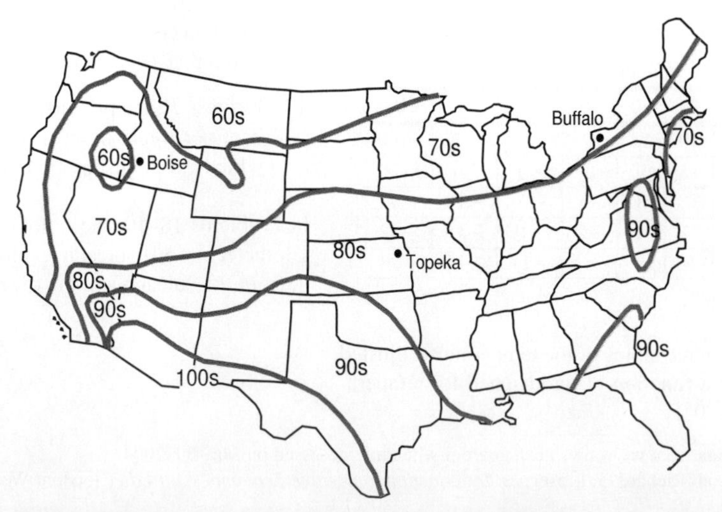

Figure 9.5: Weather map showing predicted high temperatures, T, on a summer day

Example 1 Estimate the predicted value of T in Boise, Idaho; Topeka, Kansas; and Buffalo, New York.

Solution Boise and Buffalo are in the 70s region, and Topeka is in the 80s region. Thus, the predicted temperature in Boise and Buffalo is between 70 and 80 while the predicted temperature in Topeka is between 80 and 90.

In fact, we can say more. Although both Boise and Buffalo are in the 70s, Boise is quite close to the $T = 70$ isotherm, whereas Buffalo is quite close to the $T = 80$ isotherm. So we estimate that the temperature will be in the low 70s in Boise and in the high 70s in Buffalo. Topeka is about halfway between the $T = 80$ isotherm and the $T = 90$ isotherm. Thus, we guess that the temperature in Topeka will be in the mid 80s. In fact, the actual high temperatures for that day were 71°F for Boise, 79°F for Buffalo, and 86°F for Topeka.

Topographical Maps

Another common example of a contour diagram is a topographical map like that shown in Figure 9.6. Here, the contours separate regions of lower elevation from regions of higher elevation, and give an overall picture of the nature of the terrain. Such topographical maps are frequently colored green at the lower elevations and brown, red, or even white at the higher elevations.

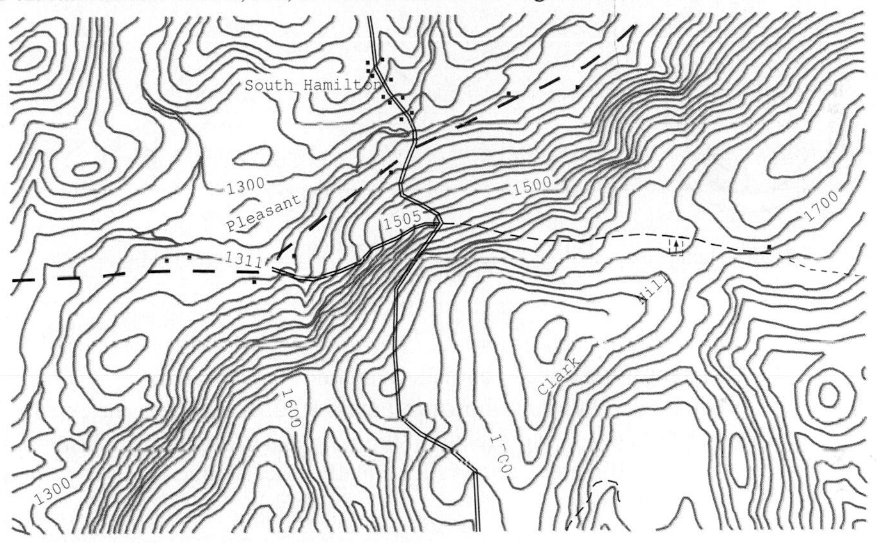

Figure 9.6: A topographical map showing the region around South Hamilton, NY

Example 2 Explain why the topographical map shown in Figure 9.7 corresponds to terrain such as that shown in Figure 9.8.

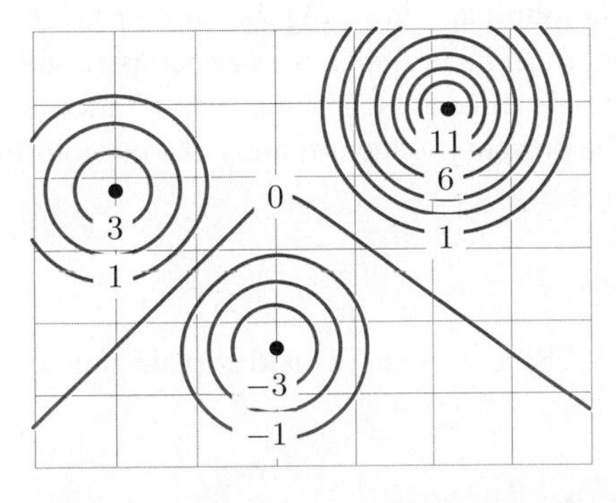

Figure 9.7: A topographical map

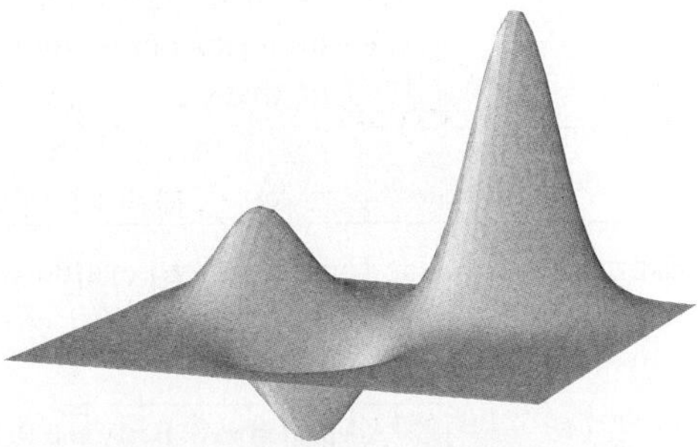

Figure 9.8: Terrain corresponding to the topographical map in Figure 9.7

Solution We see from the topographical map in Figure 9.7 that there are two hills, one with height about 12, and the other with height about 4. Most of the terrain is around height 0, and there is one valley with height about -4. This matches the terrain in Figure 9.8 since there are two hills (one taller than the other) and one valley.

The contours on a topographical map outline the contour or shape of the land. Because every point along the same contour has the same elevation, contours are also called *level curves* or *level sets*. The more closely spaced the contours, the steeper the terrain; the more widely spaced the contours, the flatter the terrain (provided, of course, that the elevation between contours varies by a constant amount). Certain features have distinctive characteristics. A mountain peak is typically surrounded by contours like those in Figure 9.9. A pass in a range of mountains may have contours that look like Figure 9.10. A long valley has parallel contours indicating the rising elevations on both sides of the valley (see Figure 9.11); a long ridge of mountains has the same type of contours, only the elevations decrease on both sides of the ridge. Notice that the elevation numbers on the contours are as important as the curves themselves.

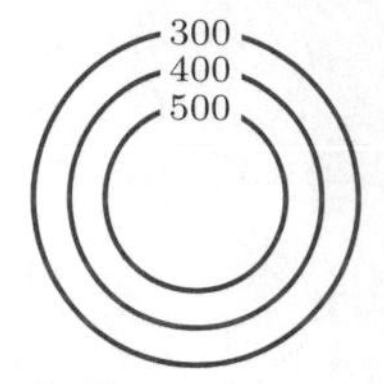

Figure 9.9: Mountain peak

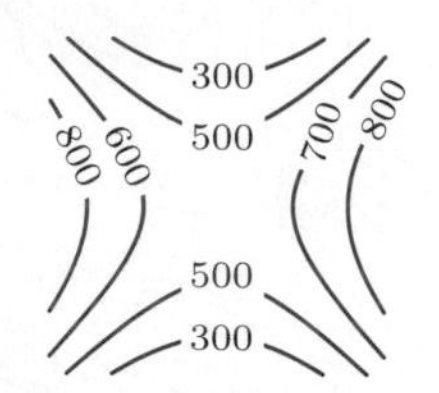

Figure 9.10: Pass between two mountains

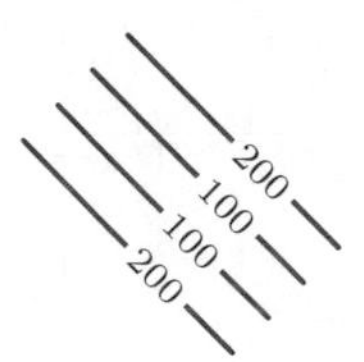

Figure 9.11: Long valley

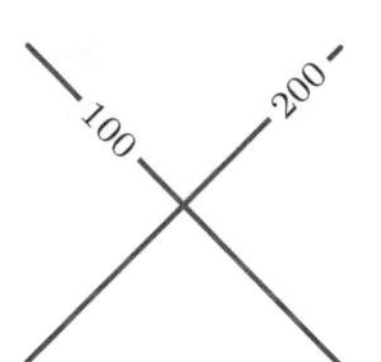

Figure 9.12: Impossible contour lines

There are some things contours cannot do. Two contours corresponding to different elevations cannot cross each other as shown in Figure 9.12. If they did, the point of intersection of the two curves would have two different elevations, which is impossible (assuming the terrain has no overhangs). We usually draw contours for equally spaced values of the function.

Using Contour Diagrams

Consider the effect of different weather conditions on US corn production. What would happen if the average temperature were to increase (due to global warming, for example) or if the rainfall were to decrease (due to a drought)? One way of estimating the effect of these climatic changes is to use Figure 9.13. This map is a contour diagram giving the corn production $C = f(R, T)$ in the US as a function of the total rainfall, R, in inches, and average temperature, T, in degrees Fahrenheit, during the growing season.[3] Suppose at the present time, $R = 15$ inches and $T = 76$°F. Production is measured as a percentage of the present production; thus, the contour through $R = 15, T = 76$ is $C = 100$, that is, $C = f(15, 76) = 100$.

Example 3 Use Figure 9.13 to evaluate $f(18, 78)$ and $f(12, 76)$ and explain the answers in terms of corn production.

[3] Adapted from S. Beaty and R. Healy, *The Future of American Agriculture*, Scientific American, Vol. 248, No. 2, February, 1983.

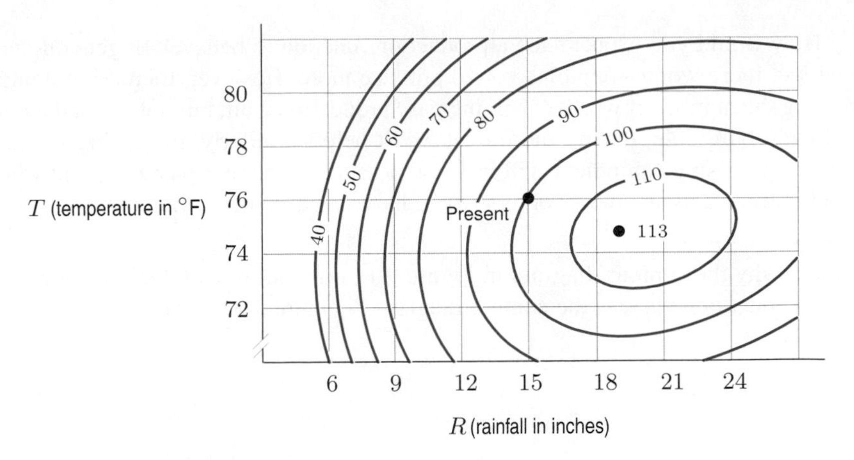

Figure 9.13: Corn production, C, as a function of rainfall and temperature

Solution The point with R-coordinate 18 and T-coordinate 78 is on the contour with value $C = 100$, so $f(18, 78) = 100$. This means that if the annual rainfall were 18 inches and the temperature were 78°F, the country would produce about the same amount of corn as at present, although it would be wetter and warmer than it is now. The point with R-coordinate 12 and T-coordinate 76 is about halfway between the $C = 80$ and $C = 90$ contours, so $f(12, 76) \approx 85$. This means that if the rainfall dropped to 12 inches and the temperature stayed at 76°F, then corn production would drop to about 85% of what it is now.

Example 4 Describe how corn production changes as a function of rainfall if temperature is fixed at the present value in Figure 9.13. Describe how corn production changes as a function of temperature if rainfall is held constant at the present value. Give common sense explanations for your answers.

Solution To see what happens to corn production if the temperature stays fixed at 76°F but the rainfall changes, look along the horizontal line $T = 76$. Starting from the present and moving left along the line $T = 76$, the values on the contours decrease. In other words, if there is a drought, corn production decreases. Conversely, as rainfall increases, that is, as we move from the present to the right along the line $T = 76$, corn production increases, reaching a maximum of more than 110% when $R = 21$, and then decreases (too much rainfall floods the fields). If, instead, rainfall remains at the present value and temperature increases, we move up the vertical line $R = 15$. Under these circumstances corn production decreases; a 2° increase causes a 10% drop in production. This makes sense since hotter temperatures lead to greater evaporation and hence drier conditions, even with rainfall constant at 15 inches. Similarly, a decrease in temperature leads to a very slight increase in production, reaching a maximum of around 102% when $T = 74$, followed by a decrease (the corn won't grow if it is too cold).

Cobb-Douglas Production Functions

Suppose you are running a small printing business, and decide to expand because you have more orders than you can handle. How should you expand? Should you start a night shift and hire more workers? Should you buy more expensive but faster computers which will enable the current staff to keep up with the work? Or should you do some combination of the two?

Obviously, the way such a decision is made in practice involves many other considerations—such as whether you could get a suitably trained night shift, or whether there are any faster computers available. Nevertheless, you might model the quantity, P, of work produced by your business as a function of two variables: your total number, N, of workers, and the total value, V, of your equipment.

How would you expect such a production function to behave? In general, having more equipment and more workers enables you to produce more. However, increasing equipment without increasing the number of workers will increase production a bit, but not beyond a point. (If equipment is already lying idle, having more of it won't help.) Similarly, increasing the number of workers without increasing equipment will increase production, but not past the point where the equipment is fully utilized, as any new workers would have no equipment available to them.

Example 5 Explain why the contour diagram in Figure 9.14 does not model the behavior expected of the production function, whereas the contour diagram in Figure 9.15 does.

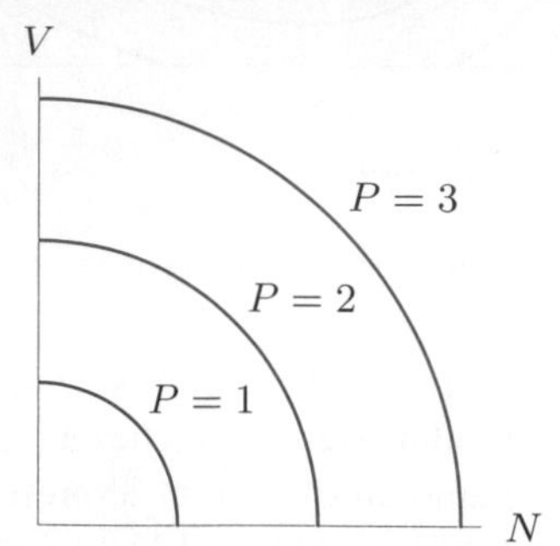

Figure 9.14: Incorrect contours for printing production

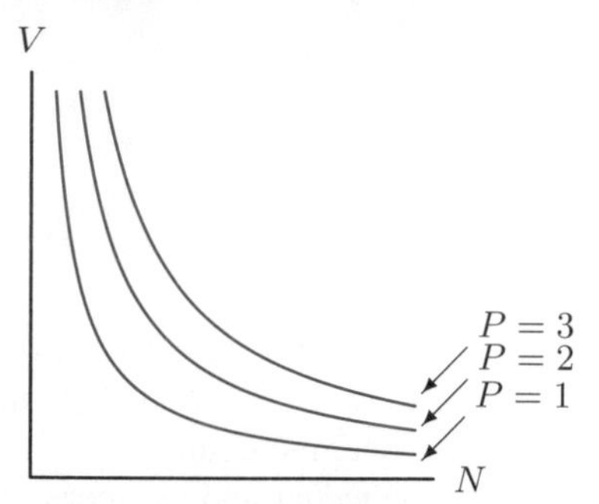

Figure 9.15: Correct contours for printing production

Solution Production P should be an increasing function of N and an increasing function of V. We see that both contour diagrams (in Figure 9.14 and Figure 9.15) satisfy this condition. Which of the contour diagrams has production increasing in the correct way? First look at the contour diagram in Figure 9.14. Fixing V at a particular value and letting N increase means moving to the right on the contour diagram. As we do so, we cross contours with larger and larger P values, meaning that production increases indefinitely. On the other hand, in Figure 9.15, as we move in the same direction we eventually find ourselves moving nearly parallel to the contours, crossing them less and less frequently. Therefore, production increases more and more slowly as N increases while V is held fixed. Similarly, if we hold N fixed and let V increase, the contour diagram in Figure 9.14 shows production increasing at a steady rate, whereas Figure 9.15 shows production increasing, but at a decreasing rate. Thus, Figure 9.15 fits the expected behavior of the production function best.

The Cobb-Douglas Production Model

In 1928, Cobb and Douglas used a simple formula to model the production of the entire US economy in the first quarter of the last century. Using government estimates of P, the total yearly production between 1899 and 1922, and of K, the total capital investment over the same period, and of L, the total labor force, they found that P was well approximated by the function:

$$P = 1.01L^{0.75}K^{0.25}.$$

This function turned out to model the US economy surprisingly accurately, both for the period on which it was based, and for some time afterward. The contour diagram of this function is similar to that in Figure 9.15. In general, production is often modeled by a function of the following form:

> **Cobb-Douglas Production Function**
>
> $$P = f(N, V) = cN^{\alpha}V^{\beta}$$
>
> where P is the total quantity produced and c, α, and β are positive constants with $0 < \alpha < 1$ and $0 < \beta < 1$.

Contour Diagrams and Tables

Table 9.5 shows the heat index as a function of temperature and humidity. The heat index is a temperature which tells you how hot it feels as a result of the combination of the two. We can also display this function using a contour diagram. Scales for the two independent variables (temperature and humidity) go on the axes. The heat indices shown range from 64 to 151, so we will draw contours at values of 70, 80, 90, 100, 110, 120, 130, 140, and 150. How do we know where the contour for 70 goes? Table 9.5 shows that, when humidity is 0%, a heat index of 70 occurs between 75°F and 80°F, so the contour will go approximately through the point $(76, 0)$. It also goes through the point $(75, 10)$. Continuing in this way, we can approximate the 70 contour. See Figure 9.16. You can construct all the contours in Figure 9.17 in a similar way.

Table 9.5 *Heat index (°F)*

Humidity (%) \ Temperature (°F)	70	75	80	85	90	95	100	105	110	115
0	64	69	73	78	83	87	91	95	99	103
10	65	70	75	80	85	90	95	100	105	111
20	66	72	77	82	87	93	99	105	112	120
30	67	73	78	84	90	96	104	113	123	135
40	68	74	79	86	93	101	110	123	137	151
50	69	75	81	88	96	107	120	135	150	
60	70	76	82	90	100	114	132	149		

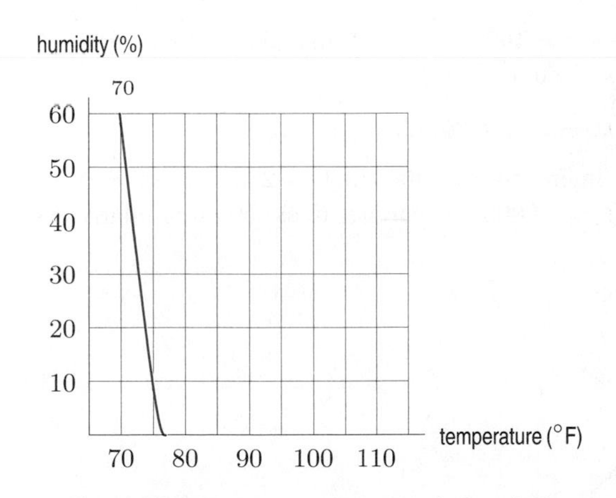

Figure 9.16: The contour for a heat index of 70

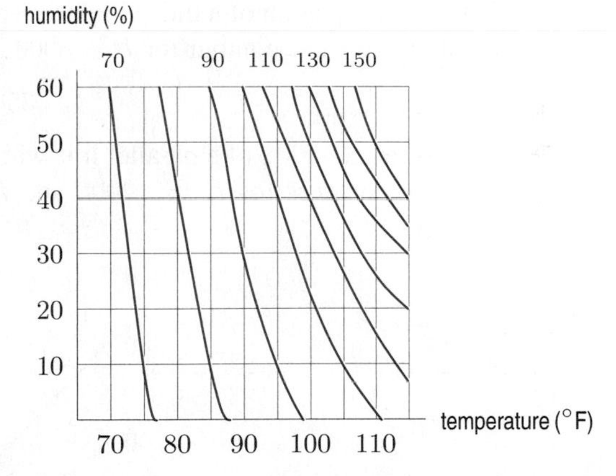

Figure 9.17: Contour diagram for the heat index

Example 6 Heat exhaustion is likely to occur where the heat index is 105 or higher. On the contour diagram in Figure 9.17, shade in the region where heat exhaustion is likely to occur.

Solution The shaded region in Figure 9.18 shows the values of temperature and humidity at which the heat index is above 105.

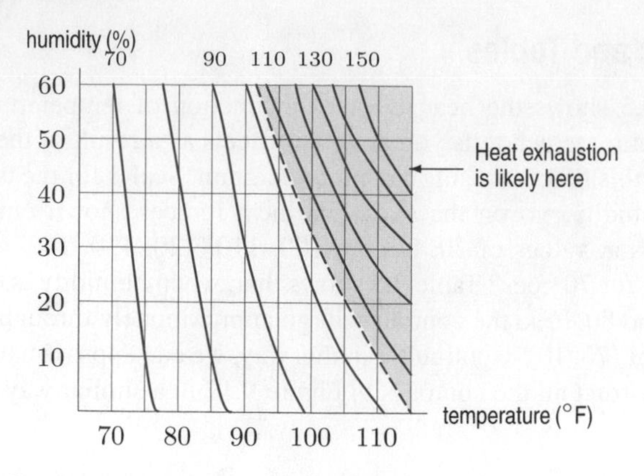

Figure 9.18: Shaded region shows conditions under which heat exhaustion is likely

Finding Contours Algebraically

Algebraic equations for the contours of a function f are easy to find if we have a formula for $f(x, y)$. A contour consists of all the points (x, y) where $f(x, y)$ has a constant value, c. Its equation is

$$f(x, y) = c.$$

Example 7 Draw a contour diagram for the airline revenue function $R = 350x + 200y$. Include contours for $R = 4000, 8000, 12000, 16000$.

Solution The contour for $R = 4000$ is given by

$$350x + 200y = 4000.$$

This is the equation of a line with intercepts $x = 4000/350 = 11.43$ and $y = 4000/200 = 20$. (See Figure 9.19.) The contour for $R = 8000$ is given by

$$350x + 200y = 8000.$$

This is the equation of a parallel line with intercepts $x = 8000/350 = 22.86$ and $y = 8000/200 = 40$. The contours for $R = 12000$ and $R = 16000$ are parallel lines drawn similarly. (See Figure 9.19.)

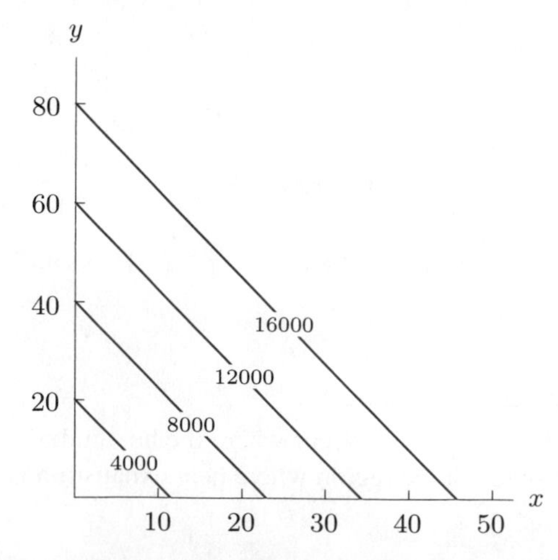

Figure 9.19: A contour diagram for $R = 350x + 200y$

Problems for Section 9.2

1. Figure 9.20 shows contour diagrams of temperature in °C in a room at three different times. Describe the heat flow in the room. What could be causing this?

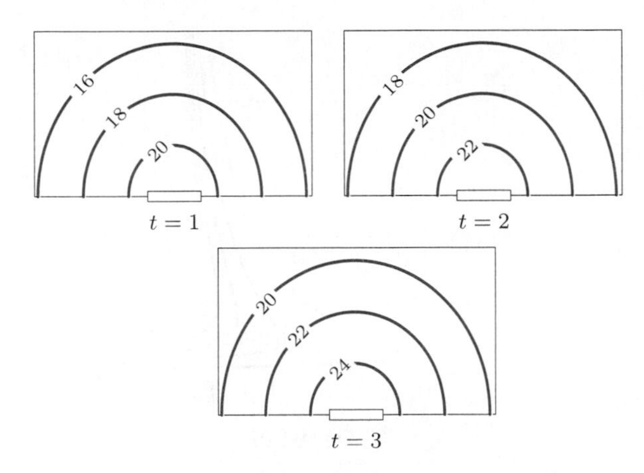

Figure 9.20

2. Figure 9.21 shows contours for the function $z = f(x, y)$. Is z an increasing or a decreasing function of x? Is z an increasing or a decreasing function of y?

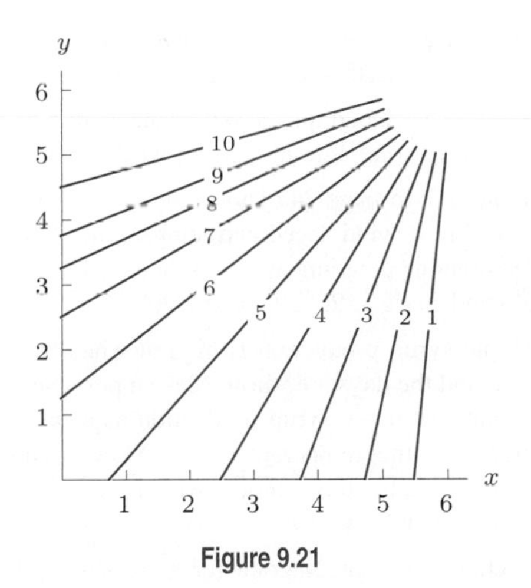

Figure 9.21

3. A manufacturer sells two products, one at a price of \$3000 a unit and the other at a price of \$12,000 a unit. A quantity q_1 of the first product and q_2 of the second product are sold at a total cost of \$4000 to the manufacturer.

   (a) Express the manufacturer's profit, π, as a function of q_1 and q_2.

   (b) Sketch contours of π for $\pi = 10{,}000$, $\pi = 20{,}000$, and $\pi = 30{,}000$ and the break-even curve $\pi = 0$.

4. Figure 9.22 is a contour diagram for the demand for orange juice as a function of the price of orange juice and the price of apple juice. Which axis corresponds to orange juice? Which to apple juice? Explain.

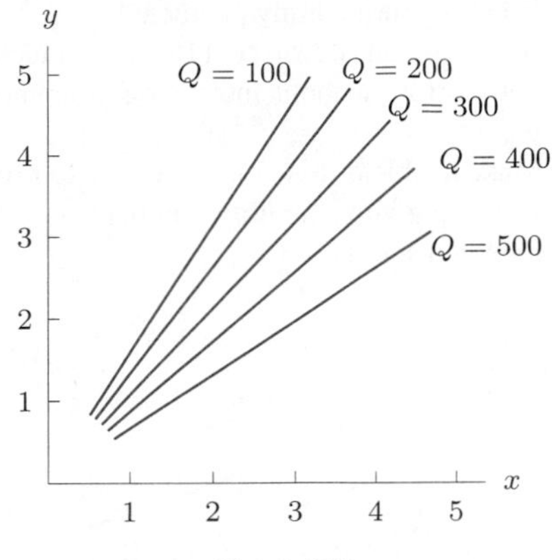

Figure 9.22

5. Figure 9.23 is a contour diagram for the sales of a product as a function of the price of the product and the amount spent on advertising. Which axis corresponds to the amount spent on advertising? Explain.

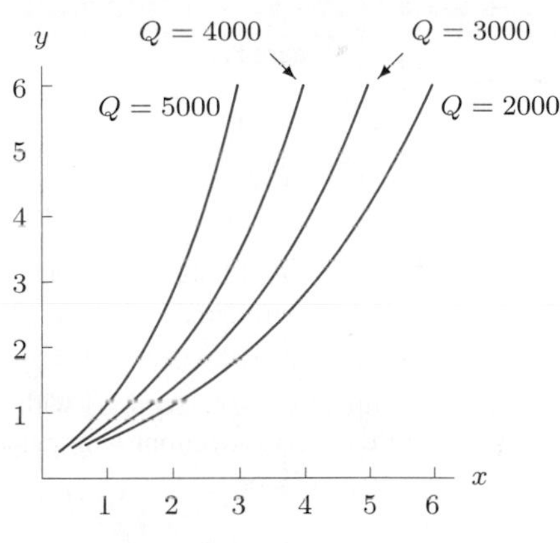

Figure 9.23

6. A topographic map is given in Figure 9.24. How many hills are there? Estimate the x- and y-coordinates of the tops of the hills. Which hill is the highest? A river runs through the valley; in which direction is it flowing?

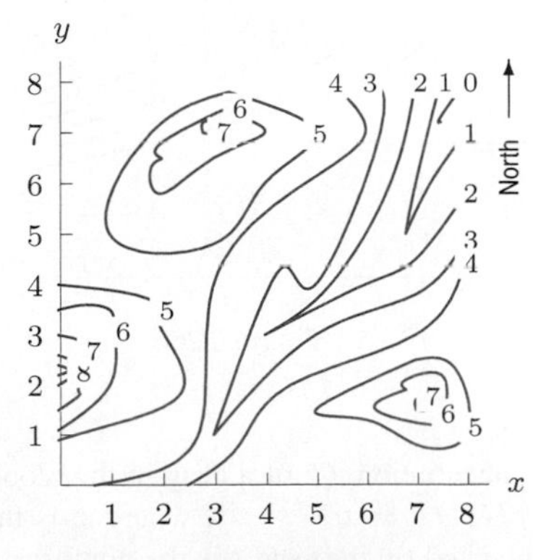

Figure 9.24

7. Figure 9.25 is a contour diagram of the monthly payment on a 5-year car loan as a function of the interest rate and the amount you borrow. The interest rate is 13% and you borrow \$6000.

(a) What is your monthly payment?
(b) If interest rates drop to 11%, how much more can you borrow without increasing your monthly payment?
(c) Make a table of how much you can borrow, without increasing your monthly payment, as a function of the interest rate.

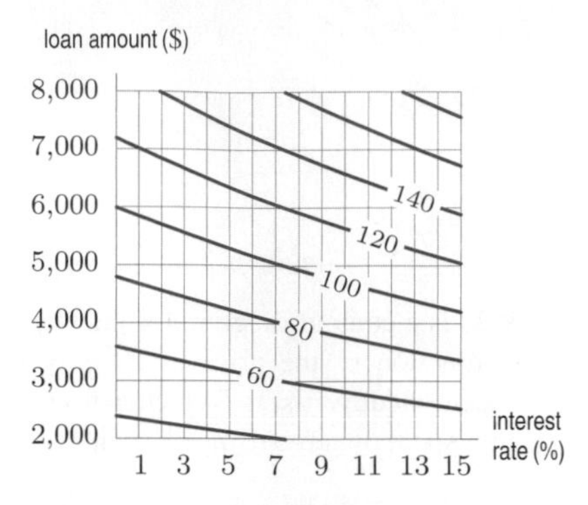

Figure 9.25

8. The contour diagram in Figure 9.26 shows your happiness as a function of love and money.

(a) Describe in words your happiness as a function of:
(i) Money, with love fixed.
(ii) Love, with money fixed.
(b) Graph two different cross-sections with love fixed and two different cross-sections with money fixed.

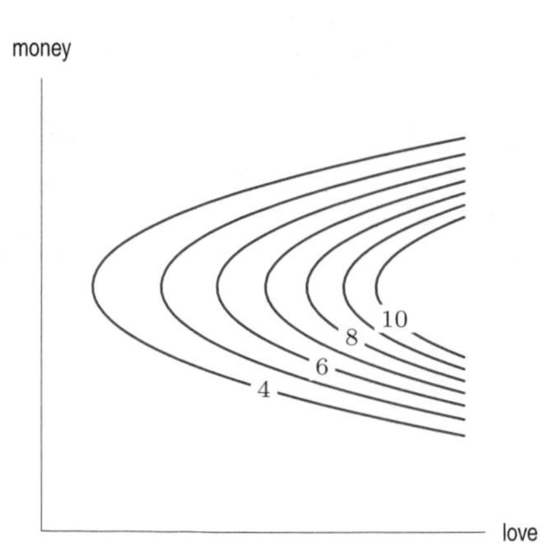

Figure 9.26

9. The concentration, C, of a drug in the blood is given by $C = f(x,t) = te^{-t(5-x)}$, where x is the amount of drug injected (in mg) and t is the number of hours since the injection. The contour diagram of $f(x,t)$ is given in Figure 9.27. Explain the diagram by varying one variable at a time: describe f as a function of x if t is held fixed, and then describe f as a function of t if x is held fixed.

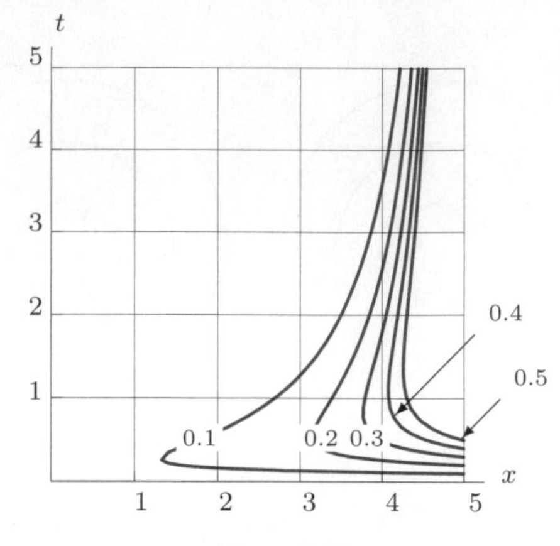

Figure 9.27

Problems 10–12 refer to the map in Figure 9.5 on page 348.

10. Give the range of daily high temperatures for

(a) Pennsylvania **(b)** North Dakota
(c) California.

11. Sketch a possible graph of the predicted high temperature T on a line north-south through Topeka.

12. Sketch possible graphs of the predicted high temperature on a north-south line and an east-west line through Boise.

13. Table 9.3 on page 347 shows the wind-chill factor as a function of wind speed and temperature. Draw a possible contour diagram for this function. Include contours at wind chills of 20°, 0°, and −20°.

14. Maple syrup production is highest when the nights are cold and the days are warm. Make a possible contour diagram for maple syrup production as a function of the high (daytime) temperature and the low (nighttime) temperature. Label the contours with 10, 20, 30, and 40 (in liters of maple syrup.)

15. Sketch a contour diagram for $z = y - \sin x$. Include at least four labeled contours. Describe the contours in words and how they are spaced.

In Problems 16–21, sketch a contour diagram for the function with at least four labeled contours. Describe in words the contours and how they are spaced.

16. $f(x,y) = x + y$ **17.** $f(x,y) = x + y + 1$

18. $f(x,y) = 3x + 3y$ **19.** $f(x,y) = -x - y$

20. $f(x,y) = 2x - y$ **21.** $f(x,y) = y - x^2$

22. Figure 9.28 shows contours of the function giving the species density of breeding birds at each point in the US, Canada, and Mexico.[4] Are the following statements true or false? Explain your answers.

(a) Moving from south to north across Canada, the species density increases.
(b) In general, peninsulas (for example, Florida, Baja California, the Yucatan) have lower species densities than the areas around them.
(c) The species density around Miami is over 100.
(d) The greatest rate of change in species density with distance is in Mexico. If this is true, mark the point and direction giving the maximum rate.

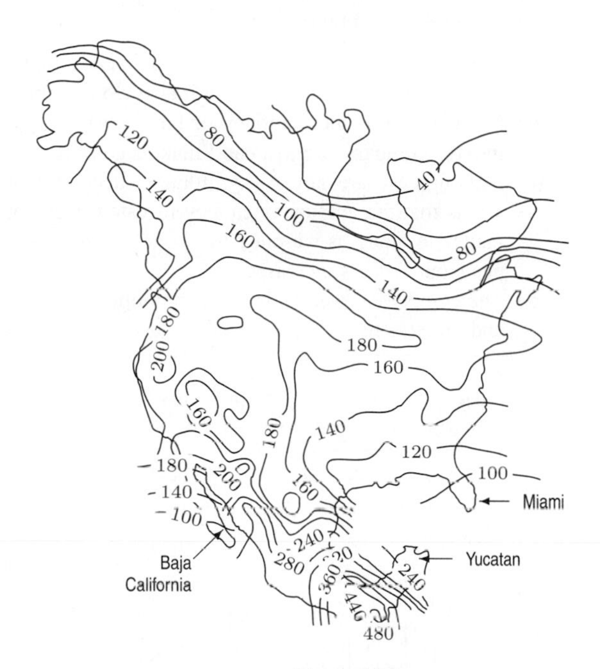

Figure 9.28

23. In a small printing business, $P = 2N^{0.6}V^{0.4}$, where N is the number of workers, V is the value of the equipment, and P is production, in thousands of pages per day.

(a) If this company has a labor force of 300 workers and 200 units worth of equipment, what is production?
(b) If the labor force is doubled (to 600 workers), how does production change?
(c) If the company purchases enough equipment to double the value of its equipment (to 400 units), how does production change?
(d) If both N and V are doubled from the values given in part (a), how does production change?

24. The quantity, Q, of a certain item produced depends on the number of units of labor, L, and of capital, K, according to the function $Q = 900L^{1/2}K^{2/3}$.

(a) If $L = 70$ and $K = 50$, what quantity is produced?
(b) Find Q when $L = 140$ and $K = 100$. In general, discuss the effects on Q of doubling both L and K.

25. Figure 9.29 gives contour diagrams for different Cobb-Douglas production functions $F(L, K)$. Match each contour diagram with the correct statement.

(A) Tripling each input triples output.
(B) Quadrupling each input doubles output.
(C) Doubling each input almost triples output.

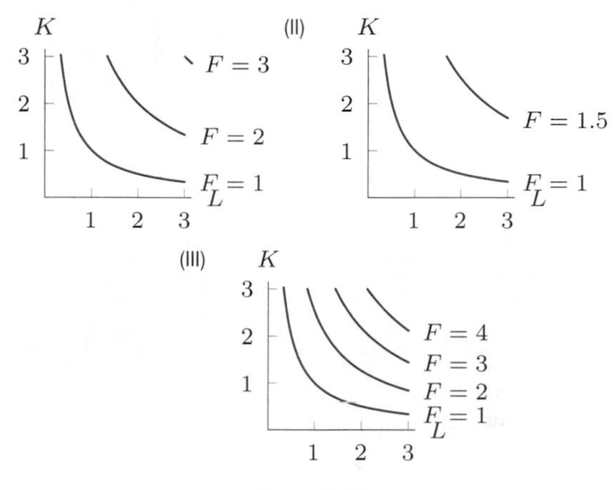

Figure 9.29

26. Each of the contour diagrams in Figure 9.30 shows population density in a certain region of a city. Choose the contour diagram that best corresponds to each of the following situations. Many different matchings are possible. Pick a reasonable one and justify your choice.

(a) The middle contour is a highway.
(b) The middle contour is an open sewage canal.
(c) The middle contour is a railroad line.

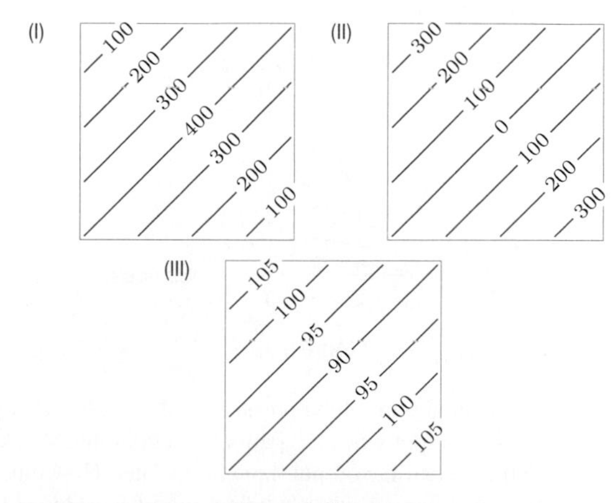

Figure 9.30

[4]From the undergraduate senior thesis of Professor Robert Cook, Director of Harvard's Arnold Arboretum.

27. Match tables (a)–(d) with the contour diagrams (I)–(IV) in Figure 9.31.

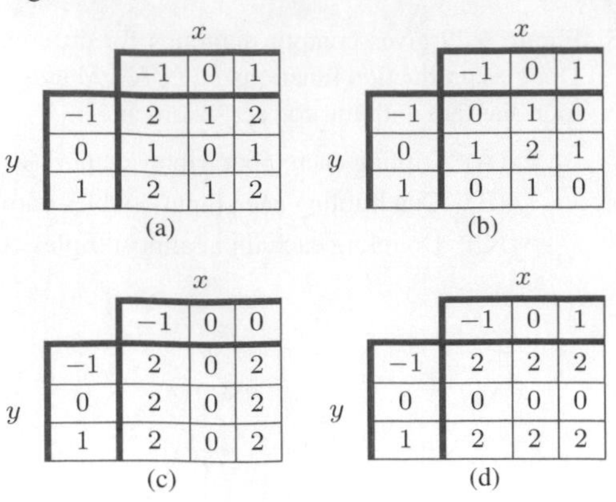

(a)

y \ x	−1	0	1
−1	2	1	2
0	1	0	1
1	2	1	2

(b)

y \ x	−1	0	1
−1	0	1	0
0	1	2	1
1	0	1	0

(c)

y \ x	−1	0	0
−1	2	0	2
0	2	0	2
1	2	0	2

(d)

y \ x	−1	0	1
−1	2	2	2
0	0	0	0
1	2	2	2

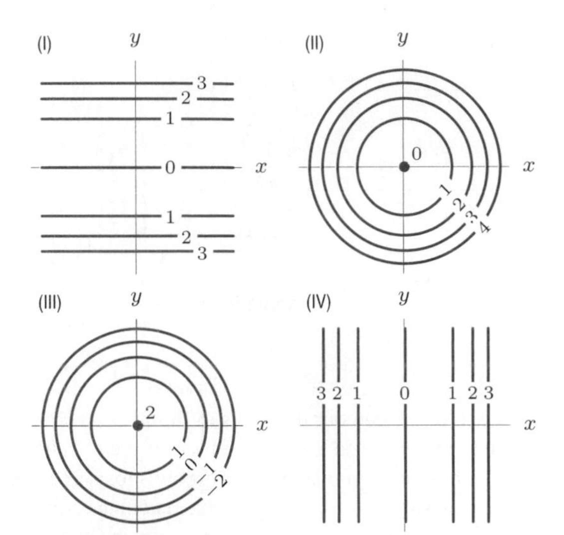

Figure 9.31

28. Figure 9.32 shows cardiac output (in liters per minute) in patients suffering from shock as a function of blood pressure in the central veins (in mm Hg) and the time in hours since the onset of shock.[5]

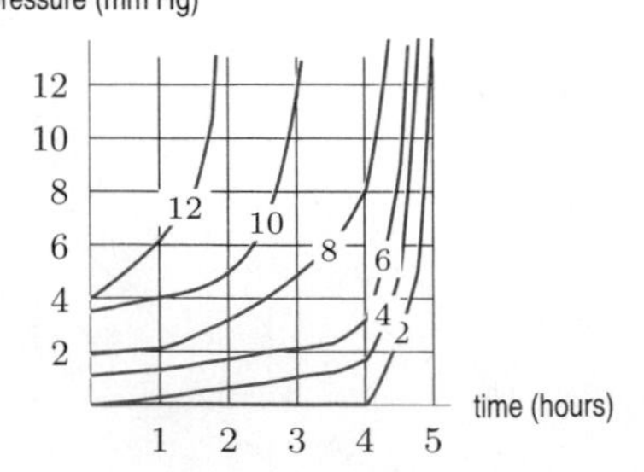

Figure 9.32

(a) In a patient with blood pressure of 4 mm Hg, what is cardiac output when the patient first goes into shock? Estimate cardiac output three hours later. How much time has passed when cardiac output is reduced to 50% of the initial value?

(b) In patients suffering from shock, is cardiac output an increasing or decreasing function of blood pressure?

(c) Is cardiac output an increasing or decreasing function of time, t, where t represents the elapsed time since the patient went into shock?

(d) If blood pressure is 3 mm Hg, explain how cardiac output changes as a function of time. In particular, does it change rapidly or slowly during the first two hours of shock? During hours 2 to 4? During the last hour of the study? Explain why this information is useful to a physician treating a patient for shock.

29. Match the following descriptions of a company's success with the contour diagrams of success as a function of money and work in Figure 9.33.

(a) Our success is measured in dollars, plain and simple. More hard work won't hurt, but it also won't help.

(b) No matter how much money or hard work we put into the company, we just can't make a go of it.

(c) Although we are not always totally successful, it seems that the amount of money invested does not matter. As long as we put hard work into the company, our success increases.

(d) The company's success is based on both hard work and investment.

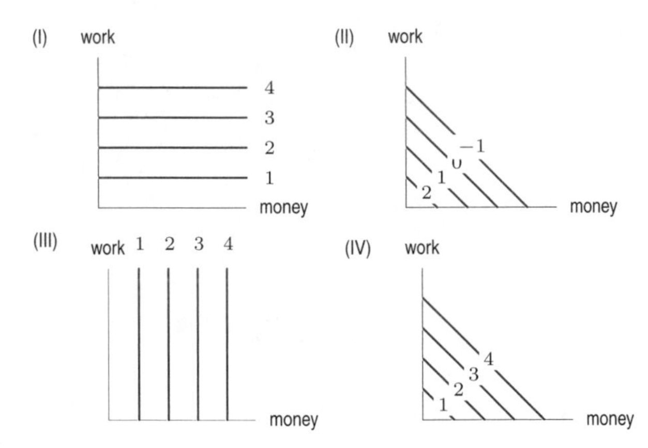

Figure 9.33

30. Figure 9.34 shows a contour map of a hill with two paths, A and B.

(a) On which path, A or B, will you have to climb more steeply?

(b) On which path, A or B, will you probably have a better view of the surrounding countryside? (Assuming trees do not block your view.)

(c) Alongside which path is there more likely to be a stream?

[5] Arthur C. Guyton and John E. Hall, *Textbook of Medical Physiology, Ninth Edition*, p. 288, (Philadelphia: W. B. Saunders, 1996).

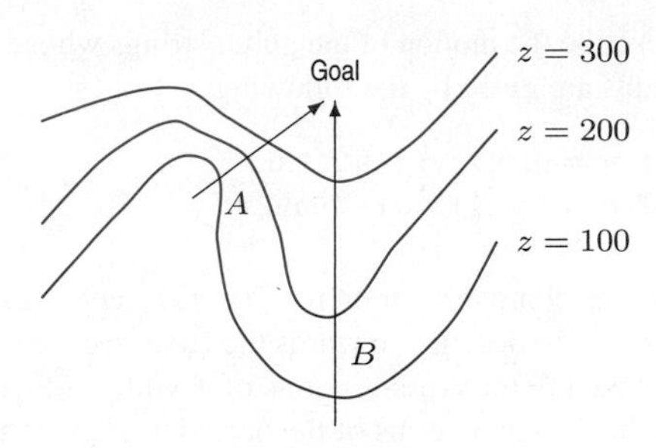

Figure 9.34

31. Antibiotics can be toxic in large doses. If repeated doses of an antibiotic are to be given, the rate at which the medicine is excreted through the kidneys should be monitored by a physician. One measure of kidney function is the glomerular filtration rate, or GFR, which measures the amount of material crossing the outer (or glomerular) membrane of the kidney, in milliliters per minute. A normal GFR is about 125 ml/min. Figure 9.35 gives a contour diagram of the percent, P, of a dose of mezlocillin (an antibiotic) excreted, as a function of the patient's GFR and the time, t, in hours since the dose was administered.[6]

(a) In a patient with a GFR of 50, approximately how long will it take for 30% of the dose to be excreted?
(b) In a patient with a GFR of 60, approximately what percent of the dose has been excreted after 5 hours?
(c) Explain how we can tell from the graph that, for a patient with a fixed GFR, the amount excreted changes very little after 12 hours.
(d) Is the percent excreted an increasing or decreasing function of time? Explain why this makes sense.
(e) Is the percent excreted an increasing or decreasing function of GFR? Explain what this means to a physician giving antibiotics to a patient with kidney disease.

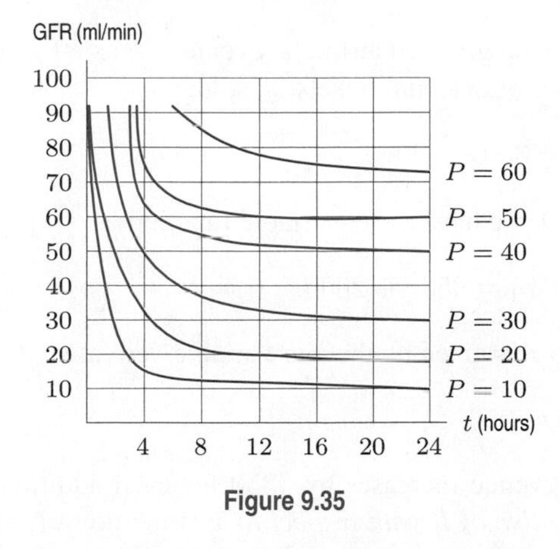

Figure 9.35

32. Match the pairs of functions (a)–(d) with the contour diagrams (I)–(IV). In each case, show which contours represent f and which represent g. (The x- and y-scales are equal.)

(a) $f(x, y) = x + y$, $g(x, y) = x - y$
(b) $f(x, y) = 2x + 3y$, $g(x, y) = 2x - 3y$
(c) $f(x, y) = x^2 - y$, $g(x, y) = 2y + \ln |x|$
(d) $f(x, y) = x^2 - y^2$, $g(x, y) = xy$

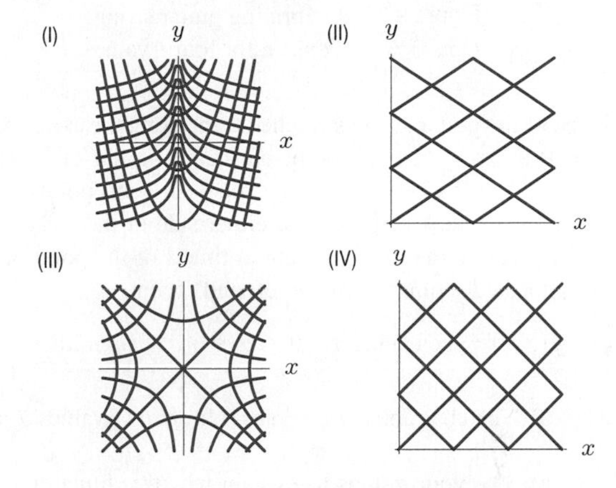

33. You like pizza and you like cola. The contour diagrams in Figure 9.36 show your happiness as a function of the number of pizzas and the number of colas you have. Which diagram represents your happiness if:

(a) There is no such thing as too many pizzas and too much cola?
(b) There is such a thing as too many pizzas or too much cola?
(c) There is such a thing as too much cola but no such thing as too many pizzas?

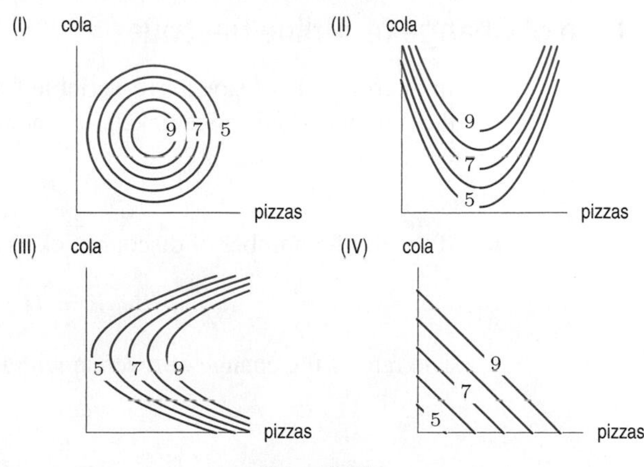

Figure 9.36

[6]Peter G. Welling and Francis L. S. Tse, *Pharmacokinetics of Cardiovascular, Central Nervous System, and Antimicrobial Drugs*, The Royal Society of Chemistry, 1985, p. 316.

Problems 34–38 concern a vibrating guitar string. Snapshots of the string at millisecond intervals are shown in Figure 9.37.

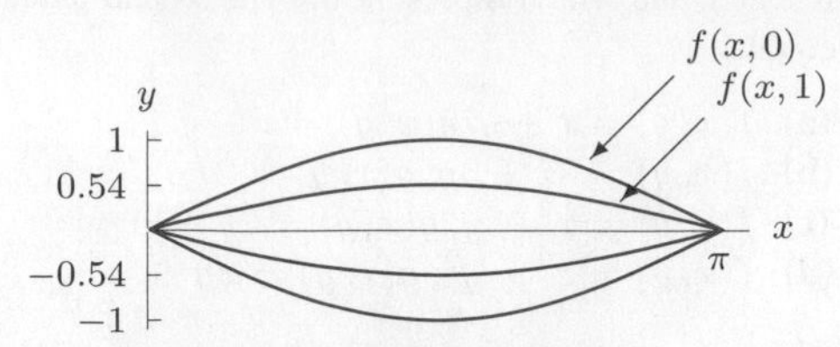

Figure 9.37: A vibrating guitar string: $f(x,t) = \cos t \sin x$ for four t values

Think of the guitar string stretched tight along the x-axis from $x = 0$ to $x = \pi$. Each point on the string has an x-value, $0 \leq x \leq \pi$. After it has been plucked, each point on the string moves back and forth on either side of the x-axis. Let $y = f(x,t)$ be the displacement at time t of the point on the string located x units from the left end. Then

$$y = f(x,t) = \cos t \sin x, \quad 0 \leq x \leq \pi, \quad t \text{ in milliseconds.}$$

34. **(a)** Sketch graphs of y versus x for fixed t values, $t = 0$, $\pi/4$, $\pi/2$, $3\pi/4$, π.

(b) Use your graphs to explain why this function could represent a vibrating guitar string.

35. Explain what the functions $f(x,0)$ and $f(x,1)$ represent in terms of the vibrating string.

36. Explain what the functions $f(0,t)$ and $f(1,t)$ represent in terms of the vibrating string.

37. Describe the motion of the guitar strings whose displacements are given by the following:

(a) $y = g(x,t) = \cos 2t \sin x$
(b) $y = h(x,t) = \cos t \sin 2x$

38. Use the contour diagram for $f(x,t) = \cos t \sin x$ in Figure 9.38 to describe in words the cross-sections of f with t fixed and the cross-sections of f with x fixed. Explain what you see in terms of the behavior of the string.

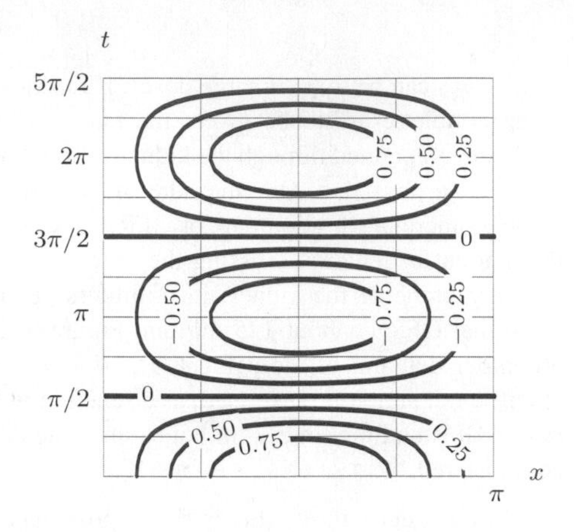

Figure 9.38

9.3 PARTIAL DERIVATIVES

In one-variable calculus we saw how the derivative measures the rate of change of a function. We begin by reviewing this idea.

Rate of Change of Airline Revenue

In Section 9.1 we saw a two-variable function which gives an airline's revenue, R, as a function of the number of full price tickets, x, and the number of discount tickets, y, sold:

$$R = f(x,y) = 350x + 200y.$$

If we fix the number of discount tickets at $y = 10$, we have a one-variable function

$$R = f(x,10) = g(x) = 350x + 2000.$$

The rate of the change of revenue with respect to x is given by the one-variable derivative

$$g'(x) = 350.$$

This tells us that, if y is fixed at 10, then the revenue increases by \$350 for each additional full price ticket sold. We call $g'(x)$ the *partial derivative of R with respect to x* at the point $(x, 10)$. If $R = f(x,y)$, we write

$$\frac{\partial R}{\partial x} = f_x(x,10) = g'(x) = 350.$$

Example 1 Find the rate of change of revenue, R, as y increases with x fixed at $x = 20$.

Solution Substituting $x = 20$ into $R = 350x + 200y$ gives the one-variable function

$$R = h(y) = 350(20) + 200y = 7000 + 200y$$

The rate of change of R as y increases with x fixed is

$$\frac{\partial R}{\partial y} = f_y(20, y) = h'(y) = 200.$$

We call $\partial R/\partial y = f_y(20, y)$ the *partial derivative of R with respect to y* at the point $(20, y)$. The fact that both partial derivatives of R are positive corresponds to the fact that the revenue is increasing as more of either type of ticket is sold.

Definition of the Partial Derivative

For any function $f(x, y)$ we study the influence of x and y separately on the value $f(x, y)$ by keeping one fixed and letting the other vary. The method of the previous example allows us to calculate the rates of change of $f(x, y)$ with respect to x and y. For all points (a, b) at which the limits exist, we make the following definitions:

Partial Derivatives of f With Respect to x and y

The *partial derivative of f with respect to x* at (a, b) is the derivative of f with y constant:

$$f_x(a,b) = \begin{array}{c}\text{Rate of change of } f \text{ with } y \text{ fixed}\\ \text{at } b, \text{ at the point } (a,b)\end{array} = \lim_{h\to 0} \frac{f(a+h,b) - f(a,b)}{h}.$$

The *partial derivative of f with respect to y* at (a, b) is the derivative of f with x constant:

$$f_y(a,b) = \begin{array}{c}\text{Rate of change of } f \text{ with } x \text{ fixed}\\ \text{at } a, \text{ at the point } (a,b)\end{array} = \lim_{h\to 0} \frac{f(a,b+h) - f(a,b)}{h}.$$

If we think of a and b as variables, $a = x$ and $b = y$, we have the **partial derivative functions** $f_x(x, y)$ and $f_y(x, y)$.

Just as with ordinary derivatives, there is an alternative notation:

Alternative Notation for Partial Derivatives

If $z = f(x, y)$ we can write

$$f_x(x,y) = \frac{\partial z}{\partial x} \quad \text{and} \quad f_y(x,y) = \frac{\partial z}{\partial y}$$

$$f_x(a,b) = \left.\frac{\partial z}{\partial x}\right|_{(a,b)} \quad \text{and} \quad f_y(a,b) = \left.\frac{\partial z}{\partial y}\right|_{(a,b)}$$

We use the symbol ∂ to distinguish partial derivatives from ordinary derivatives. In cases where the independent variables have names different from x and y, we adjust the notation accordingly. For example, the partial derivatives of $f(u, v)$ are denoted by f_u and f_v.

Estimating Partial Derivatives from a Table

Example 2 An experiment[7] done on rats to measure the toxicity of formaldehyde yielded the data shown in Table 9.6. The values in the table show the percent, P, of rats that survived an exposure with concentration c (in parts per million) after t months, so $P = f(t, c)$. Using Table 9.6, estimate $f_t(18, 6)$ and $f_c(18, 6)$. Interpret your answers in terms of formaldehyde toxicity.

Table 9.6 *Percent, P, of rat population surviving after exposure to formaldehyde vapor*

Conc. c (ppm) \ Time t (months)	0	2	4	6	8	10	12	14	16	18	20	22	24
0	100	100	100	100	100	100	100	100	100	100	99	97	95
2	100	100	100	100	100	100	100	100	99	98	97	95	92
6	100	100	100	99	99	98	96	96	95	93	90	86	80
15	100	100	100	99	99	99	99	96	93	82	70	58	36

Solution For $f_t(18, 6)$, we fix c at 6 ppm, and find the rate of change of percent surviving, P, with respect to t. We have

$$f_t(18,6) \approx \frac{\Delta P}{\Delta t} = \frac{f(20,6) - f(18,6)}{20 - 18} = \frac{90 - 93}{20 - 18} \approx -1.5\ \%\ \text{per month}.$$

This is the rate of change of percent surviving, P, *in the time t direction* at the point $(18, 6)$.The fact that it is negative means that P is decreasing as we read across the $c = 6$ row of the table in the direction of increasing t (that is, horizontally from left to right in Table 9.6). For $f_c(18, 6)$, we fix t at 18, and calculate the rate of change of P as we move in the direction of increasing c (that is, from top to bottom in Table 9.6.) We have

$$f_c(18,6) \approx \frac{\Delta P}{\Delta c} = \frac{f(18,15) - f(18,6)}{15 - 6} = \frac{82 - 93}{15 - 6} = -1.22\%\ \text{per ppm}.$$

The rate of change of P as c increases is about -1.22% per ppm. This means that as the concentration increases by 1 ppm from 6 ppm, the percent surviving 18 months decreases by about 1.22% per unit increase ppm. The partial derivative is negative because fewer rats survive this long when the concentration of formaldehyde increases. (That is, P goes down as c goes up.)

Using Partial Derivatives to Estimate Values of the Function

Example 3 Use Table 9.6 and partial derivatives to estimate the percent of rats surviving if they are exposed to formaldehyde with a concentration of

(a) 6 ppm for 18.5 months (b) 18 ppm for 24 months (c) 9 ppm for 20.5 months

[7] James E. Gibson, *Formaldehyde Toxicity*, p. 125, (Hemisphere Publishing Company, McGraw-Hill, 1983).

Solution

(a) Since $t = 18.5$ and $c = 6$, we want to evaluate $P = f(18.5, 6)$. Table 9.6 tells us that $f(18, 6) = 93\%$ and we have just calculated

$$\left.\frac{\partial P}{\partial t}\right|_{(18,6)} = f_t(18, 6) = -1.5\% \text{ per month.}$$

This partial derivative tells us that after 18 months of exposure to formaldehyde at a concentration of 6 ppm, P decreases by 1.5% for every additional month of exposure. Therefore after an additional 0.5 month, we have

$$P \approx 93 - 1.5(0.5) = 92.25\%.$$

(b) Now we wish to evaluate $f(24, 18)$. The closest entry to this in Table 9.6 is $f(24, 15) = 36$. We keep t fixed at 24 and increase c from 15 to 18. We estimate the rate of change in P as c changes; this is $\partial P/\partial c$. We see from Table 9.6 that

$$\left.\frac{\partial P}{\partial c}\right|_{(24,15)} \approx \frac{\Delta P}{\Delta c} = \frac{36 - 80}{15 - 6} = -4.89\% \text{ per ppm.}$$

The percent surviving 24 months goes down from 36% by about 4.89% for every unit increase in the formaldehyde concentration above 15 ppm. We have:

$$f(24, 18) \approx 36 - 4.89(3) = 21.33\%.$$

We estimate that only about 21% of the rats would survive for 24 months if they were exposed to formaldehyde as strong as 18 ppm. Since this figure is an extrapolation from the available data, we should use it with caution.

(c) To estimate $f(20.5, 9)$, we use the closest entry $f(20, 6) = 90$. As we move from $(20, 6)$ to $(20.5, 9)$, the percentage, P, changes both due to the change in t and due to the change in c. We estimate the two partial derivatives at $t = 20$, $c = 6$:

$$\left.\frac{\partial P}{\partial t}\right|_{(20,6)} \approx \frac{\Delta P}{\Delta t} = \frac{86 - 90}{22 - 20} = -2\% \text{ per month}$$

$$\left.\frac{\partial P}{\partial c}\right|_{(20,6)} \approx \frac{\Delta P}{\Delta c} = \frac{70 - 90}{15 - 6} = 2.22\% \text{ per month}$$

The change in P due to a change of $\Delta t = 0.5$ month and $\Delta c = 3$ ppm is

$$\begin{aligned}\Delta P &\approx \text{Change due to } \Delta t + \text{Change due to } \Delta c\\ &= -2(0.5) - 2.22(3)\\ &= -7.66\end{aligned}$$

So for $t = 20.5$, $c = 9$ we have

$$f(20.5, 9) \approx f(20, 6) - 7.66 = 82.34\%.$$

In Example 3 part (c), we estimated changes in the function using the relationship between ΔP, Δt, and Δc. The general form of this relationship is called *local linearity*:

$$\begin{matrix}\text{Change} \\ \text{in } f\end{matrix} \approx \begin{matrix}\text{Rate of change} \\ \text{in } x\text{-direction}\end{matrix} \cdot \Delta x + \begin{matrix}\text{Rate of change} \\ \text{in } y\text{-direction}\end{matrix} \cdot \Delta y$$

$$\Delta f \approx f_x \cdot \Delta x + f_y \cdot \Delta y$$

Estimating Partial Derivatives from a Contour Diagram

If we move parallel to one of the axes on a contour diagram, the partial derivative is the rate of change of the value of the function on the contours. For example, if the values on the contours are increasing in the direction of positive change, then the partial derivative must be positive.

Example 4 Figure 9.39 shows the contour diagram for the temperature $H(x,t)$ (in °F) in a room as a function of distance x (in feet) from a heater and time t (in minutes) after the heater has been turned on. What are the signs of $H_x(10,20)$ and $H_t(10,20)$? Estimate these partial derivatives and explain the answers in practical terms.

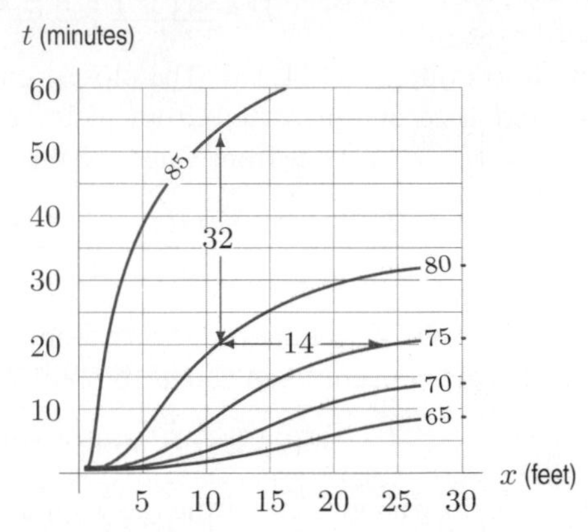

Figure 9.39: Temperature in a heated room

Solution The point $(10,20)$ is on the $H = 80$ contour. As x increases, we move toward the $H = 75$ contour, so H is decreasing and $H_x(10,20)$ is negative. This makes sense because as we move further from the heater, the temperature drops. On the other hand, as t increases, we move toward the $H = 85$ contour, so H is increasing and $H_t(10,20)$ is positive. This also makes sense, because it says that as time passes, the room warms up.

To estimate the partial derivatives, use a difference quotient. Looking at the contour diagram, we see there is a point on the $H = 75$ contour about 14 units to the right of $(10,20)$. Hence, H decreases by 5 when x increases by 14, so the rate of change of H with respect to x is about $\Delta H/\Delta x = -5/14 \approx -0.36$. Thus, we find

$$H_x(10,20) \approx -0.36°\text{F/ft}.$$

This means that near the point 10 feet from the heater, after 20 minutes the temperature drops about 1/3 of a degree for each foot we move away from the heater.

To estimate $H_t(10,20)$, we look again at the contour diagram and notice that the $H = 85$ contour is about 32 units directly above the point $(10,20)$. So H increases by 5 when t increases by 32. Hence,

$$H_t(10,20) \approx \frac{\Delta H}{\Delta t} = \frac{5}{32} \approx 0.16°\text{F/min}.$$

This means that after 20 minutes the temperature is going up about 1/6 of a degree each minute at the point 10 ft from the heater.

Using Units to Interpret Partial Derivatives

The units of the independent and dependent variables can often be helpful in explaining the meaning of a partial derivative.

Example 5 Suppose that your weight w in pounds is a function $f(c,n)$ of the number c of calories you consume daily and the number n of minutes you exercise daily. Using the units for w, c and n, interpret in everyday terms the statements

$$\left.\frac{\partial w}{\partial c}\right|_{(2000,15)} = 0.02 \quad \text{and} \quad \left.\frac{\partial w}{\partial n}\right|_{(2000,15)} = -0.025.$$

Solution The units of $\partial w/\partial c$ are pounds per calorie. The statement

$$\left.\frac{\partial w}{\partial c}\right|_{(2000,15)} = 0.02$$

means that if you are presently consuming 2000 calories daily and exercising 15 minutes daily, you will weigh 0.02 pounds more for each extra calorie you consume daily, or about 2 pounds for each extra 100 calories per day. The units of $\partial w/\partial n$ are pounds per minute. The statement

$$\left.\frac{\partial w}{\partial n}\right|_{(2000,15)} = -0.025$$

means that for the same calorie consumption and number of minutes of exercise, you will weigh 0.025 pounds less for each extra minute you exercise daily, or about 1 pound less for each extra 40 minutes per day. So if you eat an extra 100 calories each day and exercise about 80 minutes more each day, your weight should remain roughly steady.

Problems for Section 9.3

1. Using the contour diagram for $f(x, y)$ in Figure 9.40, decide whether each of these partial derivatives is positive, negative, or approximately zero.

(a) $f_x(4, 1)$ **(b)** $f_y(4, 1)$
(c) $f_x(5, 2)$ **(d)** $f_y(5, 2)$

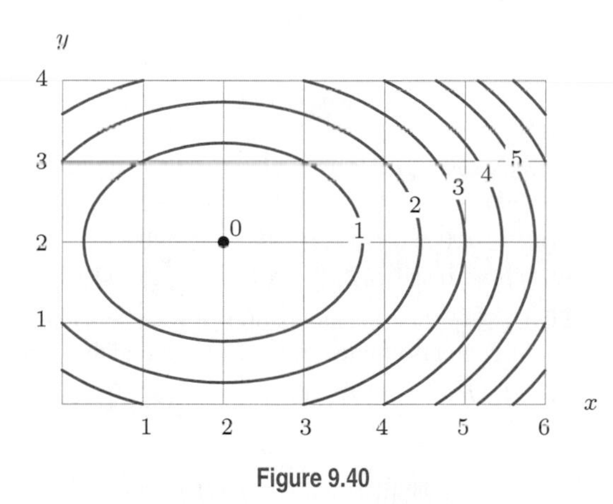

Figure 9.40

2. According to the contour diagram for $f(x, y)$ in Figure 9.40, which is larger: $f_x(3, 1)$ or $f_x(5, 2)$? Explain.

3. The quantity Q (in pounds) of beef that a certain community buys during a week is a function $Q = f(b, c)$ of the prices of beef, b, and chicken, c, during the week. Do you expect $\partial Q/\partial b$ to be positive or negative? What about $\partial Q/\partial c$?

4. A drug is injected into a patient's blood vessel. The function $c = f(x, t)$ represents the concentration of the drug at a distance x mm in the direction of the blood flow measured from the point of injection and at time t seconds since the injection. What are the units of the following partial derivatives? What are their practical interpretations? What do you expect their signs to be?

(a) $\partial c/\partial x$ **(b)** $\partial c/\partial t$

5. The demand for coffee, Q, in pounds sold per week, is a function of the price of coffee, c, in dollars per pound and the price of tea, t, in dollars per pound, so $Q = f(c, t)$.

(a) Do you expect f_c to be positive or negative? What about f_t? Explain.
(b) Interpret each of the following statements in terms of the demand for coffee:

$f(3, 2) = 780$ $f_c(3, 2) = -60$ $f_t(3, 2) = 20$

6. Figure 9.41 is a contour diagram for $z = f(x, y)$. Is f_x positive or negative? Is f_y positive or negative? Estimate $f(2, 1)$, $f_x(2, 1)$, and $f_y(2, 1)$.

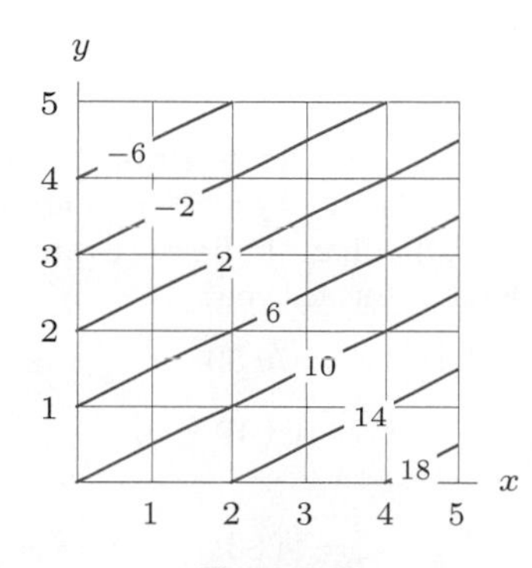

Figure 9.41

7. The monthly mortgage payment in dollars, P, for a house is a function of three variables

$$P = f(A, r, N),$$

where A is the amount borrowed in dollars, r is the interest rate, and N is the number of years before the mortgage is paid off.

(a) $f(92000, 14, 30) = 1090.08$. What does this tell you, in financial terms?

(b) $\left.\frac{\partial P}{\partial r}\right|_{(92000,14,30)} = 72.82$. What is the financial significance of the number 72.82?

(c) Would you expect $\partial P/\partial A$ to be positive or negative? Why?

(d) Would you expect $\partial P/\partial N$ to be positive or negative? Why?

8. Table 9.7 gives the number of calories burned per minute, $B = f(s, w)$, for someone roller-blading,[8] as a function of the person's weight, w, and speed, s.

(a) Is f_w positive or negative? Is f_s positive or negative? What do your answers tell us about the effect of weight and speed on calories burned per minute?

(b) Estimate $f_w(160, 10)$ and $f_s(160, 10)$. Interpret your answers.

Table 9.7 *Calories burned per minute*

$w\backslash s$	8 mph	9 mph	10 mph	11 mph
120 lbs	4.2	5.8	7.4	8.9
140 lbs	5.1	6.7	8.3	9.9
160 lbs	6.1	7.7	9.2	10.8
180 lbs	7.0	8.6	10.2	11.7
200 lbs	7.9	9.5	11.1	12.6

9. The sales of a product, $S = f(p, a)$, is a function of the price, p, of the product (in dollars per unit) and the amount, a, spent on advertising (in thousands of dollars).

(a) Do you expect f_p to be positive or negative? Why?

(b) Explain the meaning of the statement $f_a(8, 12) = 150$ in terms of sales.

10. You borrow \$$A$ at an interest rate of r% (per month) and pay it off over t months by making monthly payments of $P = g(A, r, t)$ dollars. In financial terms, what do the following statements tell you?

(a) $g(8000, 1, 24) = 376.59$

(b) $\left.\frac{\partial g}{\partial A}\right|_{(8000,1,24)} = 0.047$

(c) $\left.\frac{\partial g}{\partial r}\right|_{(8000,1,24)} = 44.83$

[8]From the August 28, 1994, issue of *Parade Magazine*.

11. Figure 9.42 shows a contour diagram for the monthly payment P as a function of the interest rate, r%, and the amount, L, of a 5-year loan. Estimate $\partial P/\partial r$ and $\partial P/\partial L$ at the point where $r = 8$ and $L = 5000$. Give the units and the financial meaning of your answers.

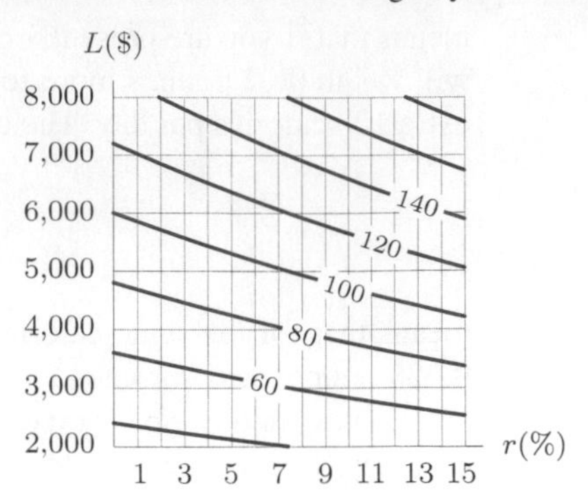

Figure 9.42

12. Estimate $z_x(1, 0)$ and $z_x(0, 1)$ and $z_y(0, 1)$ from the contour diagram for $z(x, y)$ in Figure 9.43.

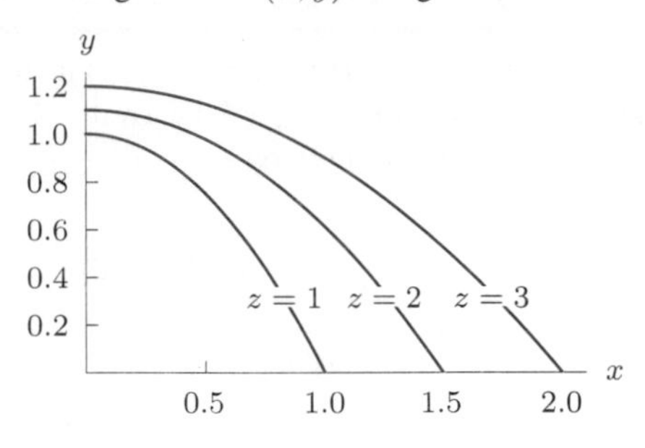

Figure 9.43

13. In each case, give a possible contour diagram for the function $f(x, y)$ if

(a) $f_x > 0$ and $f_y > 0$ **(b)** $f_x > 0$ and $f_y < 0$

(c) $f_x < 0$ and $f_y > 0$ **(d)** $f_x < 0$ and $f_y < 0$

14. Figure 9.44 shows contours of $f(x, y)$ with values of f on the contours omitted. If $f_x(P) > 0$, find the sign of

(a) $f_y(P)$ **(b)** $f_y(Q)$ **(c)** $f_x(Q)$

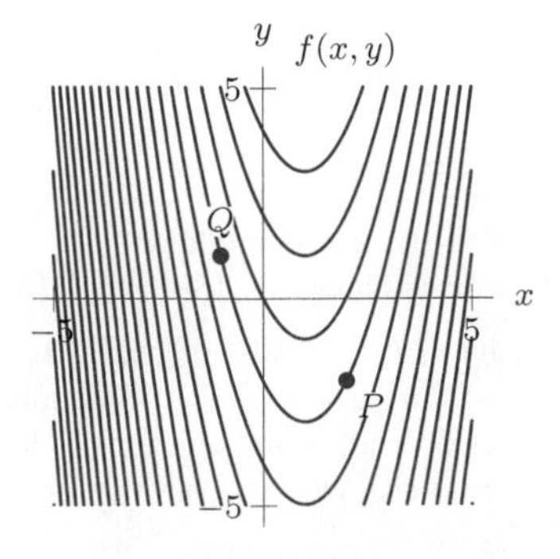

Figure 9.44

15. Figure 9.13 on page 351 gives a contour diagram of corn production as a function of rainfall, R, in inches and temperature, T, in °F. Corn production, C, is measured as a percentage of the present production, and $C = f(R, T)$. Estimate the following quantities. Give units and interpret your answers in terms of corn production:

(a) $f_R(15, 76)$ **(b)** $f_T(15, 76)$

16. For a function $f(x, y)$, we are given $f(100, 20) = 2750$, and $f_x(100, 20) = 4$, and $f_y(100, 20) = 7$. Estimate $f(105, 21)$.

17. For a function $f(r, s)$, we are given $f(50, 100) = 5.67$, and $f_r(50, 100) = 0.60$, and $f_s(50, 100) = -0.15$. Estimate $f(52, 108)$.

18. People commuting to a city can choose to go either by bus or by train. The number of people who choose either method depends in part upon the price of each. Let $f(P_1, P_2)$ be the number of people who take the bus when P_1 is the price of a bus ride and P_2 is the price of a train ride. What can you say about the signs of $\partial f/\partial P_1$ and $\partial f/\partial P_2$? Explain your answers.

19. Suppose that x is the price of one brand of gasoline and y is the price of a competing brand. Then q_1, the quantity of the first brand sold in a fixed time period, depends on both x and y, so $q_1 = f(x, y)$. Similarly, if q_2 is the quantity of the second brand sold during the same period, $q_2 = g(x, y)$. What do you expect the signs of the following quantities to be? Explain.

(a) $\partial q_1/\partial x$ and $\partial q_2/\partial y$
(b) $\partial q_1/\partial y$ and $\partial q_2/\partial x$

20. In the 1940s the quantity, q, of beer sold each year in Britain was found to depend on I (the aggregate personal income, adjusted for taxes and inflation), p_1 (the average price of beer), and p_2 (the average price of all other goods and services). Would you expect $\partial q/\partial I, \partial q/\partial p_1, \partial q/\partial p_2$ to be positive or negative? Give reasons for your answers.

For Problems 21–23, refer to Table 9.3 on page 347 giving the temperature adjusted for wind-chill, C, in °F, as a function $f(w, T)$ of the wind speed, w, in mph, and the temperature, T, in °F. The temperature adjusted for wind-chill tells you how cold it feels, as a result of the combination of wind and temperature.

21. Estimate $f_w(10, 25)$. What does your answer mean in practical terms?

22. Estimate $f_T(5, 20)$. What does your answer mean in practical terms?

23. From Table 9.3 you can see that when the temperature is 20°F, the temperature adjusted for wind-chill drops by an average of about 0.8°F with every 1 mph increase in wind speed from 5 mph to 10 mph. Which partial derivative is this telling you about?

24. An airline's revenue, R, is a function of the number of full price tickets, x, and the number of discount tickets, y, sold. Values of $R = f(x, y)$ are in Table 9.1 on page 344.

(a) Evaluate $f(200, 400)$, and interpret your answer.
(b) Is $f_x(200, 400)$ positive or negative? Is $f_y(200, 400)$ positive or negative? Explain.
(c) Estimate the partial derivatives in part (b). Give units and interpret your answers in terms of revenue.

25. In Problem 24 the revenue is \$150,000 when 200 full-price tickets and 400 discount tickets are sold; that is, $f(200, 400) = 150{,}000$. Use this fact and the partial derivatives $f_x(200, 400) = 350$ and $f_y(200, 400) = 200$ to estimate the revenue when

(a) $x = 201$ and $y = 400$ **(b)** $x = 200$ and $y = 405$
(c) $x = 203$ and $y = 406$

26. Table 9.6 on page 362 gives the percent of rats surviving, P, as a function of time, t, in months and concentration of formaldehyde, c, in ppm, so $P = f(t, c)$. Use partial derivatives to estimate the percent surviving after 26 months when the concentration is 15.

27. The cardiac output, represented by c, is the volume of blood flowing through a person's heart, per unit time. The systemic vascular resistance (SVR), represented by s, is the resistance to blood flowing through veins and arteries. Let p be a person's blood pressure. Then p is a function of c and s, so $p = f(c, s)$.

(a) What does $\partial p/\partial c$ represent?

Suppose now that $p = kcs$, where k is a constant.

(b) Sketch the level curves of p. What do they represent? Label your axes.
(c) For a person with a weak heart, it is desirable to have the heart pumping against less resistance, while maintaining the same blood pressure. Such a person may be given the drug nitroglycerine to decrease the SVR and the drug Dopamine to increase the cardiac output. Represent this on a graph showing level curves. Put a point A on the graph representing the person's state before drugs are given and a point B for after.
(d) Right after a heart attack, a patient's cardiac output drops, thereby causing the blood pressure to drop. A common mistake made by medical residents is to get the patient's blood pressure back to normal by using drugs to increase the SVR, rather than by increasing the cardiac output. On a graph of the level curves of p, put a point D representing the patient before the heart attack, a point E representing the patient right after the heart attack, and a third point F representing the patient after the resident has given the drugs to increase the SVR.

9.4 COMPUTING PARTIAL DERIVATIVES ALGEBRAICALLY

The partial derivative $f_x(x,y)$ is the ordinary derivative of the function $f(x,y)$ with respect to x with y fixed, and the partial derivative $f_y(x,y)$ is the ordinary derivative of $f(x,y)$ with respect to y with x fixed. Thus, we can use all the techniques for differentiation from single-variable calculus to find partial derivatives.

Example 1 Let $f(x,y) = x^2 + 5y^2$. Find $f_x(3,2)$ and $f_y(3,2)$ algebraically.

Solution We use the fact that $f_x(3,2)$ is the derivative of $f(x,2)$ at $x = 3$. To find f_x, we fix y at 2:

$$f(x,2) = x^2 + 5(2^2) = x^2 + 20.$$

Differentiating with respect to x gives

$$f_x(x,2) = 2x \qquad \text{so} \qquad f_x(3,2) = 2(3) = 6.$$

Similarly, $f_y(3,2)$ is the derivative of $f(3,y)$ at $y = 2$. To find f_y, we fix x at 3:

$$f(3,y) = 3^2 + 5y^2 = 9 + 5y^2.$$

Differentiating with respect to y, we have

$$f_y(3,y) = 10y \qquad \text{so} \quad f_y(3,2) = 10(2) = 20.$$

Example 2 Let $f(x,y) = x^2 + 5y^2$ as in Example 1. Find f_x and f_y as functions of x and y.

Solution To find f_x, we treat y as a constant. Thus $5y^2$ is a constant and the derivative with respect to x of this term is 0. We have

$$f_x(x,y) = 2x + 0 = 2x.$$

To find f_y, we treat x as a constant and so the derivative of x^2 with respect to y is zero. We have

$$f_y(x,y) = 0 + 10y = 10y.$$

Example 3 Find both partial derivatives of each of the following functions:
(a) $f(x,y) = 3x + e^{-5y}$ (b) $f(x,y) = x^2y$ (c) $f(u,v) = u^2e^{2v}$

Solution

(a) To find f_x, we treat y as a constant, so the term e^{-5y} is a constant, and the derivative of this term is zero. Likewise, to find f_y, we treat x as a constant. We have

$$f_x(x,y) = 3 + 0 = 3 \qquad \text{and} \qquad f_y(x,y) = 0 + (-5)e^{-5y} = -5e^{-5y}.$$

(b) To find f_x, we treat y as a constant, so the function is treated as a constant times x^2. The derivative of a constant times x^2 is the constant times $2x$, and so we have

$$f_x(x,y) = (2x)y = 2xy \qquad \text{Similarly,} \qquad f_y(x,y) = (x^2)(1) = x^2.$$

(c) To find f_u, we treat v as a constant, and to find f_v, we treat u as a constant. We have

$$f_u(u,v) = (2u)(e^{2v}) = 2ue^{2v} \qquad \text{and} \qquad f_v(u,v) = u^2(2e^{2v}) = 2u^2e^{2v}.$$

Example 4 The concentration C of bacteria in the blood (in millions of bacteria/ml) following the injection of an antibiotic is a function of the dose x (in gm) injected and the time t (in hours) since the injection. Suppose we are told that $C = f(x,t) = te^{-xt}$. Evaluate the following quantities and explain what each one means in practical terms: (a) $f_x(1,2)$ (b) $f_t(1,2)$

Solution

(a) To find f_x, we treat t as a constant and differentiate with respect to x, giving

$$f_x(x,t) = -t^2e^{-xt}.$$

Substituting $x = 1, t = 2$ gives

$$f_x(1,2) = -4e^{-2} \approx -0.54.$$

To see what $f_x(1,2)$ means, think about the function $f(x,2)$ of which it is the derivative. The graph of $f(x,2)$ in Figure 9.45 gives the concentration of bacteria as a function of the dose two hours after the injection. The derivative $f_x(1,2)$ is the slope of this graph at the point $x = 1$; it is negative because a larger dose reduces the bacteria population. More precisely, the partial derivative $f_x(1,2)$ gives the rate of change of bacteria concentration with respect to the dose injected, namely a decrease in bacteria concentration of 0.54 million/ml per gram of additional antibiotic injected.

(b) To find f_t, treat x as a constant and differentiate using the product rule:

$$f_t(x,t) = 1 \cdot e^{-xt} - xte^{-xt}.$$

Substituting $x = 1, t = 2$ gives

$$f_t(1,2) = e^{-2} - 2e^{-2} \approx -0.14.$$

To see what $f_t(1,2)$ means, think about the function $f(1,t)$ of which it is the derivative. The graph of $f(1,t)$ in Figure 9.46 gives the concentration of bacteria at time t if the dose of antibiotic is 1 gm. The derivative $f_t(1,2)$ is the slope of the graph at the point $t = 2$; it is negative because after 2 hours the concentration of bacteria is decreasing. More precisely, the partial derivative $f_t(1,2)$ gives the rate at which the bacteria concentration is changing with respect to time, namely a decrease in bacteria concentration of 0.14 million/ml per hour.

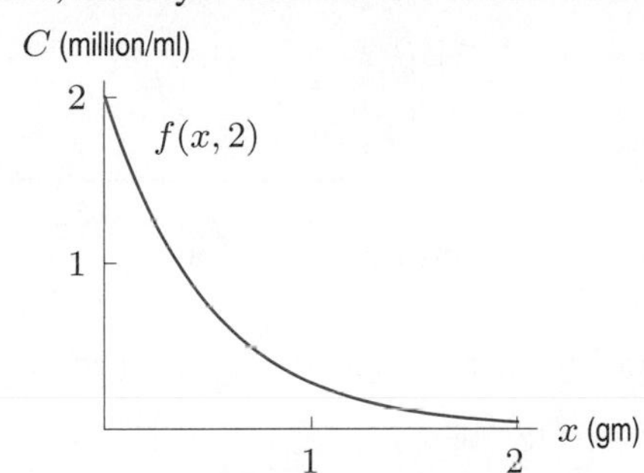

Figure 9.45: Bacteria concentration after 2 hours as a function of the quantity of antibiotic injected

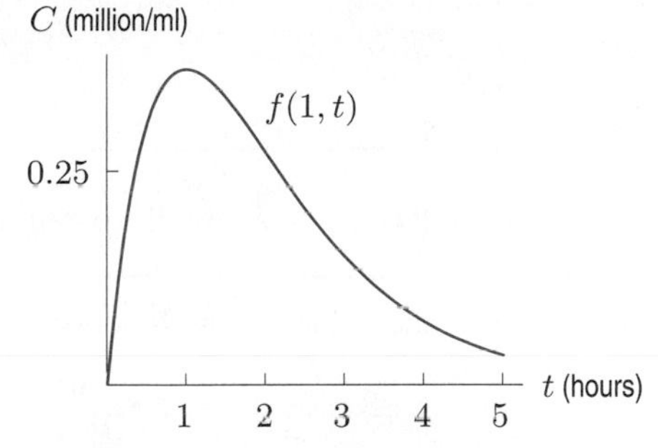

Figure 9.46: Bacteria concentration as a function of time if 1 unit of antibiotic is injected

---

Example 5 Let's consider a small printing business where N is the number of workers, V is the value of the equipment (in units of \$25,000), and P is the production, measured in thousands of pages per day. Suppose the production function for this company is given by

$$P = f(N,V) = 2N^{0.6}V^{0.4}.$$

(a) If this company has a labor force of 100 workers and 200 units worth of equipment, what is the production output of the company?
(b) Find $f_N(100,200)$ and $f_V(100,200)$. Interpret your answers in terms of production.

Solution (a) We have $N = 100$ and $V = 200$, so

$$\text{Production} = 2(100)^{0.6}(200)^{0.4} = 263.9 \text{ thousand pages per day.}$$

(b) To find f_N, we treat V as a constant and differentiate with respect to N:

$$f_N(N,V) = 2(0.6)N^{-0.4}V^{0.4}.$$

Substituting $N = 100$, $V = 200$ gives

$$f_N(100,200) = 1.2(100^{-0.4})(200^{0.4}) \approx 1.583 \text{ thousand pages/worker.}$$

This tells us that if we have 200 units of equipment and increase the number of workers by 1 from 100 to 101, the production output will go up by about 1.58 units, or 1580 pages per day.

Similarly, to find $f_V(100, 200)$, we treat N as a constant and differentiate with respect to V:

$$f_V(N, V) = 2(0.4)N^{0.6}V^{-0.6}.$$

Substituting $N = 100$, $V = 200$ gives

$$f_V(100, 200) = 0.8(100^{0.6})(200^{-0.6}) \approx 0.53 \text{ thousand pages/unit of equipment.}$$

This tells us that if we have 100 workers and increase the value of the equipment by 1 unit ($25,000) from 200 units to 201 units, the production goes up by about 0.53 units, or 530 pages per day.

Second-Order Partial Derivatives

Since the partial derivatives of a function are themselves functions, we can usually differentiate them, giving *second-order partial derivatives*. A function $z = f(x, y)$ has two first-order partial derivatives, f_x and f_y, and four second-order partial derivatives.

The Second-Order Partial Derivatives of $z = f(x, y)$

$$\frac{\partial^2 z}{\partial x^2} = f_{xx} = (f_x)_x, \qquad \frac{\partial^2 z}{\partial x \partial y} = f_{yx} = (f_y)_x,$$

$$\frac{\partial^2 z}{\partial y \partial x} = f_{xy} = (f_x)_y, \qquad \frac{\partial^2 z}{\partial y^2} = f_{yy} = (f_y)_y.$$

It is usual to omit the parentheses, writing f_{xy} instead of $(f_x)_y$ and $\frac{\partial^2 z}{\partial y\, \partial x}$ instead of $\frac{\partial}{\partial y}\left(\frac{\partial z}{\partial x}\right)$.

Example 6 Use the values of the function $f(x, y)$ in Table 9.8 to estimate $f_{xy}(1, 2)$ and $f_{yx}(1, 2)$.

Table 9.8 *Values of $f(x, y)$*

y \ x	0.9	1.0	1.1
1.8	4.72	5.83	7.06
2.0	6.48	8.00	9.60
2.2	8.62	10.65	12.88

Solution Since $f_{xy} = (f_x)_y$, we first estimate f_x:

$$f_x(1, 2) \approx \frac{f(1.1, 2) - f(1, 2)}{0.1} = \frac{9.60 - 8.00}{0.1} = 16.0,$$

$$f_x(1, 2.2) \approx \frac{f(1.1, 2.2) - f(1, 2.2)}{0.1} = \frac{12.88 - 10.65}{0.1} = 22.3.$$

Thus,

$$f_{xy}(1, 2) \approx \frac{f_x(1, 2.2) - f_x(1, 2)}{0.2} = \frac{22.3 - 16.0}{0.2} = 31.5.$$

Similarly,

$$f_{yx}(1, 2) \approx \frac{f_y(1.1, 2) - f_y(1, 2)}{0.1} \approx \frac{1}{0.1}\left(\frac{f(1.1, 2.2) - f(1.1, 2)}{0.2} - \frac{f(1, 2.2) - f(1, 2)}{0.2}\right)$$

$$= \frac{1}{0.1}\left(\frac{12.88 - 9.60}{0.2} - \frac{10.65 - 8.00}{0.2}\right) = 31.5.$$

Observe that in this example, $f_{xy} = f_{yx}$ at the point $(1, 2)$.

Example 7 Compute the four second-order partial derivatives of $f(x, y) = xy^2 + 3x^2e^y$.

Solution From $f_x(x, y) = y^2 + 6xe^y$ we get

$$f_{xx}(x,y) = \frac{\partial}{\partial x}(y^2 + 6xe^y) = 6e^y \quad \text{and} \quad f_{xy}(x,y) = \frac{\partial}{\partial y}(y^2 + 6xe^y) = 2y + 6xe^y.$$

From $f_y(x, y) = 2xy + 3x^2e^y$ we get

$$f_{yx}(x,y) = \frac{\partial}{\partial x}(2xy+3x^2e^y) = 2y+6xe^y \quad \text{and} \quad f_{yy}(x,y) = \frac{\partial}{\partial y}(2xy+3x^2e^y) = 2x+3x^2e^y.$$

Observe that $f_{xy} = f_{yx}$ in this example.

The Mixed Partial Derivatives Are Equal

It is not an accident that the estimates for $f_{xy}(1, 2)$ and $f_{yx}(1, 2)$ are equal in Example 6, because the same values of the function are used to calculate each one. The fact that $f_{xy} = f_{yx}$ in Example 7 corroborates the following general result:

> If f_{xy} and f_{yx} are continuous at (a, b), then
>
> $$f_{xy}(a, b) = f_{yx}(a, b).$$

Most of the functions we will encounter not only have f_{xy} and f_{yx} continuous, but all their higher order partial derivatives (such as f_{xxy} or f_{xyyy}) will be continuous. We call such functions *smooth*.

Problems for Section 9.4

Find the partial derivatives in Problems 1–15. The variables are restricted to a domain on which the function is defined.

1. f_x and f_y if $f(x, y) = x^2 + 2xy + y^3$
2. $\dfrac{\partial z}{\partial x}$ if $z = x^2e^y$
3. f_x and f_y if $f(x, y) = 2x^2 + 3y^2$
4. $\dfrac{\partial Q}{\partial p}$ if $Q = 5a^2p - 3ap^3$
5. $\dfrac{\partial P}{\partial r}$ if $P = 100e^{rt}$
6. f_t if $f(t, a) = 5a^2t^3$
7. f_x and f_y if $f(x, y) = 100x^2y$
8. f_x and f_y if $f(x, y) = 10x^2e^{3y}$
9. z_x if $z = x^2y + 2x^5y$
10. f_u and f_v if $f(u, v) = u^2 + 5uv + v^2$
11. $\dfrac{\partial A}{\partial h}$ if $A = \frac{1}{2}(a + b)h$
12. $\dfrac{\partial}{\partial m}\left(\dfrac{1}{2}mv^2\right)$
13. f_x and f_y if $f(x, y) = 5x^2y^3 + 8xy^2 - 3x^2$
14. $\dfrac{\partial V}{\partial r}$ and $\dfrac{\partial V}{\partial h}$ if $V = \frac{4}{3}\pi r^2h$
15. $f_x(1, 2)$ and $f_y(1, 2)$ if $f(x, y) = x^3 + 3x^2y - 2y^2$
16. If $f(x, y) = x^3 + 3y^2$, find $f(1, 2)$, $f_x(1, 2)$, and $f_y(1, 2)$.
17. If $f(u, v) = 5uv^2$, find $f(3, 1)$, $f_u(3, 1)$, and $f_v(3, 1)$.
18. A manufacturing company produces two items in quantities q_1 and q_2, respectively. Total production costs are given by

$$\text{Cost} = f(q_1, q_2) = 16 + 1.2q_1 + 1.5q_2 + 0.2q_1q_2.$$

Find $f(500, 1000)$, $f_{q_1}(500, 1000)$, and $f_{q_2}(500, 1000)$. Give units with your answers and interpret each of your answers in terms of production cost.

19. **(a)** Let $f(x, y) = x^2 + y^2$. Estimate $f_x(2, 1)$ and $f_y(2, 1)$ using the contour diagram for f in Figure 9.47.
(b) Estimate $f_x(2, 1)$ and $f_y(2, 1)$ from a table of values for f with $x = 1.9, 2, 2.1$ and $y = 0.9, 1, 1.1$.
(c) Compare your estimates in parts (a) and (b) with the exact values of $f_x(2, 1)$ and $f_y(2, 1)$ found algebraically.

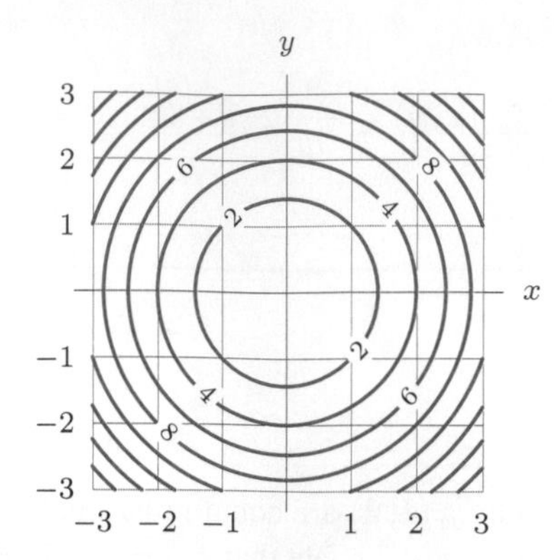

Figure 9.47

20. The amount of money, $\$B$, in a bank account earning interest at a continuous rate, r, depends on the amount deposited, $\$P$, and the time, t, it has been in the bank, where

$$B = Pe^{rt}.$$

Find $\partial B/\partial t$, $\partial B/\partial r$ and $\partial B/\partial P$ and interpret each in financial terms.

21. The DuBois formula relates a person's surface area, s, in m^2, to weight, w, in kg, and height, h, in cm, by

$$s = f(w, h) = 0.01w^{0.25}h^{0.75}.$$

Find $f(65, 160)$, $f_w(65, 160)$, and $f_h(65, 160)$. Interpret your answers in terms of surface area, height, and weight.

22. The cost of renting a car from a certain company is \$40 per day plus 15 cents per mile, and so we have

$$C = 40d + 0.15m.$$

Find $\partial C/\partial d$ and $\partial C/\partial m$. Give units and explain why your answers make sense.

23. The Cobb-Douglas production function for a product is given by

$$Q = 25K^{0.75}L^{0.25},$$

where Q is the quantity produced for a capital investment of K dollars and a labor investment of L.
(a) Find Q_K and Q_L.
(b) Find the values of Q, Q_K and Q_L given that $K = 60$ and $L = 100$.
(c) Interpret each of the values you found in part (b) in terms of production.

24. Show that the Cobb-Douglas function

$$Q = bK^{\alpha}L^{1-\alpha} \quad \text{where} \quad 0 < \alpha < 1$$

satisfies the equation

$$K\frac{\partial Q}{\partial K} + L\frac{\partial Q}{\partial L} = Q.$$

25. Figure 9.48 is a contour diagram of $f(x, y)$. In each of the following cases, list the marked points in the diagram (there may be none or more than one) at which
(a) $f_x < 0$ **(b)** $f_y > 0$
(c) $f_{xx} > 0$ **(d)** $f_{yy} < 0$

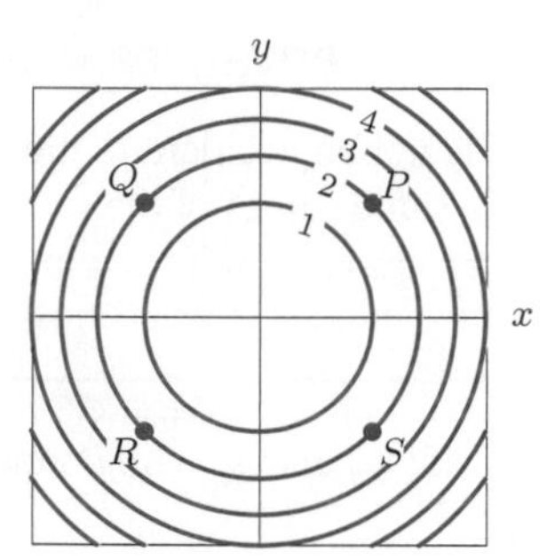

Figure 9.48

For Problems 26–37, calculate all four second order partial derivatives and confirm that the mixed partials are equal.

26. $f(x, y) = x^2y$
27. $f(x, y) = x^2+2xy+y^2$
28. $f(x, y) = xe^y$
29. $f(x, y) = \dfrac{2x}{y}, \quad y \neq 0$
30. $f = 5 + x^2y^2$
31. $f = e^{xy}$
32. $Q = 5p_1^2p_2^{-1}, \quad p_2 \neq 0$
33. $V = \pi r^2 h$
34. $P = 2KL^2$
35. $B = 5xe^{-2t}$
36. $f(x, t) = t^3 - 4x^2t$
37. $f = 100e^{rt}$

38. Is there a function f which has the following partial derivatives? If so what is it? Are there any others?

$$f_x(x, y) = 4x^3y^2 - 3y^4,$$
$$f_y(x, y) = 2x^4y - 12xy^3.$$

9.5 CRITICAL POINTS AND OPTIMIZATION

To optimize a function means to find the largest or smallest value of the function. If the function represents profit, we may want to find the conditions that maximize profit. On the other hand, if the function represents cost, we may want to find the conditions that minimize cost. In Chapter 4, we saw how to optimize a function of one variable by investigating critical points. In this section, we see how to extend the notions of critical points and local extrema to a function of more than one variable.

Local and Global Maxima and Minima for Functions of Two Variables

Functions of several variables, like functions of one variable, can have *local and global extrema.* (That is, local and global maxima and minima.) A function has a local extremum at a point where it takes on the largest or smallest value in a small region around the point. Global extrema are the largest or smallest value anywhere. For a function f defined on a domain R, we say:

- f has a **local maximum** at P_0 if $f(P_0) \geq f(P)$ for all points P near P_0
- f has a **local minimum** at P_0 if $f(P_0) \leq f(P)$ for all points P near P_0
- f has a **global maximum** at P_0 if $f(P_0) \geq f(P)$ for all points P in R
- f has a **global minimum** at P_0 if $f(P_0) \leq f(P)$ for all points P in R

Example 1 Table 9.9 gives a table of values for a function $f(x, y)$. Estimate the location and value of any global maxima or minima for $0 \leq x \leq 1$ and $0 \leq y \leq 20$.

Table 9.9 *Where are the extreme points of this function $f(x, y)$?*

y \ x	0	0.2	0.4	0.6	0.8	1.0
0	80	84	82	76	71	65
5	86	90	88	73	77	71
10	91	95	93	88	82	76
15	87	91	89	84	78	72
20	82	86	84	79	73	67

Solution The global maximum value of the function appears to be 95 at the point $(0.2, 10)$. Since the table only gives certain values, we cannot be sure that this is exactly the maximum. (The function might have a larger value at, for example, $(0.3, 11)$.) The global minimum value of this function on the points given is 65 at the point $(1, 0)$.

Example 2 Figure 9.49 gives a contour diagram for a function $f(x, y)$. Estimate the location and value of any local maxima or minima. Are any of these global maxima or minima on the square shown?

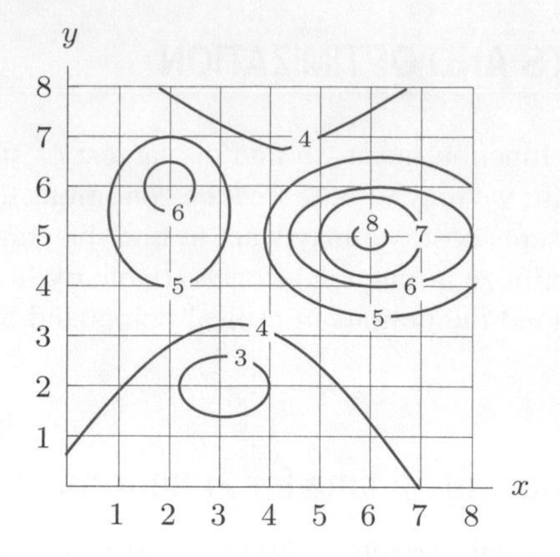

Figure 9.49: Where are the local and global extreme points of this function?

Solution

There is a local maximum of above 8 near the point $(6, 5)$, a local maximum of above 6 near the point $(2, 6)$, and a local minimum of below 3 near the point $(3, 2)$. The value above 8 is the global maximum and the value below 3 is the global minimum on the given domain.

In Example 1 and Example 2, we can estimate the location and value of extreme points, but we do not have enough information to find them exactly. This is usually true when we are given a table of values or a contour diagram. To find local or global extrema exactly, we usually need to have a formula for the function.

Finding a Local Maximum or Minimum Analytically

In one-variable calculus, the local extrema of a function occur at points where the derivative is zero or undefined. How does this generalize to the case of functions of two or more variables? Suppose that a function $f(x, y)$ has a local maximum at a point (x_0, y_0) which is not on the boundary of the domain of f. If the partial derivative $f_x(x_0, y_0)$ were defined and positive, then we could increase f by increasing x. If $f_x(x_0, y_0) < 0$, then we could increase f by decreasing x. Since f has a local maximum at (x_0, y_0), there can be no direction in which f is increasing, so we must have $f_x(x_0, y_0) = 0$. Similarly, if $f_y(x_0, y_0)$ is defined, then $f_y(x_0, y_0) = 0$. The case in which $f(x, y)$ has a local minimum is similar. Therefore, we arrive at the following conclusion:

If a function $f(x, y)$ has a local maximum or minimum at a point (x_0, y_0) not on the boundary of the domain of f, then either

$$f_x(x_0, y_0) = 0 \quad \text{and} \quad f_y(x_0, y_0) = 0$$

or (at least) one partial derivative is undefined at the point (x_0, y_0). Points where each of the partial derivatives is either zero or undefined are called **critical points**.

As in the single variable case, the fact that (x_0, y_0) is a critical point for f does not necessarily mean that f has a maximum or a minimum there.

How Do We Find Critical Points?

To find critical points of a function f, we find the points where both partial derivatives of f are zero or undefined.

Example 3 Find and analyze the critical points of $f(x, y) = x^2 - 2x + y^2 - 4y + 5$.

Solution To find the critical points, we set both partial derivatives equal to zero:

$$f_x(x, y) = 2x - 2 = 0,$$
$$f_y(x, y) = 2y - 4 = 0.$$

Solving these equations gives $x = 1$ and $y = 2$. Hence, f has only one critical point, namely $(1, 2)$. What is the behavior of f near $(1, 2)$? The values of the function in Table 9.10 suggest that the function has a local minimum value of 0 at the point $(1, 2)$.

Table 9.10 *Values of $f(x, y)$ near the point $(1, 2)$*

y \ x	0.8	0.9	1.0	1.1	1.2
1.8	0.08	0.05	0.04	0.05	0.08
1.9	0.05	0.02	0.01	0.02	0.05
2.0	0.04	0.01	0.00	0.01	0.04
2.1	0.05	0.02	0.01	0.02	0.05
2.2	0.08	0.05	0.04	0.05	0.08

Example 4 A manufacturing company produces two products which are sold in two separate markets. The company's economists analyze the two markets and determine that the quantities, q_1 and q_2, demanded by consumers and the prices, p_1 and p_2 (in dollars), of each item are related by the equations

$$p_1 = 600 - 0.3q_1 \quad \text{and} \quad p_2 = 500 - 0.2q_2.$$

Thus, if the price for either item increases, the demand for it decreases. The company's total production cost is given by

$$C = 16 + 1.2q_1 + 1.5q_2 + 0.2q_1q_2.$$

If the company wants to maximize its total profits, how much of each product should it produce? What is the maximum profit? [9]

Solution The total revenue R is the sum of the revenues, p_1q_1 and p_2q_2, from each market. Substituting for p_1 and p_2, we get

$$\begin{aligned} R &= p_1q_1 + p_2q_2 \\ &= (600 - 0.3q_1)q_1 + (500 - 0.2q_2)q_2 \\ &= 600q_1 - 0.3q_1^2 + 500q_2 - 0.2q_2^2. \end{aligned}$$

Thus the total profit π is given by

$$\begin{aligned} \pi &= R - C \\ &= 600q_1 - 0.3q_1^2 + 500q_2 - 0.2q_2^2 - (16 + 1.2q_1 + 1.5q_2 + 0.2q_1q_2) \\ &= -16 + 598.8q_1 - 0.3q_1^2 + 498.5q_2 - 0.2q_2^2 - 0.2q_1q_2. \end{aligned}$$

To maximize π, we compute partial derivatives:

$$\frac{\partial \pi}{\partial q_1} = 598.8 - 0.6q_1 - 0.2q_2,$$
$$\frac{\partial \pi}{\partial q_2} = 498.5 - 0.4q_2 - 0.2q_1.$$

[9] Adapted from M. Rosser, *Basic Mathematics for Economists*, p. 316, (New York: Routledge, 1993).

Since the partial derivatives are defined everywhere, the only critical points of π are those where the partial derivatives of π are both equal to zero. Thus, we solve the equations for q_1 and q_2,

$$598.8 - 0.6q_1 - 0.2q_2 = 0,$$
$$498.5 - 0.4q_2 - 0.2q_1 = 0,$$

giving

$$q_1 = 699.1 \approx 699 \quad \text{and} \quad q_2 = 896.7 \approx 897.$$

To see whether this is a maximum, we look at a table of values of profit π around this point. Table 9.11 suggests that profit is greatest at $(699, 897)$. So the company should produce 699 units of the first product priced at \$390.30 per unit, and 897 units of the second product priced at \$320.60 per unit. The maximum profit is then $\pi(699, 897) = \$432{,}797$.

Table 9.11 *Does this profit function have a maximum at* $(699, 897)$*?*

		Quantity, q_1		
		698	699	700
Quantity, q_2	896	432,796.4	432,796.9	432,796.8
	897	432,796.7	432,797.0	432,796.7
	898	432,796.6	432,796.7	432,796.2

Is a Critical Point a Local Maximum or a Local Minimum?

We can often see whether a critical point is a local maximum or minimum or neither by looking at a table or contour diagram. The following analytic method may also be useful in distinguishing between local maxima and minima.[10] It is analogous to the Second Derivative Test in Chapter 4.

Second Derivative Test for Functions of Two Variables

Suppose (x_0, y_0) is a critical point where $f_x(x_0, y_0) = f_y(x_0, y_0) = 0$. Let

$$D = f_{xx}(x_0, y_0)f_{yy}(x_0, y_0) - f_{xy}(x_0, y_0)^2.$$

- If $D > 0$ and $f_{xx}(x_0, y_0) > 0$, then f has a local minimum at (x_0, y_0).
- If $D > 0$ and $f_{xx}(x_0, y_0) < 0$, then f has a local maximum at (x_0, y_0).
- If $D < 0$, then f has neither a local maximum or minimum at (x_0, y_0).
- If $D = 0$, the test is inconclusive.

Example 5 Use the second derivative test to confirm that the critical point $q_1 = 699.1$, $q_2 = 896.7$ gives a local maximum of the profit function π of Example 4.

Solution To see whether or not we have found a maximum point, we compute the second-order partial derivatives:

$$\frac{\partial^2 \pi}{\partial q_1^2} = -0.6, \quad \frac{\partial^2 \pi}{\partial q_2^2} = -0.4, \quad \frac{\partial^2 \pi}{\partial q_1 \partial q_2} = -0.2.$$

Since

$$D = \frac{\partial^2 \pi}{\partial q_1^2}\frac{\partial^2 \pi}{\partial q_2^2} - \left(\frac{\partial^2 \pi}{\partial q_1 \partial q_2}\right)^2 = (-0.6)(-0.4) - (-0.2)^2 = 0.2 > 0,$$

the second derivative test implies that we have found a local maximum point.

[10] An explanation of this test can be found, for example, in *Multivariable Calculus* by W. McCallum et al. (New York: John Wiley, 1997).

Problems for Section 9.5

1. Figure 9.50 shows contours of $f(x, y)$. List the x- and y-coordinates and the value of the function at any local maximum and local minimum points, and identify which is which. Are any of these local extrema also global extrema on the region shown? If so, which ones?

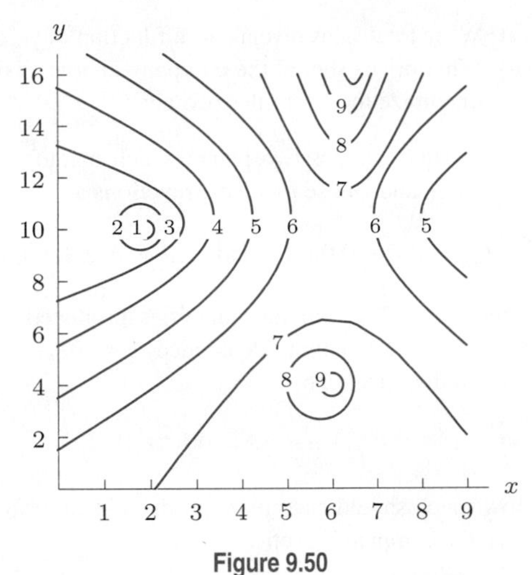

Figure 9.50

2. Figure 9.51 shows contours of $f(x, y)$. List x- and y-coordinates and the value of the function at any local maximum and local minimum points, and identify which is which. Are any of these local extrema also global extrema on the region shown? If so, which ones?

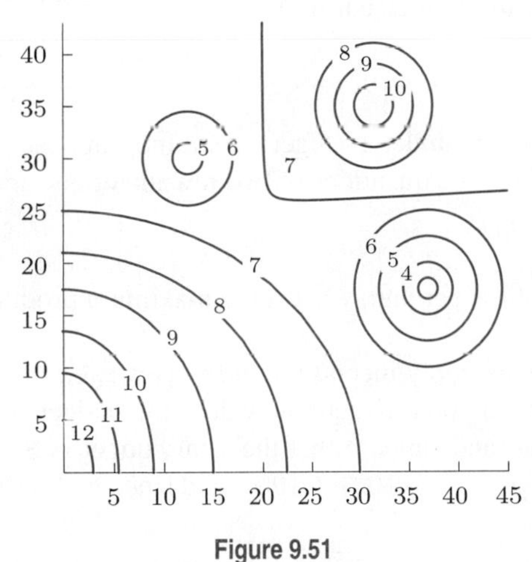

Figure 9.51

In Problems 3–11, find all the critical points and determine whether each is a local maximum, local minimum, or neither.

3. $f(x, y) = x^2 + y^2 + 6x - 10y + 8$

4. $f(x, y) = x^2 + 4x + y^2$

5. $f(x, y) = x^2 + xy + 3y$

6. $f(x, y) = y^3 - 3xy + 6x$

7. $f(x, y) = x^3 - 3x + y^3 - 3y$

8. $f(x, y) = x^3 + y^2 - 3x^2 + 10y + 6$

9. $f(x, y) = x^2 - 2xy + 3y^2 - 8y$

10. $f(x, y) = x^3 + y^3 - 3x^2 - 3y + 10$

11. $f(x, y) = x^3 + y^3 - 6y^2 - 3x + 9$

12. Find the values of x and y which maximize $f(x, y) = 400 - 3x^2 - 4x + 2xy - 5y^2 + 48y$

13. By looking at the weather map in Figure 9.5 on page 348, find the maximum and minimum daily high temperatures in the states of Mississippi, Alabama, Pennsylvania, New York, California, Arizona, and Massachusetts.

In Problems 14–16, estimate the position and approximate value of the global maxima and minima on the region shown.

14.

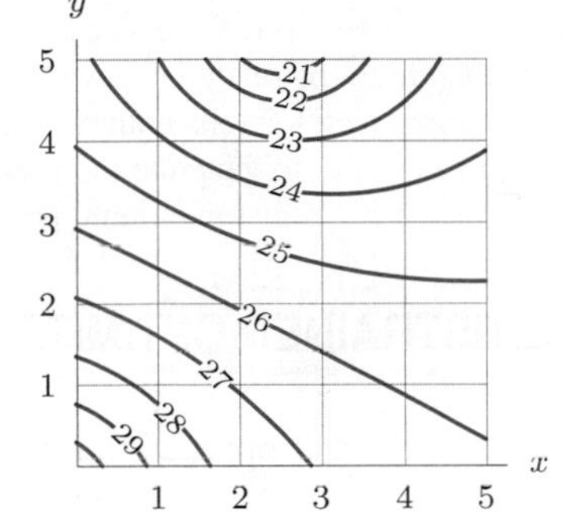

15.

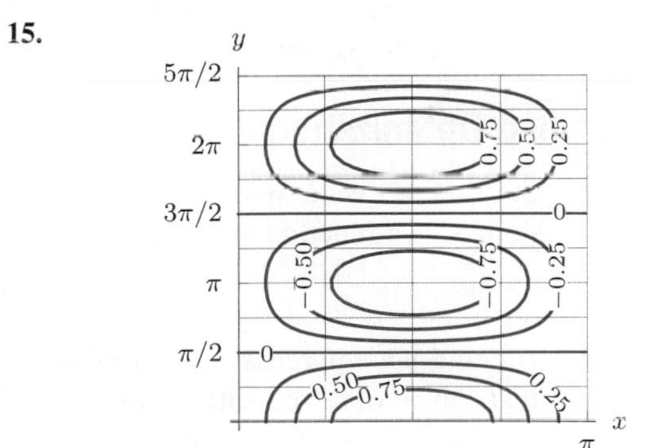

16.

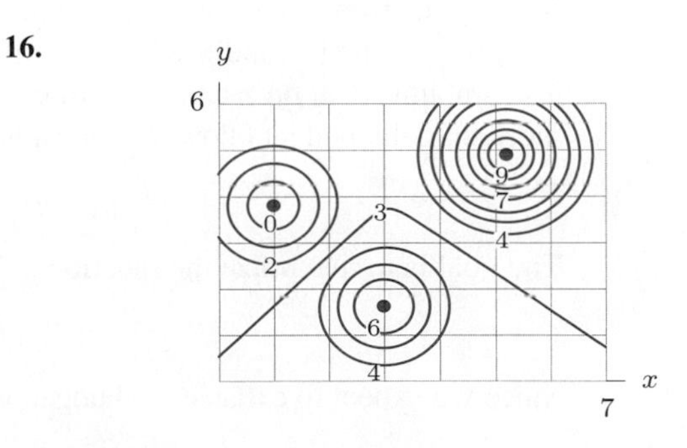

17. For $f(x, y) = A - (x^2 + Bx + y^2 + Cy)$, what values of A, B, and C give f a local maximum value of 15 at the point $(-2, 1)$?

18. A missile has a guidance device which is sensitive to both temperature, $t°$C, and humidity, h. The range in km over which the missile can be controlled is given by

$$\text{Range} = 27{,}800 - 5t^2 - 6ht - 3h^2 + 400t + 300h.$$

What are the optimal atmospheric conditions for controlling the missile?

19. The quantity of a product demanded by consumers is a function of its price. The quantity of one product demanded may also depend on the price of other products. For example, the demand for tea is affected by the price of coffee; the demand for cars is affected by the price of gas. The quantities demanded, q_1 and q_2, of two products depend on their prices, p_1 and p_2, as follows

$$q_1 = 150 - 2p_1 - p_2$$

$$q_2 = 200 - p_1 - 3p_2.$$

(a) What does the fact that the coefficients of p_1 and p_2 are negative tell you? Give an example of two products that might be related this way.
(b) If one manufacturer sells both products, how should the prices be set to generate the maximum possible revenue? What is that maximum possible revenue?

20. A company sells two products which are partial substitutes for each other, such as coffee and tea. If the price of one product rises, then the demand for the other product rises. The quantities demanded, q_1 and q_2, are given as a function of the prices, p_1 and p_2, by

$$q_1 = 517 - 3.5p_1 + 0.8p_2, \quad q_2 = 770 - 4.4p_2 + 1.4p_1.$$

(a) Write total sales revenue as a function of p_1 and p_2.
(b) What prices should the company charge in order to maximize the total sales revenue? [11]

21. A company operates two plants which manufacture the same item and whose total cost functions are

$$C_1 = 8.5 + 0.03q_1^2 \quad \text{and} \quad C_2 = 5.2 + 0.04q_2^2,$$

where q_1 and q_2 are the quantities produced by each plant. The total quantity demanded, $q = q_1 + q_2$, is related to the price, p, by

$$p = 60 - 0.04q.$$

How much should each plant produce in order to maximize the company's profit? [12]

9.6 CONSTRAINED OPTIMIZATION

Many real optimization problems are constrained by external circumstances. For example, a city wanting to build a public transportation system has a limited number of tax dollars available. A nation trying to maintain its balance of trade must spend less on imports than it earns on exports. In this section, we see how to find an optimum value under such constraints.

A Constrained Optimization Problem

Suppose we want to maximize the production of a firm under a budget constraint. Suppose production, f, is a function of two variables, x and y, which are quantities of two raw materials, and

$$f(x, y) = x^{2/3}y^{1/3}.$$

If x and y are purchased at prices of p_1 and p_2 dollars per unit, what is the maximum production f that can be obtained with a budget of c dollars?

To increase f without regard to the budget, we simply increase x and y. However, the budget prevents us from increasing x and y beyond a certain point. Exactly how does the budget constrain us? Suppose that x and y each cost \$100 per unit, and suppose that the total budget is \$378,000. The amount spent on x and y together is given by $g(x, y) = 100x + 100y$, and since we can't spend more than the budget allows, we must have:

$$g(x, y) = 100x + 100y \leq 378{,}000.$$

The goal is to maximize the function

$$f(x, y) = x^{2/3}y^{1/3}.$$

Since we expect to exhaust the budget, we have

$$100x + 100y = 378{,}000.$$

[11] Adapted from M. Rosser, *Basic Mathematics for Economists*, p. 318 (New York: Routledge, 1993).
[12] Adapted from M. Rosser, *Basic Mathematics for Economists*, p. 318 (New York: Routledge, 1993).

Example 1 A company has production function $f(x, y) = x^{2/3}y^{1/3}$ and budget constraint $100x + 100y = 378{,}000$.

(a) If \$100,000 is spent on x, how much can be spent on y? What is the production in this case?
(b) If \$200,000 is spent on x, how much can be spent on y? What is the production in this case?
(c) Which of the two options above is the better choice for the company? Do you think this is the best of all possible options?

Solution

(a) If the company spends \$100,000 on x, then it has \$278,000 left to spend on y. In this case, we have $100x = 100{,}000$, so $x = 1000$, and $100y = 278{,}000$, so $y = 2780$. Therefore,

$$\text{Production} = f(1000, 2780) = (1000)^{2/3}(2780)^{1/3} = 1406 \text{ units.}$$

(b) If the company spends \$200,000 on x, then it has \$178,000 left to spend on y. Therefore, $x = 2000$ and $y = 1780$, and so

$$\text{Production} = f(2000, 1780) = (2000)^{2/3}(1780)^{1/3} = 1924 \text{ units.}$$

(c) Of these two options, (b) is better since production is larger in this case. This is probably not optimal, since there are many other combinations of x and y that we have not checked.

Graphical Approach: Maximizing Production Subject to a Budget Constraint

How can we find the maximum value of production? We want to maximize the *objective function*

$$f(x, y) = x^{2/3}y^{1/3}$$

subject to $x \geq 0$ and $y \geq 0$ and the budget constraint

$$g(x, y) = 100x + 100y = 378{,}000.$$

The constraint is represented by the line in Figure 9.52. Any point on or below the line represents a pair of values of x and y that we can afford. A point on the line completely exhausts the budget, while a point below the line represents values of x and y which can be bought without using up the budget. Any point above the line represents a pair of values that we cannot afford.

Figure 9.52 also shows some contours of the production function f. Since we want to maximize f, we want to find the point which lies on the contour with the largest possible f value *and* which

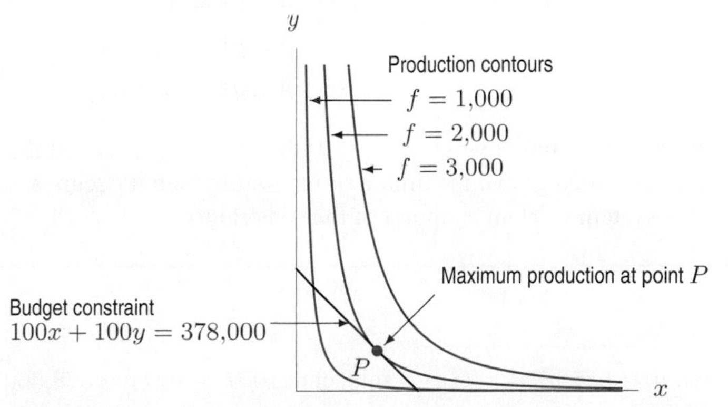

Figure 9.52: At the optimal point P the budget constraint is tangent to a production contour

lies within the budget. The point we are looking for must lie on the budget constraint because we should spend all the available money. The key observation is this: The maximum occurs at a point P where the budget constraint is tangent to a production contour. (See Figure 9.52.) The reason is that if we are on the constraint line to the left of P, moving right on the constraint increases f; if we are on the line to the right of P, moving left increases f. Thus, the maximum value of f on the budget constraint line occurs at the point P.

In general, provided f and g are smooth, we have the following result:

If $f(x, y)$ has a global maximum or minimum on the constraint $g(x, y) = c$, it occurs at a point where the graph of the constraint is tangent to a contour of f, or at an endpoint of the constraint.[13]

Analytical Approach: The Method of Lagrange Multipliers

Suppose we want to optimize $f(x, y)$ subject to the constraint $g(x, y) = c$. We make the following definition.

Suppose P_0 is a point satisfying the constraint $g(x, y) = c$.

- f has a **local maximum** at P_0 **subject to the constraint** if $f(P_0) \geq f(P)$ for all points P near P_0 satisfying the constraint.
- f has a **global maximum** at P_0 **subject to the constraint** if $f(P_0) \geq f(P)$ for all points P satisfying the constraint.

Local and global minima are defined similarly.

It can be shown[14] that the constraint is tangent to a contour of f at the point which satisfies the equations laid out in the following method.

Method of Lagrange Multipliers To optimize $f(x, y)$ subject to the constraint $g(x, y) = c$, solve the following system of three equations

$$\begin{aligned} f_x(x, y) &= \lambda g_x(x, y), \\ f_y(x, y) &= \lambda g_y(x, y), \\ g(x, y) &= c, \end{aligned}$$

for the three unknowns x, y, and λ; the number λ is called the *Lagrange multiplier*. If f has a constrained global maximum or minimum, then it occurs at one of the solutions (x_0, y_0) to this system or at an endpoint of the constraint.

Example 2 Maximize $f(x, y) = x^{2/3}y^{1/3}$ subject to $100x + 100y = 378{,}000$ and $x \geq 0$, $y \geq 0$.

[13] If the constraint has endpoints.

[14] See W. McCallum, et al., *Multivariable Calculus*, (New York: John Wiley, 2005).

Solution Differentiating gives

$$f_x(x,y) = \frac{2}{3}x^{-1/3}y^{1/3} \quad \text{and} \quad f_y(x,y) = \frac{1}{3}x^{2/3}y^{-2/3}$$

and

$$g_x(x,y) = 100 \quad \text{and} \quad g_y(x,y) = 100,$$

leading to the equations

$$\begin{aligned} \frac{2}{3}x^{-1/3}y^{1/3} &= \lambda(100) \\ \frac{1}{3}x^{2/3}y^{-2/3} &= \lambda(100) \\ 100x + 100y &= 378{,}000. \end{aligned}$$

The first two equations show that we must have

$$\frac{2}{3}x^{-1/3}y^{1/3} = \frac{1}{3}x^{2/3}y^{-2/3}.$$

Using the fact that $x^{-1/3} = 1/x^{1/3}$, we can rewrite this as

$$\frac{2y^{1/3}}{3x^{1/3}} = \frac{x^{2/3}}{3y^{2/3}}.$$

Multiplying through by the denominators gives

$$2y^{1/3}(3y^{2/3}) = x^{2/3}(3x^{1/3}),$$

and simplifying using $y^{1/3} \cdot y^{2/3} = y^1$ gives

$$\begin{aligned} 6y &= 3x \\ 2y &= x. \end{aligned}$$

Since we must also satisfy the constraint $100x + 100y = 378{,}000$, we substitute $x = 2y$ and get

$$\begin{aligned} 100(2y) + 100y &= 378{,}000 \\ 300y &= 378{,}000 \\ y &= 1260. \end{aligned}$$

Since $x = 2y$, we have $x = 2520$. The optimum value occurs at $x = 2520$ and $y = 1260$. For these values,

$$f(2520, 1260) = (2520)^{2/3}(1260)^{1/3} \approx 2000.1.$$

The endpoints of the constraint are the points $(3780, 0)$ and $(0, 3780)$. Since

$$f(3780, 0) = f(0, 3780) = 0,$$

we see that the maximum value of f is approximately 2000 and that it occurs at $x = 2520$ and $y = 1260$.

The Meaning of λ

In the previous example, we never found (or needed) the value of λ. However, λ does have a practical interpretation. In the production problem we maximized

$$f(x, y) = x^{2/3}y^{1/3}$$

subject to the constraint

$$g(x, y) = 100x + 100y = 378{,}000.$$

We solved the equations

$$\frac{2}{3}x^{-1/3}y^{1/3} = 100\lambda,$$
$$\frac{1}{3}x^{2/3}y^{-2/3} = 100\lambda,$$
$$100x + 100y = 378{,}000,$$

to get $x = 2520, y = 1260$. Continuing to find λ gives us

$$\lambda \approx 0.0053.$$

Suppose now we do another, apparently unrelated calculation. Suppose our budget is increased by \$1000, from \$378,000 to \$379,000. The new budget constraint is

$$100x + 100y = 379{,}000.$$

The corresponding solution is at $x = 2527, y = 1263$ and the new maximum value (instead of $f = 2000.1$) is

$$f = (2527)^{2/3}(1263)^{1/3} \approx 2005.4.$$

The additional \$1000 in the budget increased the production level f by 5.3 units. Notice that production increased by $5.3/1000 = 0.0053$ units per dollar, which is our value of λ. The value of λ represents the extra production achieved by increasing the budget by one dollar—in other words, the extra "bang" you get for an extra "buck" of budget.

Solving for λ in either of the equations $f_x = \lambda g_x$ or $f_y = \lambda g_y$ suggests that the Lagrange multiplier is given by the ratio of the changes:

$$\lambda \approx \frac{\Delta f}{\Delta g} = \frac{\text{Change in optimum value of } f}{\text{Change in } g}.$$

These results suggest the following interpretations of the Lagrange multiplier λ:

- The value of λ is approximately the change in the optimum value of f when the value of the constraint is increased by 1 unit.
- The value of λ represents the rate of change of the optimum value of f as the constraint increases.

Example 3 The quantity of goods produced according to the function $f(x, y) = x^{2/3}y^{1/3}$ is maximized subject to the budget constraint $100x + 100y = 378{,}000$. Suppose the budget is increased to allow a small increase in production. What price must the product sell for if it is to be worth the increased budget?

Solution We know that $\lambda = 0.0053$. Therefore, increasing the budget by \$1 increases production by about 0.0053 unit. In order to make the increase in budget profitable, the extra goods produced must sell for more than \$1. If the price is p in dollars, we must have $0.0053p > 1$. Thus, we need $p > 1/0.0053 \approx \$189$.

Example 4 If x and y are the amounts of raw materials used, the quantity, Q, of a product manufactured is

$$Q = xy.$$

Assume that x costs \$20 per unit, y costs \$10 per unit, and the production budget is \$10,000.

(a) How many units of x and y should be purchased in order to maximize production? How many units are produced at that point?

(b) Find the value of λ and interpret it.

Solution (a) We maximize $f(x, y) = xy$ subject to the constraint $g(x, y) = 20x + 10y = 10{,}000$ and $x \geq 0$, $y \geq 0$. We have the following partial derivatives:

$$f_x = y, \qquad f_y = x, \qquad \text{and} \qquad g_x = 20, \qquad g_y = 10.$$

The method of Lagrange multipliers gives the following equations:

$$\begin{aligned} y &= 20\lambda \\ x &= 10\lambda \\ 20x + 10y &= 10{,}000. \end{aligned}$$

Substituting values of x and y from the first two equations into the third gives

$$\begin{aligned} 20(10\lambda) + 10(20\lambda) &= 10{,}000 \\ 400\lambda &= 10{,}000 \\ \lambda &= 25. \end{aligned}$$

Substituting $\lambda = 25$ in the first two equations above gives $x = 250$ and $y = 500$.

The endpoints of the constraint are the point $(500, 0)$ and $(0, 1000)$. Since

$$f(500, 0) = 0 \quad \text{and} \quad f(0, 1000) = 0,$$

the maximum value of the production function is

$$f(250, 500) = 250 \cdot 500 = 125{,}000 \text{ units.}$$

The company should purchase 250 units of x and 500 units of y, giving 125,000 units of products.

(b) We have $\lambda = 25$. This tells us that if the budget is increased by \$1, we expect production to go up by about 25 units. If the budget goes up by \$1000, maximum production will increase about 25,000 to a total of roughly 150,000 units.

The Lagrangian Function

Constrained optimization problems are frequently solved using a *Lagrangian function*, $\mathcal{L}$. For example, to optimize the function $f(x, y)$ subject to the constraint $g(x, y) = c$, we use the Lagrangian function

$$\mathcal{L}(x, y, \lambda) = f(x, y) - \lambda(g(x, y) - c).$$

To see why the function $\mathcal{L}$ is useful, compute the partial derivatives of $\mathcal{L}$:

$$\begin{aligned} \frac{\partial \mathcal{L}}{\partial x} &= \frac{\partial f}{\partial x} - \lambda \frac{\partial g}{\partial x}, \\ \frac{\partial \mathcal{L}}{\partial y} &= \frac{\partial f}{\partial y} - \lambda \frac{\partial g}{\partial y}, \\ \frac{\partial \mathcal{L}}{\partial \lambda} &= -(g(x, y) - c). \end{aligned}$$

Notice that if (x_0, y_0) gives a maximum or minimum value of $f(x, y)$ subject to the constraint $g(x, y) = c$ and λ_0 is the corresponding Lagrange multiplier, then at the point (x_0, y_0, λ_0), we have

$$\frac{\partial \mathcal{L}}{\partial x} = 0 \quad \text{and} \quad \frac{\partial \mathcal{L}}{\partial y} = 0 \quad \text{and} \quad \frac{\partial \mathcal{L}}{\partial \lambda} = 0.$$

In other words, (x_0, y_0, λ_0) is a critical point for the unconstrained problem of optimization of the Lagrangian, $\mathcal{L}(x, y, \lambda)$.

We can therefore attack constrained optimization problems in two steps. First, write down the Lagrangian function $\mathcal{L}$. Second, find the critical points of $\mathcal{L}$.

Problems for Section 9.6

In Problems 1–10, use Lagrange multipliers to find the maximum or minimum values of $f(x, y)$ subject to the constraint.

1. $f(x, y) = xy, \quad 5x + 2y = 100$
2. $f(x, y) = x^2 + 3y^2 + 100, \quad 8x + 6y = 88$
3. $f(x, y) = x^2 + 4xy, \quad x + y = 100$
4. $f(x, y) = 5xy, \quad x + 3y = 24$
5. $f(x, y) = x + y, \quad x^2 + y^2 = 1$
6. $f(x, y) = 3x - 2y, \quad x^2 + 2y^2 = 44$
7. $f(x, y) = x^2 + y^2, \quad 4x - 2y = 15$
8. $f(x, y) = x^2 + y, \quad x^2 - y^2 = 1$
9. $f(x, y) = xy, \quad 4x^2 + y^2 = 8$
10. $f(x, y) = x^2 + y^2, \quad x^4 + y^4 = 2$
11. Figure 9.53 shows contours of $f(x, y)$ and the constraint $g(x, y) = c$. Approximately what values of x and y maximize $f(x, y)$ subject to the constraint? What is the approximate value of f at this maximum?

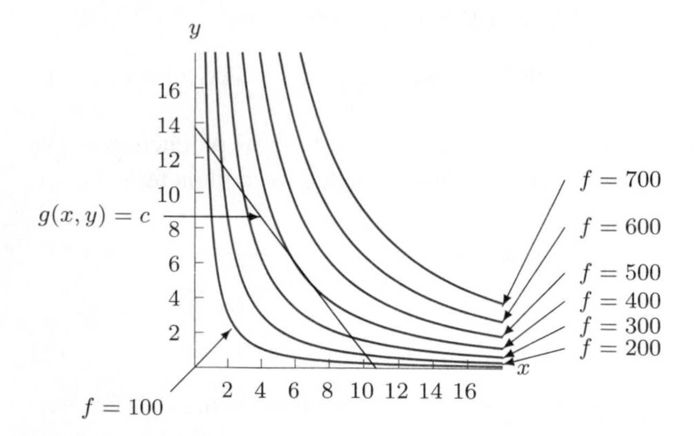

Figure 9.53

12. The quantity, Q, of a certain product manufactured depends on the quantity of labor, L, and of capital, K, used according to the function

$$Q = 900L^{1/2}K^{2/3}.$$

Labor costs \$100 per unit and capital costs \$200 per unit. What combination of labor and capital should be used to produce 36,000 units of the goods at minimum cost? What is that minimum cost?

13. The quantity, Q, of a good produced depends on the quantities x_1 and x_2 of two raw materials used:

$$Q = x_1^{0.3}x_2^{0.7}.$$

A unit of x_1 costs \$10, and a unit of x_2 costs \$25. We want to maximize production with a budget of \$50 thousand for raw materials.

(a) What is the objective function?
(b) What is the constraint?

14. The quantity, Q, of a good produced depends on the quantities x_1 and x_2 of two raw materials used:

$$Q = x_1^{0.6}x_2^{0.4}.$$

A unit of x_1 costs \$127, and a unit of x_2 costs \$92. We want to minimize the cost, C, of producing 500 units of the good.

(a) What is the objective function?
(b) What is the constraint?

15. Figure 9.54 shows level curves for production $f(x, y)$ as a function of the quantities x and y of two raw materials utilized. The cost of the materials is $15x + 20y$ thousand dollars. What is the maximum production possible with a budget of 300 thousand dollars? How much of each raw material should be purchased to achieve this maximum?

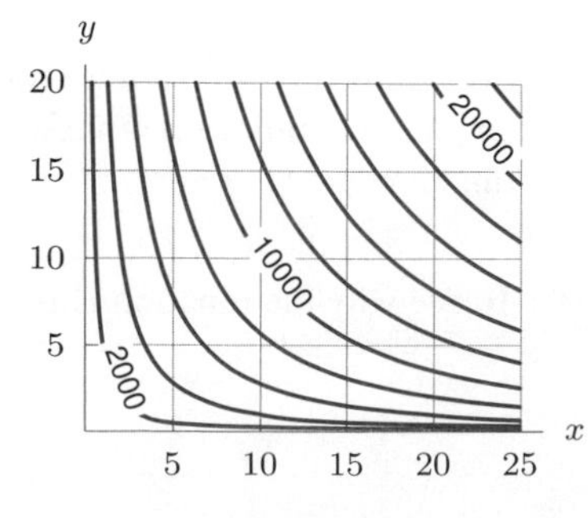

Figure 9.54

16. The quantity, Q, of a product manufactured by a company is given by

$$Q = aK^{0.6}L^{0.4},$$

where a is a positive constant, K is the quantity of capital and L is the quantity of labor used. Capital costs are \$20 per unit, labor costs are \$10 per unit, and the company wants costs for capital and labor combined to be no higher than \$150. Suppose you are asked to consult for the company, and learn that 5 units each of capital and labor are being used.

(a) What do you advise? Should the company use more or less labor? More or less capital? If so, by how much?

(b) Write a one sentence summary that could be used to sell your advice to the board of directors.

17. A firm manufactures a commodity at two different factories. The total cost of manufacturing depends on the quantities, q_1 and q_2, supplied by each factory, and is expressed by the *joint cost function*,

$$C = f(q_1, q_2) = 2q_1^2 + q_1q_2 + q_2^2 + 500.$$

The company's objective is to produce 200 units, while minimizing production costs. How many units should be supplied by each factory?

18. The Cobb-Douglas production function for a product is

$$P = 5L^{0.8}K^{0.2},$$

where P is the quantity produced, L is the size of the labor force, and K is the amount of total equipment. Each unit of labor costs \$300, each unit of equipment costs \$100, and the total budget is \$15,000.

(a) Make a table of L and K values which exhaust the budget. Find the production level, P, for each.

(b) Use the method of Lagrange multipliers to find the optimal way to spend the budget.

19. For a cost function, $f(x, y)$, the minimum cost for a production of 50 is given by $f(33, 87) = 1200$, with $\lambda = 15$. Estimate the cost if the production quota is:

(a) Raised to 51 **(b)** Lowered to 49

20. A company has the production function $P(x, y)$, which gives the number of units that can be produced for given values of x and y; the cost function $C(x, y)$ gives the cost of production for given values of x and y.

(a) If the company wishes to maximize production at a cost of \$50,000, what is the objective function f? What is the constraint equation? What is the meaning of λ in this situation?

(b) If instead the company wishes to minimize the costs at a fixed production level of 2000 units, what is the objective function f? What is the constraint equation? What is the meaning of λ in this situation?

21. A steel manufacturer can produce $P(K, L)$ tons of steel using K units of capital and L units of labor, with production costs $C(K, L)$ dollars. With a budget of \$600,000, the maximum production is 2,500,000 tons, using \$400,000 of capital and \$200,000 of labor. The Lagrange multiplier is $\lambda = 3.17$.

(a) What is the objective function?

(b) What is the constraint?

(c) What are the units for λ?

(d) What is the practical meaning of the statement $\lambda = 3.17$?

22. You have set aside 20 hours to work on two class projects. You want to maximize your grade (measured in points) which depends on how you divide your time between the two projects.

(a) What is the objective function for this optimization problem and what are its units?

(b) What is the constraint?

(c) Suppose you solve the problem by the method of Lagrange multipliers. What are the units for λ?

(d) What is the practical meaning of the statement $\lambda = 5$?

23. The terminal velocity (meters/second) that a two-stage rocket achieves is a function of the amount of fuel x_1 and x_2 (measured in liters) loaded into the two stages. You wish to minimize the total quantity of fuel required to achieve a specified terminal velocity, v_0.

(a) What is the objective function and what are its units?

(b) What is the constraint?

(c) Suppose you solve the problem by the method of Lagrange multipliers. What are the units for λ?

(d) What is the practical meaning of the statement $\lambda = 8$ when the terminal velocity is 50 meters/second?

Figure 9.55 shows contours of f. In Problems 24–26 give an approximate maximum or minimum value of f for $0 \le x \le 300$, $0 \le y \le 300$ subject to the given constraint.

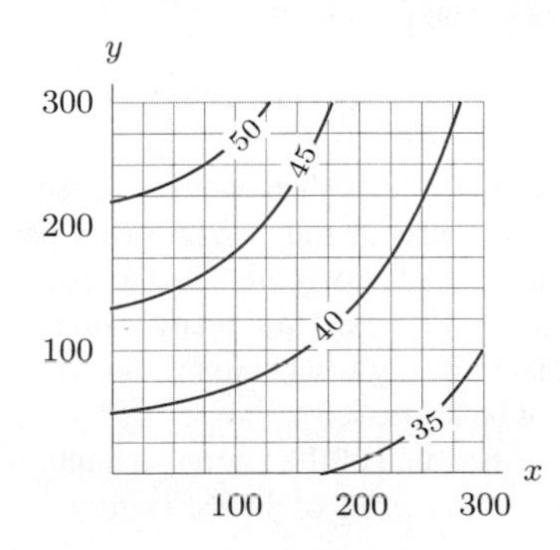

Figure 9.55

24. Minimum, constraint is $y = 100$

25. Maximum, constraint is $y = 100$

26. Maximum, constraint is $y = x$

27. A company manufactures x units of one item and y units of another. The total cost in dollars, C, of producing these two items is approximated by the function

$$C = 5x^2 + 2xy + 3y^2 + 800.$$

(a) If the production quota for the total number of items (both types combined) is 39, find the minimum production cost.
(b) Estimate the additional production cost or savings if the production quota is raised to 40 or lowered to 38.

28. The quantity, q, of a product manufactured depends on the number of workers, W, and the amount of capital invested, K, and is given by the Cobb-Douglas function

$$q = 6W^{3/4}K^{1/4}.$$

In addition, labor costs are \$10 per worker and capital costs are \$20 per unit and the budget is \$3000.

(a) What are the optimum number of workers and the optimum number of units of capital?
(b) Recompute the optimum values of W and K when the budget is increased by \$1. Check that increasing the budget by \$1 allows the production of λ extra units of the product, where λ is the Lagrange multiplier.

29. An automobile manufacturing plant currently employs 1500 workers and has capital investment of 4 million dollars per month. The production function is

$$Q = x^{0.4}y^{0.6},$$

where Q is the number of cars produced per month, x is the number of workers and y is capital investment. Each worker's salary is \$2100 per month and each unit of capital costs \$1000 per month.

(a) How many cars is the factory currently assembling each month?
(b) Due to the sluggish economy, the factory decides to decrease production to 2000 cars per month. The factory wants to minimize the cost of producing these 2000 cars. How many workers will need to be laid off? By what amount should monthly investment be decreased?
(c) Give the value of the Lagrange multiplier, λ, and interpret it in terms of the car factory.

30. Figure 9.56 shows contours labeled with values of $f(x, y)$ and a constraint $g(x, y) = c$. Mark the approximate points at which:

(a) f has a maximum
(b) f has a maximum on the constraint $g = c$.

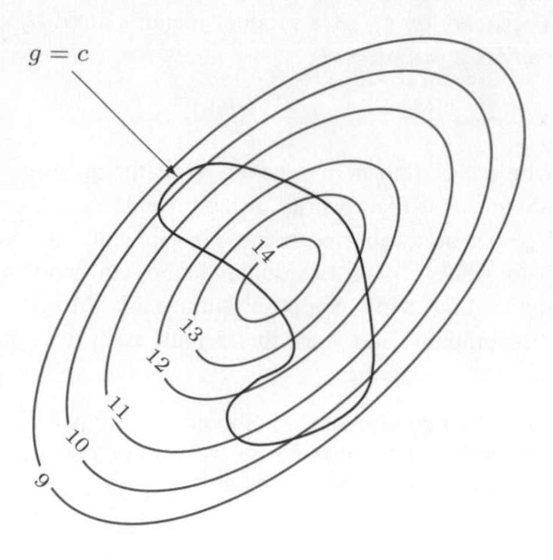

Figure 9.56

31. A monopolistic producer of two goods A and B has a joint total cost function

$$C = 10q_1 + q_1q_2 + 10q_2,$$

where q_1 and q_2 denote the quantities of A and B respectively. The demand curves for the corresponding prices p_1 and p_2 are

$$p_1 = 50 - q_1 + q_2$$
$$p_2 = 30 + 2q_1 - q_2.$$

(a) Find the maximum profit if the firm produces a total of 15.
(b) Estimate the new optimal profit if the production quota increases by one unit.

32. If x_1 and x_2 are the number of items of two goods bought, a customer's utility is

$$U(x_1, x_2) = 2x_1x_2 + 3x_1.$$

The unit cost is \$1 for the first good and \$3 for the second. Use Lagrange multipliers to find the maximum value of U if the consumer's disposable income is \$100. Estimate the new optimal utility if the consumer's disposable income increases by \$6.

33. Each person tries to balance his or her time between leisure and work. The tradeoff is that as you work less your income falls. Therefore each person has *indifference curves* which connect the number of hours of leisure, l, and income, s. If, for example, you are indifferent between 0 hours of leisure and an income of \$1125 a week on the one hand, and 10 hours of leisure and an income of \$750 a week on the other hand, then the points $l = 0$, $s = 1125$, and $l = 10$, $s = 750$ both lie on the same indifference curve. Table 9.12 gives information on three indifference curves, I, II, and III.

Table 9.12

Weekly income			Weekly leisure hours		
I	II	III	I	II	III
1125	1250	1375	0	20	40
750	875	1000	10	30	50
500	625	750	20	40	60
375	500	625	30	50	70
250	375	500	50	70	90

(a) Graph the three indifference curves.
(b) You have 100 hours a week available for work and leisure combined, and you earn \$10/hour. Write an equation in terms of l and s which represents this constraint.
(c) On the same axes, graph this constraint.
(d) Estimate from the graph what combination of leisure hours and income you would choose under these circumstances. Give the corresponding number of hours per week you would work.

CHAPTER SUMMARY

- **Functions of two variables**
Represented by: tables, graphs, formulas, cross-sections (one variable fixed), contours (function value fixed).
- **Partial derivatives**
Definition as a difference quotient, interpreting using units, estimating from a contour diagram or a table, computing from a formula, second-order partial derivatives.
- **Optimization**
Critical points, local and global maxima and minima.
- **Constrained optimization**
Geometric interpretation of Lagrange multiplier method, solving Lagrange multiplier problems algebraically, interpreting λ, Lagrangian function.

REVIEW PROBLEMS FOR CHAPTER NINE

1. The heat index is a temperature which tells you how hot it feels as a result of the combination of temperature and humidity. See Table 9.13. Heat exhaustion is likely to occur when the heat index reaches 105°F.

(a) If the temperature is 80°F and the humidity is 50%, how hot does it feel?
(b) At what humidity does 90°F feel like 90°F?
(c) Make a table showing the approximate temperature at which heat exhaustion becomes a danger, as a function of humidity.
(d) Explain why the heat index is sometimes above the actual temperature and sometimes below it.

Table 9.13 *Heat index (°F) as a function of humidity (H%) and temperature (T°F)*

H \ T	70	75	80	85	90	95	100	105	110	115
0	64	69	73	78	83	87	91	95	99	103
10	65	70	75	80	85	90	95	100	105	111
20	66	72	77	82	87	93	99	105	112	120
30	67	73	78	84	90	96	104	113	123	135
40	68	74	79	86	93	101	110	123	137	151
50	69	75	81	88	96	107	120	135	150	
60	70	76	82	90	100	114	132	149		

2. Using Table 9.13, graph heat index as a function of humidity with temperature fixed at 70°F and at 100°F. Explain the features of each graph and the difference between them in common sense terms.

For each of the functions in Problems 3–4, make a contour plot in the region $-2 < x < 2$ and $-2 < y < 2$. In each case, what is the equation and the shape of the contour lines?

3. $z = 3x - 5y + 1$ **4.** $z = 2x^2 + y^2$

5. You are an anthropologist observing a native ritual. Sixteen people arrange themselves with their backs to you along a bench; all but the three on the far left side are seated. The first person on the far left is standing with her hands at her side, the second is standing with his hands raised and the third is standing with her hands at her side. At some unseen signal, the first one sits down, and everyone else copies what his neighbor to the left was doing one second earlier. Every second that passes, this behavior is repeated until all are once again seated.

(a) Draw graphs at several different times showing how the height depends upon the distance along the bench.
(b) Graph the location of the raised hands as a function of time.
(c) What US ritual is most closely related to what you have observed?

6. Draw a contour diagram for the function $C = 40d + 0.15m$. Include contours for $C = 50, 100, 150, 200$.

7. Each of the contour diagrams in Figure 9.57 shows population density in a certain region. Choose the contour diagram that best corresponds to each of the following situations. Many different matchings are possible. Pick any reasonable one and justify your choice.

(a) The center of the diagram is a city.

(b) The center of the diagram is a lake.
(c) The center of the diagram is a power plant.

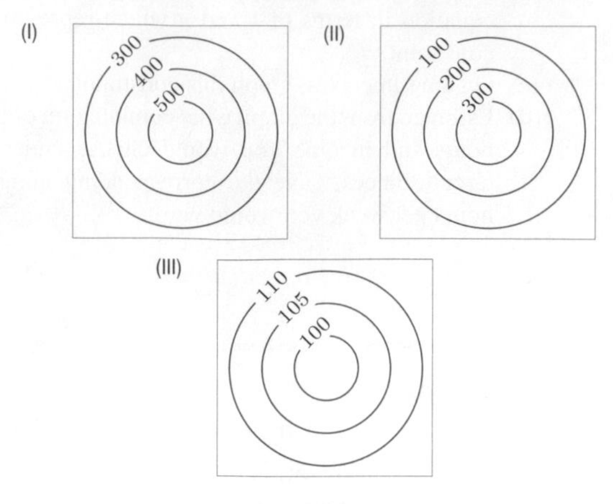

Figure 9.57

8. Figure 9.58 shows the contours of the temperature H in a room near a recently opened window. Label the three contours with reasonable values of H if the house is in the following locations.

(a) Minnesota in winter (where winters are harsh).
(b) San Francisco in winter (where winters are mild).
(c) Houston in summer (where summers are hot).
(d) Oregon in summer (where summers are mild).

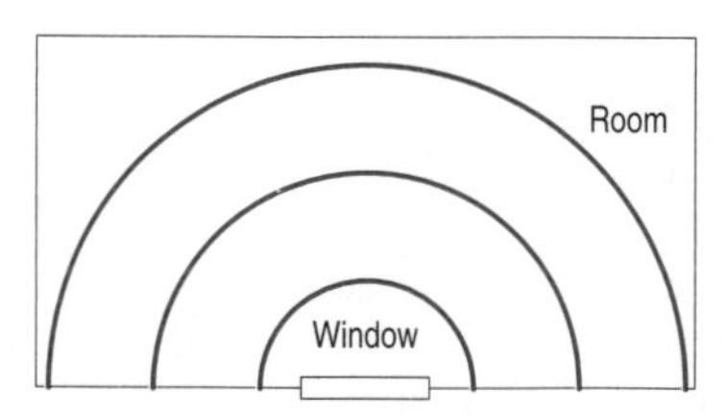

Figure 9.58

9. You are in a room 30 feet long with a heater at one end. In the morning the room is 65°F. You turn on the heater, which quickly warms up to 85°F. Let $H(x,t)$ be the temperature x feet from the heater, t minutes after the heater is turned on. Figure 9.39 on page 364 shows the contour diagram for H. How warm is it 10 feet from the heater 5 minutes after it was turned on? 10 minutes after it was turned on?

10. Using the contour diagram in Figure 9.39 on page 364, sketch the graphs of the one-variable functions $H(x,5)$ and $H(x,20)$. Interpret the two graphs in practical terms, and explain the difference between them.

11. The following table gives values of a function $f(x,y)$.

(a) Is f_x positive or negative? Is f_y positive or negative?
(b) Estimate $f_x(10,6)$ and $f_y(10,6)$.
(c) Use partial derivatives to estimate $f(30,9)$ and $f(34,8)$.

	$x=0$	$x=10$	$x=20$	$x=30$
$y=0$	89	80	74	71
$y=2$	93	85	80	76
$y=4$	98	91	85	81
$y=6$	104	98	92	88
$y=8$	112	105	99	94

Problems 12–16 investigate how the demand for coffee depends on the price of coffee and the price of tea. Suppose that the demand for coffee, Q, in thousands of pounds per week, depends on the price of coffee, c, and the price of tea, t, both in dollars per pound, according to the formula

$$Q = f(c,t) = 100\frac{t}{c}.$$

12. **(a)** If the price of tea is \$1 per pound, the demand for coffee is $Q = f(c,1) = 100/c$. Graph this demand curve, with Q as a function of c, for $0 \le c \le 6$.
(b) On the same axes, sketch demand curves for coffee with $t = 1, 2, 3, 4, 5$, and label each curve with the corresponding value of t. What do you observe? How does the demand curve for coffee change as the price of tea increases? Explain in terms of the demand for coffee why this is reasonable.

13. Is Q an increasing or decreasing function of c? Is Q an increasing or decreasing function of t? Explain in terms of the demand for coffee why this is so.

14. Make a table showing the value of Q when $t = 1, 2, 3, 4$, and $c = 1, 2, 3, 4$. (You should have 16 values of Q in your table.) Use your table to check your answers to Problem 13.

15. We can also think of the demand for tea as a function of the price of coffee and the price of tea. Give a possible formula for the demand for tea as a function of c and t.

16. **(a)** Draw contours for $Q = 25$, $Q = 50$, $Q = 100$, $Q = 200$.
(b) Determine from the contour diagram whether Q is an increasing or a decreasing function of c. Is Q an increasing or a decreasing function of t?
(c) Demand is called *elastic* if small increases in the price of the product cause large changes in the demand for the product. If the price of tea is fixed at $t = 1$, is the demand for coffee more elastic at low or high prices? Explain using the contour diagram.

17. The cornea is the front surface of the eye. Corneal specialists use a TMS, or Topographical Modeling System, to produce a "map" of the curvature of the eye's surface. A computer analyzes light reflected off the eye and draws level curves joining points of constant curvature. The regions between these curves are colored different colors.

The first two pictures in Figure 9.59 are cross-sections of eyes with constant curvature, the smaller being about 38 units and the larger about 50 units. For contrast, the third eye has varying curvature.

(a) Describe in words how the TMS map of an eye of constant curvature will look.

(b) Draw the TMS map of an eye with the cross-section in Figure 9.60. Assume the eye is circular when viewed from the front, and the cross-section is the same in every direction. Put reasonable numeric labels on your level curves.

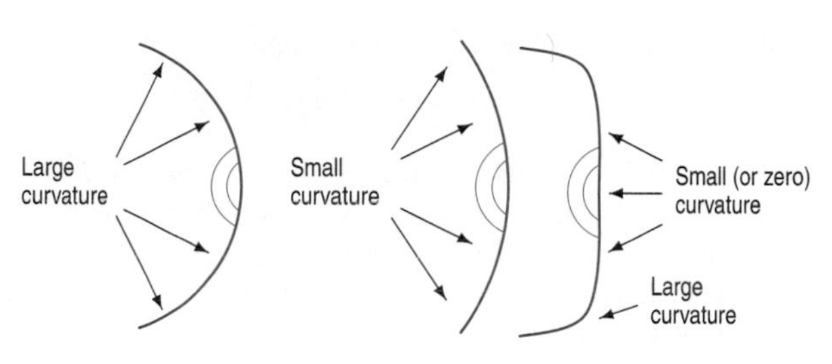

Figure 9.59: Pictures of eyes with different curvature

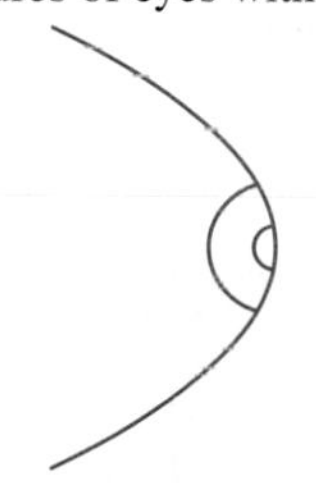

Figure 9.60

18. The cost of producing one unit of a product is given by

$$c = a + bx + ky,$$

where x is the amount of labor used (in man hours) and y is the amount of raw material used (by weight) and a and b and k are constants. What does $\partial c/\partial x = b$ mean? What is the practical interpretation of b?

19. Your monthly car payment in dollars is $P = f(P_0, t, r)$, where $\$P_0$ is the amount you borrowed, t is the number of months it takes to pay off the loan, and $r\%$ is the interest rate. What are the units, the financial meanings, and the signs of $\partial P/\partial t$ and $\partial P/\partial r$?

20. Table 9.4 on page 347 gives the quantity of beef bought, C, as a function of household income, I, and the price of beef, p. Thus we have $C = f(I, p)$.

(a) Find $f_p(80, 4.0)$ and interpret it in terms of beef consumption.

(b) Find $f_I(80, 4.0)$ and interpret it in terms of beef consumption.

(c) Use partial derivatives to estimate the quantity of beef bought by a family with a household income of $110,000 if the price of beef is $4.00 per pound.

For Problems 21–22 refer to Table 9.5 on page 353 giving the heat index, I, in °F, as a function $f(H, T)$ of the relative humidity, H, and the temperature, T, in °F. The heat index is a temperature which tells you how hot it feels as a result of the combination of humidity and temperature.

21. Estimate $\partial I/\partial H$ and $\partial I/\partial T$ for typical weather conditions in Tucson in summer ($H = 10, T = 100$). What do your answers mean in practical terms for the residents of Tucson?

22. Answer the question in Problem 21 for Boston in summer ($H = 50, T = 80$).

23. Suppose that x is the average price of a new car and that y is the average price of a gallon of gasoline. Then q_1, the number of new cars bought in a year, depends on both x and y, so $q_1 = f(x, y)$. Similarly, if q_2 is the quantity of gas bought in a year, then $q_2 = g(x, y)$.

(a) What do you expect the signs of $\partial q_1/\partial x$ and $\partial q_2/\partial y$ to be? Explain.

(b) What do you expect the signs of $\partial q_1/\partial y$ and $\partial q_2/\partial x$ to be? Explain.

24. Figure 9.61 shows the density of the fox population P (in foxes per square kilometer) for southern England. Draw two different graphs of the fox population as a function of kilometers north, with kilometers east fixed at two different values, and draw two different graphs of the fox population as a function of kilometers east, with kilometers north fixed at two different values.

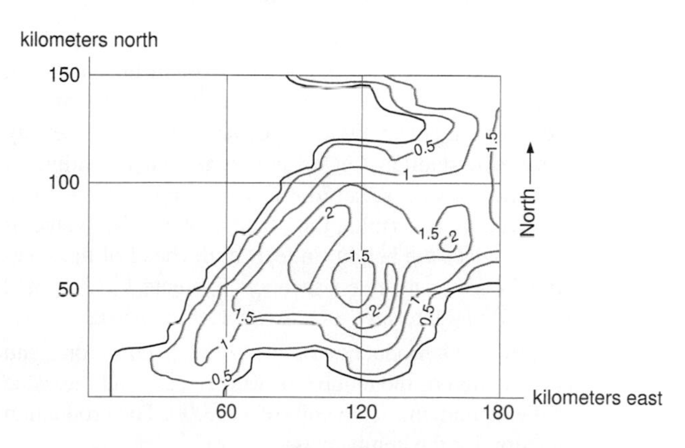

Figure 9.61

25. Figure 9.61 gives a contour diagram for the number n of foxes per square kilometer in southwestern England. Estimate $\partial n/\partial x$ and $\partial n/\partial y$ at the points A, B, and C, where x is kilometers east and y is kilometers north.

26. Use the diagram from Problem 20, page 348 to estimate $H_T(T,w)$ for $T = 10, 20, 30$ and $w = 0.1, 0.2, 0.3$. What is the practical meaning of these partial derivatives?

27. Repeat Problem 26 for $H_w(T,w)$ at $T = 10$, 20, and 30 and $w = 0.1$, 0.2, and 0.3. What is the practical meaning of these partial derivatives?

Find the partial derivatives in Problems 28–33. The variables are restricted to a domain on which the function is defined.

28. f_x and f_y if $f(x,y) = x^2 + xy + y^2$

29. P_a and P_b if $P = a^2 - 2ab^2$

30. $\dfrac{\partial Q}{\partial p_1}$ and $\dfrac{\partial Q}{\partial p_2}$ if $Q = 50p_1p_2 - p_2^2$

31. $\dfrac{\partial f}{\partial x}$ and $\dfrac{\partial f}{\partial t}$ if $f = 5xe^{-2t}$

32. $\dfrac{\partial P}{\partial K}$ and $\dfrac{\partial P}{\partial L}$ if $P = 10K^{0.7}L^{0.3}$

33. f_x and f_y if $f(x,y) = \sqrt{x^2 + y^2}$

34. A soft drink company is interested in seeing how the demand for its products is affected by prices. The company believes that the quantity, q, of its soft drinks sold depends on p_1, the average price of the company's soft drinks, p_2, the average price of competing soft drinks, and p_3, the average amount of money spent by the company on advertising:

$$q = C - 8\cdot 10^6 p_1 + 4\cdot 10^6 p_2 + 2p_3.$$

(a) What does the constant C represent in terms of soft drink sales?

(b) Find the marginal demand for soft drinks with respect to changes in p_1, p_2, and p_3. Explain why the signs and relative magnitudes of your answers are reasonable. [Note: The marginal demand is the rate of change of quantity demanded with price.]

35. You are in a stadium doing the wave. This is a ritual in which members of the audience stand up and down in such a way as to create a wave that moves around the stadium. Normally a single wave travels all the way around the stadium, but we assume there is a continuous sequence of waves. Let $h(x,t) = 5 + \cos(0.5x - t)$ be the function describing this stadium wave. The value of $h(x,t)$ gives the height (in feet) of the head of the spectator in seat x at time t seconds. Evaluate $h_x(2,5)$ and $h_t(2,5)$ and interpret each in terms of the wave.

36. A company's production output, P, is given in tons, and is a function of the number of workers, N, and the value of the equipment, V, in units of \$25,000. The production function for the company is

$$P = f(N,V) = 5N^{0.75}V^{0.25}.$$

The company currently employs 80 workers, and has equipment worth \$750,000. What are N and V? Find the values of f, f_N, and f_V at these values of N and V. Give units and explain what each answer means in terms of production.

37. Find all critical points of $f(x,y) = x^2 + 3y^2 - 4x + 6y + 10$.

38. Find all critical points of $f(x,y) = x^3 - 3x + y^2$. Make a table of values to determine if each critical point is a local minimum, a local maximum, or neither.

39. Two products are manufactured in quantities q_1 and q_2 and sold at prices of p_1 and p_2, respectively. The cost of producing them is given by

$$C = 2q_1^2 + 2q_2^2 + 10.$$

(a) Find the maximum profit that can be made, assuming the prices are fixed.

(b) Find the rate of change of that maximum profit as p_1 increases.

40. The production function for a company is given by

$$P = 24K^{0.6}L^{0.4},$$

where P is the amount produced by the company, L is the amount spent on labor, and K is the value of the equipment, or capital. The total amount spent on L and K together cannot exceed \$1000. If the goal is to maximize production, how much should be spent on labor and how much on capital?

(a) Set up the equations to solve this problem using Lagrange multipliers.

(b) What are the optimal values of L and K?

(c) How many units are produced at this level of production?

(d) Find the value of λ and interpret it.

41. The director of a neighborhood health clinic has an annual budget of \$600,000. How should the budget be allocated to maximize the number of patient visits, V, which is a function of the number of doctors, D, and the number of nurses, N, and given by

$$V = 1000D^{0.6}N^{0.3}.$$

Doctors receive an annual salary of \$40,000, while nurses get \$10,000.

(a) Set up the director's constrained optimization problem.

(b) Solve the problem formulated in part (a).

(c) Find the value of the Lagrange multiplier and interpret its meaning in this problem.

42. Find expressions for K and L which maximize output given by the Cobb-Douglas production function

$$Q = AK^{\alpha}L^{\beta},$$

where A, α and β are constants, subject to the cost constraint

$$c_K K + c_L L = M.$$

PROJECTS FOR CHAPTER NINE

1. A Heater in a Room

Figure 9.62 shows the contours of the temperature along one wall of a heated room through one winter day, with time indicated as on a 24-hour clock. The room has a heater located at the left-most corner of the wall and one window in the wall. The heater is controlled by a thermostat about 2 feet from the window.

(a) Where is the window? **(b)** When is the window open?
(c) When is the heat on?
(d) Draw graphs of the temperature along the wall of the room at 6 am, at 11 am, at 3 pm (15 hours) and at 5 pm (17 hours).
(e) Draw a graph of the temperature as a function of time at the heater, at the window and midway between them.
(f) The temperature at the window at 5 pm (17 hours) is less than at 11 am. Why do you think this might be?
(g) To what temperature do you think the thermostat is set? How do you know?
(h) Where is the thermostat?

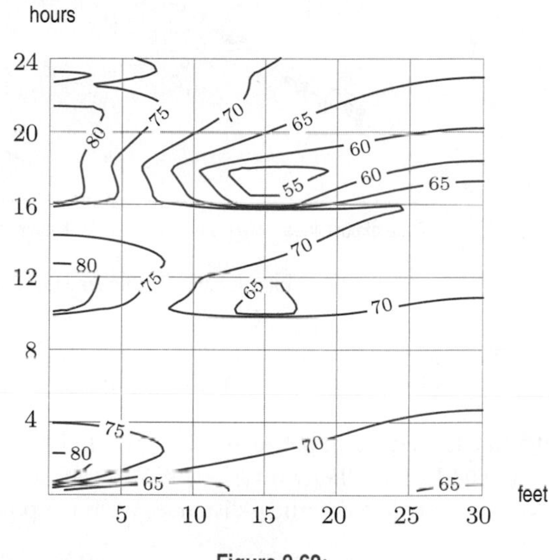

Figure 9.62:

2. Optimizing Relative Prices for Adults and Children

Some items are sold at a discount to senior citizens or children. The reason is that these groups are more sensitive to price, so a discount has greater impact on their purchasing decisions. The seller faces an optimization problem: How large a discount to offer in order to maximize profits? Suppose a theater can sell q_c child tickets and q_a adult tickets at prices p_c and p_a, according to the demand functions:

$$q_c = r{p_c}^{-4} \quad \text{and} \quad q_a = s{p_a}^{-2},$$

and has operating costs proportional to the total number of tickets sold. What should be the relative price of children's and adults' tickets?

3. Maximizing Production and Minimizing Cost: "Duality"

A company's production function is $P = 270x_1^{1/3}x_2^{2/3}$ for quantities x_1 and x_2 of two raw materials, costing \$4 per unit and \$27 per unit, respectively.

(a) How much of each raw material should be used to maximize production if the budget for raw materials is \$324? What is the maximum production achieved, P_0?
(b) What is the minimum cost at which a production level of P_0 can be achieved? How much of each raw material is used at this minimum?
(c) Comment on the relationship between your answers to parts (a) and (b).

FOCUS ON THEORY

DERIVING THE FORMULA FOR A REGRESSION LINE

Suppose we want to find the "best fitting" line for some experimental data. In the Focus on Modeling section beginning on page 78, we used a computer or calculator to find the formula for this line. In this section, we derive this formula.

We decide which line fits the data best by using the following criterion. The data is plotted in the plane. The distance from a line to the data points is measured by adding the squares of the vertical distances from each point to the line. The smaller this sum of squares is, the better the line fits the data. The line with the minimum sum of square distances is called the *least squares line*, or the *regression line*. If the data is nearly linear, the least squares line will be a good fit; otherwise it may not be. (See Figure 9.63.)

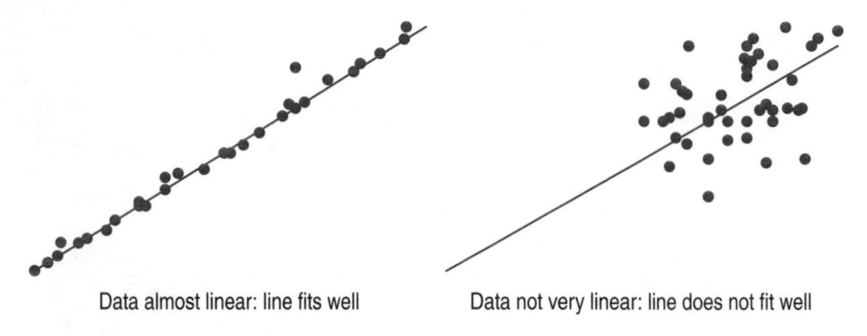

Figure 9.63: Fitting lines to data points

Example 1 Find a least squares line for the following data points: $(1,1)$, $(2,1)$, and $(3,3)$.

Solution Suppose the line has equation $y = b + mx$. If we find b and m then we have found the line. So, for this problem, b and m are the two variables. We want to minimize the function $f(b,m)$ that gives the sum of the three squared vertical distances from the points to the line in Figure 9.64.

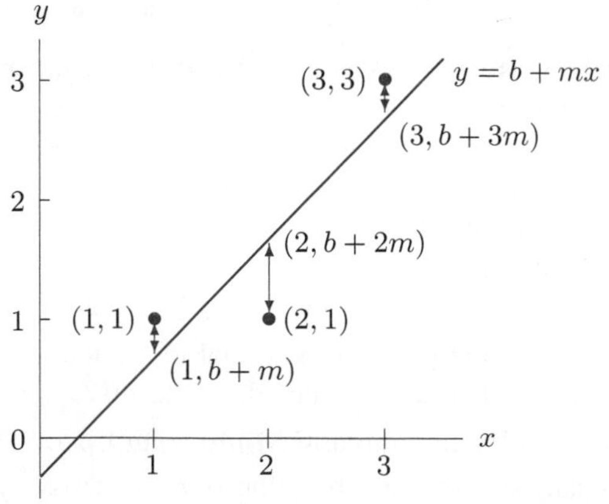

Figure 9.64: The least squares line minimizes the sum of the squares of these vertical distances

The vertical distance from the point $(1,1)$ to the line is the difference in the y-coordinates $1-(b+m)$; similarly for the other points. Thus, the sum of squares is

$$f(b,m) = (1-(b+m))^2 + (1-(b+2m))^2 + (3-(b+3m))^2.$$

To minimize f we look for critical points. First we differentiate f with respect to b:

$$\begin{aligned} f_b(b,m) &= -2(1-(b+m)) - 2(1-(b+2m)) - 2(3-(b+3m)) \\ &= -2+2b+2m-2+2b+4m-6+2b+6m \\ &= -10+6b+12m. \end{aligned}$$

Now we differentiate with respect to m:

$$\begin{aligned} f_m(b,m) &= 2(1-(b+m))(-1) + 2(1-(b+2m))(-2) + 2(3-(b+3m))(-3) \\ &= -2+2b+2m-4+4b+8m-18+6b+18m \\ &= -24+12b+28m. \end{aligned}$$

The equations $f_b = 0$ and $f_m = 0$ give a system of two linear equations in two unknowns:

$$\begin{aligned} -10+6b+12m &= 0, \\ -24+12b+28m &= 0. \end{aligned}$$

The solution to this pair of equations is the critical point $b = -1/3$ and $m = 1$. Since

$$D = f_{bb}f_{mm} - (f_{mb})^2 = (6)(28) - 12^2 = 24 \quad \text{and} \quad f_{bb} = 6 > 0,$$

we have found a local minimum. This local minimum is also the global minimum of f. Thus, the least squares line is

$$y = x - \frac{1}{3}.$$

As a check, notice that the line $y = x$ passes through the points $(1,1)$ and $(3,3)$. It is reasonable that introducing the point $(2,1)$ moves the y-intercept down from 0 to $-1/3$.

Derivation of the Formulas for the Regression Line

We use the method of Example 1 to derive the formulas for the least-squares line $y = b + mx$ generated by data points (x_1, y_1), (x_2, y_2), $\cdots$, (x_n, y_n). Notice that we are looking for the slope and y-intercept, so we think of m and b as the variables.

For each data point (x_i, y_i), the corresponding point directly above the line or below it on the line has the y-coordinate $b + mx_i$. Thus, the squares of the vertical distances from the point to the line is $(y_i - (b + mx_i))^2$. (See Figure 9.65.)

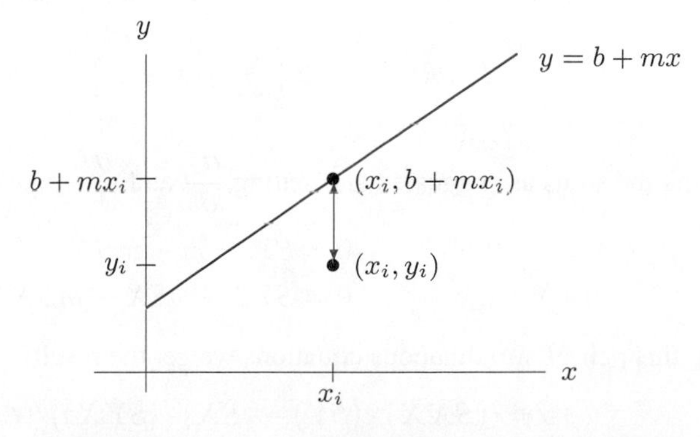

Figure 9.65: The vertical distance from a point to the line

We find the sum of the n squared distances from points to the line, and think of the sum as a function of m and b:

$$f(b, m) = \sum_{i=1}^{n}(y_i - (b + mx_i))^2.$$

To minimize this function, we first find the two partial derivatives, f_b and f_m. We use the chain rule and the properties of sums.

$$\begin{aligned}
f_b(b, m) &= \frac{\partial}{\partial b}\left(\sum_{i=1}^{n}(y_i - (b + mx_i))^2\right) = \sum_{i=1}^{n}\frac{\partial}{\partial b}(y_i - (b + mx_i))^2 \\
&= \sum_{i=1}^{n} 2(y_i - (b + mx_i)) \cdot \frac{\partial}{\partial b}(y_i - (b + mx_i)) \\
&= \sum_{i=1}^{n} 2(y_i - (b + mx_i)) \cdot (-1) \\
&= -2\sum_{i=1}^{n}(y_i - (b + mx_i))
\end{aligned}$$

$$\begin{aligned}
f_m(b, m) &= \frac{\partial}{\partial m}\left(\sum_{i=1}^{n}(y_i - (b + mx_i))^2\right) = \sum_{i=1}^{n}\frac{\partial}{\partial m}(y_i - (b + mx_i))^2 \\
&= \sum_{i=1}^{n} 2(y_i - (b + mx_i)) \cdot \frac{\partial}{\partial m}(y_i - (b + mx_i)) \\
&= \sum_{i=1}^{n} 2(y_i - (b + mx_i)) \cdot (-x_i) \\
&= -2\sum_{i=1}^{n}(y_i - (b + mx_i)) \cdot x_i
\end{aligned}$$

We now set the partial derivatives equal to zero and solve for m and b. This is easier than it looks: we simplify the appearance of the equations by temporarily substituting other symbols for the sums: write SY for $\sum y_i$, SX for $\sum x_i$, SXY for $\sum y_i x_i$ and SXX for $\sum x_i^2$. Remember, the x_i and y_i are all constants. We get a pair of simultaneous linear equations in m and b; solving for m and b gives us formulas in terms of SX, SY, SXY, and SXX. We separate $f_b(b, m)$ into three sums as shown:

$$f_b(b, m) = -2\left(\sum_{i=1}^{n} y_i - b\sum_{i=1}^{n} 1 - m\sum_{i=1}^{n} x_i\right).$$

Similarly we can separate $f_m(b, m)$ after multiplying through by x_i:

$$f_m(b, m) = -2\left(\sum_{i=1}^{n} y_i x_i - b\sum_{i=1}^{n} x_i - m\sum_{i=1}^{n} x_i^2\right).$$

Rewriting the sums as suggested and setting $\dfrac{\partial f}{\partial b}$ and $\dfrac{\partial f}{\partial m}$ equal to zero we have:

$$\begin{aligned}
0 &= SY - bn - mSX \\
0 &= SYX - bSX - mSXX
\end{aligned}$$

Solving this pair of simultaneous equations we get the result:

$$\begin{aligned}
b &= ((SXX) \cdot (SY) - (SX) \cdot (SYX))/(n(SXX) - (SX)^2) \\
m &= (n(SYX) - (SX) \cdot (SY))/(n(SXX) - (SX)^2)
\end{aligned}$$

Writing these expressions with summation notation we arrive at the following result:

The least squares line for data points (x_1, y_1), (x_2, y_2), $\cdots$, (x_n, y_n) is the line $y = b + mx$ where

$$b = \left(\sum_{i=1}^{n} x_i^2 \sum_{i=1}^{n} y_i - \sum_{i=1}^{n} x_i \sum_{i=1}^{n} y_i x_i\right) \Big/ \left(n \sum_{i=1}^{n} x_i^2 - \left(\sum_{i=1}^{n} x_i\right)^2\right)$$

$$m = \left(n \sum_{i=1}^{n} y_i x_i - \sum_{i=1}^{n} x_i \sum_{i=1}^{n} y_i\right) \Big/ \left(n \sum_{i=1}^{n} x_i^2 - \left(\sum_{i=1}^{n} x_i\right)^2\right).$$

Example 2 Use these formulas to find the best fitting line for the data point $(1, 5)$, $(2, 4)$, $(4, 3)$.

Solution We compute the sums needed in the formulas:

$$\sum_{i=1}^{3} x_i = 1 + 2 + 4 = 7$$

$$\sum_{i=1}^{3} y_i = 5 + 4 + 3 = 12$$

$$\sum_{i=1}^{3} x_i^2 = 1^2 + 2^2 + 4^2 = 1 + 4 + 16 = 21$$

$$\sum_{i=1}^{3} y_i x_i = (5)(1) + (4)(2) + (3)(4) = 5 + 8 + 12 = 25$$

Since $n = 3$, we have:

$$\begin{aligned} b &= \left(\sum_{i=1}^{3} x_i^2 \sum_{i=1}^{3} y_i - \sum_{i=1}^{3} x_i \sum_{i=1}^{3} y_i x_i\right) \Big/ \left(3 \sum_{i=1}^{3} x_i^2 - \left(\sum_{i=1}^{3} x_i\right)^2\right) \\ &= ((21)(12) - (7)(25)) \,/\, (3(21) - (7^2)) \\ &= 77/14 = 5.5 \end{aligned}$$

and

$$\begin{aligned} m &= \left(3 \sum_{i=1}^{3} y_i x_i - \sum_{i=1}^{3} x_i \sum_{i=1}^{3} y_i\right) \Big/ \left(3 \sum_{i=1}^{3} x_i^2 - \left(\sum_{i=1}^{3} x_i\right)^2\right) \\ &= (3(25) - (7)(12)) \,/\, (3(21) - (7^2)) \\ &= -9/14 = -0.64 \end{aligned}$$

The least squares line for these three points is

$$y = 5.5 - 0.64x.$$

To check this equation, plot the line and the three points together.

Many calculators have the formulas for the least squares line built in, so that when you enter the data, out come the values of b and m. At the same time, you get the *correlation coefficient*, which measures how close the data points actually come to fitting the least squares line.

Problems on Deriving the Formula for Regression Lines

In Problems 1–2, use the method of Example 1 to find the least squares line. Check by graphing the points with the line.

1. $(-1, 2), (0, -1), (1, 1)$ **2.** $(0, 2), (1, 4), (2, 5)$

In Problems 3–5, use the formulas for b and m to check that you get the same result as in the problem or example specified.

3. $(-1, 2), (0, -1), (1, 1)$. See Problem 1.

4. $(0, 2), (1, 4), (2, 5)$. See Problem 2.

5. $(1, 1), (2, 1), (3, 3)$. See Example 1.

In Problems 6–7, we transform nonlinear data so that it looks more linear. For example, suppose the data points (x, y) fit the exponential equation,

$$y = Ce^{ax},$$

where a and C are constants. Taking the natural log of both sides, we get

$$\ln y = ax + \ln C.$$

Thus, $\ln y$ is a linear function of x. To find a and C, we can use least squares for the graph of $\ln y$ against x.

6. The population of the US was about 180 million in 1960, grew to 206 million in 1970, and 226 million in 1980.

(a) Assuming that the population was growing exponentially, use logarithms and the method of least squares to estimate the population in 1990.

(b) According to the national census, the 1990 population was 249 million. What does this say about the assumption of exponential growth?

(c) Predict the population in the year 2010.

7. A biological rule of thumb states that as the area A of an island increases tenfold, the number of animal species, N, living on it doubles. The table contains data for islands in the West Indies. Assume that N is a power function of A.

(a) Use the biological rule of thumb to find

(i) N as a function of A

(ii) $\ln N$ as a function of $\ln A$

(b) Using the data given, tabulate $\ln N$ against $\ln A$ and find the line of best fit. Does your answer agree with the biological rule of thumb?

Island	Area (sq km)	Number of species
Redonda	3	5
Saba	20	9
Montserrat	192	15
Puerto Rico	8858	75
Jamaica	10854	70
Hispaniola	75571	130
Cuba	113715	125

Chapter Ten

MATHEMATICAL MODELING USING DIFFERENTIAL EQUATIONS

Mathematical modeling means using mathematics to represent a situation. The equations or formulas which make up the model may be used to make predictions.

One particular type of model is a *differential equation*. If we know the rate of change, or derivative, of an unknown function, we may be able to use this information to write an equation involving the derivative. Such an equation is called a *differential equation*.

In this chapter, we use differential equations to model money in a bank account, pollution in the Great Lakes, the quantity of drug in the body, and net worth of a company. We use systems of differential equations to model the interaction of two populations (such as two species or two businesses) and the spread of a disease.

10.1 MATHEMATICAL MODELING: SETTING UP A DIFFERENTIAL EQUATION

Sometimes we do not know a key function, but we do have information about its rate of change, or its derivative. Then we may be able to write a new type of equation, called a *differential equation*, from which we can get information about the original function. For example, we may use what we know about the derivative of a population function (its rate of change) to predict the population in the future.

In this section, we start from a verbal description and use it to write a differential equation.

Marine Harvesting

We begin by investigating the effect of fishing on a fish population. Suppose that, left alone, a fish population increases at a continuous rate of 20% per year. Suppose that fish are also being harvested (caught) by fishermen at a constant rate of 10 million fish per year. How does the fish population change over time?

Notice that we have been given information about the rate of change, or derivative, of the fish population. Combined with information about the initial population, we can use this to predict the population in the future. We know that

$$\begin{matrix}\text{Rate of change} \\ \text{of fish population}\end{matrix} = \begin{matrix}\text{Rate of increase} \\ \text{due to breeding}\end{matrix} - \begin{matrix}\text{Rate fish removed} \\ \text{due to harvesting}\end{matrix}.$$

Suppose the fish population, in millions, is P and its derivative is dP/dt, where t is time in years. If left alone, the fish population increases at a continuous rate of 20% per year, so we have

$$\begin{aligned}\text{Rate of increase due to breeding} &= 20\% \cdot \text{Current population} \\ &= 0.20P \text{ million fish/year.}\end{aligned}$$

In addition,

$$\text{Rate fish removed by harvesting} = 10 \text{ million fish/year.}$$

Since the rate of change of the fish population is dP/dt, we have

$$\frac{dP}{dt} = 0.20P - 10.$$

This is a differential equation that models how the fish population changes. The unknown quantity in the equation is the function giving P in terms of t.

Net Worth of a Company

A company earns revenue (income) and also makes payroll payments. Assume that revenue is earned continuously, that payroll payments are made continuously, and that the only factors affecting net worth are revenue and payroll. The company's revenue is earned at a continuous annual rate of 5% times its net worth. At the same time, the company's payroll obligations are paid out at a constant rate of 200 million dollars a year.

We use this information to write a differential equation to model the net worth of the company, W, in millions of dollars, as a function of time, t, in years. We know that

$$\begin{matrix}\text{Rate at which} \\ \text{net worth is changing}\end{matrix} = \begin{matrix}\text{Rate revenue} \\ \text{is earned}\end{matrix} - \begin{matrix}\text{Rate payroll payments} \\ \text{are made}\end{matrix}.$$

Since the company's revenue is earned at a rate of 5% of its net worth, we have

$$\text{Rate revenue is earned} = 5\% \cdot \text{ Net worth } = 0.05W \text{ million dollars/year.}$$

Since payroll payments are made at a rate of 200 million dollars a year, we have

$$\text{Rate payroll payments are made} = 200 \text{ million dollars/year.}$$

Putting these two together, since the rate at which net worth is changing is dW/dt, we have

$$\frac{dW}{dt} = 0.05W - 200.$$

This is a differential equation that models how the net worth of the company changes. The unknown quantity in the equation is the function giving net worth W as a function of time t.

Pollution in a Lake

If clean water flows into a polluted lake and a stream takes water out, the level of pollution in the lake will decrease (assuming no new pollutants are added).

Example 1 The quantity of pollutant in the lake decreases at a rate proportional to the quantity present. Write a differential equation to model the quantity of pollutant in the lake. Is the constant of proportionality positive or negative? Use the differential equation to explain why the graph of the quantity of pollutant against time is decreasing and concave up, as in Figure 10.1.

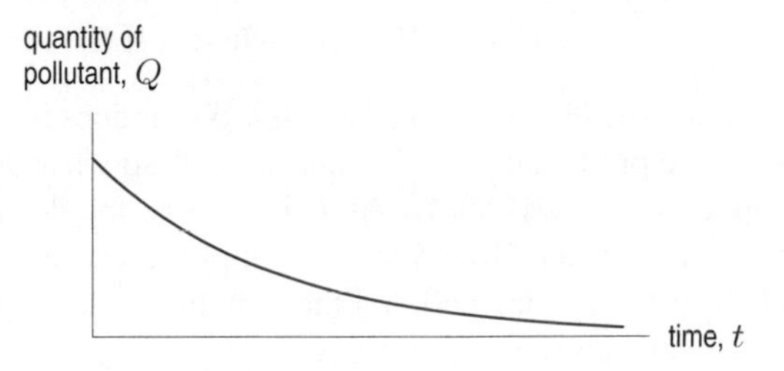

Figure 10.1: Quantity of pollutant in a lake

Solution Let Q denote the quantity of pollutant present in the lake at time t. The rate of change of Q is proportional to Q, so dQ/dt is proportional to Q. Thus, the differential equation is

$$\frac{dQ}{dt} = kQ.$$

Since no new pollutants are being added to the lake, the quantity Q is decreasing over time, so dQ/dt is negative. Thus, the constant of proportionality k is negative.

Why does the differential equation $dQ/dt = kQ$, with k negative, give us the graph shown in Figure 10.1? Since k is negative and Q is positive, we know kQ is negative. Thus, dQ/dt is negative, so the graph of Q against t is decreasing as in Figure 10.1. Why is it concave up? Since Q is getting smaller and k is fixed, as t increases, the product kQ is getting smaller in magnitude so the derivative dQ/dt is getting smaller in magnitude. Thus, the graph of Q is more horizontal as t increases. Therefore, the graph is concave up. See Figure 10.1.

The Quantity of a Drug in the Body

In the previous example, the rate at which pollutants leave a lake is proportional to the quantity of pollutants in the lake. This model works for any contaminants flowing in or out of a fluid system with complete mixing. Another example is the quantity of drug in a person's body.

Example 2 A patient having major surgery is given the antibiotic vancomycin intravenously at a rate of 85 mg per hour. The rate at which the drug is excreted from the body is proportional to the quantity present, with proportionality constant 0.1 if time is in hours. Write a differential equation for the quantity, Q in mg, of vancomycin in the body after t hours.

Solution The quantity of vancomycin, Q, is increasing at a constant rate of 85 mg/hour and is decreasing at a rate of 0.1 times Q. The administration of 85 mg/hour makes a positive contribution to the rate of change dQ/dt. The excretion at a rate of $0.1Q$ makes a negative contribution to dQ/dt. Putting these together, we have

$$\text{Rate of change of a quantity} = \text{Rate in} - \text{Rate out},$$

so

$$\frac{dQ}{dt} = 85 - 0.1Q.$$

The Logistic Model

A population in a confined space grows proportionally to the product of the current population, P, and the difference between the *carrying capacity*, L, and the current population. (The carrying capacity is the maximum population the environment can sustain.) We use this information to write a differential equation for the population P.

The rate of change of P is proportional to the product of P and $L - P$, so

$$\frac{dP}{dt} = kP(L - P), \qquad \text{where } k \text{ is the constant of proportionality.}$$

This is called a *logistic differential equation.* What does it tell us about the graph of P? The derivative dP/dt is the product of k and P and $L - P$, so when P is small, the derivative dP/dt is small and the population grows slowly. As P increases, the derivative dP/dt increases and the population grows more rapidly. However, as P approaches the carrying capacity L, the term $L - P$ is small, and dP/dt is again small and the population grows more slowly. The *logistic growth curve* in Figure 10.2 satisfies these conditions.

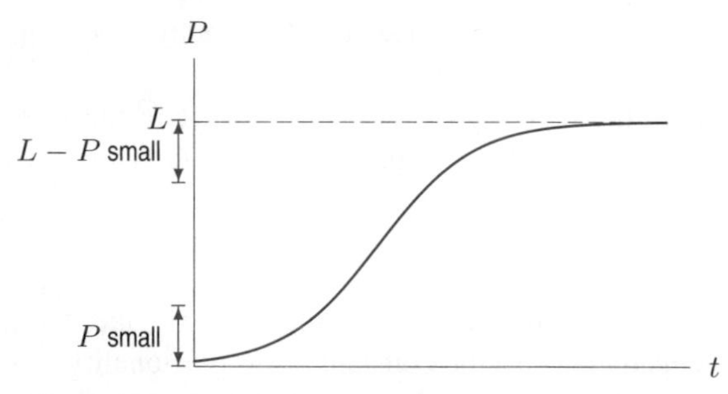

Figure 10.2: The logistic growth curve is a solution to $dP/dt = kP(L - P)$

Problems for Section 10.1

1. A population of insects grows at a rate proportional to the size of the population. Write a differential equation for the size of the population, P, as a function of time, t. Is the constant of proportionality positive or negative?

2. Match the graphs in Figure 10.3 with the following descriptions.
   - **(a)** The population of a new species introduced onto a tropical island
   - **(b)** The temperature of a metal ingot placed in a furnace and then removed
   - **(c)** The speed of a car traveling at uniform speed and then braking uniformly
   - **(d)** The mass of carbon-14 in a historical specimen
   - **(e)** The concentration of tree pollen in the air over the course of a year.

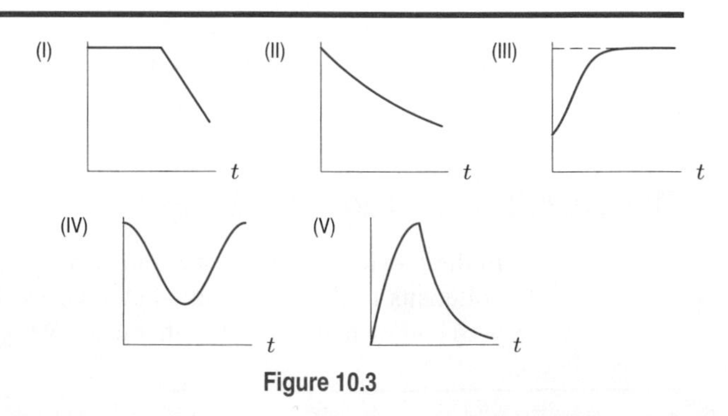

Figure 10.3

3. Money in a bank account earns interest at a continuous annual rate of 5% times the current balance. Write a differential equation for the balance, B, in the account as a function of time, t, in years.

4. Radioactive substances decay at a rate proportional to the quantity present. Write a differential equation for the quantity, Q, of a radioactive substance present at time t. Is the constant of proportionality positive or negative?

5. Match the graphs in Figure 10.4 with the following descriptions.

(a) The temperature of a glass of ice water left on the kitchen table.
(b) The amount of money in an interest-bearing bank account into which $50 is deposited.
(c) The speed of a constantly decelerating car.
(d) The temperature of a piece of steel heated in a furnace and left outside to cool.

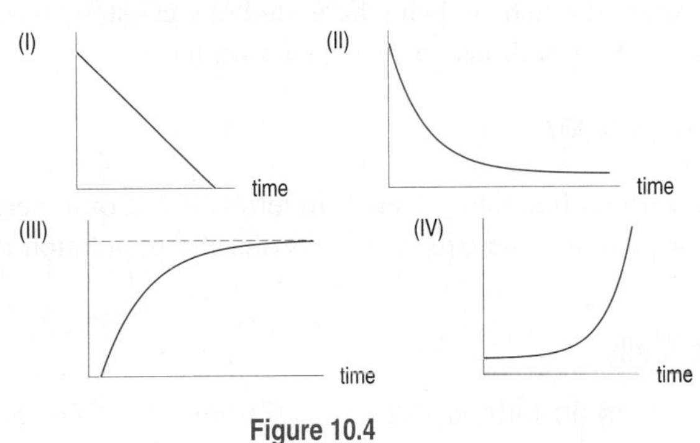

Figure 10.4

6. A cup of coffee contains about 100 mg of caffeine. Caffeine is metabolized and leaves the body at a continuous rate of about 17% every hour.

(a) Write a differential equation for the amount, A, of caffeine in the body as a function of the number of hours, t, since the coffee was consumed.
(b) Use the differential equation to find dA/dt at the start of the first hour (right after the coffee is consumed.) Use your answer to estimate the change in the amount of caffeine during the first hour.

7. Alcohol is metabolized and excreted from the body at a rate of about one ounce of alcohol every hour. If some alcohol is consumed, write a differential equation for the amount of alcohol, A (in ounces), remaining in the body as a function of t, the number of hours since the alcohol was consumed.

8. A bank account that initially contains $25,000 earns interest at a continuous rate of 4% per year. Withdrawals are made out of the account at a constant rate of $\$2,000$ per year. Write a differential equation for the balance, B, in the account as a function of the number of years, t.

9. Morphine is administered to a patient intravenously at a rate of 2.5 mg per hour. About 34.7% of the morphine is metabolized and leaves the body each hour. Write a differential equation for the amount of morphine, M, in milligrams, in the body as a function of time, t, in hours.

10. A pollutant spilled on the ground decays at a rate of 8% a day. In addition, clean-up crews remove the pollutant at a rate of 30 gallons a day. Write a differential equation for the amount of pollutant, P, in gallons, left after t days.

11. Toxins in pesticides can get into the food chain and accumulate in the body. A person consumes 10 micrograms a day of a toxin, ingested throughout the day. The toxin leaves the body at a continuous rate of 3% every day. Write a differential equation for the amount of toxin, A, in micrograms, in the person's body as a function of the number of days, t.

12. A person deposits money into an account at a continuous rate of $6000 a year, and the account earns interest at a continuous rate of 7% per year.

(a) Write a differential equation for the balance in the account, B, in dollars, as a function of years, t.
(b) Use the differential equation to calculate dB/dt if $B = 10{,}000$ and if $B = 100{,}000$. Interpret your answers.

13. The graphs in Figure 10.5 represent the temperature, $H(°\text{C})$, of four eggs as a function of time, t, in minutes. Match three of the graphs with the descriptions (a)–(c). Write a similar description for the fourth graph, including an interpretation of any intercepts and asymptotes.

(a) An egg is taken out of the refrigerator (just above 0°C) and put into boiling water.
(b) Twenty minutes after the egg in part (a) is taken out of the fridge and put into boiling water, the same thing is done with another egg.
(c) An egg is taken out of the refrigerator at the same time as the egg in part (a) and left to sit on the kitchen table.

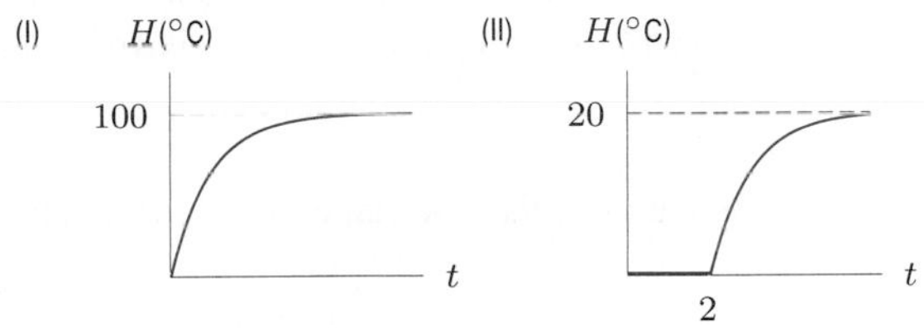

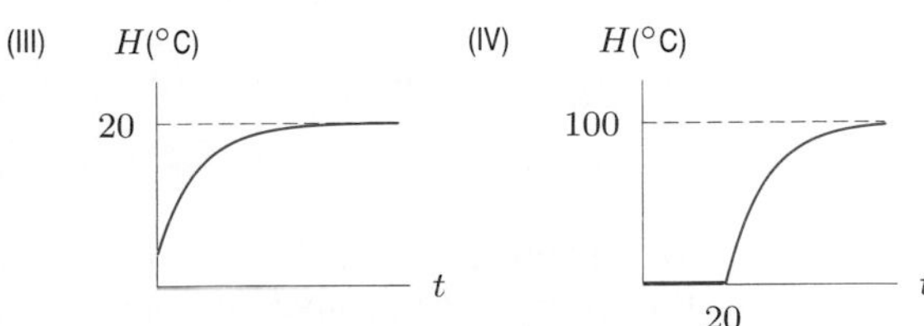

Figure 10.5

14. A quantity W satisfies the differential equation

$$\frac{dW}{dt} = 5W - 20.$$

(a) Is W increasing or decreasing at $W = 10$? $W = 2$?
(b) For what values of W is the rate of change of W equal to zero?

15. A quantity y satisfies the differential equation

$$\frac{dy}{dt} = -0.5y.$$

Under what conditions is y increasing? Decreasing?

10.2 SOLUTIONS OF DIFFERENTIAL EQUATIONS

What does it mean to "solve" a differential equation? A differential equation is an equation involving the derivative of an unknown function. The unknown is not a number but a function. A *solution* to a differential equation is any function that satisfies the differential equation.

In this section, we see how to solve a differential equation numerically and how to check whether or not a function is a solution to a differential equation. In the next section, we see how to visualize a solution.

Another Look at Marine Harvesting

Let's take another look at the fish population discussed in Section 10.1. Left alone, the population increases at a continuous rate of 20% per year. The fish are being harvested at a constant rate of 10 million fish per year. If P is the fish population, in millions, in year t, then we have

$$\frac{dP}{dt} = 0.20P - 10.$$

Solving this differential equation means to find a function giving P in terms of t. Combined with information about the initial population, we can use the equation to predict the population at any time in the future.

Solving the Differential Equation Numerically

Suppose at time $t = 0$, the fish population is 60 million. We can substitute $P = 60$ into the differential equation to compute the derivative, dP/dt:

$$\text{At time } t = 0, \qquad \frac{dP}{dt} = 0.20P - 10 = 0.20(60) - 10 = 12 - 10 = 2.$$

Since at $t = 0$, the fish population is changing at a rate of 2 million fish a year, at the end of the first year, the fish population will have increased by about 2 million fish. So:

$$\text{At } t = 1, \qquad \text{we estimate} \quad P = 60 + 2 = 62.$$

We use this new value of P to estimate dP/dt during the second year:

$$\text{At time } t = 1, \qquad \frac{dP}{dt} = 0.20P - 10 = 12.4 - 10 = 2.4.$$

During the second year, the fish population increased by about 2.4 million fish, so:

$$\text{At } t = 2, \qquad \text{we estimate} \quad P = 62 + 2.4 = 64.4.$$

We use this value of P to estimate the rate of change during the third year, and so on. Continuing in this fashion, we compute the approximate values of P in Table 10.1. This table gives approximate numerical values for P at future times.

Table 10.1 *Approximate values of the fish population, as a function of time*

t (years)	0	1	2	3	4	5	...
P (millions)	60	62	64.4	67.28	70.74	74.89	...

A Formula for the Solution to the Differential Equation

A function $P = f(t)$ which satisfies the differential equation

$$\frac{dP}{dt} = 0.20P - 10$$

is called a *solution* of the differential equation. Table 10.1 shows approximate numerical values of a solution. It is sometimes (but not always) possible to find a formula for the solution. In this particular case, there is a formula; it is

$$P = 50 + Ce^{0.20t},$$

where C is any constant. We check that this is a solution to the differential equation by substituting it into the left and right sides of the differential equation separately. We find

$$\begin{aligned}\text{Left side} &= \frac{dP}{dt} = 0.20Ce^{0.20t} \\ \text{Right side} = 0.20P - 10 &= 0.20(50 + Ce^{0.20t}) - 10 \\ &= 10 + 0.20Ce^{0.20t} - 10 \\ &= 0.20Ce^{0.20t}.\end{aligned}$$

Since we get the same expression on both sides, we say that $P = 50 + Ce^{0.20t}$ is a solution of this differential equation. Any choice of C works, so the solutions form a family of functions with parameter C. Several members of the family of solutions are graphed in Figure 10.6.

Finding the Arbitrary Constant: Initial Conditions

To find a value for the constant C—in other words, to select a single solution from the family of solutions—we need an additional piece of information, usually the initial population. In this case, we know that $P = 60$ when $t = 0$, so substituting into

$$P = 50 + Ce^{0.20t}$$

gives

$$\begin{aligned}60 &= 50 + Ce^{0.20(0)} \\ 60 &= 50 + C \cdot 1 \\ C &= 10.\end{aligned}$$

The function $P = 50 + 10e^{0.20t}$ satisfies the differential equation *and* the initial condition that $P = 60$ when $t = 0$.

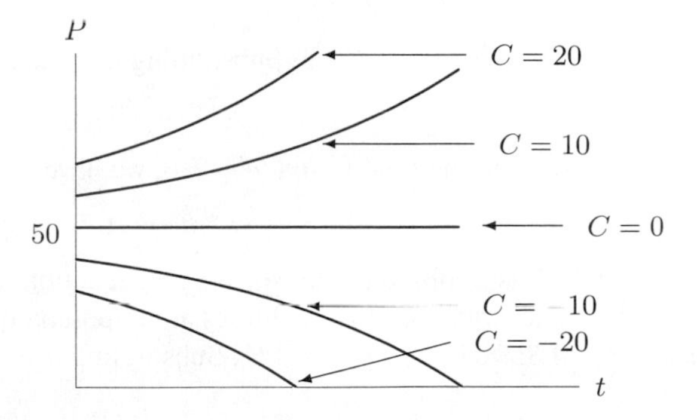

Figure 10.6: Solution curves for $dP/dt = 0.20P - 10$: Members of the family $P = 50 + Ce^{0.20t}$

General Solutions and Particular Solutions

For the differential equation $dP/dt = 0.20P - 10$, it can be shown that every solution is of the form $P = 50 + Ce^{0.20t}$ for some value of C. We say that the *general solution* of the differential equation $dP/dt = 0.20P - 10$ is the family of functions $P = 50 + Ce^{0.20t}$. The solution $P = 50 + 10e^{0.20t}$ that satisfies the differential equation together with the initial condition that $P = 60$ when $t = 0$ is called a *particular solution*. The differential equation and the initial condition together is called an *initial-value problem*.

Example 1 (a) Check that $P = Ce^{2t}$ is a solution to the differential equation

$$\frac{dP}{dt} = 2P.$$

(b) Find the particular solution satisfying the initial condition $P = 100$ when $t = 0$.

Solution (a) Since $P = Ce^{2t}$ where C is a constant, we find expressions for each side:

$$\text{Left side} = \frac{dP}{dt} = Ce^{2t} \cdot 2 = 2Ce^{2t}$$
$$\text{Right side} = 2P = 2Ce^{2t}.$$

Since the two expressions are equal, $P = Ce^{2t}$ is a solution to the differential equation.

(b) We substitute $P = 100$ and $t = 0$ into the general solution $P = Ce^{2t}$, and solve for C:

$$100 = Ce^{2(0)}$$
$$100 = C \cdot 1$$
$$100 = C.$$

The particular solution for this initial-value problem is $P = 100e^{2t}$.

Example 2 Decide whether or not $y = e^{-2x}$ is a solution of the differential equation $y' - 2y = 0$.

Solution Note that $y' = dy/dx$. Differentiating $y = e^{-2x}$ gives $y' = -2e^{-2x}$. Substituting, we have

$$y' - 2y = -2e^{-2x} - 2e^{-2x} = -4e^{-2x} \neq 0,$$

and so $y = e^{-2x}$ is not a solution to this differential equation.

Example 3 (a) What conditions must be imposed on the constants C and k if $y = Ce^{kt}$ is a solution to the differential equation

$$\frac{dy}{dt} = -0.5y?$$

(b) What additional conditions must be imposed on C and k if $y = Ce^{kt}$ also satisfies the initial condition that $y = 10$ when $t = 0$?

Solution (a) If $y = Ce^{kt}$, then $dy/dt = Cke^{kt}$. Substituting into the equation $dy/dt = -0.5y$ gives

$$Cke^{kt} = -0.5(Ce^{kt}),$$

and therefore, assuming $C \neq 0$, so $Ce^{kt} \neq 0$, we have

$$k = -0.5.$$

So $y = Ce^{-0.5t}$ is a solution to the differential equation. If $C = 0$, then $Ce^{kt} = 0$ is a solution to the differential equation. No conditions are imposed on C.

(b) Since $k = -0.5$, we have $y = Ce^{-0.5t}$. Substituting $y = 10$ when $t = 0$ gives

$$10 = Ce^0$$

so

$$10 = C.$$

So $y = 10e^{-0.5t}$ is a solution to the differential equation together with the initial condition.

Problems for Section 10.2

1. Check that $y = t^4$ is a solution to the differential equation $t\dfrac{dy}{dt} = 4y$.

2. Decide whether or not each of the following is a solution to the differential equation $xy' - 2y = 0$.

(a) $y = x^2$ **(b)** $y = x^3$

3. **(a)** Determine which of the following functions is a solution to the differential equation

$$x\frac{dy}{dx} = 3y.$$

(i) $y = Cx^2$ (ii) $y = Cx^3$
(iii) $y = x^3 + C$

(b) For any function which is a solution, find C if $y = 40$ when $x = 2$.

4. If the initial population of fish is 70 million, use the differential equation $dP/dt = 0.2P - 10$ to estimate the fish population after 1, 2, 3 years.

5. Fill in the missing values in Table 10.2 given that $dy/dt = 0.5y$. Assume the rate of growth, given by dy/dt, is approximately constant over each unit time interval.

Table 10.2

t	0	1	2	3	4
y	8				

6. Fill in the missing values in Table 10.3 given that $dy/dt = 0.5t$. Assume the rate of growth, given by dy/dt, is approximately constant over each unit time interval.

Table 10.3

t	0	1	2	3	4
y	8				

7. For a certain quantity y, assume that $dy/dt = -0.20y$. Fill in the values of y in Table 10.4. Assume that the rate of growth given by dy/dt is approximately constant over each unit time interval.

Table 10.4

t	0	1	2	3	4
y	125				

8. Fill in the missing values in Table 10.5 given that $dy/dt = 4 - y$. Assume the rate of growth, given by dy/dt, is approximately constant over each unit time interval.

Table 10.5

t	0	1	2	3	4
y	8				

9. Show that, for any constant P_0, the function $P = P_0e^t$ satisfies the differential equation

$$\frac{dP}{dt} = P.$$

10. Suppose $Q = Ce^{kt}$ satisfies the differential equation

$$\frac{dQ}{dt} = -0.03Q.$$

What (if anything) does this tell you about the values of C and k?

11. Find the values of k for which $y = x^2 + k$ is a solution to the differential equation $2y - xy' = 10$.

12. Is there a value of n which makes $y = x^n$ a solution to the equation $13x(dy/dx) = y$? If so, what value?

In Problems 13–21, use the fact that the derivative gives the slope of a curve to decide which of the graphs (A)–(F) in Figure 10.7 represents a possible solution to the differential equation.

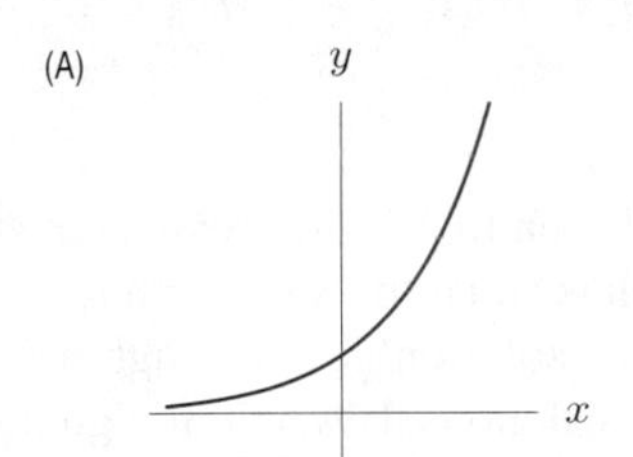

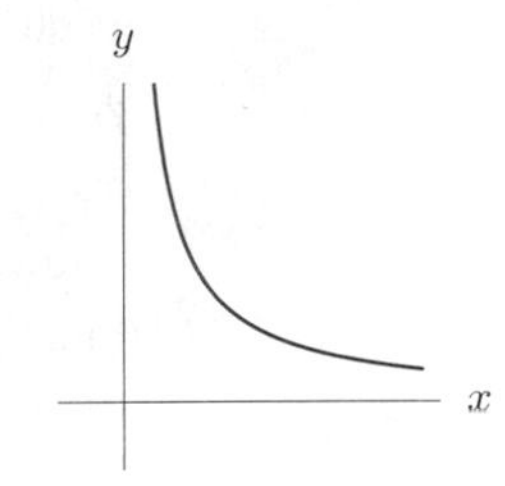

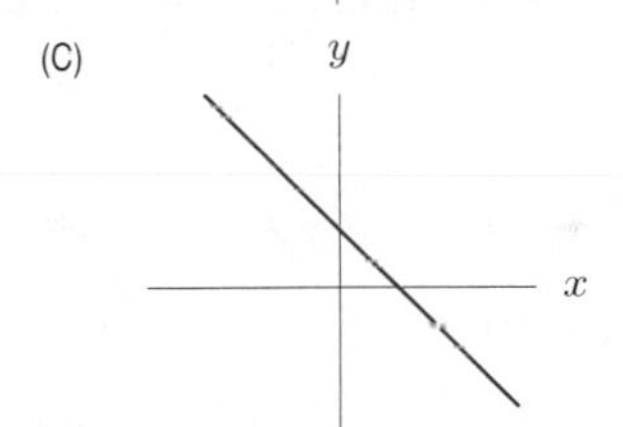

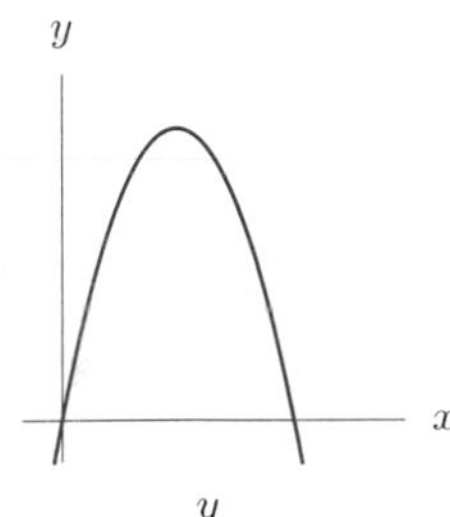

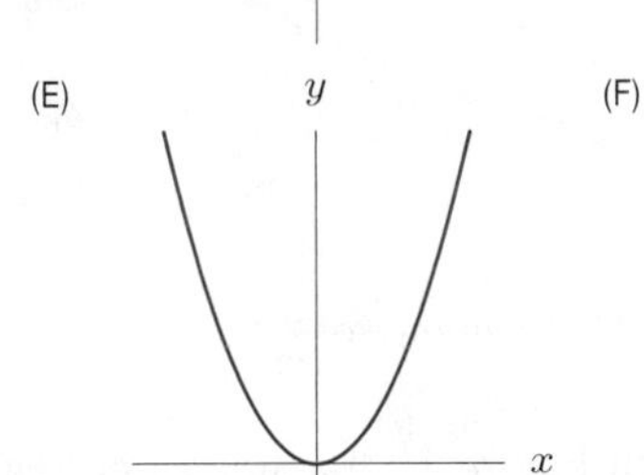

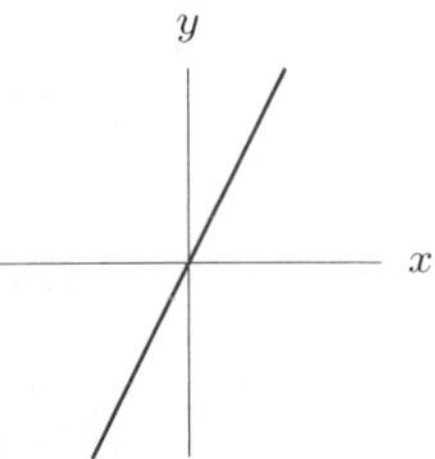

Figure 10.7

13. $\frac{dy}{dx} = -1$ **14.** $\frac{dy}{dx} = 0.1$ **15.** $\frac{dy}{dx} = -y^2$

16. $\frac{dy}{dx} = 2x$ **17.** $\frac{dy}{dx} = 2$ **18.** $\frac{dy}{dx} = y$

19. $\frac{dy}{dx} = -\frac{1}{x^2}$ **20.** $\frac{dy}{dx} = 1 - x$ **21.** $\frac{dy}{dx} = 2y$

22. Match solutions and differential equations. (Note: Each equation may have more than one solution, or no solutions.)

(a) $\frac{dy}{dx} = \frac{y}{x}$ (I) $y = x^3$

(b) $\frac{dy}{dx} = 3\frac{y}{x}$ (II) $y = 3x$

(c) $\frac{dy}{dx} = 3x$ (III) $y = e^{3x}$

(d) $\frac{dy}{dx} = y$ (IV) $y = 3e^x$

(e) $\frac{dy}{dx} = 3y$ (V) $y = x$

23. Pick out which functions are solutions to which differential equations. (Note: Functions may be solutions to more than one equation or to none; an equation may have more than one solution.)

(a) $\frac{dy}{dx} = -2y$ (I) $y = 2\sin x$

(b) $\frac{dy}{dx} = 2y$ (II) $y = \sin 2x$

(c) $\frac{d^2y}{dx^2} = 4y$ (III) $y = e^{2x}$

(d) $\frac{d^2y}{dx^2} = -4y$ (IV) $y = e^{-2x}$

10.3 SLOPE FIELDS

In this section, we see how to visualize a differential equation and its solutions. Let's start with the equation

$$\frac{dy}{dx} = y.$$

Any solution to this differential equation has the property that at any point in the plane, the slope of its graph is equal to its y coordinate. (That's what the equation $dy/dx = y$ is telling us!) This means that if the solution goes through the point $(0, 1)$, its slope there is 1; if it goes through a point with $y = 4$ its slope is 4. A solution going through $(0, 2)$ has slope 2 there; at the point where $y = 8$ the slope of this solution is 8. (See Figure 10.8.)

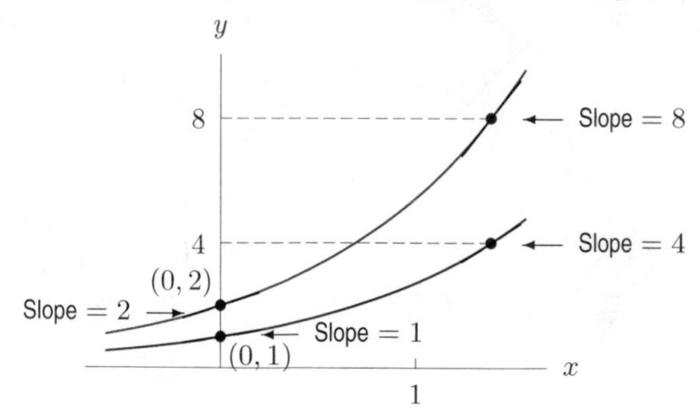

Figure 10.8: Solutions to $\frac{dy}{dx} = y$

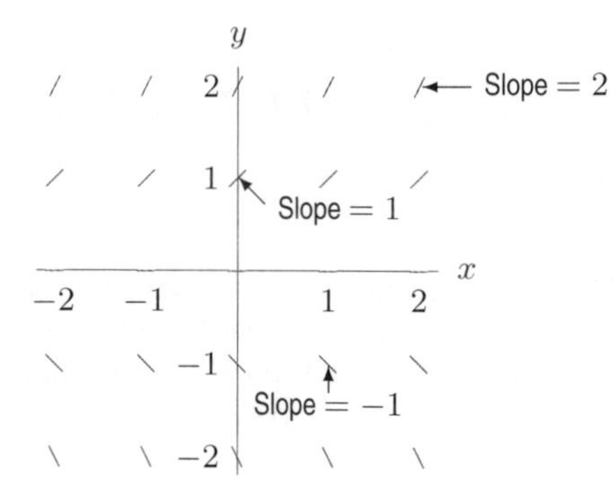

Figure 10.9: Visualizing the slope of y, if $\frac{dy}{dx} = y$

In Figure 10.9 a small line segment is drawn at the marked points showing the slope of the solution curve there. Since $dy/dx = y$, the slope at the point $(1, 2)$ is 2 (the y-coordinate), and so we draw a line segment there with slope 2. We draw a line segment at the point $(0, -1)$ with slope -1, and so on. If we draw many of these line segments, we have the *slope field* for the equation $dy/dx = y$ shown in Figure 10.10. Above the x-axis, the slopes are positive (because y is positive there), and the slopes increase as we move upward (as y increases). Below the x-axis, the slopes are negative, and get more so as we move downward. Notice that on any horizontal line (where y is constant) the slopes are constant. In the slope field you can see the ghost of the solution curve lurking. Start anywhere on the plane and move so that the slope lines are tangent to your path; you will trace out one of the solution curves. Try penciling in some solution curves on Figure 10.10, some above the x-axis and some below. The curves you draw should have the shape of exponential functions. By substituting $y = Ce^x$ into the differential equation, you can check that each curve in the family of exponentials, $y = Ce^x$, is a solution to this differential equation.

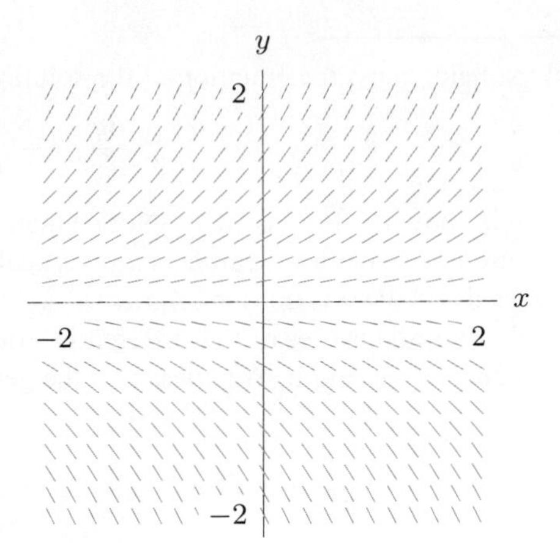

Figure 10.10: Slope field for $\frac{dy}{dx} = y$

In most problems, we are interested in getting the solution curves from the slope field. Think of the slope field as a set of signposts pointing in the direction you should go at each point. Imagine starting anywhere in the plane: look at the slope field at that point and start to move in that direction. After a small step, look at the slope field again, and alter your direction if necessary. Continue to move across the plane in the direction the slope field points, and you'll trace out a solution curve. Notice that the solution curve is not necessarily the graph of a function, and even if it is, we may not have a formula for the function. Geometrically, solving a differential equation means finding the family of solution curves.

Example 1 Figure 10.11 shows the slope field of the differential equation $\dfrac{dy}{dx} = 2x$.

(a) What do you notice about the slope field?
(b) Compare the solution curves in Figure 10.12 with the formula $y = x^2 + C$ for the solutions to this differential equation.

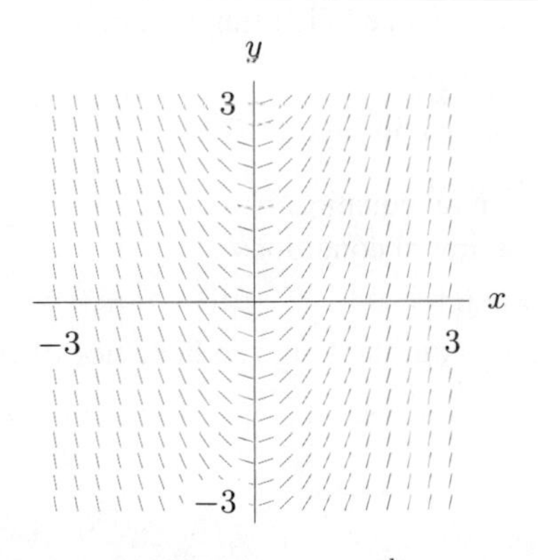

Figure 10.11: Slope field for $\frac{dy}{dx} = 2x$

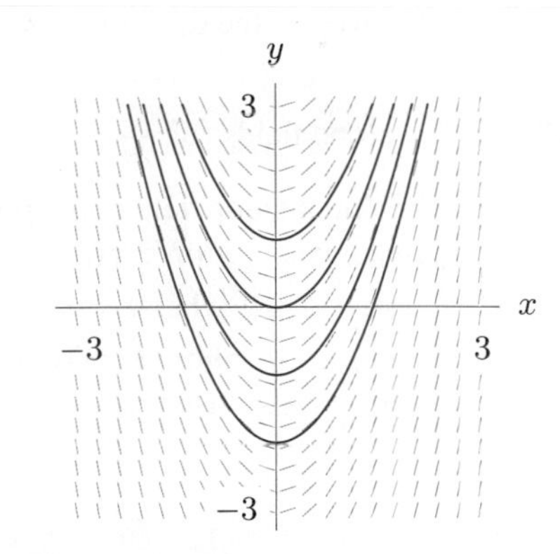

Figure 10.12: Some solutions to $\frac{dy}{dx} = 2x$

Solution

(a) In Figure 10.11, notice that on any vertical line (where x is constant) the slopes are all the same. This is because in this differential equation dy/dx depends on x only. (In the previous example, $dy/dx = y$, the slopes depended on y only.)
(b) The solution curves in Figure 10.12 look like parabolas. It is easy to check by substitution that

$$y = x^2 + C \qquad \text{is a solution to} \qquad \frac{dy}{dx} = 2x,$$

so the parabolas $y = x^2 + C$ are solution curves.

Example 2 Using the slope field, guess the equation of the solution curves of the differential equation

$$\frac{dy}{dx} = -\frac{x}{y}.$$

Solution The slope field is shown in Figure 10.13. Notice that on the y-axis, where x is 0, the slope is 0. On the x-axis, where y is 0, the line segments are vertical and the slope is undefined. At the origin the slope is undefined and there is no line segment.

What do the solution curves of this differential equation look like? The slope field suggests they are circles centered at the origin. We guess that the general solution to this differential equation is

$$x^2 + y^2 = r^2.$$

This solution is derived in the Focus on Theory section on page 441.

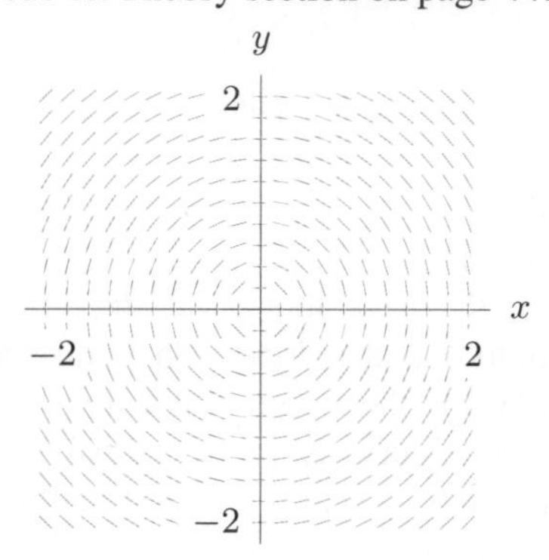

Figure 10.13: Slope field for $\frac{dy}{dx} = -\frac{x}{y}$

The previous example shows that the solutions to differential equations may sometimes be expressed as *implicit functions*. Implicit functions are ones which have not been "solved" for y; in other words, the dependent variable is not expressed as an explicit function of x.

Example 3 The slope fields for $\frac{dy}{dt} = 2 - y$ and $\frac{dy}{dt} = \frac{t}{y}$ are shown in Figure 10.14.

(a) Which slope field corresponds to which differential equation?

(b) Sketch solution curves on each slope field with initial conditions

(i) $y = 1$ when $t = 0$ (ii) $y = 3$ when $t = 0$ (iii) $y = 0$ when $t = 1$

(c) For each solution curve, can you say anything about the long-run behavior of y? In particular, as $t \to \infty$, what happens to the value of y?

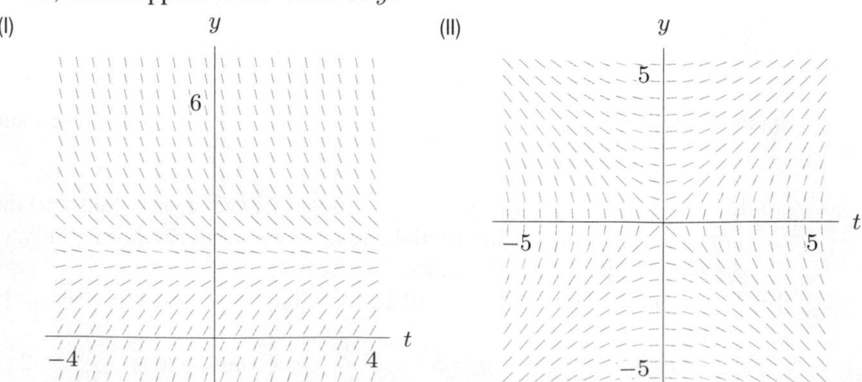

Figure 10.14: Slope fields for $\frac{dy}{dt} = 2 - y$ and $\frac{dy}{dt} = \frac{t}{y}$: Which is which?

Solution

(a) Consider the slopes at different points for the two differential equations. In particular, look at the line $y = 2$ in Figure 10.14. The equation $dy/dt = 2 - y$ has slope 0 all along this line, whereas the line $dy/dt = t/2$ has slope $t/2$. Since slope field (I) looks horizontal at $y = 2$, slope field (I) corresponds to $dy/dt = 2 - y$ and slope field (II) corresponds to $dy/dt = t/y$.

(b) The initial conditions (i) and (ii) give the value of y when t is 0; that is, the y-intercept. To draw the solution curve satisfying the condition (i), draw the solution curve with y-intercept 1. For (ii), draw the solution curve with y-intercept 3. For (iii), the solution goes through the point $(1, 0)$, so draw the solution curve passing through this point. See Figures 10.15 and 10.16.

(c) For $dy/dt = 2 - y$, all solution curves have $y = 2$ as a horizontal asymptote, so $y \to 2$ as $t \to \infty$. For $dy/dt = t/y$ with initial conditions $(0, 1)$ and $(0, 3)$, we see that $y \to \infty$ as $t \to \infty$. The graph has asymptotes which appear to be diagonal lines. In fact, they are $y = t$ and $y = -t$, so $y \to \pm\infty$ as $t \to \infty$.

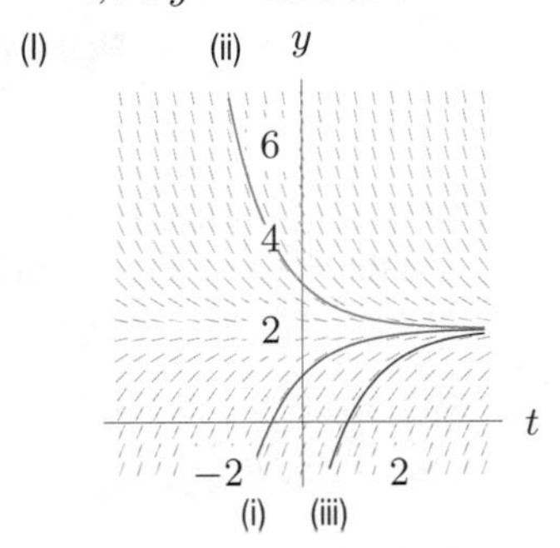

Figure 10.15: Solution curves for $\dfrac{dy}{dt} = 2 - y$

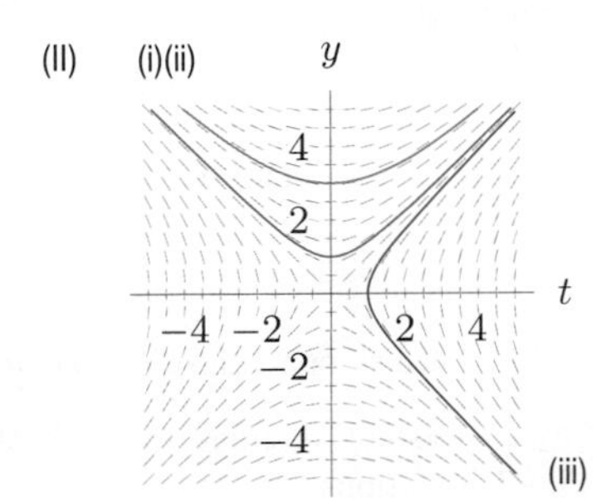

Figure 10.16: Solution curves for $\dfrac{dy}{dt} = \dfrac{t}{y}$

Existence and Uniqueness of Solutions

Since differential equations are used to model many real situations, the question of whether a solution exists and is unique can have great practical importance. If we know how the velocity of a satellite is changing, can we know its velocity for all future time? If we know the initial population of a city, and we know how the population is changing, can we predict the population in the future? Common sense says yes: if we know the initial value of some quantity and we know exactly how it is changing, we should be able to figure out the future value of the quantity.

In the language of differential equations, an initial-value problem (that is, a differential equation and an initial condition) representing a real situation almost always has a unique solution. One way to see this is by looking at the slope field. Imagine starting at the point representing the initial condition. Through that point there will usually be a line segment pointing in the direction the solution curve must go. By following the line segments in the slope field, we trace out the solution curve. Several examples with different starting points are shown in Figure 10.17. In general, at each point there is one line segment and therefore only one direction for the solution curve to go. Thus the solution curve *exists* and is *unique* provided we are given an initial point.

It can be shown that if the slope field is continuous as we move from point to point in the plane, we can be sure that the solution curve exists around every point. Ensuring that each point has only one solution curve through it requires a slightly stronger condition.

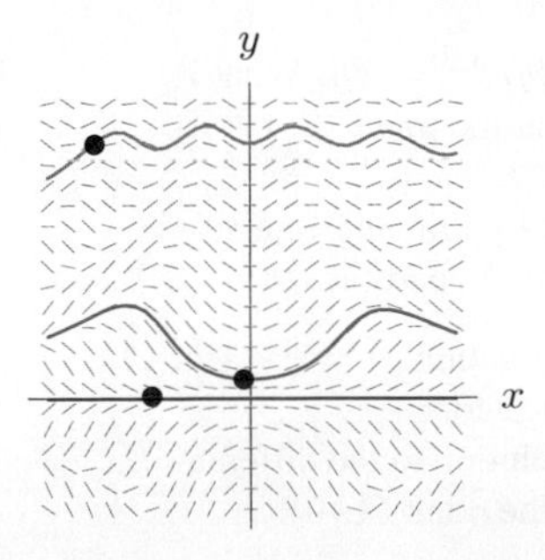

Figure 10.17: There is one and only one solution curve through each point in the plane for this slope field

Problems for Section 10.3

1. Figure 10.18 displays sketches of two slope fields. Sketch three solution curves for each of these fields.

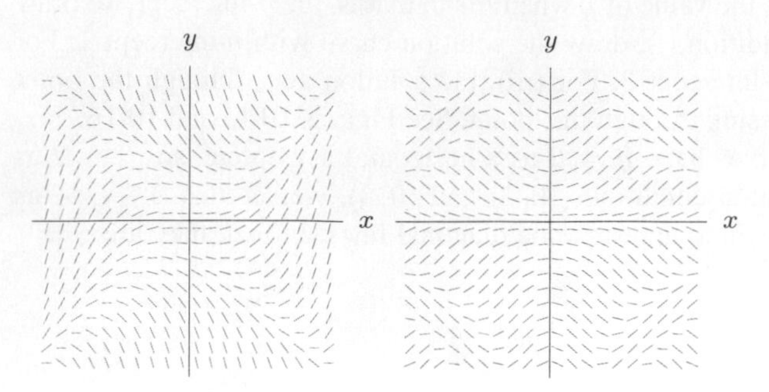

Figure 10.18

2. Figure 10.19 is a slope field for $dy/dx = y - 10$.
   (a) Draw the solution curve for each of the following initial conditions:
   (i) $y = 8$ when $x = 0$ (ii) $y = 12$ when $x = 0$
   (iii) $y = 10$ when $x = 0$
   (b) Since $dy/dx = y - 10$, when $y = 10$, we have $dy/dx = 10 - 10 = 0$. Explain why this matches your answer to part (iii).

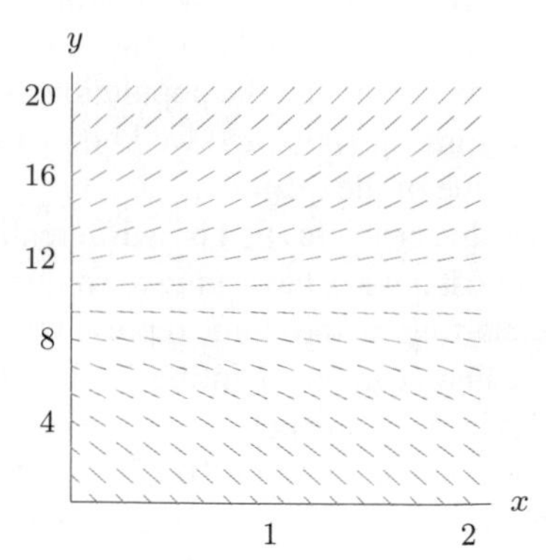

Figure 10.19: Slope field for $dy/dx = y - 10$

3. **(a)** Consider the slope field for $dy/dx = xy$. What is the slope of the line segment at the point $(2, 1)$? At $(0, 2)$? At $(-1, 1)$? At $(2, -2)$?
   (b) Sketch part of the slope field by drawing line segments with the slopes calculated in part (a).

4. **(a)** Sketch the slope field for the equation $y' = x - y$ in Figure 10.20 at the points indicated.
   (b) Check that $y = x - 1$ is the solution to the differential equation passing through the point $(1, 0)$.

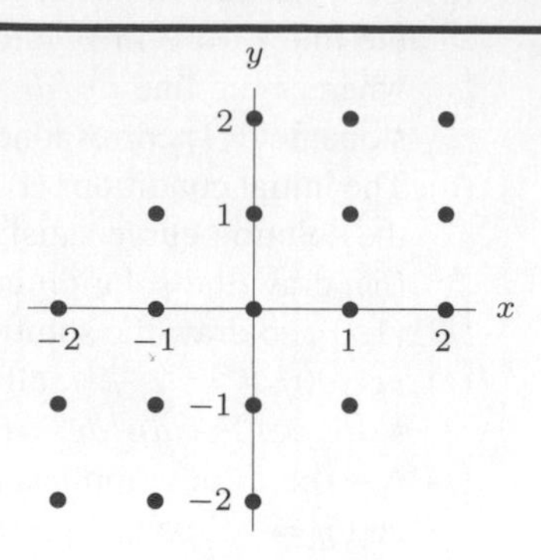

Figure 10.20

5. Which one of the following differential equations best fits the slope field shown in Figure 10.21? Explain.

   I. $y' = 1 + y$ II. $y' = 2 - y$
   III. $y' = (1 + y)(2 - y)$ IV. $y' = 1 + x$
   V. $y' = xy$

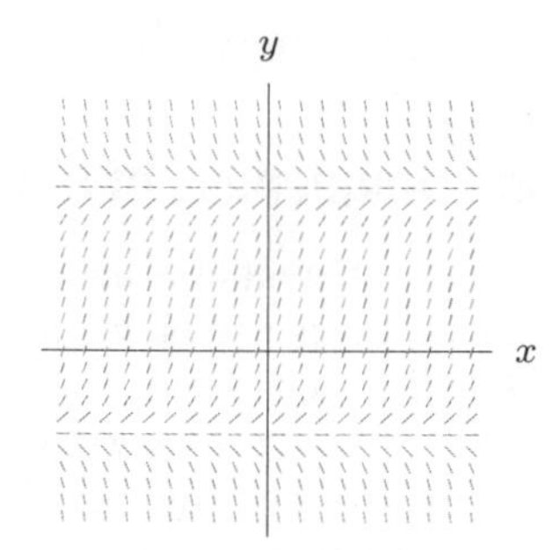

Figure 10.21

6. Which one of the following differential equations best fits the slope field shown in Figure 10.22? Explain.

   I. $dP/dt = P - 1$ II. $dP/dt = P(P - 1)$
   III. $dP/dt = 3P(1 - P)$ IV. $dP/dt = 1/3P(1-P)$

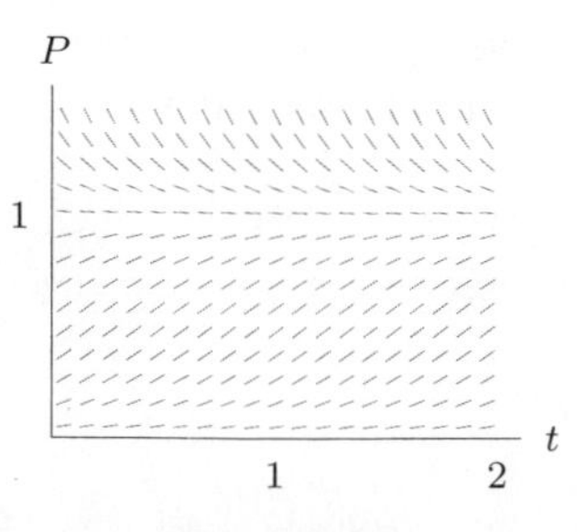

Figure 10.22

7. Match each of the slope field segments in (I)–(VI) with one or more of the differential equations in (a)–(f).

(a) $y' = e^{-x^2}$ **(b)** $y' = \cos y$ **(c)** $y' = \cos(4-y)$
(d) $y' = y(4-y)$ **(e)** $y' = y(3-y)$ **(f)** $y' = x(3-x)$

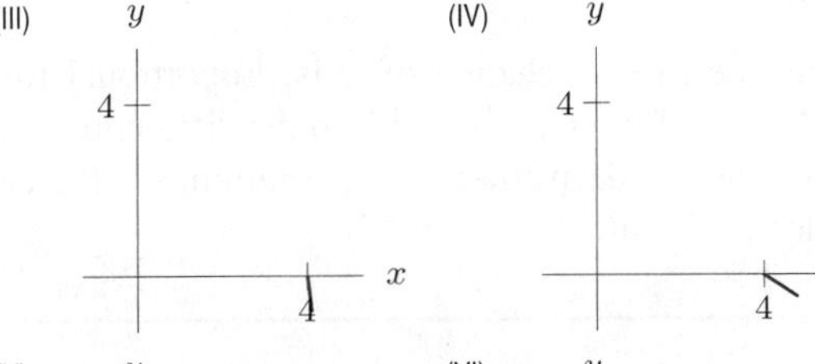

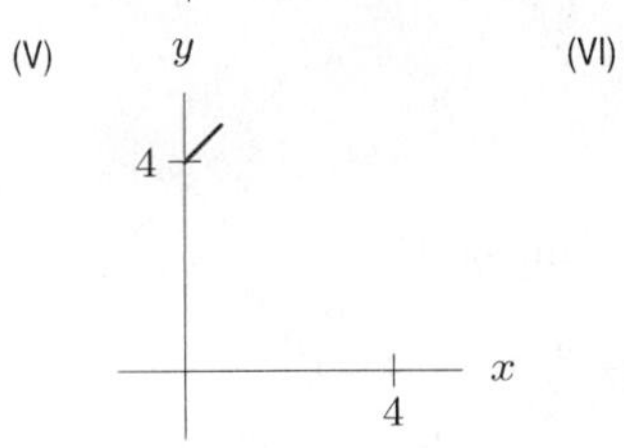

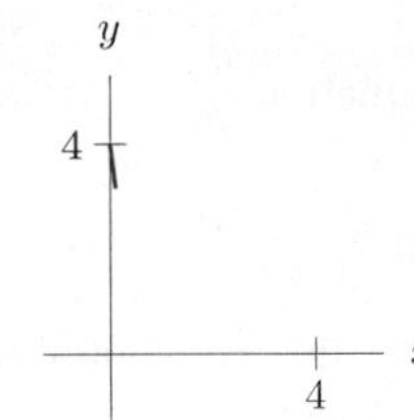

8. Match the slope fields in Figure 10.23 with their differential equations:

(a) $y' = -y$ **(b)** $y' = y$ **(c)** $y' = x$
(d) $y' = 1/y$ **(e)** $y' = y^2$

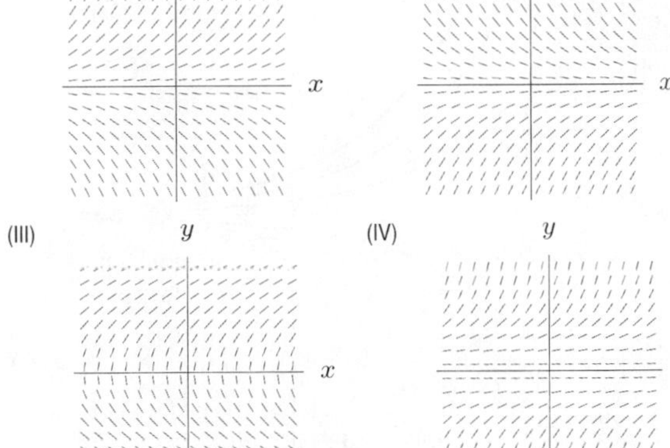

Figure 10.23

9. Match the slope fields in Figure 10.24 with their differential equations:

(a) $y' = 1 + y^2$ **(b)** $y' = x$ **(c)** $y' = \sin x$
(d) $y' = y$ **(e)** $y' = x - y$ **(f)** $y' = 4 - y$

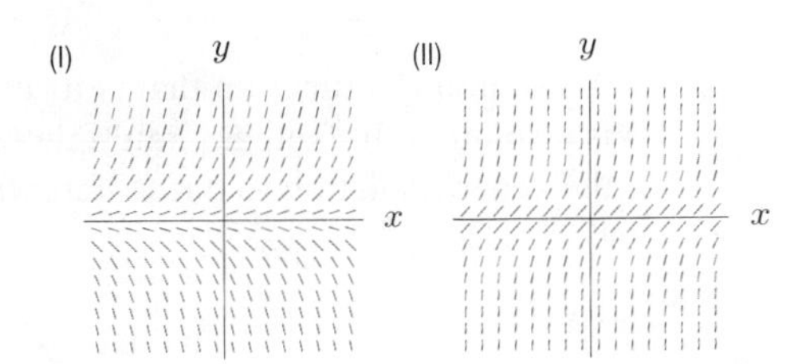

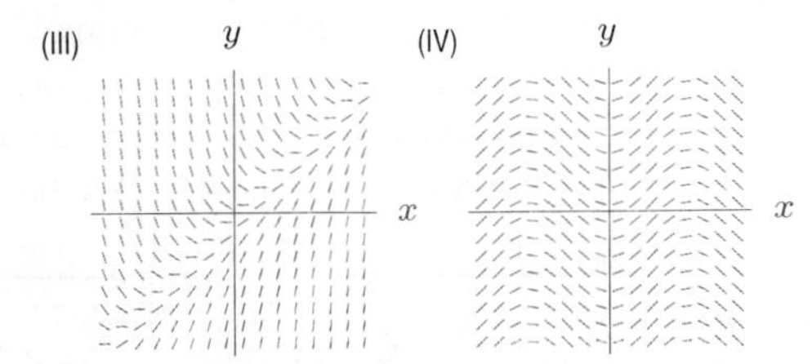

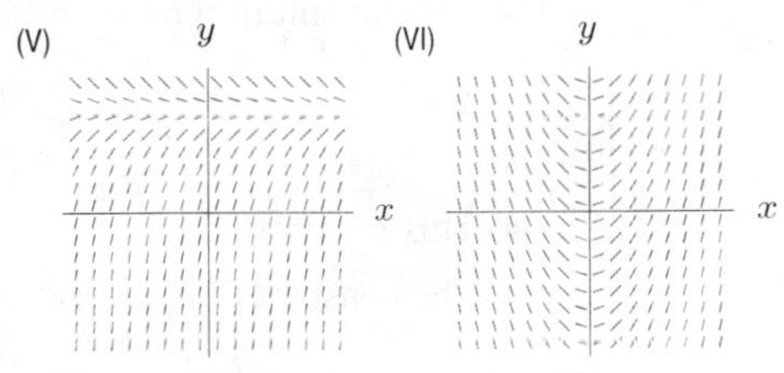

Figure 10.24: Each slope field is graphed for $-5 \le x \le 5$, $-5 \le y \le 5$

For Problems 10–15, consider a solution curve for each of the slope fields in Problem 9. Write one or two sentences describing qualitatively the long-run behavior of y. For example, as x increases, does $y \to \infty$, or does y remain finite? You may get different limiting behavior for different starting points. In each case, your answer should discuss how the limiting behavior depends on the starting point.

10. Slope field (I)

11. Slope field (II)

12. Slope field (III)

13. Slope field (IV)

14. Slope field (V)

15. Slope field (VI)

10.4 EXPONENTIAL GROWTH AND DECAY

What is a solution to the differential equation

$$\frac{dy}{dt} = y?$$

A solution is a function that is its own derivative. The function $y = e^t$ has this property, so $y = e^t$ is a solution. In fact, any multiple of e^t also has this property. The family of functions $y = Ce^t$ is the general solution to this differential equation. If k is a constant, the differential equation

$$\frac{dy}{dt} = ky,$$

is similar. This differential equation says that the rate of change of y is proportional to y. The constant k is the constant of proportionality. By substituting $y = Ce^{kt}$ into the differential equation, you can check that $y = Ce^{kt}$ is a solution. For another derivation of the solution, see the Focus on Theory section on page 441. We have the following result:

> The general solution to the differential equation $\dfrac{dy}{dt} = ky$ is
>
> $$y = Ce^{kt} \qquad \text{for any constant C.}$$
>
> - This is exponential growth for $k > 0$, and exponential decay for $k < 0$.
> - The constant C is the value of y when t is 0.

Graphs of solution curves for some $k > 0$ are in Figure 10.25. For $k < 0$, the graphs are reflected about the y-axis. See Figure 10.26.

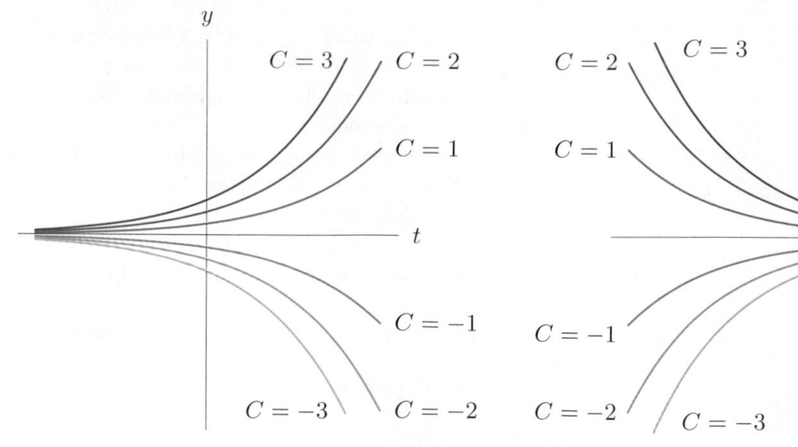

Figure 10.25: Graphs of $y = Ce^{kt}$, which are solutions to $\dfrac{dy}{dt} = ky$ for some fixed $k > 0$

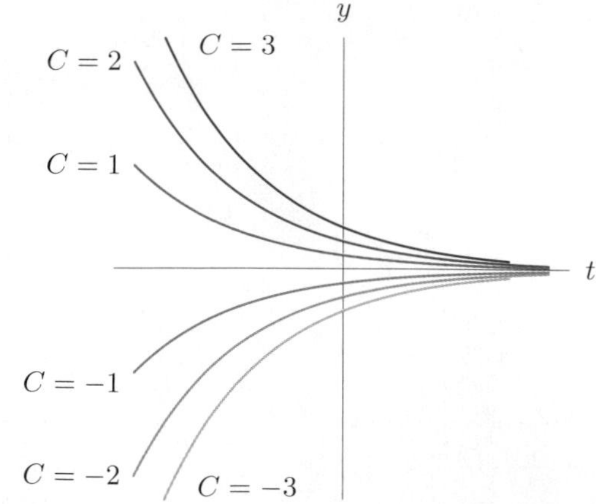

Figure 10.26: Graphs of $y = Ce^{kt}$, which are solutions to $\dfrac{dy}{dt} = ky$ for some fixed $k < 0$

Example 1

(a) Find the general solution to each of the following differential equations:

(i) $\dfrac{dy}{dt} = 0.05y$ (ii) $\dfrac{dP}{dt} = -0.3P$ (iii) $\dfrac{dw}{dz} = 2z$ (iv) $\dfrac{dw}{dz} = 2w$

(b) For differential equation (i), find the particular solution satisfying $y = 50$ when $t = 0$.

Solution (a) The differential equations given in (i), (ii), and (iv) are all examples of exponential growth or decay, since each is in the form

$$\text{Derivative} = \text{Constant} \cdot \text{Dependent variable}.$$

Notice that differential equation (iii) is not in this form. Example 1 on page 407 showed that the solution to (iii) is $w = z^2 + C$. The general solutions are

(i) $y = Ce^{0.05t}$ (ii) $P = Ce^{-0.3t}$ (iii) $w = z^2 + C$ (iv) $w = Ce^{2z}$

(b) The general solution to (i) is $y = Ce^{0.05t}$. Substituting $y = 50$ and $t = 0$ gives

$$50 = Ce^{0.05(0)}$$
$$50 = C \cdot 1.$$

So $C = 50$, and the particular solution to this initial value problem is $y = 50e^{0.05t}$.

Population Growth

Consider the population P of a region where there is no immigration or emigration. The rate at which the population is growing is often proportional to the size of the population. This means larger populations grow faster, as we expect since there are more people to have babies. If the population has a continuous growth rate of 2% per unit time, then we know

$$\text{Rate of growth of population} = 2\% \text{ of Current population},$$

so

$$\frac{dP}{dt} = 0.02P.$$

This equation is of the form $dP/dt = kP$ for $k = 0.02$ and has the general solution $P = Ce^{0.02t}$. If the initial population at time $t = 0$ is P_0, then $P_0 = Ce^{0.02(0)} = C$. So $C = P_0$ and we have

$$P = P_0e^{0.02t}.$$

Continuously Compounded Interest

In Chapter 1 we introduced continuous compounding as the limiting case in which interest was added more and more often. Here we approach continuous compounding from a different point of view. We imagine interest being accrued at a rate proportional to the balance at that moment. Thus, the larger the balance, the faster interest is earned and the faster the balance grows.

Example 2 A bank account earns interest continuously at a rate of 5% of the current balance per year. Assume that the initial deposit is \$1000 and that no other deposits or withdrawals are made.

(a) Write a differential equation satisfied by the balance in the account.
(b) Solve the differential equation and graph the solution.

Solution (a) We are looking for B, the balance in the account in dollars, as a function of t, time in years. Interest is being added continuously to the account at a rate of 5% of the balance at that moment, so

$$\text{Rate at which balance is increasing} = 5\% \text{ of Current balance}.$$

Thus, a differential equation that describes the process is

$$\frac{dB}{dt} = 0.05B.$$

Notice that it does not involve the \$1000, the initial condition, because the initial deposit does not affect the process by which interest is earned.

(b) Since $B_0 = 1000$ is the initial value of B, the solution to this differential equation is

$$B = B_0 e^{0.05t} = 1000e^{0.05t}.$$

This function is graphed in Figure 10.27.

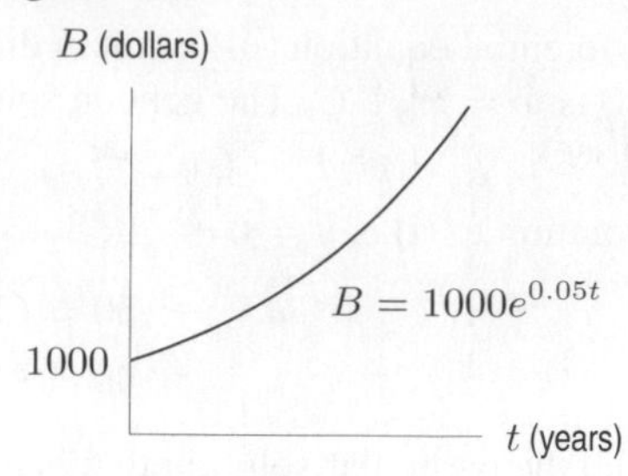

Figure 10.27: Bank balance against time

You may wonder how we can represent an amount of money by a differential equation, since money can only take on discrete values (you can't have fractions of a cent). In fact, the differential equation is only an approximation, but for large amounts of money, it is a pretty good approximation.

Pollution in the Great Lakes

In the 1960s pollution in the Great Lakes became an issue of public concern. We will set up a model for how long it would take the lakes to flush themselves clean, assuming no further pollutants were being dumped in the lake.

Let Q be the total quantity of pollutant in a lake of volume V at time t. Suppose that clean water is flowing into the lake at a constant rate r and that water flows out at the same rate. Assume that the pollutant is evenly spread throughout the lake, and that the clean water coming into the lake immediately mixes with the rest of the water.

How does Q vary with time? First, notice that since pollutants are being taken out of the lake but not added, Q decreases, and the water leaving the lake becomes less polluted, so the rate at which the pollutants leave decreases. This tells us that Q is decreasing and concave up. In addition, the pollutants will never be completely removed from the lake though the quantity remaining will become arbitrarily small. In other words, Q is asymptotic to the t-axis. (See Figure 10.28.)

Setting Up a Differential Equation for the Pollution

To understand how Q changes with time, we write a differential equation for Q. We know that

$$\begin{matrix}\text{Rate } Q \\ \text{changes}\end{matrix} = -\left(\begin{matrix}\text{Rate pollutants} \\ \text{leave in outflow}\end{matrix}\right)$$

where the negative sign represents the fact that Q is decreasing. At time t, the concentration of pollutants is Q/V and water containing this concentration is leaving at rate r. Thus,

$$\begin{matrix}\text{Rate pollutants} \\ \text{leave in outflow}\end{matrix} = \begin{matrix}\text{Rate of} \\ \text{outflow}\end{matrix} \times \text{Concentration} = r \cdot \frac{Q}{V}.$$

So the differential equation is

$$\frac{dQ}{dt} = -\frac{r}{V}Q$$

and its general solution is

$$Q = Q_0 e^{-rt/V}.$$

Table 10.6 contains values of r and V for four of the Great Lakes.[1] We use this data to calculate how long it would take for certain fractions of the pollution to be removed from Lake Erie.

[1]Data from William E. Boyce and Richard C. DiPrima, *Elementary Differential Equations* (New York: Wiley, 1977).

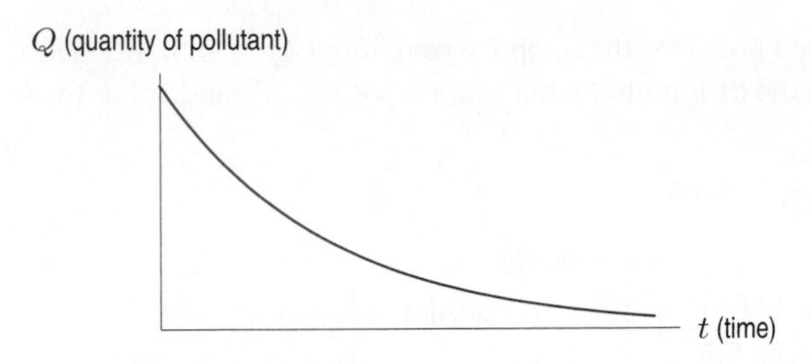

Figure 10.28: Pollutant in lake versus time

Table 10.6 *Volume and outflow in Great Lakes*

	V (km^3)	r (km^3/year)
Superior	12,200	65.2
Michigan	4900	158
Erie	460	175
Ontario	1600	209

Example 3 How long will it take for 90% of the pollution to be removed from Lake Erie? For 99% to be removed?

Solution For Lake Erie, $r/V = 175/460 = 0.38$, so at time t we have

$$Q = Q_0 e^{-0.38t}.$$

When 90% of the pollution has been removed, 10% remains, so $Q = 0.1Q_0$. Substituting gives

$$0.1Q_0 = Q_0 e^{-0.38t}.$$

Canceling Q_0 and solving for t gives

$$t = \frac{-\ln(0.1)}{0.38} \approx 6 \text{ years.}$$

Similarly, when 99% of the pollution has been removed, $Q = 0.01Q_0$, so we solve

$$0.01Q_0 = Q_0 e^{-0.38t},$$

giving

$$t = \frac{-\ln(0.01)}{0.38} \approx 12 \text{ years.}$$

The Quantity of a Drug in the Body

As we saw in Section 10.1, the rate at which a drug leaves a patient's body is proportional to the quantity of the drug left in the body. If we let Q represent the quantity of drug left, then

$$\frac{dQ}{dt} = -kQ.$$

The negative sign indicates the quantity of drug in the body is decreasing. The solution to this differential equation is $Q = Q_0 e^{-kt}$; the quantity decreases exponentially. The constant k depends on the drug and Q_0 is the amount of drug in the body at time zero. Sometimes physicians convey information about the relative decay rate with a *half life*, which is the time it takes for Q to decrease by a factor of $1/2$.

Example 4 Valproic acid is a drug used to control epilepsy; its half-life in the human body is about 15 hours.

(a) Use the half-life to find the constant k in the differential equation $dQ/dt = -kQ$, where Q represents the quantity of drug in the body t hours after the drug is administered.

(b) At what time will 10% of the original dose remain?

Solution (a) Since the half-life is 15 hours, we know that the quantity remaining $Q = 0.5Q_0$ when $t = 15$. We substitute into the solution to the differential equation, $Q = Q_0e^{-kt}$, and solve for k:

$$\begin{aligned}
Q &= Q_0e^{-kt} \\
0.5Q_0 &= Q_0e^{-k(15)} \\
0.5 &= e^{-15k} && \text{(Divide through by } Q_0) \\
\ln 0.5 &= -15k && \text{(Take the natural logarithm of both sides)} \\
k &= \frac{-\ln 0.5}{15} = 0.0462. && \text{(Solve for } k)
\end{aligned}$$

(b) To find the time when 10% of the original dose remains in the body, we substitute $0.10Q_0$ for the quantity remaining, Q, and solve for the time, t.

$$\begin{aligned}
0.10Q_0 &= Q_0e^{-0.0462t} \\
0.10 &= e^{-0.0462t} \\
\ln 0.10 &= -0.0462t \\
t &= \frac{\ln 0.10}{-0.0462} = 49.84.
\end{aligned}$$

There will be 10% of the drug still in the body at $t = 49.84$, or after about 50 hours.

Problems for Section 10.4

Find solutions to the differential equations in Problems 1–6, subject to the given initial condition.

1. $\dfrac{dy}{dx} = -0.14y, \quad y = 5.6 \text{ when } x = 0$
2. $\dfrac{dw}{dr} = 3w, \quad w = 30 \text{ when } r = 0$
3. $\dfrac{dP}{dt} = 0.02P, \quad P(0) = 20$
4. $\dfrac{dQ}{dt} = \dfrac{Q}{5}, \quad Q = 50 \text{ when } t = 0$
5. $\dfrac{dy}{dx} + \dfrac{y}{3} = 0, \quad y(0) = 10$
6. $\dfrac{dp}{dq} = -0.1p, \quad p = 100 \text{ when } q = 5$
7. A deposit of \$5000 is made to a bank account paying 1.5% annual interest, compounded continuously.
   - **(a)** Write a differential equation for the balance in the account, B, as a function of time, t, in years.
   - **(b)** Solve the differential equation.
   - **(c)** How much money is in the account in 10 years?
8. A deposit is made to a bank account paying an annual interest rate of 7% compounded continuously. No other deposits or withdrawals are made to the account.
   - **(a)** Write a differential equation satisfied by B, the balance in the account after t years.
   - **(b)** Solve the differential equation given in part (a).
   - **(c)** If the initial deposit is \$5000, give the particular solution satisfying this initial condition.
   - **(d)** How much is in the account after 10 years?
9. The amount of ozone, Q, in the atmosphere is decreasing at a rate proportional to the amount of ozone present. If time t is measured in years, the constant of proportionality is -0.0025. Write a differential equation for Q as a function of t, and give the general solution for the differential equation. If this rate continues, approximately what percent of the ozone in the atmosphere now will decay in the next 20 years?
10. Radioactive iodine decays at a continuous rate of about 9% per day. Write a differential equation to model this behavior. Find the general solution.
11. Using the model in the text and the data in Table 10.6 on page 415, find how long it would take for 90% of the pollution to be removed from Lake Michigan and from Lake Ontario, assuming no new pollutants are added. Explain how you can tell which lake will take longer to be purified just by looking at the data in the table.
12. Use the model in the text and the data in Table 10.6 on page 415 to determine which of the Great Lakes would require the longest time and which would require the shortest time for 80% of the pollution to be removed, assuming no new pollutants are being added. Find the ratio of these two times.
13. The rate of growth of a tumor is proportional to the size of the tumor.
   - **(a)** Write a differential equation satisfied by S, the size of the tumor, in mm, as a function of time, t.
   - **(b)** Find the general solution to the differential equation.
   - **(c)** If the tumor is 5 mm across at time $t = 0$, what does that tell you about the solution?
   - **(d)** If, in addition, the tumor is 8 mm across at time $t = 3$, what does that tell you about the solution?

14. The rate at which a drug leaves the bloodstream and passes into the urine is proportional to the quantity of the drug in the blood at that time. If an initial dose of Q_0 is injected directly into the blood, 20% is left in the blood after 3 hours.

(a) Write and solve a differential equation for the quantity, Q, of the drug in the blood after t hours.
(b) How much of this drug is in a patient's body after 6 hours if the patient is given 100 mg initially?

15. In some chemical reactions, the rate at which the amount of a substance changes with time is proportional to the amount present. For example, this is the case as δ-glucono-lactone changes into gluconic acid.

(a) Write a differential equation satisfied by y, the quantity of δ-glucono-lactone present at time t.
(b) If 100 grams of δ-glucono-lactone is reduced to 54.9 grams in one hour, how many grams will remain after 10 hours?

16. Oil is pumped continuously from a well at a rate proportional to the amount of oil left in the well. Initially there were 1 million barrels of oil in the well; six years later 500,000 barrels remain.

(a) At what rate was the amount of oil in the well decreasing when there were 600,000 barrels remaining?
(b) When will there be 50,000 barrels remaining?

17. Hydrocodone bitartrate is used as a cough suppressant. After the drug is fully absorbed, the quantity of drug in the body decreases at a rate proportional to the amount left in the body. The half-life of hydrocodone bitartrate in the body is 3.8 hours and the dose is 10 mg.

(a) Write a differential equation for the quantity, Q, of hydrocodone bitartrate in the body at time t, in hours since the drug was fully absorbed.
(b) Solve the differential equation given in part (a).
(c) Use the half-life to find the constant of proportionality, k.
(d) How much of the 10 mg dose is still in the body after 12 hours?

18. Warfarin is a drug used as an anticoagulant. After administration of the drug is stopped, the quantity remaining in a patient's body decreases at a rate proportional to the quantity remaining. The half-life of warfarin in the body is 37 hours.

(a) Sketch the quantity, Q, of warfarin in a patient's body as a function of the time, t, since stopping administration of the drug. Mark the 37 hours on your graph.
(b) Write a differential equation satisfied by Q.
(c) How many days does it take for the drug level in the body to be reduced to 25% of the original level?

10.5 APPLICATIONS AND MODELING

In the last section, we considered several situations modeled by the differential equation

$$\frac{dy}{dt} = ky.$$

In this section, we consider situations where the rate of change of y is a linear function of y of the form

$$\frac{dy}{dt} = k(y - A), \qquad \text{where } k \text{ and } A \text{ are constants.}$$

The Quantity of a Drug in the Body

A patient is given the drug warfarin, an anticoagulant, intravenously at the rate of 0.5 mg/hour. Warfarin is metabolized and leaves the body at the rate of about 2% per hour. A differential equation for the quantity, Q (in mg), of warfarin in the body after t hours is given by

$$\text{Rate of change} = \text{Rate in} - \text{Rate out}$$

$$\frac{dQ}{dt} = 0.5 - 0.02Q.$$

What does this tell us about the quantity of warfarin in the body for different initial values of Q?

If Q is small, then $0.02Q$ is also small and the rate the drug is excreted is less than the rate at which the drug is entering the body. Since the rate in is greater than the rate out, the rate of change is positive and the quantity of drug in the body is increasing. If Q is large enough that $0.02Q$ is greater than 0.5, then $0.5 - 0.02Q$ is negative, so dQ/dt is negative and the quantity is decreasing.

For small Q, the quantity will increase until the rate in equals the rate out. For large Q, the quantity will decrease until the rate in equals the rate out. What is the value of Q at which the rate in exactly matches the rate out? We have

$$\begin{aligned}\text{Rate in} &= \text{Rate out}\\ 0.5 &= 0.02Q\\ Q &= 25.\end{aligned}$$

If the amount of warfarin in the body is initially 25 mg, then the amount being excreted exactly matches the amount being added. The quantity of drug Q will stay constant at 25 mg. Notice also that when $Q = 25$, the derivative dQ/dt is zero, since

$$\frac{dQ}{dt} = 0.5 - 0.02(25) = 0.5 - 0.5 = 0.$$

If the initial quantity is 25, then the solution is the horizontal line $Q = 25$. This solution is called an *equilibrium solution*.

The slope field for this differential equation is shown in Figure 10.29, with solution curves drawn for $Q_0 = 20$, $Q_0 = 25$, and $Q_0 = 30$. In each case, we see that the quantity of drug in the body is approaching the equilibrium solution of 25 mg. The solution curve with $Q_0 = 30$ should remind you of an exponential decay function. It is, in fact, an exponential decay function that has been shifted up 25 units.

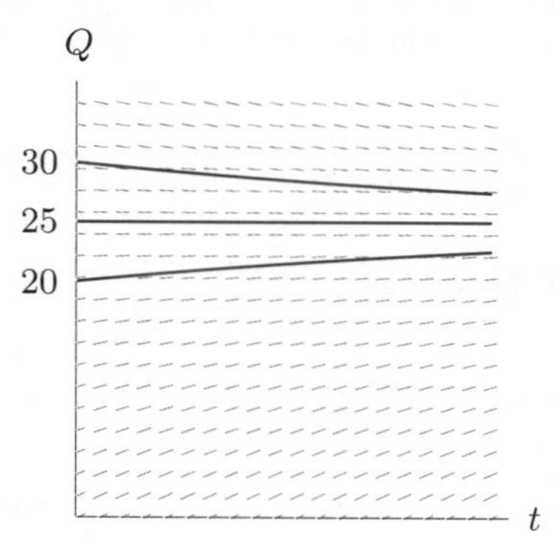

Figure 10.29: Slope field for $dQ/dt = 0.5 - 0.02Q$

Solving the Differential Equation $dy/dt = k(y - A)$

The drug concentration in the previous example satisfies a differential equation of the form

$$\frac{dy}{dt} = k(y - A).$$

Let us find the general solution to this equation. Since A is a constant, $dA/dt = 0$ so that we have

$$\frac{d}{dt}(y - A) = \frac{dy}{dt} - \frac{dA}{dt} = \frac{dy}{dt} - 0 = k(y - A).$$

Thus $y - A$ satisfies an exponential differential equation, so $y - A$ must be of the form

$$y - A = Ce^{kt}.$$

For an alternative derivation of the solution, see the Focus on Theory section on page 441.

> The general solution to the differential equation
>
> $$\frac{dy}{dt} = k(y - A)$$
>
> is
>
> $$y = A + Ce^{kt}, \qquad \text{for any constant } C.$$

Warning: Notice that, for differential equations of this form, the arbitrary constant C is *not* the initial value of the variable, but rather the initial value of $y - A$.

Example 1 Give the solution to each of the following differential equations:

(a) $\frac{dy}{dt} = 0.02(y - 50)$ (b) $\frac{dP}{dt} = 5(P - 10), \quad P = 8$ when $t = 0$

(c) $\frac{dy}{dt} = 3y - 300$ (d) $\frac{dW}{dt} = 500 - 0.1W$

Solution (a) The general solution is $y = 50 + Ce^{0.02t}$.

(b) The general solution is $P = 10 + Ce^{5t}$. Use the initial condition to solve for C:

$$8 = 10 + C(e^0)$$
$$8 = 10 + C$$

So $C = -2$, and the particular solution is $P = 10 - 2e^{5t}$.

(c) First rewrite the right-hand side of the equation in the form $k(y - A)$ by factoring out a 3:

$$\frac{dy}{dt} = 3(y - 100).$$

The general solution to this differential equation is $y = 100 + Ce^{3t}$.

(d) We begin by factoring out the coefficient of W:

$$\frac{dW}{dt} = 500 - 0.1W = -0.1\left(W - \frac{500}{0.1}\right) = -0.1(W - 5000).$$

The general solution to this differential equation is $W = 5000 + Ce^{-0.1t}$.

Example 2 At the start of this section, we gave the following differential equation for the quantity of warfarin in the body:

$$\frac{dQ}{dt} = 0.5 - 0.02Q.$$

Write the general solution to this differential equation. Find particular solutions for $Q_0 = 20$, $Q_0 = 25$, and $Q_0 = 30$.

Solution We first rewrite the differential equation in the form $dQ/dt = k(Q - A)$ by factoring out -0.02:

$$\frac{dQ}{dt} = 0.5 - 0.02Q = -0.02(Q - 25).$$

The general solution to this differential equation is

$$Q = 25 + Ce^{-0.02t}.$$

To find the particular solution when $Q_0 = 20$, we use the initial condition to solve for C:

$$20 = 25 + C(e^0)$$
$$C = -5.$$

The particular solution when $Q_0 = 20$ is $Q = 25 - 5e^{-0.02t}$.

When $Q_0 = 25$, we have $C = 0$ and the particular solution is the horizontal line $Q = 25$. When $Q_0 = 30$, we have $C = 5$ and the particular solution is $Q = 25 + 5e^{-0.02t}$. These three solutions are the three we saw earlier in Figure 10.29.

Example 3 A company's revenue is earned at a continuous annual rate of 5% of its net worth. At the same time, the company's payroll obligations are paid out at a constant rate of 200 million dollars a year. We saw in Section 10.1 that the differential equation governing the net worth, W (in millions of dollars), of this company in year t is given by

$$\begin{aligned} \text{Rate of change of } W &= \text{Rate in} - \text{Rate out} \\ \frac{dW}{dt} &= 0.05W - 200. \end{aligned}$$

(a) Solve the differential equation, assuming an initial net worth of W_0 million dollars.
(b) Sketch the solution for $W_0 = 3000$, 4000 and 5000. For which of these values of W_0 does the company go bankrupt? In which year?

Solution (a) Factor out 0.05 to get

$$\frac{dW}{dt} = 0.05(W - 4000).$$

The general solution is

$$W = 4000 + Ce^{0.05t}.$$

To find C we use the initial condition that $W = W_0$ when $t = 0$.

$$\begin{aligned} W_0 &= 4000 + Ce^0 \\ W_0 - 4000 &= C \end{aligned}$$

Substituting this value for C into $W = 4000 + Ce^{0.05t}$ gives

$$W = 4000 + (W_0 - 4000)e^{0.05t}.$$

(b) If $W_0 = 4000$, then $W = 4000$, the equilibrium solution.
If $W_0 = 5000$, then $W = 4000 + 1000e^{0.05t}$.
If $W_0 = 3000$, then $W = 4000 - 1000e^{0.05t}$. The graphs of these functions are in Figure 10.30. Notice that if the net worth starts with W_0 near, but not equal to, \$4000 million, then W moves further away. We see that if $W_0 = 3000$, the value of W goes to 0, and the company goes bankrupt. Solving $W = 0$ gives $t \approx 27.7$, so the company goes bankrupt in its twenty-eighth year.

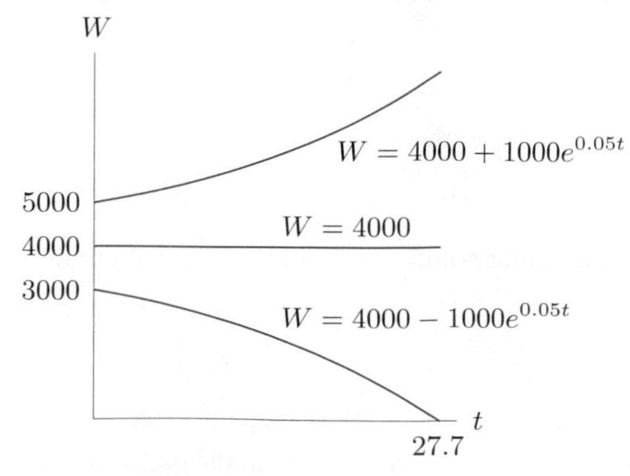

Figure 10.30: Solutions to $\frac{dW}{dt} = 0.05W - 200$

Equilibrium Solutions

Figure 10.29 shows the quantity of warfarin in the body for several different initial quantities. All these curves are solutions to the differential equation

$$\frac{dQ}{dt} = 0.5 - 0.02Q = -0.02(Q - 25),$$

and all the solutions have the form

$$Q = 25 + Ce^{-0.02t}$$

for some C. Notice that $Q \to 25$ as $t \to \infty$ for all solutions because $e^{-0.02t} \to 0$ as $t \to \infty$. In other words, in the long run, the quantity approaches the *equilibrium solution* of $Q = 25$ no matter what the initial quantity.

Notice that the equilibrium solution can be found directly from the differential equation by solving $dQ/dt = 0$:

$$\frac{dQ}{dt} = -0.02(Q - 25) = 0,$$

giving $Q = 25$. Because Q always gets closer and closer to the equilibrium value of 25 as $t \to \infty$, we call $Q = 25$ a *stable* equilibrium for Q.

A different situation is shown in Figure 10.30 with the solutions to the differential equation $dW/dt = 0.05W - 200$. We find the equilibrium by looking at the solution curves or by setting $dW/dt = 0$:

$$\frac{dW}{dt} = 0.05W - 200 = 0.05(W - 4000) = 0,$$

giving $W = 4000$ as the equilibrium solution. This equilibrium solution is called *unstable* because if W starts near, but not equal to, 4000, the net worth W moves further away from 4000 as $t \to \infty$.

- An **equilibrium solution** is constant for all values of the independent variable. The graph is a horizontal line. Equilibrium solutions can be identified by setting the derivative of the function to zero.
- An equilibrium solution is **stable** if a small change in the initial conditions gives a solution which tends toward the equilibrium as the independent variable tends to positive infinity.
- An equilibrium solution is **unstable** if a small change in the initial conditions gives a solution curve which veers away from the equilibrium as the independent variable tends to positive infinity.

In general, a differential equation may have more than one equilibrium solution or no equilibrium solution.

Example 4 Find the equilibrium solution for each of the following differential equations. Determine whether the equilibrium solution is stable or unstable.

(a) $\dfrac{dH}{dt} = -2(H - 20)$ (b) $\dfrac{dB}{dt} = 2(B - 10)$

Solution (a) To find equilibrium solutions, we set $dH/dt = 0$:

$$\frac{dH}{dt} = -2(H - 20) = 0,$$

giving $H = 20$ as the equilibrium solution. The general solution to this differential equation is $H = 20 + Ce^{-2t}$. The solution curves for $H_0 = 10$, $H_0 = 20$, and $H_0 = 30$ are shown in Figure 10.31. We see that the equilibrium solution is stable.

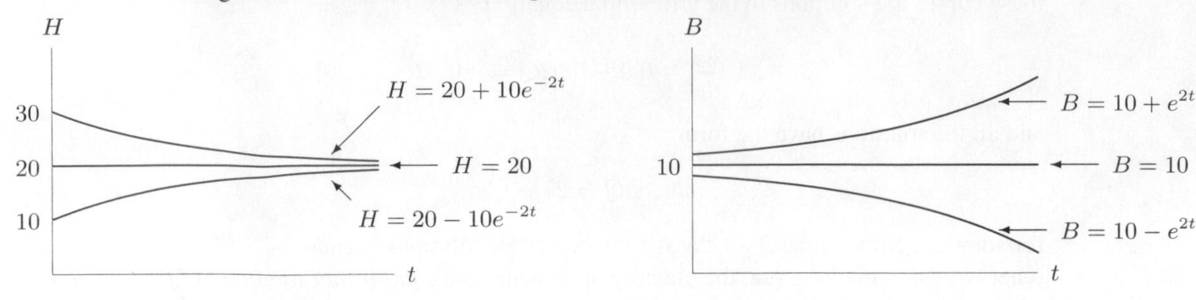

Figure 10.31: $H = 20$ is stable equilibrium

Figure 10.32: $B = 10$ is unstable equilibrium

(b) To find equilibrium solutions, we set $dB/dt = 0$:

$$\frac{dB}{dt} = 2(B - 10) = 0,$$

giving $B = 10$ as the equilibrium solution. The general solution to this differential equation is $B = 10 + Ce^{2t}$. The solution curves for $B_0 = 9$, $B_0 = 10$, and $B_0 = 11$ are shown in Figure 10.32. We see that the equilibrium solution is unstable.

Newton's Law of Heating and Cooling

Newton proposed that the temperature of a hot object decreases at a rate proportional to the difference between its temperature and that of its surroundings. Similarly, a cold object heats up at a rate proportional to the temperature difference between the object and its surroundings.

For example, a hot cup of coffee standing on a table cools at a rate proportional to the temperature difference between the coffee and the surrounding air. As the coffee cools, the rate at which it cools decreases because the temperature difference between the coffee and the air decreases. In the long run, the rate of cooling tends to zero and the temperature of the coffee approaches room temperature. Figure 10.33 shows the temperature of two cups of coffee against time, one starting at a higher temperature than the other, but both tending toward room temperature in the long run.

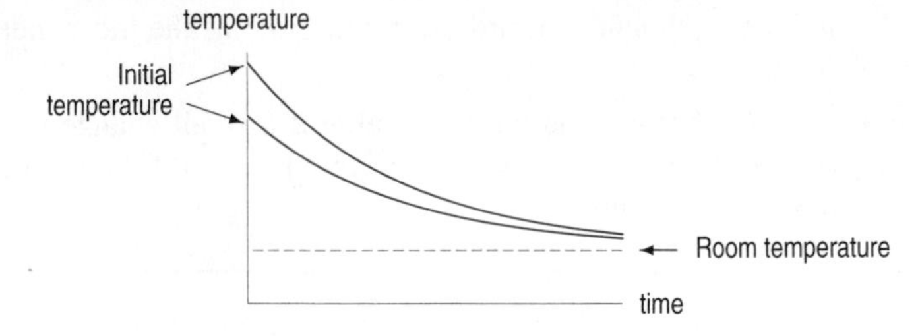

Figure 10.33: Temperature of coffee versus time

Let H be the temperature at time t of a cup of coffee in a 70°F room. Newton's Law says that the rate of change of H is proportional to the temperature difference between the coffee and the room:

$$\text{Rate of change of temperature} = \text{Constant} \cdot \text{Temperature difference}.$$

The rate of change of temperature is dH/dt. The temperature difference between the coffee and the room is $(H - 70)$, so

$$\frac{dH}{dt} = \text{Constant} \cdot (H - 70).$$

What about the sign of the constant? If the coffee starts out hotter than 70° (that is, $H - 70 > 0$), then the temperature of the coffee decreases (i.e., $dH/dt < 0$) and so the constant must be negative:

$$\frac{dH}{dt} = -k(H - 70) \qquad k > 0.$$

What can we learn from this differential equation? Suppose we take $k = 1$. The slope field for this differential equation in Figure 10.34 shows several solution curves. Notice that, as we expect, the temperature of the coffee is approaching the temperature of the room. The general solution to this differential equation is

$$H = 70 + Ce^{-t},$$

where C is an arbitrary constant.

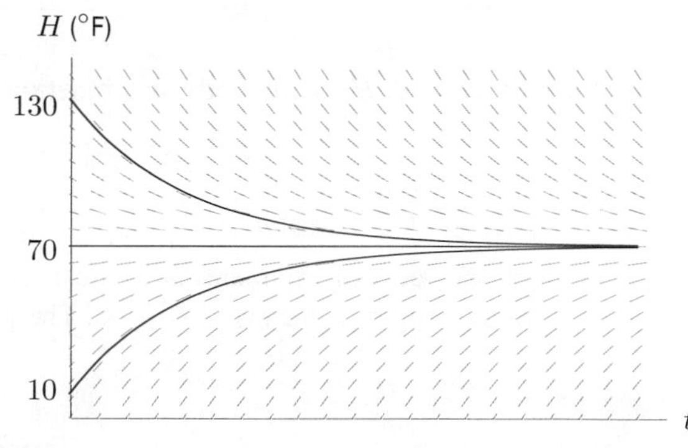

Figure 10.34: Slope field for $\dfrac{dH}{dt} = -(H - 70)$

Example 5 The body of a murder victim is found at noon in a room with a constant temperature of 20°C. At noon the temperature of the body is 35°C; two hours later the temperature of the body is 33°C.

(a) Find the temperature, H, of the body as a function of t, the time in hours since it was found.
(b) Graph H against t. What happens to the temperature in the long run?
(c) At the time of the murder, the victim's body had the normal body temperature, 37°C. When did the murder occur?

Solution (a) Newton's Law of Cooling says that

$$\text{Rate of change of temperature} = \text{Constant} \cdot \text{Temperature difference}.$$

Since the temperature difference is $H - 20$, we have for some constant k

$$\frac{dH}{dt} = -k(H - 20).$$

The general solution is

$$H = 20 + Ce^{-kt}.$$

To determine C, we use the fact that $H = 35$ at $t = 0$:

$$35 = 20 + C \cdot e^0$$
$$35 = 20 + C$$

So $C = 15$ and we have

$$H = 20 + 15e^{-kt}.$$

To find k, we use the fact that $H = 33$ when $t = 2$:

$$33 = 20 + 15e^{-k(2)}.$$

We isolate the exponential and solve for k:

$$\begin{aligned}13 &= 15e^{-2k}\\ \frac{13}{15} &= e^{-2k}\\ \ln\left(\frac{13}{15}\right) &= -2k\\ k &= -\frac{\ln(13/15)}{2} = 0.072.\end{aligned}$$

Therefore, the temperature, H, of the body as a function of time, t, is given by

$$H = 20 + 15e^{-0.072t}.$$

(b) The graph of $H = 20 + 15e^{-0.072t}$ has a vertical intercept of $H = 35$, the initial temperature. The temperature decays exponentially with a horizontal asymptote of $H = 20$. (See Figure 10.35.) "In the long run" means as $t \to \infty$. The graph shows that $H \to 20$ as $t \to \infty$.

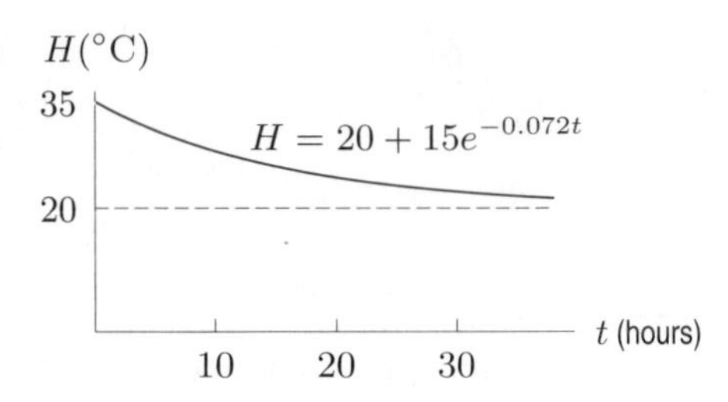

Figure 10.35: Temperature of dead body

(c) We want to know when the temperature was 37°C. We substitute $H = 37$ and solve for t:

$$\begin{aligned}37 &= 20 + 15e^{-0.072t}\\ \frac{17}{15} &= e^{-0.072t}.\end{aligned}$$

Taking natural logs on both sides gives

$$\ln\left(\frac{17}{15}\right) = -0.072t$$

so

$$t = -\frac{\ln(17/15)}{0.072} = -1.74 \text{ hours.}$$

The murder occurred about 1.74 hours before noon, that is, about 10:15 am.

Problems for Section 10.5

Find particular solutions in Problems 1–8.

1. $\dfrac{dy}{dt} = 0.5(y - 200), \quad y = 50$ when $t = 0$
2. $\dfrac{dH}{dt} = 3(H - 75), \quad H = 0$ when $t = 0$
3. $\dfrac{dB}{dt} = 4B - 100, \quad B = 20$ when $t = 0$
4. $\dfrac{dP}{dt} = P + 4, \quad P = 100$ when $t = 0$
5. $\dfrac{dm}{dt} = 0.1m + 200, \quad m(0) = 1000$
6. $\dfrac{dQ}{dt} = 0.3Q - 120, \quad Q = 50$ when $t = 0$
7. $\dfrac{dB}{dt} + 2B = 50, \quad B(1) = 100$
8. $\dfrac{dB}{dt} + 0.1B - 10 = 0 \qquad B(2) = 3$

9. Check that $y = A + Ce^{kt}$ is a solution to the differential equation
$$\frac{dy}{dt} = k(y - A).$$

10. A company earns 2% per month on its assets, paid continuously, and its expenses are paid out continuously at a rate of $\$80,000$ per month.

(a) Write a differential equation for the value, V, of the company as a function of time, t, in months.
(b) What is the equilibrium solution for the differential equation? What is the significance of this value for the company?
(c) Solve the differential equation found in part (a).
(d) If the company has assets worth $\$3$ million at time $t = 0$, what are its assets worth one year later?

11. Money in an account earns interest at a continuous rate of 8% per year, and payments are made continuously out of the account at the rate of $\$5000$ a year. The account initially contains $\$50,000$. Write a differential equation for the amount of money in the account, B, in t years. Solve the differential equation. Does the account ever run out of money? If so, when?

12. A bank account earns 7% annual interest compounded continuously. You deposit $\$10,000$ in the account, and withdraw money continuously from the account at a rate of $\$1000$ per year.

(a) Write a differential equation for the balance, B, in the account after t years.
(b) What is the equilibrium solution to the differential equation? (This is the amount that must be deposited now for the balance to stay the same over the years.)
(c) Find the solution to the differential equation.
(d) How much is in the account after 5 years?
(e) Graph the solution. What happens to the balance in the long run?

13. A patient is given the drug theophylline intravenously at a rate of 43.2 mg/hour to relieve acute asthma. The rate at which the drug leaves the patient's body is proportional to the quantity there, with proportionality constant 0.082 if time, t, is in hours. The patient's body contains none of the drug initially.

(a) Describe in words how you expect the quantity of theophylline in the patient to vary with time.
(b) Write a differential equation satisfied by the quantity of theophylline in the body, $Q(t)$.
(c) Solve the differential equation and graph the solution. What happens to the quantity in the long run?

14. A bank account earns 10% annual interest, compounded continuously. Money is deposited in a continuous cash flow at a rate of $\$1200$ per year into the account.

(a) Write a differential equation that describes the rate at which the balance $B = f(t)$ is changing.
(b) Solve the differential equation given an initial balance $B_0 = 0$.
(c) Find the balance after 5 years.

15. Suppose $\$1000$ is put into a bank account and earns interest continuously at a rate of i per year, and in addition, continuous payments are made out of the account at a rate of $\$100$ a year. Sketch the amount of money in the account as a function of time if the interest rate is

(a) 5% **(b)** 10% **(c)** 15%

In each case, first find an expression for the amount of money in the account at time t in years.

16. Morphine is often used as a pain-relieving drug. The half-life of morphine in the body is 2 hours. Suppose morphine is administered to a patient intravenously at a rate of 2.5 mg per hour, and the rate at which the morphine is eliminated is proportional to the amount present.

(a) Show that, to three decimal places, the constant of proportionality for the rate at which morphine leaves the body (in mg/hour) is $k = -0.347$.
(b) Write a differential equation for the quantity, Q, of morphine in the blood after t hours.
(c) Use the differential equation to find the equilibrium solution. (This is the long-term amount of morphine in the body, once the system has stabilized.)

17. Dead leaves accumulate on the ground in a forest at a rate of 3 grams per square centimeter per year. At the same time, these leaves decompose at a continuous rate of 75% per year. Write a differential equation for the total quantity of dead leaves (per square centimeter) at time t. Sketch a solution showing that the quantity of dead leaves tends toward an equilibrium level. What is that equilibrium level?

18. A chain smoker smokes five cigarettes every hour. From each cigarette, 0.4 mg of nicotine is absorbed into the person's bloodstream. Nicotine leaves the body at a rate proportional to the amount present, with constant of proportionality -0.346 if t is in hours.

(a) Write a differential equation for the level of nicotine in the body, N, in mg, as a function of time, t, in hours.
(b) Solve the differential equation from part (a). Initially there is no nicotine in the blood.
(c) The person wakes up at 7 am and begins smoking. How much nicotine is in the blood when the person goes to sleep at 11 pm (16 hours later)?

19. A bank account earns 5% annual interest compounded continuously. Continuous payments are made out of the account at a rate of $\$12,000$ per year for 20 years.

(a) Write a differential equation describing the balance $B = f(t)$, where t is in years.
(b) Solve the differential equation given an initial balance of B_0.
(c) What should the initial balance be such that the account has zero balance after precisely 20 years?

20. **(a)** Find all equilibrium solutions for the equation

$$\frac{dy}{dx} = 0.5y(y-4)(2+y).$$

(b) With a calculator or computer, draw a slope field for this differential equation. Use it to determine whether each equilibrium solution is stable or unstable.

21. **(a)** Find the equilibrium solution of the equation

$$\frac{dy}{dt} = 0.5y - 250.$$

(b) Find the general solution of this equation.
(c) Graph several solutions with different initial values.
(d) Is the equilibrium solution stable or unstable?

22. Figure 10.36 gives the slope field for a differential equation. Estimate all equilibrium solutions and indicate whether each is stable or unstable.

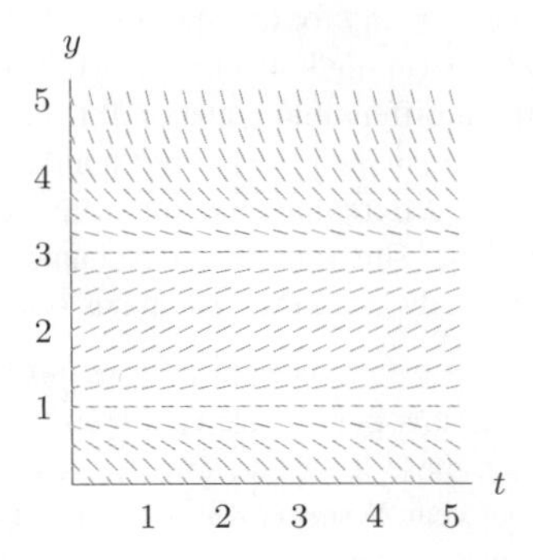

Figure 10.36

23. According to a simple physiological model, an athletic adult male needs 20 calories per day per pound of body weight to maintain his weight. If he consumes more or fewer calories than those required to maintain his weight, his weight changes at a rate proportional to the difference between the number of calories consumed and the number needed to maintain his current weight; the constant of proportionality is $1/3500$ pounds per calorie. Suppose that a particular person has a constant caloric intake of I calories per day. Let $W(t)$ be the person's weight in pounds at time t (measured in days).

(a) What differential equation has solution $W(t)$?
(b) Solve this differential equation.
(c) Graph $W(t)$ if the person starts out weighing 160 pounds and consumes 3000 calories a day.

24. An item is initially sold at a price of $\$p$ per unit. Over time, market forces push the price toward the equilibrium price, $\$p^*$, at which supply balances demand. The Evans Price Adjustment model says that the rate of change in the market price, $\$p$, is proportional to the difference between the market price and the equilibrium price.

(a) Write a differential equation for p as a function of t.
(b) Solve for p.
(c) Sketch solutions for various different initial prices, both above and below the equilibrium price.
(d) What happens to p as $t \to \infty$?

25. A yam is put in a 200°C oven and heats up according to the differential equation

$$\frac{dH}{dt} = -k(H-200), \quad \text{for } k \text{ a positive constant.}$$

(a) If the yam is at 20°C when it is put in the oven, solve the differential equation.
(b) Find k using the fact that after 30 minutes the temperature of the yam is 120°C.

26. At 1:00 pm one winter afternoon, there is a power failure at your house in Wisconsin, and your heat does not work without electricity. When the power goes out, it is 68°F in your house. At 10:00 pm, it is 57°F in the house, and you notice that it is 10°F outside.

(a) Assuming that the temperature, T, in your home obeys Newton's Law of Cooling, write the differential equation satisfied by T.
(b) Solve the differential equation to estimate the temperature in the house when you get up at 7:00 am the next morning. Should you worry about your water pipes freezing?
(c) What assumption did you make in part (a) about the temperature outside? Given this (probably incorrect) assumption, would you revise your estimate up or down? Why?

27. A detective finds a murder victim at 9 am. The temperature of the body is measured at 90.3°F. One hour later, the temperature of the body is 89.0°F. The temperature of the room has been maintained at a constant 68°F.

(a) Assuming the temperature, T, of the body obeys Newton's Law of Cooling, write a differential equation for T.
(b) Solve the differential equation to estimate the time the murder occurred.

28. A drug is administered intravenously at a constant rate of r mg/hour and is excreted at a rate proportional to the quantity present, with constant of proportionality $\alpha > 0$.

(a) Solve a differential equation for the quantity, Q, in milligrams, of the drug in the body at time t hours. Assume there is no drug in the body initially. Your answer will contain r and α. Graph Q against t. What is Q_∞, the limiting long-run value of Q?
(b) What effect does doubling r have on Q_∞? What effect does doubling r have on the time to reach half the limiting value, $\frac{1}{2}Q_\infty$?
(c) What effect does doubling α have on Q_∞? On the time to reach $\frac{1}{2}Q_\infty$?

10.6 MODELING THE INTERACTION OF TWO POPULATIONS

So far we have used a differential equation to model the growth of a single quantity. We now consider the growth of two interacting populations, a situation which requires a system of two differential equations. Examples include two species competing for food, one species preying on another, or two species helping each other (symbiosis).

A Predator-Prey Model: Robins and Worms

We model a predator-prey system using what are called the Lotka-Volterra equations. Let's look at a simplified and idealized case in which robins are the predators and worms the prey.[2] Suppose there are r thousand robins and w million worms. If there were no robins, the worms would increase exponentially according to the equation

$$\frac{dw}{dt} = aw \qquad \text{where } a \text{ is a positive constant.}$$

If there were no worms, the robins would have no food and so their population would decrease according to the equation[3]

$$\frac{dr}{dt} = -br \qquad \text{where } b \text{ is a positive constant.}$$

Now imagine the effect of the two populations on one another. Clearly, the presence of the robins is bad for the worms, so

$$\frac{dw}{dt} = aw - \text{Effect of robins on worms.}$$

On the other hand, the robins do better with the worms around, so

$$\frac{dr}{dt} = -br + \text{Effect of worms on robins.}$$

How exactly do the two populations interact? Let's assume the effect of one population on the other is proportional to the number of "encounters." (An encounter is when a robin eats a worm.) The number of encounters is likely to be proportional to the product of the populations because if one population is held fixed, the number of encounters should be directly proportional to the other population. So we assume

$$\frac{dw}{dt} = aw - cwr \qquad \text{and} \qquad \frac{dr}{dt} = -br + kwr,$$

where c and k are positive constants.

To analyze this system of equations, let's look at the specific example with $a = b = c = k = 1$:

$$\frac{dw}{dt} = w - wr \qquad \text{and} \qquad \frac{dr}{dt} = -r + wr.$$

The Phase Plane

To see the growth of the populations, we want graphs of r and w against t. However, it is easier to obtain a graph of r against w first. If we plot a point (w, r) representing the number of worms and robins at any moment, then, as the populations change, the point moves. The wr-plane on which the point moves is called the *phase plane* and the path of the point is called the *phase trajectory*.

To find the phase trajectory, we need a differential equation relating w and r directly. We have the two differential equations

$$\frac{dw}{dt} = w - wr \qquad \text{and} \qquad \frac{dr}{dt} = -r + wr.$$

Thinking of r as a function of w and w as a function of t, the chain rule gives

$$\frac{dr}{dt} = \frac{dr}{dw} \cdot \frac{dw}{dt}.$$

[2]Based on work by Thomas A. McMahon.

[3]This assumption unrealistically predicts that the robin population will decay exponentially, rather than die out in finite time.

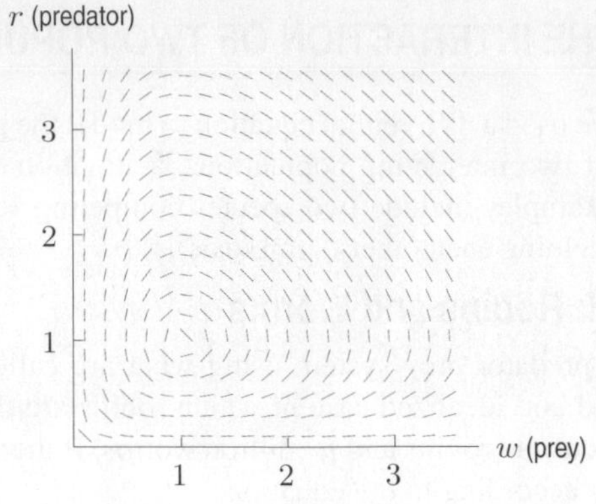

Figure 10.37: Slope field for $\dfrac{dr}{dw} = \dfrac{-r + wr}{w - wr}$

This tells us that

$$\frac{dr}{dw} = \frac{dr/dt}{dw/dt},$$

so we have

$$\frac{dr}{dw} = \frac{-r + wr}{w - wr}.$$

Figure 10.37 shows the slope field of this differential equation in the phase plane.

The Slope Field and Equilibrium Points

We can get an idea of what solutions of this equation look like from the slope field. At the point $(1, 1)$ there is no slope drawn because dr/dw is undefined there since the rates of change of both populations with respect to time are zero:

$$\frac{dw}{dt} = 1 - 1 \cdot 1 = 0, \qquad \text{and} \qquad \frac{dr}{dt} = -1 + 1 \cdot 1 = 0.$$

In terms of worms and robins, this means that if at some moment $w = 1$ and $r = 1$ (that is, there are 1 million worms and 1 thousand robins), then w and r remain constant forever. The point $w = 1$, $r = 1$ is therefore an equilibrium solution. The origin is also an equilibrium point, since if $w = 0$ and $r = 0$, then w and r remain constant. The slope field suggests that there are no other equilibrium points. We check this by solving

$$\frac{dw}{dt} = w - wr = 0 \qquad \text{and} \qquad \frac{dr}{dt} = -r + rw = 0,$$

which yields only $w = 0$, $r = 0$ and $w = 1$, $r = 1$ as solutions.

Trajectories in the wr-Phase Plane

Let's look at the trajectories in the phase plane. A point on a curve represents a pair of populations (w, r) existing at the same time t (though t is not shown on the graph). A short time later, the pair of populations is represented by a nearby point. As time passes, the point traces out a trajectory. It can be shown that the trajectory is a closed curve. See Figure 10.38.

In which direction does the point move on the trajectory? Look at the original pair of differential equations. They tell us how w and r change with time. Imagine, for example, that we are at the point P_0 in Figure 10.39, where $w = 2.2$ and $r = 1$; then

$$\frac{dr}{dt} = -r + wr = -1 + (2.2)(1) = 1.2 > 0.$$

Therefore, r is increasing, so the point is moving in the direction shown by the arrow in Figure 10.39.

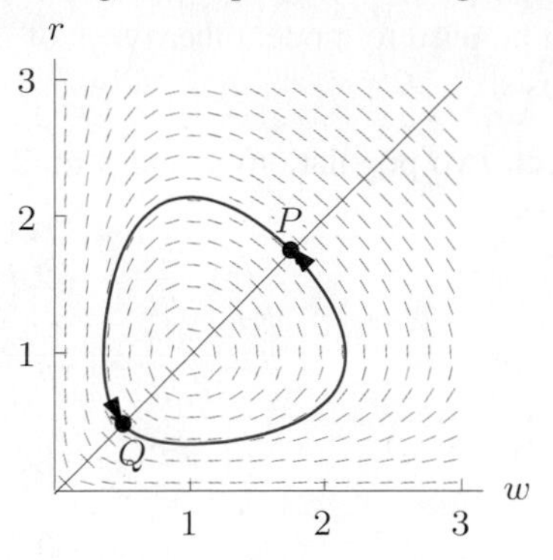

Figure 10.38: Solution curve is closed

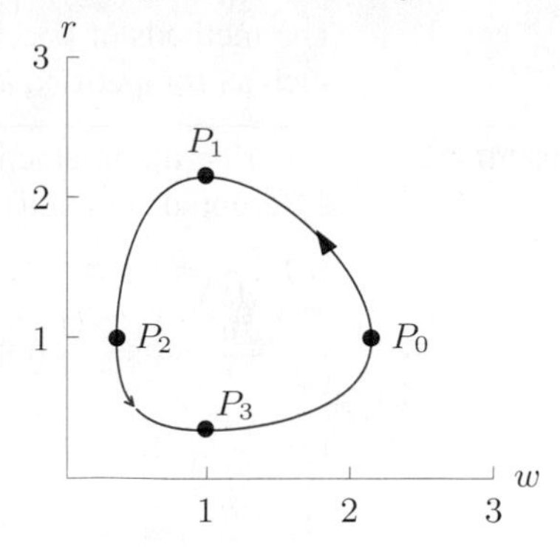

Figure 10.39: A trajectory

Example 1 Suppose that at time $t = 0$, there are 2.2 million worms and 1 thousand robins. Describe how the robin and worm populations change over time.

Solution The trajectory through the point P_0 where $w = 2.2$ and $r = 1$ is shown on the slope field in Figure 10.38 and by itself in Figure 10.39.

Initially there are lots of worms so the robin population does well. The robin population is increasing and the worm population is decreasing until there are about 2.2 thousand robins and 1 million worms (point P_1 in Figure 10.39). At this point, there are too few worms to sustain the robin population; it begins to decrease and the worm population continues to fall as well. The robin population falls dramatically until there are about 1 thousand robins and 0.4 million worms (point P_2 in Figure 10.39). With so few robins, the worm population starts to recover, but the robin population is still decreasing. The worm population increases until there are about 0.4 thousand robins and 1 million worms (P_3 in Figure 10.39). Now there are lots of worms for the small population of robins so both populations increase. The populations return to the starting values (since the trajectory forms a closed curve) and the cycle starts over.

Problem 17 on page 431 shows how to calculate approximate coordinates of points on the curve.

The Populations as Functions of Time

The shape of a trajectory tells us how the populations vary with time. We use this information to graph each population against time, as in Figure 10.40. The fact that the trajectory is a closed curve means that both populations oscillate periodically. Both populations have the same period, and the worms (the prey) are at their maximum a quarter of a cycle before the robins.

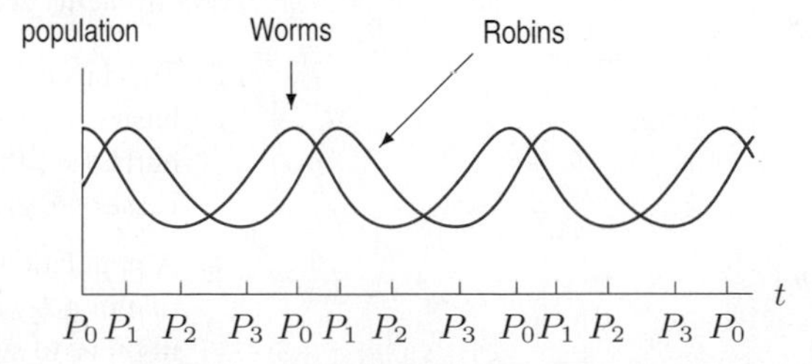

Figure 10.40: Populations of robins (in thousands) and worms (in millions) over time

Lynxes and Hares

A predator-prey system for which there are long-term data is the Canadian lynx and the hare. Both animals were of interest to fur trappers and the records of the Hudson Bay Company shed some light on their populations through much of the 20th century. These records show that both populations oscillated up and down, quite regularly, with a period of about ten years. This is the behavior predicted by Lotka-Volterra equations.

Other Forms of Species Interaction

The methods of this section can be used to model other types of interactions between two species, such as competition and symbiosis.

Example 2 Describe the interactions between two populations x and y modeled by the following systems of differential equations.

(a) $$\frac{dx}{dt} = 0.2x - 0.5xy \qquad \frac{dy}{dt} = 0.6y - 0.8xy$$

(b) $$\frac{dx}{dt} = -2x + 5xy \qquad \frac{dy}{dt} = -y + 0.2xy$$

(c) $$\frac{dx}{dt} = 0.5x \qquad \frac{dy}{dt} = -1.6y + 2xy$$

(d) $$\frac{dx}{dt} = 0.3x - 1.2xy \qquad \frac{dy}{dt} = -0.7y + 2.5xy$$

Solution

(a) If we ignore the interaction terms with xy, we have $dx/dt = 0.2x$ and $dy/dt = 0.6y$, so both populations grow exponentially. Since both interaction terms are negative, each species inhibits the other's growth, such as when deer and elk compete for food.

(b) If we ignore the interaction terms, the populations of both species decrease exponentially. However, both interaction terms are positive, meaning each species benefits from the other, so the relationship is symbiotic. An example is the pollination of plants by insects.

(c) Ignoring interaction, x grows and y decays. But the interaction term means that y benefits from x, in the way birds that build nests benefit from trees.

(d) Without the interaction, x grows and y decays. The interaction terms show that y hurts x while x benefits y. This is a predator-prey model where y is the predator and x is the prey.

Problems for Section 10.6

Problems 1–3 give the rates of growth of two populations, x and y, measured in thousands.

(a) Describe in words what happens to the population of each species in the absence of the other.

(b) Describe in words how the species interact with one another. Give reasons why the populations might behave as described by the equations. Suggest species that might interact that way.

1. $$\frac{dx}{dt} = 0.01x - 0.05xy \qquad \frac{dy}{dt} = -0.2y + 0.08xy$$

2. $$\frac{dx}{dt} = 0.01x - 0.05xy \qquad \frac{dy}{dt} = 0.2y - 0.08xy$$

3. $$\frac{dx}{dt} = 0.2x \qquad \frac{dy}{dt} = 0.4xy - 0.1y$$

4. The following system of differential equations represents the interaction between two populations, x and y.

$$\frac{dx}{dt} = -3x + 2xy$$
$$\frac{dy}{dt} = -y + 5xy$$

(a) Describe how the species interact. How would each species do in the absence of the other? Are they helpful or harmful to each other?

(b) If $x = 2$ and $y = 1$, does x increase or decrease? Does y increase or decrease? Justify your answers.

(c) Write a differential equation involving dy/dx.

(d) Use a computer or calculator to draw the slope field for the differential equation in part (c).

(e) Draw the trajectory starting at point $x = 2$, $y = 1$ on your slope field, and describe how the populations change as time increases.

Create a system of differential equations to model the situations in Problems 5–7. You may assume that all constants of proportionality are 1.

5. Two businesses are in competition with each other. Both businesses would do well without the other one, but each hurts the other's business. The values of the two businesses are given by x and y.

6. A population of fleas is represented by x, and a population of dogs is represented by y. The fleas need the dogs in order to survive. The dog population, however, is unaffected by the fleas.

7. The concentrations of two chemicals are denoted by x and y, respectively. Alone, each decays at a rate proportional to its concentration. Together, they interact to form a third substance. As the third substance is created, the concentrations of the initial two populations get smaller.

8. Two companies, A and B, are in competition with each other. Let x represent the net worth (in millions of dollars) of Company A, and y represent the net worth (in millions of dollars) of Company B. Four trajectories are given in Figure 10.41. For each trajectory: Describe the initial conditions. Describe what happens initially: Do the companies gain or lose money early on? What happens in the long run?

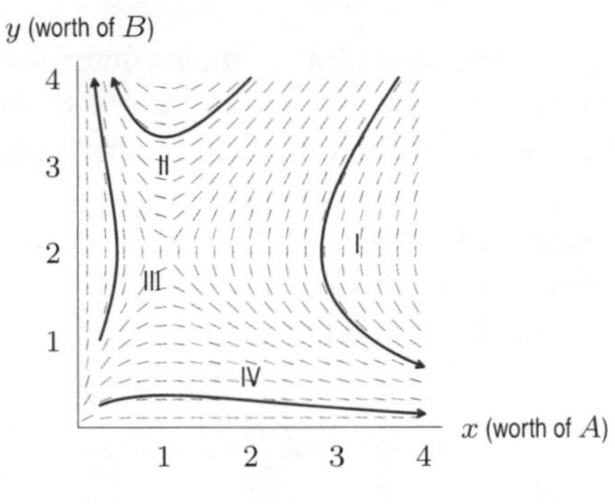

Figure 10.41

For Problems 9–19, let w be the number of worms (in millions) and r the number of robins (in thousands) living on an island. Suppose w and r satisfy the following differential equations, which correspond to the slope field in Figure 10.42.

$$\frac{dw}{dt} = w - wr, \qquad \frac{dr}{dt} = -r + wr.$$

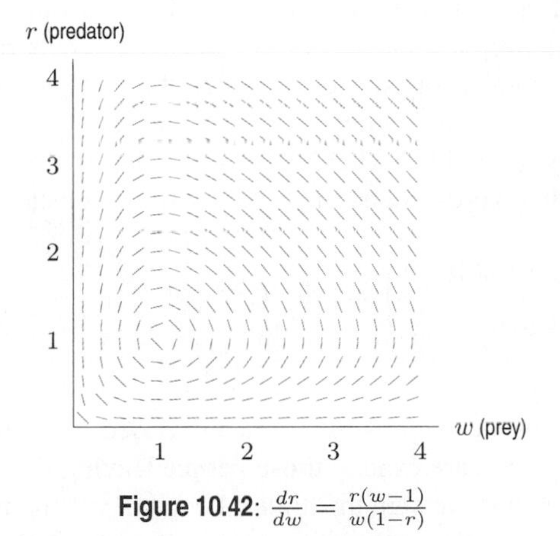

Figure 10.42: $\frac{dr}{dw} = \frac{r(w-1)}{w(1-r)}$

9. Explain why these differential equations are a reasonable model for interaction between the two populations. Why have the signs been chosen this way?

10. Solve these differential equations in the two special cases when there are no robins and when there are no worms living on the island.

11. Describe and explain the symmetry you observe in the slope field. What consequences does this symmetry have for the solution curves?

12. Assume $w = 2$ and $r = 2$ when $t = 0$. Do the numbers of robins and worms increase or decrease at first? What happens in the long run?

13. For the case discussed in Problem 12, estimate the maximum and the minimum values of the robin population. How many worms are there at the time when the robin population reaches its maximum?

14. On the same axes, graph w and r (the worm and the robin populations) against time. Use initial values of 1.5 for w and 1 for r. You may do this without units for t.

15. People on the island like robins so much that they decide to import 200 robins all the way from England, to increase the initial population to $r = 2.2$ when $t = 0$. Does this make sense? Why or why not?

16. Assume that $w = 3$ and $r = 1$ when $t = 0$. Do the numbers of robins and worms increase or decrease initially? What happens in the long run?

17. At $t = 0$ there are 2.2 million worms and 1 thousand robins.

(a) Use the differential equations to calculate the derivatives dw/dt and dr/dt at $t = 0$.
(b) Use the initial values and your answer to part (a) to estimate the number of robins and worms at $t = 0.1$.
(c) Using the method of part (a) and (b), estimate the number of robins and worms at $t - 0.2$ and 0.3.

18. **(a)** Assume that there are 3 million worms and 2 thousand robins. Locate the point corresponding to this situation on the slope field given in Figure 10.42. Draw the trajectory through this point.
(b) In which direction does the point move along this trajectory? Put an arrow on the trajectory and justify your answer using the differential equations for dr/dt and dw/dt given in this section.
(c) How large does the robin population get? What is the size of the worm population when the robin population is at its largest?
(d) How large does the worm population get? What is the size of the robin population when the worm population is at its largest?

19. Repeat Problem 18 if initially there are 0.5 million worms and 3 thousand robins.

20. For each system of differential equations in Example 2, determine whether x increases or decreases and whether y increases or decreases when $x = 2$ and $y = 2$.

21. For each system of equations in Example 2, write a differential equation involving dy/dx. Use a computer or calculator to draw the slope field for $x, y > 0$. Then draw the trajectory through the point $x = 3, y = 1$.

10.7 MODELING THE SPREAD OF A DISEASE

Differential equations can be used to predict when an outbreak of a disease becomes so severe that it is called an *epidemic*[4] and to decide what level of vaccination is necessary to prevent an epidemic. Let's consider a specific example.

Flu in a British Boarding School

In January 1978, 763 students returned to a boys' boarding school after their winter vacation. A week later, one boy developed the flu, followed immediately by two more. By the end of the month, nearly half the boys were sick. Most of the school had been affected by the time the epidemic was over in mid-February.[5]

Being able to predict how many people will get sick, and when, is an important step toward controlling an epidemic. This is one of the responsibilities of Britain's Communicable Disease Surveillance Centre and the US's Center for Disease Control and Prevention.

The S-I-R model

We apply one of the most commonly used models for an epidemic, called the S-I-R model, to the boarding school flu example. Imagine the population of the school divided into three groups:

S = the number of *susceptibles*, the people who are not yet sick but who could become sick

I = the number of *infecteds*, the people who are currently sick

R = the number of *recovered*, or *removed*, the people who have been sick and can no longer infect others or be reinfected.

In this model, the number of susceptibles decreases with time, as people become infected. We assume that the rate people become infected is proportional to the number of contacts between susceptible and infected people. We expect the number of contacts between the two groups to be proportional to both S and I. (If S doubles, we expect the number of contacts to double; similarly, if I doubles, we expect the number of contacts to double.) Thus, we assume that the number of contacts is proportional to the product, SI. In other words, we assume that for some constant $a > 0$,

$$\frac{dS}{dt} = -\left(\begin{array}{c}\text{Rate susceptibles}\\ \text{get sick}\end{array}\right) = -aSI.$$

(The negative sign is used because S is decreasing.)

The number of infecteds is changing in two ways: newly sick people are added to the infected group and others are removed. The newly sick people are exactly those people leaving the susceptible group and so accrue at a rate of aSI (with a positive sign this time). People leave the infected group either because they recover (or die), or because they are physically removed from the rest of the group and can no longer infect others. We assume that people are removed at a rate proportional to the number sick, or bI, where b is a positive constant. Thus,

$$\frac{dI}{dt} = \begin{array}{c}\text{Rate susceptibles}\\ \text{get sick}\end{array} - \begin{array}{c}\text{Rate infecteds}\\ \text{get removed}\end{array} = aSI - bI.$$

[4]Exactly when a disease should be called an epidemic is not always clear. The medical profession generally classifies a disease an epidemic when the frequency is higher than usually expected—leaving open the question of what is usually expected. See, for example, *Epidemiology in Medicine* by C. H. Hennekens and J. Buring (Boston: Little, Brown, 1987).

[5]Data from the Communicable Disease Surveillance Centre (UK); reported in "Influenza in a Boarding School," *British Medical Journal* March 4, 1978, and by J. D. Murray in *Mathematical Biology* (New York: Springer Verlag, 1990).

Assuming that those who have recovered from the disease are no longer susceptible, the recovered group increases at the rate of bI, so

$$\frac{dR}{dt} = bI.$$

We are assuming that having the flu confers immunity on a person, that is, that the person cannot get the flu again. (This is true for a given strain of flu, at least in the short run.)

We can use the fact that the total population $S+I+R$ is not changing. (The total population, the total number of boys in the school, did not change during the epidemic; see Problem 2 on page 435.) Thus, once we know S and I, we can calculate R. So we restrict our attention to the two equations

$$\frac{dS}{dt} = -aSI$$
$$\frac{dI}{dt} = aSI - bI.$$

The Constants a and b

The constant a measures how infectious the disease is—that is, how quickly it is transmitted from the infecteds to the susceptibles. In the case of the flu, we know from medical accounts that the epidemic started with one sick boy, with two more becoming sick roughly a day later. Thus, when $I = 1$ and $S = 762$, we have $dS/dt \approx -2$, enabling us to roughly[6] approximate a:

$$a = -\frac{dS/dt}{SI} = \frac{2}{762 \cdot 1} = 0.0026.$$

The constant b represents the rate at which infected people are removed from the infected population. In this case of the flu, boys were generally taken to the infirmary within one or two days of becoming sick. Assuming half the infected population was removed each day, we take $b \approx 0.5$. Thus, our equations are:

$$\frac{dS}{dt} = -0.0026SI$$
$$\frac{dI}{dt} = 0.0026SI - 0.5I.$$

The Phase Plane

As in Section 10.6, we look at trajectories in the phase plane. Thinking of I as a function of S, and S as a function of t, we use the chain rule to get

$$\frac{dI}{dt} = \frac{dI}{dS} \cdot \frac{dS}{dt},$$

so

$$\frac{dI}{dS} = \frac{dI/dt}{dS/dt}.$$

Substituting for dI/dt and dS/dt, we get

$$\frac{dI}{dS} = \frac{0.0026SI - 0.5I}{-0.0026SI}.$$

Assuming I is not zero, this equation simplifies to approximately

$$\frac{dI}{dS} = -1 + \frac{192}{S}.$$

The slope field of this differential equation is shown in Figure 10.43. The trajectory with initial condition $S_0 = 762$, $I_0 = 1$ is shown in Figure 10.44. Time is represented by the arrow showing the direction that a point moves on the trajectory. The disease starts at the point $S_0 = 762$, $I_0 = 1$. At first, more people become infected and fewer are susceptible. In other words, S decreases and I increases. Later, I decreases as S continues to decrease.

[6] The values of a and b are close to those obtained by J. D. Murray in *Mathematical Biology* (New York: Springer Verlag, 1990).

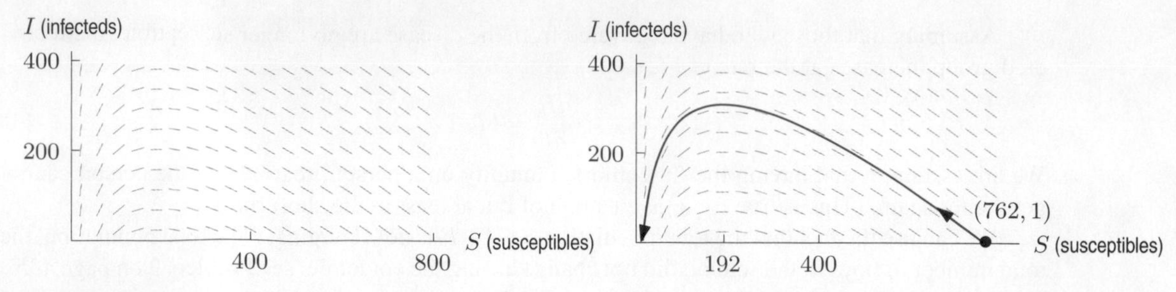

Figure 10.43: Slope field for $dI/dS = -1 + 192/S$

Figure 10.44: Trajectory for $S_0 = 762$, $I_0 = 1$

What does the SI-Phase Plane Tell Us?

To learn how the disease progresses, look at the shape of the curve in Figure 10.44. The value of I first increases, then decreases to zero. This peak value of I occurs when $S \approx 200$. We can determine exactly when the peak value occurs by solving

$$\frac{dI}{dS} = -1 + \frac{192}{S} = 0,$$

which gives

$$S = 192.$$

Notice that the peak value for I always occurs at the same value of S, namely $S = 192$. The graph shows that if a trajectory starts with $S_0 > 192$, then I first increases and then decreases to zero. On the other hand, if $S_0 < 192$, there is no peak as I decreases right away.

For this example, the value $S = 192$ is called a *threshold population*. If S_0 is around or below 192, there is no epidemic. If S_0 is significantly greater than 192, an epidemic occurs.[7]

The phase diagram makes clear that the maximum value of I is about 300, which is the maximum number infected at any one time. In addition, the point at which the trajectory crosses the S-axis represents the time when the epidemic has passed (since $I = 0$). Thus, the S-intercept shows how many boys never get the flu and, hence, how many do get sick.

Threshold Value

For the general SIR model, we have the following result:

$$\text{Threshold population} = \frac{b}{a}.$$

If S_0, the initial number of susceptibles, is above b/a, there is an epidemic; if S_0 is below b/a, there is no epidemic. See Problem 11.

How Many People Should Be Vaccinated?

Faced with an outbreak of the flu or, as happened on several US campuses in the 1980s, of the measles, many institutions consider a vaccination program. How many students must be vaccinated in order to control an outbreak? To answer this, we can think of vaccination as removing people from the S category (without increasing I), which amounts to moving the initial point on the trajectory to the left, parallel to the S-axis. To avoid an epidemic, the initial value of S_0 should be around or below the threshold value. Therefore, the boarding school epidemic would have been avoided if all but 192 students had been vaccinated.

Graphs of S and I against t

On the trajectory in Figure 10.44, the number of susceptible people decreases throughout the epidemic. This makes sense since people are getting sick and then well again, and thus are no longer susceptible to infection. The trajectory also shows that the number of infected people increases and then decreases. Graphs of S and I against time, t, are shown in Figure 10.45.

[7]Here we are using J. D. Murray's definition of an epidemic as an outbreak in which the number of infecteds increases from the initial value, I_0. See *Mathematical Biology* (New York: Springer Verlag, 1990).

To get the scale on the time axis, we would need to use numerical methods. It turns out that the number of infecteds peaked after about 6 days and then dropped. The epidemic ran its course in about 20 days.

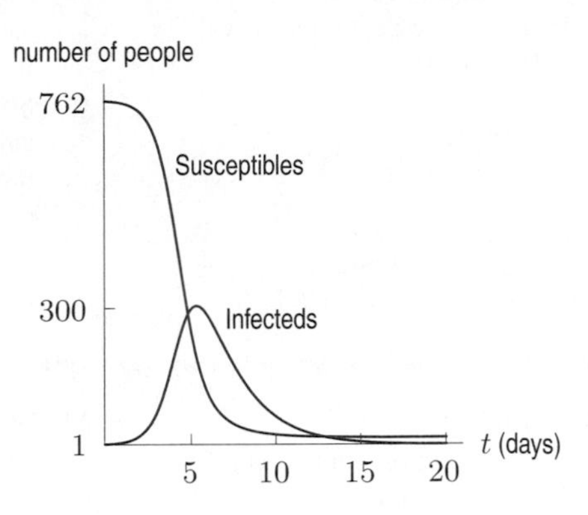

Figure 10.45: Progress of the flu over time

Problems for Section 10.7

1. Let I be the number of infected people and S be the number of susceptible people in an outbreak of a disease. Explain why it is reasonable to model the interaction between these two groups by the differential equations

$$\frac{dS}{dt} = -aSI$$

$$\frac{dI}{dt} = aSI - bI \quad \text{where } a, b \text{ are positive constants.}$$

Why have the signs been chosen this way? Why is the constant a the same in both equations?

2. Show that if S, I, and R satisfy the differential equations in Problem 1, the total population, $S+I+R$, is constant.

3. Explain how you can tell from the graph of the trajectory shown in Figure 10.44 on page 434 that most people at the British boarding school eventually got sick.

4. **(a)** In a school of 150 students, one of the students has the flu initially. What is I_0? What is S_0?
   (b) Use these values of I_0 and S_0 and the equation

$$\frac{dI}{dt} = 0.0026SI - 0.5I$$

   to determine whether the number of infected people initially increases or decreases. What does this tell you about the spread of the disease?

5. Repeat Problem 4 for a school with 350 students.

6. **(a)** On the slope field for dI/dS in Figure 10.43 on page 433, draw the trajectory through the point where $I = 1$ and $S = 400$.
   (b) How many susceptible people are there when the number of infected people is at its maximum?

7. Use Figure 10.45 to estimate the maximum number of infecteds. What does this represent? When does it occur?

8. Compare the diseases modeled by each of the following differential equations with the flu model in this section. Match each set of differential equations with one of the following statements. Write a system of differential equations corresponding to each of the unmatched statements.

(I) $\dfrac{dS}{dt} = -0.04SI$, $\quad \dfrac{dI}{dt} = 0.04SI - 0.2I$

(II) $\dfrac{dS}{dt} = -0.002SI$, $\quad \dfrac{dI}{dt} = 0.002SI - 0.3I$

(III) $\dfrac{dS}{dt} = -0.03SI$, $\quad \dfrac{dI}{dt} = 0.03SI$

   (a) More infectious; infecteds removed more slowly.
   (b) More infectious; infecteds removed more quickly.
   (c) Less infectious; infecteds removed more slowly.
   (d) Less infectious; infecteds removed more quickly.
   (e) Infecteds never removed.

9. For the equations (I) in Problem 8, what is the threshold value of S?

10. For the equations (II) in Problem 8, suppose $S_0 = 100$. Does the disease spread initially? What if $S_0 = 200$?

11. Let S and I satisfy the differential equations in Problem 1. Assume $I \neq 0$.
   (a) If $dI/dt = 0$, find S.
   (b) Show that I increases if S is greater than the value you found in part (a). Show that I decreases if S is less than the value you found in part (a).
   (c) Explain how you know that your answer to part (a) is the threshold value.

12. During World War I, a particularly lethal form of flu killed about 40 million people around the world.[8] The epidemic started in an army camp of 45,000 soldiers outside of Boston, where the first soldier fell sick on September 7, 1918. With time, t, in days since September 7, values of the constants in the SIR model,

$$\frac{dS}{dt} = -aSI$$
$$\frac{dI}{dt} = aSI - bI,$$

are estimated to be $a = 0.000267$, $b = 9.865$.

(a) What are the initial values, S_0 and I_0?
(b) Explain how you know that this model predicts an epidemic in this case.
(c) Find the differential equation for dI/dS. Sketch its slope field and estimate the total number of soldiers infected over the course of the disease.
(d) Solve the differential equation for dI/dS analytically. Use the solution to solve approximately for the number of soldiers affected over the course of the disease.

CHAPTER SUMMARY

- **Differential equations terminology**
  Family of solutions, particular solution, initial conditions, stable/unstable equilibrium solutions
- **Slope fields**
  Visualizing the solution to a differential equation
- **Solving differential equations analytically**
  Solutions to $dy/dt = ky$, and $dy/dt = k(y - A)$, the logistic model
- **Modeling with differential equations**
  Growth and decay, pollution in a lake, quantity of drug in the body, Newton's law of heating and cooling, net worth of a company
- **Systems of differential equations**
  Interaction of two species or businesses, predator-prey model, spread of a disease

REVIEW PROBLEMS FOR CHAPTER TEN

1. Is $y = x^3$ a solution to the differential equation $xy' - 3y = 0$? Justify your answer.

2. For a certain quantity y, assume that $dy/dt = \sqrt{y}$. Fill in the value of y in the following table. Assume that the rate of growth, dy/dt, is approximately constant over each unit time interval and that the initial value of y is 100.

t	0	1	2	3	4
y	100				

3. Slope fields for $dy/dx = 1 + x$ and $dy/dx = 1 + y$ are in Figures 10.46 and 10.47.

(a) Which slope field corresponds to which equation?
(b) On each slope field, draw the solution curve through the origin.
(c) For each slope field, list all equilibrium solutions and indicate whether each is stable or unstable.

(I)

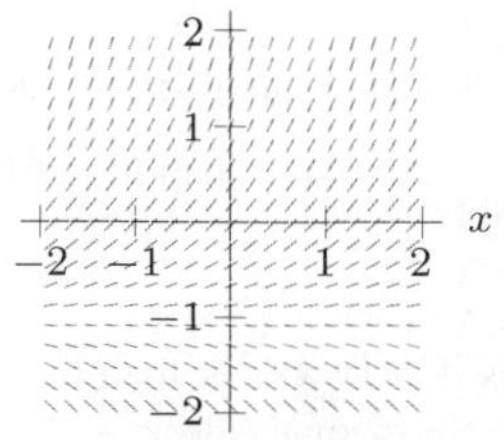

Figure 10.46

(II)

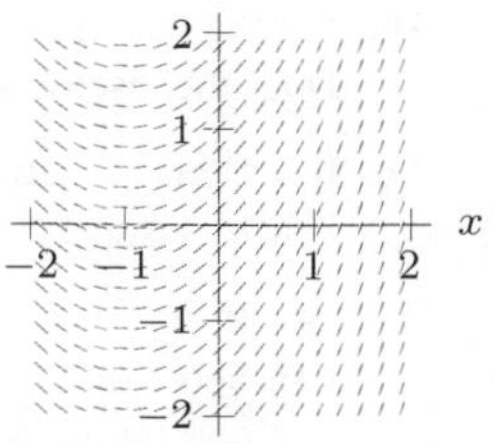

Figure 10.47

4. Figure 10.48 is the slope field for the equation $y' = x+y$.

(a) Sketch the solutions that pass through the points
(i) $(0, 0)$ (ii) $(-3, 1)$ (iii) $(-1, 0)$
(b) From your sketch, guess the equation of the solution passing through $(-1, 0)$.
(c) Check your solution to part (b) by substituting it into the differential equation.

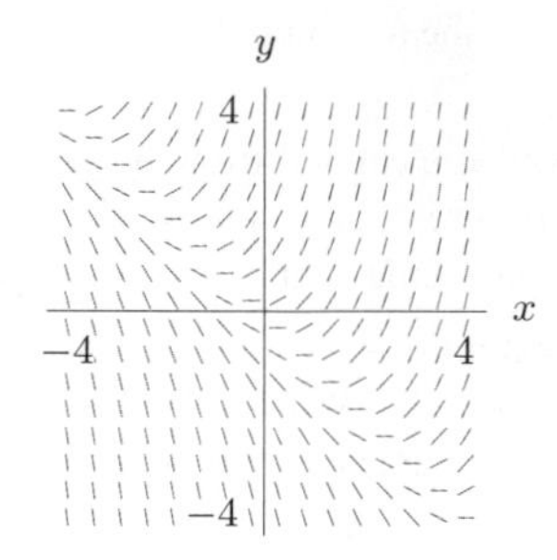

Figure 10.48: Slope field for $y' = x + y$

5. Find the general solution to the differential equation

$$\frac{dy}{dt} = 2t.$$

6. Write a differential equation whose solution is the temperature as a function of time of a bottle of orange juice taken out of a 40°F refrigerator and left in a 65°F room. Solve the equation and graph the solution.

[8] "Capturing a Killer Flu Virus," J. Taukenberger, A. Reid, T. Fanning, in *Scientific American*, Vol. 292, No. 1, Jan. 2005.

7. The Gompertz equation, which models growth of animal tumors, is $y' = -ay\ln(y/b)$, where a and b are positive constants. Write a paragraph describing the similarities and/or differences between solutions to this differential equation with $a = 1$ and $b = 2$ and solutions to the equation $y' = y(2-y)$. Use Figures 10.49 and 10.50.

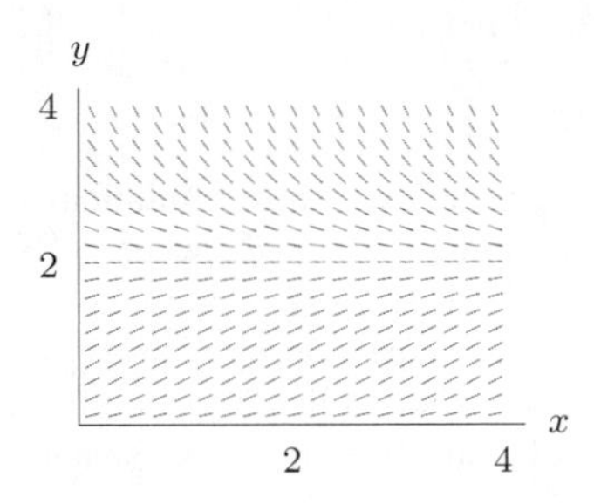

Figure 10.49: Slope field for $y' = -y\ln(y/2)$

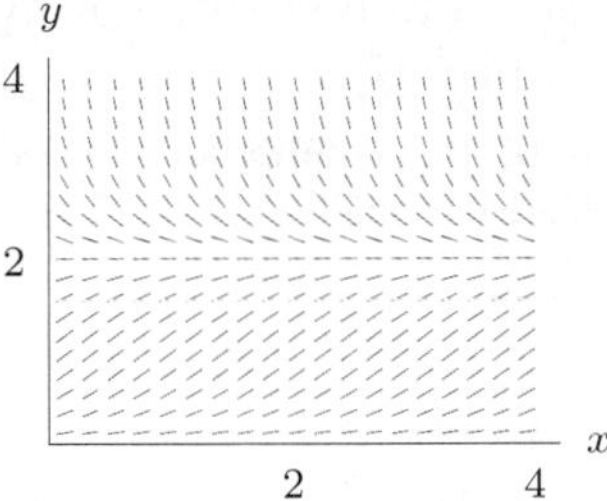

Figure 10.50: Slope field for $y' = y(2-y)$

Solve the differential equations in Problems 8–17.

8. $\dfrac{dP}{dt} = t$

9. $\dfrac{dy}{dt} = 5y$

10. $\dfrac{dy}{dt} = 5t$

11. $\dfrac{dP}{dt} = 0.03P$

12. $\dfrac{dA}{dt} = -0.07A$

13. $\dfrac{1}{Q}\dfrac{dQ}{dt} = 2$

14. $\dfrac{dP}{dt} = 10 - 2P$

15. $\dfrac{dy}{dt} = 100 - y$

16. $\dfrac{dy}{dx} = 0.2y - 8$

17. $\dfrac{dH}{dt} = 0.5H + 10$

For Problems 18–21, solve the differential equations with the given initial conditions and graph the solutions.

18. $\dfrac{dP}{dt} = 0.08P, \quad P = 5000 \text{ when } t = 0$

19. $\dfrac{dy}{dt} = -0.2y, \quad y = 25 \text{ when } t = 0$

20. $\dfrac{dP}{dt} = 0.08P - 50, \quad P(0) = 10$

21. $\dfrac{dH}{dt} = 100 - 0.5H, \quad H(0) = 40$

22. Draw the slope field for the differential equation in Problem 14, and sketch three different solution curves on the slope field.

23. Money in a bank account grows continuously at an annual rate of r (when the interest rate is 5%, $r = 0.05$, and so on). Suppose \$1000 is put into the account in 2000.

(a) Write a differential equation satisfied by M, the amount of money in the account at time t, measured in years since 2000.
(b) Solve the differential equation.
(c) Sketch the solution until the year 2030 for interest rates of 5% and 10%.

24. The amount of land in use for growing crops increases as the world's population increases. Suppose $A(t)$ represents the total number of hectares of land in use in year t. (A hectare is about $2\frac{1}{2}$ acres.)

(a) Explain why it is plausible that $A(t)$ satisfies the equation $A'(t) = kA(t)$. What assumptions are you making about the world's population and its relation to the amount of land used?
(b) In 1950 about $1 \cdot 10^9$ hectares of land were in use; in 1980 the figure was $2 \cdot 10^9$. If the total amount of land available for growing crops is thought to be $3.2 \cdot 10^9$ hectares, when does this model predict it is exhausted? (Let $t = 0$ in 1950.)

25. The radioactive isotope carbon-14 is present in small quantities in all life forms, and it is constantly replenished until the organism dies, after which it decays to stable carbon-12 at a rate proportional to the amount of carbon-14 present, with a half-life of 5730 years. Suppose $C(t)$ is the amount of carbon-14 present at time t.

(a) Find the value of the constant k in the differential equation $C' = -kC$.
(b) In 1988 three teams of scientists found that the Shroud of Turin, which was reputed to be the burial cloth of Jesus, contained 91% of the amount of carbon-14 contained in freshly made cloth of the same material.[9] How old is the Shroud of Turin, according to these data?

26. A bank account that earns 10% interest compounded continuously has an initial balance of zero. Money is deposited into the account at a continuous rate of \$1000 per year.

(a) Write a differential equation that describes the rate of change of the balance $B = f(t)$.
(b) Solve the differential equation.

27. One theory on the speed an employee learns a new task claims that the more the employee already knows, the slower he or she learns. Suppose that the rate at which a person learns is equal to the percentage of the task not yet learned. If y is the percentage learned by time t, the percentage not yet learned by that time is $100 - y$, so we can model this situation with the differential equation

[9] *The New York Times*, October 18, 1988

$$\frac{dy}{dt} = 100 - y.$$

(a) Find the general solution to this differential equation.

(b) Sketch several solutions.

(c) Find the particular solution if the employee starts learning at time $t = 0$ (so $y = 0$ when $t = 0$).

28. **(a)** What are the equilibrium solutions for the differential equation

$$\frac{dy}{dt} = 0.2(y - 3)(y + 2)?$$

(b) Use a graphing calculator or computer to sketch a slope field for this differential equation. Use the slope field to determine whether each equilibrium solution is stable or unstable.

29. As you know, when a course ends, students start to forget the material they have learned. One model (called the Ebbinghaus model) assumes that the rate at which a student forgets material is proportional to the difference between the material currently remembered and some positive constant, a.

(a) Let $y = f(t)$ be the fraction of the original material remembered t weeks after the course has ended. Set up a differential equation for y. Your equation will contain two constants; the constant a is less than y for all t.

(b) Solve the differential equation.

(c) Describe the practical meaning (in terms of the amount remembered) of the constants in the solution $y = f(t)$.

For Problems 30–33, suppose x and y are the populations of two different species. Describe in words how each population changes with time.

30.

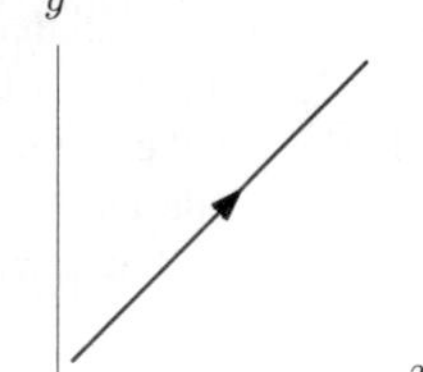

31.

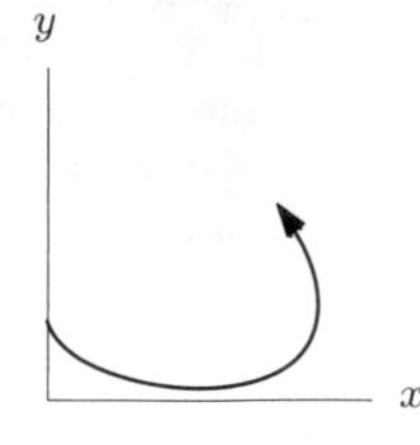

32.

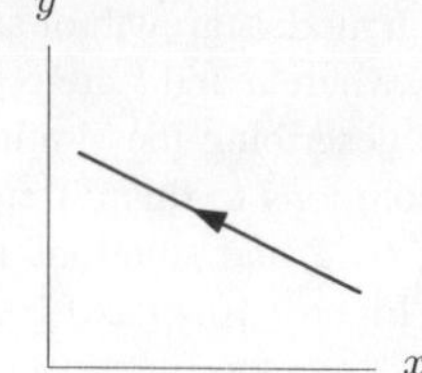

33.

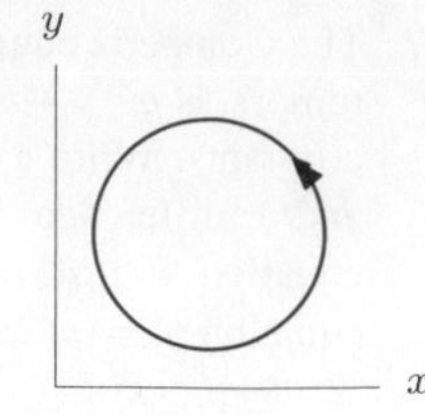

34. **(a)** What are the equilibrium solutions to the differential equation $y' = f(y)$, where $f(y)$ is graphed in Figure 10.51?

(b) Sketch the slope field for $y' = f(y)$.

(c) For each of the following initial conditions, sketch the solution curve on your slope field:

(i) $y(0) = 0$ (ii) $y(0) = 1$ (iii) $y(0) = 6$
(iv) $y(0) = 8$ (v) $y(0) = 10$ (vi) $y(0) = 16$
(vii) $y(0) = 17$

(d) Which of the equilibrium solutions are stable? Which are unstable?

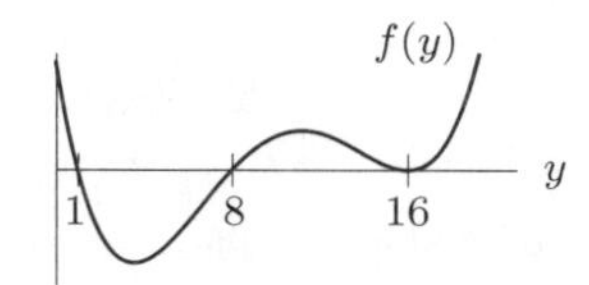

Figure 10.51

35. **(a)** Graph $f(y) = y - y^2$.

(b) Sketch the slope field of the differential equation

$$\frac{dy}{dx} = y - y^2.$$

(c) Describe how the two graphs are related. In particular, explain how you can determine the equilibrium solutions to the differential equation from the graph of $f(y)$ and how you can tell if the equilibrium solutions are stable.

PROJECTS FOR CHAPTER TEN

1. Harvesting and Logistic Growth In this project, we look at the effects of *harvesting* a population which is growing logistically. Harvesting could be, for example, fishing or logging. An important question is what level of harvesting leads to a *sustainable yield*. In other words, how much can be harvested without having the population depleted in the long run?

(a) When there is no fishing, a population of fish is governed by the differential equation

$$\frac{dN}{dt} = 2N - 0.01N^2,$$

where N is the number of fish at time t in years. Sketch a graph of dN/dt against N. Mark on your graph the equilibrium values of N.

Notice on your graph that if N is between 0 and 200, then dN/dt is positive and N increases. If N is greater than 200, then dN/dt is negative and N decreases. Check this by sketching a slope field for this differential equation. Use the slope field to sketch solutions showing N against t for various initial values. Describe what you see.

(b) Fish are now removed by fishermen at a continuous rate of 75 fish/year. Let P be the number of fish at time t with harvesting. Explain why P satisfies the differential equation

$$\frac{dP}{dt} = 2P - 0.01P^2 - 75.$$

(c) Sketch dP/dt against P. Find and label the intercepts.

(d) Sketch the slope field for the differential equation for P.

(e) Recall that if dP/dt is positive for some values of P, then P increases for these values, and if dP/dt is negative for some values of P, then P decreases for these values. The value of P, however, never goes past an equilibrium value. Use this information and the graph from part (c) to answer the following questions:

(i) What are the equilibrium values of P?

(ii) For what initial values of P does P increase? At what value does P level off?

(iii) For what initial values of P does P decrease?

(f) Use the slope field in part (d) to sketch graphs of P against t, with the initial values:

(i) $P(0) = 40$ (ii) $P(0) = 50$ (iii) $P(0) = 60$
(iv) $P(0) = 150$ (v) $P(0) = 170$

(g) Using the graphs you drew, decide what the equilibrium values of the populations are and whether or not they are stable.

(h) We now look at the effect of different levels of fishing on a fish population. If fishing takes place at a continuous rate of H fish/year, the fish population P satisfies the differential equation

$$\frac{dP}{dt} = 2P - 0.01P^2 - H.$$

(i) For each of the values $H = 75, 100, 200$, plot dP/dt against P.

(ii) For which of the three values of H that you considered in part (i) is there an initial condition such that the fish population does not die out eventually?

(iii) Looking at your answer to part (ii), decide for what values of H there is an initial value for P such that the population does not die out eventually.

(iv) Recommend a policy to ensure long-term survival of the fish population.

2. Population Genetics

Population genetics is the study of hereditary traits in a population. A specific hereditary trait has two possibilities, one dominant, such as brown eyes, and one recessive, such as blue eyes.[10] Let b denote the gene responsible for the recessive trait and B denote the gene responsible for the dominant trait. Each member of the population has a pair of these genes—either BB (dominant individuals), bb (recessive individuals), or Bb (hybrid individuals). The *gene frequency* of the b gene is the total number of b genes in the population divided by the total number of all genes (b and B) controlling this trait. The gene frequency is essentially constant when there are no mutations or outside influences on the population. In this project, we consider the effect of mutations on the gene frequency.

Let q denote the gene frequency of the b gene. Then q is between 0 and 1 (since it is a fraction of a whole) and, since b and B are the only genes influencing this trait, the gene frequency of the B gene is $1 - q$. Let time t be measured in generations. Every generation, a

[10] Adapted from C. C. Li, *Population Genetics* (Chicago: University of Chicago Press, 1995).

fraction k_1 of the b genes mutate to become B genes, and a fraction k_2 of the B genes mutate to become b genes.

(a) Explain why the gene frequency q satisfies the differential equation:

$$\frac{dq}{dt} = -k_1 q + k_2(1-q).$$

(b) If $k_1 = 0.0001$ and $k_2 = 0.0004$, simplify the differential equation for q and solve it. The initial value is q_0. Sketch the solutions with $q_0 = 0.1$ and $q_0 = 0.9$. What is the equilibrium value of q? Explain how you can tell that the gene frequency gets closer to the equilibrium value as generations pass. Explain how you can tell that the equilibrium value is completely determined by the relative mutation rates.

(c) Repeat part (b) if $k_1 = 0.00003$ and $k_2 = 0.00001$.

3. The Spread of SARS

In the spring of 2003, SARS (Severe Acute Respiratory Syndrome) spread rapidly in several Asian countries and Canada. Predicting the course of the disease—how many people would be infected, how long it would last—was important to officials trying to minimize the impact of the disease. This project analyzes the spread of SARS through interaction between infected and susceptible people.

The variables are S, the number of susceptibles, I, the number of infecteds who can infect others, and R, the number removed (this group includes those in quarantine and those who die, as well as those who have recovered and acquired immunity). Time, t, is in days since March 17, 2003, the date the World Health Organization (WHO) started to publish daily SARS reports. On March 17, Hong Kong reported 95 cases. In this model

$$\frac{dS}{dt} = -aSI$$
$$\frac{dI}{dt} = aSI - bI,$$

and $S + I + R = 6.8$ million, the population of Hong Kong in 2003.[11] Estimates based on WHO data give $a = 1.25 \cdot 10^{-8}$.

(a) What are S_0 and I_0, the initial values of S and I?

(b) During March 2003, the value of b was about 0.06. Using a calculator or computer, sketch the slope field for this system of differential equations and the solution trajectory corresponding to the initial conditions. (Use $0 \le S \le 7 \cdot 10^6, 0 \le I \le 0.4 \cdot 10^6$.)

(c) What does your graph tell you about the total number of people infected over the course of the disease if $b = 0.06$? What is the threshold value? What does this value tell you?

(d) During April, as public health officials worked to get the disease under control, people who had been in contact with the disease were quarantined. Explain why quarantining has the effect of raising the value of b.

(e) Using the April value, $b = 0.24$, sketch the slope field. (Use the same value of a and the same window.)

(f) What is the threshold value for $b = 0.24$? What does this tell you? Comment on the quarantine policy.

(g) Comment on the effectiveness of each of the following policies intended to prevent an epidemic and protect a city from an outbreak of SARS in a nearby region.

I Close off the city from contact with the infected region. Shut down roads, airports, trains, and other forms of direct contact.

II Install a quarantine policy. Isolate anyone who has been in contact with a SARS patient or anyone who shows symptoms of SARS.

[11] www.census.gov, International Data Base (IDB), accessed June 8, 2004.

FOCUS ON THEORY

SEPARATION OF VARIABLES

We have seen how to sketch solution curves of a differential equation using a slope field. Now we see how to solve certain differential equations analytically, finding an equation for the solution curve.

First, we look at a familiar example, the differential equation

$$\frac{dy}{dx} = -\frac{x}{y},$$

whose solution curves are the circles

$$x^2 + y^2 = C.$$

We can check that these circles are solutions by differentiation; the question now is how they were obtained. The method of *separation of variables* works by putting all the xs on one side of the equation and all the ys on the other, giving

$$y\,dy = -x\,dx.$$

We then integrate each side separately:

$$\int y\,dy = -\int x\,dx,$$

$$\frac{y^2}{2} = -\frac{x^2}{2} + k.$$

This gives the circles we were expecting:

$$x^2 + y^2 = C \qquad \text{where } C = 2k.$$

You might worry about whether it is legitimate to separate the dx and the dy. The reason it can be done is explained at the end of this section.

The Exponential Growth and Decay Equations

We use separation of variables to derive the general solution of the equation

$$\frac{dy}{dt} = ky.$$

Separating variables, we have

$$\frac{1}{y}dy = k\,dt,$$

and integrating,

$$\int \frac{1}{y}\,dy = \int k\,dt,$$

gives

$$\ln|y| = kt + C \quad \text{for some constant } C.$$

Solving for $|y|$ leads to

$$|y| = e^{kt+C} = e^{kt}e^C = Ae^{kt}$$

where $A = e^C$, so A is positive. Thus,

$$y = (\pm A)e^{kt} = Be^{kt}$$

where $B = \pm A$, so B is any nonzero constant. Even though there's no C leading to $B = 0$, we can have $B = 0$ because $y = 0$ is a solution to the differential equation. We lost this solution when we divided through by y at the first step. Thus we have derived the solution used earlier in the chapter:

$$y = Be^{kt} \quad \text{for any constant } B.$$

Example 1 Find all solutions of

$$\frac{dy}{dt} = k(y - A).$$

Solution We separate variables and integrate:

$$\int \frac{1}{y - A}\,dy = \int k\,dt.$$

This gives

$$\ln|y - A| = kt + D,$$

where D is a constant of integration. Solving for y leads to

$$|y - A| = e^{kt+D} = e^{kt}e^{D} = Be^{kt}$$

or

$$y - A = (\pm B)e^{kt} = Ce^{kt}$$
$$y = A + Ce^{kt}.$$

Also, $C = 0$ gives a solution. This is the same result we used earlier.

Example 2 Find and sketch the solution to

$$\frac{dP}{dt} = 2P - 2Pt \qquad \text{satisfying } P = 5 \text{ when } t = 0.$$

Solution Factoring the right-hand side gives

$$\frac{dP}{dt} = P(2 - 2t).$$

Separating variables, we get

$$\int \frac{dP}{P} = \int (2 - 2t)\,dt,$$

so

$$\ln|P| = 2t - t^2 + C.$$

Solving for P leads to

$$|P| = e^{2t-t^2+C} = e^{C}e^{2t-t^2} = Ae^{2t-t^2}$$

with $A = e^{C}$, so $A > 0$. In addition, $A = 0$ gives a solution. Thus the general solution to the differential equation is

$$P = Be^{2t-t^2} \quad \text{for any } B.$$

To find the value of B, substitute $P = 5$ and $t = 0$ into the general solution, giving

$$5 = Be^{2\cdot 0 - 0^2} = B$$

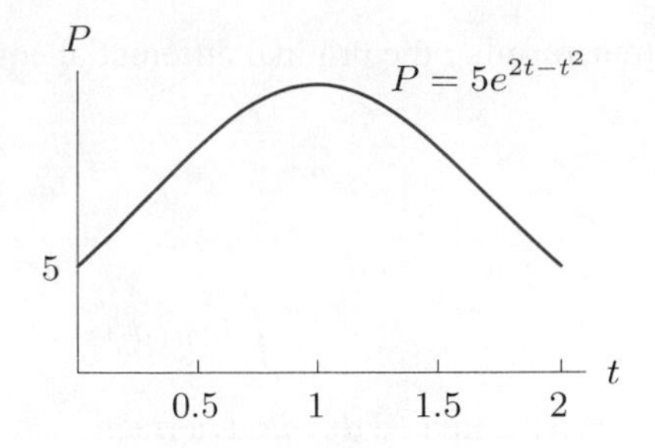

Figure 10.52: Bell-shaped solution curve

so

$$P = 5e^{2t-t^2}.$$

The graph of this function is in Figure 10.52. Since the solution can be rewritten as

$$P = 5e^{1-1+2t-t^2} = 5e^1 e^{-1+2t-t^2} = (5e)e^{-(t-1)^2},$$

the graph has the same shape as the graph of $y = e^{-t^2}$, the bell-shaped curve of statistics. Here the maximum, normally at $t = 0$, is shifted one unit to the right to $t = 1$.

Justification For Separation of Variables

Suppose a differential equation can be written in the form

$$\frac{dy}{dx} = g(x)f(y).$$

Provided $f(y) \neq 0$, we write $f(y) = 1/h(y)$ so the right-hand side can be thought of as a fraction,

$$\frac{dy}{dx} = \frac{g(x)}{h(y)}.$$

If we multiply through by $h(y)$ we get

$$h(y)\frac{dy}{dx} = g(x).$$

Thinking of y as a function of x, so $y = y(x)$, and $dy/dx = y'(x)$, we can rewrite the equation as

$$h(y(x)) \cdot y'(x) = g(x).$$

Now integrate both sides with respect to x:

$$\int h(y(x)) \cdot y'(x)\, dx = \int g(x)\, dx.$$

The form of the integral on the left suggests that we use the substitution $y = y(x)$. Since $dy = y'(x)\, dx$, we get

$$\int h(y)\, dy = \int g(x)\, dx.$$

If we can find antiderivatives of h and g, then this gives the equation of the solution curve.

Note that transforming the original differential equation,

$$\frac{dy}{dx} = \frac{g(x)}{h(y)},$$

into

$$\int h(y)\,dy = \int g(x)\,dx$$

looks as though we have treated dy/dx as a fraction, cross-multiplied and then integrated. Although that's not exactly what we have done, you may find this a helpful way of remembering the method. In fact, the dy/dx notation was introduced by Leibniz to allow shortcuts like this (more specifically, to make the chain rule look like cancellation).

Problems on Separation of Variables

Use separation of variables to find the solutions to the differential equations in Problems 1–12, subject to the given initial conditions.

1. $\frac{dP}{dt} = -2P, \quad P(0) = 1$

2. $\frac{dL}{dp} = \frac{L}{2}, \quad L(0) = 100$

3. $P\frac{dP}{dt} = 1, \quad P(0) = 1$

4. $\frac{dm}{ds} = m, \quad m(1) = 2$

5. $2\frac{du}{dt} = u^2, \quad u(0) = 1$

6. $\frac{dz}{dy} = zy, \quad z = 1$ when $y = 0$

7. $\frac{dR}{dy} + R = 1, \quad R(1) = 0.1$

8. $\frac{dy}{dt} = \frac{y}{3+t}, \quad y(0) = 1$

9. $\frac{dz}{dt} = te^z$, through the origin

10. $\frac{dy}{dx} = \frac{5y}{x}, \quad y = 3$ where $x = 1$

11. $\frac{dy}{dt} = y^2(1+t), \quad y = 2$ when $t = 1$

12. $\frac{dz}{dt} = z + zt^2, \quad z = 5$ when $t = 0$

13. Determine which of the following differential equations is separable. Do not solve the equations.

(a) $y' = y$ **(b)** $y' = x + y$
(c) $y' = xy$ **(d)** $y' = \sin(x+y)$
(e) $y' - xy = 0$ **(f)** $y' = y/x$
(g) $y' = \ln(xy)$ **(h)** $y' = (\sin x)(\cos y)$
(i) $y' = (\sin x)(\cos xy)$ **(j)** $y' = x/y$
(k) $y' = 2x$ **(l)** $y' = (x+y)/(x+2y)$

Use separation of variables to solve the differential equations in Problems 14–19. Assume a, b, and k are nonzero constants.

14. $\frac{dP}{dt} = P - a$

15. $\frac{dQ}{dt} = b - Q$

16. $\frac{dP}{dt} = k(P - a)$

17. $\frac{dR}{dt} = aR + b$

18. $\frac{dP}{dt} - aP = b$

19. $\frac{dy}{dt} = ky^2(1+t^2)$

20. **(a)** Find the general solution to the differential equation modeling how a person learns:

$$\frac{dy}{dt} = 100 - y.$$

(b) Plot the slope field of this differential equation and sketch solutions with $y(0) = 25$ and $y(0) = 110$.
(c) For each of the initial conditions in part (b), find the particular solution and add to your sketch.
(d) Which of these two particular solutions could represent how a person learns?

21. **(a)** Sketch the slope field for the differential equation $dy/dx = xy$.
(b) Sketch several solution curves.
(c) Solve the differential equation analytically.

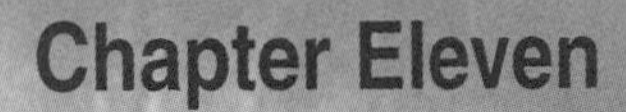

Chapter Eleven

GEOMETRIC SERIES

In this chapter, we study *geometric series* and their applications. A geometric series is a sum in which each term is a constant multiple of the preceding term. We investigate sums of geometric series with a finite number of terms and with an infinite number of terms.

Section 11.1 introduces geometric series. Section 11.2 discusses applications of geometric series in business and economics, such as annuities and the multiplier effect. In Section 11.3, we see applications in the life sciences, such as repeated doses of drugs.

11.1 GEOMETRIC SERIES

Repeated Drug Dosage

Malaria is a parasitic infection transmitted by mosquito bites, mainly in tropical areas of the world. The disease has existed since ancient times, and currently there are hundreds of millions of cases each year, with millions of deaths. Around 1630, Jesuits in Peru introduced the bark of the cinchona tree to the West as the first treatment for malaria. The drug quinine is the active ingredient in the bark, and it is still used today.

Suppose a person is given a 50 mg dose of quinine at the same time every day for the prevention of malaria. After the first dose, the person has 50 mg of quinine in the body. What about after the second dose? Each day, the person's body metabolizes some of the quinine so that, after one day, 23% of the original amount remains. After the second dose, the amount of quinine in the body is the amount from the second dose (50 mg) plus the remnants of the first dose ($50 \cdot 0.23 = 11.5$ mg) for a total of 61.5 mg.

Let Q_n represent the quantity, in mg, of quinine in the body right after the nth dose. Then

$$\begin{aligned}
Q_1 &= \text{First dose } = 50.\\
Q_2 &= \text{Second dose + Remnants of first dose } = 50 + 50(0.23) = 61.5.\\
Q_3 &= \text{Third dose + Remnants of previous doses } = 50 + 61.5(0.23) = 64.145.
\end{aligned}$$

Notice that we can multiply out the expression for Q_3 to show the contributions of the first and second dose separately:

$$\begin{aligned}
Q_3 &= 50 + 61.5(0.23) = 50 + (50 + 50(0.23))(0.23)\\
Q_3 &= 50 + 50(0.23) + 50(0.23)^2,
\end{aligned}$$

so we have

$$Q_3 = \text{Third dose + Remnants of second dose + Remnants of first dose.}$$

The multiplied out form of Q_3 enables us to guess formulas for later values of Q_n:

$$\begin{aligned}
Q_4 &= 50 + 50(0.23) + 50(0.23)^2 + 50(0.23)^3 = 64.753.\\
Q_5 &= 50 + 50(0.23) + 50(0.23)^2 + 50(0.23)^3 + 50(0.23)^4 = 64.893.\\
Q_6 &= 50 + 50(0.23) + 50(0.23)^2 + 50(0.23)^3 + 50(0.23)^4 + 50(0.23)^5 = 64.925.\\
&\vdots\\
Q_{10} &= 50 + 50(0.23) + 50(0.23)^2 + \cdots + 50(0.23)^8 + 50(0.23)^9 = 64.935.
\end{aligned}$$

Comparing the values of Q_6 and Q_{10}, we see that the quantity seems to be stabilizing at around 64.9 mg. See Figure 11.1. Notice that although we have calculated Q_6 and Q_{10} to several decimal places, there is no practical difference between values that are this close.

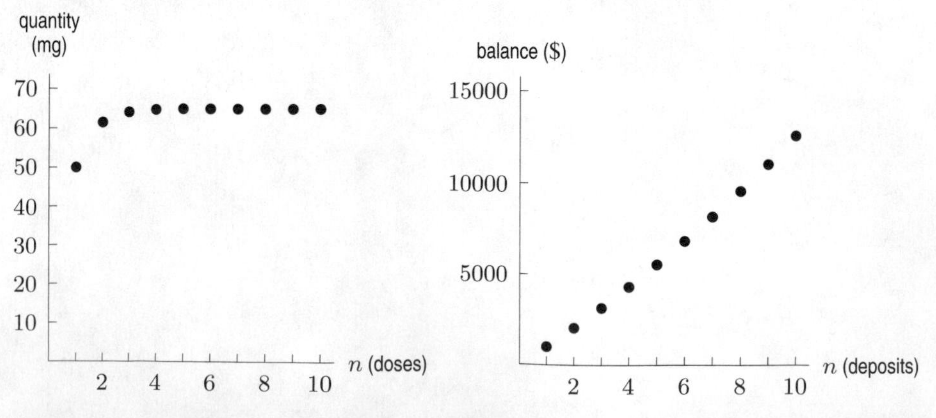

Figure 11.1: Quantity of quinine levels off

Figure 11.2: Bank balance grows without bound

Repeated Deposits into a Savings Account

People who save money often do so by putting a fixed amount aside regularly. Suppose \$1000 is deposited every year in a savings account earning 5% interest a year, compounded annually. Let B_n represent the balance, in dollars, in the account right after the nth deposit. Then

$$\begin{aligned} B_1 &= \text{First deposit } = 1000.\\ B_2 &= \text{Second deposit + Amount from first deposit } = 1000 + 1000(1.05) = 2050.\\ B_3 &= \text{Third deposit + Amount from previous deposits } = 1000 + 2050(1.05) = 3152.5.\end{aligned}$$

As before, we multiply out the expression for B_3 to show the contributions of the first and second deposits separately:

$$\begin{aligned} B_3 &= 1000 + 2050(1.05) = 1000 + (1000 + 1000(1.05))(1.05)\\ B_3 &= 1000 + 1000(1.05) + 1000(1.05)^2\\ B_3 &= \text{Third deposit + Amount from second deposit + Amount from first deposit.}\end{aligned}$$

The multiplied out formula for B_3 enables us to guess formulas for B_6 and B_{10}. Evaluating gives

$$B_6 = 1000 + 1000(1.05) + 1000(1.05)^2 + 1000(1.05)^3 + 1000(1.05)^4 + 1000(1.05)^5 = 6801.91.$$

$$B_{10} = 1000 + 1000(1.05) + 1000(1.05)^2 + \cdots + 1000(1.05)^8 + 1000(1.05)^9 = 12{,}577.89.$$

Notice that the balance is growing without bound. See Figure 11.2.

Finite Geometric Series

In the two previous examples, we encountered sums of the form $a+ar+ar^2+ar^3+\cdots+ar^8+ar^9$. Such a sum is called a *finite geometric series*. A geometric series is a sum in which each term is a constant multiple of the preceding one. The first term is a, and the constant multiplier, or common ratio, of successive terms is r.

> A **finite geometric series** with n terms has (for n a positive integer) the form
>
> $$a + ar + ar^2 + ar^3 + \cdots + ar^{n-2} + ar^{n-1}.$$

Sum of a Finite Geometric Series

In the quinine example, suppose we want to find Q_{40}, the quantity in the body after 40 doses:

$$Q_{40} = 50 + 50(0.23) + 50(0.23)^2 + \cdots + 50(0.23)^{38} + 50(0.23)^{39}.$$

To calculate Q_{40}, it appears that we have to add 40 terms. Fortunately, there is a better way.

We write S_n for the sum of the first n terms of the series, that is, up to the term ar^{n-1}:

$$S_n = a + ar + ar^2 + ar^3 + \cdots + ar^{n-2} + ar^{n-1}.$$

Multiplying both sides by r gives

$$rS_n = ar + ar^2 + ar^3 + ar^4 + \cdots + ar^{n-1} + ar^n.$$

Now subtract rS_n from S_n, which cancels out all terms except two on the right, giving

$$S_n - rS_n = a - ar^n,$$

so

$$(1-r)S_n = a(1-r^n).$$

Provided $r \neq 1$, we can solve for S_n. The result is called a *closed form* for S_n.

The **sum of a finite geometric series** is given by

$$S_n = a + ar + ar^2 + ar^3 + \cdots + ar^{n-1} = \frac{a(1-r^n)}{1-r}, \quad \text{provided } r \neq 1.$$

Note that the value of n in the formula is the number of terms in the sum S_n.

Example 1 In the quinine example, calculate and interpret Q_{40} and Q_{100}.

Solution We saw earlier that

$$Q_{40} = 50 + 50(0.23) + 50(0.23)^2 + \cdots + 50(0.23)^{39}.$$

This is a finite geometric series with $a = 50$ and $r = 0.23$. Using the formula for the sum with $n = 40$, we have

$$Q_{40} = \frac{50(1-(0.23)^{40})}{1-0.23} = 64.935.$$

The amount of quinine in the body right after the 40th dose is 64.935 mg.

Similarly, using $n = 100$, we have

$$Q_{100} = \frac{50(1-(0.23)^{100})}{1-0.23} = 64.935.$$

Right after the 100th dose, the amount of quinine in the body is still 64.935 mg. To three decimal places, the amount appears to have stabilized.

Example 2 In the bank deposit example, calculate and interpret B_{40} and B_{100}.

Solution We have

$$B_{40} = 1000 + 1000(1.05) + 1000(1.05)^2 + \cdots + 1000(1.05)^{39}.$$

This is a finite geometric series with $a = 1000$ and $r = 1.05$. The formula for the sum with $n = 40$ gives

$$B_{40} = \frac{1000(1-(1.05)^{40})}{1-1.05} = 120{,}799.77.$$

The balance in the account right after the 40th deposit is \$120,799.77.

Similarly, using $n = 100$, we have

$$B_{100} = \frac{1000(1-(1.05)^{100})}{1-1.05} = 2{,}610{,}025.16.$$

Right after the 100th deposit, the balance in the account is \$2,610,025.16. Compound interest has increased the \$100,000 investment to over \$2 million.

Infinite Geometric Series

Suppose a finite geometric series has n terms in it. What happens as $n \to \infty$? We get an infinite geometric series that goes on forever.

An **infinite geometric series** has the form

$$a + ar + ar^2 + ar^3 + \cdots + ar^{n-1} + ar^n + \cdots.$$

The "$\cdots$" at the end of the series tells us that the series is going on forever—it is infinite.

Sum of an Infinite Geometric Series

Given an infinite geometric series

$$a + ar + ar^2 + ar^3 + \cdots,$$

we call the sum of the first n terms a *partial sum*, written S_n. To calculate S_n, we use the formula

$$S_n = \frac{a(1 - r^n)}{1 - r}.$$

What happens to S_n as $n \to \infty$? It depends on the value of r. If $|r| < 1$, that is, $-1 < r < 1$, then $r^n \to 0$ as $n \to \infty$, so as $n \to \infty$,

$$S_n = \frac{a(1 - r^n)}{1 - r} \to \frac{a(1 - 0)}{1 - r} = \frac{a}{1 - r}.$$

Thus, provided $|r| < 1$, as $n \to \infty$ the partial sums S_n approach a limit of $a/(1 - r)$. When this happens, we define the sum of the infinite geometric series to be that limit and say the series *converges* to $a/(1 - r)$.

For $|r| < 1$, the **sum of the infinite geometric series** is given by

$$S = a + ar + ar^2 + ar^3 + \cdots + ar^{n-1} + ar^n + \cdots = \frac{a}{1 - r}.$$

If, on the other hand, $|r| > 1$, then r^n and the partial sums have no limit as $n \to \infty$ (if $a \neq 0$). In this case, we say the series *diverges*. If $r > 1$, the terms in the series become larger and larger in magnitude, and the partial sums diverge to $+\infty$ if $a > 0$, or to $-\infty$ if $a < 0$. If $r < -1$, the terms become larger in magnitude, the partial sums oscillate as $n \to \infty$, and the series diverges.

What happens if $r = 1$? The series is

$$a + a + a + a + \cdots,$$

so if $a \neq 0$, the partial sums grow without bound, and the series does not converge. If $r = -1$, the series is

$$a - a + a - a + a - \cdots,$$

and, if $a \neq 0$, the partial sums oscillate between a and 0, and the series does not converge.

Example 3 For each of the following infinite series, find the first three partial sums and the sum (if it exists).

(a) $10 + 10(0.75) + 10(0.75)^2 + \cdots$

(b) $250 + 250(1.2) + 250(1.2)^2 + \cdots$

Solution (a) This is an infinite geometric series with $a = 10$ and $r = 0.75$. The first three partial sums are:

$$S_1 = 10.$$
$$S_2 = 10 + 10(0.75) = 10 + 7.5 = 17.5.$$
$$S_3 = 10 + 10(0.75) + 10(0.75)^2 = 10 + 7.5 + 5.625 = 23.125.$$

Since $|r| < 1$, the series converges and the sum is

$$S = \frac{a}{1 - r} = \frac{10}{1 - 0.75} = 40.$$

If we find partial sums for larger and larger n, they get closer and closer to 40. (See Problem 15.)

(b) This is an infinite geometric series with $a = 250$ and $r = 1.2$. The first three partial sums are:

$$S_1 = 250.$$
$$S_2 = 250 + 250(1.2) = 250 + 300 = 550.$$
$$S_3 = 250 + 250(1.2) + 250(1.2)^2 = 250 + 300 + 360 = 910.$$

Since $r > 1$, the series diverges, and the partial sums grow without bound. (See Problem 16.)

Example 4 If 50 mg doses of quinine are taken daily forever, find the long-term quantity of quinine in the body right after a dose is given and right before a dose is given.

Solution Since quinine is given forever, from Example 1 we know that the long-term quantity of quinine in the body, right after a dose, is given by

$$Q = 50 + 50(0.23) + 50(0.23)^2 + \cdots.$$

This is an infinite geometric series with $a = 50$ and $r = 0.23$. Since $-1 < r < 1$, the series converges to a finite sum given by

$$Q = \frac{a}{1-r} = \frac{50}{1-0.23} = 64.935.$$

The long-term quantity of quinine in the body right after a dose is 64.935 mg. The quantity levels off to this value in Figure 11.1. In fact, after ten doses, the quantity already agrees with the long-term value to three decimal places, since $Q_{10} = 64.395$.

What is the long-term quantity of quinine right before a dose? Since a dose is 50 mg, the quantity of quinine in the body right before a dose is $64.935 - 50 = 14.935$ mg. Thus, in the long term, the quinine level oscillates between 15 mg and 65 mg.

Example 5 Suppose that \$1000 a year is deposited forever into the bank account in Example 2. Does the balance in the account stabilize at a fixed amount? Explain.

Solution Since the deposits are made forever, the quantity in the account right after a deposit is represented by the sum

$$B = 1000 + 1000(1.05) + 1000(1.05)^2 + \cdots.$$

This is an infinite geometric series with $a = 1000$ and $r = 1.05$. Since r is larger than 1, the series diverges. This makes sense, since if you keep depositing \$1000 in an account, your balance grows without bound, even if you don't earn interest. This matches Figure 11.2 and Example 2.

Problems for Section 11.1

1. Find the sum of the following series in two ways: by adding terms and by using the geometric series formula.

$$3 + 3\cdot 2 + 3\cdot 2^2$$

2. Find the sum of the following series in two ways: by adding terms and by using the geometric series formula.

$$50 + 50(0.9) + 50(0.9)^2 + 50(0.9)^3$$

In Problems 3–14, find the sum, if it exists.

3. $2 + 2^2 + 2^3 + \cdots + 2^{10}$

4. $20 + 20(1.4) + 20(1.4)^2 + \cdots + 20(1.4)^8$

5. $1000 + 1000(1.08) + 1000(1.08)^2 + 1000(1.08)^3 + \cdots$

6. $500 + 500(0.6) + 500(0.6)^2 + \cdots + 500(0.6)^{15}$

7. $30 + 30(0.85) + 30(0.85)^2 + 30(0.85)^3 + \cdots$

8. $25 + 25(0.2) + 25(0.2)^2 + 25(0.2)^3 + \cdots$

9. $1 + \frac{1}{2} + \frac{1}{2^2} + \cdots + \frac{1}{2^8}$

10. $1 + \frac{1}{3} + \frac{1}{3^2} + \frac{1}{3^3} + \cdots$

11. $3 + \frac{3}{2} + \frac{3}{4} + \frac{3}{8} + \cdots + \frac{3}{2^{10}}$

12. $1000 + 1500 + 2250 + 3375 + 5062.5 + \cdots$

13. $200 + 100 + 50 + 25 + 12.5 + \cdots$

14. $-2 + 1 - \frac{1}{2} + \frac{1}{4} - \frac{1}{8} + \frac{1}{16} - \cdots$

15. In Example 3(a), we found partial sums of the geometric series with $a = 10$ and $r = 0.75$ and showed that the sum of this series is 40. Find the partial sums S_n for $n = 5, 10, 15, 20$. As n gets larger, do the partial sums appear to be approaching 40?

16. In Example 3(b), we found partial sums for the geometric series with $a = 250$ and $r = 1.2$. Find the partial sums S_n for $n = 5, 10, 15, 20$. As n gets larger, do the partial sums appear to grow without bound, as expected if $r > 1$?

17. Every month, \$500 is deposited into an account earning 0.5% interest a month, compounded monthly.

- **(a)** How much is in the account right after the 6th deposit? Right before the 6th deposit?
- **(b)** How much is in the account right after the 12th deposit? Right before the 12th deposit?

18. Each year, a family deposits \$5000 into an account paying 8.12% interest per year, compounded annually. How much is in the account right after the 20th deposit?

19. Each morning, a patient receives a 25 mg injection of an anti-inflammatory drug, and 40% of the drug remains in the body after 24 hours. Find the quantity in the body:

- **(a)** Right after the 3rd injection.
- **(b)** Right after the 6th injection.
- **(c)** In the long term, right after an injection.

20. A drug is given in daily doses of 100 mg. After 24 hours, 82% of the previous day's dose remains in the body. What is the long term amount of drug in the body, right after and right before a dose is given?

21. In Example 4, we saw that if a 50 mg dose of quinine is given every 24 hours, the long term quantity of quinine in the body is about 65 mg right after a dose and about 15 mg right before a dose. The concentration of quinine in the body is measured in milligrams of quinine per kilogram of body weight. To be effective, the average concentration of quinine in the body must be at least 0.4 mg/kg. Concentrations above 3.0 mg/kg are not safe.

- **(a)** Estimate the average quantity of quinine in the body over the long term by averaging the long-term quantities of quinine in the body right after a dose and right before a dose.
- **(b)** Find the average concentration for a person weighing 70 kilograms. Is this treatment safe and effective for such a person?
- **(c)** For what range of weights would this treatment produce a long-term average concentration that is
  (i) Too low? (ii) Unsafe?

22. A used car costs \$15,000 and repairs are \$500 over the first year and increase by 20% each subsequent year. Find the total cost of owning the car for ten years.

11.2 APPLICATIONS TO BUSINESS AND ECONOMICS

Annuities

An *annuity* is a sequence of equal payments or deposits made at regular intervals indefinitely or over a specified period of time. We can use the sum of a geometric series to calculate the total value of an annuity.

Example 1 An annuity pays \$5000 every year into an account that earns 7% interest per year, compounded annually. What is the balance in the account right after the 10th deposit?

Solution The 10th deposit contributes \$5000 to the balance. The previous deposit has earned interest for a year, so it contributes \$5000(1.07). The deposit the year before that has earned interest for two years, so it contributes $\$5000(1.07)^2$. Continuing, we see that

$$\text{Balance after } 10^{\text{th}} \text{ deposit } = 5000 + 5000(1.07) + 5000(1.07)^2 + \cdots + 5000(1.07)^9.$$

This sum is a finite geometric series with $a = 5000$ and $r = 1.07$. We use the formula for the sum with $n = 10$:

$$\text{Balance after } 10^{\text{th}} \text{ deposit } = \frac{a(1-r^n)}{1-r} = \frac{5000(1-(1.07)^{10})}{1-1.07} = 69{,}082.24.$$

The balance in the account right after the 10th deposit is \$69,082.24.

Present Value of an Annuity

The *present value of an annuity* is the amount of money that must be deposited today to make a series of fixed payments in the future. How can we compute this present value? We begin by considering a single payment. Suppose a payment of \$1000 is to be made three years in the future from an account paying interest at a rate of 8% per year, compounded annually. The present value is the amount P such that

$$1000 = P(1.08)^3,$$

so we have

$$\text{Present value} = P = 1000(1.08)^{-3}.$$

To find the present value of four payments of \$1000 each, one made now, one in one year, one in two years, and one in three years, we add their present values. With the same 8% interest, we have

$$\text{Present value of four payments} = 1000 + 1000(1.08)^{-1} + 1000(1.08)^{-2} + 1000(1.08)^{-3}.$$

This pattern allows us to find the present value of any annuity, as the following example shows.

Example 2 An account earns 8% interest per year, compounded annually. Twenty payments of \$10,000 each, made once a year starting now, are to be made out of the account. How much must be deposited in the account now to cover these payments? In other words, what is the present value of this annuity?

Solution The present value of the payment to be made immediately is \$10,000. The present value of next year's payment is $\$10{,}000(1.08)^{-1}$. Since the 20th payment is made 19 years in the future, the present value of the 20th payment is $\$10{,}000(1.08)^{-19}$. The present value, P, of the entire annuity, in dollars, is the sum

$$P = 10{,}000 + 10{,}000(1.08)^{-1} + 10{,}000(1.08)^{-2} + \cdots + 10{,}000(1.08)^{-19}.$$

Rewriting $(1.08)^{-2} = \left((1.08)^{-1}\right)^2$ and $(1.08)^{-3} = \left((1.08)^{-1}\right)^3$, and so on, shows that P is the sum of the finite geometric series

$$P = 10{,}000 + 10{,}000(1.08)^{-1} + 10{,}000((1.08)^{-1})^2 + \cdots + 10{,}000((1.08)^{-1})^{19}.$$

We use the formula for the sum with $a = 10{,}000$ and $r = (1.08)^{-1}$ and $n = 20$, giving

$$\text{Present value} = P = \frac{a(1-r^n)}{1-r} = \frac{10{,}000(1-((1.08)^{-1})^{20})}{1-(1.08)^{-1}} = 106{,}035.99.$$

Thus, $\$106,035.99$ must be deposited now to cover the payments for this annuity. Notice that the annuity pays out a total of $20 \cdot \$10{,}000 = \$200{,}000$, so the present value is considerably less than the amount eventually paid out.

Example 3 The annuity in Example 2 now makes annual payments of \$10,000 in perpetuity (that is, forever), rather than just twenty times. What is the present value of this annuity?

Solution Since the payments are to be made forever, the present value is given by the infinite sum:

$$\text{Present value} = 10{,}000 + 10{,}000(1.08)^{-1} + 10{,}000((1.08)^{-1})^2 + 10{,}000((1.08)^{-1})^3 + \cdots.$$

This is an infinite geometric series with $a = 10{,}000$ and $r = (1.08)^{-1} = 0.925926$. Since $-1 < r < 1$, this series converges to a finite sum. We have

$$\text{Present value} = \frac{a}{1-r} = \frac{10{,}000}{1-(1.08)^{-1}} = 135{,}000.$$

The present value of this annuity in perpetuity is \$135,000. Notice that the amount needed to make annual payments forever is only about \$29,000 more than the amount needed to make 20 annual payments. This is the power of compound interest.

The Multiplier Effect

A government decides to give a tax rebate to stimulate the economy. What is the total effect of the rebate on spending? Because the portion of the rebate that is spent by one individual becomes income for another individual, and the proportion spent by that individual then becomes income for another individual, and so on, the total effect of the rebate on the economy is much larger than the size of the rebate itself. This is called the *multiplier effect*.

Example 4 A government gives tax rebates totaling 3 billion dollars. Everyone who receives money spends 75% of it and saves the other 25%. Find the total additional spending resulting from this tax rebate.

Solution The additional spending refers to all the additional money spent by consumers as a result of this tax rebate. The recipients of the 3 billion dollars spend 75% of what they receive, for a total of 2.25 billion dollars. (Notice that the initial amount spent is $3(0.75) = 2.25$ billion dollars, not 3 billion dollars.) The recipients of this 2.25 billion dollars spend 75% of that, or $2.25(0.75) = 1.6875$ billion dollars. The recipients of this money spend 75% of this amount, and so on indefinitely. Thus,

$$\text{Total additional spending} = 2.25 + 2.25(0.75) + 2.25(0.75)^2 + 2.25(0.75)^3 + \cdots \text{ billion dollars.}$$

This is an infinite geometric series with $a = 2.25$ and $r = 0.75$. Since $-1 < r < 1$, this infinite series converges to a finite sum. We have

$$\text{Total additional spending} = \frac{a}{1-r} = \frac{2.25}{1-0.75} = 9 \text{ billion dollars.}$$

Thus, under these assumptions, a tax rebate of 3 billion dollars generates 9 billion dollars of additional spending.

Market Stabilization

Suppose a manufacturer produces a fixed number of units of a product each year, and that each year a fixed percentage of these units (regardless of age) fail or go out of use. The total number of units of this product in use in the long run right after the annual production is completed is called the *market stabilization point*.

Example 5 The US Mint produces about 13 billion pennies a year and about 10% of them are removed from circulation each year.[1] Approximately how many pennies are in circulation?

Solution We assume that the Mint produces 13 billion (bn) pennies every year. In any year, there are 13 billion pennies produced that year and 13(0.9) bn pennies remaining from the previous year (since 10% went out of circulation). There are $13(0.9)^2$ bn pennies remaining from two years before, and so on. If N is the number of pennies in circulation, in billions, then

$$N = 13 + 13(0.9) + 13(0.9)^2 + \cdots.$$

This is an infinite geometric series with $a = 13$ and $r = 0.9$. Since $-1 < r < 1$, the series converges, and its sum is given by

$$N = \frac{a}{1-r} = \frac{13}{1-0.9} = 130.$$

Thus, there are about 130 billion pennies in circulation today. The market has stabilized at 130 billion pennies.

[1] From www.pennies.org/pennyfacts.html.

Problems for Section 11.2

1. Annual deposits of \$2000 are made into an account paying 6% interest per year, compounded continuously. What is the balance in the account right after and right before the 5th deposit?

2. A yearly deposit of \$1000 is made into a bank account that pays 8.5% interest per year, compounded annually. What is the balance in the account right after the 20th deposit? How much of the balance comes from the annual deposits and how much comes from interest?

3. An annuity earning 0.5% per month, compounded monthly, is to make 36 monthly payments of \$1000 each, starting now. What is the present value of this annuity?

4. An annuity makes annual payments of \$50,000, starting now, from an account paying 7.2% interest per year, compounded annually. Find the present value of the annuity if it makes

(a) Ten payments **(b)** Payments in perpetuity

5. A donor sets up an endowment to fund an annual scholarship of \$10,000. The endowment earns 6% interest per year, compounded annually. Find the amount that must be deposited now if the endowment is to fund one award each year, starting now and continuing

(a) Until twenty awards have been made
(b) Forever

6. Twenty annual payments of \$5000 each, with the first payment one year from now, are to be made from an account earning 10% per year, compounded annually. How much must be deposited now to cover the payments?

7. What is the present value of an annuity that pays \$20,000 each year, in perpetuity, starting today from an account that pays 1% interest per year, compounded annually?

8. A deposit of \$100,000 is made into an account paying interest compounded annually at 8% per year. Annual payments of \$10,000 each, starting now, will be made out of the account. How many payments can be made before the account runs out of money?

9. An employer pays you 1 penny the first day you work and doubles your wages each day after that. Find your total earnings after working 7 days a week for

(a) One week **(b)** Two weeks
(c) Three weeks **(d)** Four weeks

10. An employee accepts a job with a starting salary of \$30,000 and a cost-of-living increase of 4% every year for the next 10 years. What is the employee's salary at the start of the 11th year and what are her total earnings during the first 10 years?

11. Find the market stabilization point for a product if 10,000 new units of the product are manufactured each year and 25% of the total number of units in use fail each year.

12. The Bureau of Engraving and Printing produces about 18 million new \$1 bills a day; worn bills are removed by Federal Reserve Banks. There are about 4 billion dollar bills currently in circulation. Assuming that a fixed percentage of \$1 bills are removed from circulation each day, use a geometric series to estimate this percentage.[2]

13. Every year, a company sells 1000 units of a product while 20% of the total number in use fail.

(a) Find the market stabilization point for this product.
(b) If the stabilization point is approached very slowly, the number of units in use may not get close to this value because market conditions change first. Make a table for the number, S_n, of units in use right after annual production is completed, for $n = 5, 10, 15, 20$, to see how rapidly this market approaches the stabilization point.

14. One way of valuing a company is to calculate the present value of all its future earnings. Suppose a farm expects to sell \$1000 worth of Christmas trees once a year forever, with the first sale in the immediate future. What is the present value of this Christmas tree business? Assume that the interest rate is 4% per year, compounded continuously.

15. To stimulate the economy, the government gives a tax rebate totaling 5 billion dollars. Find the total additional spending resulting from this tax rebate if everyone who receives money spends

(a) 80% of it. **(b)** 90% of it.

16. A government gives a tax rebate of N dollars to stimulate the economy. Everyone who receives money spends a fixed fraction, k, of the money received, with $0 < k < 1$.

(a) Find a formula (in terms of N and k) for the total additional spending resulting from the tax rebate.
(b) If $k = 0.85$, what is the total additional spending as a multiple of the size, N, of the tax rebate?

[2] www.factmonster.com/ipka/A0774850.html and www.ustreas.gov/usss/index.htm?money-damaged.htm.

11.3 APPLICATIONS TO LIFE SCIENCES

Steady State Drug Levels

A patient is given discrete doses of a drug at regular time intervals. Since some of each dose is metabolized and excreted over each time interval, we saw in Section 11.1 that the quantity of drug in the body levels off to a *steady state*. At the steady state, the quantity of drug in the body will vary between a maximum level right after a dose is taken and a minimum level right before a dose is taken. At the steady state, the amount eliminated during one dosage interval is equal to the amount of one dose. (See Problem 20.)

Example 1 A person with an ear infection takes a 200 mg ampicillin tablet once every 4 hours. About 12% of the drug in the body at the start of a four hour period is still there at the end of that period. What quantity of ampicillin is in the body

(a) Right after taking the 3rd tablet? (b) Right after taking the 6th tablet?

(c) At the steady state level, right after and right before taking a tablet?

Solution Let Q_n represent the quantity of ampicillin, in mg, in the body right after taking the nth tablet. We have

$$Q_n = 200 + 200(0.12) + 200(0.12)^2 + \cdots + 200(0.12)^{n-1}.$$

We use the formula for the sum of a finite geometric series with $a = 200$ and $r = 0.12$.

(a) Using $n = 3$, we have

$$Q_3 = \frac{a(1-r^n)}{1-r} = \frac{200(1-(0.12)^3)}{1-0.12} = 226.88 \text{ mg}.$$

(b) Using $n = 6$, we have

$$Q_6 = \frac{a(1-r^n)}{1-r} = \frac{200(1-(0.12)^6)}{1-0.12} = 227.2720 \text{ mg}.$$

(c) The steady state level, the quantity Q in mg, right after a tablet is taken is the sum of the infinite geometric series

$$Q = 200 + 200(0.12) + 200(0.12)^2 + 200(0.12)^3 + \cdots.$$

Since $r = 0.12$, and $-1 < r < 1$, the series converges to a finite sum:

$$\text{Steady state level right after a dose} = Q = \frac{a}{1-r} = \frac{200}{1-0.12} = 227.2727 \text{ mg}.$$

We see from the answer to part (b) that the quantity of ampicillin in the body is almost at the steady state level after only 6 tablets have been taken.

At steady state, the quantity of ampicillin in the body right before a dose is exactly one dose less than the quantity right after the dose, so

$$\begin{aligned}\text{Steady state level before a dose} &= \text{Steady state level after a dose} - \text{Size of one dose}\\ &= 227.2727 - 200\\ &= 27.2727 \text{ mg}.\end{aligned}$$

Problem 5 shows how this relationship can be used to calculate Q, the long-term ampicillin level.

Example 2 Valproic acid, a drug used to control epilepsy, has a half-life of 15 hours. If D mg of valproic acid is taken every 12 hours, what is the steady state level of the drug right after taking a tablet?

Solution After a single dose D, the quantity, Q, of valproic acid in the body decays exponentially, so $Q = Db^t$, where t is time in hours. Since the half-life is 15 hours, we solve for b in the equation

$$\begin{aligned} 0.5D &= Db^{15} \\ 0.5 &= b^{15} \\ b &= (0.5)^{1/15}. \end{aligned}$$

Since doses are given every 12 hours, we want to know what fraction of the drug remains after 12 hours. Using $t = 12$, we have

$$\text{Fraction remaining after 12 hours} = b^{12} = ((0.5)^{1/15})^{12} = (0.5)^{12/15} = (0.5)^{0.8}.$$

If a dose of D mg of valproic acid is given every 12 hours, the steady state level right after taking a tablet is given by

$$\text{Steady state level} = D + D(0.5)^{0.8} + D\left((0.5)^{0.8}\right)^2 + \cdots.$$

This is an infinite geometric series with $r = (0.5)^{0.8} = 0.57435$. Since $-1 < r < 1$, the series converges and its sum is

$$\text{Steady state level} = \frac{D}{1 - (0.5)^{0.8}} = 2.35D.$$

Thus, the steady state level right after taking a dose is about 2.35 times the dose amount.

Accumulation of Toxins in the Body

Toxins or poisons found in herbicides or pesticides can get into the food chain and accumulate in people's bodies through the food they eat. We can use a geometric series to calculate the total accumulation of a poison in the body.

Example 3 Every day, a person consumes 5 micrograms of a toxin, which leaves the body at a continuous rate of 2% per day. In the long run, how much toxin has accumulated in the body at the end of each day?

Solution Since the toxin leaves the body at a continuous rate of 2% every day, the 5 mg amount consumed one day earlier has decayed to $5e^{-0.02}$, the 5 mg consumed two days earlier has decayed to $5e^{-0.02(2)} = 5(e^{-0.02})^2$, and so on. In the long term, at the end of each day we have

$$\text{Total accumulation of toxin} = 5 + 5(e^{-0.02}) + 5(e^{-0.02})^2 + 5(e^{-0.02})^3 + \cdots.$$

This is an infinite geometric series with $a = 5$ and $r = e^{-0.02} = 0.9802$. Since $-1 < r < 1$, the series converges and its sum is

$$\text{Total accumulation of toxin} = \frac{a}{1-r} = \frac{5}{1 - e^{-0.02}} = 252.5 \text{ micrograms}.$$

Over time, 252.5 micrograms of toxin have accumulated in the body at the end of each day.

Depletion of Natural Resources

Geometric series can be used to estimate how long a finite natural resource (such as oil) will last, assuming that current usage levels increase at a constant percentage rate.

Example 4 At the end of the year 2003, world oil reserves were about 1148 billion barrels. During 2003, about 27 billion barrels of oil were consumed. Over the past decade, oil consumption has been increasing at about 1% per year.[3] Assuming oil consumption increases at this rate in the future, how long will the reserves last?

Solution Under these assumptions, the oil used in 2004, in billions of barrels, is predicted to be $27(1.01)$. In 2005, we predict $27(1.01)^2$ billion barrels to be used, and $27(1.01)^3$ the next year, and so on. Thus, Q_n, total quantity of oil used in n years, in billions of barrels, is

$$Q_n = 27(1.01) + 27(1.01)^2 + 27(1.01)^3 + \cdots + 27(1.01)^n.$$

This is a finite geometric series with n terms where $a = 27(1.01)$ and $r = 1.01$. The formula for the sum gives

$$Q_n = \frac{a(1-r^n)}{1-r} = 27(1.01)\left(\frac{1-(1.01)^n}{1-1.01}\right) = 2727((1.01)^n - 1).$$

Since the total reserves are 1148 billion barrels, we want to find the value of n for which Q_n reaches 1148. We can estimate n numerically or graphically, or we can find n analytically:

$$\begin{aligned} 2727((1.01)^n - 1) &= 1148 \\ (1.01)^n - 1 &= \frac{1148}{2727} = 0.4210 \\ (1.01)^n &= 1.4210. \end{aligned}$$

Taking logarithms and using $\ln(A^p) = p \ln A$, we have

$$\begin{aligned} n \ln(1.01) &= \ln(1.4210) \\ n &= \frac{\ln(1.4210)}{\ln(1.01)} = 35.3 \text{ years.} \end{aligned}$$

Thus, if present consumption patterns are maintained, the world's oil supply will be exhausted in just over 35 years. However, if consumption patterns change, the length of time until the reserves run out can be very different. Problem 2 concerns predictions using a 2.5% yearly increase and a 0.5% yearly decrease, the maximum and minimum figures for the past decade.

Geometric Series and Differential Equations

In this section, we have used geometric series formulas to model drug levels; in Chapter 10, we used differential equations to model drug levels. The question now arises: When should we use a geometric series, and when should we use a differential equation? The answer depends on whether the drug is given in *discrete* doses (such as a tablet each morning) or *continuously* (such as intravenously).

Example 5 A patient receives 25 mg of a drug each day, and the drug is metabolized and eliminated at a continuous rate of 10% per day. Find the quantity of drug in the patient's body in the long run:

(a) Using a geometric series, assuming the 25 mg dose of the drug is administered in a single injection each morning. (Find the quantity both before and after an injection is given.)

(b) Using a differential equation, assuming the 25 mg dose of the drug is administered intravenously throughout the day.

[3] www.bp.com/liveassets/bp_internet/globalbp/globalbp_uk_english/publications/energy_reviews/STAGING/local_assets/downloads/pdf/natural_gas_section_2004.pdf, accessed May 15, 2005.

Solution (a) A 25 mg injection is given each day. Since the drug is metabolized at a continuous rate of 10% per day, the quantity remaining a day later is $25e^{-0.1}$ mg. The quantity remaining two days later is $25(e^{-0.1})^2$ mg. The long-term quantity is given by

$$\text{Steady state level after injection} = 25 + 25(e^{-0.1}) + 25(e^{-0.1})^2 + 25(e^{-0.1})^3 + \cdots.$$

We use the formula for the sum of an infinite geometric series with $a = 25$ and $r = e^{-0.1}$:

$$\text{Steady state level after injection} = \frac{a}{1-r} = \frac{25}{1-e^{-0.1}} = 262.7 \text{ mg}.$$

Since each injection is 25mg,

$$\text{Steady state level before injection} = 262.7 - 25 = 237.7 \text{ mg}.$$

(b) The drug is entering the body at the continuous rate of 25 mg per day and leaving at the continuous rate of 0.1 times the current level in the body. If Q represents the amount of drug in the body after t days, then the quantity Q satisfies the differential equation

$$\frac{dQ}{dt} = 25 - 0.1Q.$$

As we saw in Chapter 10, the long-term amount of drug in the body is the equilibrium solution of the differential equation. The equilibrium solution occurs when Q is not changing, that is, when $dQ/dt = 0$. We have

$$\frac{dQ}{dt} = 25 - 0.1Q = 0$$
$$Q = 250 \text{ mg}.$$

The long-term amount of the drug in the body when the drug is given continuously is 250 mg. Notice that this is between the upper and lower levels of the drug when it is given in discrete doses.

Problems for Section 11.3

1. In 2003, world oil consumption was 28.5 billion barrels,[4] an increase of 2.1% over 2002. Assuming that consumption continues to increase at the same percentage rate, make a table showing yearly consumption between 2003 and 2012, inclusive. Find the total quantity of oil consumed during this decade.

2. As in Example 4, assume that oil reserves at the end of 2003 were 1148 billion barrels and that consumption in 2003 was 27 billion barrels. Estimate how long these reserves will last assuming consumption

- **(a)** Decreases by 0.5% per year
- **(b)** Increases by 2.5% per year

3. Every morning, a patient receives a 50 mg injection of a drug. At the end of a 24-hour period, 60% of the drug remains in the body. What quantity of drug is in the body

- **(a)** Right after the 3rd injection?
- **(b)** Right after the 7th injection?
- **(c)** Right after receiving an injection, at the steady state?

4. The half-life of warfarin, an anticoagulant, is 37 hours. A patient receives an injection of 5 mg of warfarin at the same time every day for 10 days.

- **(a)** How much warfarin is in the body after the 10th injection?
- **(b)** How much warfarin is in the body one day after the 10th injection? Ten days after the 10th injection?

5. This problem shows another way of deriving the long-run ampicillin level. (See page 455.) In the long run the ampicillin levels off to Q mg right after each tablet is taken. Four hours later, right before the next dose, there will be less ampicillin in the body. However, if stability has been reached, the amount of ampicillin that has been excreted is exactly 200 mg because taking one more tablet raises the level back to Q mg. Use this information to solve for Q.

[4] *Statistical Review of World Energy 2004*, www.bp.com.

6. A single dose of 120 mg is taken by a patient at the same time every day. In one day, 30% of the drug is excreted.

(a) Use an infinite geometric series to find the steady state level of the drug in the body right after a dose.
(b) Show that at the steady state level, the amount excreted in one day is equal to the daily dose.

7. A patient takes a single oral 50 mg dose of the antidepressant fluoxetine at the same time every day. The half-life of fluoxetine is 3 days.

(a) What fraction of a dose remains in the body after a 24-hour period?
(b) What is the level of fluoxetine in the body right after taking the 7th dose?
(c) What is the steady state level of fluoxetine in the body right after taking a dose?

8. A person with chronic pain takes a 30 mg tablet of morphine every 4 hours. The half-life of morphine is 2 hours.

(a) How much morphine is in the body right after and right before taking the 6th tablet?
(b) Find the steady state levels of morphine in the body right after and right before taking a tablet.

9. **(a)** An allergy drug with a half-life of 18 weeks is given in 100 mg doses once a week. Find the steady state level of the drug in the body right after a dose.
(b) The drug does not become effective until the quantity of drug in the body, right after a dose is given, reaches 2000 mg. How many weeks does it take until the drug becomes effective?

10. Smoking a certain brand of cigarette puts 1.2 mg of nicotine into the body. Nicotine leaves the body at a continuous rate of 34.65% per hour, but it can be lethal if the quantity in the body reaches 60 mg. If a person smokes a cigarette with each of the following frequencies, find the steady state amount of nicotine in the body right after a cigarette. Does the nicotine reach the lethal level?

(a) Every hour. **(b)** Every half hour.
(c) Every 15 minutes. **(d)** Every 6 minutes.
(e) Every 3 minutes.

11. Each day at lunch a person consumes 8 micrograms of a toxin found in a pesticide and the toxin decays at a continuous rate of 0.5% per day. In the long run, how much of this toxin accumulates in the person's body? Give the long term amount found in the body right after and right before the person eats lunch.

12. At the end of the year 2004, the total reserve of a mineral was 350,000 m^3. In the year 2005, about 5000 m^3 of this mineral was used. Each year, consumption of the mineral is expected to increase by 8%. Under these assumptions, when will all reserves of the mineral be depleted?

13. We use 1500 kg of a mineral this year and consumption of the mineral is increasing annually by 4%. The total reserves of the mineral are estimated to be 120,000 kg. Approximately when will the reserves run out?

14. At the end of 2003, natural gas reserves were 175.78 trillion m^3; during 2003, about 2591 billion m^3 of natural gas were consumed.[5] Estimate how long natural gas reserves will last if consumption increases at 2% per year. [Note: One trillion $= 10^{12}$, one billion $= 10^9$.]

15. Over the past decade, natural gas consumption has been increasing at between 0% and 5% a year. Using the data from Problem 14, estimate how long the natural gas reserves will last assuming the rate of increase is

(a) 0% **(b)** 5%

Problems 16–18 concern how long reserves of the mineral in Problem 12 last if usage patterns change. For example, as reserves get lower, substitutes may be developed.

16. How long will the reserves last if the annual increase in usage is 4%?

17. How long will the reserves last if the annual usage stays constant at 5000 m^3 per year?

18. How long will the reserves last if the usage decreases each year by 4%?

19. **(a)** A drug is administered at dosage intervals equal to the half-life. (That is, the second dose is given when half the first dose remains.) Find the steady state level of drug in the body, right after a dose is administered, as a function of the dose, D.
(b) If the desirable long-term amount of a drug in the body is 300 mg and if doses are to be given every half-life, what should be the size of each dose?

20. A dose, D, of a drug is taken at regular time intervals, and a fraction r remains after one time interval. Show that at the steady state level, the amount of the drug excreted between doses is equal to the size of one dose.

CHAPTER SUMMARY

- **Geometric series**
  Finite and infinite.
- **Sums of geometric series**
  Partial sums, convergence of infinite series.
- **Applications to business and economics**
  Annuities, present value, multiplier effect, market stabilization.
- **Applications to life sciences**
  Repeated drug doses, accumulation of toxins, depletion of natural resources.

[5] www.bp.com/liveassets/bp_internet/globalbp/globalbp_uk_english/publications/energy_reviews/STAGING/local_assets/downloads/pdf/oil_section_2004.pdf.

REVIEW PROBLEMS FOR CHAPTER ELEVEN

In Problems 1–8, find the sum, if it exists.

1. $5+5\cdot 3+5\cdot 3^2+\cdots+5\cdot 3^{12}$

2. $20+20(1.45)+20(1.45)^2+\cdots+20(1.45)^{14}$

3. $100+100(0.85)+100(0.85)^2+\cdots+100(0.85)^{10}$

4. $1000+1000(1.05)+1000(1.05)^2+\cdots$

5. $75+75(0.22)+75(0.22)^2+\cdots$

6. $500(0.4)+500(0.4)^2+500(0.4)^3+\cdots$

7. $31500+6300+1260+252+\cdots$

8. $65+\dfrac{65}{1.02}+\dfrac{65}{(1.02)^2}+\cdots+\dfrac{65}{(1.02)^{18}}$

9. Around January 1, 1993, Barbra Streisand signed a contract with Sony Corporation for $2 million a year for 10 years. Suppose the first payment was made on the day of signing and that all other payments are made on the first day of the year. Suppose also that all payments are made into a bank account earning 4% a year, compounded annually.

(a) How much money was in the account
  (i) On the night of December 31, 1999?
  (ii) On the day the last payment was made?

(b) What was the present value of the contract on the day it was signed?

10. A person inhaling the smoke from a cigarette ingests 0.4 mg of nicotine. After one hour, 71% of the nicotine remains in the body. If a person smokes one cigarette every hour beginning at 7 am, how much nicotine is in the body right after smoking the cigarette at 11 pm?

11. Figure 11.3 shows the quantity of the drug atenolol in the blood as a function of time, with the first dose at time $t=0$. Atenolol is taken in 50 mg doses once a day to lower blood pressure.

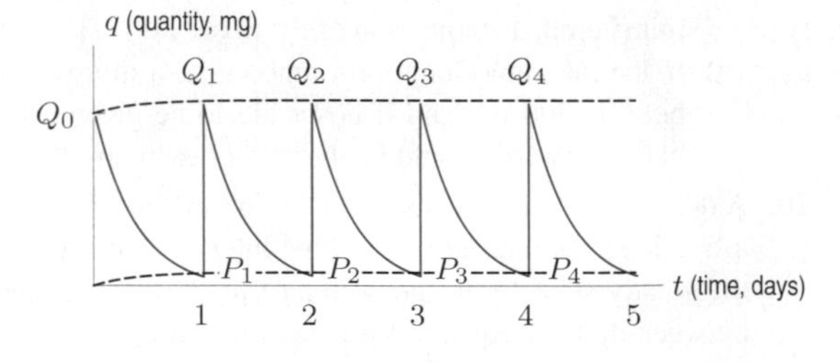

Figure 11.3

(a) If the half-life of atenolol in the blood is 6.3 hours, what percentage of the atenolol present at the start of a 24-hour period is still there at the end?

(b) Find expressions for the quantities $Q_0, Q_1, Q_2, Q_3, \ldots$, and Q_n shown in Figure 11.3. Write the expression for Q_n in closed-form.

(c) Find expressions for the quantities $P_1, P_2, P_3, \ldots$, and P_n shown in Figure 11.3. Write the expression for P_n in closed-form.

12. Cephalexin is an antibiotic with a half-life in the body of 0.9 hours, taken in tablets of 250 mg every six hours.

(a) What percentage of the cephalexin in the body at the start of a six-hour period is still there at the end (assuming no tablets are taken during that time)?

(b) Write an expression for Q_1, Q_2, Q_3, Q_4, where Q_n mg, is the amount of cephalexin in the body right after the nth tablet is taken.

(c) Express Q_3, Q_4 in closed-form and evaluate them.

(d) Write an expression for Q_n and put it in closed-form.

(e) If the patient keeps taking the tablets, use your answer to part (d) to find the quantity of cephalexin in the body in the long run, right after taking a pill.

13. This problem deals with the question of estimating the cumulative effect of a tax cut on a country's economy. Suppose the government proposes a tax cut totaling $100 million. We assume that all the people who have extra money to spend would spend 80% of it and save 20%. Thus, of the extra income generated by the tax cut, $\$100(0.8)$ million $=$ $80 million would be spent and so become extra income to someone else. Assume that these people also spend 80% of their additional income, or $\$80(0.8)$ million, and so on. Calculate the total additional spending created by such a tax cut.

14. This problem illustrates how banks create credit and can thereby lend out more money than has been deposited. Suppose that initially $100 is deposited in a bank. Experience has shown bankers that on average only 8% of the money deposited is withdrawn by the owner at any time. Consequently, bankers feel free to lend out 92% of their deposits. Thus $92 of the original $100 is loaned out to other customers (to start a business, for example). This $92 will become someone else's income and, sooner or later, will be redeposited in the bank. Then 92% of $92, or $\$92(0.92)=\$84.64$, is loaned out again and eventually redeposited. Of the $84.64, the bank again loans out 92%, and so on.

(a) Find the total amount of money deposited in the bank as a result of these transactions.

(b) The total amount of money deposited divided by the original deposit is called the *credit multiplier*. Calculate the credit multiplier for this example and explain what this number tells us.

15. Before World War I, the British government issued what are called *consols*, which pay the owner or his heirs a fixed amount of money every year forever. (Cartoonists of the time described aristocrats living off such payments as "pickled in consols.") What should a person expect to pay for consols which pay £10 a year forever? Assume the first payment is one year from the date of purchase and that interest remains 4% per year, compounded annually. (£ denotes pounds, the British unit of currency.)

16. A repeating decimal can always be expressed as a fraction. This problem shows how writing a repeating decimal as a geometric series enables you to find the fraction. Consider the decimal $0.232323\ldots$.

(a) Use the fact that $0.232323\ldots = 0.23 + 0.0023 + 0.000023 + \cdots$ to write $0.232323\ldots$ as a geometric series.

(b) Use the formula for the sum of a geometric series to show that $0.232323\ldots = 23/99$.

17. A ball is dropped from a height of 10 feet and bounces. Each bounce is $\frac{3}{4}$ of the height of the bounce before. Thus, after the ball hits the floor for the first time, the ball rises to a height of $10(\frac{3}{4}) = 7.5$ feet, and after it hits the floor for the second time, it rises to a height of $7.5(\frac{3}{4}) = 10(\frac{3}{4})^2 = 5.625$ feet.

(a) Find an expression for the height to which the ball rises after it hits the floor for the nth time.

(b) Find an expression for the total vertical distance the ball has traveled when it hits the floor for the first, second, third, and fourth times.

(c) Find an expression for the total vertical distance the ball has traveled when it hits the floor for the nth time. Express your answer in closed-form.

Problems 18–20 are about *bonds*, which are issued by a government to raise money. An individual who buys a $1000 bond gives the government $1000 and in return receives a fixed sum of money, called the *coupon*, every six months or every year for the life of the bond. At the time of the last coupon, the individual also gets the $1000, or *principal* back.

18. What is the present value of a $1000 bond which pays $50 a year for 10 years, starting one year from now? Assume the interest rate is 6% per year, compounded annually.

19. What is the present value of a $1000 bond which pays $50 a year for 10 years, starting one year from now? Assume the interest rate is 4% per year, compounded annually.

20. **(a)** What is the present value of a $1000 bond which pays $50 a year for 10 years, starting one year from now? Assume the interest rate is 5% per year, compounded annually.

(b) Since $50 is 5% of $1000, this bond is often called a 5% bond. What does your answer to part (a) tell you about the relationship between the principal and the present value of this bond when the interest rate is 5%?

(c) If the interest rate is more than 5% per year, compounded annually, which is larger: the principal or the present value of the bond? Why do you think the bond is then described as *trading at a discount*?

(d) If the interest rate is less than 5% per year, compounded annually, why is the bond described as *trading at a premium*?

PROJECTS FOR CHAPTER ELEVEN

1. Do you have any common ancestors?

In this project, we estimate the number of ancestors you have and determine whether you have any common ancestors. (A common ancestor is one who appears on two sides of your family tree. For example, if your great-grandmother on your mother's mother's side is also your grandmother on your father's side, then she would be a common ancestor.)

(a) In general, each person has 2 biological parents, 4 biological grandparents, 8 biological great-grandparents, and so on. Write a formula for the number of ancestors you have, going back n generations.

(b) How long is one generation? Estimate the age of typical parents when a baby is born. This is the length of time for a generation. How many generations are included if we go back 100 years? 500 years? 1000 years? 2000 years?

(c) Use your answers to parts (a) and (b) to estimate the number of ancestors you have if we go back 100 years, 500 years, 1000 years, or 2000 years.

(d) In parts (a) and (c), we counted every ancestor separately, so we assumed that you have no common ancestors. Use the fact that the population of the world was about 6 billion people in 1999 and was about 200 million people in the year 1 AD to determine whether this is a reasonable assumption. Explain your reasoning.

2. Harrod-Hicks model of an expanding national economy

The Harrod-Hicks model predicts that if a national economy is growing, then the national income in one year is related to the national income in the preceding year. If we let $f(n)$ be the national income in year n, then the model predicts, for some constants k and h with $k > 1$ and $h > 0$, that

$$f(n+1) = kf(n) - h.$$

(a) Let $C = f(0)$. Write $f(1)$, $f(2)$, and $f(3)$ in terms of k, h, and C.

(b) Show that

$$f(1) = kC - h,$$
$$f(2) = k^2C - (1 + k)h,$$
$$f(3) = k^3C - (1 + k + k^2)h.$$

Use these formulas to guess a formula for $f(n)$.

(c) Use the formula for the sum of a finite geometric series to rewrite the formula for $f(n)$ in closed form.

3. Probability of Winning in Sports

In certain sports, winning a game requires a lead of two points. That is, if the score is tied you have to score two points in a row to win.

(a) For some sports (e.g. tennis), a point is scored every play. Suppose your probability of scoring the next point is always p. Then, your opponent's probability of scoring the next point is always $1 - p$.

(i) What is the probability that you win the next two points?

(ii) What is the probability that you and your opponent split the next two points, that is, that neither of you wins both points?

(iii) What is the probability that you split the next two points but you win the two after that?

(iv) What is the probability that you either win the next two points or split the next two and then win the next two after that?

(v) Give a formula for your probability w of winning a tied game.

(vi) Compute your probability of winning a tied game when $p = 0.5$; when $p = 0.6$; when $p = 0.7$; when $p = 0.4$. Comment on your answers.

(b) In other sports (e.g. volleyball), you can score a point only if it is your turn, with turns alternating until a point is scored. Suppose your probability of scoring a point when it is your turn is p, and your opponent's probability of scoring a point when it is her turn is q.

(i) Find a formula for the probability S that you are the first to score the next point, assuming it is currently your turn.

(ii) Suppose that if you score a point, the next turn is yours. Using your answers to part (a) and your formula for S, compute the probability of winning a tied game (if you need two points in a row to win).

- Assume $p = 0.5$ and $q = 0.5$ and it is your turn.
- Assume $p = 0.6$ and $q = 0.5$ and it is your turn.

APPENDIX

The following projects require the use of a spreadsheet. They can all be done using only the ideas in Chapter 1. In addition, Project 8 (Verhulst: The Logistic Model) and Project 9 (The Spread of Information) give another perspective on material in Chapters 4 and 10. Project 4 (Comparing Home Mortgages) uses a geometric series but can be done before Chapter 11.

SPREADSHEET PROJECTS

1. MALTHUS: POPULATION OUTSTRIPS FOOD SUPPLY

In this project, we compare exponential and linear growth. We see the eventual dominance of exponential functions over linear functions.

One of the most famous models of population growth was made by Thomas Malthus in the early 19th century. Malthus believed that while human population increased exponentially, its means of subsistence increased linearly. The gloomy conclusion that Malthus drew from this observation was that the population of the earth would inevitably outstrip its means of subsistence, resulting in an inadequate supply of food. (Malthus went on to note that this state of affairs could only be averted by war, famine, epidemic disease, wide-scale sexual restraint, or other such drastic checks on population growth.)

The following table shows part of a spreadsheet showing such as a scenario.[1] The starting population is 1 million, while the available food feeds 2 million people. The population grows at an annual rate of 3% per year and the food production increases by 100,000 per year. These population and food growth rates are in cells on the right of the spreadsheet. The fourth column contains the ratio of available food per person in the population. The spreadsheet includes a safety ratio—so long as the food-to-population ratio is above this figure of 1.5, the fifth column displays "Yes"; whenever the ratio drops below this figure the fifth column displays "No" (as it does by the end of the 21st century).

We see that at first there is plenty of food—the ratio of food to population is 2, which means that there is twice as much food as is necessary to feed the population. For the first few years the ratio increases, but at a certain point it starts to decrease, and eventually the ratio drops below one.

Year	Population	Food supply	Ratio	Above safety ratio?	
1999	1000000	2000000	2.00	Yes	Annual pop growth rate 3.00%
2000	1030000	2100000	2.04	Yes	
2001	1060900	2200000	2.07	Yes	
2002	1092727	2300000	2.10	Yes	Annual food growth rate 100,000
2003	1125509	2400000	2.13	Yes	
2004	1159274	2500000	2.16	Yes	
2005	1194052	2600000	2.18	Yes	Safety ratio 1.5
2006	1229874	2700000	2.20	Yes	
2007	1266770	2800000	2.21	Yes	
2008	1304773	2900000	2.22	Yes	
⋮	⋮	⋮	⋮	⋮	
2098	18658866	11900000	0.64	No	
2099	19218632	12000000	0.62	No	
2100	19795191	12100000	0.61	No	

1. Set up your spreadsheet to look like the one shown in the table, extending it to the year 2100. Virtually every cell must contain a formula — the exceptions being the six cells containing "1999," "1,000,000," "2,000,000," "3.00%," "100,000," and "1.5."

[1]From Graeme Bird.

2. (a) About what year is the food-to-population ratio the highest?
   (b) In which year does this ratio reach 1?
3. There are at least two ways to improve upon the current situation: we can lower the population growth rate, or we can increase the food supply.
   (a) What would the population growth rate have to be lowered to, in order to have the food-to-population ratio not reach 1 until the year 2100? (Keep the food supply increasing at 100,000 per year.)
   (b) What would the food supply rate have to be increased to, in order to achieve this same goal, of the ratio not reaching 1 until the year 2100? (Keep the population growth rate at its original 3%.)
4. Using the original scenario, create each of the following charts (both line and column). Extend the charts to the year 2100, so that the point where the population outstrips the food supply is clearly evident.
   (a) Showing population and food, with the years on the horizontal axis.
   (b) Showing only the ratio, with the years on the horizontal axis.

2. CREDIT CARD DEBT

You have a credit card on which you owe $2000. Your credit card company charges a monthly interest rate of 1.5% and requires a minimum monthly payment of 2.5% of your current balance. (This payment scheme is similar to ones used by many credit card companies, but see Question 8.)

[Note: For Questions 1–2, you will not need a spreadsheet, although you will need a calculator.]

1. If the monthly interest rate is 1.5%, what is the effective annual interest rate?
2. As a rule, the minimum monthly payment required exceeds the interest accrued in a month. (For example, here the minimum monthly payment of 2.5% exceeds the monthly interest charges of 1.5%.) Explain why this should be the case. What would happen to the card balance if the minimum required payment was less than the accrued interest?

Suppose that you decide to pay off your $2000 credit card debt by making only the minimum required payment every month. Assume that you make no further charges to the card, since you're trying to pay it off.

3. Since 2.5% of $2000 is $50, your first monthly payment is $50. Before you do any spreadsheet calculations, guess how long it will take to bring your total balance down from $2000 to less than $50, assuming that you make only the minimum required payment every month. A rough guess is fine; use common sense and explain your reasoning.
4. Over time, your monthly payments, which start at $50, will decrease. Explain why this happens. Does the fact that your monthly payments decrease affect the answer you gave to Question 3?
5. Although you only owe the credit card company $2000, you will end up paying quite a bit more than $2000, due to interest charges. Make your best guess (before making any specific calculations) as to how much, roughly, you will end up having paid the credit card company for your initial $2000 debt.

Set up a spreadsheet showing the number of months since you began paying off your debt, your current balance, the interest due that month, and the payment you make, each in a separate column. Each quantity should be calculated by a formula.

Example: To figure out what formulas you need, recall that your initial balance is $2000, the monthly interest charged is 1.5%, and the minimum required payment is 2.5%. Thus, at the beginning of Month 1, your balance is $2000, since you have paid off nothing yet. Therefore, at the end of Month 1, the interest that you owe is 1.5% of the $2000 balance, or $30. Your minimum payment is 2.5% of the $2000 balance, or $50. Thus, at the beginning of Month 2, your new balance is the old balance of \$2000 plus the \$30 in interest minus the \$50 payment, or \$1980. Notice that the figures for Month 2 depend on the figures for Month 1; similarly, the figures for Month 3 depend on Month 2 figures, and so on. Follow this procedure to figure out what formulas you need in each column of your spreadsheet.

Once you have set up a working spreadsheet, answer the following questions.

6. How good was your guess in Question 3? Using your spreadsheet, find out how many months it takes to bring your balance down to less than \$50. Was your guess close, or were you surprised by how long it really takes?
7. How good was your guess in Question 5? Using your spreadsheet, figure out exactly how much you pay the credit card company to bring your balance down to less than \$50. How does this figure compare to the original debt of \$2000?
8. Use your spreadsheet to find out how long it takes to bring your balance down to \$0. Or can't you tell? Does there ever come a point when you have exactly paid off your debt? [Hint: Eventually, the minimum monthly payments and the interest charges become unrealistic. In what way are they unrealistic? How do real credit card companies avoid this problem?]
9. Now let's try experimenting with the numbers and see what happens. In each of the following cases, make the appropriate changes to your spreadsheet. Assume that as soon as your balance is under \$50, you pay it off in a lump sum.
   (a) If every month you pay \$1 more than the minimum required payment, how long does it take to bring your debt down to less than \$50? How much do you end up paying to your creditors in total? How much money do you save by using this payment scheme instead of the one in Question 5?
   (b) Your first monthly payment is \$50. If you paid \$50 every month, instead of the minimum required payment, how long does it take to bring your debt down to less than \$50? How much money do you save by using this payment scheme instead of the one in Question 5?
   (c) Recently, many credit card companies have made offers similar to the following: if you transfer your debt from a competitor's card to their card, they will charge you a lower interest rate. Suppose you find a credit card company willing to make this transaction, and that their monthly interest rate is 1%, not 1.5%. Leaving all the other original assumptions unchanged, how long does it take to bring your debt down to less than \$50, and how much do you pay your creditors in total? How much money do you save compared to what you would have paid your original card company?
10. Comparing the scheme used in Questions 3–5 with each of the schemes in Question 9, what conclusions can you reach about paying off a credit card debt?

3. CHOOSING A BANK LOAN

A local bank offers the following loan packages. Use a spreadsheet to decide which option is the best. The packages are as follows:

- A loan of \$2000, at an annual rate of 9%, payable in 24 monthly installments.
- A loan of \$2000, at an annual rate of 10%, payable in 36 monthly installments.
- A loan of \$2000, at an annual rate of 9.25%, payable in 52 biweekly installments.

Interest is compounded with the same frequency as payments are made. Notice that the first and last loans have two-year payoff periods; the middle loan is for three years.

1. Use a spreadsheet to decide which loan is cheapest in terms of total payoff to the bank. (See the following hint.)
2. Use a spreadsheet to decide which loan is easiest to afford in terms of lowest monthly payment. (See the following hint.)

Hint: The difficult part of Questions 1 and 2 is figuring out your monthly (or biweekly) payments. There are formulas that give the payment based on the period and amount of the loan and the interest charged, but instead of using them, we will use a spreadsheet. The idea is that you can make an educated guess as to what the payment ought to be, and then use a spreadsheet to check your answer. By looking at the spreadsheet, you can decide whether your guess was too high or too low, and thus improve upon your original guess. It's surprising how quickly you can zero in on the required monthly payment, down to the nearest penny, by this guess-and-check method.

For example, consider the first loan, the two-year \$2000 loan at 9%. Set up a spreadsheet with the initial balance of \$2000, the interest for the first month, which is $(9\%/12) \cdot \$2000 = \$15$, and a guess at the monthly payment. There are lots of ways to make a guess at the monthly payment. One way is to say that if you were to borrow \$2000 for 2 years at 9%, you would owe about $\$2000(1.09)^2 = \$2376$. (Never mind the monthly compounding—this is just a rough approximation.) To pay off this amount in 24 equal monthly payments would require $\$2376/24 = \$99$. So, we guess a monthly payment of \$100. Using this guess, the second month's balance will be

$$(\$2000) + (\$15 \text{ in interest for month } 1) - (\text{Payment of } \$100) = \$1915.$$

Thus, the next month's interest will be $(9\%/12) \cdot \$1915$, and the next month's payment should be the same as the first month's payment, or \$100. Continue this process until 24 months' (two years') worth of payments have been made. You will see that the final balance is negative, meaning you paid the bank more than you really owed. This means that \$100 is too high a monthly payment to pay off your \$2000 loan. (We could have predicted that this was the case when we made our estimate above. Do you see why?) So, since \$100 is too high, you might guess that an \$80 monthly payment would be right. If you do, you'll see that you'd still owe the bank some money after 24 months had passed. This tells you that \$80 is too low a monthly payment, and that the actual payment is somewhere between \$80 and \$100. This procedure can be repeated until the exact monthly payment is reached.

4. COMPARING HOME MORTGAGES

To do this project, first go to any bank and ask for a fact sheet of their most recent *mortgage loan rates*. Banks are happy to provide them.

Obtain rates for a thirty-year loan, a fifteen-year loan, a thirty-year biweekly loan, and a twenty-year loan (if available) for \$100,000 with zero points. (Note: Some loans include points. A point is an additional fee paid to the lender at the time of the loan equal to 1% of the amount borrowed. Typically, you get lower interest rates by paying a point or two. We will only consider loans with zero points.)

The following formula can be used to determine your payment, x:

$$x = \frac{Pr^n(r-1)}{r^n - 1},$$

where P is the amount of the loan—in this case \$100,000—and n is the number of payments. For a thirty-year loan with monthly payments, $n = 360$; for a thirty-year biweekly loan, $n = 780$ (there are 26 payments every year). Finally, r is the interest rate per period plus 1. (For example, if the interest rate is 2%, then $r = 1.02$.) In Question 4, you will derive this formula for x using the following formula for the sum of a geometric series:[2]

$$1 + r + r^2 + \cdots + r^{n-2} + r^{n-1} = \frac{1 - r^n}{1 - r}.$$

[2]Geometric series are discussed in detail in Chapter 11.

1. Using the information given, as well as the fact sheet you got from the bank, determine which loan (30-year, 30-year biweekly, 20-year or 15-year) is best if you intend to live in your house for the full term of the loan. Assume the best mortgage is the one that ends up costing you the least overall. (The situation in real life can be more complicated when points and taxes are considered.) Although it's possible to work this problem without a spreadsheet, you might want to set one up anyway.
2. Banks usually require that the monthly payments not exceed some stated fraction of the applicant's monthly income. For this reason, it is generally easier to qualify for loans with smaller monthly payments. Thus, it may be that the "best loan"—the one you found in Question 1—is not the "easiest" loan to qualify for. Which of the loans on your fact sheet has the lowest monthly payment? The highest?
3. Suppose that you expect to sell your house for $145,000 in five years. In this case, which loan should you take? [Hint: The goal here is to maximize profit. Figure out how much money you have paid to the bank after five years, and your remaining debt at that time. When you sell your house, the remaining debt is paid to the bank immediately, so your total profit is (Selling price of house)−(Loan payoff to bank)−(Amount paid to bank during first five years).]
4. Derive the formula for the monthly payment, x. [Hint: Set up a geometric series in terms of x and r to give your loan balance after n months. The loan balance equals 0 when you have paid off your loan; use this fact to solve for x. Simplify the resulting expression (by summing a geometric series) to get the formula given for x.]

5. PRESENT VALUE OF LOTTERY WINNINGS

On Thursday, February 24th, 1993, Bruce Hegarty of Dennis Port, MA, received the first installment of the $26,680,940 prize he won in the Mass Millions state lottery. Mr. Hegarty was scheduled to receive 19 more such installments on a yearly basis. Each check written by the Lottery Commission is for one twentieth of the total prize, or $1,334,047. Why doesn't the Lottery Commission pay all of Mr. Hegarty's prize up front, instead of making him wait for twenty years?

1. Compute the present value of the money paid out by the Lottery Commission, assuming annual discount rates (interest rates) of 5%, 10% and 15%. In each case, what percent does the present value represent of the face value of the prize, $26,680,940?
2. What discount rate would result in a present value of the payments worth only half the face value of the prize?
3. Graph the present value of the payments against the discount rate, ranging from a rate of 0% up to 15%. Describe the graph. What does it tell you about why the Lottery Commission does not pay the prize money up front?

6. COMPARING INVESTMENTS

Consider two investment projects. Project A is built in one year at an initial cost of $10,000. It then yields the following decreasing stream of benefits over a five year period: $5000, $4000, $3000, $2000, $1000. Project B is built in two years. Initial costs are $10,000 in the first year and $5000 in the second year. It then yields yearly profits of $6000 for the next four years. Which of these investment projects is preferable?

1. Compute the present values of both projects assuming an annual discount rate (interest rate) of 4%. Which project seems preferable? [Hint: Treat expenditures as negative and income as positive.]
2. Compute the present values of both projects assuming an annual discount rate of 16%. Which project seems preferable now?

3. Describe in complete sentences why one of the investment projects is favored by a low discount rate, whereas the other is favored by a high discount rate.
4. The discount rate at which the present value of a project becomes zero is known as the *internal rate of return*. What is the internal rate of return of project A? Of project B? [Hint: Guess different discount rates until you find the one that brings the present value down to $0.]
5. Make a chart of the present value of the two investments against discount rates ranging from 0% to 30%. What features of this chart correspond to the internal rates of return of the two projects?

7. INVESTING FOR THE FUTURE: TUITION PAYMENTS

Parents of two teenagers, ages 13 and 17, deposit a sum of money into an account earning interest at the rate of 7% per year compounded annually. The deposit will be used for a series of eight annual college tuition payments of $10,000 each. Payments out of the account will begin one year after the initial deposit.

1. Use a spreadsheet to model the savings account that the parents opened. At the end of every year, the account earns 7% interest, and then there is a $10,000 withdrawal. Determine what initial deposit provides just enough money to make the eight yearly payments of $10,000. Do this by guessing different values, and seeing which value leaves you with nothing exactly nine years later.
2. Having answered Question 1, use a spreadsheet to compute the present value of eight yearly payments of $10,000 each, beginning one year in the future, at a discount rate of 7%.
3. Compare your answer to Question 1 with your answer to Question 2. Is this a coincidence? Discuss.
4. Suppose the parents have only $50,000 to deposit into the savings account. What annual interest rate must the account earn if the eight payments of $10,000 are to be made? [Hint: Compute the present value of the payments for various discount rates.]

8. NEW OR USED?

You are deciding whether to buy a new or used car (of the same make) and how many years to keep the car. You want to minimize your total costs, which consist of two parts: the loss in value of the car and the repairs. A new car costs $20,000 and loses 20% of its value each year. Repairs are $400 the first year and increase by 25% each subsequent year.

1. Set up a spreadsheet that gives, for each year, the value of the car, its loss in value, its repair costs, and the total cost for that year. The first two lines look like this, rounded to the nearest dollar:

Year	Value	Loss	Repair	Cost
1	20,000	4000	400	4400
2	16,000	3200	500	3700
⋮	⋮	⋮	⋮	⋮

2. Which year has the lowest cost?
3. You intend to keep your car 5 years. Compare your total costs for a new car and for a two-year-old car.
4. How old a car should you buy if you plan to keep it for 5 years?

less

5. Add two columns to your spreadsheet showing the average yearly cost for second-hand cars of different ages kept for 4 years and 5 years. Which is the best buy?
6. A new car costs \$30,000 and loses 25% of its value each year; repairs start at \$500 and increase at 10% per year. If you buy a seven-year-old car, should you keep it for 4 or 5 years? What is the average yearly cost in each case?
7. A car costs \$30,000 when new. You buy a four-year-old car and keep it for 5 years; repairs are as in Problem 6. Find the rate at which it loses value if the average yearly cost is \$2300.

9. VERHULST: THE LOGISTIC MODEL

The relative growth rate of a population, P, over a time interval, Δt, is given by

$$\text{Relative growth rate} = \frac{1}{P} \cdot \frac{\Delta P}{\Delta t}.$$

In exponential growth, the relative growth rate is a constant. Although exponential growth is often used to model populations, this model predicts that a population will increase without limit, which is unrealistic. In the 1830s, a Belgian mathematician, P. F. Verhulst, suggested the *logistic model*, in which the relative growth rate of a population decreases to 0 linearly as the population increases. Verhulst's model predicts that the population size eventually levels off to a value known as the *carrying capacity*.

To see how Verhulst's logistic model works, assume that a pair of breeding rabbits is introduced onto a small island with no rabbits. At the outset the rabbit population doubles every month. This means that initially the relative growth rate is 100% per month. Eventually, though, as the population grows, the relative growth rate drops down to 0% per month. Suppose that the growth rate reaches 0 when the population reaches 10,000 rabbits. (Thus, 10,000 is the carrying capacity of the island.) Using spreadsheets, we will model the rabbit population over time.

1. Let P be the population and r be the relative growth rate per month. Verhulst assumed that the relative growth rate decreases linearly as population increases. This means that r goes from 100% to 0% as P goes from 0 to 10,000 rabbits. Explain why the following formula for r corresponds to Verhulst's assumptions: $r = 0.0001(10{,}000 - P)$.
2. Use a spreadsheet to model the monthly rabbit population on the island for the first two years (24 months). Graph the rabbit population over time. Describe the behavior of the rabbit population. [Hint: Start with two rabbits, and compute the growth rate using the formula in Question 1. Then, for each month, update the rabbit population as well as the relative growth rate.]
3. Draw a graph comparing your logistic model of the rabbit population to a population growing exponentially at a constant relative growth rate of 100% per month. Both models should start out with two rabbits. Describe the similarities and the differences between the two charts. What advantages does the logistic model have over the exponential model? (You'll have to be careful when setting the chart parameters; otherwise, all you'll be able to see is the exponential population which climbs so quickly that the logistic one will not be visible at all.)
4. The key to the logistic model is that the relative growth rate is decreasing linearly as the population increases. However, this does not mean that the relative growth rate is decreasing linearly over time. Make a chart of the relative growth rate against time for the first two years. Describe the behavior of the relative growth rate over time.
5. (a) Different assumptions about the growing rabbit population lead to different logistic curves. In Question 1, we assumed that the relative growth rate was 100% initially, dropping to 0% when the population reached 10,000. This led to the formula $r = 0.0001(10000 - P)$. Now assume that the initial relative growth rate is 10% (instead of 100%). What is the new formula relating r and P? (The carrying capacity is still 10,000, so $r = 10\%$ when $P = 0$, and r decreases to 0% as P increases to 10,000.)

(b) Using your new formula for the relative growth rate, r, let's see how different initial populations of rabbits lead to different logistic curves. Model the following scenarios, over a 5-year (60-month) period: a population starting at 100 rabbits, a population starting at 5,000 rabbits, a population starting at 12,500 rabbits, and a population starting at 17,500 rabbits. Place all of your data on the same chart. What happens to the rabbit population when it starts out above the island's rabbit carrying capacity? Why does this make sense?

10. THE SPREAD OF INFORMATION: A COMPARISON OF TWO MODELS

The spread of information through a population is important to policy makers. For example, agricultural ministries use mathematical models to understand the spread of technical innovations or new seed types through their countries.

In this project, you will compare two different models—one of them logistic —for the spread of information. In both cases, assume that the population is 10,000 and that initially only 100 people have the information. Let N be the number of people who have the information at time t.

Model 1: If the information is spread by mass media (TV, radio, newspapers), the absolute rate, $\Delta N/\Delta t$, at which the information is spread is assumed to be proportional to the number of people *not* having the information at that time. If t is in days, the constant of proportionality is 10%. For example, on the first day, the number of people not having the information is $10{,}000 - 100 = 9900$. Since 10% of 9900 is 990, the rate of spread of information is 990 people per day on the first day. This means that on the second day, the number of people not having the information is 8910 and that the rate of spread is 10% of 8910, or 891 people per day, and so on.

Model 2: If instead the information is spread by word of mouth, the absolute rate at which information is spread is assumed to be proportional to the product of the number of people who know and the number of people who do not know. If t is in days, the constant of proportionality is 0.002%. For example, on the first day, the product of the number of people who know and the number of people who do not know is $100 \cdot 9900 = 990{,}000$. Since 0.002% of 990,000 is about 20, the rate of spread of information is 20 people per day on the first day. This means that on the second day, the product of the number of people who know and the number who do not know is $120 \cdot 9880 = 1{,}188{,}600$, giving a rate of spread of 0.002% of 1,188,600 or about 24 people per day, and so on.

1. Using spreadsheets, compare the spread of information throughout the population using both models. Make a chart comparing both models' predictions of the number of people over time who have the information. Describe the similarities and differences between the two models. Why is Model 1 used when mass media is present? Why is Model 2 used when mass media is absent?
2. Which of the two models is logistic? How can you tell? What type of growth does the other model exhibit? How can you tell?
3. By definition, a population exhibits logistic growth if its relative rate of change is a decreasing linear function of the current population. Explain why Model 2 leads to logistic growth although it was defined in terms of absolute growth rates.
4. The solutions from our spreadsheets are only approximations. Discuss why this is the case. [Hint: There's more going on here than rounding error.]

11. THE FLU IN WORLD WAR I

During World War I, a particularly lethal form of flu killed about 40 million people around the world.[3] The epidemic started in an army camp of 45,000 soldiers outside of Boston, where the first soldier fell sick on September 7, 1918. In this problem you will make a spreadsheet for the SIR

[3] "Capturing a Killer Flu Virus," J. Taukenberger, A. Reid, T. Fanning, in *Scientific American*, Vol. 292, No. 1, Jan. 2005.

model of the 1918 flu outbreak. Starting from the initial values S_0 and I_0, for any increment in time Δt, the changes in the number of susceptibles and infecteds are approximated by

$$\Delta S \approx -aSI\Delta t$$
$$\Delta I \approx aSI\Delta t - bI\Delta t.$$

1. Choosing $\Delta t = 0.1$, $a = 0.0003, b = 10$, make a spreadsheet whose first few lines look like this:

t	S	I	ΔS	ΔI
0	44999	1	-1.3500	0.3500
0.1	44997.65	1.3500	-1.82240	0.4724
0.2	44995.828	1.8224	...	...

2. How many soldiers got sick on the fifth day? How many were susceptible on this day?
3. Alter the spreadsheet so that it accepts any values of $\Delta t, a, b$ input by the user.
4. Using the values $a = 0.000267, b = 9.865$ for the 1918 epidemic, decrease the value of Δt until a stable estimate is reached for the number of soldiers sick on September 16th. How many soldiers had been infected by this date?
5. Approximately how long did it take for the 1918 epidemic to run its course?

PRETEST

The pretest in this section covers skills that are used in this book. Doing it at the start of a course will help you identify areas in algebra where reviewing early on will help you succeed.

Problems for a Pretest

1. What is the equation of the line in Figure .1?

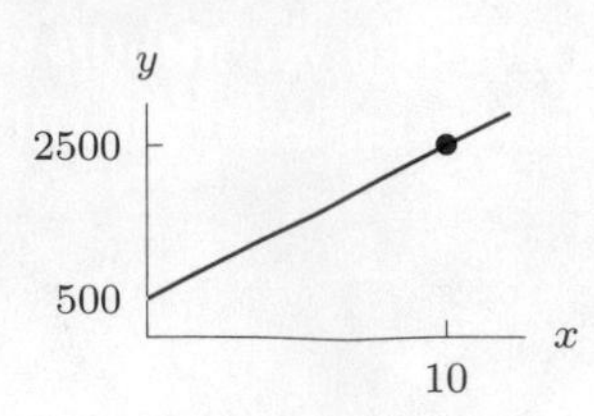

Figure .1

2. What is the value of a in Figure .2?

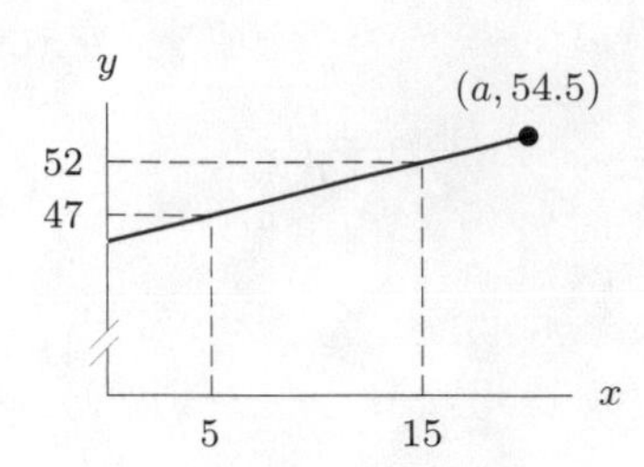

Figure .2

3. Find a_1 and a_2 from Figure .3.

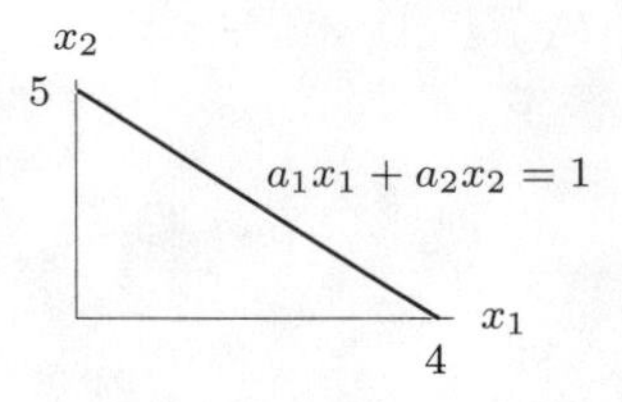

Figure .3

4. Match each of the four lines in Figure .4 with one of the slopes 0, 1, 2, -1, -2.

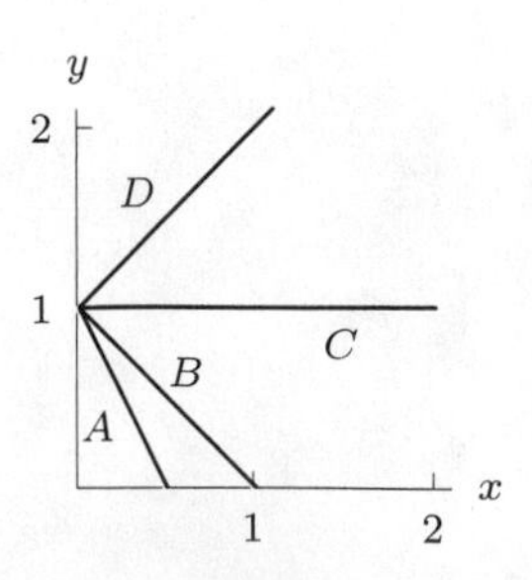

Figure .4

5. Solve for P_0:

$$P = P_0\left(1 + \frac{r}{n}\right)^t.$$

6. Solve for r:

$$P = \left(1 + \frac{r}{n}\right)^t.$$

7. If $f(t) = 4t + (t+3)^2$, find $f(0)$.

8. If $f(x) = 4x + 28$, solve $f(x) = 0$.

9. Which of the following is equivalent to $16 - x^2 + 6x$?

(a) $7 + (3 - x)^2$
(b) $16 - (x - 3)^2$
(c) $25 - (x - 3)^2$
(d) $52 - (x - 6)^2$
(e) $25 - (x + 3)^2$
(f) None of the above

10. What are the coordinates of the point Q in Figure .5?

11. What are the coordinates of the point P in Figure .5?

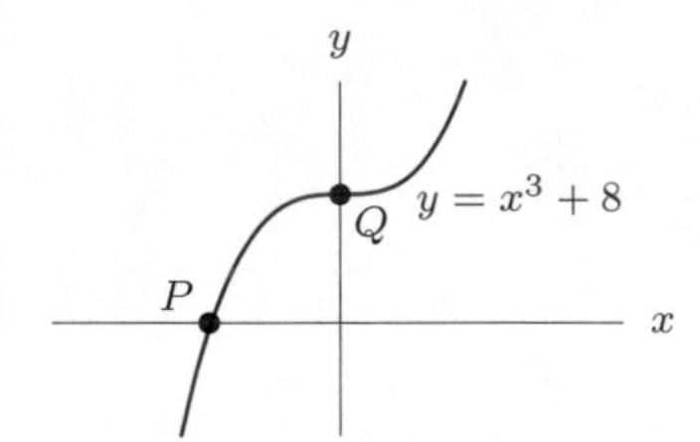

Figure .5

12. Find $|-10| + |10|$.

13. Which of the following is equivalent to $\frac{T}{E} + E$?

(a) $\frac{E^2 + T}{E}$
(b) $\frac{E}{E^2 + T}$
(c) $E + T$
(d) $\frac{T + E}{E + 1}$
(e) $\frac{T + E}{E}$

14. Which of the following is equivalent to $\frac{1}{y} + \frac{1}{u}$?

(a) $\frac{1}{uy}$
(b) $\frac{1}{u + y}$
(c) $\frac{u + y}{uy}$
(d) $\frac{u}{y}$
(e) $\frac{2}{y + u}$

15. Which of the following is equivalent to $x^{-1}y + y^{-1}x$?

(a) $\dfrac{xy}{x^2+y^2}$
(b) $\dfrac{x^2+y^2}{xy}$
(c) $\dfrac{xy}{x+y}$
(d) $\dfrac{x+y}{xy}$
(e) $\dfrac{y+x}{x+y}$

16. If $\dfrac{\sqrt[3]{t}}{t} = t^a$, what is a?

17. Simplify $(x^2x^5)^4$.

18. Simplify $W^{2/3}W^{3/2}$.

19. Expand and collect like terms: $-5u\left(\dfrac{3}{u} - 2v\right)$.

20. Solve $3p + 4 = 9$ for p.

21. Solve $2x + a = 5(1 - x)$ for x.

22. Solve $2\sqrt{w} - 3 = 9$ for w.

23. Solve $\dfrac{2}{3} = \dfrac{5}{n}$ for n.

24. Solve $Cq + Dq = A$ for q.

25. With $m = 2, x_0 = 3, y_0 = 10$, solve $y - y_0 = m(x - x_0)$ for x.

26. Find the value of x when $y = 0$ if $-2x + 9y - 9 = 0$.

27. Find the value of s when $t = s + 1$ if $3s - 8t = 7$.

28. Factor $12x^2 - 6a^2x$.

29. Factor $C^2a^3b - Cab^3$.

30. Which of the following is the factored form of $x^2 + 2x - 15$?

(a) $(x+1)^2 - 16$
(b) $(x-3)(x+5)$
(c) $x(x+2) - 15$
(d) $(x+2)(x-15)$
(e) None of the above

31. What are the zeroes of $(x-2)(x+1)(2x-5)$?

32. What is the y-intercept of $y = (x-2)(x+1)(2x-5)$?

33. If $W^2 - 2W - 8 = 4(W - 2)$, find all possible values of W.

34. If $s = d^2 + d + 1$ and $d = g + 1$ express s in terms of g.

35. If $\theta = x^2 - 1$ express $(\theta + 1)^2$ in terms of x.

36. If $b = -2z$ express $b^2 - 2b$ in terms of z.

37. The graph of the equation $a^x = y$ passes through the point $(-3, \frac{1}{8})$. What is a?

38. If $s = 2$ and $t = 0$ is a solution to $as^5 = s^7 + t$, what is a?

39. Using Table .1, what value of w gives $q = 7$?

Table .1

w	-4	-1	1	6	10
q	4	2	7	-2	2

40. Using Table .2, what is the temperature at 6 pm?

Table .2

Time (hours after midnight)	0	6	12	18
Temperature (°F)	36	48	65	55

41. For some constant k, let $M = kN$. If $M = 0.84$ when $N = 12$, what is M when $N = 6$?

42. What is the area of the shaded rectangle in Figure .6?

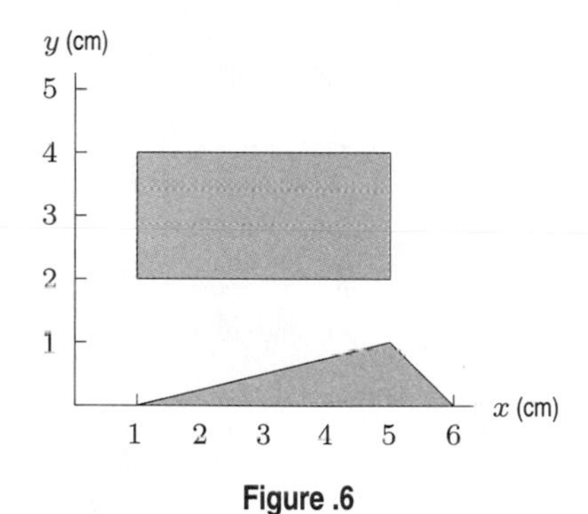

Figure .6

43. What is the area of the shaded triangle in Figure .6?

44. A price increases by a factor of 1.042. By what percent has the price increased?

45. A car, initially valued at $21,000 is now worth 32% less. What is the car's current value?

46. An item priced at $45 is on sale for 25% off. The customer pays the sale price plus 8% sales tax. How much does the customer pay?

Problems on Trigonometry (Optional)

1. What is the vertical intercept of $y = \cos x$?

2. What are the zeroes of $y = \sin x$ between $x = 0$ and $x = 2\pi$ (inclusive)?

3. How many degrees is 3π radians?

4. What are the coordinates of the highest point on the sine graph in Figure .7?

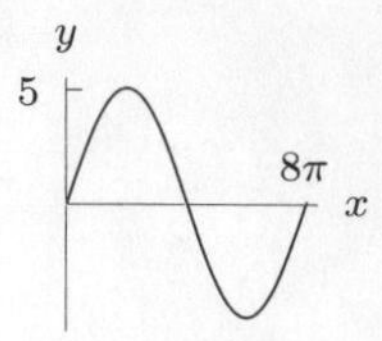

Figure .7

ANSWERS TO ODD NUMBERED PROBLEMS

Section 1.1

1 5 kilometers, 23 minutes

3 Argentina produced 9 million metric tons of wheat in 2002

7 $f(5) = 25$

9 $f(5) = 2$

11 SHORT ANSWER NOT WRITTEN

13 Greatest number of species at intermediate number of snails; yes

15 (a) 100 species at 500 ft
(b) k: value of N at sea level
c: lowest elevation with no bats

17 (a) 0.14 mg nicotine
(b) 4 hours
(c) 0.4
(d) Time nicotine level zero

19 (a) \$1000
(b) \$2200
(c) 20 years

21 (a) 13 million tons, 2 million tons, 10 million tons
(b) US: rose, then steady
India: rose
Former SU: rose, then fell

23

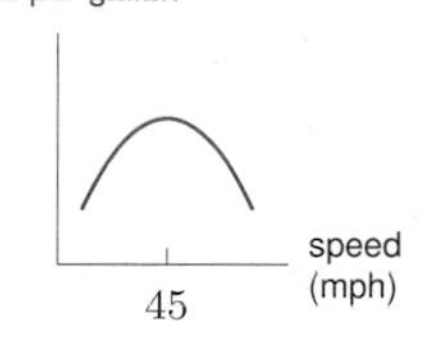

Section 1.2

1 Slope: $-12/7$
Vertical intercept: $2/7$

3 Slope: 2
Vertical intercept: $-2/3$

5 $y = (1/2)x + 2$

7 $y = (1/2)x + 2$

9 (a) l_1
(b) l_3
(c) l_2
(d) l_4

11 (a) (V)
(b) (IV)
(c) (I)
(d) (VI)
(e) (II)
(f) (III)

13 $C = 25 + 0.05m$

15 (a) Linear
(b) Linear
(c) Not linear

17 (a) $q = -(1/3)p + 8$
(b) $p = -3q + 24$

19 (a) 300 miles
(b) 50 mph
(c) $D = 300 + 50t$

21 (a) $\Delta w/\Delta h$ constant
(b) $w = 5h - 174$; 5 lbs/in
(c) $h = 0.2w + 34.8$; 0.2 in/lb

23 (a) $C = 12 + 0.2w$
(b) 0.2 \$/kg
(c) \$12

25 (a) $S = 113 - 0.94t$
(b) During 2038

27 (c)

29 (a) 60, 40 years
(b) (ii)
(c) 6.375 beats/minute more under new formula

31 No

Section 1.3

1 Concave down

3 Concave up

5 Decreasing
Concave up

7 (a)

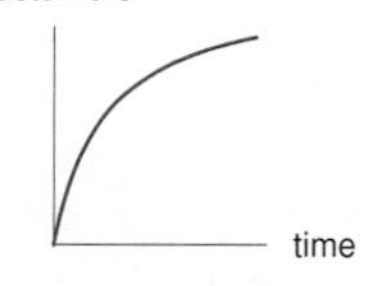

(b) Concave down

9 8

11 (a) Approximately -98 million pounds/year
(b) Yes, between 1996 and 1997

13 (a) Increasing, concave down
(b) 6 cm/year

15 1.61 million tons/year

17 (a) \$8894 million
(b) \$1778.8 million per year

19 2,353,747.3 households/year

21 (a) (i) -0.019 (mg/ml)/min
(ii) -0.016 (mg/ml)/min
(b) Decreasing; magnitude of rate decreasing

23 (a) -11 cm/sec
(b) -5.5 (cm/sec)/kg

25 (a) (i) Positive
(ii) Positive
(iii) Negative
(iv) Positive
(b) (i) $0 \le t \le 5$
(ii) $0 \le t \le 20$
(c) 25 m^3/week

27 (a) Concave up; no
(b) 2.6 m/sec

29 Decreasing, concave down

31

Section 1.4

1 (a) Between 20 and 60 units
(b) About 40 units

3 (a) \$4000
(b) \$2
(c) \$10

(d)

\$
$R(q)$
$C(q)$
5000
q
$q_0 = 500$

(e) 500

5 (a) \$75; \$7.50 per unit
(b) \$150

7 (a) When more than roughly 335 items are produced and sold
(b) About \$650

9 (a) When there are are more than 1000 customers
(b)

\$
14000
10000
5000
R
C
q
1000 2000

11 More than 875 students

13 (a) Answers vary, consulting company
(b) Answers vary, computer software company

15 (a) $V(t) = -1500t + 15{,}000$
(b) $V(3) = \$10{,}500$

17 (a) Price \$12, sell 60
(b) Decreasing

19 5500: Quantity demanded at price 0
100: Drop in quantity demanded if price increases \$1

21 (a) $C = 5q + 7000$
$R = 12q$
(b) $q = 1520$, $\pi(12) = \$3640$
(c) $C = 17{,}000 - 200p$
$R = 2000p - 40p^2$
$\pi(p) = -40p^2 + 2200p - 17{,}000$
(d) At \$27.50 per shirt the profit is \$13,250

23 (a) First: demand curve;
Second: supply curve
(b) Roughly 14
(c) Roughly 24
(d) Lower
(e) Any price less than or equal to \$143
(f) Any price greater than or equal to \$110

25 (a) $q = 820 - 20p$
(b) $p = 41 - 0.05q$

27

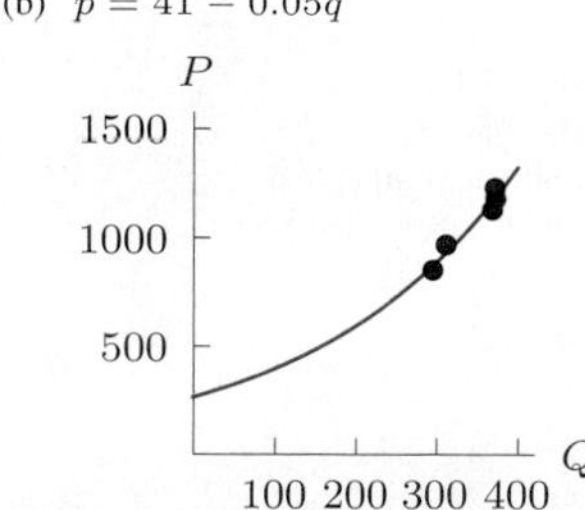

29 (a) $40b + 10s = 1000$

(b)

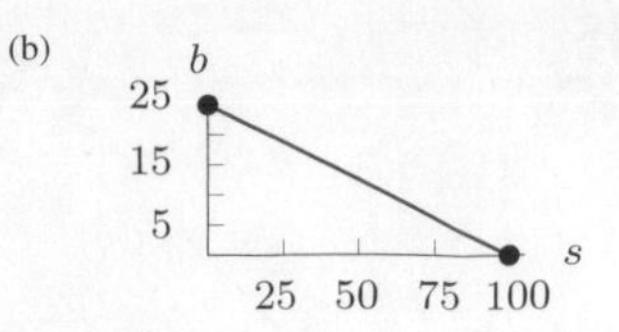

(c) The intercepts are $(0, 25)$ and $(100, 0)$

31 (a)

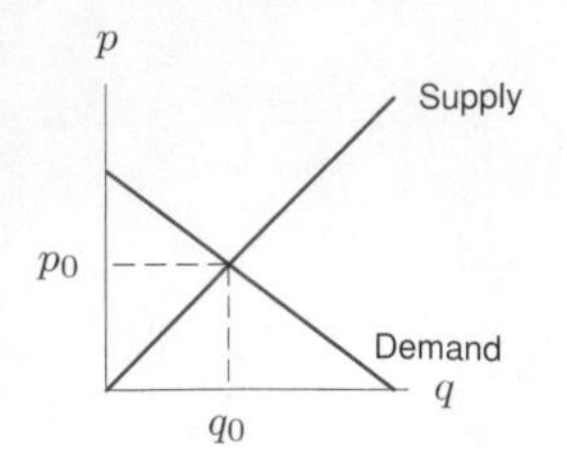

(b) Equilibrium price will increase; equilibrium quantity will decrease
(c) Equilibrium price and quantity will decrease

33 $q = 4p - 28$

35 (a) $p = 100, q = 500$
(b) $p = 102, q = 460$
(c) Consumer pays \$2
Producer pays \$4
(d) \$2760

37 (a) Demand: $q = 100 - 2p$
Supply: $q = 2.85p - 50$
(b) New equilibrium price $p \approx \$30.93$
New equilibrium quantity $q \approx 38.14$ units
(c) Consumer pays \$0.93
Producer pays \$0.62
Total \$1.55
(d) \$59.12

Section 1.5

1 (a) (i), 12%
(b) (ii), 1000
(c) Yes, (iv)

3 (a) $P = 1000 + 50t$
(b) $P = 1000(1.05)^t$

5 (a) $80 - 4t$
(b) $80(0.95)^t$

7 (a) $W = 32.4(1.036)^t$
(b) \$44.54 trillion

(c)

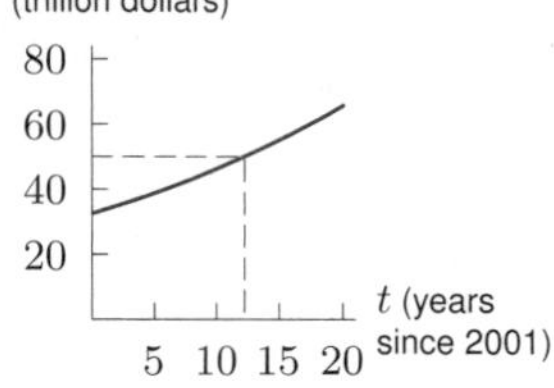

(d) During 2013

9 (a) 1.26% per year
(b) 6.4 billion, 6.90 billion
(c) 83 million people per year

11 $f(x) = 4.30(1.4)^x$

13 $y = 500(1.59)^t$

15 1.7%

17 (a) $h(x) = 31 - 3x$
(b) $g(x) = 36(1.5)^x$

19 (a) $C = 6.03 + 0.094t$,
0.094 bn tons/yr
(b) $C = 6.03(1.015)^t$,
1.5 %/yr

21 $d = 670(1.096)^{h/1000}$

23 Decreasing; concave up

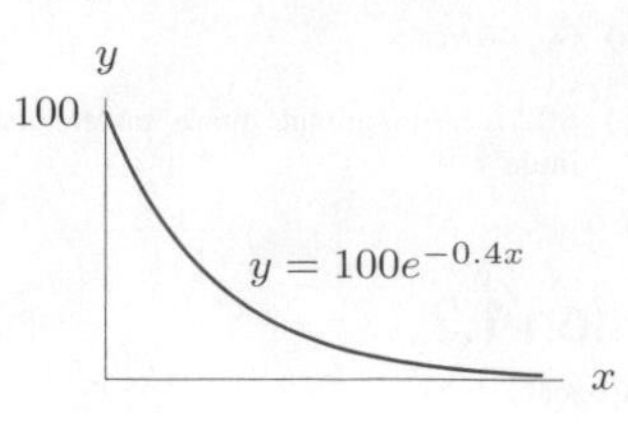

25 (a) 125%
(b) 9 times

27 Min. wage grew 4.69% per year

Section 1.6

1 $t = (\ln 7)/(\ln 5) \approx 1.209$

3 $t = (\ln 2)/(\ln 1.02) \approx 35.003$

5 $t = (\ln 5)/(\ln 3) \approx 1.465$

7 $t = (\ln a)/(\ln b)$

9 $t = \ln 2.5 \approx 0.9163$

11 $t = 2(\ln 5 - \ln 3) \approx 1.0217$

13 $t = (\ln B - \ln P)/r$

15 $t = \ln 8 - \ln 5 \approx 0.47$

17 5; 7%

19 15; −6% (continuous)

21 (a) (i) $P = 1000(1.05)^t$
(ii) $P = 1000e^{0.05t}$
(b) (i) 1629 (ii) 1649

23 $P = 15(1.2840)^t$; growth

25 $P = P_0(1.2214)^t$; growth

27 $P = 15e^{0.4055t}$

29 $P = 174e^{-0.1054t}$

31 (a) $P(0.5) \approx 779$;
$P(1) \approx 607$
(b) 223
(c) Approximately 4.6
(d)

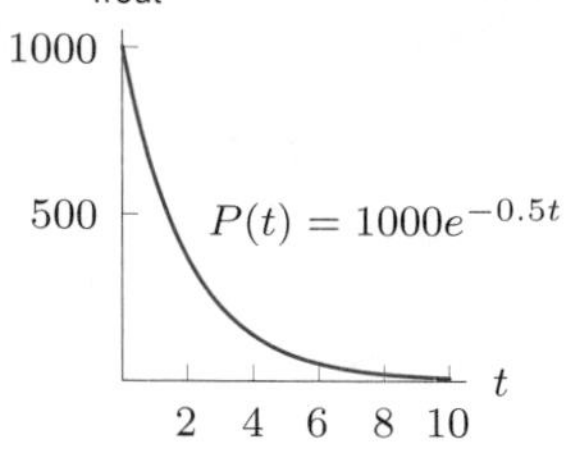

33 (a) 6%
(b) $P = 100(1.0618)^t$, 6.18%

35 $W = 32.4e^{0.0354t}$

37 8.33%

39 $P = 6.4e^{0.01252t}$

41 1990

Section 1.7

1 Just over 6 hours

3 \$10,976.23

5 (a) \$6885.64
(b) 5.9%

7 About 10.24 years

9 (a) $P(t) = 200(1.05)^t$
(b)

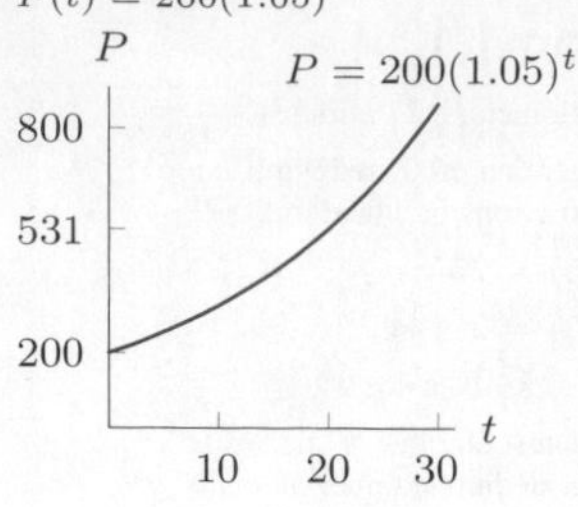

(c) 326
(d) ≈ 15 years

11 \$7866.28

13 (a) 15,678.7 years
(b) 5728.5 years

15 $P = 500e^{0.5493t}$,
7794

17 A: continuous
B: annual
\$20

19 96.34 years

21 (a) \$5068.93
(b) \$4878.84

23 0.0187, or 18.7%
$5000e^{0.0187t}$
448,766

25 (a) 8.75 years
(b) About 9.01 years

27 About 6.58 years

29 It is a fake

31 (a) Option 1
(b) Option 1: \$10.929 million;
Option 2: \$10.530 million

33 Loan

35 No

Section 1.8

1 (a) $h^2 + 6h + 11$
(b) 11
(c) $h^2 + 6h$

3 (a) 4
(b) 2
(c) $(x + 1)^2$
(d) $x^2 + 1$
(e) $t^2(t + 1)$

5 (a) e
(b) e^2
(c) e^{x^2}
(d) e^{2x}
(e) $e^t t^2$

7 (a) 9
(b) 20
(c) 25
(d) 11

9 (a) $10x^2 + 3$
(b) $20x^2 + 60x + 45$
(c) $4x + 9$

11 (a) $y = 2^u, u = 3x - 1$
(b) $P = \sqrt{u}, u = 5t^2 + 10$
(c) $w = 2 \ln u, u = 3r + 4$

13

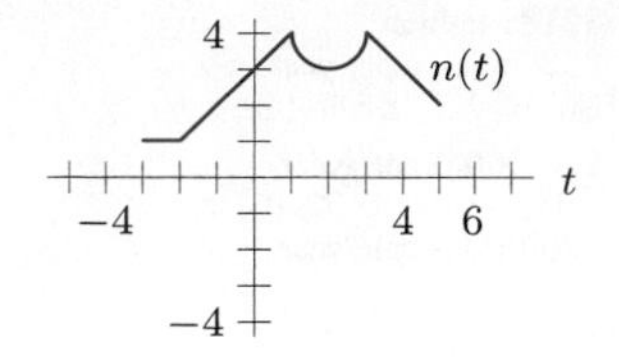

15

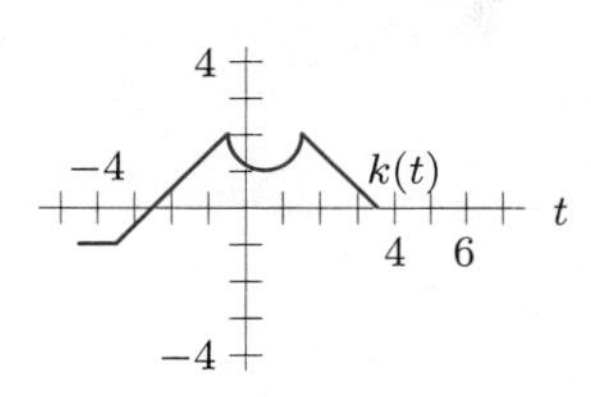

17

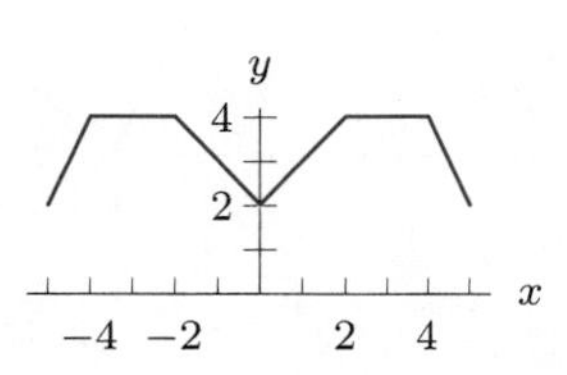

19

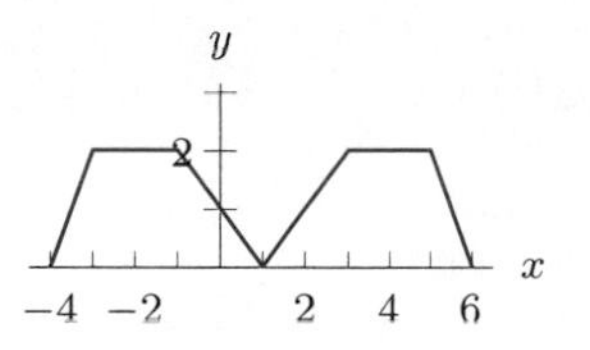

21

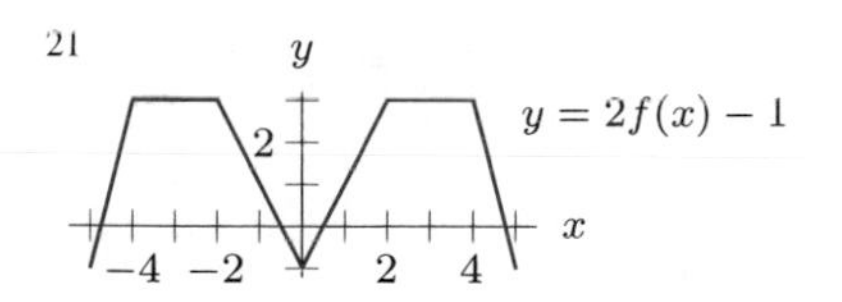

23 (a)

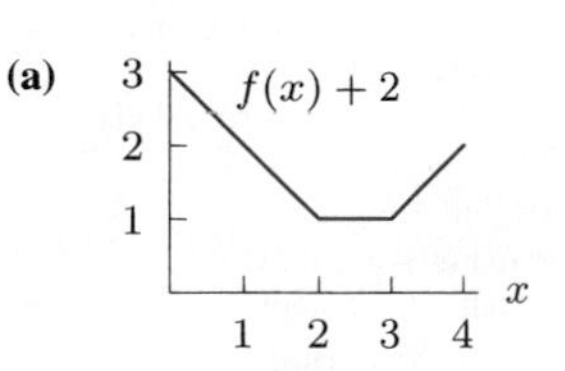

(b) $f(x - 1)$

(c) $3f(x)$

(d)

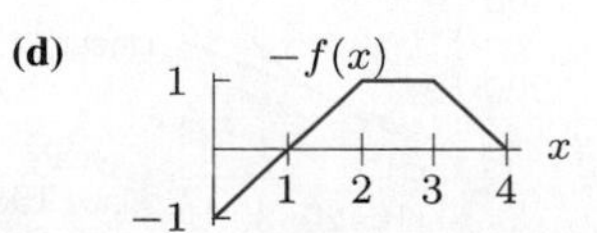

25 (a) $f(x) + 2$

(b) $f(x - 1)$

(c) $3f(x)$

(d) $-f(x)$

27

29

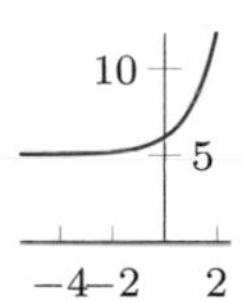

31 (a)

x	$f(x) + 3$
0	13
1	9
2	6
3	7
4	10
5	14

(b)

x	$f(x - 2)$
2	10
3	6
4	3
5	4
6	7
7	11

(c)

x	$5g(x)$
0	10
1	15
2	25
3	40
4	60
5	75

(d)

x	$-f(x) + 2$
0	-8
1	-4
2	-1
3	-2
4	-5
5	-9

(e)

x	$g(x - 3)$
3	2
4	3
5	5
6	8
7	12
8	15

(f)

x	$f(x) + g(x)$
0	12
1	9
2	8
3	12
4	19
5	26

33 1.1

35 (a) $y = 2x^2 + 1$

$y = 2x^2 + 1$ $y = x^2$

(b) $y = 2(x^2 + 1)$
(c) No

37 (a) A, B positive; C negative
(b) $A + B$
(c) A

Section 1.9

1 $y = 5x^{1/2}$

3 Not a power function.

5 $y = 9x^{10}$

7 Not a power function

9 $y = 8x^{-1}$

11 Not a power function

13 $S = kh^2$

15 $v = d/t$

17 Inversely proportional: $s \approx 6/w$

19 Blood mass $= 0.05$ (Body mass); 3.5 kilograms

21 $k = 1095$
16,782 cm^2

23 (a) $T = kB^{1/4}$
(b) $k = 17.4$
(c) 50.3 seconds

25 (a) $1.3\ \text{m}^2$
(b) 86.8 kg
(c) $h = 112.6s^{4/3}$

27 (a) 0.5125
(b) 0.3162
(c) $201{,}583$ dynes/cm^2

29 (a) Degree 2; leading coefficient positive
(b)

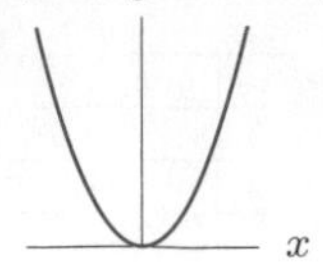

(c) One turning point

31 (a) Degree 3; negative
(b)

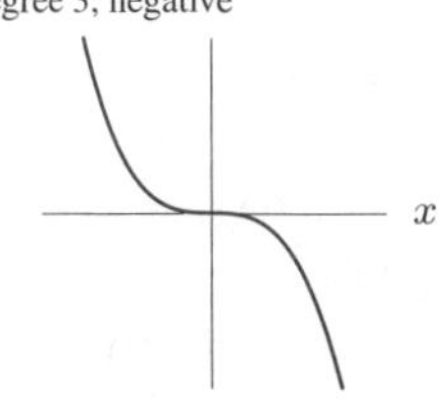

(c) Two turning points

33 (a) Degree 4; negative
(b)

x

(c) One turning point

35 (a) Degree 7; positive
(b)

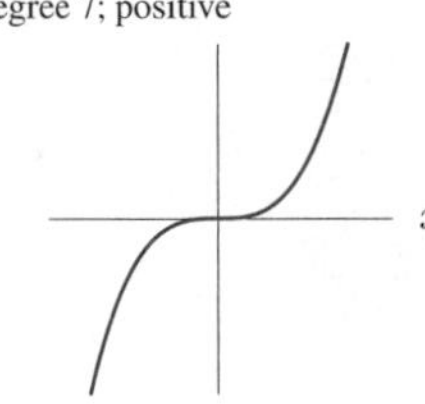

(c) Two turning points

37 (a) $q = -8p + 700$
(b) $R = -8p^2 + 700p$
(c) At roughly \$44 per unit, revenue $\approx \$15{,}300$

39 (a) $y = (x-2)^3 + 1$
(b) $y = -(x+3)^2 - 2$

41 No

Section 1.10

1 (b) Max at 7 pm; min at 9 am
(c) Period = one day; amplitude $\approx 2^\circ$C

3 $8\pi; 3$

5 (b)

7 Amplitude = 3; Period = 2π

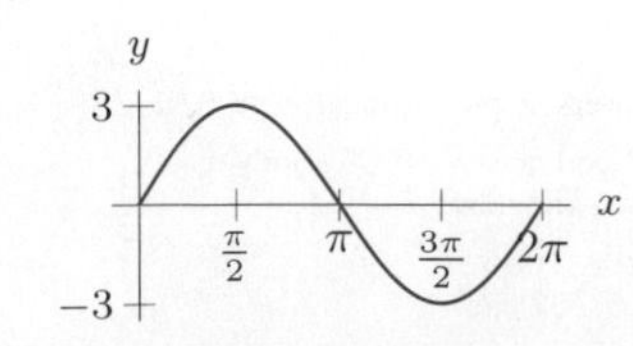

9 Amplitude = 3; Period = π

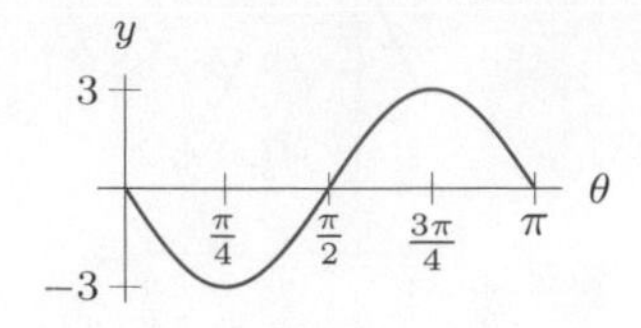

11 Amplitude = 4; Period = 4π

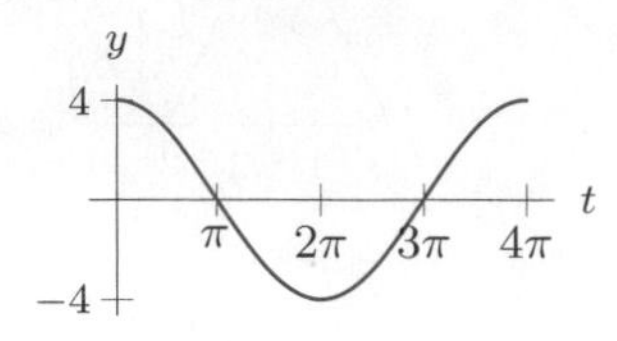

13 $0.35\cos(2\pi t/5.4) + 4$

15 (a) 5
(b) 8
(c) $f(x) = 5\cos((\pi/4)x)$

17 (a) Period = 12 months;
Amplitude = 4,500 cases
(b) About 2000 cases and 2000 cases

19 $y = 7\sin(\pi t/5)$

21 $f(x) = 5\cos(x/3)$

23 $f(x) = -4\sin(2x)$

25 $f(x) = -8\cos(x/10)$

27 $f(x) = 5\sin((\pi/3)x)$

29 $f(x) = 3 + 3\sin((\pi/4)x)$

31 Depth = $7 + 1.5\sin(\pi t/3)$

33 (a) Period is 12; amplitude is 4
(b) $g(34) = 11$; $g(60) = 14$

Chapter 1 Review

1 (a) (IV)
(b) (II)
(c) (III)

3

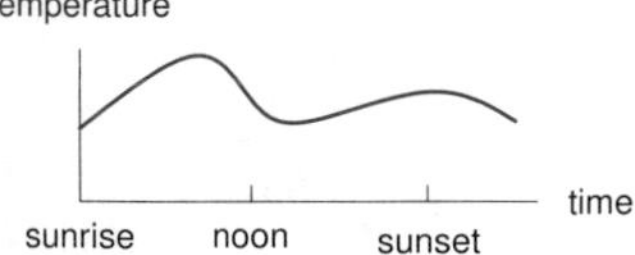

5 (a) 8; 7
(b) 10

7 (a) -2
(b) 10
(c) 3
(d) 3

9 $y = 8/3 - x/3$

11 $x = -1$

13 $P = 11.3 + 0.4t$

15 (a) Slope = 1.8
(b) $^\circ\text{F} = 1.8(^\circ\text{C}) + 32$
(c) 68°F
(d) -40°

17

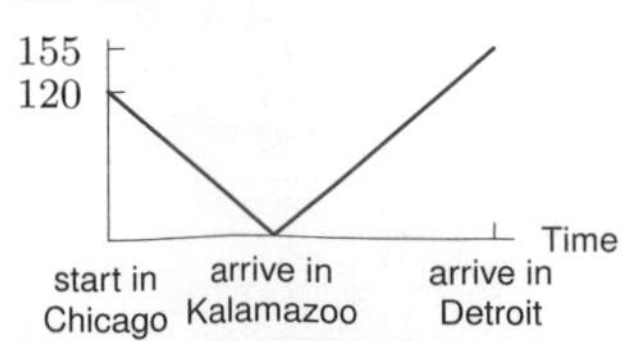

19 (a) \$2181 million
(b) \$727 million per year
(c) No

21 (a) 220,000 people/year
(b) $0.19, -0.36, 0.92, 0.13$
(c) 220,000 people/year

23 (a) $k(t)$
(b) $h(t)$
(c) $g(t)$

25 $y = (-3/7)x + 3$

27 $y = 3e^{0.2197t}$
or $y = 3(1.2457)^t$

29 $z = 1 - \cos\theta$

31 (a) $R(P) = kP(L-P)$
$(k > 0)$
(b)

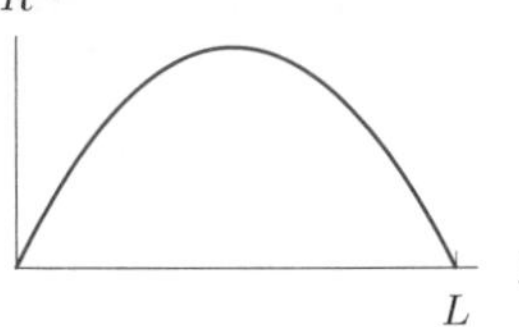

33 $y = -(x+3)^2 + 2$

35 $(\log 11)/(\log 3) = 2.2$

37 $\ln(100)/5 = 0.921$

39 $P = (1.0833)^t$; $Q = (0.741)^t$.

41 (a) $\ln(2x+3)$
(b) $2\ln x + 3$
(c) $4x + 9$

43 $2z + 1$

45 $2zh - h^2$

47 $f(x) = x^3$
$g(x) = x + 1$

49 (a) $R(n) = 7 + 1.5n$
(b) $R(2) = \$10$ and $R(8) = \$19$

51 (a) First price list:
$C_1(q) = 100 + 0.03q$ dollars
Second price list:
$C_2(q) = 200 + 0.02q$ dollars
(b) First price list
(c) 10,000

53 (a) Roughly 360 scoops
(b) Roughly 120 scoops

55 (a) $y = 2700 + 486t$
486 zebra mussels per year
(b) $y = 2700(1.18)^t$
18% per year

57 About 6.80 billion

59 (a) $P = 6t + 60$
(b) $P = 60(1.056)^t$

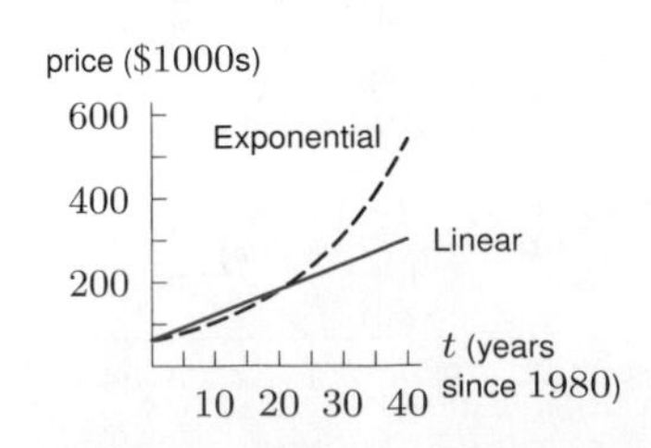

(d) Exponential

61 (a) 15%
(b) $P = 10(1.162)^t$
(c) 16.2%
(d) Graphs are the same since functions are equal

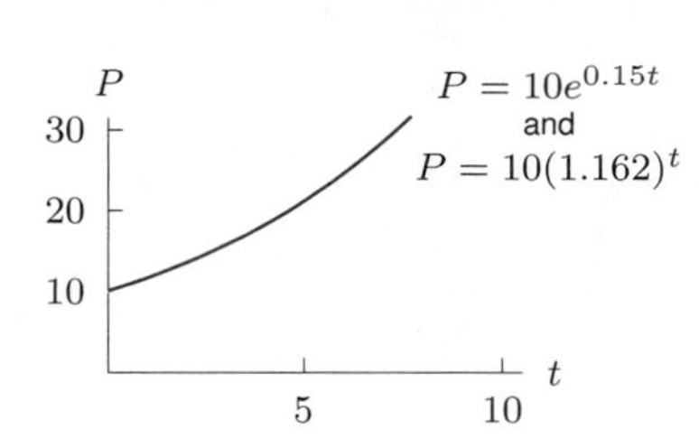

63 (a) 47.6%
(b) 23.7%

65 (a) $P(t) = (0.975)^t$
(b)

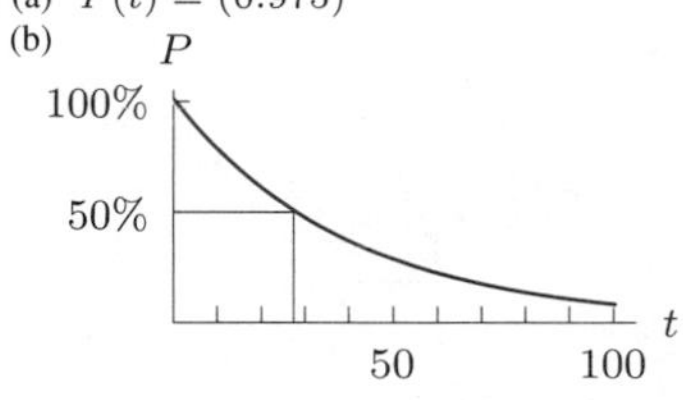

(c) About 27 years
(d) About 8%

67 7.925 hours

69 (a) Option 1
(b) $2102.54, $2051.27, $2000
(c) $2000, $1951.23, $1902.46

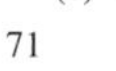
71 (a)

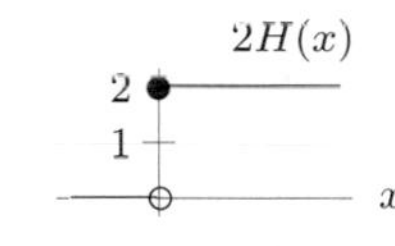

(b)

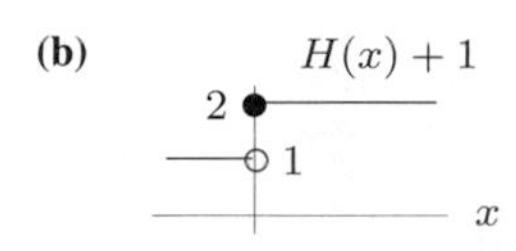

(c)

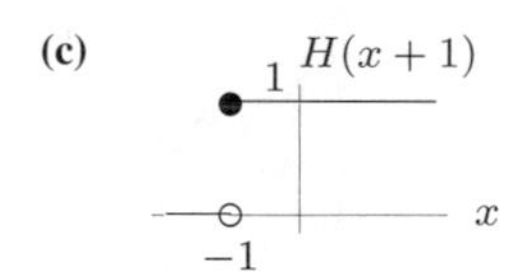

(d)

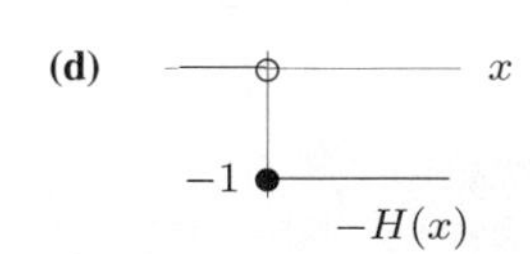

(e)

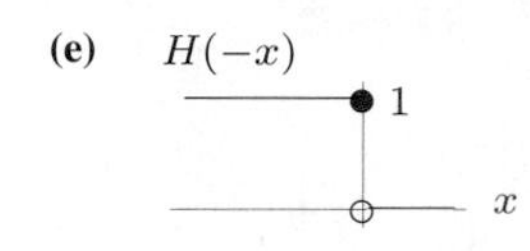

73 0.3 seconds

75 $f(x) = 2 - \sin x$

77 US: 156 volts max, 60 cycles/sec
Eur: 339 volts max, 50 cycles/sec

Modeling: Fitting Data

1 (a)

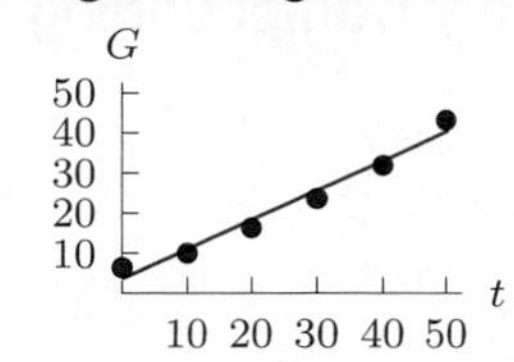

(b) Increasing at a rate of $734 billion/year
(c) $43.9 trillion, $54.9 trillion,
More confidence in the 2005 prediction

3 (a)

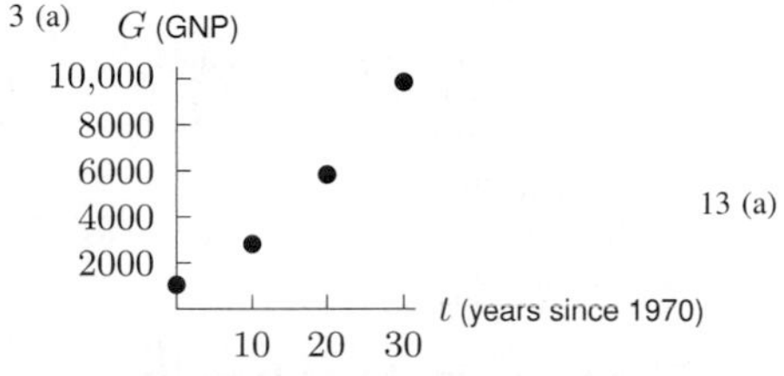

Yes

(b)

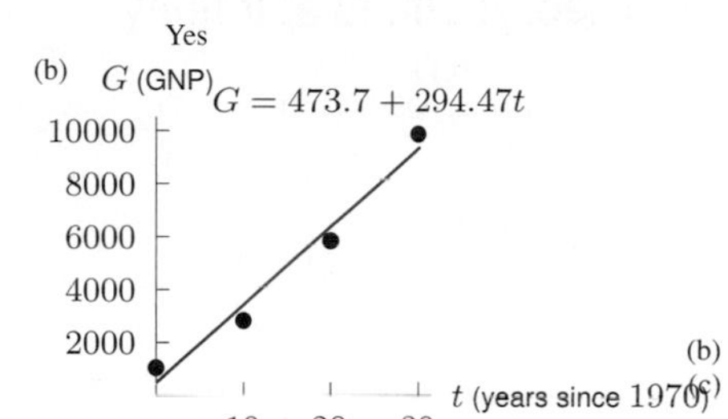

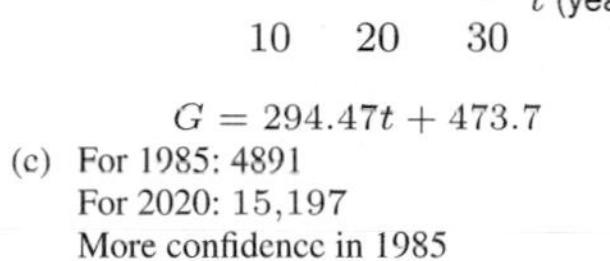
$G = 294.47t + 473.7$
(c) For 1985: 4891
For 2020: 15,197
More confidence in 1985

5 (a) $S = 0.08v + 1.77$
(b)

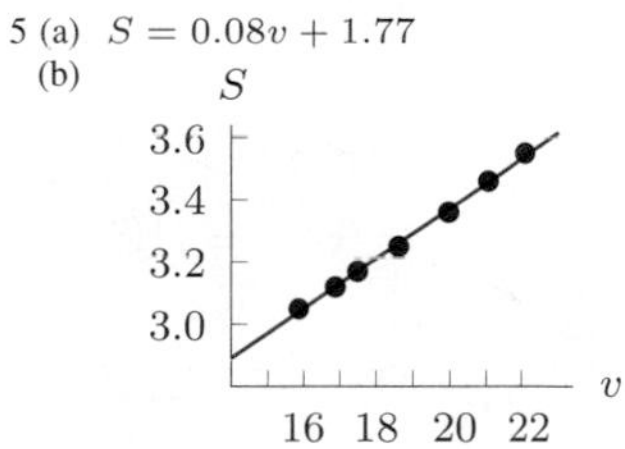

Yes

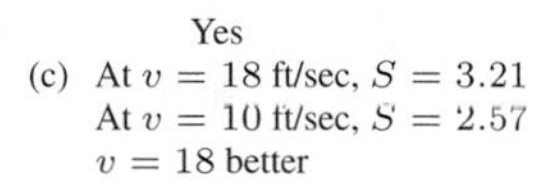
(c) At $v = 18$ ft/sec, $S = 3.21$
At $v = 10$ ft/sec, $S = 2.57$
$v = 18$ better

7 (a) $0.0026 = 0.26\%$
(b) For 1900, 272.27
For 1980, 335.1

9 (a) $P = 180.5(1.011)^t$

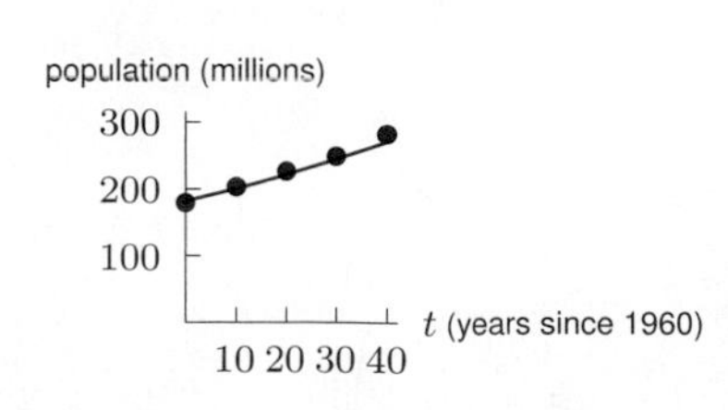

(b) 1.1% per year
(c) 348.0 million

11 $C = 418.9 + 3.612q$
Marginal cost $3.612 per unit
$r = 0.984$

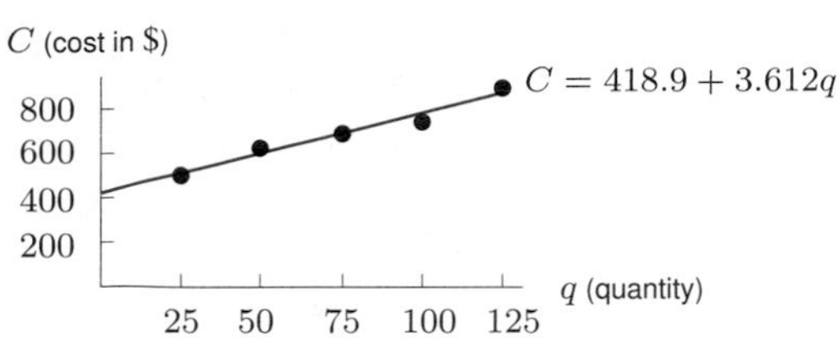

13 (a)

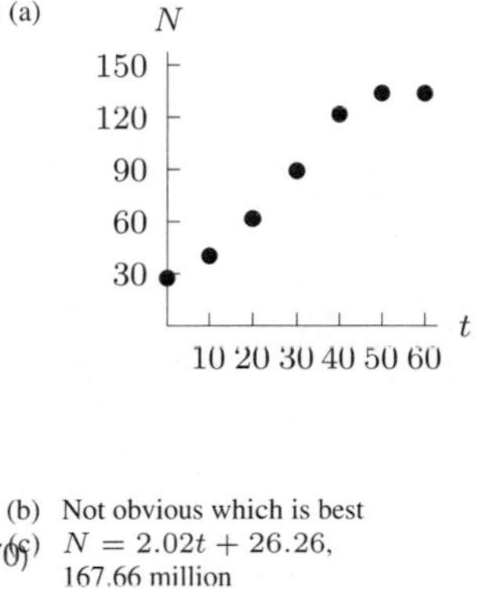

(b) Not obvious which is best
(c) $N = 2.02t + 26.26$,
167.66 million

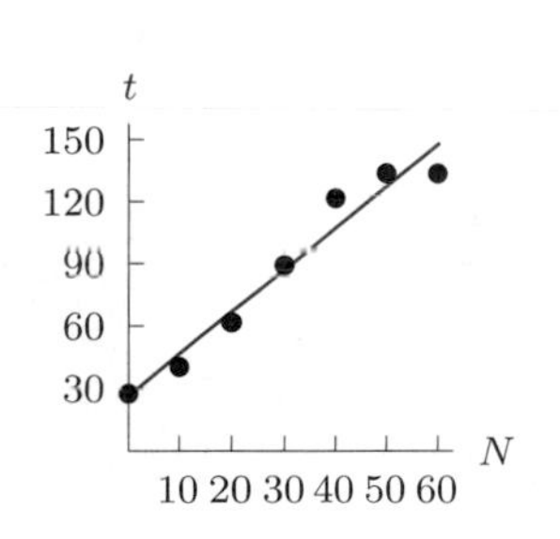

(e) $N = 32.4(1.028)^t$, Answers may vary
223.9 million

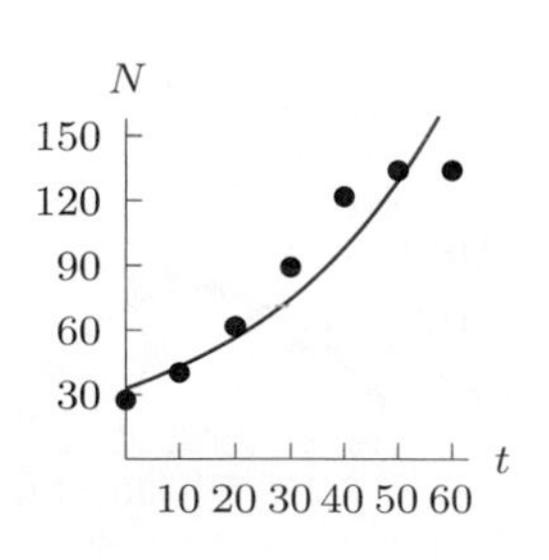

(f) 2.8%

15 (a)

(b)

$Y = -0.021x + 1.23$. Success decreases 21%/yard

(c)

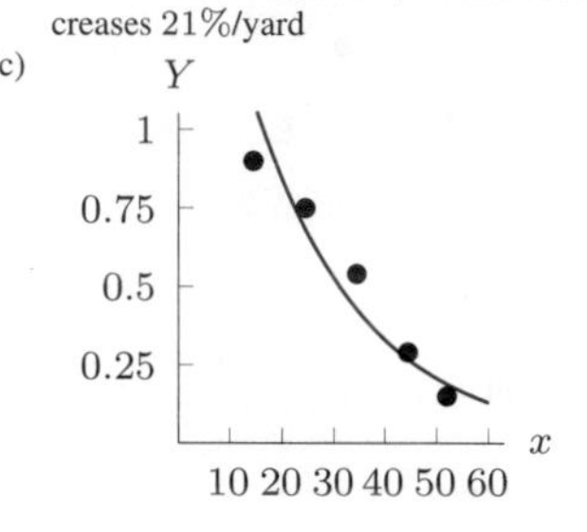

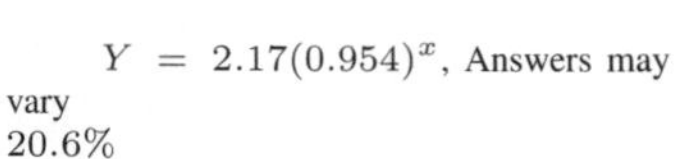

$Y = 2.17(0.954)^x$, Answers may vary
20.6%

(d) Linear

17 Leading coefficient > 0

19 (a)

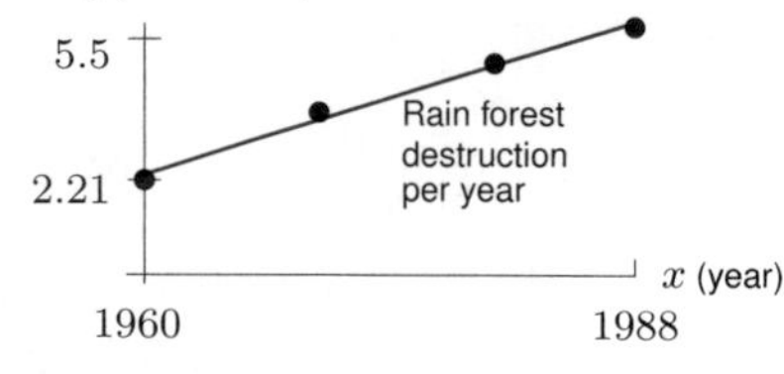

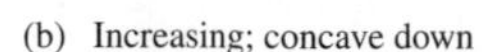

(b) Increasing; concave down
(c) $y \approx -1884.66 + 248.92 \ln x$
(d) About 8.6 million hectares

21 (a) Quadratic regression
(b) $N = -0.0886t^2 + 3.93t + 17.7$, Answers may vary
(c) 15,135 warheads
(d)

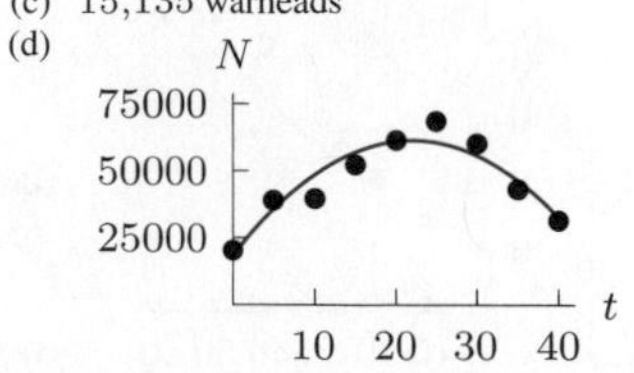

23 (a) Polynomial (Cubic)
(b) Polynomial (Quadratic)
(c) Exponential
(d) Linear
(e) Exponential
(f) Linear

Modeling: Compounding

1 27%

3 (a) \$160,356.77
(b) \$165,510.22
(c) \$165,891.05
(d) \$165,989.48
(e) \$166,005.85

5 6.18%

7 8.33%

9 (a) 1.0408107
1.0408108
1.0408108
4% compounded continuously $\approx 4.08108\%$
(b) $e^{0.04} \approx 1.048108$

11 (a) V, (b) III, (c) IV, (d) I, (e) II

13 (a) 13,900 cruzados
(b) 24.52%

Theory: Limits at Infinity

1 Largest: $0.1x^4$
Smallest: $1000x^2$

3 $y = x^{2/3}$ is larger as $x \to \infty$

5 As $x \to \infty$, x^5 largest positive;
As $x \to -\infty$, $-x^3$ largest positive

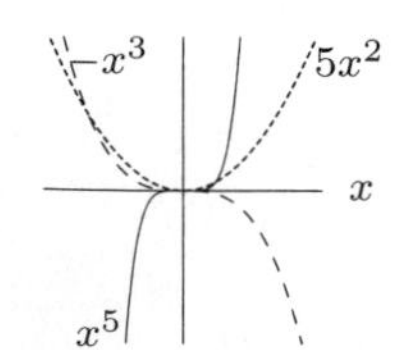

7

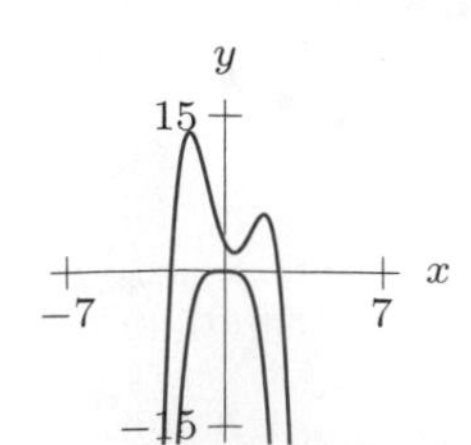

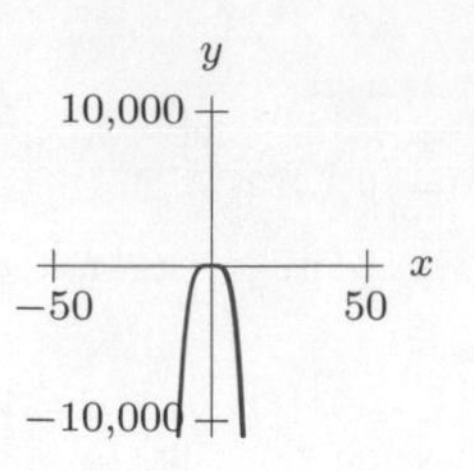

9

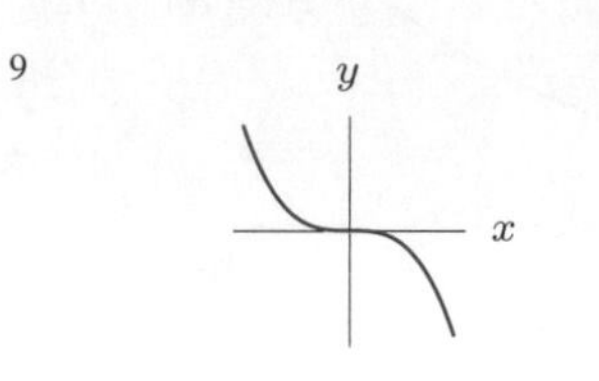

11 (b) (i)

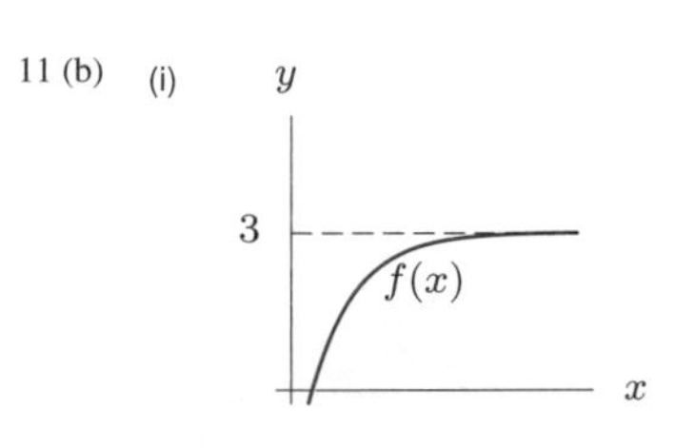

(ii)

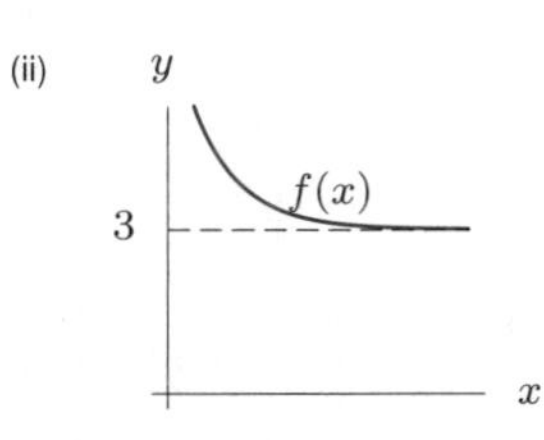

(iii)

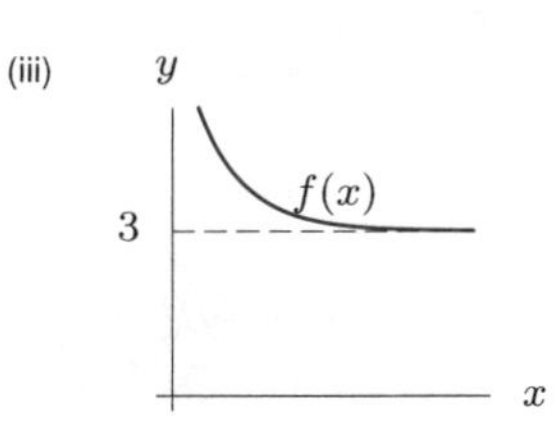

(iv)

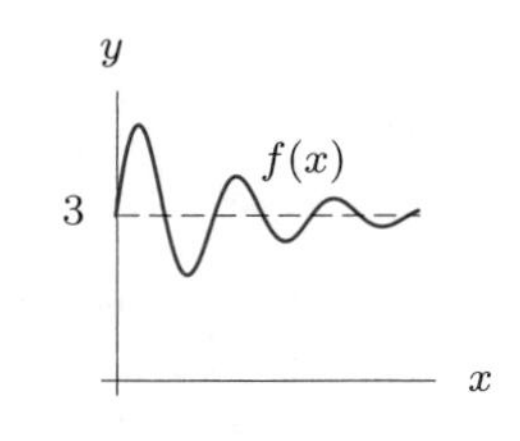

13 0

15 $-\infty, -\infty$

17 $8, -\infty$

19 $3x^5$

21 $x^{1/2}$

23 $0.5x^4$

25 A: $0.2x^5$,
B: x^4,
C: $5x^3$,
D: $70x^2$

27 A is $100x^2$
B is x^5
C is 3^x

29

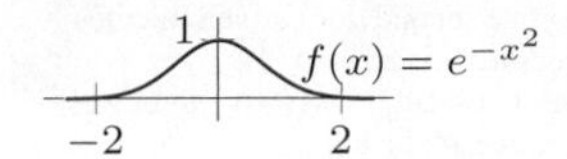

(a) Increasing for $x < 0$; decreasing for $x > 0$
(b) Concave down
(c) As $x \to \infty$, $f(x) \to 0$; as $x \to -\infty$, $f(x) \to 0$

Section 2.1

1 (a) 8 ft/sec
(b) 6 ft/sec

3 (a) 63 cubic millimeters
(b) 10.5 cubic millimeters/month
(c) 44.4 cubic millimeters/month

5 $P'(6) = 25{,}754$ people/year

7 $v_{\text{avg}} = 8.125$ ft/sec
$v(0.2) \approx 4.5$ ft/sec

9 (a) Positive at C and G
Negative at A and E. Zero at B, D, and F
(b) Largest at G
Most negative at A

11 $f'(2) \approx 40.268$

13 (a) Negative
(b) $f'(1) = -3$

15 $P'(0) = 10$

17 (a) Positive
(b) $f'(2) \approx 0.95$, $f'(10) \approx 0.55$

19 $f'(1) \approx 1.0005$;
$f'(2) \approx 1.6934$;
f is concave up on $[1, 2]$

21 (a) $g(2) = 5$
(b) $g'(2) = -0.4$

23

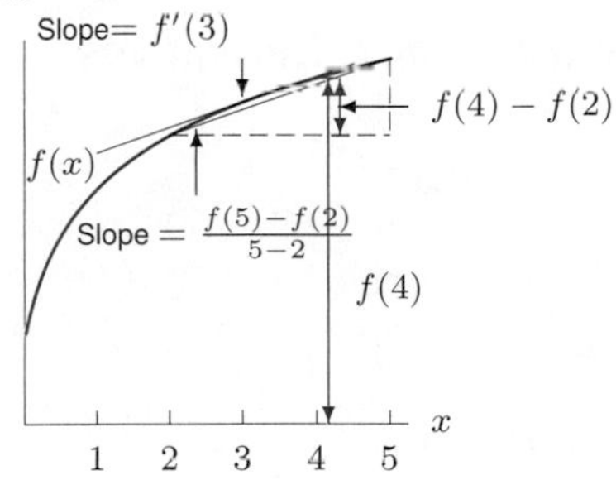
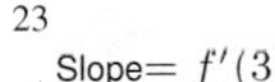

25 (a) The slopes of the two tangent lines at $x = a$ are equal for all a
(b) A vertical shift does not change the slope

Section 2.2

1 1.0, 0.3, −0.5, −1

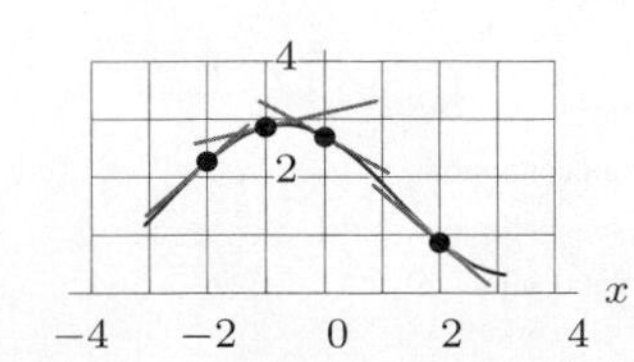

3

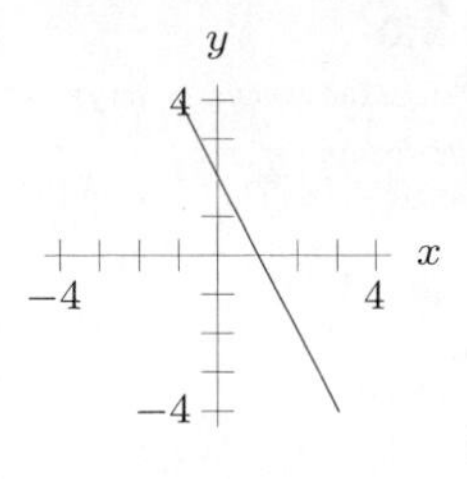

5

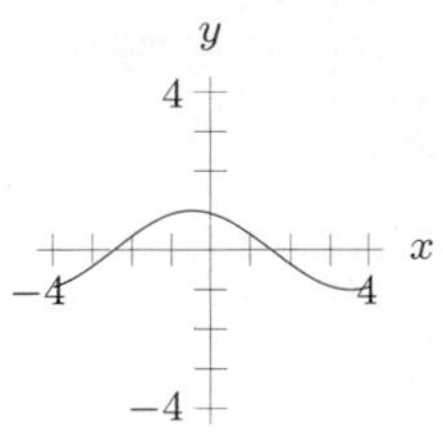

7

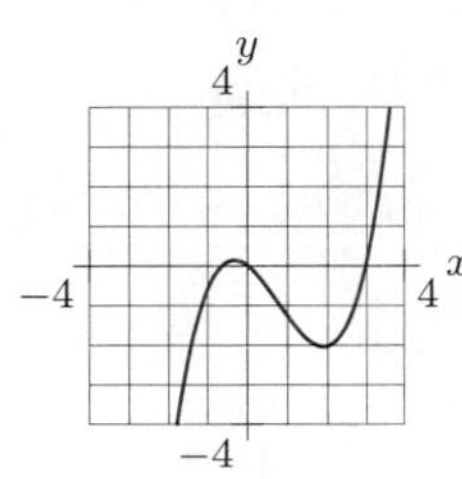

9 VIII

11 II

13 (a) x_3
(b) x_4
(c) x_5
(d) x_3

15 $-6, -3, -1.8, -1.2, -1.2$

17

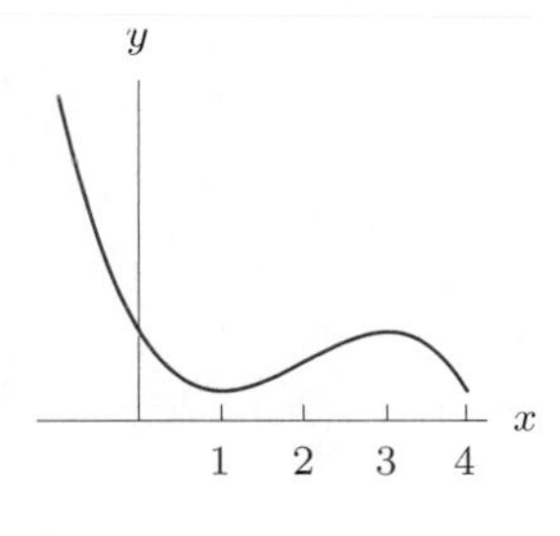

19

x
-1.5
$f'(x)$

21

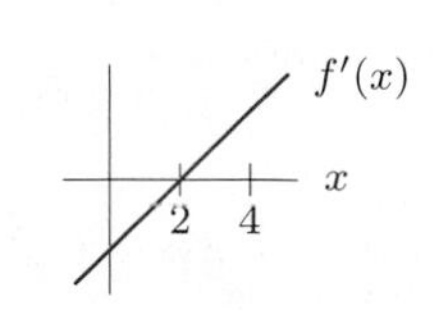

23

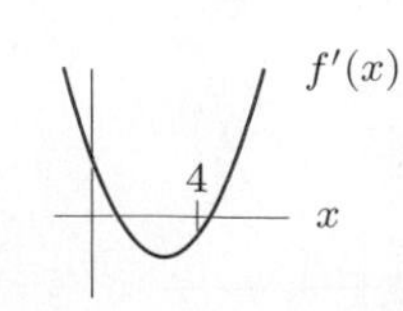

25

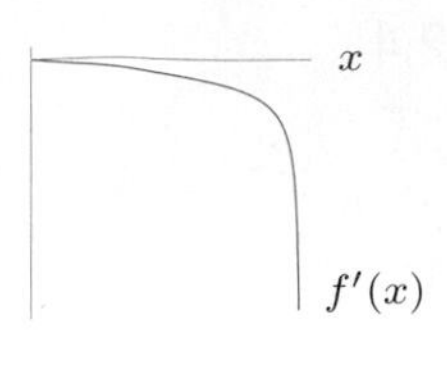

27 (a) $f'(1) \approx 0.95$
$f'(2) \approx 0.49$
$f'(3) \approx 0.33$
$f'(4) \approx 0.25$
$f'(5) \approx 0.20$
(b) $1/x$

29 (a) $x_1 < x < x_3$
(b) $0 < x < x_1$; $x_3 < x < x_5$

Section 2.3

1 kilograms/meter

3 (a) 12 pounds, 5 dollars
(b) Positive
(c) 12 pounds, 0.4 dollars/pound

5 (a) ml; minutes
(b) ml; minutes/ml

7 (a) Negative
(b) °F/min

9 (a) Positive
(b) Child weighs 45 pounds at 8 years
(c) lbs/year
(d) The child is growing at a rate of 4 lbs/year at 8 years of age
(e) Decrease

11 (a) kg/week
(b) Growing at 96 gm/wk in week 24

13 (a) Less
(b) Greater

15 wind stronger at 15.1 km than at 15 km

17 1.338 billion people; growing at 8 million people per year

19 Dollars/year

21 $f(21) \approx 65$
$f(19) \approx 71$
$f(25) \approx 53$

23 (a) 20 minutes,
0.36 mg,
-0.002 mg/minute
(b) $f(21) \approx 0.358$
$f(30) \approx 0.34$

25 (a) -5 (cm/sec)/kg
(b) -0.25cm/sec
(c) $v'(2) = -5$

27 8001.5 meters

29 (a) 1.7 (liters/minute)/hour
(b) 0.028 liter/minute
(c) $g'(2) = 1.7$

31 (a) Fat
(b) Protein

33 (a) 2.0 kg/week
(b) 0.6 kg/week
(c) 0.3 kg/week

35 I-fat, II-protein

37 (a) Liters per centimeter
(b) About 0.042 liters per centimeter
(c) Cannot expand much more

Section 2.4

1 (a) Negative
  (b) Negative
  (c) Positive

3 $f'(x) = 0$
  $f''(x) = 0$

5 $f'(x) < 0$
  $f''(x) > 0$

7 $f'(x) < 0$
  $f''(x) < 0$

9 $s'(t)$: positive
  $s''(t)$: positive or zero

11 Derivative:
  Pos. $-2.3 < t < -0.5$
  Neg. $-0.5 < t < 4$
  Second Derivative:
  Pos. $0.5 < t < 4$
  Neg. $-2.3 < t < 0.5$

13

Point	f	f'	f''
A	$-$	0	$+$
B	$+$	0	$-$
C	$+$	$-$	$-$
D	$-$	$+$	$+$

15 (b)

17 (a) Both negative
  (b) $f'(2) \approx -4$, $f'(8) \approx -21$

19

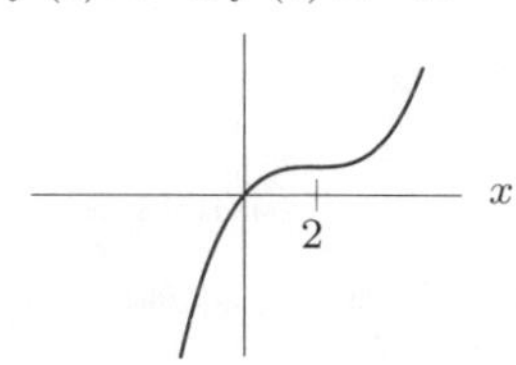

21 22 only possible value

23 (a)

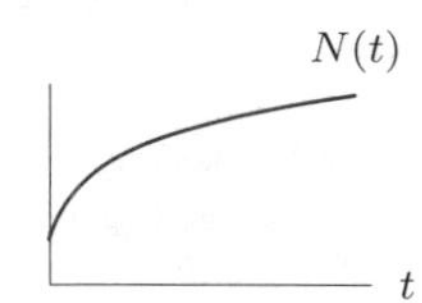

  (b) dN/dt is positive.
  d^2N/dt^2 is negative.

25 (a)

  (b) Derivative of utility is positive
  2nd derivative of utility is negative

27

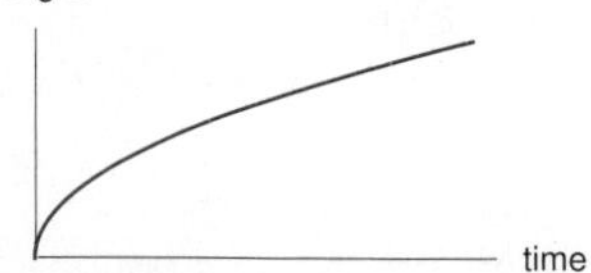

29 (a) IV, (b) III, (c) II, (d) I,
  (e) IV, (f) II

Section 2.5

1 About $\$0.42$ (answers may vary)

3 (a) Dollars/barrel
  (b) 101 barrels cost \$3 more than 100 barrels

5 At $q = 5$;
  At $q = 40$

7 (a) \$2408
  (b) \$2192

9 (a) \$4348
  (b) \$11 profit
  (c) No, company will lose money

11 (a) \$2200 profit
  (b) \$5 increase

13 (a) $\$1.8$ million
  (b) \$28,000 increase
  (c) \$35,000 decrease
  (d) \$4000 increase
  \$5000 decrease

15 (a) Fixed costs
  (b) Decreases slowly, then increases

Chapter 2 Review

1 Positive: A and D
  Negative: C and F
  Most positive: A
  Most negative: F

3 $f'(2) \approx 9.89$

5 $15.47, 57.65, 135.90,$
  $146.35, 158.55$ people/min

7 $f'(x)$ positive: $4 \leq x \leq 8$
  $f'(x)$ negative: $0 \leq x \leq 3$
  $f'(x)$ greatest: at $x \approx 8$

9

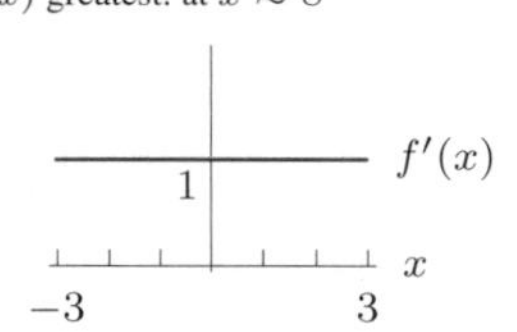

11

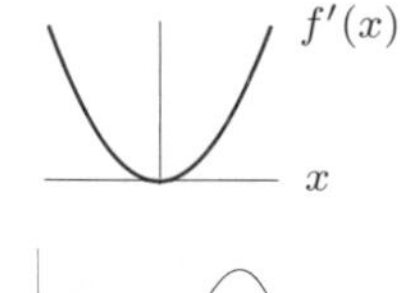

13

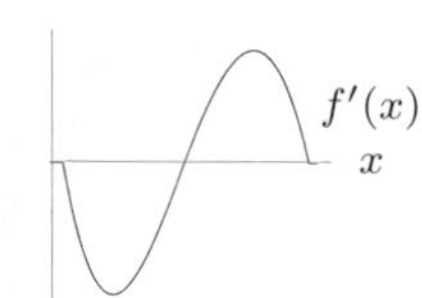

15 (a) Negative
  (b) Degrees/min

17 41

19 (a) $f(0) = 80$; $f'(0) = 0.50$
  (b) $f(10) = 85$.

21 (a) Investing the \$1000 at 5%
  would yield \$1649 after 10 years
  (b) Dollars/%

23 Feet/mile; negative

25 (a) $f'(a)$ is always positive
  (c) $f'(100) = 2$: more
  $f'(100) = 0.5$: less

27

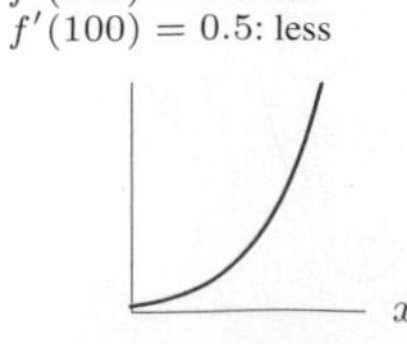

29 A positive second derivative indicates
  a successful campaign
  A negative second derivative indicates
  an unsuccessful campaign

31 (a)

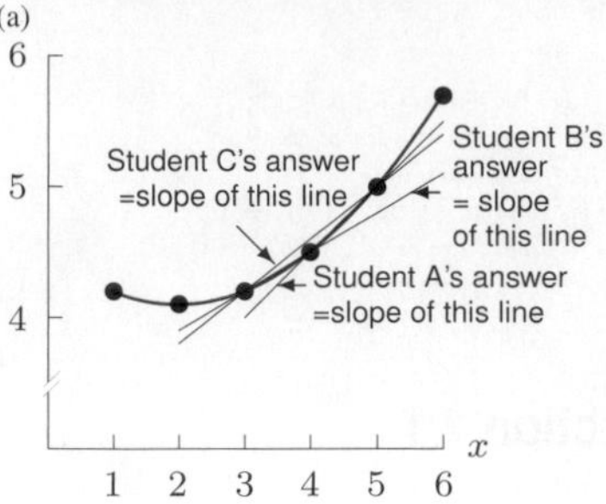

  (b) Student C's
  (c) $f'(x) = \dfrac{f(x+h) - f(x-h)}{2h}$

33 (a) About 1.8 joules/sec
  (b) Rate of energy consumption needed decreases,
  then increases at faster and faster rate
  (c) Possible answer:

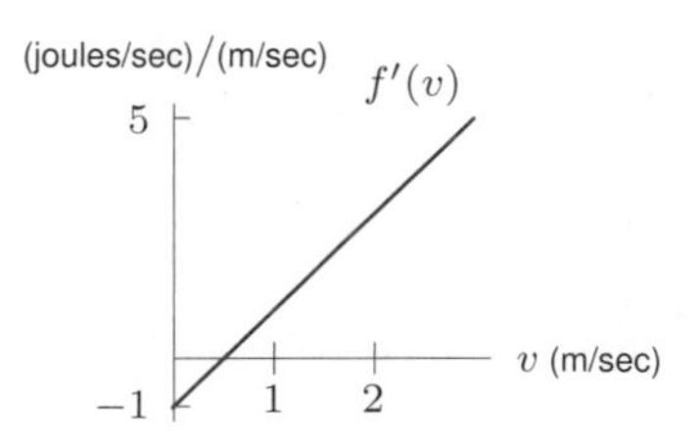

Theory: Limits, Derivatives

1

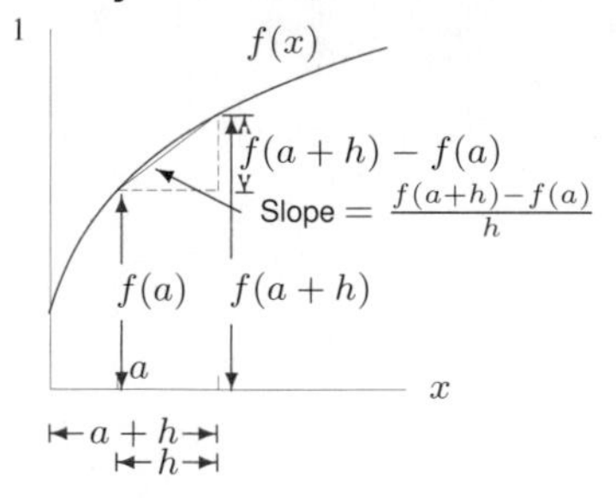

3 1

5 27

7 2.7

9 Yes

11 No; yes

13 Yes

15 Yes

17 No

19 Not continuous

21 Not continuous

23 Not continuous

Section 3.1

1 0

3 $12x^{11}$

5 $\frac{4}{3}x^{1/3}$

7 $12t^3 - 4t$

9 $-4x^{-5}$

11 $2x+5$

13 $6x+7$

15 $8.4q - 0.5$

17 $-5t^{-6}$

19 $-(7/2)r^{-9/2}$

21 $-(1/3)\theta^{-4/3}$

23 $15t^4 - \frac{5}{2}t^{-1/2} - 7t^{-2}$

25 $6t - 6/t^{3/2} + 2/t^3$

27 $(3/2)x^{1/2} + (1/2)x^{-1/2}$

29 $2kx$

31 $2aP + 3bP^2$

33 $b/(2\sqrt{t})$

35 $3ab^2$

37 (a) $2t-4$
(b) $f'(1) = -2, f'(2) = 0$

39 $f'(0) = 3, f'(3) = 9$, and $f'(-2) = -1$

41 $P'(2) = 16$

43 $f'(t) = 4t^3 - 6t + 5, f''(t) = 12t^2 - 6$

45 (a) $y = 12x - 16$
(b) Underestimates

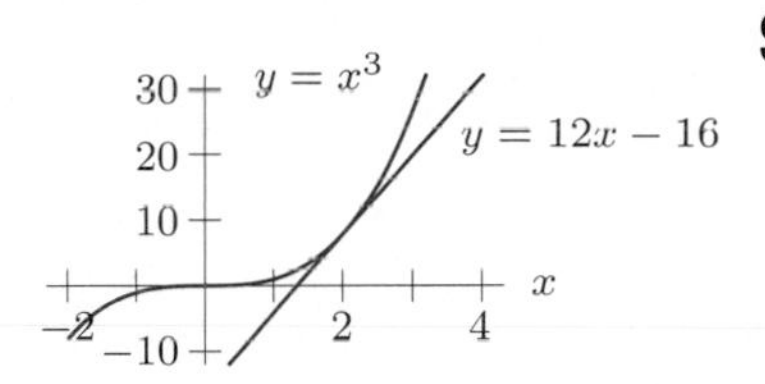

47 $y = -4 - x$

49 \$100

51 (a) 770 bushels per acre
(b) 40 bushels per acre per pound of fertilizer
(c) Use more fertilizer

53 (a) $dC/dq = 0.24q^2 + 75$
(b) $C(50) = \$14{,}750$; $C'(50) = \$675$ per item

55 $f'(x) = 3x^2 - 12x - 15$,
$x = -1$ and $x = 5$

59 (a) $v(t) = -32t$
$v \le 0$ because the height is decreasing
(b) $t \approx 8.84$ seconds
$v = -282.88$ feet/sec $\approx$-192.87 mph

Section 3.2

1 $2e^x + 2x$

3 $10t + 4e^t$

5 $(\ln 2)2^x - 6x^{-4}$

7 $(\ln 2)2^x + 2(\ln 3)3^x$

9 $3 - 2(\ln 4)4^x$

11 $3000(\ln 1.02)(1.02)^t$

13 Ce^t

15 $Ae^x - 2Bx$

17 $3/q$

19 $2t + 5/t$

21 $2x + 4 - 3/x$

23 $f'(-1) \approx -0.736$
$f'(0) = -2$
$f'(1) \approx -5.437$

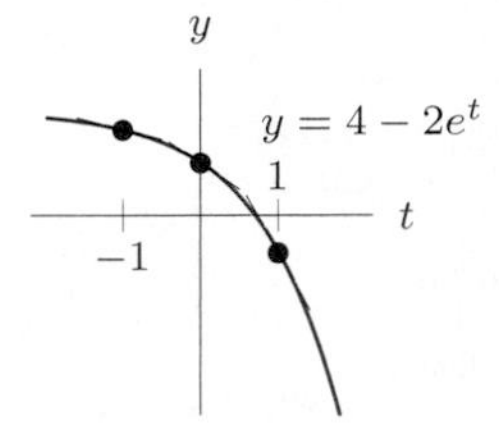

25 (a) $f'(0) = -1$
(b) $y = -x$

27 (a) 9.58 million
(b) 0.058 million/year

29 $\approx$ 7.95 cents/year

31 (a) $P = 4.1(1.02)^t$
(b) $\frac{dP}{dt} = 4.1(1.02)^t(\ln 1.02)$
$\frac{dP}{dt}|_{t=0} = 0.0812$
$\frac{dP}{dt}|_{t=25} = 0.1332$

33 (a) $y = x - 1$
(b) 0.1, 1
(c) Yes

35 $c = -1/\ln 2$

Section 3.3

1 $56x(4x^2+1)^6$

3 $8q(q^2+1)^3$

5 $300t^2(t^3+1)^{99}$

7 $3s^2/(2\sqrt{s^3+1})$

9 $0.7e^{0.7t}$

11 $-0.2e^{-0.2t}$

13 $24e^{0.12t}$

15 $216q(3q^2-5)^2$

17 $25e^{5t+1}$

19 $(e^{\sqrt{s}})/(2\sqrt{s})$

21 $1/(x-1)$

23 $e^{-x}/(1-e^{-x})$

25 $25/(5t+1)$

27 $3/(3t+2)$

29 $5 + 1/(x+2)$

31 $0.5/(x(1+\ln x)^{0.5})$

33 $-x/\sqrt{1-x^2}$

35 $y = 3x - 5$

37 $y = -0.899x + 8.089$

39 69.8 m/yr

41 $f(2) \approx 6065, f'(2) \approx -1516$

43 \$596.73/yr

45 (a) $13,394$ fish
(b) 8037 fish/month

47 (a) $P(1+r/100)^t \ln(1+r/100)$
(b) $Pt(1+r/100)^{t-1}/100$

49 $1/2$

51 -1

53 0.8

55 -0.4

Section 3.4

1 $f'(x) = 12x - 1$

3 $e^x(x+1)$

5 $5e^{x^2} + 10x^2e^{x^2}$

7 $\ln x + 1$

9 $30t + 11$

11 $t + 2t\ln t$

13 $1 - 3/x^2$

15 $e^{-t^2}(1-2t^2)$

17 $2p/(2p+1) + \ln(2p+1)$

19 $2we^{w^2}(5w^2+8)$

21 $(3t^2+5)(t^2-7t+2) + (t^3+5t)(2t-7)$

23 $(1-x)/e^x$

25 $-2/(1+t)^2$

27 $(15+10y+y^2)/(5+y)^2$

29 abe^{bt}

31 $ae^{-bx} - abxe^{-bx}$

33 $(1-2\alpha)e^{-2\alpha}e^{\alpha e^{-2\alpha}}$

35 $y = 0$

37 (a) About 2247 units
(b) $q'(10) \approx -180$

39 60.65 mg, 30.33 mg/hr,
41.04 mg, -12.31 mg/hr

41 (a) $f'(15) > 0, f'(45) < 0$

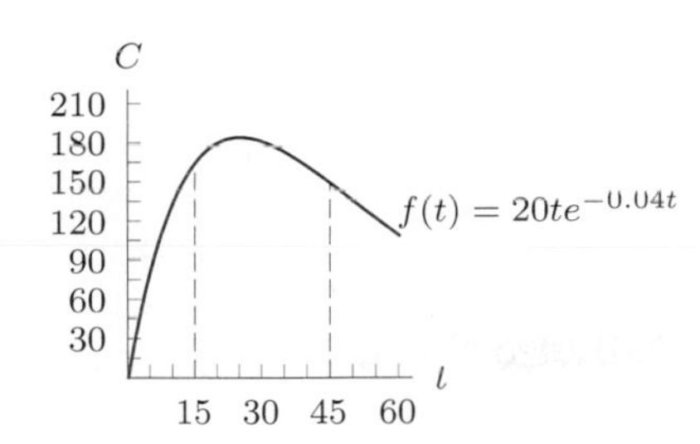

(b) $f(30) \approx 181$ mg/ml, $f'(30) \approx -1.2$ mg/ml/min

43 (a) -6
(b) 0
(c) -2

Section 3.5

1 $5\cos x$

3 $2t - 5\sin t$

5 $2q + 2\sin q$

7 $3\cos(3x)$

9 $-8t\sin(t^2)$

11 $2x\cos(x^2)$

13 $-4\sin(4\theta)$

15 $2x\cos x - x^2\sin x$

17 $3\theta^2\cos\theta - \theta^3\sin\theta$

19 $(2t\cos t + t^2\sin t)/(\cos t)^2$

21 $y = -x + \pi$

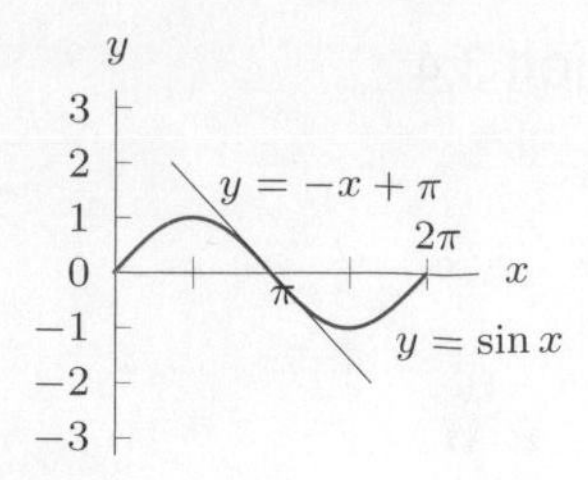

23 (a)

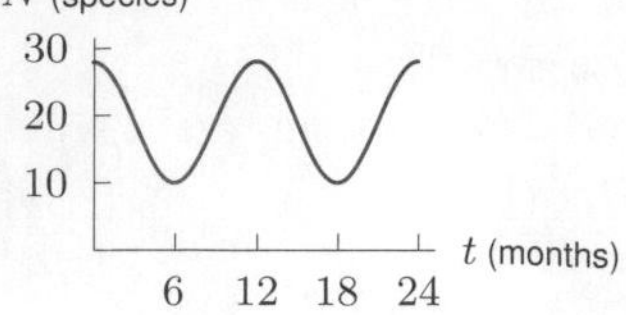

(b) $-4.712\sin(\pi t/6)$
(c) 26.8 species, -2.36 species/month, 23.5 species, 4.08 species/month

25 (a) max \$2600; min \$1400; April 1

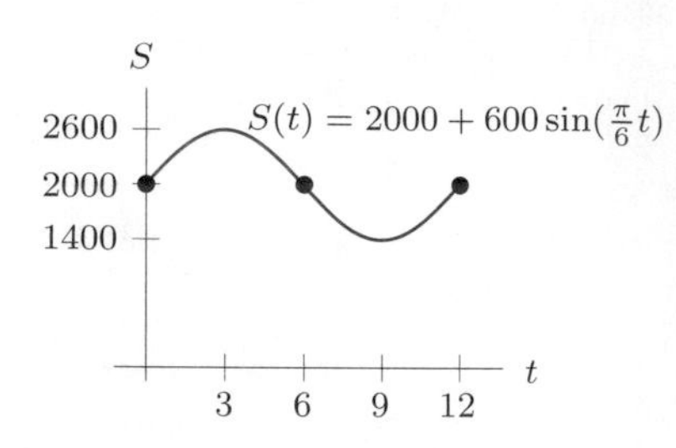

(b) $S(2) \approx 2519.62$; $S'(2) \approx 157.08$

Chapter 3 Review

1 $24t^3$
3 $2e^{2t}$
5 $0.08e^{0.08q}$
7 $e^{3x}(1+3x)$
9 $6(1+3t)e^{(1+3t)^2}$
11 $5+3.6e^{1.2t}$
13 $90(5x-1)^2$
15 $10e^x(1+e^x)^9$
17 $x(2\ln x+1)$
19 $8t+7\cos t$
21 $(-e^{-t}-1)/(e^{-t}-t)$
23 $(50x-25x^2)/(e^x)$
25 $2x\cos x - x^2\sin x$
27 e
29 $(e^x-2-e^{-x})/(1-e^{-x})^2$
31 xe^x+e^x
33 $-3x^2\sin(x^3)$
35 $(t^2+6t+13)/(t+3)^2$
37 $((\ln 3)3^x)/3-(33x^{-3/2})/2$
39 $6y^2e^{(y^3)}e^{2e^{(y^3)}}$
41 $f'(0)=7$, $f'(2)=3$, and $f'(-1)=18$
43 $y=2x-1$
45 About -285 people/year
47 6 billion people, 0.078 billion people per year, 6.833 billion people, 0.089 billion people per year
49 (a) $2, 1, -3$
(b) $f(2.1)\approx 0.7$, overestimate
$f(1.98)\approx 1.06$, underestimate
Second better
51 (a) $dH/dt = -60e^{-2t}$
(b) $dH/dt < 0$
(c) At $t=0$
53 (a) \$13,050
(b) $V'(t) = -4.063(0.85)^t$; thousands of dollars/year
(c) $V'(4) = -\$2121$ per year
55 (a) 2
(b) 15
(c) 11
(d) $-1/4$
57 At $x=0$:
$y=x$, $\sin(\pi/6)\approx 0.524$
At $x=\pi/3$:
$y = x/2+(3\sqrt{3}-\pi)/6$,
$\sin(\pi/6)\approx 0.604$
59 0.35
61 0.75
63 -21.22
65 (a) $x<1/2$
(b) $x>1/4$
(c) $x<0$ or $x>0$
67 (c) $B'(10)\approx -3776.63$
69 $y=x$ and $y=\ln(x+1)$ look the same (like the line $y=x$);
$y=\sqrt{x}$ and $y=\sqrt{2x-x^2}$ look like the line $x=0$;
$y=x^2$, $y=x^3+\frac{1}{2}x^2$, $y=x^3$, and $y=\frac{1}{2}\ln(x^2+1)$ all look like the line $y=0$
71 $b=1/40$ and $a=169.36$
73 (a) $Y(0)=105°$
(b) $350°$
(c) After about 42 minutes
(d) About 1.67 degrees/minute
75 (a) $(3C/2)D^2-(4/3)D^3$
(b) $D<9C/8$

Practice: Differentiation

1 $2t+4t^3$
3 $15x^2+14x-3$
5 $-2/x^3+5/(2\sqrt{x})$
7 $10e^{2x}-2\cdot 3^x(\ln 3)$
9 $2pe^{p^2}+10p$
11 $2x\sqrt{x^2+1}+x^3/\sqrt{x^2+1}$
13 $16/(2t+1)$
15 $2^x(\ln 2)+2x$
17 $6(2q+1)^2$
19 bke^{kt}
21 $2x\ln(2x+1)+2x^2/(2x+1)$
23 $10\cos(2x)$
25 $15\cos(5t)$
27 $2e^x+3\cos x$
29 $17+12x^{-1/2}$
31 $20x^3-2/x^3$
33 $1, x\neq -1$
35 $-3\cos(2-3x)$
37 $6/(5r+2)^2$
39 $ae^{ax}/(e^{ax}+b)$
41 $5(w^4-2w)^4(4w^3-2)$
43 $(\cos x-\sin x)/(\sin x+\cos x)$
45 $6/w^4+3/(2\sqrt{w})$
47 $(2t-ct^2)e^{-ct}$
49 $(\cos\theta)e^{\sin\theta}$
51 $3x^2/a+2ax/b-c$
53 $2r(r+1)/(2r+1)^2$
55 $2e^t+2te^t+1/(2t^{3/2})$
57 $x^2\ln x$
59 $6x\left(x^2+5\right)^2\left(3x^3-2\right)\left(6x^3+15x-2\right)$
61 $(2abr-ar^4)/(b+r^3)^2$
63 $20w/(a^2-w^2)^3$

Section 4.1

1 Four

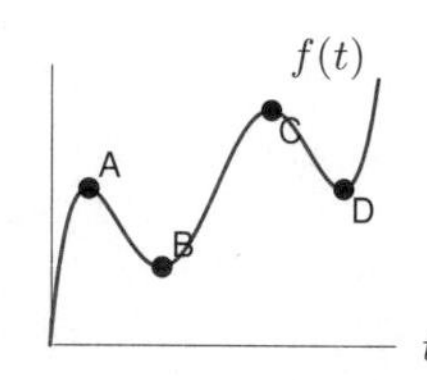

3 One

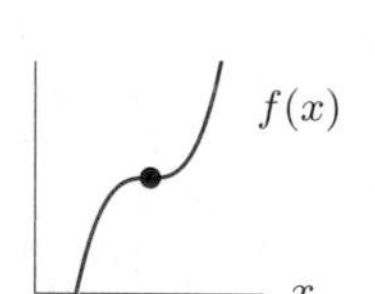

5 After 18 hours

7 (a)

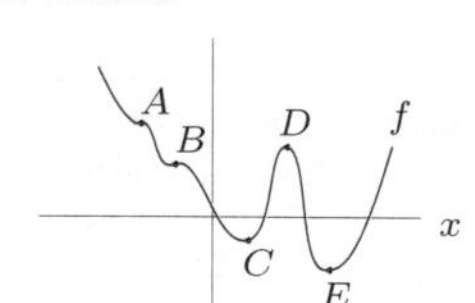

(b)

9 Increasing for all x, no critical point
11 Local min: $(2.3, -13.0)$
13 Alternately incr/decr
15 A: local max
B: local min
C: neither

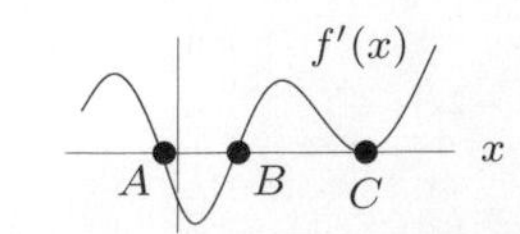

17

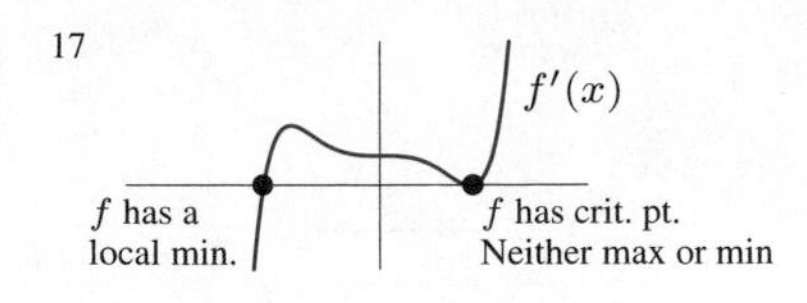

21 Local max

23 $a = -6$; $b = 14$

25

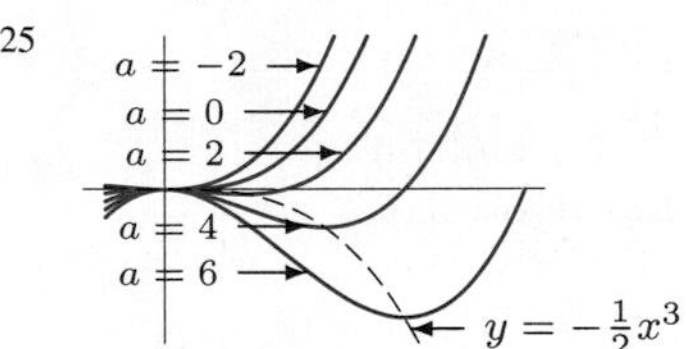

27 $a = -1/3$

29 (a) $f(\theta) = 0$ at $\theta = 0$
(b) $f'(\theta) = 1 - \cos\theta > 0$ for $0 < \theta \leq 1$. Only zero is at origin

Section 4.2

1 Three

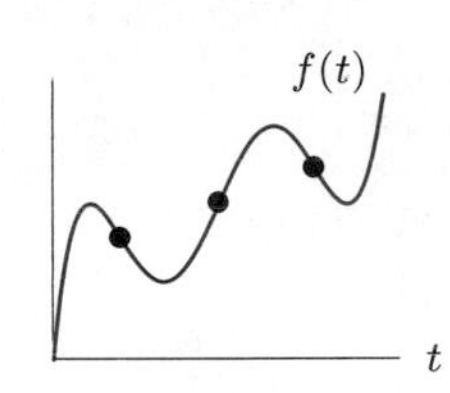

3 One

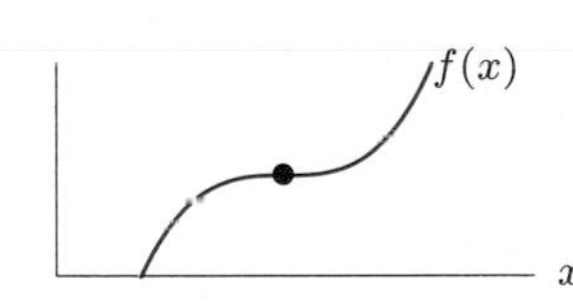

5

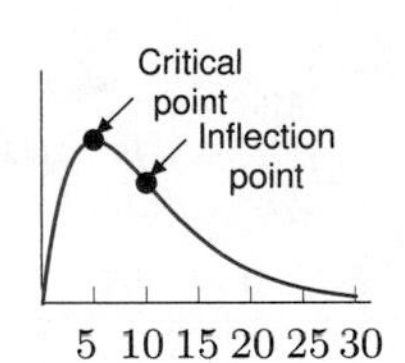

7

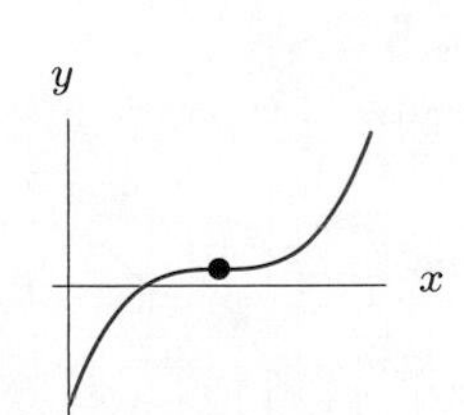

9 (a) Critical
(b) Inflection

11 Critical points: $x = \frac{5}{2}$, local minimum
Inflection points: None

13 Critical points:
$x = -3$ (local max) and $x = 2$ (local min)
Inflection point: $x = -1/2$

15 Critical points: $x = -1$, local min; $x = 0$ local max; $x = 1$, local min
Inflection points: $x = -1/\sqrt{3}$, $x = 1/\sqrt{3}$

17 Critical points:
$x = 0$ (local max) and $x = \pm 2$ (local minima)
Inflection points: $x = \pm 2/\sqrt{3}$

19 Critical points: $x = 0$, local max; $x = 4$, local min
Inflection point: $x = 3$

21 $x = -1, 1/2$

23

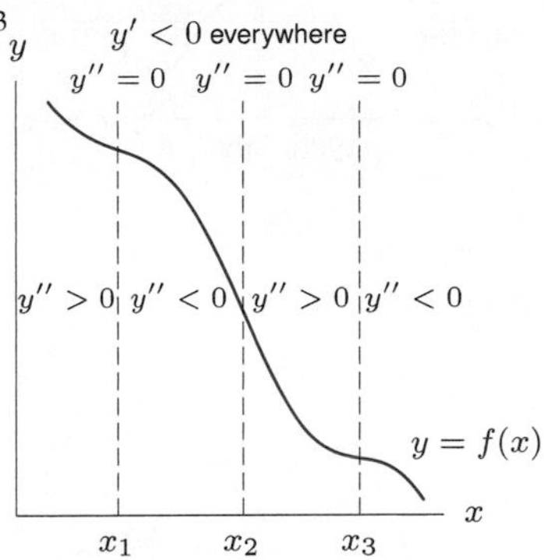

25

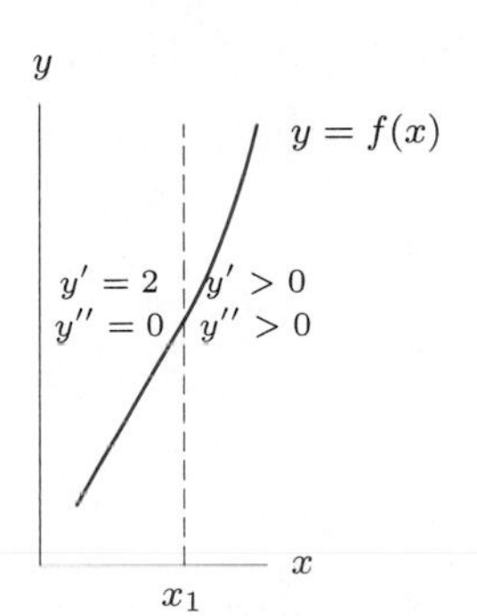

27 (a)

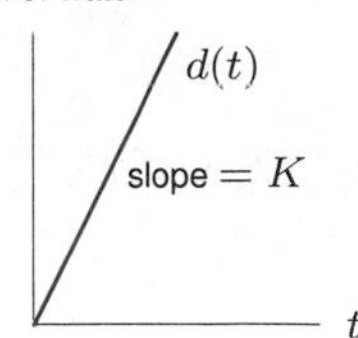

(b)

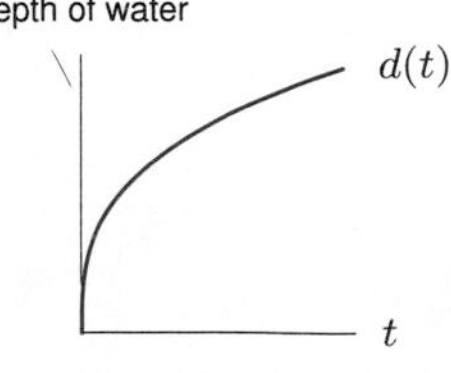

29

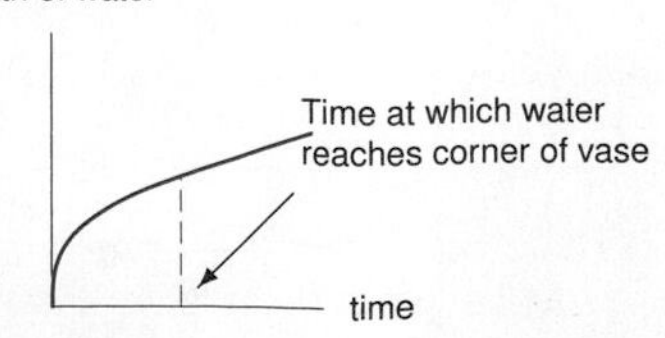

31 (a) Concavity changes at y_1 and y_3

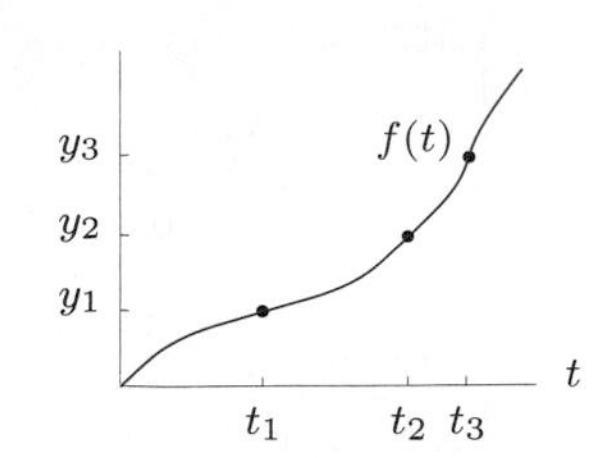

(b) $f(t)$ grows most quickly where vase is skinniest and most slowly where vase is widest. Ratio is about $16 : 1$

33 (a)

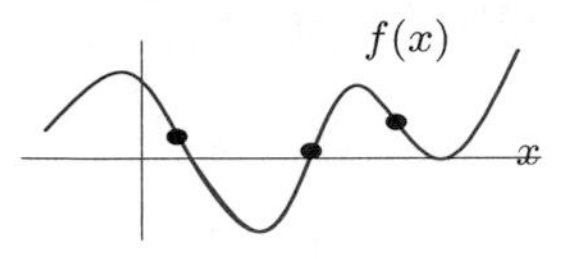

(b)

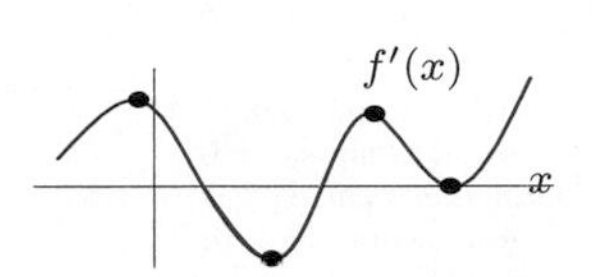

(c)

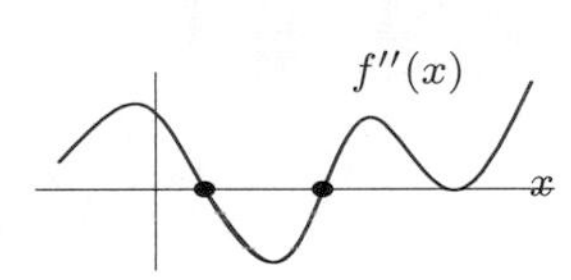

Section 4.3

1

Local and global max
Local max
$f(x)$
Local max
Local min
Local and global min
50 40 30 20 10
1 2 3 4 5 6 x

3 44.1 feet

5

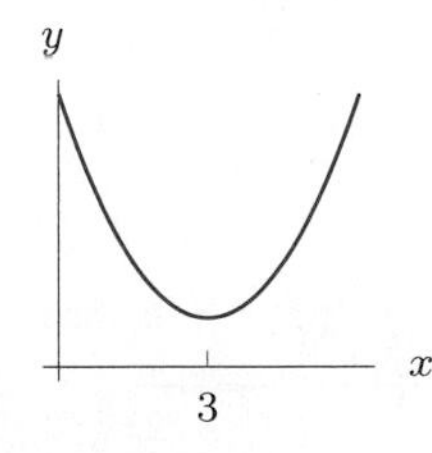

7

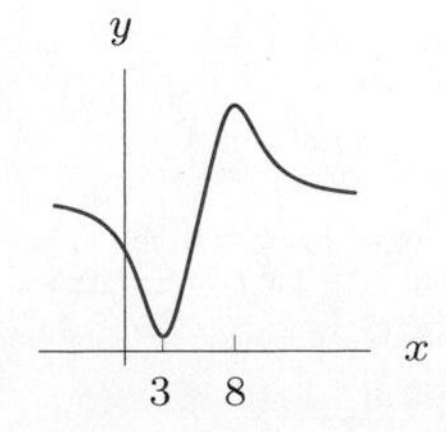

9 True

11

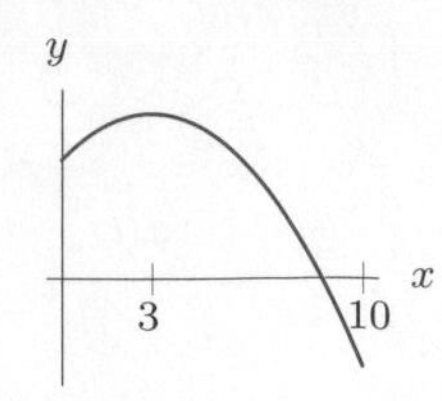

13

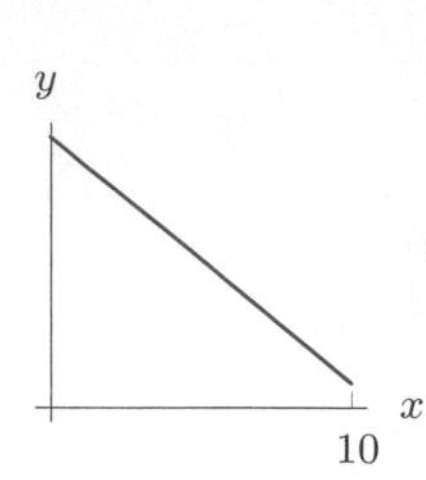

15 (a) $D = C$
(b) $D = C/2$

17 (a) $f'(x) = 3x^2 - 6x$, $f''(x) = 6x - 6$
(b) $x = 0, 2$
(c) $x = 1$
(d) Local maximum: $x = 0$
Local minimum: $x = 2$
Global maxima: $x = 0, 3$
Global minima: $x = -1, 2$
(e)

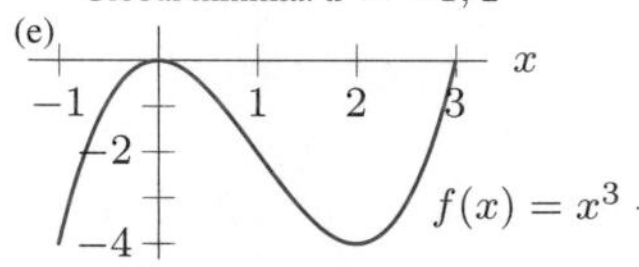

19 (a) $f'(x) = 3x^2 - 6x - 9$, $f''(x) = 6x - 6$
(b) $x = -1, 3$
(c) $x = 1$
(d) Local and global maximum : $x = -1$
Local minimum: $x = 3$
Global maximum: $x = -5$
(e)

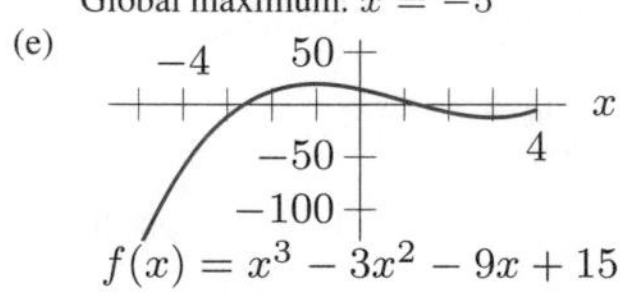

21 (a) $f'(x) = e^{-x}(\cos x - \sin x)$
$f''(x) = -2e^{-x} \cos x$
(b) $x = \pi/4, 5\pi/4$
(c) $x = \pi/2, 3\pi/2$
(d) Global maximum: $x = \pi/4$;
Global minimum: $x = 5\pi/4$
(e)

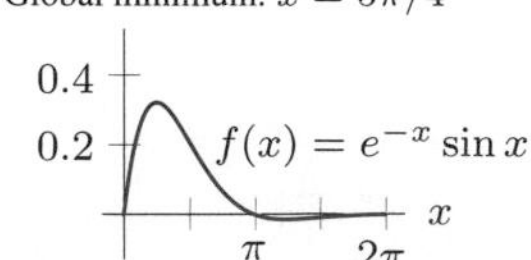

23 Global min = 2 at $x = 1$
No global max

25 Global min = 1 at $x = 1$
No global max

27 Global max = 2 at $t = 0, \pm 2\pi, \ldots$
Global min = −2 at $t = \pm\pi, \pm 3\pi, \ldots$

29 $r = (2/3)R$

31 50 m by 50 m

33 about $4583

35 $x = y = \sqrt[3]{V}$

37 306 children

39 (a) $y = \begin{cases} 200x & x \leq 200 \\ 600x - x^2 & x > 200 \end{cases}$

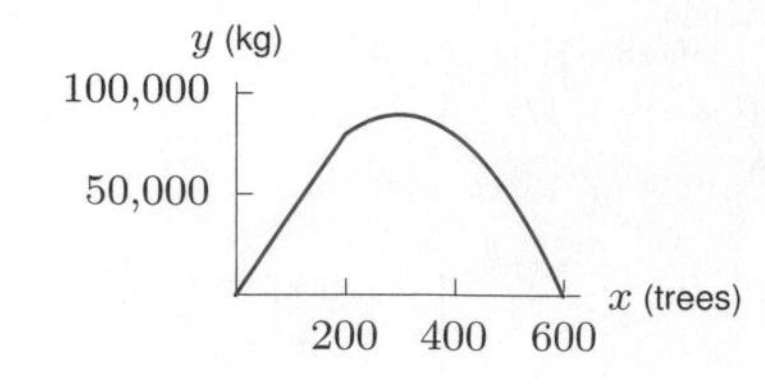

(b) 300 trees/km^2

41 (a) $0 \leq y \leq a$

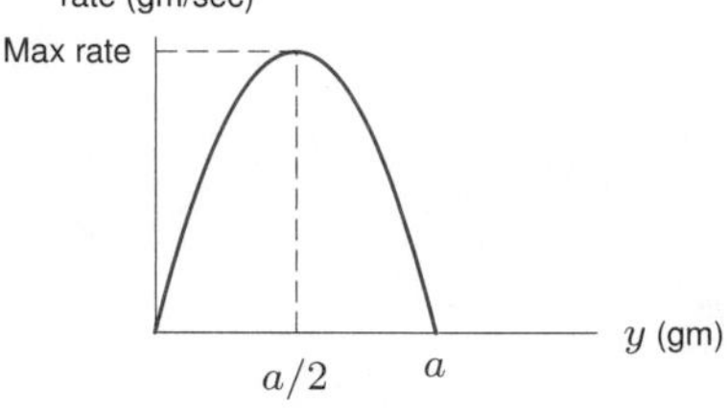

(b) $y = a/2$

43 (a) $H = 1/b$, $S = ae^{-1}/b$
(b) a: Increases
b: Decreases

45 (a) 10
(b) 9

47 (a) $k(\ln k - \ln S_0) - k + S_0 + I_0$
(b) Both

49 Max: 0.552; Min: 0.358

51 (a) $E = 500e \left(\dfrac{200 - \cos\theta}{\sin\theta} \right) + 2000e$
$\left(\arctan\left(\dfrac{500}{2000}\right) \leq \theta \leq \pi/2 \right)$
(b) $\theta = \pi/3$
(c) Independent of e, but dependent on $\overline{AB}/\overline{AL}$

Section 4.4

1 (a) $q = 2500$
(b) $3 per unit
(c) $3000

3

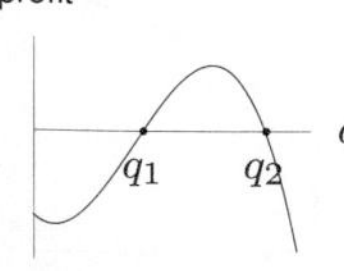

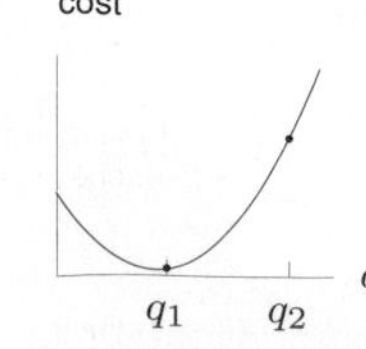

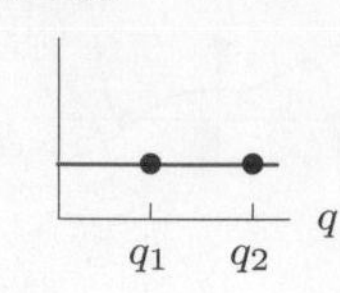

5 (a) $9
(b) −$3
(c) $C'(78) = R'(78)$

7 (a) Increase production
(b) $q = 8000$

9 (a) $q = 350$
(b)

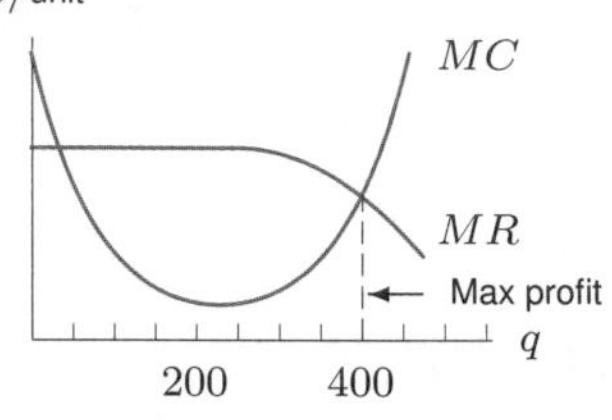

11 (a) Increase
(b) Decrease
(c) Decrease

13 $q = 4000$

15 Above 2000

17 $R(q) = 45q - 0.01q^2$
Max revenue at $q = 2250$,
$p = \$22.50$; Revenue = $50,625

19 (a) $q = 400$
(b) $5 per unit
(c) $700.

21 $14.

23 (a) $-5q^2 + 3994q - 5$
(b) 399.4
(c) $797,596.80

25 Maximum revenue = $27,225
Minimum = $0

27 (a) q/r months
(b) $(ra/q) + rb$ dollars
(c) $C = (ra/q) + rb + kq/2$ dollars
(d) $q = \sqrt{2ra/k}$

29 $L = [\beta pcK^{\alpha}/w]^{1/(1-\beta)}$

Section 4.5

1 (a) $a(1000) \approx \$1.60$ per unit
(b)

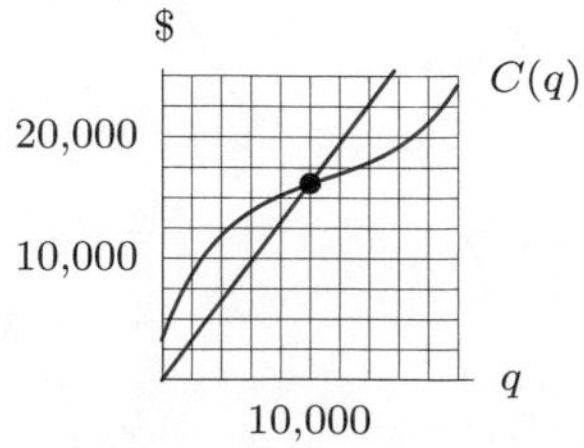

(c) 18,000 units

3 $MC = \$20$; $a(q) = \$25$

5

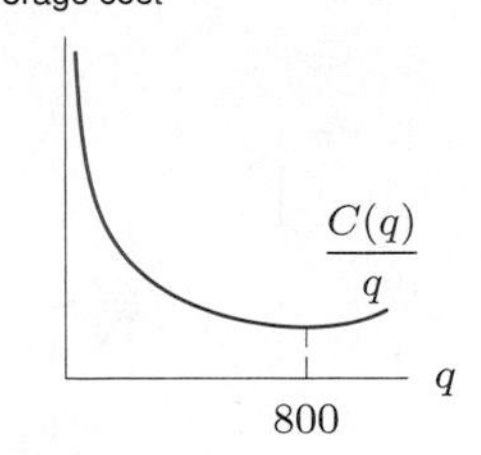

7 (a) Making money
(b) Increase; increase
(c) Increase

9 (a) $C(q) = 0.01q^3 - 0.6q^2 + 13q$
(b) $1
(c) $q = 30$, $a(30) = 4$
(d) Marginal cost is 4

11 (a)

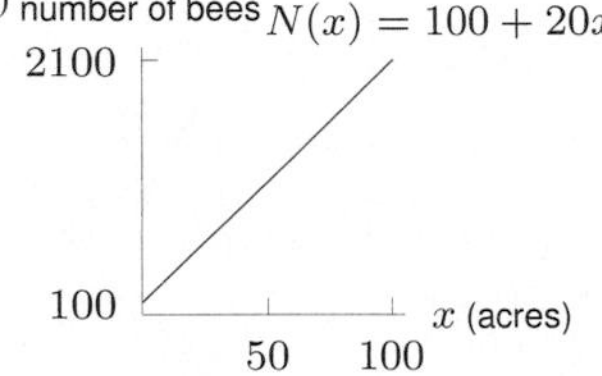

(b) (i) $N'(x) = 20$
(ii) $N(x)/x = (100/x) + 20$

13

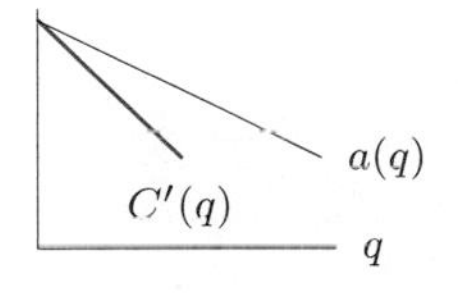

Section 4.6

1 (a) 1.5% decrease
(b) 1.5% increase

3 Elastic

5 Elastic

7 Low

9 $E = 1.25$, 2.5% decrease

15 $12.91

Section 4.7

1 (a) 40 billion
(b)

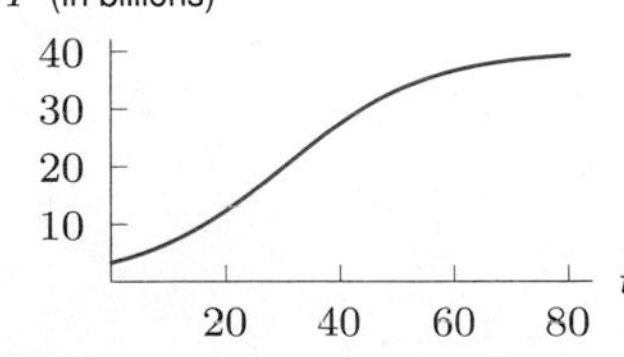

(c) 2020; 2095

3 (a) 4th month
(b) 2090 sales
(c) 4180 sales

7 (a) 36 thousand; total number infected
(b) $t \approx 16$, $n \approx 18$ thousand
(c) Virus spreading fastest
(d) Number infected half total

9 (a) About 0.252
(c) 1975
(d) 1975

13 (a) 5000
(b) 499
(c) $P(t) = 5000/(1 + 499e^{-1.78t})$

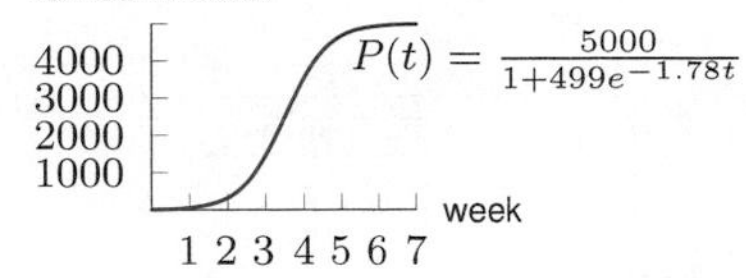

(d) About 3.5 weeks; 2500

15 (a) C: largest
B: smallest
(b) A: largest
B: smallest
(c) C: safest

19 Effective for 85%,
lethal for 6%.

Section 4.8

1 (a)

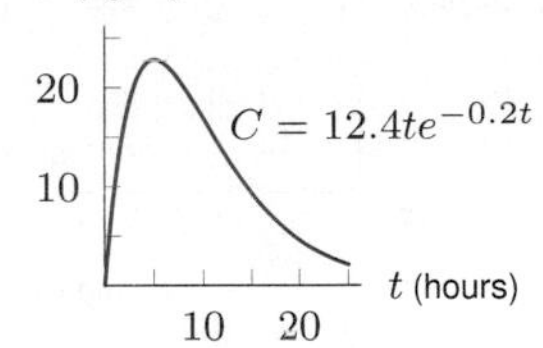

(b) 5 hrs, 22.8 ng/ml
(c) 1 to 14.4 hrs
(d) At least 20.8 hrs

3 (a) After about 6 hours
(b) After about 5 hours

5

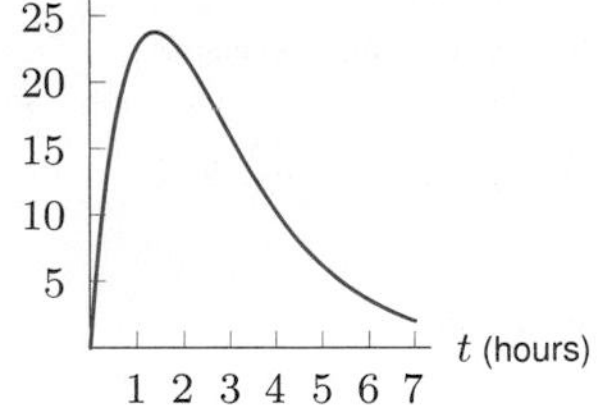

Chapter 4 Review

1

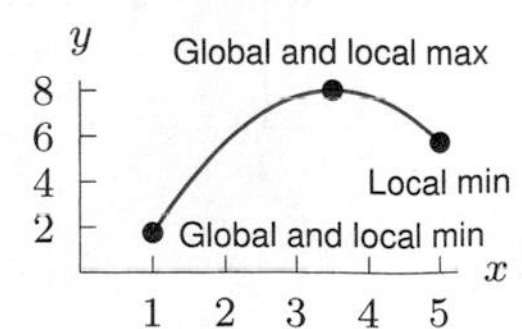

3 (a) $f(1)$ local minimum;
$f(0)$, $f(2)$ local maxima
(b) $f(1)$ global minimum
$f(2)$ global maximum

5 Local maximum, minimum, or neither
Price relatively constant around July 1

7 (a) Increasing for $x > 0$
Decreasing for $x < 0$
(b) Local and global min: $f(0)$

9 (a) Increasing for $0 < x < 4$
Decreasing for $x < 0$ and $x > 4$
(b) Local max: $f(4)$
Local min: $f(0)$

11 (a) cm/week
(b) Growing at 1.6 cm/wk in week 24

13 (a) Week 14
(b) Point of fastest growth

15 $a = 3e$, $b = -3$

17 One

21 (a) Polynomial; negative leading coefficient; degree 2
(b) Exponential.
(c) Logistic
(d) Logarithmic.
(e) Polynomial; positive leading coefficient; degree 2
(f) Exponential.
(g) Surge
(h) Periodic
(i) Polynomial; negative leading coefficient; degree 3

23 (a) $0
(b) $96.56
(c) Raise the price by $5

25 (a) $\pi(q)$ max when
$R(q) > C(q)$ and R and Q are farthest apart

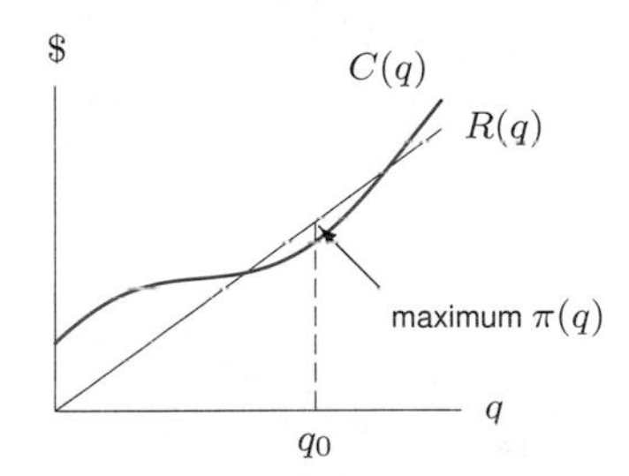

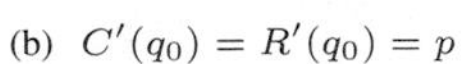

(b) $C'(q_0) = R'(q_0) = p$
(c)

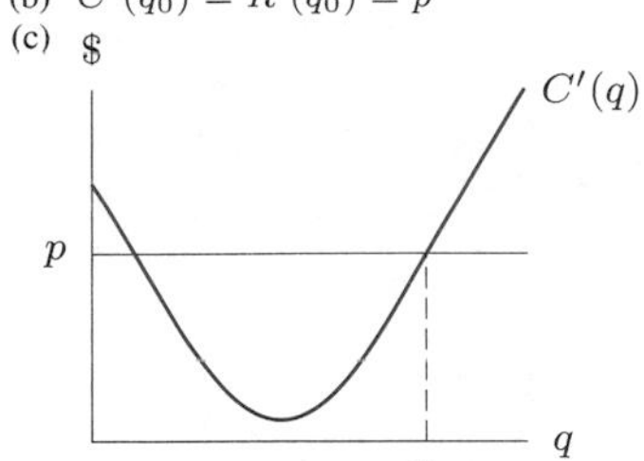

27

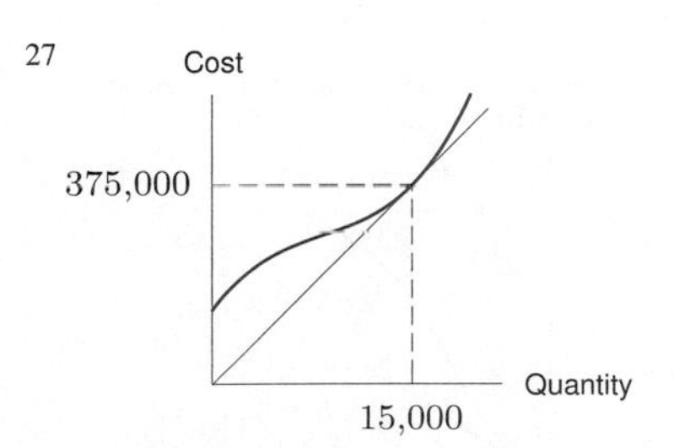

29 $C'(2)$

31 $(C(75) - C(50))/25$

33 $C(3)/3$

35 $E = 0.05$, demand is inelastic.

39 (a) $k(1 - 2P/L) \cdot dP/dt$

41 Global min $= 3$ at $z = 1/2$
No global max

43 Global max $= 1/2$ at $x = 1$
No global min

45 (a)

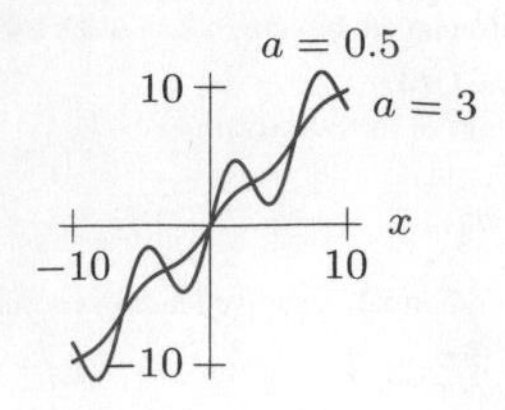

(b) $-1 \leq a \leq 1$

47 (b) 2 hours
(c) Equal

49 (a) 200; Just before 8:30

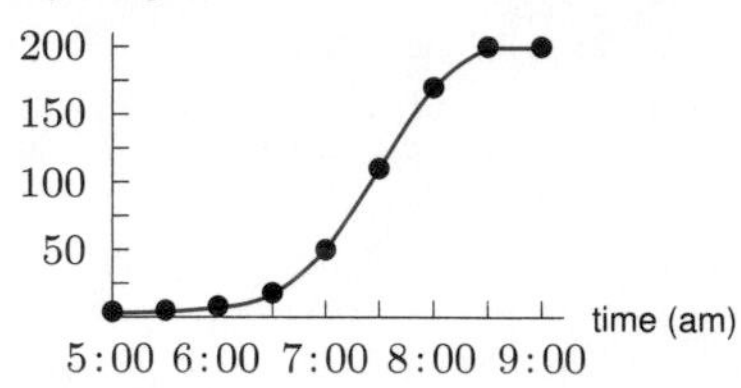

(b)

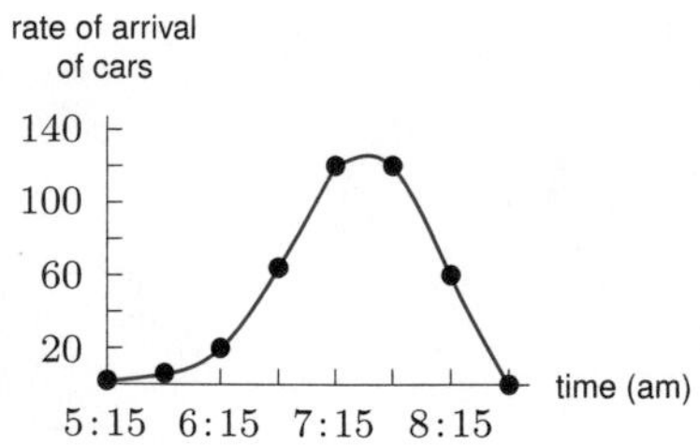

(c) About 7:30

51 (a) $1/(2e)$
(b) $(\ln 2) + 1$

53 (a) Vertical intercept: $W = Ae^{-e^b}$, Horizontal asymptote $W = A$
(b) No critical points, inflection point at $t = b/c$, $W = Ae^{-1}$
(c)

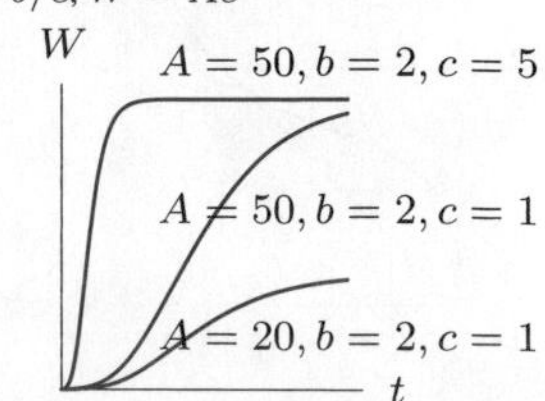

(d) Yes

Section 5.1

1 (a) Lower estimate $= 122$ ft
Upper estimate $= 298$ ft
(b)

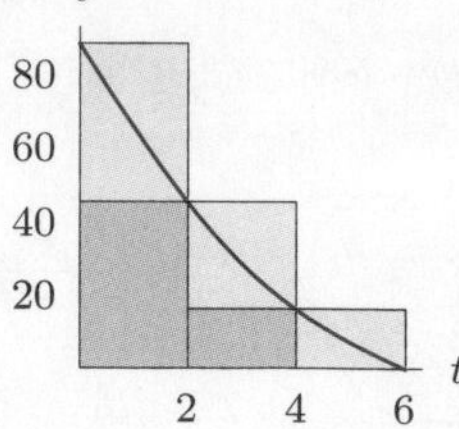

3 250 meters

5 (a) A: 8 hrs
B: 4 hrs
(b) 100 km/hr
(c) A: 400 km
B: 100 km

7 ≈ 455 feet or 0.086 miles

9 (a) About 420 kg
(b) 336 and 504 kg

11 (a) 570 m^3
(b) Every 2 minutes

13

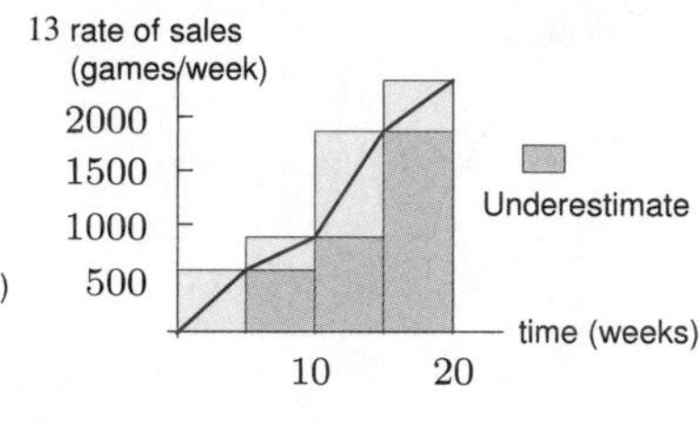

15 (a)

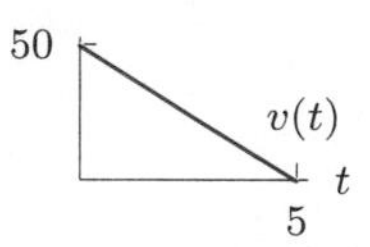

(b) 125 feet
(c) 4 times as far

17 60 m (Other answers possible.)

19 (b) \$6151

Section 5.2

1 1692.5

3 About 543

5 (a) 224

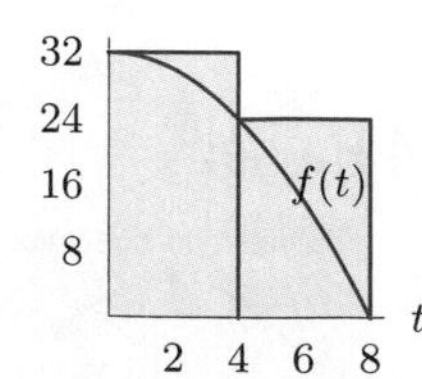

(b) 96

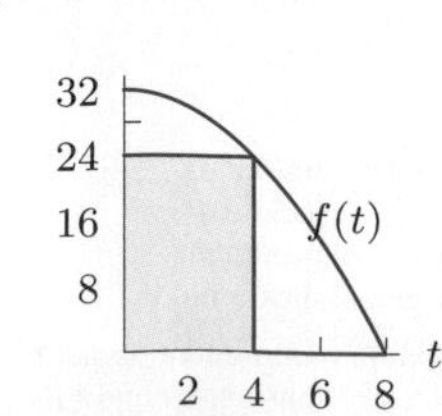

(c) 200

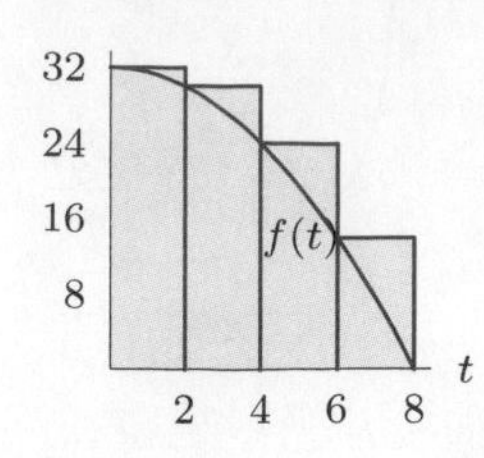

(d) 136

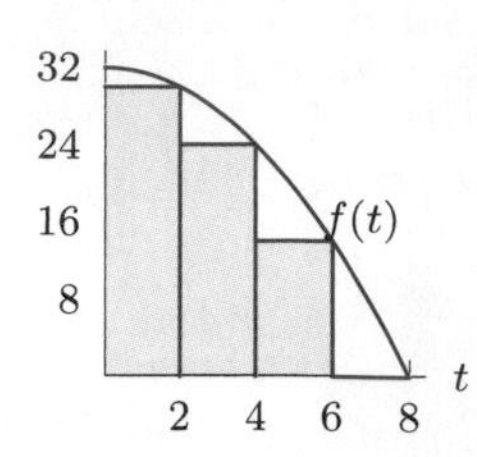

7 350

9 About 20

11 (a) 3.6

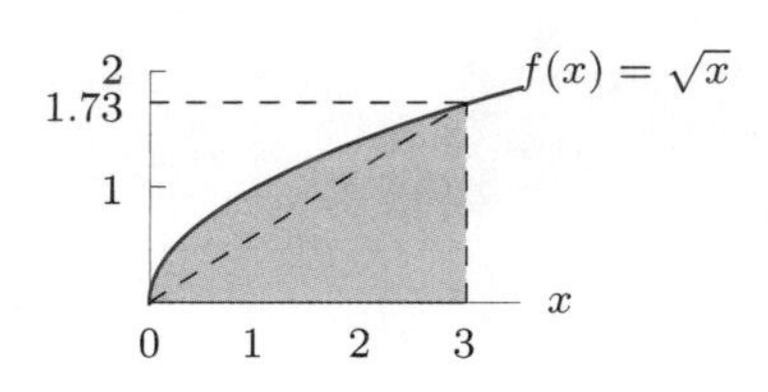

(b) 3.4641

13 (a) 2

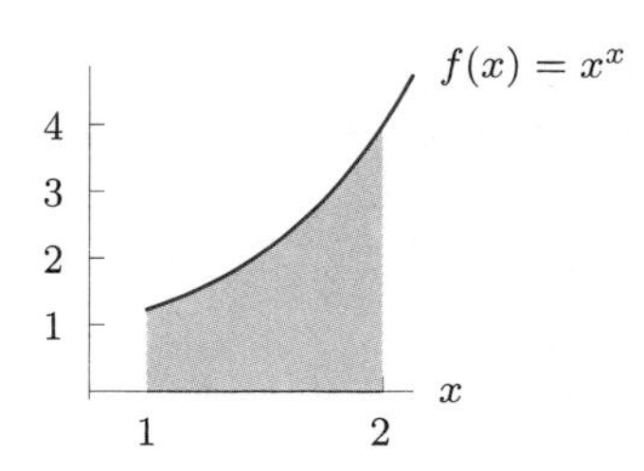

(b) 2.05045

15 (a) 78

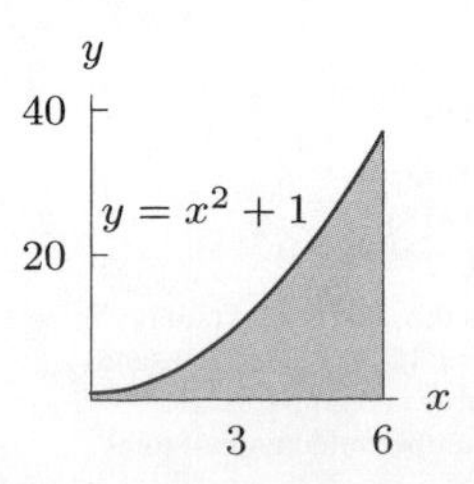

(b) 46; underestimate

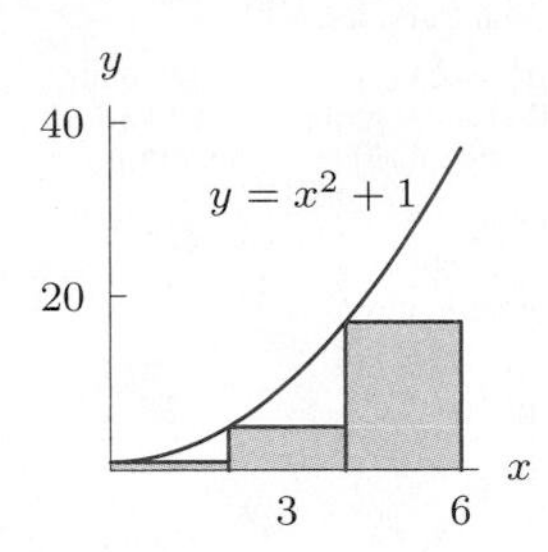

(c) 118; overestimate

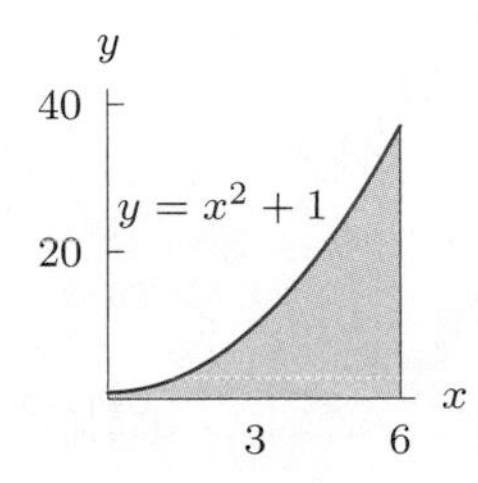

17 93.47

19 448.0

21 2.350

23 2.9

25 1.30

27 1.728, 1.816, 1.772, $n = 140$

29 (a) 4; 0, 4, 8, 12, 16; 25, 23, 22, 20, 17
(b) 360; 328
(c) 8; 0, 8, 16; 25, 22, 17
(d) 376; 312

Section 5.3

1 About 192

3 (a) 13
(b) 1

5 Positive

7 Positive

9 -40

11 (a) -4
(b) 0
(c) 8

13 II

15 III

17 V < IV < II < III < I
I, II, III positive
IV, V negative

19 6.77

21 (a) -2
(b) $-A/2$

23 About 3.34

25 14,052.5

27 14.688

29 13.457

31 2.762

Section 5.4

1 (b) 772.8 million metric tons

3 Change in position; meters

5 Change in world pop; bn people

7 Total amount $= \int_0^{60} f(t)\,dt.$

9 1417 antibodies

11 (a) Concave up
(b) 3.1 kg

13 741.6 liters

15 6 months: A more
First year: B more
Same: roughly 9 months
A roughly 170 sales
B roughly 250 sales

17 (b) $t = 6$ hours
(c) $t = 11$ hrs

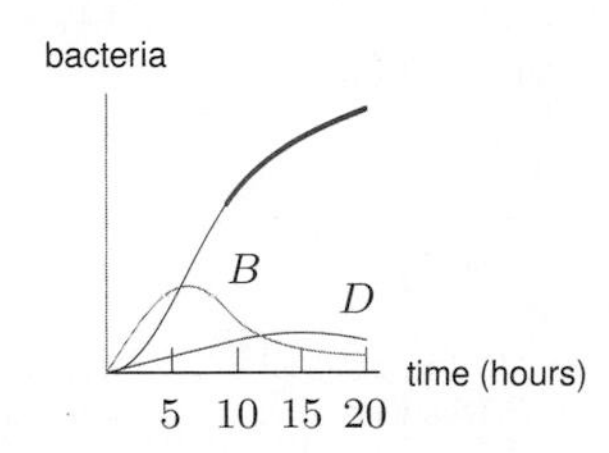

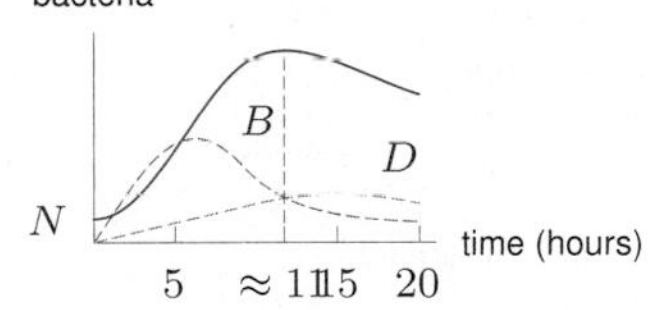

19 (a) About 430 liters
(b) $\int_0^3 60f(t)\,dt$
(c) About 470 liters

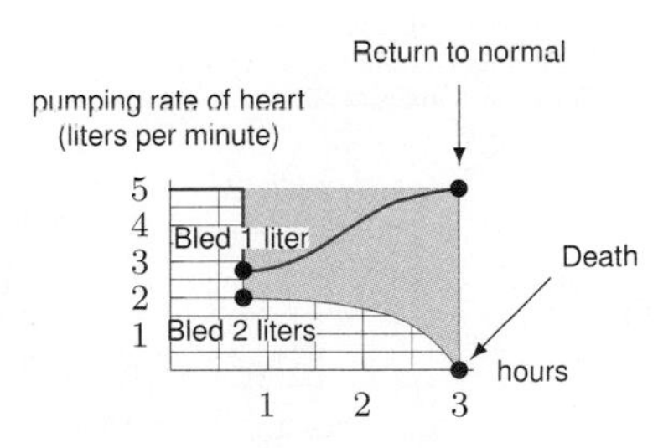

21 \$2392.87

23 (a) $\int_0^3 (40t - 10t^2)dt$
(b)

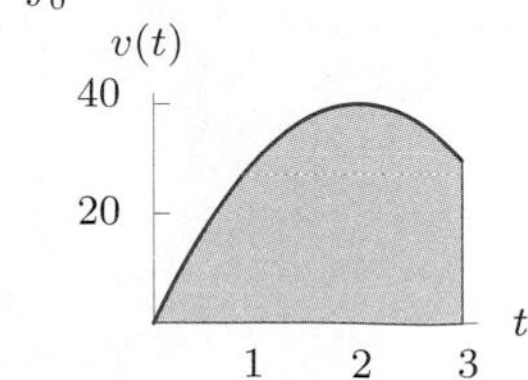

(c) 90 miles

25 15 cm to the left

27 25 cm to the right

29 65 km from home
3 hours
90 km

31 Product B has a greater peak concentration
Product A peaks sooner
Product B has a greater overall bioavailability
Product A should be used

33

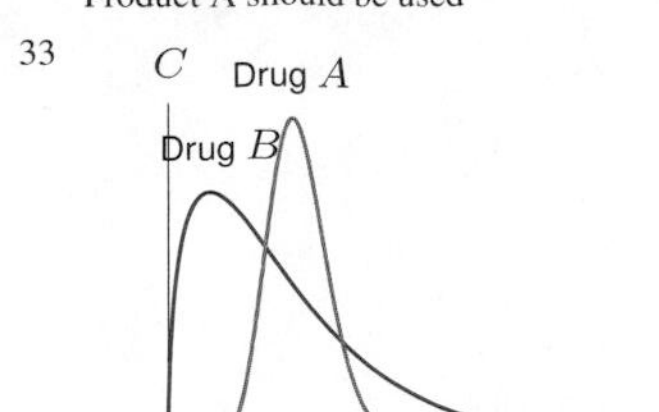

35 About \$13,800

37 (a) $\int_0^T 49(1 - (0.8187)^t)\,dt$
(meters)
(b) $T \approx 107$ seconds

Section 5.5

1 Dollars; cost of increasing production from 800 tons to 900 tons

3 (a)

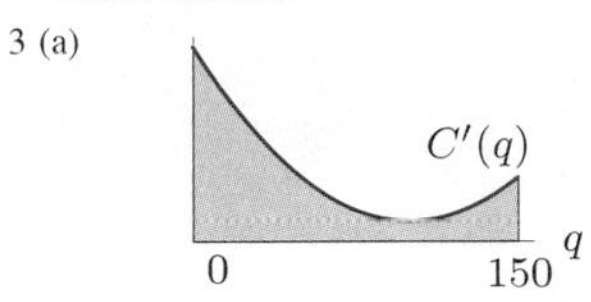

(b) \$22,775
(c) $C'(150) = 18.5$
(d) $C(151) \approx \$22{,}793.50$

5 7.65 million people

7 (a) \$18,650
(b) $C'(400) = 28$

9 (a)

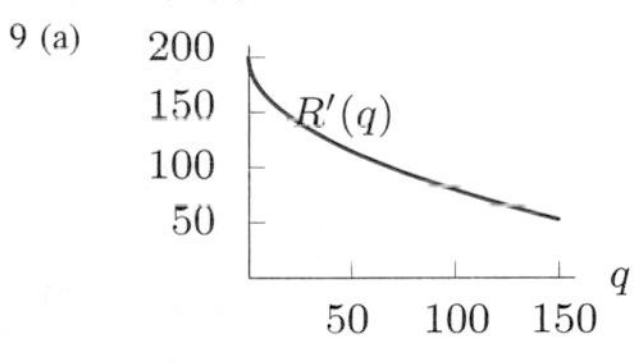

(b) \$12,000
(c) Marginal revenue is \$80/unit
Total revenue is \$12,080

11 (a) 7.54 inches after 8 hours
(b) 1.41 inches/hour

Chapter 5 Review

1 (a) 408
(b) 390

3 1096

5 Meters per second

7 Foot-pounds

9 1.44

11 1.15

13 -0.083

15 84

17 0.0833

19 About 0.1667

21 (a) 430 ft
(b) (ii)

23 (a)

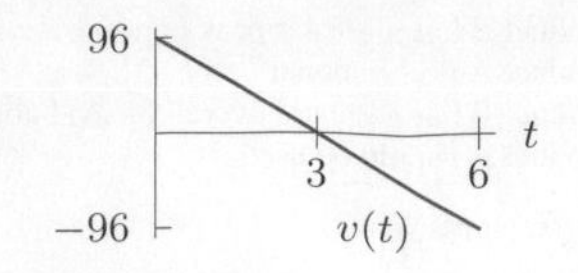

(b) 3 sec, 144 feet
(c) 80 feet

25 (a) Species B for both
(b) Species A

27 (a) Inflection point
(b) About 50 cm

29 (a) Negative
(b) Positive
(c) Negative
(d) Positive

33 (a) $\int_0^5 (10 + 8t - t^2)\, dt$
(b) 100
(c) 108.33

35 (a) Forward for $t < 3$, backward for $t > 3$
(b) Farthest forward: $t = 3$
Farthest backward:
no upper bound.

37 $300,000

39 (a)

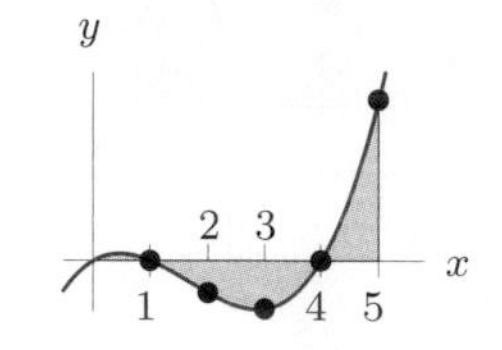

(b) I_1 is largest.
I_4 is smallest.
I_1 is the only positive value.

41 9 years

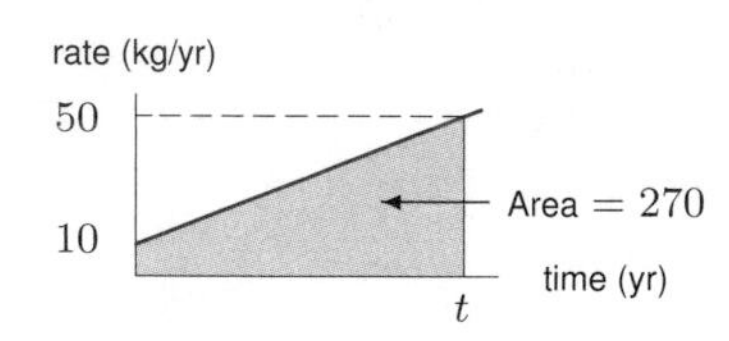

43 $F(x)$ is decreasing for
$x < -2$ and $x > 2$;
$F(x)$ is increasing for
$-2 < x < 2$

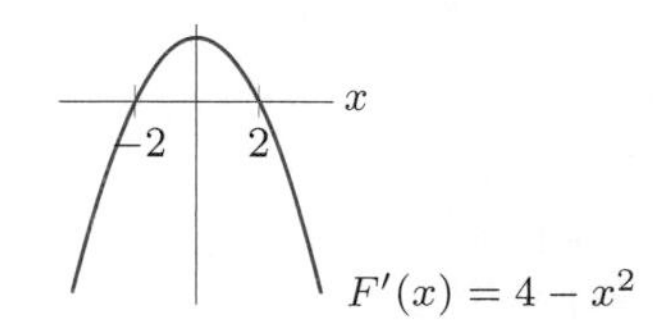

Theory: Second Fund. Thm.

1 x^3

3 xe^x

5 (a) 0
(b) F increases
(c) $F(1) \approx 1.4$, $F(2) \approx 4.3$, $F(3) \approx 10.1$

9 9

11 $8c$

Section 6.1

1 (a) $\int_0^5 f(x)\, dx \approx 25$
(b) About 5
(c) $(\int_0^5 f(x)\, dx)/5 \approx 5$

3 (a) 20
(b) $10/3$

5 (a) (i) 1/4
(ii) 1/4
(iii) 0
(b) Not true

7 -3

9 About 2.5

11 (a) 100 cases; 5 cases
(b) 32 cases

13 (a) 120 mm Hg
(b) 80 mm Hg
(c) 100 mm Hg
(d) Less

15 (a) 0.375 thousand/hour
(b) 1.75 thousands

17 (a) 9.9 hours
(b) 14.4 hours
(c) 12.0 hours

19 (a) 687,000
(b) 694,857

21 (a) $E(t) = 1.4e^{0.07t}$
(b) $0.2(e^7 - 1) \approx 219$
million megawatt-hours
(c) 1972
(d) Graph $E(t)$ and estimate t such that
$E(t) = 219$

23 $(a) < (c) < (b) < (d)$

Section 6.2

1 Equilibrium price $\approx$ $8 per unit.
Equilibrium quantity ≈ 345 units.
Consumer surplus $\approx$ $2000

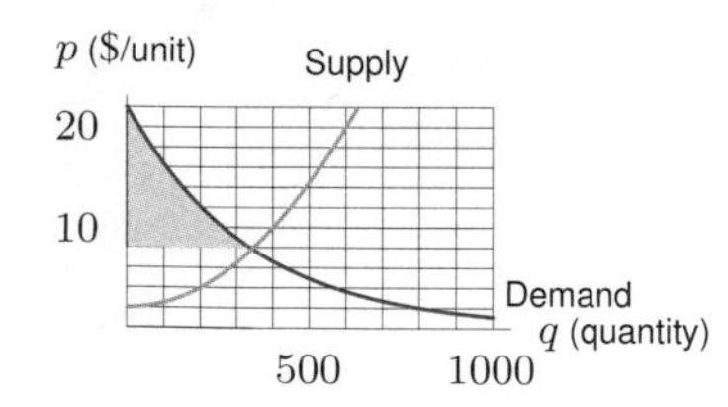

Producer surplus $\approx$ $1400

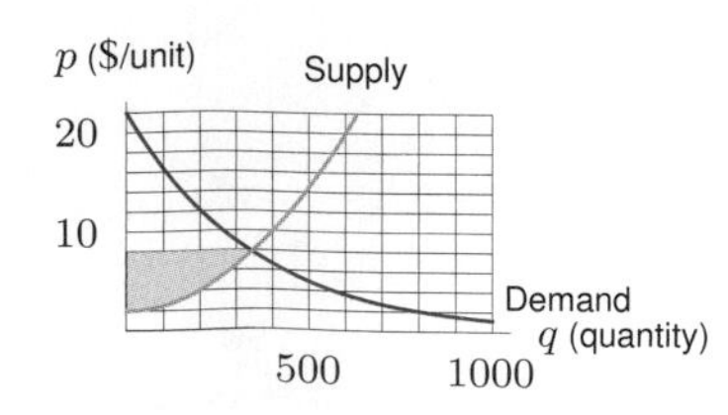

3 42.667

5 200

7 (a) $2250, $2625, $4875
(b) Consumer surplus: less
Producer surplus: greater
Total gains: less

9 (a) $p^* = \$11.43$, $q^* = 322$ units
(b) Consumer surplus = $2513.52;
Producer surplus = $1033.62

11 (a) Less
(b) Can't tell
(c) Less

15 (a)

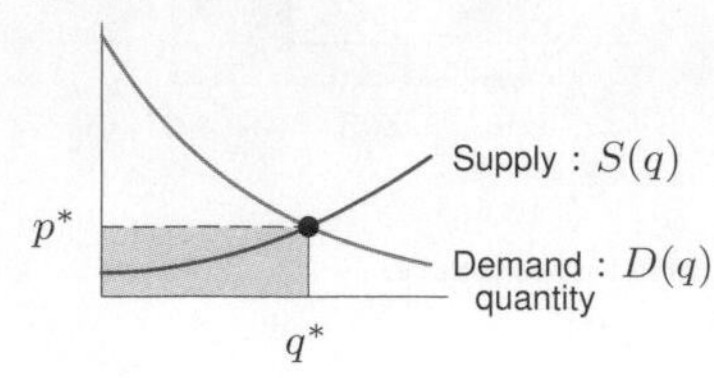

(b)

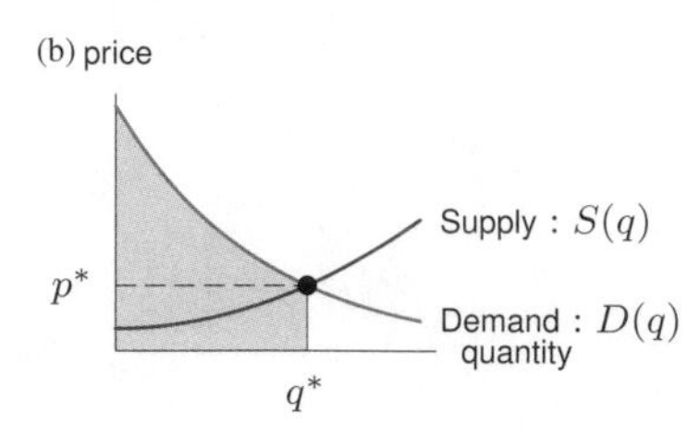

(c)

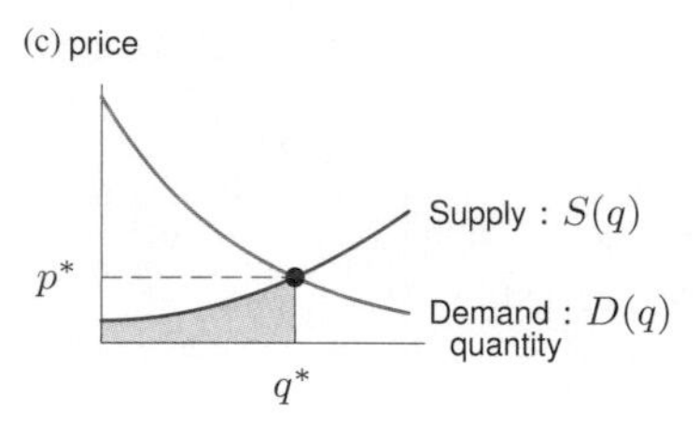

(d)

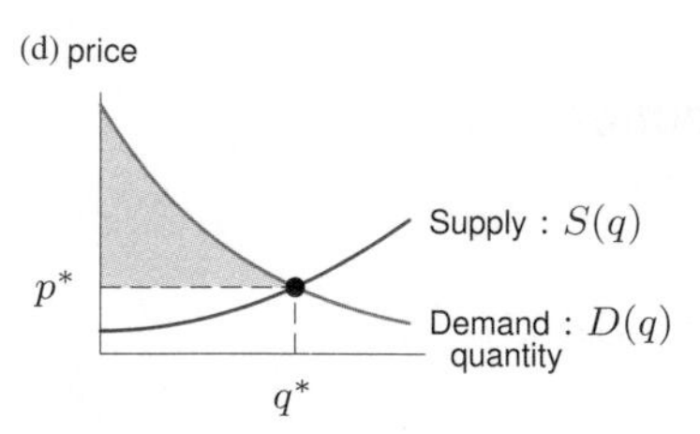

(e)

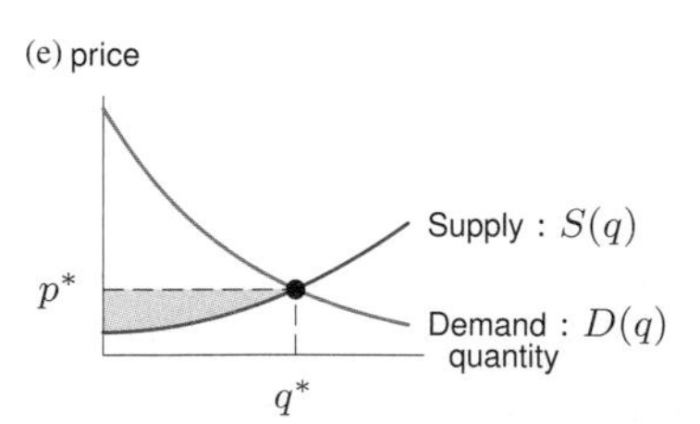

(f)

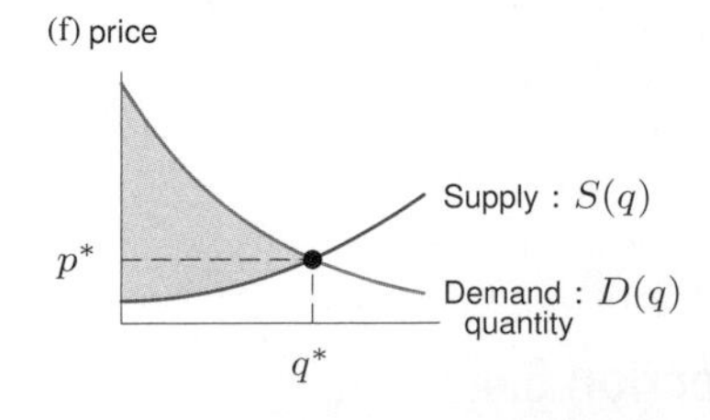

Section 6.3

1 $\$4026.35$

3 $P = \$139{,}761.16$
$F = \$464{,}023.39$

5 (a) (i) $\$18{,}846.59$
(ii) $\$16{,}484.00$
(b) (i) $\$21{,}249.47$
(ii) $\$24{,}591.24$

7 $\$635.37$ per year

9 No; present value $= \$306{,}279$

11 (a) Lump sum better at 6%
Continuous payments better at 5%
(b) Interest rates above about 5.5%

13 $\$41{,}508$

15 (a) 42,143.85 and 57,270.86
(b) 104,042.07 and 141,386.68

17 (a) 10.6 years
(b) 624.9 million dollars

Section 6.4

1 (a) $P = 100 + 10t$
(b) $P = 100(1.10)^t$
(c)

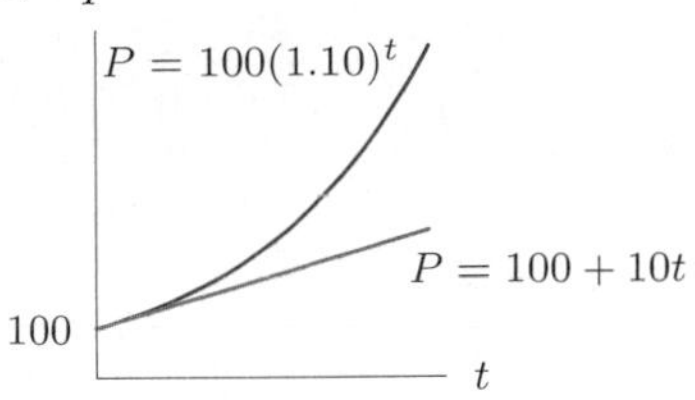

3 (a) $P = 4000 - 100t$
(b) $P = 4000(0.95)^t$
Case (a)

5 49% increase

7 18% decrease

9 Increasing $0 \le t \le 10$

11 Decreasing: $0 < t < 5$
Increasing: $5 < t < 10$

13 (a) $0 \le t \le 10$, 10.5%
(b) $10 \le t \le 15$, 2.5%
(c) 7.8% increase

15 Increases by about 5.13%

17 (a) About -0.36
(b) Decrease of about 30%

Chapter 6 Review

1 1.7

3 (a) 0.79

5 (a) $Q(10) \approx 2.7$,
$Q(20) \approx 1.8$
(b) 2.21
(c) 2.18

7 About 45

9 (a) III
(b) I
(c) II and IV

13 (a) Consumer: 87 units
Suppliers: 100 units
Down
(b) $p^* = \$48$
$q^* = 91$ units
(c) Consumer surplus: \$2096
Producer surplus: \$1143

15 Future value $= \$72{,}980.16$
Present value $= \$29{,}671.52$

17 \$12,402 million.

19 In 10 years

23 11%

Section 7.1

1 $5x$

3 $x^3/3$

5 $x^5/5$

7 $5q^3/3$

9 $y^3 - y^4/4$

11 $x^3 + 5x$

13 $t^3 + (7t^2/2) + t$

15 $5x^2/2 - 2x^{3/2}/3$

17 $t^4/4 - t^3/6 - t^2/2$

19 $-1/t$

21 $F(x) = x^7/7 + x^{-5}/35 + C$

23 $-e^{-3t}/3$

25 $G(t) = 5t + \sin t + C$

27 $F(x) = 3x$
(only possibility)

29 $F(x) = x^3/3$
(only possibility)

31 $F(x) = 2x + 2x^2 + (5/3)x^3$
(only possibility)

33 $3x^2/2 + C$

35 $2x^3 + C$

37 $(x^4/4) - (x^2/2) + C$

39 $(x^4/4) + 2x^2 + 8x + C$

41 $q^3/3 + 5q^2/2 + 2q + C$

43 $4x^{3/2} + C$

45 $-5/t - 3/t^2 + C$

47 $x^2/2 + 2x^{1/2} + C$

49 $-20e^{-0.05t} + C$

51 $-e^{-3t}/3 + C$

53 $-\cos t + C$

55 $25e^{4x} + C$

57 $5\sin x + 3\cos x + C$

59 $\sin(4x)/4 + C$

61 $10x - 4\cos(2x) + C$

63 $F(x) = x^3/3 + x + 5$

65 $F(x) = 2e^{3x} + 3$

67 (a) $20q - 2q^2$
(b) $p = 20 - 2q$

Section 7.2

1 $\frac{1}{6}(x^2+1)^6 + C$

3 $\frac{1}{4}(x+10)^4 + C$

5 $2\sqrt{x^2+1} + C$

7 $-(1/2)e^{-x^2} + C$

9 $(1/33)(t^3-3)^{11} + C$

11 $(1/9)(x^2-4)^{9/2} + C$

13 $-2\sqrt{4-x} + C$

15 $0.5\sin(t^2) + C$

17 $(1/5)x^5 + 2x^3 + 9x + C$

19 $(1/5)y^5 + (1/2)y^4 + (1/3)y^3 + C$

21 $-(2/9)(\cos 3t)^{3/2} + C$

23 $(1/7)\sin^7\theta + C$

25 $(1/35)\sin^7 5\theta + C$

27 $-\frac{1}{2}\cos(x^2) + C$

29 $\frac{1}{6}e^{3x^2} + C$

31 $\frac{1}{10}\ln(5q^2+8) + C$

33 $\ln|e^t + t| + C$

35 $2\sin\sqrt{x} + C$

37 $2\sqrt{x+e^x} + C$

39 $(1/2)\ln(x^2+2x+19) + C$

41 (a) Yes; $-0.5\cos(x^2) + C$
(b) No
(c) No
(d) Yes; $-1/(2(1+x^2)) + C$
(e) No
(f) Yes; $-\ln|2+\cos x| + C$

43 (a) (i) $x^3/3 + 5x^2 + 25x + C$
(ii) $(x+5)^3/3 + C$
(b) No; differ by a constant

Section 7.3

1 10

3 6

5 $1/2$

7 $81/4$

9 52

11 $8/15$

13 $\ln 2$

15 $2e - 2 \approx 3.437$

17 $1 - \cos 1 \approx 0.460$

19 6

21 (a) $(\ln 2)/2$
(b) $(\ln 2)/2$

23 $1 - e^{-1}$

25 2

27 21

29 2.32

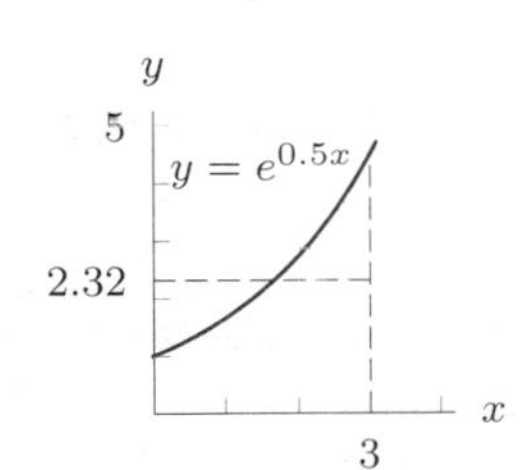

31 $\ln 3$

33 $(300)^{1/3} = 6.694$

35 (a) 5.3 billion, 6.1 billion
(b) 5.7 billion

37 (a)

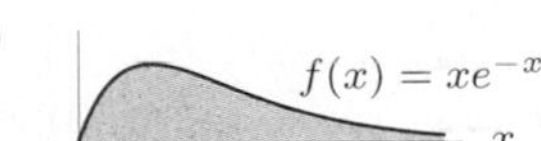

(b) 0.9596, 0.9995,
0.99999996
(c) Converges to 1

39 (a) $-(1/b) + 1$
(b) Converges to 1

41 (a) 26.42, 95.96, 99.95, 100
(b) 100

43 (a) 18, 61.2, 198

(b) $2\sqrt{b} - 2$
(c) Does not converge

45 $1/20$

47 (a) No
(b) $\int_0^\infty (60/50^t)\,dt$
(c) Yes; 15.34 miles

Section 7.4

1 $1, 0, -1/2, 0, 1$

3 Maximum at $(4, 3)$

5

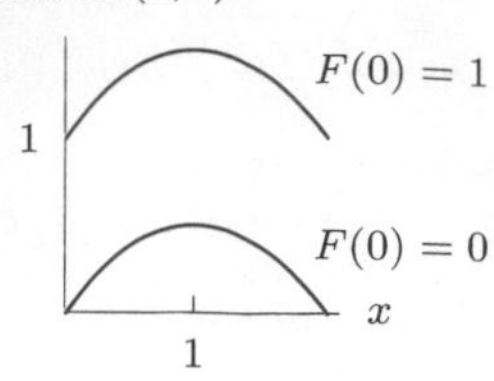

7

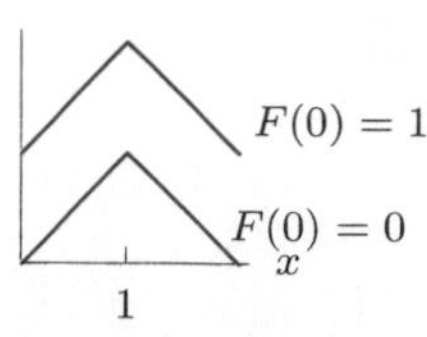

9 (a) Increasing for $x < -2$, $x > 2$
Decreasing for $-2 < x < 2$
Local maximum at $x = -2$
Local minimum at $x = 2$
(b)

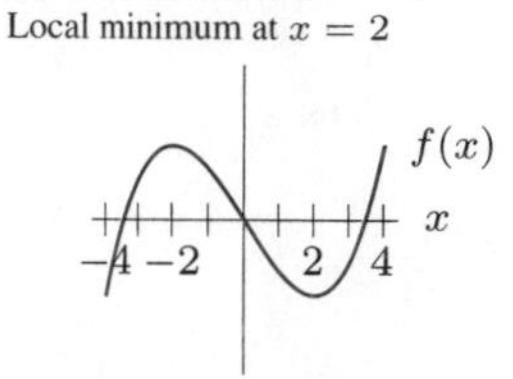

11 (a) $f(x)$ increasing when $2 < x < 5$
$f(x)$ decreasing when $x < 2$ or $x > 5$
f has a local minimum at $x = 2$
f has a local maximum at $x = 5$
(b)

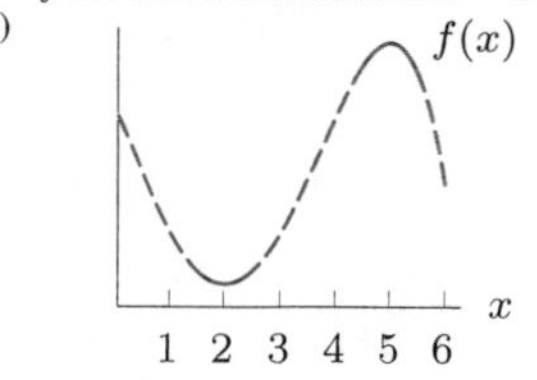

13

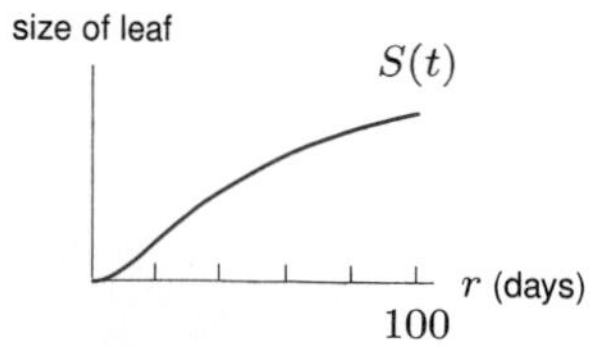

15 Largest: $F(2)$
Smallest: $F(4)$
None negative

17 Critical points: $(0, 5)$, $(2, 21)$, $(4, 13)$, $(5, 15)$

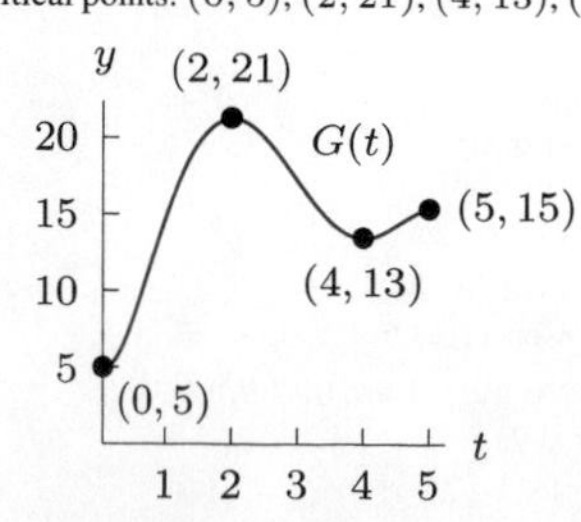

19 $F(2) = 8$; $F(5) = -8$; $F(6) = 2$

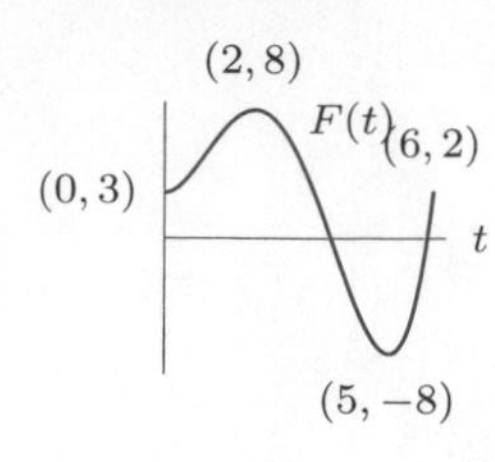

21

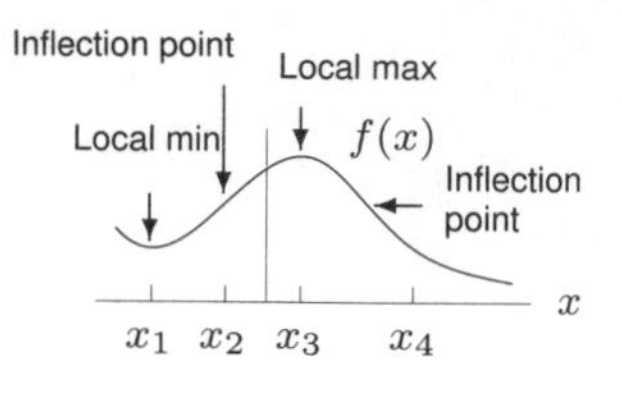

23 $f(3) - f(2)$,
$[f(4) - f(2)]/2$,
$f(4) - f(3)$

25

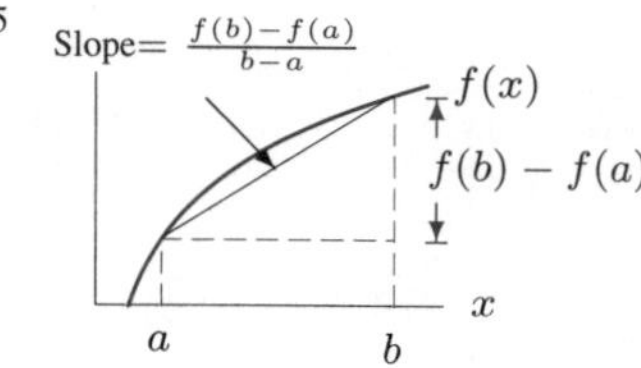

27

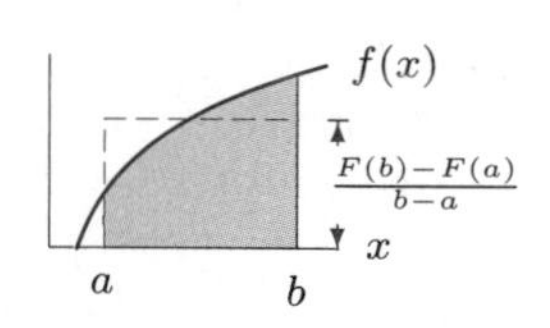

Chapter 7 Review

1 $10x + 2x^4$

3 $F(x) = (x^2/2) + (x^6/6) - (x^{-4}/4) + C$

5 $P(r) = \pi r^2 + C$

7 $-1/2z^2$

9 $G(x) = (x+1)^4/4 + C$

11 $3x^3 + C$

13 $(x+1)^3/3 + C$

15 $3\ln|t| + \frac{2}{t} + C$

17 $4t^2 + 3t + C$

19 $(x^3/3) + \ln|x| + C$

21 $2\ln|x| - \pi\cos x + C$

23 22

25 $\sin 1 - \sin(-1) = 2\sin 1$

27 $1/2$

29 $F(x) = x^2$
(only possibility)

31 $F(x) = -\cos x + 1$
(only possibility)

33 $e^{q^2+1} + C$

35 $-1/(3(3x+1)) + C$

37 $\frac{1}{55}(5x-7)^{11} + C$

39 $\frac{1}{9}(3x^2+4)^{3/2} + C$

41 $4\sin(x^3) + C$

43 (a) $\int_0^5 155e^{0.015t}\,dt$
(b) About 800 quadrillion BTUs

45 (a) $x = 1, x = 3$
(b) Local min at $x = 1$, local max at $x = 3$
(c)

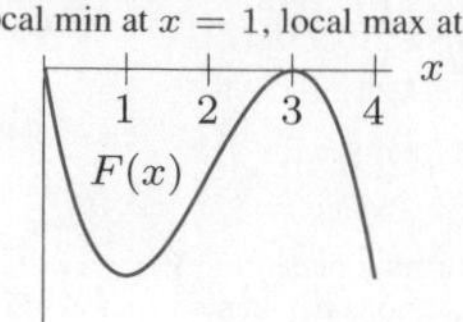

47 $(0, 1)$; $(2, 3)$; $(6, -4)$; $(8, 0)$

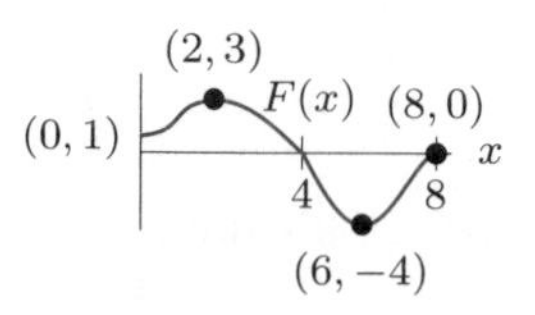

49 (a)

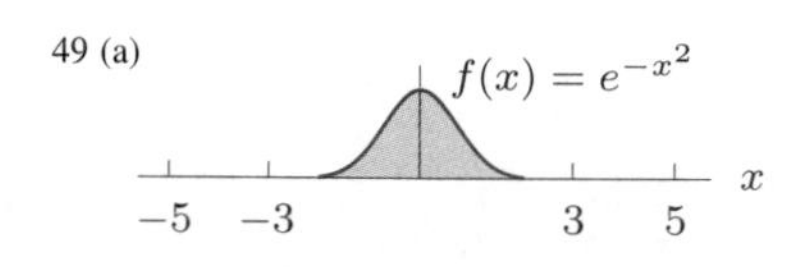

(b) 1.494; 1.764;
1.772; 1.772
(c) 1.772

Practice: Integration

1 $t^4/4 + 2t^3 + C$

3 $x^3/3 - 1/x + C$

5 $2w^{3/2} + C$

7 $t^3/3 + 5t^2/2 + t + C$

9 $w^5/5 - 3w^4 + 2w^3 - 10w + C$

11 $2q^{1/2} + C$

13 $4\ln|x| - 5/x + C$

15 $q^4/4 + 4q^2 + 15q + C$

17 $-5\cos x + 3\sin x + C$

19 $\pi hr^3/3 + C$

21 $5p^3q^4 + C$

23 $x^3 + 3e^{2x} + C$

25 $2.5e^{2q} + C$

27 $Ax^4/4 + Bx^2/2 + C$

29 $x^3/3 + 8x + e^x + C$

31 $x^4/4 + 5x^3/3 + 6x + C$

33 $Aq^2/2 + Bq + C$

35 $\frac{1}{2}e^{2t} + 5t + C$

37 $3\sin(4x) + C$

39 $\frac{1}{14}(x^2+9)^7 + C$

41 $\ln|y+2|+C$

43 $\ln(e^t+1)+C$

45 $2\sqrt{1+\sin x}+C$

Section 8.1

1

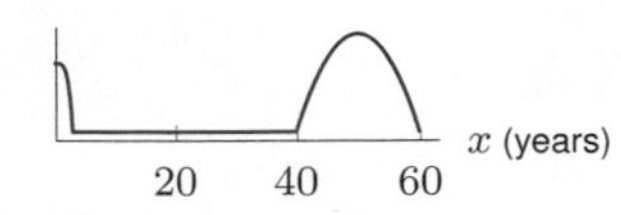

3 $1/15$

5 $1/5$

7 (b) About $3/4$

9 (a) 0.0625
(b) 0.91
(c) 0.0525

11 (a) 0.5
(b) 0.4
(c) 0.3

13 10–12: 27%
<8: 12%
>12 :45%
12–13 days

15

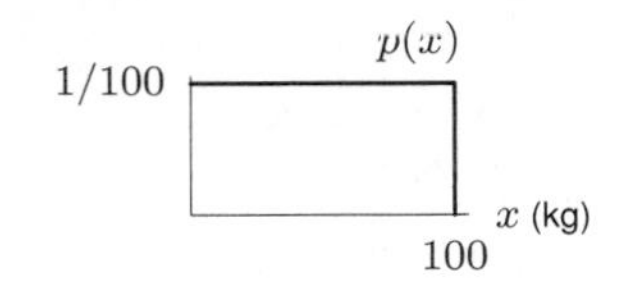

17

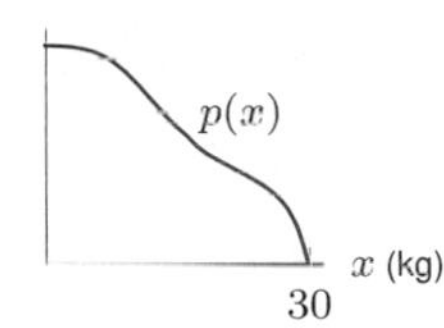

Section 8.2

1 (a) Cumulative distribution

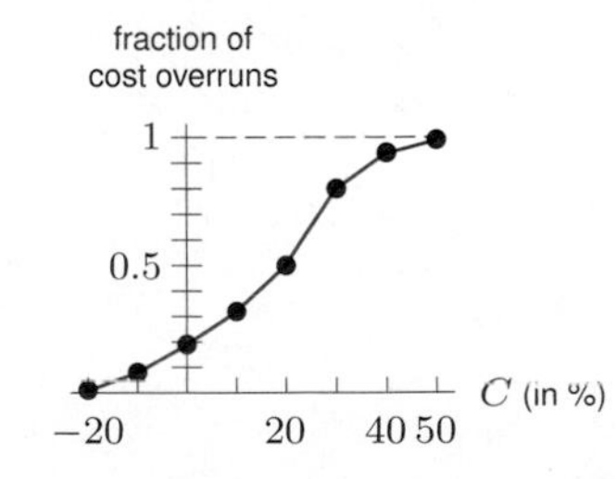

(b)

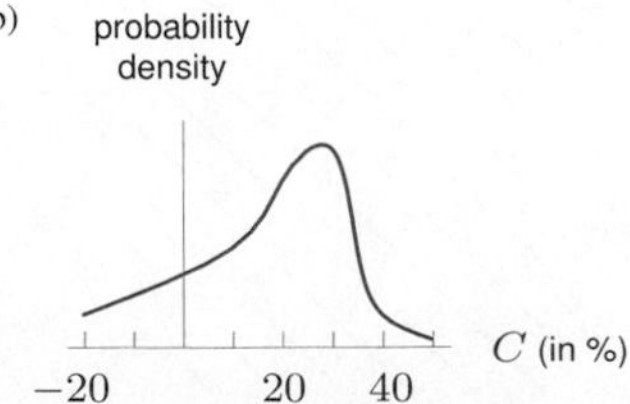

(c) More than 50%: 1%
Between 20% and 50%: 49%
Most likely: $C \approx 28\%$

3 (a)

t	0	1	2	3	4	$\cdots$
$P(t)$	0	0.60	0.975	1	1	$\cdots$

(b)

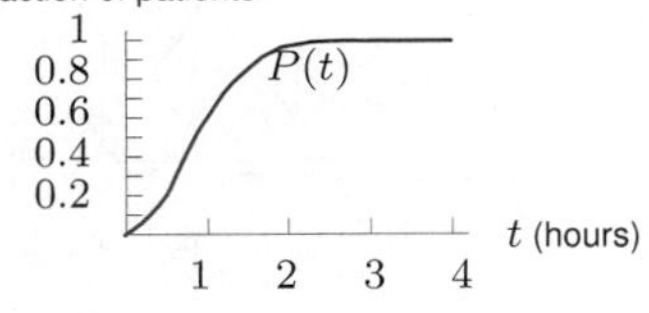

7

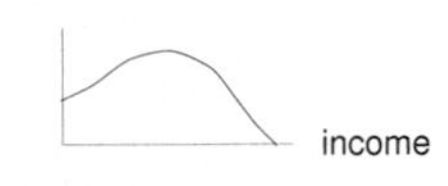

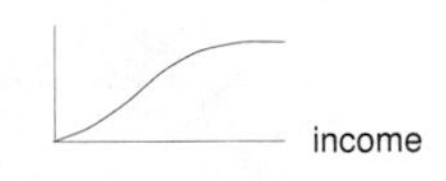

9

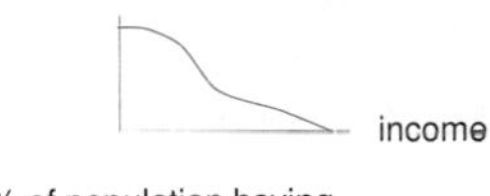

11 D

13

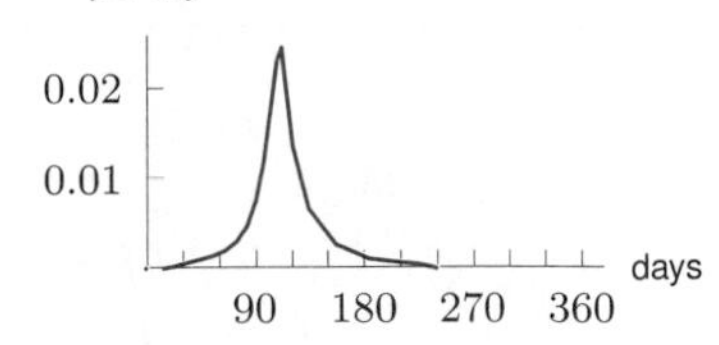

15 (a) 22.1%
(b) 33.0%
(c) 30.1%
(d) $C(h) = 1 - e^{0.4h}$

17 (a) 25%
(b) 32.5%
(c) 0

Section 8.3

1 5.35 tons

3 2.48 weeks

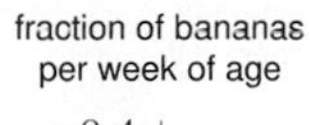

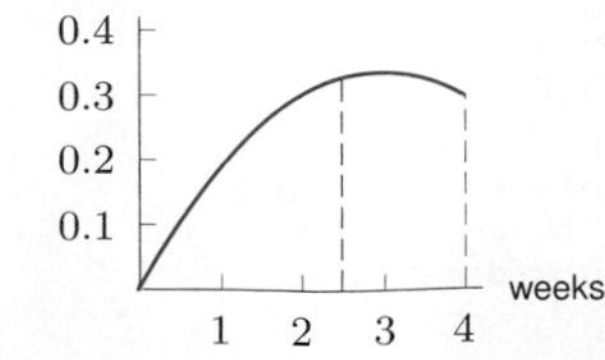

5 (b) Median ≈ 6.9
Mean ≈ 9.83

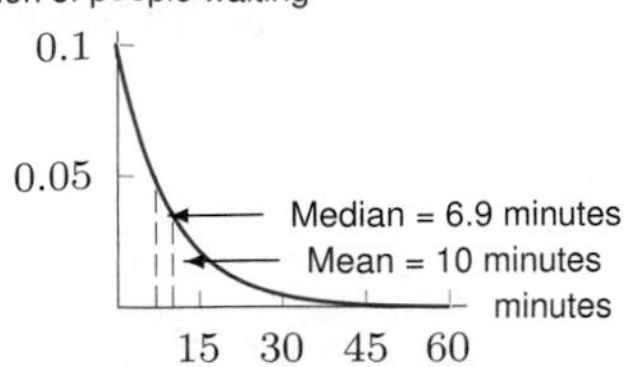

7 (a) $0.684:1$
(b) 1.6 hours
(c) 1.682 hours

9 (a) $p(x) = \frac{1}{15\sqrt{2\pi}} e^{-\frac{1}{2}\left(\frac{x-100}{15}\right)^2}$
(b) 6.7% of the population

11 14.6 days

Chapter 8 Review

1 0.04

3 0.008

7 (a) 0.19
(b) Tenth; both same
(c) $0.02, 0.38, 0.21$

9 (a)

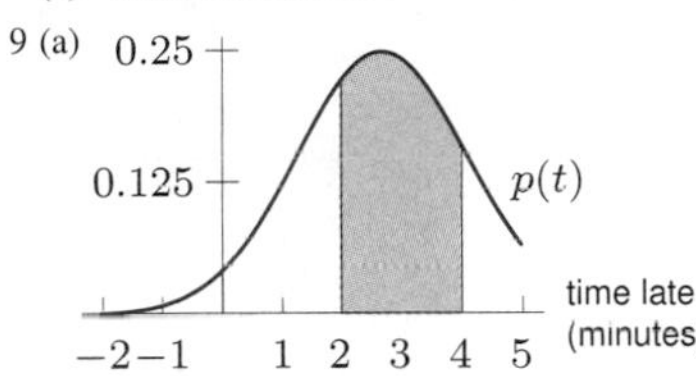

(b)

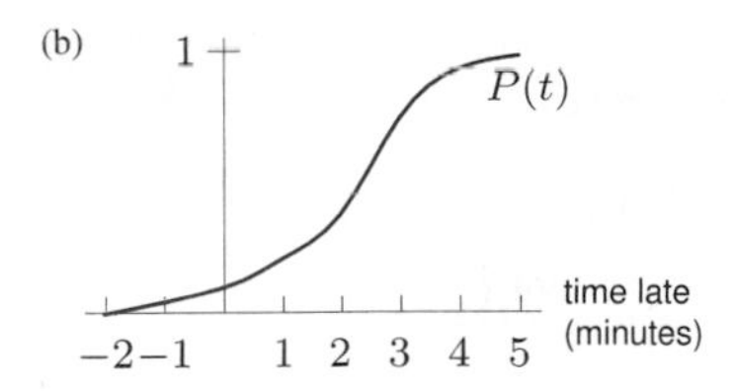

11 (a) $2/3$
(b) $1/3$
(c) Possibly many work just to pass
(d)

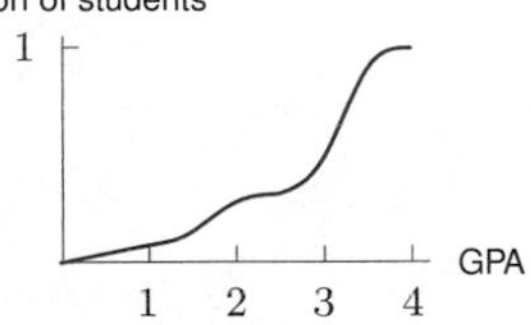

13 (a) $c = 0.0176$
(b) 9%

15 False

17 True

19 True

Section 9.1

1 Decreasing function of p
Increasing function of a

5 (a) All increasing
(b) Deposit of \$1250 gives balance \$1276 after 25 mos at 1% int

9 (a) $-19°$F
(b) 20 mph
(c) About 17.5 mph
(d) About $16.7°$F

13 $f(20, p)$: 2.65, 2.59, 2.51, 2.43
$f(100, p)$: 5.79, 5.77, 5.60, 5.53
$f(I, 3.00)$: 2.65, 4.14, 5.11, 5.35, 5.79
$f(I, 4.00)$ 2.51, 3.94, 4.97, 5.19, 5.60

17 $P = 0.052pf(I, p)/I$

19

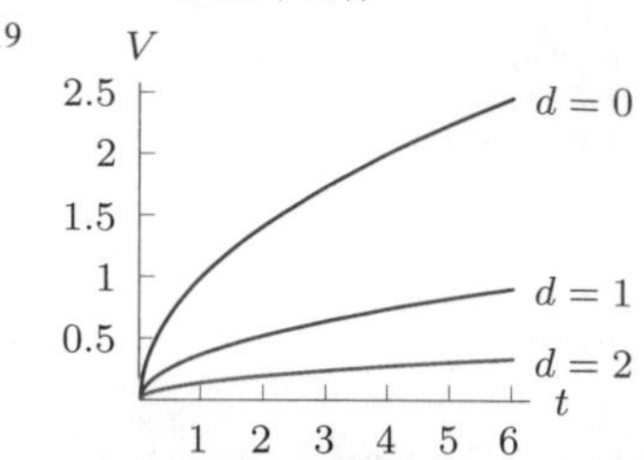

Section 9.2

3 (a) $\pi = 3q_1 + 12q_2 - 4$
(thousand dollars)

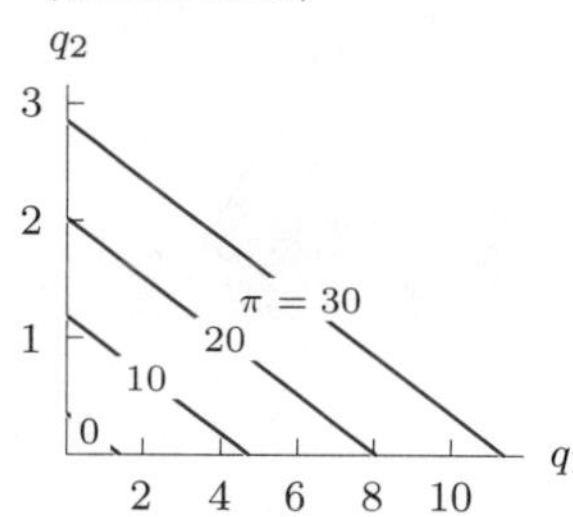

5 x-axis: price
y-axis: advertising

7 (a) About \$137
(b) About \$250

11

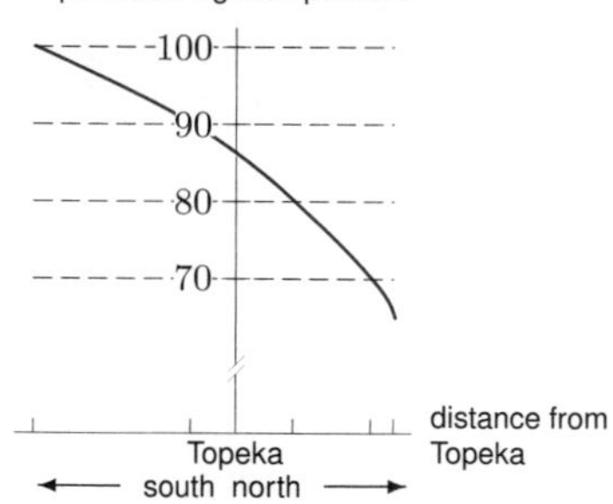

13

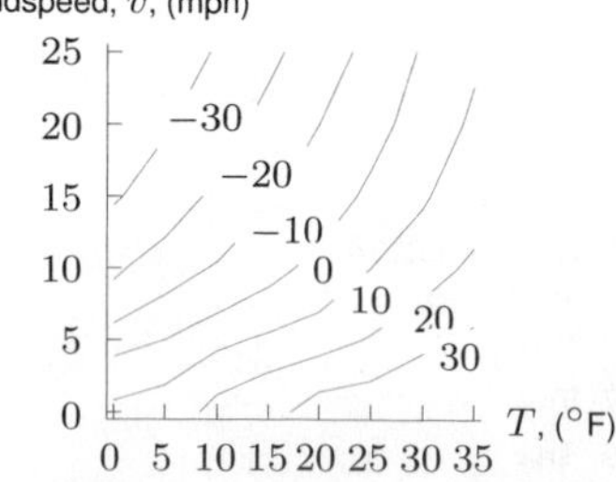

15 Contours evenly spaced

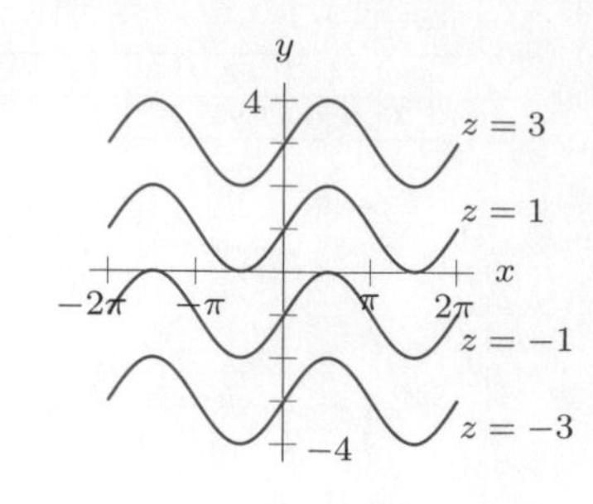

17 Contours evenly spaced

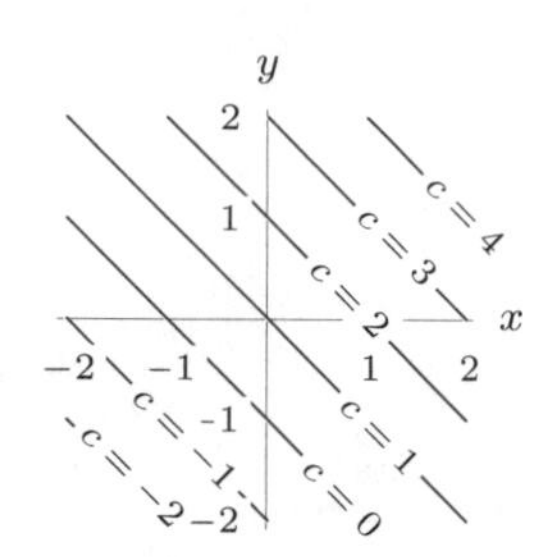

19 Contours evenly spaced

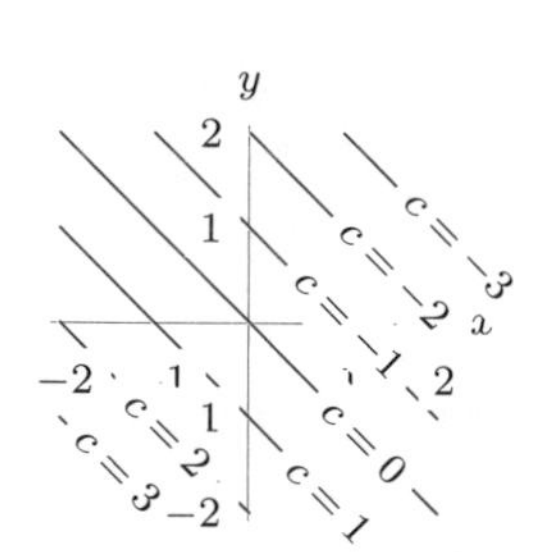

21

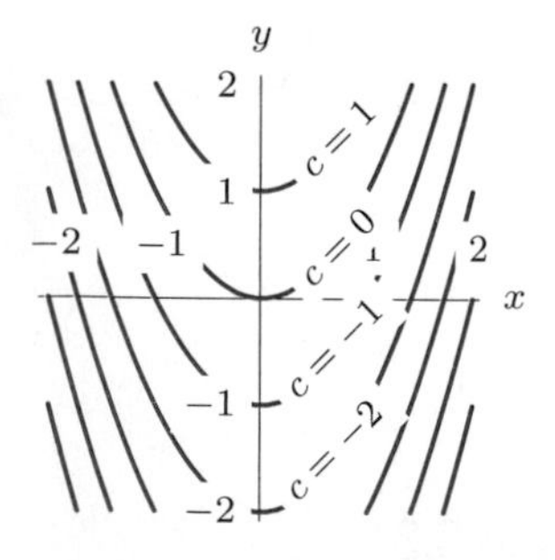

23 (a) 510.17 thousand pages/day
(b) 773.27 thousand pages/day
(c) 673.17 thousand pages/day
(d) 1020.34 thousand pages/day

25 (A) I
(B) II
(C) III

27 Table (a) matches (II)
Table (b) matches (III)
Table (c) matches (IV)
Table (d) matches (I)

29 (a) III
(b) II
(c) I
(d) IV

31 (a) 4 hours
(b) 40%
(c) Contours approx horizontal
(d) Increasing
(e) Increasing

33 (a) (IV)
(b) (I)
(c) (III)

Section 9.3

1 (a) Positive
(b) Negative
(c) Positive
(d) Zero

3 $\partial Q/\partial b < 0$
$\partial Q/\partial c > 0$

5 (a) f_c is negative
f_t is positive

7 (c) Positive
(d) Negative

9 (a) Negative

11 2.9; 0.02

13 (a)

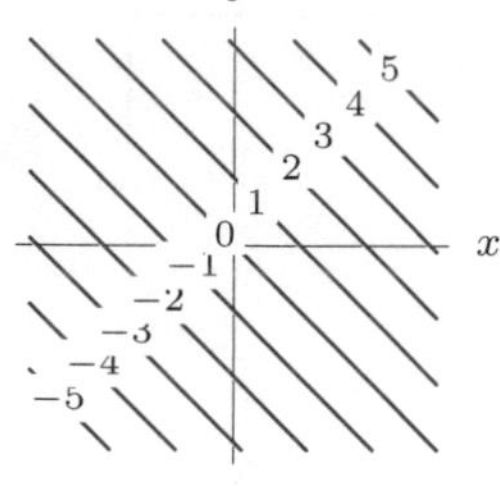

(b)

(c)

(d)

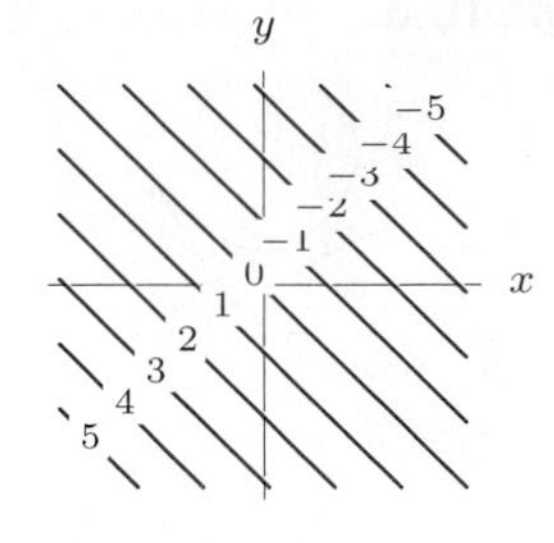

15 (a) 3.3 % / in
(b) −5% / °F

17 5.67

19 (a) Both negative
(b) Both positive

21 $f_w(10, 25) \approx -0.4°\text{F/mph}$

23 $f_w(5, 20) \approx -0.8$

25 (a) 150,350
(b) 151,000
(c) 152,250

27 (b)

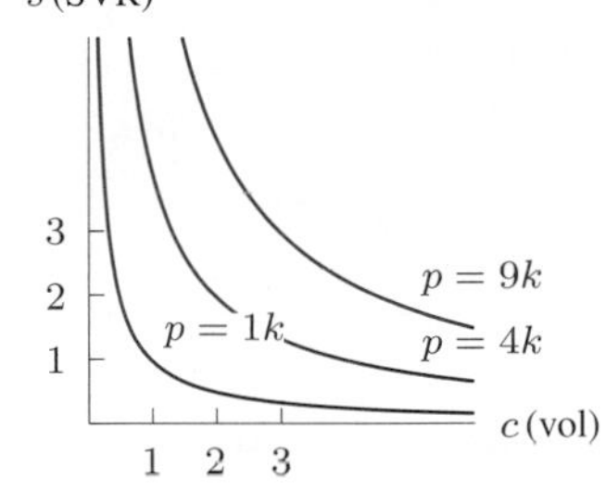

(c)

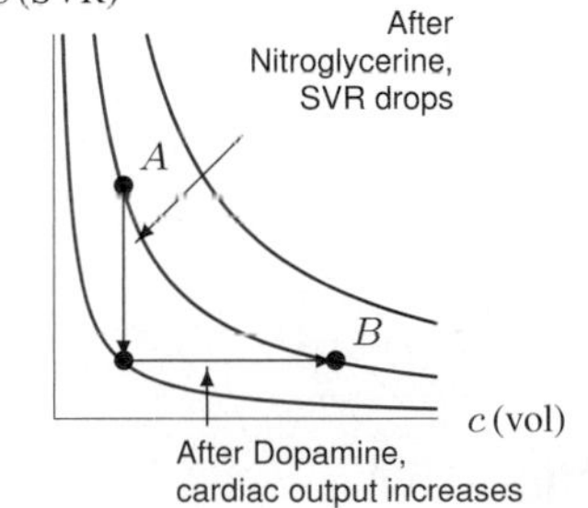

(d)

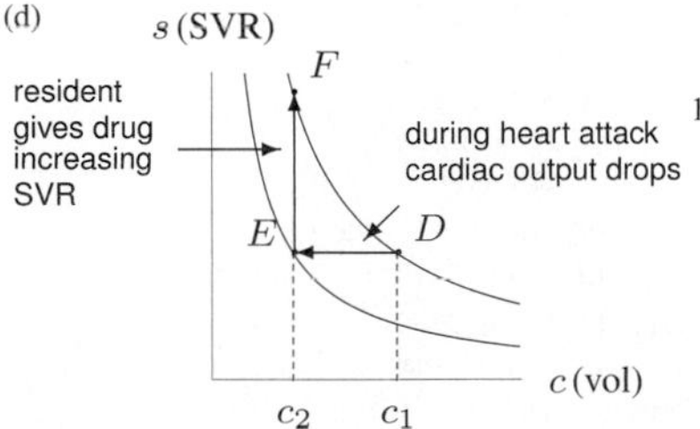

Section 9.4

1 $f_x = 2x + 2y$
$f_y = 2x + 3y^2$

3 $4x; 6y$

5 $100te^{rt}$

7 $200xy; 100x^2$

9 $2xy + 10x^4y$

11 $(a + b)/2$

13 $f_x = 10xy^3 + 8y^2 - 6x,$
$f_y = 15x^2y^2 + 16xy$

15 $f_x(1, 2) = 15,$
$f_y(1, 2) = -5$

17 15; 5; 30

19 (a) 3.3, 2.5
(b) 4.1, 2.1
(c) 4, 2

21 $1.277\ \text{m}^2, 0.005\ \text{m}^2/\text{kg}, 0.006\ \text{m}^2/\text{cm}$

23 (a) $Q_K = 18.75K^{-0.25}L^{0.25},$
$Q_L = 6.25K^{0.75}L^{-0.75}$
(b) $Q = 1704.33$
$Q_K = 21.3$
$Q_L = 4.26$

25 (a) Q, R
(b) Q, P
(c) P, Q, R, S
(d) None

27 $f_{xx} = 2, f_{xy} = 2,$
$f_{yy} = 2, f_{yx} = 2$

29 $f_{xx} = 0, f_{xy} = -2/y^2,$
$f_{yy} = 4x/y^3, f_{yx} = -2/y^2$

31 $f_{xx} = y^2e^{xy},$
$f_{xy} = (xy + 1)e^{xy},$
$f_{yy} = x^2e^{xy},$
$f_{yx} = (xy + 1)e^{xy}$

33 $V_{rr} = 2\pi h, V_{hh} = 0,$
$V_{rh} = V_{hr} = 2\pi r$

35 $B_{xx} = 0, B_{tt} = 20xe^{-2t},$
$B_{xt} = B_{tx} = -10e^{-2t}$

37 $f_{rr} = 100t^2e^{rt}, f_{tt} = 100r^2e^{rt},$
$f_{tr} = f_{rt} = 100(rt + 1)e^{rt}$

Section 9.5

1 $f(2, 10) \approx 0.5$ local and global min
$f(6, 4) \approx 9.5$ local max
$f(6.5, 16) \approx 10$ local and global max
$f(9, 10) \approx 4$ local min

3 $(-3, 5)$, minimum

5 $(-3, 6)$, neither

7 Saddle pts: $(1, -1), (-1, 1)$
local max $(-1, -1)$
local min $(1, 1)$.

9 Local minimum at $(2, 2)$

11 Local maximum at $(-1, 0)$,
Saddle points at $(1, 0)$ and $(-1, 4)$,
Local minimum at $(1, 4)$

13 Mississippi:
87 − 88 (max), 83 − 87 (min)
Alabama:
88 − 89 (max), 83 − 87 (min)
Pennsylvania:
89 − 90 (max), 80 (min)
New York:
81 − 84 (max), 74 − 76 (min)
California:
100 − 101 (max), 65 − 68 (min)
Arizona:
102 − 107 (max), 85 − 87 (min)
Massachusetts:
81 − 84 (max), 70 (min)

15 Max: 1 at $(\pi/2, 0)$; $(\pi/2, 2\pi)$
Min: −1 at $(\pi/2, \pi)$

17 $A = 10, B = 4, C = -2$

19 (b) $p_1 = p_2 = 25$
Max revenue is 4375

21 $q_1 = 300, q_2 = 225.$

Section 9.6

1 $f(10, 25) = 250$

3 $f(66.7, 33.3) = 13{,}333$

5 Min $= -\sqrt{2}$, max $= \sqrt{2}$

7 Min $= 11.25$; no max

9 Min $= -2$, max $= 2$

11 $x = 6; y = 6; f(6, 6) = 400$

13 (a) $Q = x_1^{0.3}x_2^{0.7}$
(b) $10x_1 + 25x_2 = 50{,}000$

15 $f(x, y) = 9000, x = 12, y = 6$

17 $q_1 = 50$ units
$q_2 = 150$ units

19 (a) 1215
(b) 1185

21 (a) $P(K, L)$
(b) $C(K, L) = 600{,}000$
(c) Tons/dollar
(d) Extra dollar produces approximately extra 3.17 tons

23 (a) Quantity of fuel, $x_1 + x_2$
(b) Terminal velocity (as function of x_1 and x_2) $= v_0$
(c) Liters per meter/sec
(d) 51 meters/sec requires about 8 more liters than 50 meters/sec

25 43

27 (a) $C = \$4349$
(b) \$182

29 (a) 2702
(b) 495; \$835,142
(c) \$2637.4 per car

31 (a) 475 units
(b) 505 units

33 (a)

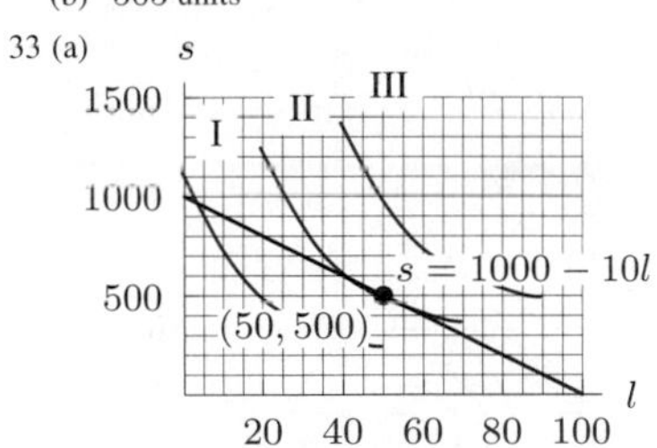

(b) $s = 1000 - 10l$

Chapter 9 Review

1 (a) 81°F
(b) 30%

3 Lines with slope 3/5, evenly spaced

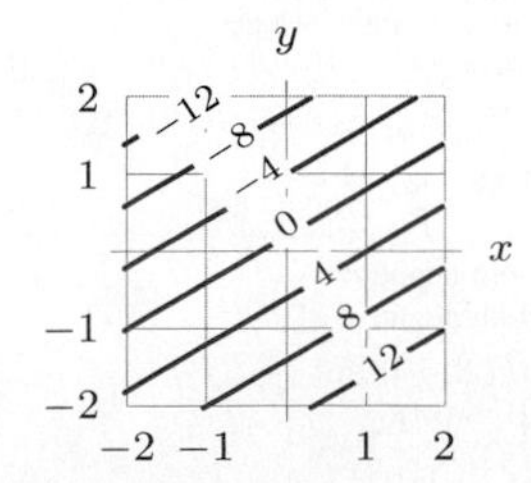

5 (a)

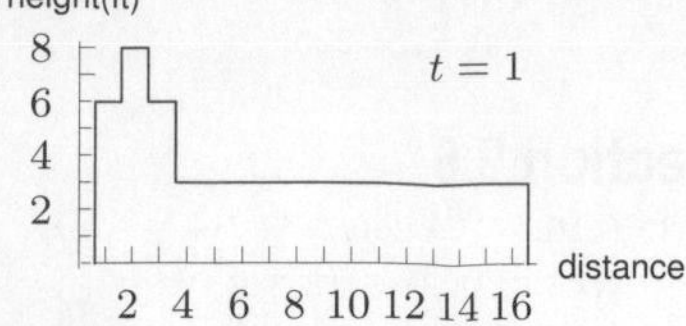

height(ft)
8
6
4
2
0
$t = 5$
distance
2 4 6 8 10 12 14 16

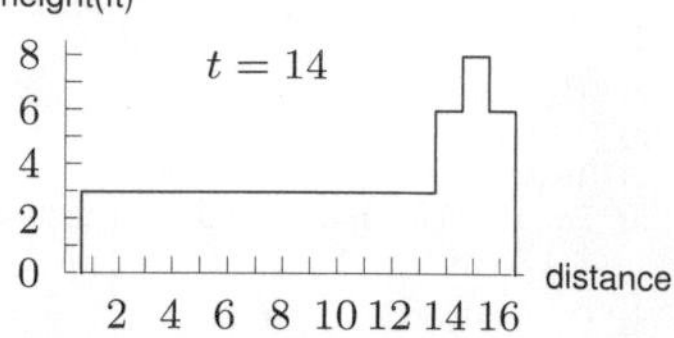

(b)

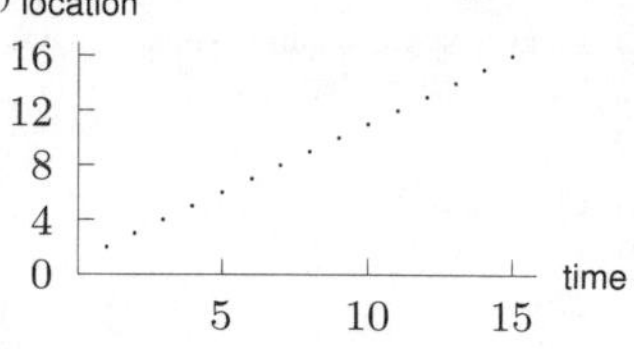

(c) The "wave" at a sports arena

9 72°F; 76°F

11 (a) $f_x < 0; f_y > 0$
(b) $-0.6; 3.5$
(c) $97; 92$

13 c: decreasing
t: increasing

15 $D = 100(c/t)$

17

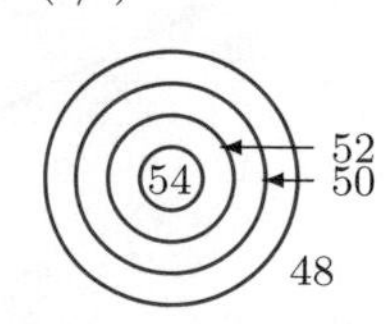

19 $\partial P/\partial t$:
dollars/month
Rate of change in payments with time
negative
$\partial P/\partial r$:
dollars/percentage point
Rate of change in payments with interest rate
positive

21 $\frac{\partial I}{\partial H}|_{(10,100)} \approx 0.4$
$\frac{\partial I}{\partial T}|_{(10,100)} \approx 1$

23 (a) Both negative
(b) Both negative

25 $(A)\, 0.06, -0.06$
$(B)\, 0, -0.05$
$(C)\, 0, 0$

27

T (°C) \ w (gm/m^3)	0.1	0.2	0.3
10	1300	900	1200
20	800	800	900
30	800	700	800

29 $P_a = 2a - 2b^2$, $P_b = -4ab$

31 $\frac{\partial f}{\partial x} = 5e^{-2t}$, $\frac{\partial f}{\partial t} = -10xe^{-2t}$

33 $f_x = x/\sqrt{x^2 + y^2}$
$f_y = y/\sqrt{x^2 + y^2}$

35 $h_x(2, 5) = -0.38$ ft/seat
$h_t(2, 5) = 0.76$ ft/sec

37 $(2, -1)$

39 (a) $P = p_1^2/8 + p_2^2/8 - 10$
(b) $\partial(\max P)/\partial p_1 = p_1/4$

41 (b) $D = 10,\ N = 20$
$V \approx 9{,}779$
(c) $\lambda \approx 14.67$, if cost in thousands

Theory: Least Squares

1 $y = 2/3 - (1/2)x$

3 $y = 2/3 - (1/2)x$

5 $y = x - 1/3$

7 (a) (i) Power function
(ii) Linear function
(b) $\ln N = 1.20 + 0.32 \ln A$
Agrees with biological rule

Section 10.1

1 $dP/dt = kP, k > 0$

3 $dB/dt = 0.05B$

5 (a) (III)
(b) (IV)
(c) (I)
(d) (II)

7 $dA/dt = -1$

9 $dM/dt = 2.5 - 0.347M$

11 $dA/dt = 10 - 0.03A$

13 (a) (I)
(b) (IV)
(c) (III)

15 y negative, y positive

Section 10.2

3 (a) $y = Cx^3$
(b) $C = 5$

5 12, 18, 27, 40.5

7 $y = 100, 80, 64, 51.2$

11 $k = 5$

13 C

15 B

17 F

19 B

21 A

23 (a) (IV)
(b) (III)
(c) (III), (IV)
(d) (II)

Section 10.3

1

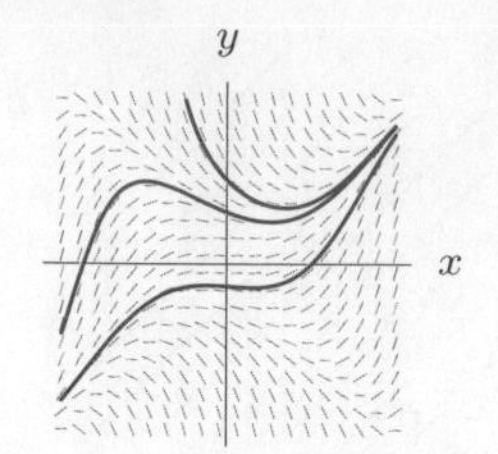

y
x

3 (a) Slopes $= 2, 0, -1, -4$
(b)

Slope $= 0$
Slope $= -1$
Slope $= 2$
Slope $= -4$

5 III: $y' = (1 + y)(2 - y)$

7 (I) = (b); (II) = (d) and (f); (III) = (f); (IV) = (c); (V) = (a) and (c); (VI) = (e)

9 (a) II
(b) VI
(c) IV
(d) I
(e) III
(f) V

11 As x increases, $y \to \infty$.

13 As $x \to \infty$, y oscillates within a certain range

15 $y \to \infty$ as $x \to \infty$

Section 10.4

1 $y = 5.6e^{-0.14x}$

3 $P = 20e^{0.02t}$

5 $y = 10e^{-x/3}$

7 (a) $dB/dt = 0.015B$
(b) $B = 5000e^{0.015t}$
(c) \$5809.17

9 $dQ/dt = -0.0025Q$,
$Q = Ce^{-0.0025t}$, About 5%

11 Michigan: 72 years
Ontario: 18 years

13 (a) $dS/dt = kS$
(b) $S = Ce^{kt}$
(c) $C = 5$
(d) $S = 5e^{0.1576t}$

15 (a) $dy/dt = ky$
(b) 0.2486 grams

17 (a) $dQ/dt = -kQ$
(b) $Q = Ce^{-kt}$
(c) $k \approx 0.182$
(d) $Q(12) \approx 1.126$ mg

Section 10.5

1 $y = 200 - 150e^{0.5t}$

3 $B = 25 - 5e^{4t}$

5 $m = 3000e^{0.1t} - 2000$

7 $B = 25 + 75e^{2-2t}$

11 $dB/dt = 0.08B - 5000$,
$B = 62{,}500 - 12{,}500e^{0.08t}$,
Yes, in 20.1 years

13 (b) $dQ/dt = 43.2 - 0.082Q$
(c) $Q = 526.8 - 526.8e^{-0.082t}$
$Q \to 526.8$ mg as $t \to \infty$

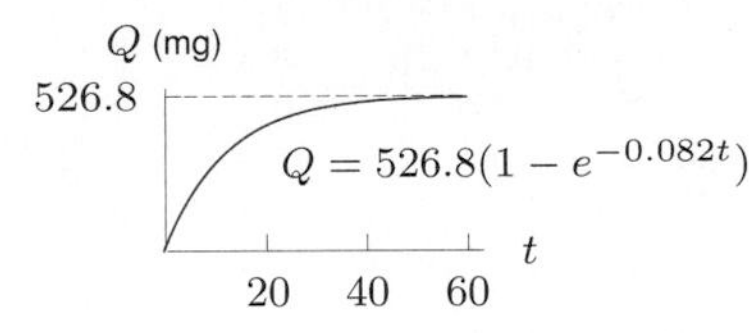

15

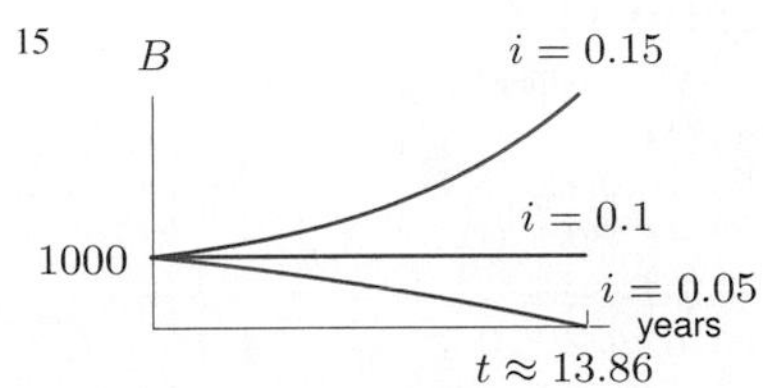

17 $dD/dt = -0.75(D - 4)$
Equilibrium = 4 g/cm^2

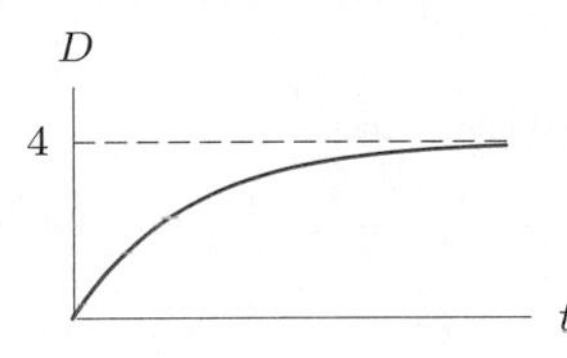

19 (a) $dB/dt = 0.05B - 12{,}000$
(b) $B = (B_0 - 240{,}000)e^{0.05t} + 240{,}000$
(c) \$151,708.93

21 (a) $y = 500$
(b) $y = 500 + Ce^{0.5t}$
(c)

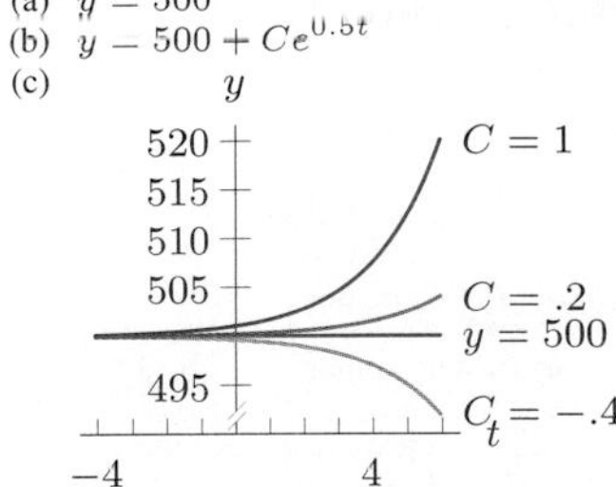

(d) Unstable

23 (a) $dW/dt = (1/3500)(I - 20W)$
(b) $W = I/20 + (W_0 - I/20)\, e^{-(2/350)t}$
(c)

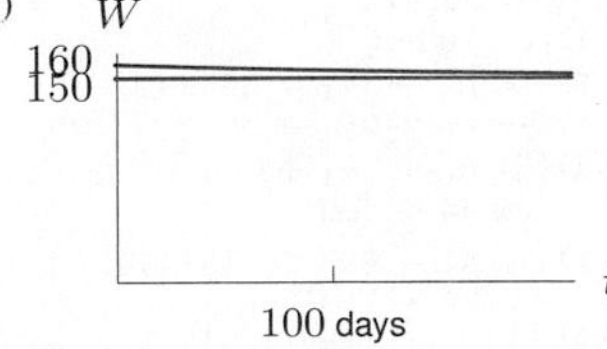

25 (a) $H = 200 - 180e^{-kt}$
(b) $k \approx 0.027$ (if t is in minutes)

27 (a) $dT/dt = -k(T - 68)$
(b) $T = 68 + 22.3e^{-0.06t}$;
3:45 am.

Section 10.6

1 (a) $x \to \infty$ exponentially
$y \to 0$ exponentially
(b) Predator-prey

3 (a) $x \to \infty$ exponentially
$y \to 0$ exponentially
(b) y is helped by the presence of x

5 $dx/dt = x - xy$,
$dy/dt = y - xy$

7 $dx/dt = -x - xy$,
$dy/dt = -y - xy$

11 Symmetric about the line $r = w$;
solutions closed curves

13 Robins:
Max ≈ 2500
Min ≈ 500
When robins are at a max,
the worm population is about 1 million

17 (a) $dw/dt = 0$
$dr/dt = 1.2$
(b) $w \approx 2.2, r \approx 1.1$
(c) At $t = 0.2$:
$w \approx 2.2, r \approx 1.3$
At $t = 0.3$
$w \approx 2.1, r \approx 1.4$

19 (a)

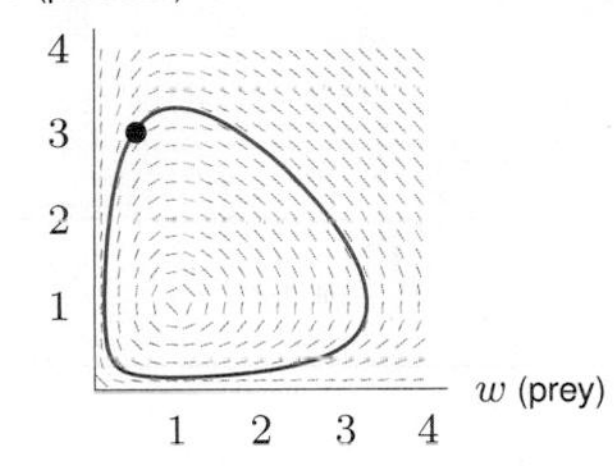

(b) Down and left
(c) $r = 3.3, w = 1$
(d) $w = 3.3, r = 1$

21 (a)

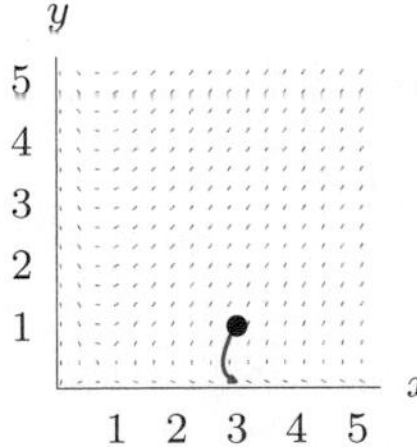

(b)

(c)

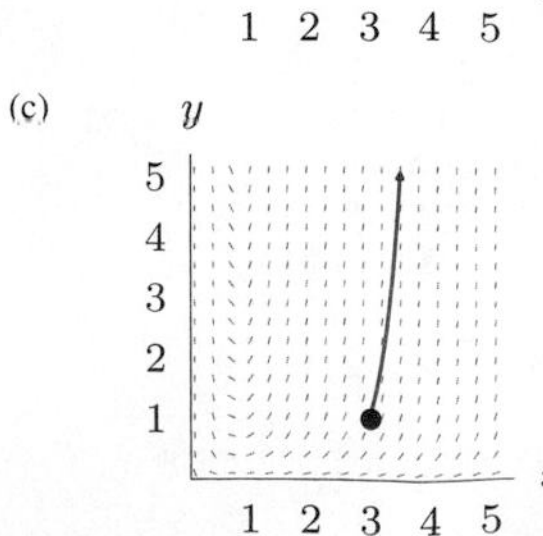

(d)

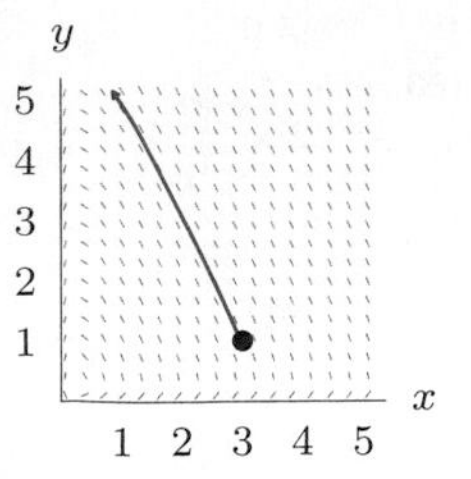

Section 10.7

5 (a) $I_0 = 1, S_0 = 349$
(b) Increases; spreads

7 About 300 boys;
$t \approx 6$ days

9 5

11 (a) b/a

Chapter 10 Review

1 Yes

3 (a) I is $y' = 1 + y$;
II is $y' = 1 + x$
(b) I

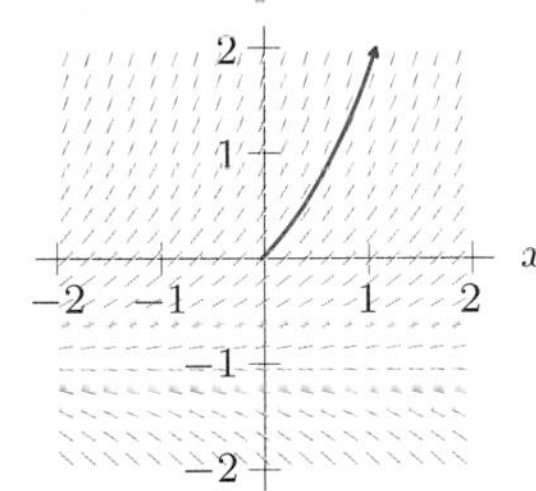

II

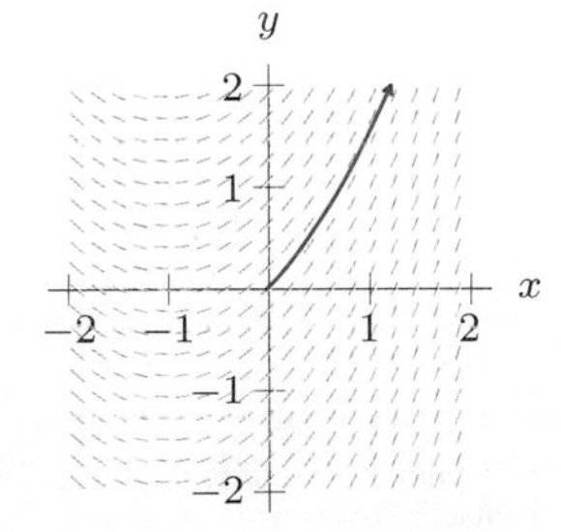

(c) I: $y' = -1$, unstable;
II: None

5 $y = t^2 + C$

9 $y = Ce^{5t}$

11 $P = Ce^{0.03t}$

13 $Q = Ce^{2t}, C \neq 0$

15 $y = 100 + Ce^{-t}$

17 $H = Ce^{0.5t} - 20$

19 $y = 25e^{-0.2t}$

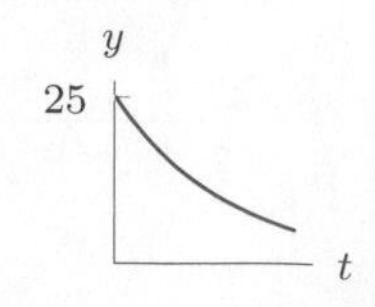

21 $H = 240e^{-0.5t} - 200$

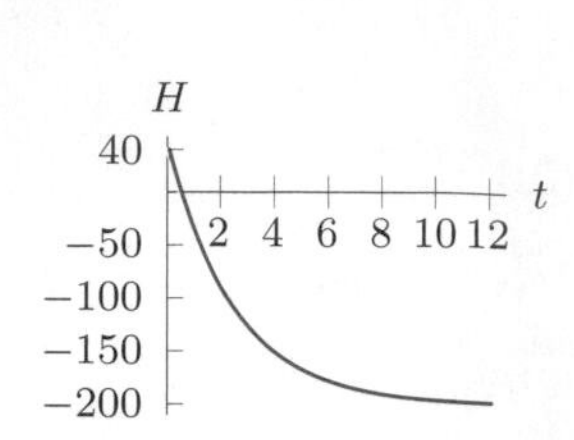

23 (a) $dM/dt = rM$
(b) $M = 1000e^{rt}$
(c)

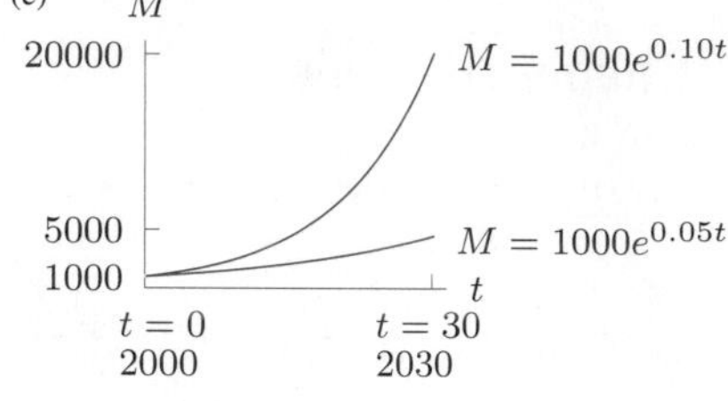
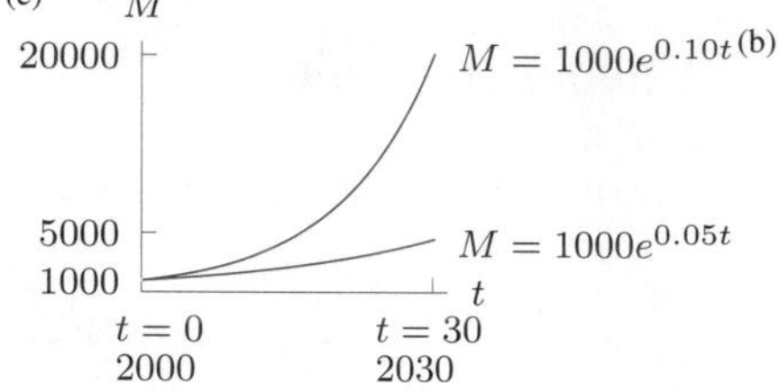

25 (a) $k \approx 0.000121$
(b) 779.4 years

27 (a) $y = ce^{-t} + 100$
(b)

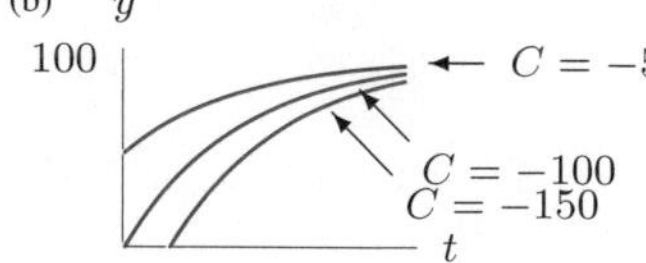

(c) $y = 100 - 100e^{-t}$

29 (a) $dy/dt = -k(y - a)$
(b) $y = (1 - a)e^{-kt} + a$
(c) a: fraction remembered in the long run
k: rate material is forgotten

31 Initially $x = 0$; y decreases, x increases. Then x increases, y increases. Finally y increases, x decreases

33 Populations oscillate

35 (c) Equilib: $y = 0, y = 1$
Stable: $y = 1$
Unstable: $y = 0$

Theory: Separation of Vars

1 $P = e^{-2t}$

3 $P = \sqrt{2t + 1}$

5 $u = 1/(1 - (1/2)t)$

7 $R = 1 - 0.9e^{1-y}$

9 $z = -\ln(1 - t^2/2)$

11 $y = -2/(t^2 + 2t - 4)$

13 (a) Yes (b) No (c) Yes
(d) No (e) Yes (f) Yes
(g) No (h) Yes (i) No
(j) Yes (k) Yes (l) No

15 $Q = b - Ae^{-t}$

17 $R = -(b/a) + Ae^{at}$

19 $y = -1/\left(k(t + t^3/3) + C\right)$

21 (a)

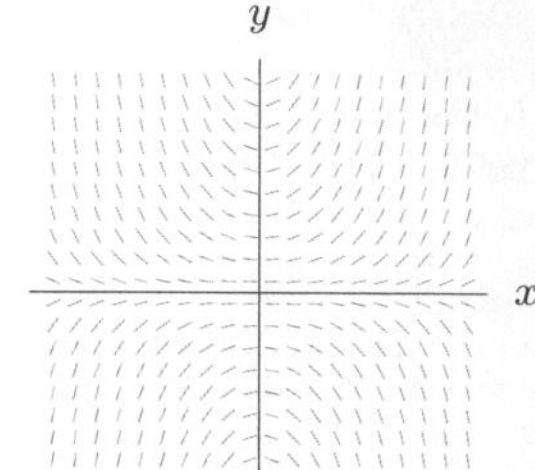
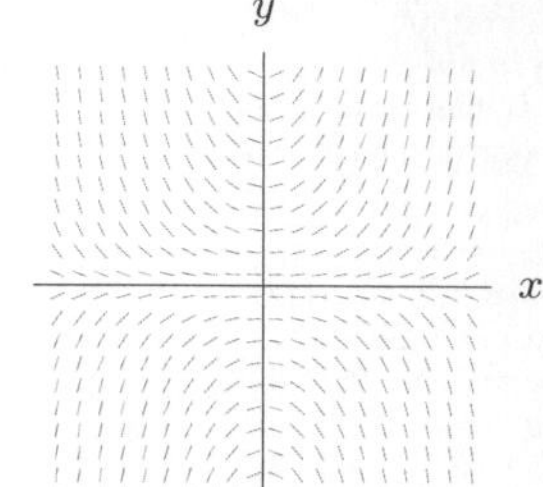

(b)

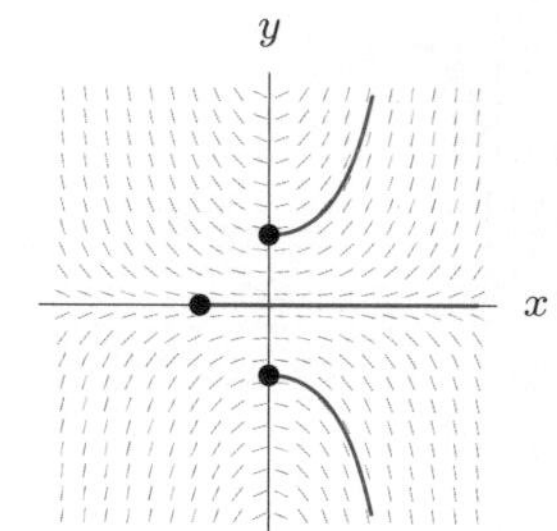

(c) $y(x) = Ae^{x^2/2}$

Section 11.1

1 21

3 2046

5 Does not exist

7 200

9 1.9961

11 $3(2^{11} - 1)/2^{10}$

13 400

15 30.51, 37.75, 39.47, 39.87, Yes

17 (a) $3037.75, $2537.75
(b) $6167.78, $5667.78

19 (a) 39 mg
(b) 41.496 mg
(c) 41.667 mg

21 (a) 40 mg
(b) 0.57 mg/kg, Yes
(c) (i) Greater than 100 kg
(ii) Less than 13.3 kg

Section 11.2

1 $11,315.60, $9,315.60

3 $33,035.37

5 (a) $121,581.16
(b) $176,666.67

7 $2.02 million

9 (a) $1.27
(b) $163.83
(c) $20,971.51
(d) $2,684,354.55

11 40,000 units

13 (a) 5000 units
(b) $S_5 = 3362$, $S_{10} = 4463$, $S_{15} = 4824$, $S_{20} = 4942$

15 (a) $20 billion
(b) $45 billion

Section 11.3

1 313.498 billion barrels

3 (a) 98 mg
(b) 121.5 mg
(c) 125 mg

5 227.42 mg

7 (a) 0.7937
(b) 194.27 mg
(c) 242.37 mg

9 (a) 2647.17 mg
(b) 37 weeks

11 1604.0 micrograms
1596.0 micrograms

13 In 36.59 years

15 (a) 67.8 years
(b) 29.6 years

17 70 years

19 (a) $2D$
(b) 150 mg

Chapter 11 Review

1 3,985,805

3 555.10

5 96.154

7 39,375

9 (a) (i) $16.43 million
(ii) $24.01 million
(b) $16.87 million

11 (a) $\approx 7\%$
(b) $Q_n = 50(1 - (0.07)^{n+1})/(1 - 0.07)$
(c) $P_n = 0.07(50)(1-(0.07)^n)/(1-0.07)$

13 $400 million

15 £250

17 (a) $h_n = 10(3/4)^n$
(b) $D_1 = 10$ feet
$D_2 = h_0 + 2h_1 = 25$ feet
$D_3 = h_0 + 2h_1 + 2h_2 = 36.25$ feet
$D_4 = h_0 + 2h_1 + 2h_2 + 2h_3$
≈ 44.69 feet
(c) $D_n = 10 + 60\left(1 - (3/4)^{n-1}\right)$

19 $1081.11

INDEX

FORMULA SUMMARY: ALGEBRA

Lines

Slope of line through (x_1, y_1) and (x_2, y_2):

$$m = \frac{y_2 - y_1}{x_2 - x_1}$$

Point-slope equation of line through (x_1, y_1) with slope m:

$$y - y_1 = m(x - x_1)$$

Slope-intercept equation of line with slope m and y-intercept b:

$$y = b + mx$$

Definition of Zero, Negative, and Fractional Exponents

$$a^0 = 1, \quad a^{-1} = \frac{1}{a}, \quad \text{and, in general, } a^{-x} = \frac{1}{a^x}$$

$$a^{1/2} = \sqrt{a}, \quad a^{1/3} = \sqrt[3]{a}, \quad \text{and, in general, } a^{1/n} = \sqrt[n]{a}.$$

$$\text{Also, } a^{m/n} = \sqrt[n]{a^m} = (\sqrt[n]{a})^m.$$

Rules of Exponents

1. $a^x \cdot a^t = a^{x+t}$ For example, $2^4 \cdot 2^3 = (2 \cdot 2 \cdot 2 \cdot 2) \cdot (2 \cdot 2 \cdot 2) = 2^7$.
2. $\dfrac{a^x}{a^t} = a^{x-t}$ For example, $\dfrac{2^4}{2^3} = \dfrac{2 \cdot 2 \cdot 2 \cdot 2}{2 \cdot 2 \cdot 2} = 2^1$.
3. $(a^x)^t = a^{xt}$ For example, $(2^3)^2 = 2^3 \cdot 2^3 = 2^6$.

Definition of Natural Log

$y = \ln x$ means $e^y = x$; *for example:* $\ln 1 = 0$, since $e^0 = 1$.

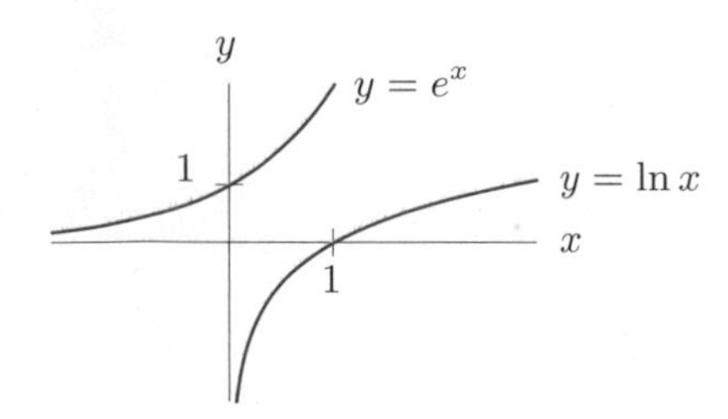

Rules of Natural Logarithms

$$\ln(AB) = \ln A + \ln B$$

$$\ln\left(\frac{A}{B}\right) = \ln A - \ln B$$

$$\ln A^p = p \ln A$$

Identities

$$\ln e^x = x$$

$$e^{\ln x} = x$$